Lecture Notes in Electrical Engineering

Volume 683

The book series *Lecture Notes in Electrical Engineering* (LNEE) publishes the latest developments in Electrical Engineering - quickly, informally and in high quality. While original research reported in proceedings and monographs has traditionally formed the core of LNEE, we also encourage authors to submit books devoted to supporting student education and professional training in the various fields and applications areas of electrical engineering. The series cover classical and emerging topics concerning:

- Communication Engineering, Information Theory and Networks
- Electronics Engineering and Microelectronics
- Signal, Image and Speech Processing
- Wireless and Mobile Communication
- Circuits and Systems
- Energy Systems, Power Electronics and Electrical Machines
- Electro-optical Engineering
- Instrumentation Engineering
- Avionics Engineering
- Control Systems
- Internet-of-Things and Cybersecurity
- Biomedical Devices, MEMS and NEMS

For general information about this book series, comments or suggestions, please contact leontina.dicecco@springer.com.

To submit a proposal or request further information, please contact the Publishing Editor in your country:

China

Jasmine Dou, Associate Editor (jasmine.dou@springer.com)

India, Japan, Rest of Asia

Swati Meherishi, Executive Editor (Swati.Meherishi@springer.com)

Southeast Asia, Australia, New Zealand

Ramesh Nath Premnath, Editor (ramesh.premnath@springernature.com)

USA, Canada:

Michael Luby, Senior Editor (michael.luby@springer.com)

All other Countries:

Leontina Di Cecco, Senior Editor (leontina.dicecco@springer.com)

**** Indexing: The books of this series are submitted to ISI Proceedings, EI-Compendex, SCOPUS, MetaPress, Web of Science and Springerlink ****

More information about this series at http://www.springer.com/series/7818

David Harvey · Haranath Kar ·
Shekhar Verma · Vijaya Bhadauria
Editors

Advances in VLSI, Communication, and Signal Processing

Select Proceedings of VCAS 2019

Springer

Editors
David Harvey
Faculty of Engineering and Technology
General Engineering Research Institute
Liverpool John Moorse University
Liverpool, UK

Haranath Kar
Department of Electronics
and Communication Engineering
Motilal Nehru National Institute
of Technology Allahabad
Prayagraj, India

Shekhar Verma
Department of Information Technology
Indian Institute of Information Technology
Allahabad, India

Vijaya Bhadauria
Department of Electronics
and Communication Engineering
Motilal Nehru National Institute
of Technology Allahabad
Prayagraj, India

ISSN 1876-1100 ISSN 1876-1119 (electronic)
Lecture Notes in Electrical Engineering
ISBN 978-981-15-6839-8 ISBN 978-981-15-6840-4 (eBook)
https://doi.org/10.1007/978-981-15-6840-4

This Springer imprint is published by the registered company Springer Nature Singapore Pte Ltd.
The registered company address is: 152 Beach Road, #21-01/04 Gateway East, Singapore 189721, Singapore

Contents

Controller Design According to Right/Left Coprime Factorization 1
A. Karimpour and D. K. Chaturvedi

Fault Detection and Classification in Microgrid Using Wavelet Transform and Artificial Neural Network 17
Priya Singh, Nitin Singh, and Niraj Kumar Choudhary

Performance of 4H-SiC IMPATT Diode at Ka- and W-Band with Temperature Variation 33
Girish Chandra Ghivela, Prince Kumar, Sourabh Tiwari, and Joydeep Sengupta

Design and Analysis of Low-Power SRAM 41
Pawar Dhiraj Kumar, Ritesh Kumar Kushwaha, and P. Karuppanan

A Unified Approach for Calculating Outage Performance of Multi-hop Regenerative Relay Network in Nakagami-*m* Fading Channel 57
Himanshu Katiyar, P. K. Verma, Arun Kumar Singh, and Saurabh Dixit

832.5 Gb/s PM-8QAM Superchannel with 5 b/s/Hz Spectral Efficiency 67
Divya Sharma, Shikha Devi, and Y. K. Prajapati

Design and Performance of High-Speed CMOS Double-Tail Dynamic Comparator Suitable for Mixed-Signal ICs 75
Avaneesh K. Dubey, Vikrant Varshney, Ankur Kumar, Pratosh K. Pal, and R. K. Nagaria

Hybrid Forecasting Model Based on Nonlinear Auto-Regressive Exogenous Network, Fourier Transform, Self-organizing Map and Pattern Recognition Model for Hour Ahead Electricity Load Forecasting 89
Nida-e-Falak and M. M. Tripathi

Structural and Optical Characterization of EZO Thin Film for Application in Optical Waveguide 109
Lucky Agarwal, Richa Singh, and Shweta Tripathi

Predictive Analysis of Wind Turbine Output Power Using Support Vector Machine(SVM) Based on Genetic Algorithm(GA). 117
Tushar Srivastava and M. M. Tripathi

Dual-Band Dual-Mode Orthogonally Placed CDRA-Based MIMO Antenna for Wi-fi/WLAN Applications 135
Ajay Kumar Dwivedi, Anand Sharma, Pramod Narayan Tripathi, and Ashutosh Kumar Singh

An Energy-Efficient Localization Scheme Using Beacon Nodes for Wireless Sensor Networks 145
Sunil Kumar, Prateek Raj Gautam, Swati Verma, and Arvind Kumar

Rain Streaks Elimination Using Hybrid Median Filter and Contrast Stretching 157
Rais Ahmad and Som Pal Gangwar

Design of Efficient Ternary Subtractor. 179
Yogesh Shrivastava and Tarun Kumar Gupta

A Multichannel Link-Layer Cooperation Protocol (MLCP) for Cognitive Radio Ad Hoc Network 191
Jahnvi Tiwari, Arun Prakash, and Rajeev Tripathi

Far Field Analysis of Defected Ground Structured Wideband Antenna for RF Energy Harvesting Applications 201
Rashmi Pandey, A. K. Shankhwar, and Ashutosh Singh

New LMI Criterion to the Robust Stability of Discrete-Time Systems with Time-Varying Delays and Generalized Overflow Nonlinearities 213
Pushpendra Kumar Gupta and V. Krishna Rao Kandanvli

PAPR Reduction in OFDM for VLC System 229
Shilpi, M. Shukla, and Arvind Kumar

Performance Evaluation of Energy-Efficient Adiabatic Logic Circuit-Based Multiplexer for Low Power Applications 239
Shivangi Jaiswal, Prashasti, Anam Khan, and Subodh Wairya

Design and Performance Evaluation of Highly Efficient Adders in Nanometer Technology. 251
Prashasti, Shivangi Jaiswal, Anam Khan, and Subodh Wairya

$\alpha - \eta - \mu/IG$ Composite Fading Model for Body-Centric Communication 263
Prasoon Raghuwanshi and Krishan Kumar

DTMOS-Based Low-Voltage Low-Power CCII+ and Biquad Filter Using ±0.25 V Supply 271
Anil Kumar Gautam, Tripurari Sharan, and P. Nemthianhoi Zou

Structural and Optical Analysis of Bulk-Hetero Interface Between MoS_2: Pentacene 291
Anshika Srivastava and Shweta Tripathi

An Efficient Watermarking Process Based on Three-Level DWT and FFT Technique 303
Jamal Arif and Som Pal Gangwar

Blockchain-Enabled Traceable, Transparent Transportation System for Blood Bank 313
Nihar Ranjan Pradhan, Akhilendra Pratap Singh, and Vinay Kumar

A Compendious Analysis of Advances in HE Methods for Contrast Enhancement 325
D. Vijayalakshmi and Malaya Kumar Nath

A Survey of Semantic Segmentation on Biomedical Images Using Deep Learning 347
Yellamelli Prajna and Malaya Kumar Nath

HT-IWT-DCT-Based Hybrid Technique of Robust Image Watermarking 359
Priyank Khare and Vinay Kumar Srivastava

Reliability and Circuit Timing Analysis with HCI and NBTI 371
S. Udaya Shankar and P. Kalpana

Realization of a Low Profile, Wideband Omni-directional Antenna for Ku-band Airborne Applications 393
Pramendra Kumar Verma and Raj Kumar

Effect of Structural Metal on Metamaterial-Based Absorber Performance 403
Sajal Agarwal and Y. K. Prajapati

Effect of Various Parameter Variations on Electrical Characteristics of Rectangular Gate All Around Junctionless Transistor 409
Manish Kumar Rai, Sanjeev Rai, and Abhinav Gupta

Slotted I-Patch with Capacitive Probe Fed Microstrip Antenna for Wideband Applications 417
Santosh Kumar Gupta and Sangaraju Varun

Development of Cloud-Based Multi-Modal m-Cardiac Management System 429
Naveen Kumar, Arun Kumar, Prabal Gupta, Rohini Srivastava, R. P. Tewari, Nitin Sahai, and Basant Kumar

PDP Analysis of CNTFET Full Adders for Single and Multiple Threshold Voltages 439
M. Elangovan, R. Ranjith, and S. Devika

A Linear OTA Using Series-Connected Source-Degenerated Bulk-Driven Floating Gate Differential Pairs 447
Tanmay Dubey and Vijaya Bhadauria

Compressive Sensing-Based Continuous EEG Monitoring: Seizure Detection Performance Comparison of Different Classifiers 459
Meenu Rani, S. B. Dhok, and R. B. Deshmukh

A Low Harmonic High Gain Subthreshold Flipped Voltage Follower-Based Bulk-Driven OTA Suitable for Low-Frequency Applications 469
Sougata Ghosh, Saumya Tripathi, and Vijaya Bhadauria

Performance Analysis of MoS_2FET for Electronic and Spintronic Application 489
Kavindra Kumar Kavi, R. A. Mishra, and Shweta Tripathi

Split Behavior of Supervised Machine Learning Algorithms for Phishing URL Detection 497
Madhurendra Purbay and Divya Kumar

LMI and YALMIP: Modeling and Optimization Toolbox in MATLAB 507
Akhilesh Kumar Ravat, Amit Dhawan, and Manish Tiwari

Improved Noise Margin and Reduced Power Consumption in Subthreshold Adiabatic Logic Using Dual Rail Power Supply 517
Sanjay Kumar Prajapati, Devarshi Shukla, and Santosh Kumar Gupta

Human Action Recognition Using a New Hybrid Descriptor 527
Om Mishra, Pranoti S. Kavimandan, M. M. Tripathi, Rajiv Kapoor, and Kalpana Yadav

Analog and Radio-Frequency Performance of Hetero-Gate-Dielectric FD SOI MOSFET in Re-S/D Technology 537
Nilesh Anand Srivastava, Anjali Priya, and Ram Awadh Mishra

Theoretical Analysis of Defected Ground Multiband Rectangular Shape Microstrip Patch Antenna 549
Salah Boubkar Salah Abdou, Anil Kumar, and Arvind Kumar

New Resistorless FDNR Simulation Configuration Employing CDDITAs 561
Kapil Bhardwaj and Mayank Srivastava

Novel CDDITA-Based-Grounded Inductance Simulation Circuits 571
Kapil Bhardwaj and Mayank Srivastava

New FDNR and FDNC Simulation Configurations Using Inverted VDDIBAs 583
Kapil Bhardwaj and Mayank Srivastava

A Low Power Approach for Designing 12-Bit Current Steering DAC 595
Abhishek Kumar, Santosh Kumar Gupta, and Vijaya Bhadauria

Design and Implementation of an Efficient Mixed Parallel-Pipeline SAD Architecture for HEVC Motion Estimation 605
Mamidi Nagaraju, Santosh Kumar Gupta, Vijaya Bhadauria, and Devarshi Shukla

Minimization of Peak-to-Average Power Ratio in DHT Precoded OFDM System by A-Law Companding 623
Ajay Kumar Yadav, Dheeraj Dubey, and Y. K. Prajapati

Leakage-Tolerant Low-Power Wide Fan-in OR Logic Domino Circuit 631
Ankur Kumar, Pratosh K. Pal, Vikrant Varshney, Avaneesh K. Dubey, and R. K. Nagaria

Virtually Doped Silicon-on-Insulator Junctionless Transistor for Reduced OFF-State Leakage Current 643
Prateek Kishor Verma and Santosh Kumar Gupta

Improved Store-Carry-Forward Scheme for Information Dissemination in Unfavorable Vehicular Distribution 657
Ankita Srivastava, Arun Prakash, and Rajeev Tripathi

Information Theory-Based Defense Mechanism Against DDOS Attacks for WSAN 667
Jyoti Bhola and Surender Soni

Comparative Analysis of Channel Estimation Techniques in Vehicular Communication 679
Devesh Shukla, Arun Prakash, and Rajeev Tripathi

A Survey Study of Diseases Diagnosed Through Imaging Methodology Using Ultrasonography 689
Kumar Mohit, Jojo Johnson, Kumari Simran, Rajeev Gupta, and Basant Kumar

Sensor Localization in WSNs Using Rotating Directional-Antenna at the Base Station 705
Prateek Raj Gautam, Sunil Kumar, and Arvind Kumar

A Survey on Proactive and Reactive Channel Switching Techniques in Cognitive Radios 719
Anushree Srivastava, Arun Prakash, and Rajeev Tripathi

Controlling GIDL Using Core–Shell Technique in Conventional Nano-Wire 731
Abhishek Kumar, Sushmita Jaiswal, and Santosh Kumar Gupta

About the Editors

Dr. David Harvey is a Professor in the Department of Electronics Engineering, Liverpool John Moores University, UK. Prof. Harvey got his doctorate degree on Real-time Microprocessor-based Analysis of Optoelectronic Data in 1984. He has been working as a Principal Electronics Design Engineer, Kratos, Analytical Instruments. He has also been working as Director Electronic Design and Manufacturing (EDAM) Centre, LJMU. He has also been Director Engineering Development Centre (EDC), LJMU. He has been consultant of over 250 companies. He has an exhaustive experience in Metrology and advanced optical/acoustic/X-ray/electronic/AFM measurement systems at nano-and micro-scales. He has guided more than 19 Ph.D scholars, has more than 139 publications in reputed journals and conferences and more than 618 citations. He has peer reviewed many journals and is actively involved in research.

Dr. Haranath Kar received the B.E. Degree from Bengal Engineering College in 1989, the M.Tech. Degree from the Banaras Hindu University, Varanasi, India, in 1992 and the Ph.D. Degree from the University of Allahabad, Allahabad, India, in 2000. After spending a brief period at the Defence Research and Development Organization as a Scientist B, he joined Motilal Nehru National Institute of Technology (MNNIT), Allahabad, India, as a Lecturer in 1991, where he became an Assistant Professor in 2001, Associate Professor in 2006 and Professor in 2007. He spent two years with the Atilim University, Turkey (2002–2004) as an Assistant Professor. He served as the Chairman of the Senate Post-Graduate Committee at MNNIT from 2009 to 2010. He was Head of Electronics and Communication Engineering Department at MNNIT during 2013–2015. His current research interests are in digital signal processing, nonlinear dynamical systems, delayed systems, robust stability, guaranteed cost control and multidimensional systems. He is a recipient of the 2002–2003 IEE Heaviside Premium Award. He was conferred with the D.N. Agrawal Award of excellence and the Bharat Vikas Award in 2005 and 2017, respectively. He was a member of editorial board of the Mathematical Problems in Engineering.

Dr. Shekhar Verma has received his BTech, MTech and PhD from IIT BHU, Varanasi. He is currently working as professor in Information Technology at Indian Institute of Information Technology Allahabad. He has published more than 100 research papers in reputed refereed International Journals and more than 80 papers in international conferences. He has supervised 20 PhD scholars and handled many R&D projects. He is a member of the "Machine Learning and Optimization Group" at IIIT Allahabad. His research interests include dimensionality reduction, Manifold regularization, Privacy Preserving Machine Learning and Deep Learning techniques.

Prof. Vijaya Bhadauria received her B.E. (Electronics) and M.E. (Control & Instrumentation) from MNREC Allahabad. She received her Ph.D. in Electronics Engineering from MNNIT Allahabad. She was Head in Department of Electronics and Communication Engineering, MNNIT Allahabad, India from Aug. 2017 to Aug. 2019. Her research interests are in the area of VLSI Circuit and System, Digital Integrated Circuit Design, Advanced Analog Integrated Circuit Design, VLSI Technology and Semiconductor Device and Modeling. She has published many papers in international journals and conferences of repute and supervised several Ph.D. students. Dr. Bhadauria worked as an ad hoc reviewer of many international journals and served as program committee member of several international conferences of repute in the area of Microelectronics and VLSI Design.

Controller Design According to Right/Left Coprime Factorization

A. Karimpour and D. K. Chaturvedi

Abstract This paper proposes a method to design a controller based on the right/left coprime factorization. In this method, the designer chooses the controller from the set of all stabilizing controllers based on some performance measures on the closed-loop transfer function. In comparison with other methods that work on closed-loop transfer functions such as H_∞ or H_2, the proposed method has a better and clearer insight into the system design procedure. The applicability of the method was shown in some case studies.

Keywords Right/left coprime factorization · All stabilizing controller · Controller design

1 Introduction

Despite significant advances in control science and the growing speed of high-speed computers, the design of a controller for LTI systems still is a challenging task. Designing controller for linear systems can be done through open-loop or closed-loop criteria [1, 2].

In designing the controller according to open-loop criteria, the designer tries to consider the requested performance requirement on the open loop transfer function. So, shaping the open-loop transfer matrix is the main idea of this method. In this method, some performance measures are gain margin (GM), phase margin (PM), steady-state error, open-loop bandwidth, etc. Mainly, root locus, Bode plot, and

A. Karimpour
Department of Electrical Engineering, Faculty of Engineering, Ferdowsi University of Mashhad, Mashhad, Iran
e-mail: karimpor@um.ac.ir

D. K. Chaturvedi (✉)
Department of Electrical Engineering, Faculty of Engineering, Dayalbagh Educational Institute, Dayalbagh, Agra, India
e-mail: dkchaturvedi@dei.ac.in

D. Harvey et al. (eds.), *Advances in VLSI, Communication, and Signal Processing*,
Lecture Notes in Electrical Engineering 683,
https://doi.org/10.1007/978-981-15-6840-4_1

Nichols chart are used in designing the controller. Nichols chart helps the designer to know about closed-loop criteria through open-loop analysis.

On the other, another approach is to consider closed-loop transfer matrices or closed-loop behavior directly. For example, a linear quadratic regulator (LQR) tries to minimize a performance measure that, directly considers some optimal performance on the states and input signal. Reference [3] applied LQR to the power system. Model predictive control (MPC) is another controller that has grown increasingly popular over the past two decades, it tries to match the closed-loop system with a desired one. MPC has been used in many areas and also some works available in their estimation part and also some works on fuzzy models used in MPC [4–7]. Others have investigated methods using historical data to design a controller for traffic systems [8]. The computational effort for real-time analysis is found in [9].

Other approaches try to shape the closed-loop transfer matrix directly. In this area, some important controllers are robust controllers such as H_∞/H_2 [10, 11]. These controllers suffer from the lack of a precise method to find the exact closed-loop shape.

Some researchers use the right/left coprime factorization and then try to find a suitable controller from the set of all stabilizing controllers [12]. Reference [13] uses the genetic algorithm to choose the suitable controller from the set of all stabilizing controller.

In this paper, right/left coprime factorization used to find the set of all stabilizing controllers. Then by choosing the exact shape for some transfer function and using some optimization procedure, one can derive a suitable controller from the set of all stabilizing controller to control the system. Section 2 of this paper explains about right/left coprime factorization and stabilizing controllers, Sect. 3 deals briefly with the procedure to consider control requirements on closed-loop transfer functions. Section 4 explains the controller design by coprime factorization and limitation on closed-loop transfer matrices, and Sect. 5 shows the applicability of the method by some case studies. Section 6 summarizes the paper with a conclusion.

2 Right/Left Coprime Factorization and Stabilizing Controller

Any transfer matrix $G(s)$ with m output and q input can be decomposed to right/left coprime factorization as Eq. 1 or Eq. 2.

$$G(s) = N(s)M^{-1}(s) \tag{1}$$

$$G(s) = \widetilde{M}^{-1}(s)\widetilde{N}(s) \tag{2}$$

where $N(s)$ is $m \times q$ transfer matrix, $M(s)$ is $q \times q$ transfer matrix, $\widetilde{N}(s)$ is $m \times q$ transfer matrix, and $\widetilde{M}(s)$ is $m \times m$ transfer matrix in the set of H_∞ space (stable

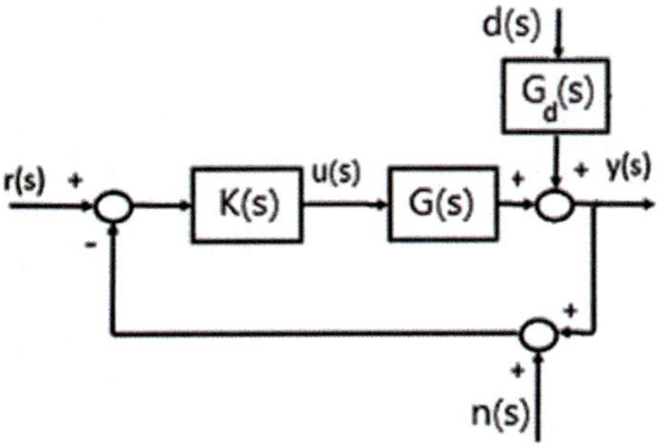

Fig. 1 System under study

transfer matrix). Meanwhile, $M(s)$ and $N(s)$ are right coprime and also $\widetilde{M}(s)$ and $\widetilde{N}(s)$ are left coprime so there exist Bezout pairs X_r, Y_r, X_l and Y_l, respectively. Equation 3 shows the relation between right/left coprime factorization and their Bezout counterparts.

$$\begin{bmatrix} M & -Y_l \\ N & X_l \end{bmatrix} \begin{bmatrix} X_r & Y_r \\ -\widetilde{N} & \widetilde{M} \end{bmatrix} = \begin{bmatrix} I & 0 \\ 0 & I \end{bmatrix} \tag{3}$$

To consider the importance of coprime factorization considers Fig. 1. In Fig. 1, $G(s)$ is the plant under control and $K(s)$ is the controller of the system. The set of all stabilizing controller can be derived from Theorem 1 [1].

Theorem 1 [1] *The set of all $K(s)$ for which the feedback system in* Fig. 1 *is internally stable equals*

$$K(s) = \left(X_r - Q_r\widetilde{N}\right)^{-1}\left(Y_r + Q_r\widetilde{M}\right) \tag{4}$$

$$K(s) = (Y_l + MQ_l)(X_l - NQ_l)^{-1} \tag{5}$$

Remark 1 If $G(s)$ is stable transfer matrix so one can clearly choose $M = \widetilde{M} = X_l = X_r = I$ and $N = \widetilde{N} = G$ and $Y_l = Y_r = 0$ so the set of all stabilizing controller are

$$K(s) = (I - Q_rG)^{-1}Q_r \tag{6}$$

$$K(s) = Q_l(I - GQ_l)^{-1} \tag{7}$$

Now, finding the right/left coprime factorization and their Bezout counterpart is the main challenge and reference [2] shows the procedure to derive them. The following algorithm is used to derive right/left coprime factorization and their Bezout counterpart [1, 2].

Step 1: Get a detectable and stabilizable $(A, B, C,$ and $D)$ of $G(s)$.

Step 2: Compute matrices F and L so that $[A + BF]$ and $[A + LC]$ are stable.

Step 3: Derive right/left coprime factorization and their Bezout counterpart in Eqs. 8 and 9.

$$\begin{bmatrix} M & -Y_l \\ N & X_l \end{bmatrix} = \begin{bmatrix} A+BF & B & -L \\ F & I & 0 \\ C+DF & D & I \end{bmatrix} \tag{8}$$

$$\begin{bmatrix} X_r & Y_r \\ -\tilde{N} & \tilde{M} \end{bmatrix} = \begin{bmatrix} A+LC & -(B+LD) & L \\ F & I & 0 \\ C & -D & I \end{bmatrix} \tag{9}$$

Next section will provide the procedure to consider performance measures into closed-loop transfer matrix.

3 Performance in Terms of Closed-Loop Criteria

There are two main approaches to design controller analytically. First one is according to shaping the open-loop transfer matrix $L(s) = G(s)K(s)$. The performance measures in this situation are, gain margin (GM), phase margin (PM), steady-state error, open-loop bandwidth, etc. Second approach is according to the shaping of the closed-loop transfer matrix. To explain this procedure, consider the output of the system and output of the controller of Fig. 1 as:

$$y(s) = T(s)r(s) + S(s)G_d(s)d(s) - T(s)n(s) \tag{10}$$

$$u(s) = K(s)S(s)r(s) + K(s)S(s)G_d(s)d(s) - K(s)S(s)n(s) \tag{11}$$

where $S(s)$ is sensitivity transfer matrix, $T(s)$ is complement sensitivity transfer matrix, $r(s)$ is reference signal, $d(s)$ is disturbance, and, $n(s)$ is the measurement noise. $S(s)$ and $T(s)$ are closed-loop transfer matrix and complement sensitivity transfer matrix, respectively, and derived by:

$$T(s) = G(s)K(s)(I + G(s)K(s))^{-1} \tag{12}$$

$$S(s) = (I + G(s)K(s))^{-1} \tag{13}$$

For reference tracking and disturbance rejection according to Eq. 10, $T(s)$ must be I and so $S(s)$ must be zero. For noise attenuation according to Eq. 10, $T(s)$ must be zero and so $S(s)$ must be I. So one need to compromise between $T(s)$ and $S(s)$ to derive suitable performance. Fortunately, since reference and disturbances are in

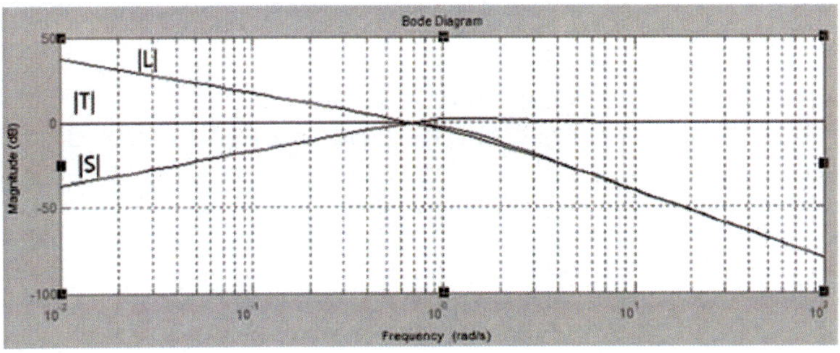

Fig. 2 Suitable shape for T, S, and L

low frequency and noise is in high frequency so suitable shapes for (s), $S(s)$, and also $L(s)$ can be shown in Fig. 2 (Note that $S(s) = (I + L(s))^{-1}$).

Some closed-loop criteria such as H_∞/H_2 try to consider an upper bound for some closed-loop transfer functions. For example, a mixed sensitivity problem considers two weighting matrix for $S(s)$ and $K(s)S(s)$ as $W_p(s)$ and $W_u(s)$, respectively, to take care of disturbance rejection and also input saturation. The H_∞ controller tries to minimize the following performance measure.

$$J = \min_{K(s)} \left\| \begin{matrix} W_p(s)S(s) \\ W_u(s)K(s)S(s) \end{matrix} \right\|_\infty \tag{14}$$

The main drawback of this kind of design is that the designer has not direct intuition into the details of design and just have an overview through chooses weight. This paper considers a procedure that the designer can have a better feel and more intuition in the design procedure. The next two secessions describe the procedure proposed by this paper.

4 Controller Design by Coprime Factorization and Limitation on Closed-Loop Transfer Matrices

To design a controller through right/left coprime factorization one must apply the performance measures on closed-loop transfer matrix. In this regard, one must know the limitation that must be considered on a closed-loop transfer matrix.

Lemma 1 *Closed-loop transfer function between the reference signal and output must have the same RHP zeros as G(s).*

Proof The transfer function between the reference signal r and the output signal y of the system in Fig. 1 is derived easily from Eq. 12 by putting controller of Eq. 4 in it:

$$T(s) = N(s)\big(Y_r(s) + Q_r(s)\widetilde{M}(s)\big) \tag{15}$$

So if there were some RHP zeros on $G(s)$, $N(S)$ losses rank at that point and so $T(s)$ also must lose rank at that point. (Note that since the term in bracket is stable so it cannot eliminate the RHP zeros of (S)).

Lemma 2 *Closed-loop transfer function between the reference signal and output must be one at the RHP poles of G(s).*

Proof Suppose $G(s)$, has an RHP pole at p so $\widetilde{M}(s)$ is zero at p and so $T(p) = N(p)\big(Y_r(p) + Q_r(p)\widetilde{M}(p)\big) = N(p)Y_r(p) = 1$ according to Eq. 3.

Lemma 3 *Poles of the closed-loop transfer function between the reference signal and output are among the eigenvalues of* $[A + BF]$, $[A + LC]$, *and poles of* $Q_r(s)$ *or* $Q_l(s)$.

Proof Proof of theorem is clear by Eq. 15.

So to choose $T_{\text{desired}}(s)/S_{\text{desired}}(s)$ one must consider the following remarks.

Remark 2 Desired transfer matrix $T_{\text{desired}}(s)$ must be zero at RHP zeros of $G(s)$ or equivalently sensitivity matrix $S_{\text{desired}}(s)$ must be one at RHP zeros of $G(s)$. (Since $T(s) + S(s) = 1.$ in the SISO case).

Remark 3 Desired transfer matrix $T_{\text{desired}}(s)$ must be one at the RHP poles of $G(s)$ or equivalently sensitivity matrix $S_{\text{desired}}(s)$ must be zero at the RHP poles of $G(s)$. (Since $T(s) + S(s) = 1.$ in the SISO case).

Remark 4 There is also some limitation on the bandwidths of the closed-loop system. For example [1] shows that bandwidth must be less than half of RHP zeros of $G(s)$. And also the bandwidth of system must be twice of RHP pole of $G(s)$.

Next session an algorithm for the design of controller based on closed-loop measures are presented.

5 Algorithm of Controller Design According to Right/Left Coprime Factorization

To design a controller through coprime factorization one must apply the performance measures on closed-loop transfer matrix. Also, limitations on the closed-loop matrix mentioned in the previous session must be considered.

So the following algorithm suggested for controller design.

1. Try to convert your performance criteria on closed-loop matrix.
   In this part, designer need to map performance measures on the closed-loop transfer matrices.

2. Analyze the system under study and find its RHP zeros and RHP poles.
3. Choose $T_{\text{desired}}/S_{\text{desired}}$ according to step 1 and consider Remarks 2–4 in the previous section.
4. Try to choose a parametric stable Q_r or Q_l with suitable order and also note to Lemma 3 of the previous section. Choose the number of parameters according to flexibility and parsimony. (High parameters are suitable for more flexibility and smaller parameters consider parsimony)
5. By use of optimization try to find parameters of Q_r or Q_l such that following performance measures are minimized.

$$J = \min_{\text{Parameters of } Q_r \text{ or } Q_l} \|T_{\text{desired}} - T_{\text{actual}}\|_2 \tag{16}$$

   By this optimization, the designer can put all of the performance measures on the closed-loop transfer matrices.
6. Check the controller on the system if it is still not acceptable to try again from 3 or 4, otherwise, it is finished.

Next section considers some case studies to clarify the mentioned procedure. The main contribution of this paper is to use the optimization part (step 5 of the algorithm) to map all of our performance requests on the closed-loop transfer function. However, here examples are very easy so the controller was found directly through analytical formulation to show the applicability of procedure.

Benefits of using right/left coprime factorization method in designing the controller is that, the designer has more insight into the controller design and better insight into the closed-loop transfer function. On the other side, mapping requested performance on the closed-loop transfer matrices is not an easy task.

6 Case Studies

In this section, two examples are considered. In the first example, the system under study has no RHP pole and zero so there no limitation according to RHP poles and zeros. In the second example, the system has two RHP poles so there is some limitation according to the existence of RHP poles.

Example 1 Consider the system in Fig. 1 and suppose

$$G(s) = \frac{200}{10s+1}\frac{1}{(0.05s+1)^2}$$

$$G_d(s) = \frac{100}{10s+1}$$

Suppose we want to design a controller such that rise time (to reach 90% of the final value) should be less than 0.3 s and the overshoot should be less than 5%. The outputs go to zero as quickly as possible if a unit disturbance applied to the system.

Reference [1] solves the problem with mixed sensitivity H_∞ approach and the step response of the system (y_1) to reference and disturbance are shown in Figs. 3 and 4 respectively. Since the response to disturbance was very sluggish so designer the weighting matrix must change. After some try and error y_2 is the step response of the system to reference and disturbance as shown in Figs. 3 and 4, respectively.

Now, we want to design a controller in the procedure mentioned in this paper. Clearly, since the plant is stable so closed-loop transfer function $T(s)$ can be found from Eq. 15 as:

$$T(s) = G(s)Q_r(s)$$

Fig. 3 Reference step response [1]

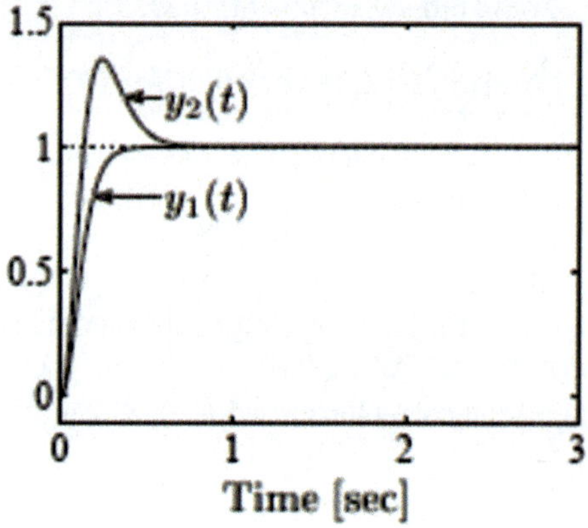

Fig. 4 Disturbance step response [1]

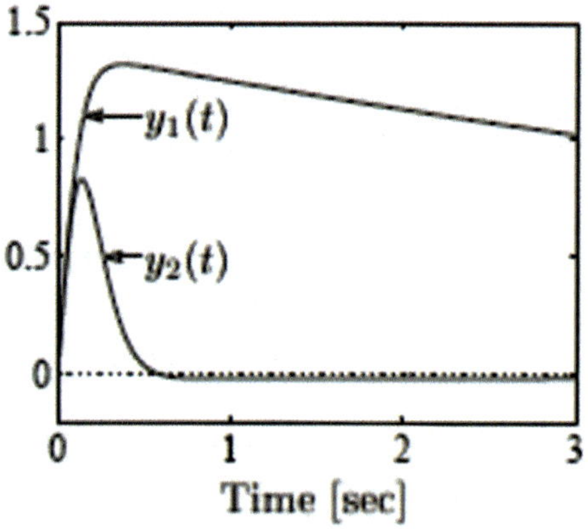

So, since, closed-loop transfer function $T(s)$ must be proper and also poles of $Q_r(s)$ will appear in poles of $T(s)$ so $Q_r(s)$ choose as:

$$Q_r(s) = \frac{as^3 + bs^2 + cs + d}{(s^2 + 14s + 100)(0.01s + 1)}$$

Now by choosing

$$T_{\text{desired}} = \frac{100}{(s^2 + 14s + 100)(0.01s + 1)}$$

Minimization of Eq. 16 leads to $a = 1.25$, $b = 50.12$, $c = 505$, $d = 50$.

Step response of the system to reference and disturbance are shown in Figs. 5 and 6 respectively. Since the response to disturbance was very sluggish so some changes must be considered to the system. To have more insight into the system problem, bode plot of $S(s)$ and $G_d(s)$ are considered in Figs. 7 and 8 correspondingly.

Figures 7 and 8 show that although $S(s)$ it is small at the frequency less than 10 rad/s but, of $G_d(s)$ is not small so we choose another $T_{\text{desired}}(S_{\text{desired}})$ to find a better situation.

$$T_{\text{desired}} = \frac{16s + 100}{(s^2 + 14s + 100)(0.01s + 1)^2}$$

Since $T(s) = G(s)Q_r(s)$ so $Q_r(s)$ must have at least four poles. So by Lemma 3, try

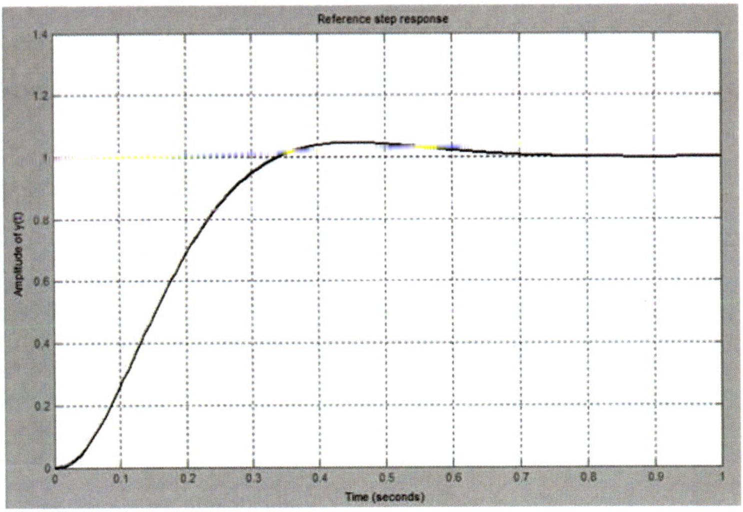

Fig. 5 Reference step response of Example 1 by proposed method

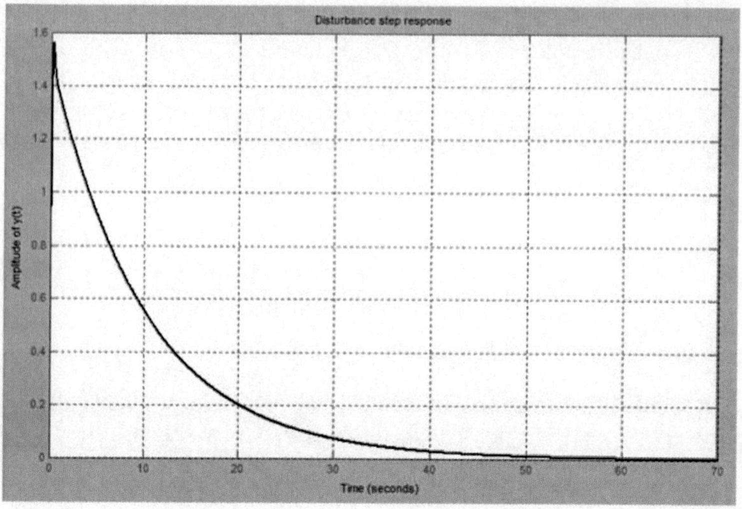

Fig. 6 Disturbance step response of Example 1 by proposed method

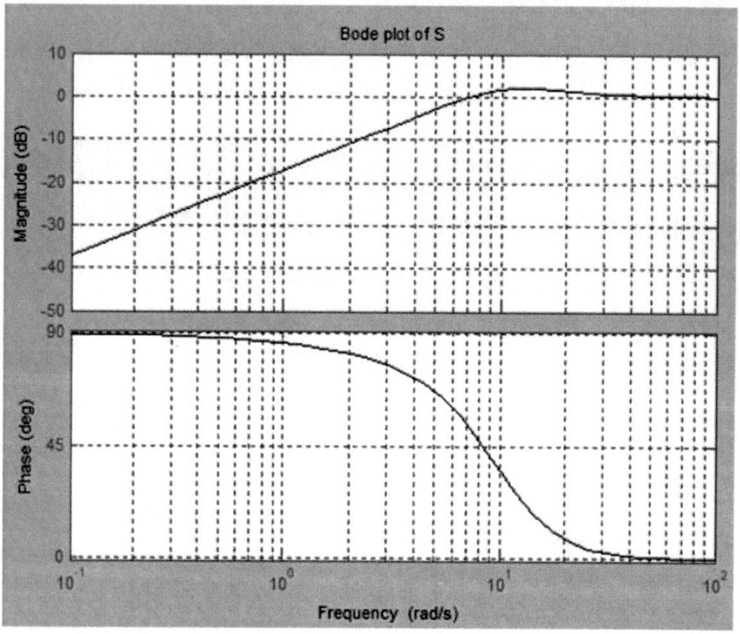

Fig. 7 Bode plot of $S(s)$

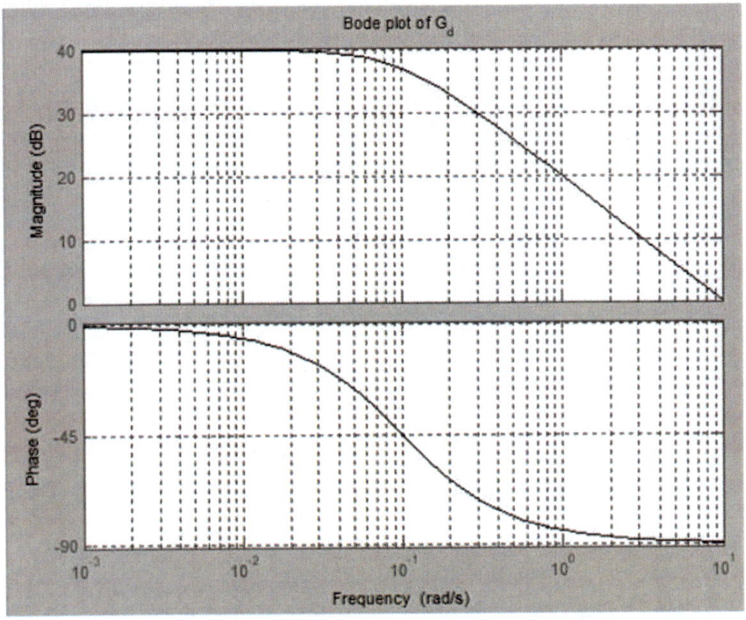

Fig. 8 Bode plot of $G_d(s)$

$$Q_r(s) = \frac{as^4 + bs^3 + cs^2 + ds + e}{\left(s^2 + 14s + 100\right)(0.01s + 1)^2}$$

Minimization of Eq. 16 leads to $a = 20, b = 927, c = 13090, d = 51300$, and $e = 5000$

Step response of the system to reference and disturbance are shown in Figs. 9 and 10 respectively.

Now, the responses of the system are acceptable. Next example is an unstable plant and derived from [2].

Example 2 Consider the system in Fig. 1 and suppose

$$G(s) = \frac{1}{(s - 1)(s - 2)}$$
$$G_d(s) = 1$$

The problem is to find a controller that the system tracks step with zero steady-state error and the final value of y equals zero when d is a sinusoid of 10 rad/s and $r = 0$.

Since this system is unstable so, first of all, a stabilizable and detectable state space of system derived as:

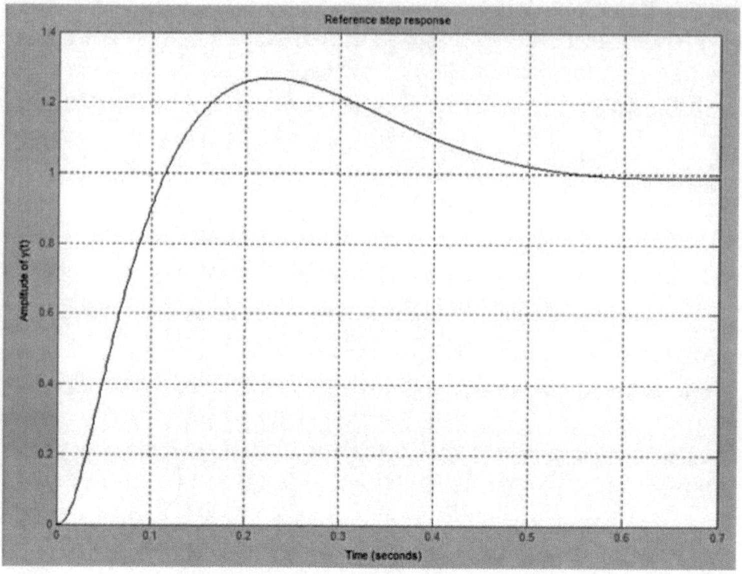

Fig. 9 Reference step response of Example 1 by the proposed method

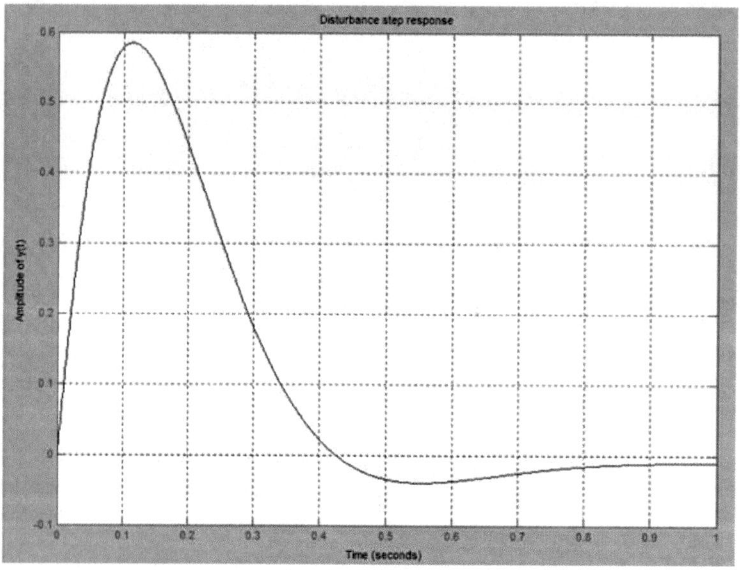

Fig. 10 Disturbance step response of Example 1 by the proposed method

$$\dot{x} = \begin{bmatrix} 0 & 1 \\ -2 & 3 \end{bmatrix} x + \begin{bmatrix} 0 \\ 1 \end{bmatrix} u$$
$$y = \begin{bmatrix} 1 & 0 \end{bmatrix} x$$

By choosing $F = \begin{bmatrix} 1 & -5 \end{bmatrix}$ and $L = \begin{bmatrix} -7 & -23 \end{bmatrix}^T$ then $A + BF$ and $A + LC$ are stable so right/left coprime factorization can be derived from Eqs. 8 and 9. Now since $G(s)$ is unstable there is some limitation on $T(s)$. Let $T_{\text{desired}} = 1 - S_{\text{desired}}$.

And let to consider the limitation on $T(s)$ and $S(s)$ let

$$S_{\text{desired}} = 1 - 2e - 6\frac{s(s-1)(s-2)\left(s^2+100\right)}{\left(s^2+0.7s+0.25\right)(0.02s+1)(0.01s+1)^2}$$

Choose a parametrized $Q_r(s)$ as

$$Q_r = \frac{as^5 + bs^4 + cs^3 + ds^2 + es + f}{\left(s^2+0.7s+0.25\right)(0.02s+1)(0.01s+1)^2}$$

After minimizing Eq. 16 the step response of the system to reference and disturbance are shown in Figs. 11 and 12 respectively.

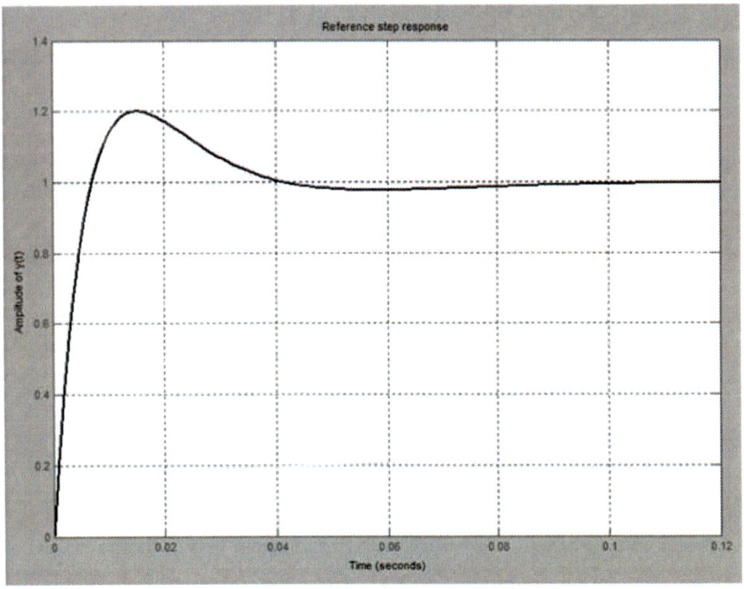

Fig. 11 Reference step response of Example 2 by the proposed method

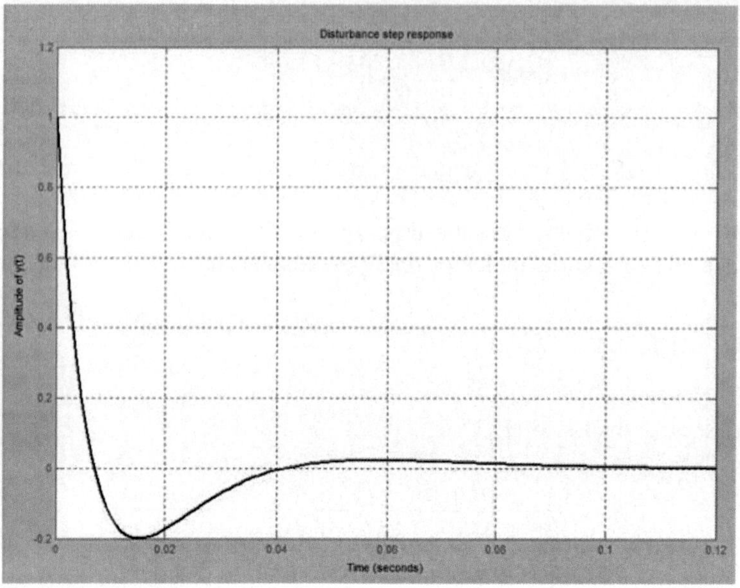

Fig. 12 Disturbance step response of Example 2 by the proposed method

7 Conclusion

In the convention method based on mixed sensitivity design (H_∞/H_2) designer must consider the performance request on the weighting function as a whole and the designer has no exact insight into the detail of the design. Also, a mixed sensitivity procedure is based on trial and error. A new method of the designing controller as described in this paper that the designer can consider his request in a more efficient way and in more detail. According to this fashion designer derive a stabilizing controller through the set of all stabilizing controller through an optimization. By choosing a high order stable Q_r or Q_l and with the use of optimization procedure one can assign a more suitable controller. So, in this method designer has more insight into designing the controller.

References

1. Skogestad S, Postlethwaite I (2007) Multivariable feedback control: analysis and design. Wiley, New York
2. Doyle J, Francis B, Tannenbaum A (1990) Feedback control theory. Macmillan Publishing Co.
3. Hasanzadeh A, Edrington CS, Mokhtari H (2011) A novel LQR based optimal tuning method for IMP-based linear controllers of power electronics/power systems. In: 2011 50th IEEE conference on decision and control and European control conference, pp 7711–7716

4. Hartley EN, Maciejowski JM (2013) Predictive control for spacecraft rendezvous in an elliptical orbit using an FPGA. In: 2013 European control conference (ECC), pp 1359–1364
5. Brunner FD, Müller MA, Allgöwer F (2018) Enhancing output-feedback MPC with set-valued moving horizon estimation. IEEE Trans Autom Control 63(9):2976–2986
6. Teng L, Wang Y, Cai W, Li H (2018) Robust fuzzy model predictive control of discrete-time Takagi-Sugeno systems with nonlinear local models. IEEE Trans Fuzzy Syst 26(5):2915–2925
7. Karimpour M, Hitihamillage L, Elkhoury N, Moridpour S, Hesami R (2018) Fuzzy approach in rail track degradation prediction. J Adv Transp
8. Karimpour M, Karimpour A, Kompany K, Karimpour A (2017) Online traffic prediction using time series: a case study. In: Integral methods in science and engineering, vol 2, Birkhäuser, Cham, pp 147–156
9. Kerrigan EC, Constantinides GA, Suardi A, Picciau A, Khusainov B (2015) Computer architectures to close the loop in real-time optimization. In: 2015 54th IEEE conference on decision and control (CDC), pp 4597–4611
10. Lin X, Liang K, Li H, Jiao Y, Nie J (2018) Robust finite-time H-infinity control with transients for dynamic positioning ship subject to input delay. Math Problems Eng
11. Ashok Kumar M, Kanthalakshmi S (2018) H_∞ tracking control for an inverted pendulum. J Vib Control 24(16):3515–3524
12. Glaria JJ, Goodwin GC (1994) A parameterization for the class of all stabilizing controllers for linear minimum phase plants. IEEE Trans Autom Control 39(2):433–434
13. Ebrahim Zadeh F, Karimpour A (2011) Designing stabilizing control with genetic algorithm, based on the transfer function. In: 3rd Iranian conference on electrical and electronics engineering (ICEEE), Gonabad, Iran pp 1–8

Fault Detection and Classification in Microgrid Using Wavelet Transform and Artificial Neural Network

Priya Singh, Nitin Singh, and Niraj Kumar Choudhary

Abstract In the proposed work, Wavelet Transform analysis and wavelet entropy methods have been used to classify various types of fault in a nine bus microgrid system. Both methods are compared and analyzed. The simulation result shows that the proposed method successfully identifies the fault type and phase involved in the fault. The proposed algorithm is validated for different locations and fault types on nine bus microgrid system. In addition to the above, wavelet analysis and wavelet coefficients are also used with the Artificial Neural Network (ANN) for detecting and classifying the faults. The different fault cases have different fault resistances and inception angles. The fault detection process is done by the summation of sixth level detail coefficients of current obtained using Discrete Wavelet Transform (DWT) based Multiresolution Analysis (MRA) technique for all the three phases while, for the classification of fault type, wavelet entropy calculations for each phase currents are acquired.

Keywords Fault classification · Wavelet analysis · Wavelet entropy (WE) · Artificial Neural Network (ANN) · Multiresolution Analysis (MRA)

1 Introduction

A fault is any abnormal condition related to the current or voltage in a power system. It can occur due to several reasons such as natural casualty, human error, tree fall, etc. The fault may be of symmetrical or unsymmetrical characteristics. Although an unsymmetrical fault is most likely to occur in a system, it is the symmetrical fault

P. Singh (✉) · N. Singh · N. K. Choudhary
Department of Electrical Engineering, MNNIT Allahabad, Prayagraj, India
e-mail: eengpriya@gmail.com

N. Singh
e-mail: nitin@mnnit.ac.in

N. K. Choudhary
e-mail: niraj@mnnit.ac.in

D. Harvey et al. (eds.), *Advances in VLSI, Communication, and Signal Processing*, Lecture Notes in Electrical Engineering 683,
https://doi.org/10.1007/978-981-15-6840-4_2

which is used to calibrate the protection devices and equipment due to its severe consequences [1, 2]. It is important to analyze extensively power system under fault conditions in order to provide system protection. It is fundamentally due to the imperative need to minimize the downtime of a line due to a fault. If the recognition of the fault took a longer time then it leads to the consequential line failures, power cuts, economic losses, wastage of maintenance worker's time and energy.

In the fault diagnosis, fault detection is a major task. In [1–3], the classification of fault techniques in prominent and hybrid techniques along with the simulation tool used, complexity levels are discussed. A wide area issues of the fault of various components of a microgrid and diagnosis methods in terms of model-based and data-driven approaches [4]. Several methods have been used in the past such as artificial neural networks [5–7], wavelet transforms [8–15], fast Fourier transform [9], decision tree [15] wavelet entropy [16–19] and hybrid method wavelet and ANN combined, etc. Wavelet multiresolution technique is considered to be one of the best methods for signal analysis in case of fault generated signal [8–10, 13]. In this Wavelet Entropy method, we decompose the signal using a predefined prototype known as mother wavelet. The wavelet entropy approach as discussed in [16–19], has been used for feature pick-up in case of fault detection. Some researchers have also used Support Vector Machine (SVM) for the detection of fault [3, 14]. Since the protection system requires a very fast algorithm to detect and locate the fault and hence it is very important to choose a method that classifies the fault as soon as possible. The purpose of the wavelet method used is to identify the type of fault by extracting datasets from the buses at nearby points of the fault location.

2 Wavelet Transform

Wavelet is the small wave whose average value is zero. Wavelet Transform (WT) is suitable for analyzing nonperiodic signal since it can locate the different components of the frequency spectrum of the signal over time, which is particularly useful for identifying transient signal components that can be used as a basis for the development of high-speed protection algorithm The criteria for wavelet is that it must be oscillatory in nature, should have zero average value and must decay to zero quickly and if the wavelet transform is discrete then it must be orthogonal to each other. We perform the translation (shifting in the time-domain) and dilation.

The DWT provides a sparse representation of transients in the fault signal. This is equivalent to the series of tree-structured discrete filter banks, where the signal x is successively downsampled when passed through it. The samples of x are first passed through a low pass filter where they are convolved with impulse response 'g' resulting $y[n]$ as shown in Eq. (1),

$$y[n] = x[n] * g[n] = \sum_{k=-\infty}^{\infty} x[k]g[n-k] \tag{1}$$

Simultaneously, the signal is also decomposed using a high pass filter 'h'. The low pass filter gives the approximate coefficient, with better resolution in time and high pass filter gives the detail coefficients with better resolution in frequency. These filters are also known as quadrature mirror filters. Half of the signals can be discarded according to Nyquist's rule. The decomposition occurs recursively at different levels for low pass filter as y_{low} and high pass filter as y_{high} as shown in Eqs. (2) and (3).

$$y_{\text{low}}[n] = \sum_{k=-\infty}^{\infty} x[k]g[2n - k] \tag{2}$$

$$y_{\text{high}}[n] = \sum_{k=-\infty}^{\infty} x[k]h[2n - k] \tag{3}$$

where the decimated DWT discretizes the scale parameter in integer power of 2, 2^j, $j = 1, 2, 3, \ldots, n$, and the translational parameter is always proportional to the scale.

3 Wavelet Entropy and ANN

3.1 Wavelet Entropy

If Wavelet Entropy (WE) takes a signal $x[n]$ which is quickly transformed at the moment k and at the scale j, then the high frequency is $D_j(k)$ and low-frequency components is $A_j(k)$ is obtained. The DWT provides a perfect reconstruction of the signal upon inversion. The coefficients $D_j(k)$ and $A_j(k)$ can be used to synthesize the reproduction signal, as given in Eqs. (4) and (5) [3].

$$D_j(k) = \left[2^{-(j+1)} f_s, 2^{-j} f_s\right] \tag{4}$$

$$A_j(k) = \left[0, 2^{-(j+1)} f_s\right] \tag{5}$$

for $j = 1, 2, \ldots, n$

where f_s is sampling frequency. Thus, original signal $x[n]$ represented as sum of all coefficients will be represented as follows

$$x[n] = D_1(n) + A_1(n) \tag{6}$$

$$= D_1(n) + D_2(n) + A_2(n) \tag{7}$$

$$= \sum_{j=1}^{j} D_j(n) + A_j(n) \tag{8}$$

In this paper, the type of entropy used is non-normalized Shannon entropy as shown in Eq. (9) using Eq. (10) to obtain the entropy of coefficients. The definition of non-normalized Shannon entropy is given by Eq. (9).

$$E_j = -\sum_k E_{j,k} \log E_{j,k} \tag{9}$$

where $E_{j,k}$ is known as wavelet energy spectrum and $D_{j,k}$ is the coefficients of E in an orthonormal basis. It is defined at scale j and instant k as shown in Eq. (10).

$$E_{j,k} = \left\| D_{j,k} \right\|^2 \tag{10}$$

3.2 *Artificial Neural Network*

An artificial neural network (ANN) is a system of interconnected simple processing units known as neurons. It mimics the model of biological neurons in the human brain. The human brain has excessively billions of neurons connected in parallel which forms the basic processing unit.

In ANN, neurons are interconnected to several other neurons with each link having some weight. These are known as synaptic weights. Thus, input fed to the network is modified by the weights and biases added to it. Bias acts as a threshold value. The weighted output thus obtained is given to nonlinear filter known as activation function or 'Transfer function' or 'Squash function' which can be sigmoid, tanh, etc. The artificial neural network investigating and modeling mechanisms specifies the actual structure and functioning of neural networks, this links the behavioral responses of the model to environmental stimuli, depending on the structure of neurons, strength of interlink connection, and processing type performed at each element or nodes. In classification tasks, the output being predicted is inarticulate variable, whereas in regression problems the output is a quantitative variable (Figs. 1 and 2).

4 Microgrid System (9 Bus)

The Simulink test model is a nine bus microgrid system with distribution feeder impedance parameters $R = 0.01273\ \Omega$, $X = 0.3519\ \Omega$ per unit length in km. These feeders are fed by the utility (X/R ratio = 6, short circuit MVA = 500 MVA and f = 60 Hz) through a transformer of rating 20 MVA, 115 kV/12.47 kV. Four DGs of

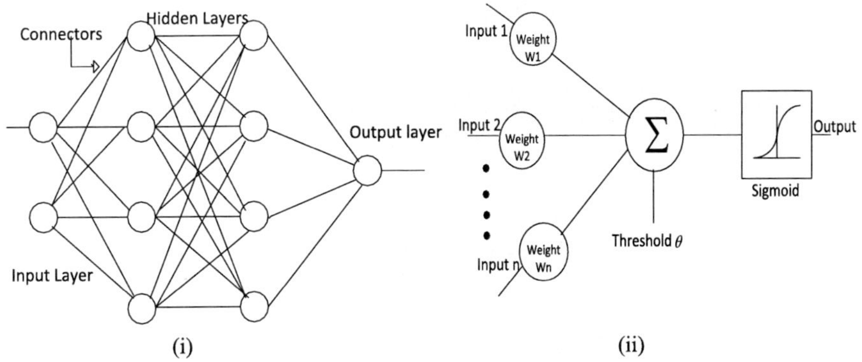

Fig. 1 (i), (ii) Simplified ANN models

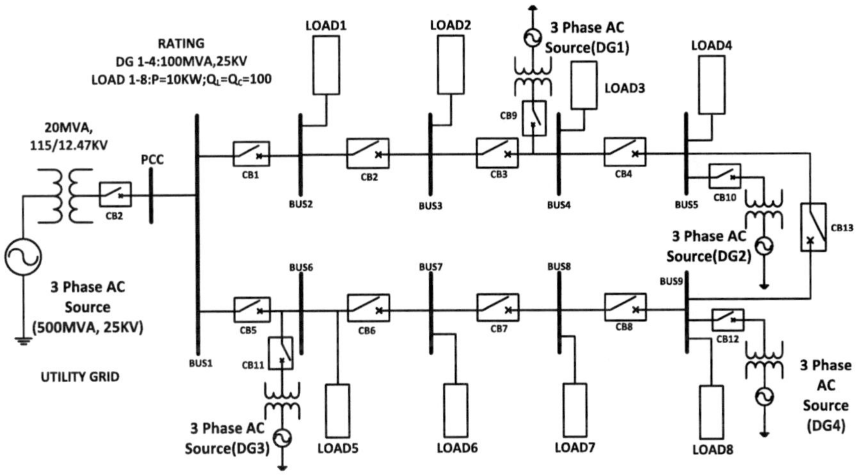

Fig. 2 9 Bus microgrid system connected to utility grid

the same rating of 100 MVA, 25 kV, $X/R = 7$ are connected at buses 4, 5, 6, and 9 through 12.47 kV/480 V transformer further the load of 10 kW is connected at each bus from 2 to 9. The system is simulated with a sampling frequency (f_s) of 12.77 kHz for which the current signals are recorded. The 6sixth level decomposition is suitable since it lies in the range of 99.80–199.60 Hz in which the most dominant harmonics lie. These time-domain generated signals are applied for the WT and ANN analysis. The simulation shows the significant effect of variable inception angles and fault resistance with variable fault locations on the current signals and therefore on the wavelet transform and ANN output. The inception angle varies from 0° to 180° for each type of fault.

5 Proposed Algorithm for Fault Classification

5.1 Selection of Mother Wavelet and Wavelet Coefficient Level

Wavelet Transform based on MRA is used for selecting the mother wavelet and wavelet level for transient analysis purposes. MRA gives flexibility in selecting the range of frequency which is of concern by removing the redundancy problem. Using MRA the approximate and detailed coefficients are obtained for level 1, A1 and D1. This process is repeated until the desired level of the detailed coefficient is regenerated for the approximation coefficient for different wavelets. In this paper, base wavelet db8 with 6th level decomposition is used here since it is compact and better localized with a time suitable to both short and fast transient analysis. It can be observed from Fig. 7 that the sixth level detailed coefficient of the current sample provides the most distinct divergence during the fault condition.

5.2 Proposed Algorithm

Let S_a, S_b, S_c be the summation of sixth level detailed coefficients for current in phase *a*, *b*, *c*, respectively, which is non-zero in faulty case if involved. By comparing the summation of the wavelet coefficient of different phases, fault detection is performed as shown in Fig. 3. The fault is classified as L-L-L fault if $S_a + S_b + S_c = 0$. In this, the magnitude of all the summation values S_a, S_b, and S_c are comparable to each other. The fault is classified as L-L fault if $S_a + S_b + S_c = 0$ also anyone summations S_a, S_b,

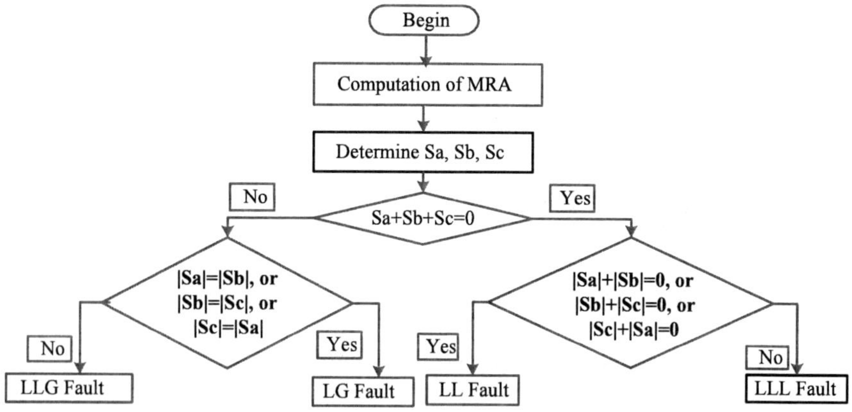

Fig. 3 Flow chart for fault classification

or S_c is equal to zero, i.e., the magnitude of one of the summation is comparatively very small with respect to other two equal in magnitude to each other.

For an unambiguous and accurate result, we applied a slight modification of the above method known as the wavelet entropy method. The transient signals of the three-phase currents are produced using the simulation model built with the power block set of the SIMULINK. A discrete wavelet transformation is performed using six levels of symmetric wavelet for the three-phase current signals (i_a, i_b, and i_c) and the ground current i_g, where $i_g = (i_a + i_b + i_c)$. Let $S_e(i_a)$, $S_e(i_b)$, $S_e(i_c)$, $S_e(i_g)$ be the entropies obtained for current in phases *a*, *b*, *c*, and ground, respectively. Shannon entropy coefficients measure the uncertainty or randomness of the signal. A fault is an unpredictable event that can occur in the system due to multiple factors in the system resulting in transients or disorders in the signal. These randomness or uncertainty are nonlinear features that can measure the divergence in the energy distribution of the signal describing the phases involved in the fault. Since entropy is an additive cost function such that $S_e i_p(0) = 0$ (for healthy phase) and $S_e i_p \neq 0$ (for faulty phase) (where i_p is the current in the phase p). The entropy of each coefficient of the four currents is then calculated and using the flow chart shown in Fig. 4 fault can be identified and detected.

Using the flow chart as shown in Fig. 5, phases in which fault is occurring can be identified. The phases involved in the fault has higher and comparatively similar uncertainty value with respect to the healthy phase, which can be observed from Tables 3 and 4. Both the flow charts mentioned in Figs. 4 and 5 are based on the wavelet entropy method for identifying and detecting the fault and the phase involved during the fault.

Further, ANN is used as a classifier with input matrix of size 2100 × 4 to build the Back Propagation Neural Network (BPNN) for training and testing with the wavelet entropy coefficients as input features $S_e(i_a)$, $S_e(i_b)$, $S_e(i_c)$, and $S_e(i_g)$. The different fault cases of (7 × 10 × 3) × 10 samples have been obtained for seven different fault locations, ten variable fault resistance values, and three different inception angles. An output matrix of the same size is prepared. The model is trained by Levenberg–Marquardt technique backpropagation with 75% of training data, 10% for validation, and 15% for testing. The further output obtained is compared with the threshold values and has been classified as either 0 or 1 depending upon their values. While extracting feature 6th level detailed coefficient was used for the current signal. Thus, the two-step process ensures a better classification of the fault type. Various performance characteristics as well as the output values, obtained from ANN has been shown in the result section.

6 Results and Discussion

We have applied the WT method and wavelet entropy on a nine bus overhead distribution line in microgrid system. The decomposition based on MRA with a db8 mother wavelet is analyzed. The different decomposition level is shown in Table 1. The sixth

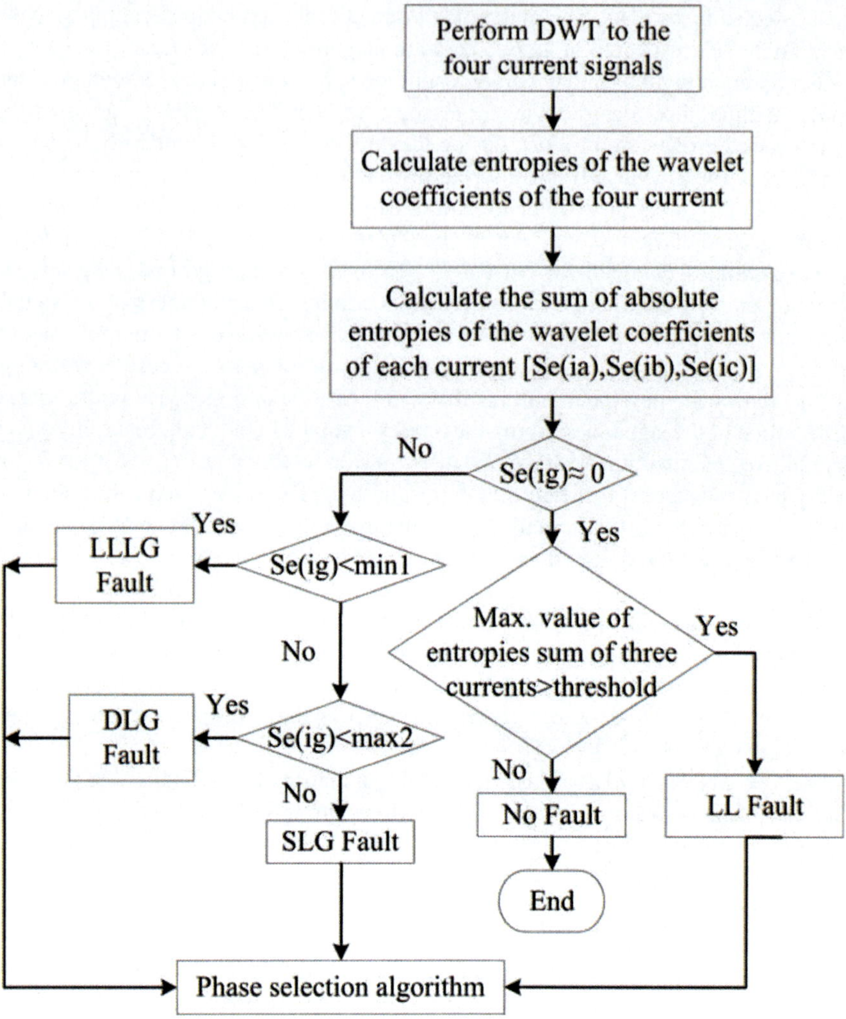

Fig. 4 Flow chart for identification and detection of a fault

level decomposition is observed as most sensitive detailed coefficient to second and third harmonics as $d_6 = 99.804–199.61$ Hz [8, 9].

As a sample, the waveform of a single line to ground fault (SLG) in phase A is shown in Fig. 6. The wavelet coefficients and detailed coefficients of phase A current are shown in Fig. 7.

In analyzing the fault by WT and modified wavelet entropy method, the wavelet coefficients of a current of different phases during the fault are compared with the threshold values obtained from the normal conditions [8, 10]. The transients of the faulty phases of current are very high hence the coefficients of the faulty phases

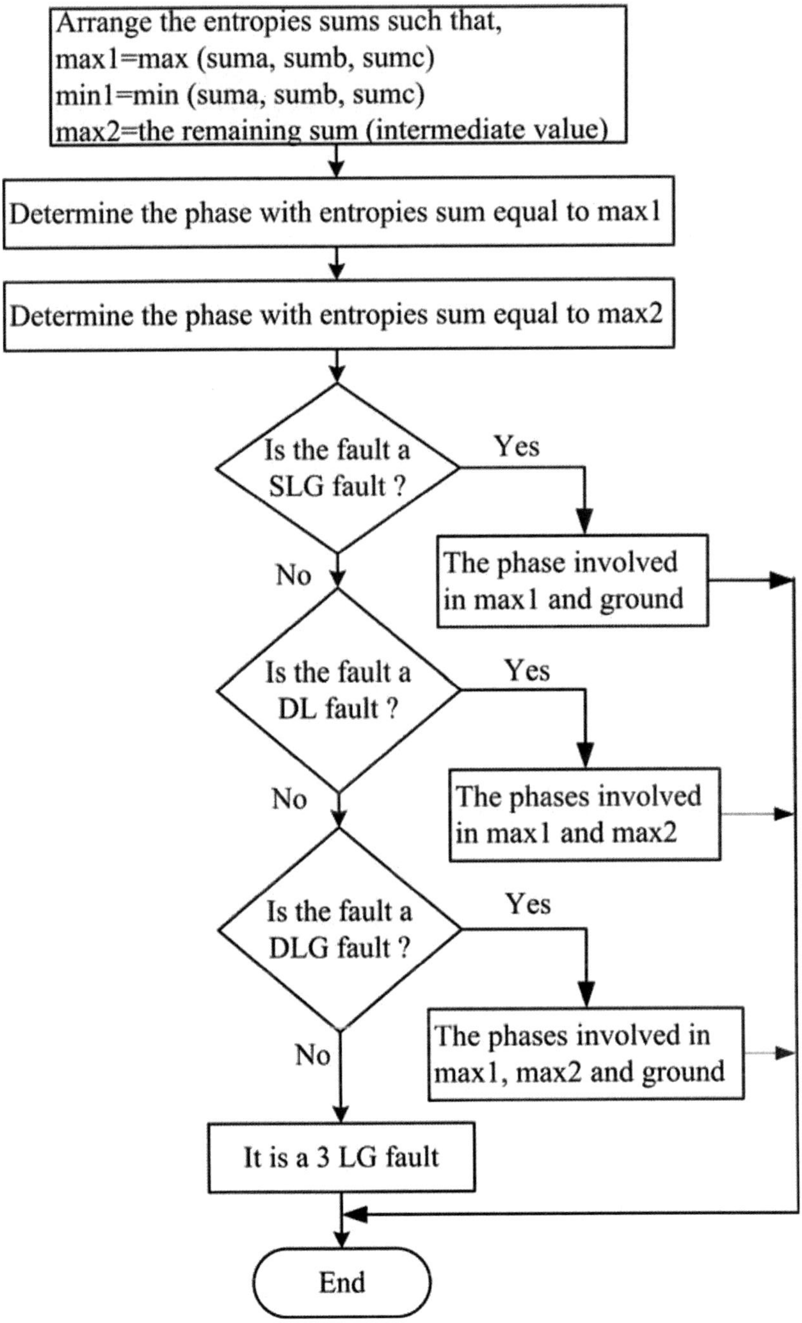

Fig. 5 Flow chart for classification of phase involved in fault

Table 1 Different decomposition levels of high pass filter as detailed coefficients

Decomposition levels	Frequency bands (Hz)
d_1	6387.5–3193.75
d_2	3193.75–1596.87
d_3	1596.87–798.43
d_4	798.43–399.21
d_5	399.21–199.60
$\mathbf{d_6}$	**199.6–99.80**
d_7	99.80–49.90
d_8	49.90–24.95

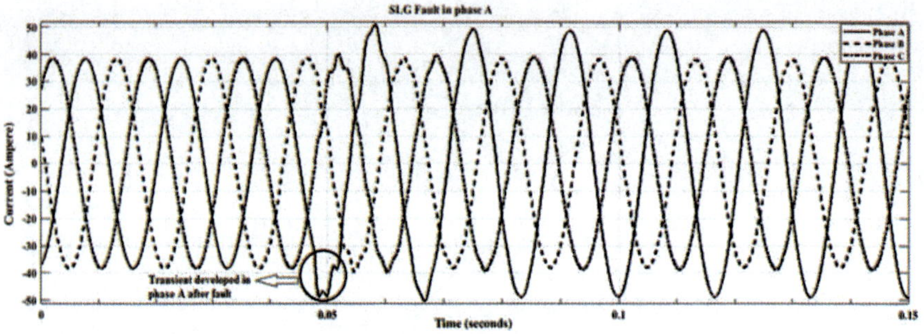

Fig. 6 Waveform of currents of SLG fault in phase A

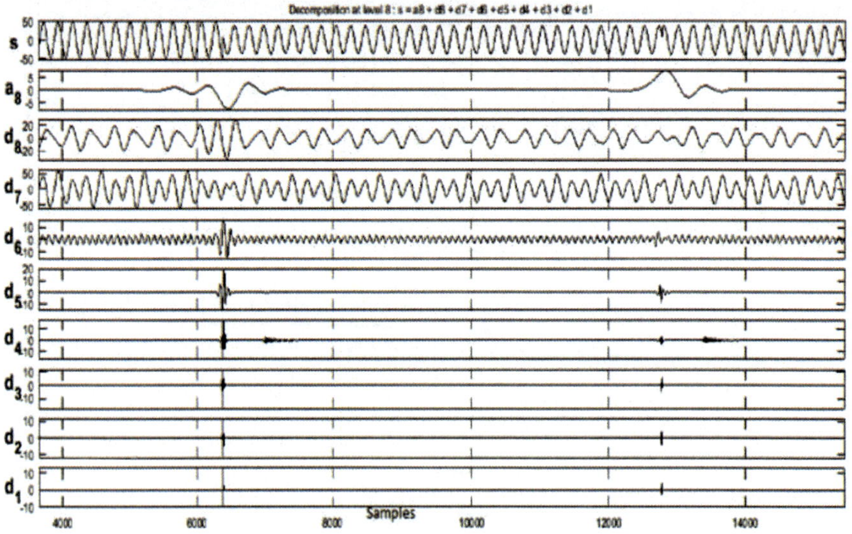

Fig. 7 The waveform of the approximate and detailed coefficients of phase A in LG fault

are comparable in values with respect to each other as shown in Tables 2, 3 and 4 predicting the fault occurred. These values along with the detection trigger the fault classification module and identification of a fault. Through the wavelet entropy method, fault detection and classification was easier and more efficient. Further, we have applied ANN along with wavelet to detect a fault in the given model which can predict the fault faster sweeping the online data. For this, as shown in Table 5 the output matrix is constructed based on the supervised learning pattern to feed in neural network and sum of detailed coefficients of all the three phases and ground to the system as the input matrix.

AI Technique (Back Propagation Neural Network) in combination with wavelet transform provides a much better classification for multiclass classification than any of the above proposed alone with less computation time. The phase involved in the

Table 2 Wavelet Transform result for a nine bus system (fault at bus 1) in grid-connected mode

Fault type	S_a	S_b	S_c	Sum
LG-A	**37.5333**	8.2129	1.5319	47.2781
LG-B	9.7448	**5.9850**	1.5319	2.2279
LG-C	9.7448	8.2129	**43.5183**	45.0501
LL-AB	**15.0180**	**16.5506**	1.5322	3.5073e−04
LL-BC	9.7452	**29.6323**	**19.8867**	4.0659e−04
LL-AC	**36.4374**	8.2130	**44.6504**	5.5863e−04
LLG-AB	**37.5348**	**5.9661**	1.5319	45.0328
LLG-BC	9.7448	**6.0023**	**43.5167**	47.2592
LLG-AC	**37.5506**	8.2129	**43.5371**	2.2264
LLLG-ABC	**37.5521**	**5.9835**	**43.5356**	3.4719e−08
LLL-ABC	**37.5521**	**5.9835**	**43.5356**	1.0373e−09

Table 3 Wavelet entropy result for a nine bus system (fault at bus 1) in grid-connected mode

Fault type	S_{ei_a}	S_{ei_b}	S_{ei_c}	S_{ei_g}
LG-A	**4.3708e+09**	2.2005e+08	4.0659e−04	**4.0659e+09**
LG-B	2.2003e+08	**5.0085e+09**	2.2013e+08	**4.2332e+09**
LG-C	2.2003e+08	2.2005e+08	**4.4666e+09**	**3.7244e+09**
LL-AB	**4.4563e+09**	**2.7037e+09**	2.2014e+08	0.1204
LL-BC	2.2004e+08	**4.5373e+09**	**2.7639e+09**	0.1236
LL-AC	**2.3086e+09**	2.2006e+08	**4.0442e+09**	0.1251
LLG-AB	**4.3815e+09**	**5.0109e+09**	2.2013e+08	**3.7159e+09**
LLG-BC	2.2003e+08	**5.0200e+09**	**3.6078e+09**	**3.6078e+09**
LLG-AC	**4.3677e+09**	2.2005e+08	**4.4716e+09**	**4.2192e+09**
LLLG-ABC	**4.3783e+09**	**5.0224e+09**	**4.4748e+09**	**1.5528e-09**
LLL-ABC	**4.3783e+09**	**5.0224e+09**	**4.4748e+09**	0

Table 4 Wavelet entropy result for a nine bus system (fault at bus 1) in Islanded mode

Fault Type	S_{ei_a}	S_{ei_b}	S_{ei_c}	S_{ei_g}
LG-A	**3.4339e+08**	3.9084e+07	3.9087e+07	**6.7319e+08**
LG-B	3.9101e+07	**3.4603e+08**	3.9087e+07	**6.7758e+08**
LG-C	3.9101e+07	3.9084e+07	**3.4539e+08**	**6.7644e+08**
LL-AB	**2.5083e+08**	**2.7785e+08**	3.9085e+07	0.0374
LL-BC	3.9100e+07	**2.5168e+08**	**2.7969e+08**	0.0374
LL-AC	**2.7799e+08**	3.9082e+07	**2.4966e+08**	0.0375
LLG-AB	**3.4384e+08**	**3.4638e+08**	3.9087e+07	**6.7528e+08**
LLG-BC	3.9101e+07	**3.4648e+08**	**3.4575e+08**	**6.7204e+08**
LLG-AC	**3.4374e+08**	3.9084e+07	**3.4584e+08**	**6.7642e+08**
LLLG-ABC	**3.4419e+08**	**3.4684e+08**	**3.4620e+08**	0
LLL-ABC	**3.4419e+08**	**3.4684e+08**	**3.4620e+08**	0

Table 5 Output matrix after comparison with threshold

Fault Type	A	B	C	G
LG-A	1	0	0	1
LG-B	0	1	0	1
LG-C	0	0	1	1
LL-AB	1	1	0	0
LL-BC	0	1	1	0
LL-AC	1	0	1	0
LLG-AB	1	1	0	1
LLG-BC	0	1	1	1
LLG-AC	1	0	1	1
LLLG-ABC	1	1	1	0

fault is represented with the BPNN index as 1 while 0 for the healthy phase. The various output waveforms that have been observed for the wavelet in combination with the artificial neural network after applying on the system are described in Fig. 8 and the mean square error has been plotted in Fig. 9.

The ANN classifier with one hidden layer and four neurons was trained until the mean square error was close to 0. The computed MSE obtained is in the order of e-2 in training and testing with a validation error of 0.039691. It can be observed that the actual output slightly deviates from the accurate result. The MSE plot (shown in Fig. 9) shows that as the training progress mean square error reduces and improves with larger number of varying parameters involved in detecting and classifying the fault further the corresponding linear regression plot provides better result when close to 1 which is possible when the errors are less distributed from the center as

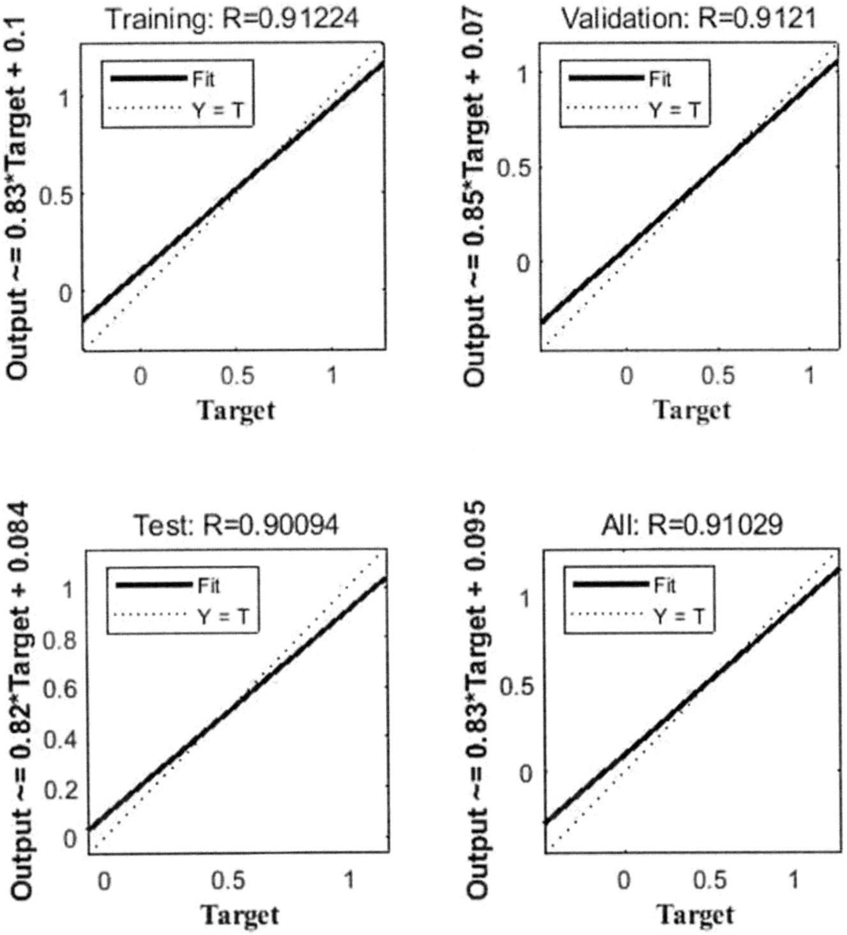

Fig. 8 Regression plot between the output versus the target of training, validating and testing

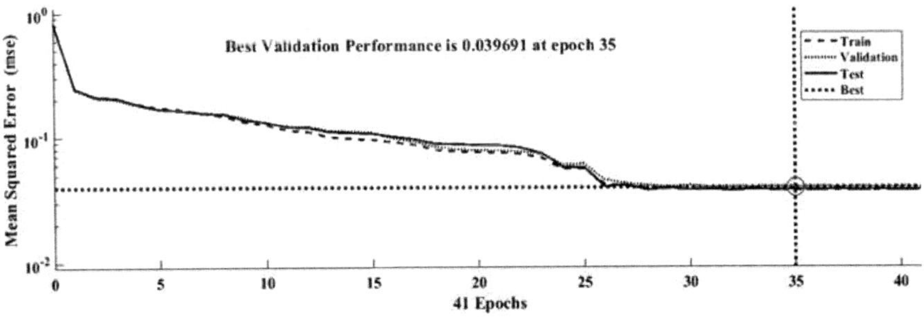

Fig. 9 Plot between the MSE and epochs

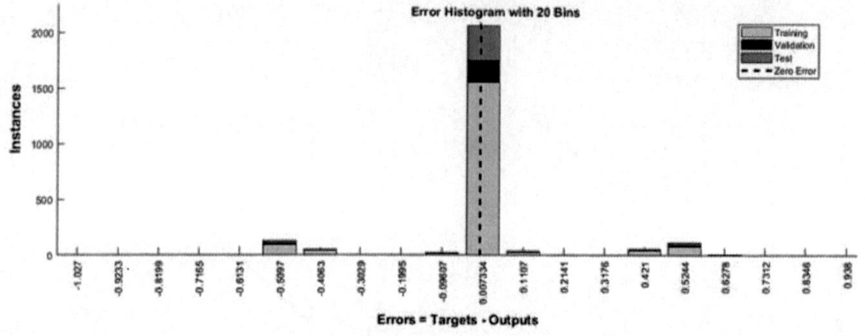

Fig. 10 Histogram plot of error values

shown in Fig. 10. The accuracy of the wavelet analysis combined with ANN has been found to be approximately 92%.

7 Conclusion

RBFNN, RNN, Hopefield NN, BPNN, etc. are different forms of NN out of which BPNN is a simplest and widely applied NN because of uncomplicated and well-developed algorithm which can perform nonlinear mapping of inputs to outputs. The proposed algorithm provides a better and simple method for detecting and classifying the fault by analyzing the transients developed in the current. The analysis indicates that the proposed scheme is efficient, reliable, and suitable for both modes of operations. In this paper, one hidden layer with four neurons is sufficient to assure the accurate result since the larger the number of neurons and hidden layers, the more complex and computational time the system will take. Further, the result can be more precise with more samples and variable parameters taken as input vectors.

References

1. Prasad A, Belwin Edward J, Ravi K (2018) A review on fault classification methodologies in power transmission systems: Part—I. J Electr Syst Inf Technol 5(1):48–60. ISSN 2314-7172
2. Prasad A, Belwin Edward J, Ravi K (2018) A review on fault classification methodologies in power transmission systems: Part-II. J Electr Syst Inf Technol 5(1):61–67. ISSN 2314-7172
3. Chan PPK, Zhu J, Qiu Z, Ng WWY, Yeung DS (2011) Comparison of different classifiers in fault detection in microgrid. In: 2011 international conference on machine learning and cybernetics, Guilin, pp 1210–1213
4. Hare J, Shi X, Gupta S, Bazzi A (2016) Fault diagnostics in smart micro-grids: a survey. Renew Sustain Energy Rev 60:1114–1124. ISSN 1364-0321
5. Sivanandam SN, Sumathi S, Deepa SN (2015) Introduction to neural networks using MATLAB 6.0. 22nd reprint, McGraw Hill

6. Upendar J, Gupta CP, Singh GK (2008) ANN based power system fault classification. IEEE Region 10 annual international conference, Proceedings/TENCON. 1–6. https://doi.org/10.1109/tencon.2008.4766623
7. Karmacharya IM, Gokaraju R (2018) Fault location in ungrounded photovoltaic system using wavelets and ANN. IEEE Trans Power Delivery 33(2):549–559
8. Srinivasa Rao P (2013) Pattern recognition approach for fault identification in power transmission lines. Int J Eng Res Appl 3:1051–1056
9. Goharrizi A, Sepehri N (2018) Application of fast fourier and wavelet transforms towards actuator leakage diagnosis: a comparative study. Int J Fluid Power 14(2):39–51. Retrieved from http://journals.riverpublishers.com/index.php/IJFP/article/view/221
10. Chanda D, Kishore NK, Sinha AK (2003) A wavelet multiresolution analysis for location of faults on transmission lines. Int J Electr Power Energy Syst 25(1):59–69. ISSN 0142-0615
11. Kirubadevi S, Sutha S (2017) Wavelet based transmission line fault identification and classification, pp 737–741. https://doi.org/10.1109/iccpeic.2017.8290461
12. Cesar TM, Pimentel SP, Marra EG, Alvarenga BP (2017) Wavelet transform analysis for grid-connected photovoltaic systems. In: 2017 6th international conference on clean electrical power (ICCEP), Santa Margherita Ligure, pp 1–6
13. Liang J, Elangovan Saikishore, Devotta JBX (1998) A wavelet multiresolution analysis approach to fault detection and classification in transmission lines. Int J Electr Power Energy Syst 20:327–332. https://doi.org/10.1016/S0142-0615(97)00076-8
14. Manohar M, Koley E, Ghosh S (2017) A reliable fault detection and classification scheme based on wavelet transform and ensemble of SVM for microgrid protection. In: 2017 3rd international conference on applied and theoretical computing and communication technology (iCATccT), Tumkur, pp 24–28
15. Kar S, Samantaray SR (2016) High impedance fault detection in microgrid using maximal overlapping discrete wavelet transform and decision tree. In: 2016 international conference on electrical power and energy systems (ICEPES), Bhopal, pp 258–263
16. Shannon CE (1948) A mathematical theory of communication. Bell Syst Tech J 27(3):379–423
17. El Safty S, El-Zonkoly A (2009) Applying wavelet entropy principle in fault classification. Int J Electr Power Energy Syst 31(10):604–607. ISSN 0142-0615
18. He Zhengyou, Chen Xiaoqin, Qian Qingquan (2007) A study of wavelet entropy measure definition and its application for fault feature pick-up and classification. J Electron 24:628–634. https://doi.org/10.1007/s11767-005-0253-0
19. Adewoleand AC, Tzoneva R (2012) Fault detection and classification in a distribution network integrated with distributed generators. In: IEEE power and energy society conference and exposition in Africa: intelligent grid integration of renewable energy resources (PowerAfrica), Johannesburg, pp 1–8

Performance of 4H-SiC IMPATT Diode at Ka- and W-Band with Temperature Variation

Girish Chandra Ghivela, Prince Kumar, Sourabh Tiwari, and Joydeep Sengupta

Abstract We have carried out the DC analysis of one-dimensional n^+-n-p-p^+ structure and its temperature dependency for 4H-SiC IMPATT diode at 36 GHz and 94 GHz. Efficiency, noise, and various dc parameters at different junction temperatures are computed and compared by taking the area of the diode as $10^{-8}m^2$ (material measurement data). This paper can be helpful to choose the best operating condition for 4H-SiC and designing the 4H-SiC-based IMPATT diode against temperature variation at Ka-band and W-band.

Keywords Ka-band · W-band · Avalanche · Drift · IMPATT · Microwave

1 Introduction

Impact avalanche transit time (IMPATT) diode is capable of generating sufficient power at microwave, millimeter wave, and sub-millimeter wave zones [1]. However, its operation is affected by temperature variation at the junction. We have preferred 36 GHz (peak frequency in Ka-band [2]) and 94 GHz (peak frequency in W-band [3]) in our analysis to study of temperature dependency because the atmospheric attenuation is relatively low at the peak frequency [3]. The DC analysis of the double drift region (DDR) structure as shown in Fig. 1 [4, 5] was carried out by solving Poisson's Eq. (1) [6, 7]

$$\frac{dE(x)}{dx} = \frac{q}{\varepsilon}[N_D - N_A + p(x) - n(x)] \tag{1}$$

where $E(x)$ is the electric field, N_D, N_A, $n(x)$, and $p(x)$ are the concentrations of donor, acceptor, electron and hole, respectively, at any point x, q is the charge of electron, and ε is the permittivity of the semiconductor.

G. C. Ghivela (✉) · P. Kumar · S. Tiwari · J. Sengupta
Electronics and Communication Engineering Department, Visvesvaraya National Institute of Technology, Nagpur 440010, India
e-mail: girishvnit2012@gmail.com

D. Harvey et al. (eds.), *Advances in VLSI, Communication, and Signal Processing*, Lecture Notes in Electrical Engineering 683,
https://doi.org/10.1007/978-981-15-6840-4_3

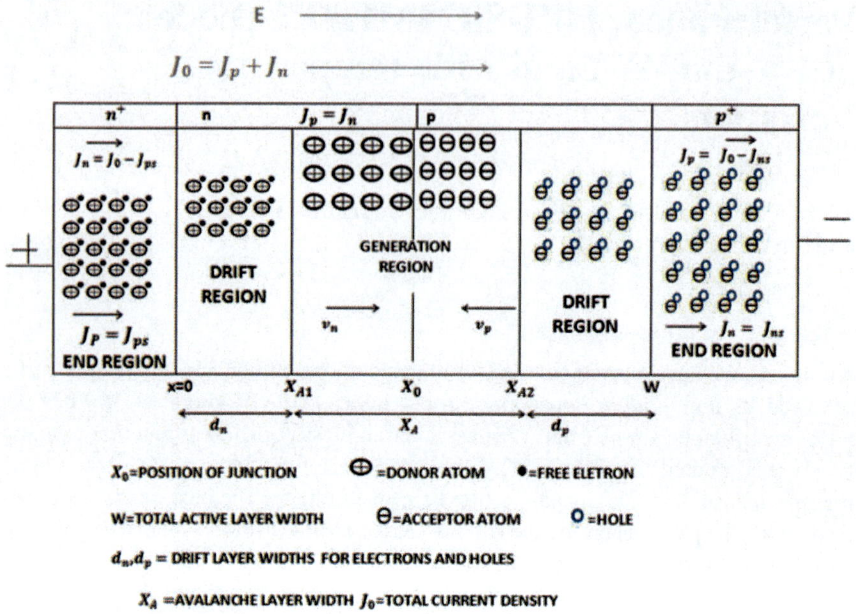

Fig. 1 Schematic of IMPATT

The conversion efficiency is calculated as [8]

$$\eta(\%) = \frac{2mV_D}{\pi V_B} \tag{2}$$

where V_D and V_B are the drop in drift region and breakdown voltage, and m is multiplication factor. V_B can be found as

$$V_B = \int_0^W E(x)\mathrm{d}x \tag{3}$$

Noise figure (M) can be found by using (4)

$$\mathrm{NM} = \frac{\langle V^2 \rangle / df}{4\mathrm{KT}(-Z_R)} \tag{4}$$

where $\langle V^2 \rangle / df$ is the noise spectral density and given by

$$\langle v^2/df \rangle = (2q/J_0 A) \cdot \left(1 + \frac{W}{X_A}\right)^2 / \alpha^2 \tag{5}$$

where A is the area of the diode, and $(-Z_R)$ is the real part of the device impedance. Total width is

$$W = W_n + W_P \tag{6}$$

where

$$W_n = 0.37V_{ns}/f \tag{7}$$

and

$$W_p = 0.37V_{ps}/f \tag{8}$$

where W_n and W_p are depletion region width in n-side and p-side, and V_{ns} and V_{ps} are saturation velocity of electrons and holes, respectively.

Mobilities of electron and hole have the form of Eqs. (9) and (10), respectively;

$$\mu_n = \frac{\mu_{n\max}(T/300)^{-\alpha}}{1 + [(N_D + N_A)/N_{n\mathrm{ref}}]^{\gamma}}(\mathrm{cm}^2/\mathrm{V\,s}) \tag{9}$$

where $\mu_{n\max}$ is the mobility for electron at 300 K. $N_{n\mathrm{ref}}$ of 1.94×10^{17} cm^{-3}, α of 2.8 and γ of 0.61 give good fitting with the measured data [9].

$$\mu_p = \frac{\mu_{p\max}(T/300)^{-\alpha}}{1 + [(N_D + N_A)/N_{p\mathrm{ref}}]^{\gamma}}(\mathrm{cm}^2/\mathrm{V\,s}) \tag{10}$$

where $\mu_{p\max}$ is the mobility for hole at 300 K. N_{pref} of 1.76×10^{19} cm^{-3}, α of 2.8 and γ of 0.34 give good match with the measured data [9].

The quality factor is measure of rate of growth of oscillation which has the form of Eq. (11),

$$Q = 1/2g \tag{11}$$

where generation rate of charge carriers (g) is

$$g = \alpha_n v_{ns} n + \alpha_p v_{ps} p \tag{12}$$

2 Material Parameters and Simulation Method

The simulation starts with DC analysis, described elsewhere [6, 10]. The Poisson's equation is simulated using MATLAB subject to boundary condition at 300 K. The

Table 1 Design parameters at 94 GHz and 300 K

Material	J_0 (A m^{-2})	N_D (m^{-3})	N_A (m^{-3})	μ_n (m^2 V s^{-1})	μ_p (m^2 V s^{-1})
4H-SiC	5×10^8	3.5×10^{23}	3.4×10^{23}	0.05	0.5

Table 2 Design parameters at 36 GHz and 300 K

Material	J_0 (A m^{-2})	N_D (m^{-3})	N_A (m^{-3})	μ_n (m^2 V s^{-1})	μ_p (m^2 V s^{-1})
4H-SiC	4×10^8	2.8×10^{23}	2.9×10^{23}	0.12	0.12

temperature dependency of the electron and hole mobility, impact ionization coefficient as well as the electron and hole saturation velocities are computed for each temperature [9]. Then, the same computer simulation method is used for the analysis as described above in case of other temperatures. The 4H-SiC material parameters at 300 K used in the simulation are given in Table 1 for 94 GHz and in Table 2 for 36 GHz. The boundary conditions are $E(x = 0) = 0$ and $E(W) = 0$; and normalized current density $P(x)$ at edges are given by Eqs. (13) and (14) [11, 12],

$$P(x = 0) = \frac{2J_{ps} - J_0}{J_0} = \left(2\frac{J_{ps}}{J_0} - 1\right) = \left(\frac{2}{M_p} - 1\right) \cong -1 \tag{13}$$

where $M_p = \frac{J_0}{J_{ps}}$ = usually a large number.

$$P(X_2) = \frac{-2J_{ns} + J_0}{J_0} = \left(-2\frac{J_{ps}}{J_0} + 1\right) = \left(-\frac{2}{M_n} + 1\right) \cong 1 \tag{14}$$

where $M_n = \frac{J_0}{J_{ns}}$ = usually a large number.

3 Simulation Results and Discussion

The simulation result of various dc parameters such as efficiency, quality factor, noise etc., are computed and compared at the peak frequencies of Ka-band and W-band in different temperature as shown in Tables 3 and 4. From the simulation result, it has been observed that the variation in breakdown voltage with corresponding temperature is more in case of 94 GHz as compared to 36 GHz. So the efficiency is very less, i.e., 18.69% in 94 GHz whereas 26.672% in 36 GHz at 300 K, and for the remaining temperature range results are shown in Tables 3 and 4. The electric field distribution at 300 K for 94 GHz and 36 GHz is shown in Figs. 2 and 3, respectively. The width of avalanche layer for 94 GHz is more than that for 36 GHz throughout the operating temperature range, which results in minimum static noise in 94 GHz shown in Fig. 4. The comparison plot of efficiency is shown in Fig. 5. Also the quality factor comparison is shown in Fig. 6.

Table 3 Results at 94 GHz in different temperature

DC parameters	300 K	330 K	350 K	400 K	450 K	500 K
V_A (V)	93.438	95.39	92.28	93.67	92.34	90.27
V_B (V)	132.31	128.07	120.81	117.4	111.8	106.4
V_D (V)	38.874	32.68	28.53	23.75	19.48	16.14
V_{ns} (10^5 m s^{-1})	1.4	1.334	1.296	1.212	1.143	1.084
V_{ps} (10^5 m s^{-1})	1.39	1.325	1.286	1.203	1.113	1.076
X_A (μm)	0.904	0.85	0.83	0.81	0.78	0.76
Noise (dbm)	−253.07	−253.15	−251.88	−252.16	−251.66	−250.99
Q (10^{-10})	1.409	1.521	1.852	2.104	2.582	3.048
η (%)	18.69	16.23	15.03	12.87	11.08	9.65

Table 4 Results at 36 GHz in different temperature

DC parameters	300 K	330 K	350 K	400 K	450 K	500 K
V_A (V)	41.477	42.765	42.32	44.117	42.552	39.667
V_B (V)	71.407	57.239	56.278	55.959	53.132	47.696
V_D (V)	29.929	14.474	13.958	11.842	10.58	8.029
V_{ns} (10^5 m s^{-1})	2.508	2.391	2.321	2.172	2.047	1.942
V_{ps} (10^5 m s^{-1})	2.5	2.383	2.314	2.165	2.041	1.936
X_A (μm)	0.523	0.643	0.638	0.636	0.62	0.587
Noise (dbm)	−252.32	−246.52	−246.27	−247.25	−245.74	−241.42
Q (10^{-10})	1.159	3.138	3.491	3.831	5.17	9.19
η (%)	26.672	16.09	15.78	13.46	12.67	10.712

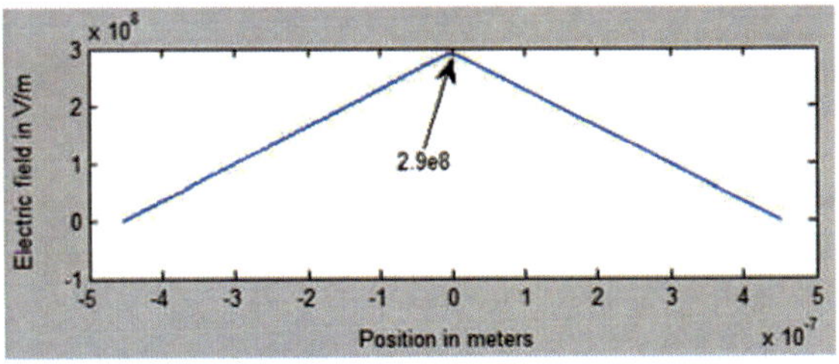

Fig. 2 Electric field distribution at 300 K for 94 GHz

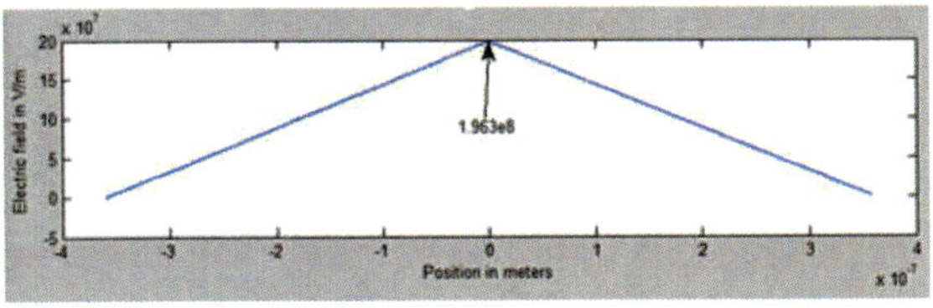

Fig. 3 Electric field distribution at 300 K for 36 GHz

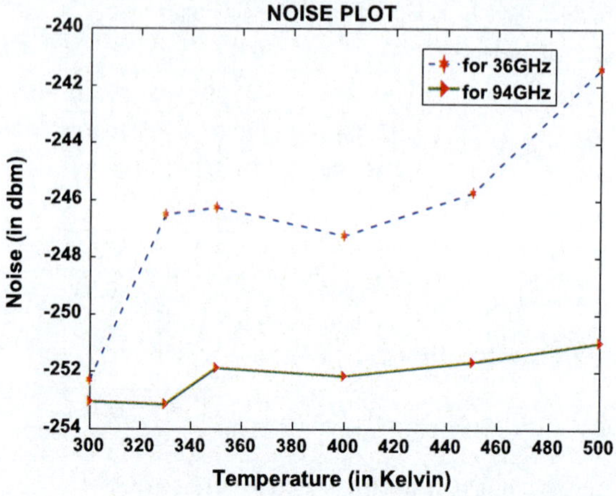

Fig. 4 Noise plot for 36 and 94 GHz at different temperatures

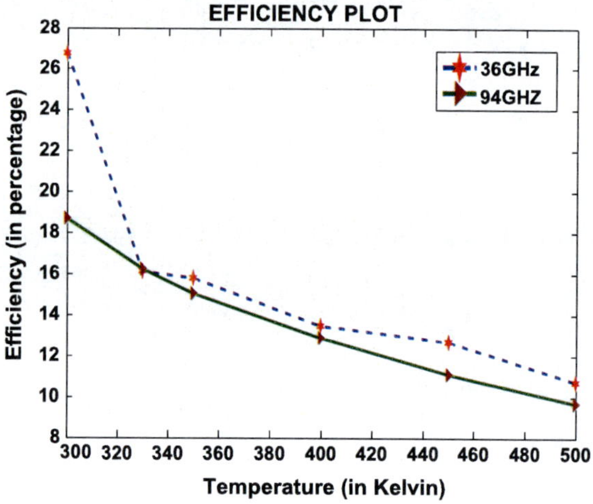

Fig. 5 Efficiency plot for 36 and 94 GHz at different temperatures

4 Conclusion

The results show that the efficiency of 4H-SiC IMPATT diode at 36 GHz is more than at 94 GHz but the noise is more in 36 GHz at same operating condition. The quality factor of the IMPATT diode at 94 GHz is more, i.e., 3.048×10^{-10} at 500 K. So, 94 GHz can provide better oscillation than 36 GHz. So the simulation result concludes that at varied temperatures, 4H-SiC IMPATT diode has better conversion (DC to rf) ability at 36 GHz than at 94 GHz. However, better oscillation at 94 GHz than at 36 GHz.

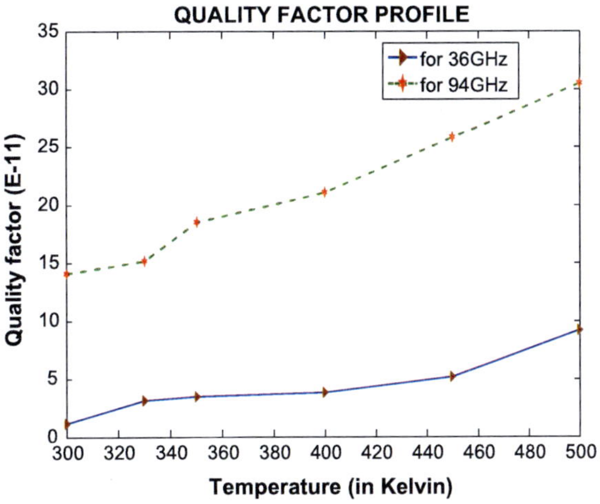

Fig. 6 Quality factor for 36 and 94 GHz at different temperatures

Acknowledgements This work was supported by Department of Electronics and Communication Engineering, VNIT, Nagpur, India.

References

1. Midford TA, Bernick RL (1979) Millimeterwave CW IMPATT diodes and oscillators. IEEE Trans Microw Theory Tech 27:483–492. https://doi.org/10.1109/TMTT.1979.1129653
2. Sengupta J, Ghivela GC, Gajbhiye A, Mitra M (2016) Measurement of noise and efficiency of 4H-SiC IMPATT diode at Ka band. Int J Electron Lett 4:134–140. https://doi.org/10.1080/21681724.2014.966774
3. Ghivela GC, Sengupta J (2019) Noise performance of avalanche transit–time devices in the presence of acoustic phonons. J Comput Electron 18:222–230. https://doi.org/10.1007/s10825-018-1289-3
4. Ghivela GC, Sengupta J (2019) Modeling and computation of double drift region transit time diode performance based on graphene-SiC. Int J Numer Model 32:1–11. https://doi.org/10.1002/jnm.2601
5. Ghivela GC, Kumar P, Sengupta J (2018) Numerical measurement of oscillating parameters of IMPATT using Group IV and Group III-V materials. In: Advances in VLSI, communication and signal processing (VCAS), MNNIT, Allahabad. https://doi.org/10.1007/978-981-32-9775-3_37
6. Ghivela GC, Sengupta J (2019) Prospects of impact avalanche transit-time diode based on chemical-vapor-deposited diamond substrate. J Electron Mater 48:1044–1053. https://doi.org/10.1007/s11664-018-6821-5
7. Ghivela GC, Sengupta J (2019) Estimation of power density in IMPATT using different materials. Int J Electron. https://doi.org/10.1080/00207217.2019.1672810
8. Ghivela GC, Sengupta J, Mitra M (2019) Ka band noise comparison for Si, Ge, GaAs, InP, WzGaN, 4H-SiC based IMPATT diode. Int J Electron Lett 7:107–116. https://doi.org/10.1080/21681724.2018.1460869
9. Sengupta J, Ghivela GC, Gajbhiye A, Jothe B, Mitra M (2014) Temperature dependence of 4H-SiC IMPATT Diode at Ka band. Int J Electr Electron Comput Syst 19:4–8

10. Ghivela GC, Sengupta J, Mitra M (2017) Space charge effect of IMPATT diode using Si, Ge, GaAs, InP, WzGaN, 4H-SiC at Ka band. IETE J Educ 58:61–66. https://doi.org/10.1080/09747338.2017.1378132
11. Datta DN et al (1982) computer analysis of DC field and current density profiles of DAR IMPATT diode. IEEE Trans Electron Devices 29:1813–1816. https://doi.org/10.1109/t-ed.1982.21032
12. Roy SK et al (1979) Computer method for the dc field and carrier current profiles in the IMPATT device starting from the field extremum in the depletion layer. In: Miller JH (ed) Proceedings of the 1st conference on numerical analysis of semiconductor devices (NASECODE I), Dublin, Ireland, pp 266–274

Design and Analysis of Low-Power SRAM

Pawar Dhiraj Kumar, Ritesh Kumar Kushwaha, and P. Karuppanan

Abstract The Static Random Access Memory (SRAM) is an integral part of memory architectures. With growing technology and scaling factor, static power consumption needs to be minimized. Therefore, an SRAM cell desires the new techniques and architecture, which can operate at very low sub-threshold voltage. In this paper, PN-based 10T SRAM cell (PPN10T) and transmission-based 8T (TG8T) have been analyzed and proposed a new 8T SRAM cell (P8T) to overwhelm the problems faced by conventional 6T SRAM (C6T) cell. The selection of W/L ratio and bit-line leakage problem has been significantly resolved for low-power SRAM design. Moreover, C6T, PPN10, TG8T cells and P10T has been compared for different performance parameters such as hold noise margin, read static noise margin (RSNM), write static noise margin (WSNM), read delay, write delay, I_{read}/I_{leak} ratio, hold power dissipation and dynamic power dissipation.

Keywords SRAM · C6T · PPN10T · TG8T · P10T · Bit-line leakage · RSNM · WSNM

1 Introduction

Memory circuits are an integral part of any system design. In order to build a processor having the capability to process instructions at a very high rate, the processing unit requires significant SRAM cache memories. These cache memories take the primary part of a silicon area in the processor and consume more leakage power [1]. With

P. D. Kumar · R. K. Kushwaha (✉) · P. Karuppanan
Department of Electronics and Communication Engineering, Motilal Nehru National Institute of Technology Allahabad, Prayagraj 211004, India
e-mail: riteshkushwaha04@gmail.com

P. D. Kumar
e-mail: dhirajdpawar@gmail.com

P. Karuppanan
e-mail: karuppanan1982@gmail.com

D. Harvey et al. (eds.), *Advances in VLSI, Communication, and Signal Processing*, Lecture Notes in Electrical Engineering 683,
https://doi.org/10.1007/978-981-15-6840-4_4

the technology advances, static power consumption in CMOS devices required to be more attention, particularly for SRAMs because of leakage power is directly dependent on the amount of transistors present in the circuit. Previous study proofs that 70% of the total memory structure in a microprocessor-based designs are occupied by transistors, and it is going to be 80% in the future [2].

Hence, for stable and energy-efficient SRAM, low-power and minimum leakage component desired. The leakage currents flowing in the circuit are different so that from various places, power can be leaked [3]. Mainly Sub-threshold current, gate leakage current and junction leakage current are taken into consideration to calculate leakage power and the sub-threshold leakage current is the most relevant component among all [2, 4].

Numerous methods have been discussed earlier to minimize the power dissipation of an SRAM cell. In the drowsy mode method [5], one extra pMOS transistor can be connected between the voltage supply and the cell, therefore, cell will remained on the standby mode. Similarly, one nMOS transistor can be connected between the cell and the ground so that it can break the path connecting-cell to the ground. Subsequently, that sleep transistors will nearly eliminated the leakage power. In the voltage scaling method [6], static (leakage) power and supply voltage has a linear relation. Thus, leakage power can be reduced by operating the cells in standby mode at low supply voltage.

Consequently, when the cell is in the standby mode, it is connected to the smaller supply voltage compared to V_{DD}. The gated-V_{DD} (gated-GND) technique and dual-threshold voltage (dual-V_t) technique have been discussed in [5] to a decrease the leakage power. In these techniques, all the transistors of SRAM cells are at low-V_t transistors so that performance is maintained and the sleep transistors are used in either supply path or ground path with high-V_t transistors.

In addition to this, power dissipation during read and write operations in SRAM is a very essential due to leakage currents. In conventional SRAM, additional energy is lost during write operation as compared to the read operation. Because bit-line voltage swing during write operation is more and content stored in the cell can get tripped very easily during the read operation that leads to very low read stability [7]. Moreover, many design methods have been reported to improve the read and write-ability of the cell operating at low voltage supply. Sharifkhani and Sachdev [8] presented a technique in which, during the write operation, the write access transistors are controlled by a more abounded voltage supply. Schmitt trigger-based SRAM cell was proposed by Kulkarni et al. [9] that suggested to overcome the problem of read disturbance. PPN-based 10T SRAM cell (PPN10T) [10], transmission-based 8T (TG8T) has been recently reported [11], which describes the energy efficiency and variation tolerance of SRAM. Schmitt-trigger-based single-ended 11T SRAM cell was reported by Ahmad et al. [12]. In this case, two nMOS are connected in series in the cell of SRAM and transistors (read buffers) are connected; hence, it takes low power and gives good write and read static noise margin (SNM).

On the other hand in order to deal with disturbance problems through the read operation, Chang et al. [13] has been proposed novel circuit which contains two word lines and two access transistors. One of the word-line was used during read operation, and the other was used in the write operation. This cell has a good read static noise margin. In addition, leakage and disturbance problems have been discussed briefly in [14, 15].

In this article detail analysis has been done for PPN-based 10T SRAM cell (PPN10T), transmission-based 8T (TG8T), conventional 6T SRAM (C6T) and projected a novel 10T SRAM cell (P10T) to tackle the problems faced by SRAM cells. The paper has been organized in the following order: in Sect. 2 transistor sizing and Bit-line leakage methods have been discussed. Section 3, explain about the proposed circuit, and performance of the circuit is analyzed in Sect. 4. Finally, in Sect. 5, conclusion of the work has been summarized.

2 Transistor Sizing and Bit-Line Leakage

SRAM cell design is sensitive to transistor sizing. The pull-up, pull-down, and access transistors sizes need to be appropriately selected to get the desired write and read operations.

The conventional 6T SRAM cell (to be called 'C6T' hereon) based on CMOS inverters is shown in Fig. 1. It consists of two back to back connected inverters and two access transistors N2 and N4. The access transistors are connected between inverters and bit-lines BL and BLB. The gates of N2 and N4 are connected to word line WL. The access transistors are turned on through the word line to enable writing and reading operation and turned off when the cell is in standby mode. The same ports are used for read and write operation.

Let us assume that the conventional 6T SRAM cells, '0' is stored at Q and '1' at Qbar ('Q'). For read operation, word-line (WL) and both bit-lines (BL and BLB) are held at '1'. Now, as Q is '0', then BL will discharge through pull-down transistor of 'N1'. For the flow of current, there must be some voltage difference between drain

Fig. 1 Conventional 6T SRAM cell

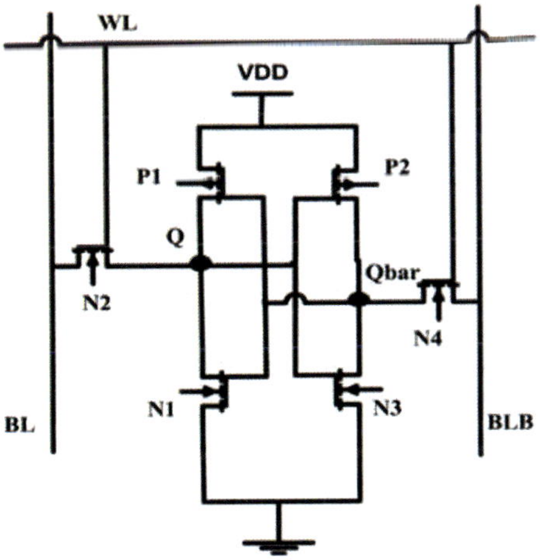

and source of transistor of 'N1'. Thus, the voltage at Q goes high by small amount for the time WL = 1. If it goes beyond this small amount, then input to right inverter (formed by P2 and N3) will become '1' and 'Q' will be '0'. So, it will flip the data during the read operation. This is called the read disturbance problem. Thus, we should ensure that the voltage at Q should not increase too much. Thus current drive strength of 'N1' must be higher than that of 'N2' to make I_{read} to flow through series connection of 'N1' and 'N2'. Drive strength of pull-down device and access device are given in Eqs. 1 and 2:

$$\beta_{\mathrm{d}} = \mu \frac{W_{\mathrm{d}}}{L_{\mathrm{d}}} (V_{\mathrm{cell}} - V_{\mathrm{th,d}})^2 (V_{\mathrm{ds,d}})^2 \tag{1}$$

$$\beta_{\mathrm{a}} = \mu \frac{W_{\mathrm{d}}}{L_{\mathrm{a}}} (V_{\mathrm{WL}} - V_{\mathrm{th,a}})^2 (V_{\mathrm{ds},a})^2 \tag{2}$$

To increase the drive strength, we cannot increase $V_{\mathrm{ds,d}}$ in Eq. (1), as it is the voltage at point Q. So, to have drive strength of pull-down device should be higher than that of access device i.e.$\beta_{\mathrm{d}} \gg \beta_{\mathrm{a}}$.

So, to avoid data flipping (to have stable read), we can (i) increase W_{d} (ii) reduce the $V_{\mathrm{th,d}}$ (iii) increase $V_{\mathrm{th,a}}$ (iv) use cell voltage V_{cell} greater than V_{WL}. However, when we don't want to go in handling voltages, we can make W_{d} higher than W_{a}. Also, increasing $V_{\mathrm{th,a}}$ will help in reducing the bit-line leakage.

In Fig. 2, during the read operation in an SRAM cell, both the bit-lines are made high. If '0' is stored in all an accessed cells of the particular column, and the threshold voltage of access transistors (connected to BL) is less, then sub-threshold leakage current I_{leak} will be flowing in each an accessed cell. As there are many cells in single column, then due to each I_{leak} the BL, so 'BL' get discharged and appear wrong output. Hence, this leakage current limits the number of cells present in one column. Consequently, increasing the threshold voltage of access transistors helps in avoiding I_{leak} and thereby increasing the number of cells present in a single column. However, increasing $V_{\mathrm{th,a}}$ will reduce the current (as in Eq. 2) I_{read} which discharges the bit-line, and it takes a longer time to read thereby increasing read time. So, there is a trade-off between leakage reduction and write noise margin or read the time.

3 Proposed Circuit

In Fig. 3, the proposed SRAM cell (to be called P8T hereon) is shown. This circuit contained two extra transistors N5 and N6 which will provide a discharging path for a read current. In conventional SRAM cell (C6T), the read current (I_{read} or Ion) flows through access transistor N2 (N4) and pull-down transistor N1 (N3). As this flow path passes through the storage nodes, C6T suffers disturbance while the read operation. Moreover, in PPN10T cell, two pull-down transistors (N5 and N6) are connected in the circuit. This forms a discharging path for read current on both sides

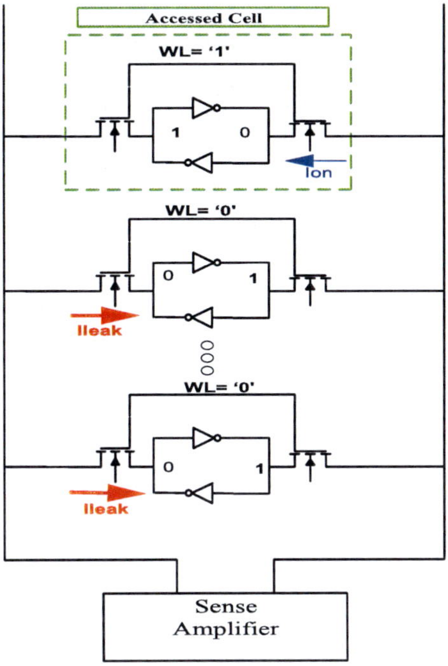

Fig. 2 Investigation of bit-line leakage problem

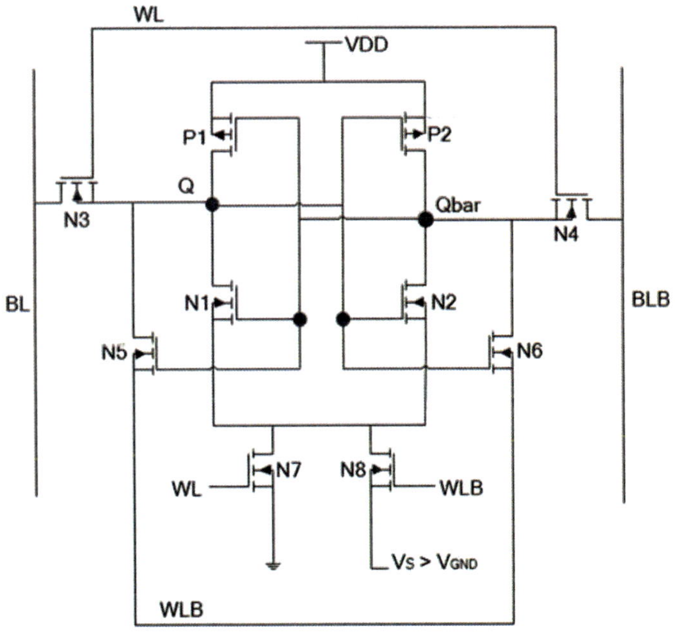

Fig. 3 Proposed SRAM cell (P10T)

of this SRAM cell, through one pull-down transistor (N5 or N6) and one access transistor (N3 or N4). Therefore, this discharging path for read current, PPN10T cell design does not suffer from disturbance. In addition to this, the pull-down transistors (N5 and N6) are connected to the pseudo storage nodes of WLB on each side of the cell. So, during a read operation, this WLB will be connected to ground, and thus the read current will flow from bit-line to ground through N5 or N6.

It is worth notice the fact that N-controlled SRAM design in which the data stored in the cell is reserved even when the cell is in the standby mode. This technique uses Dynamic Voltage Scaling (DVS) to reduce the leakage power in the cells. Due to this, the suggested design has a striking features of achieving tremendous leakage power savings. For this two-tail transistors in our proposed cell are connected. In this approach, usually a large row in the array of memory is partitioned into a number of smaller identical sub-blocks, and one pair of nMOS tail transistors controls each of these blocks.

In Fig. 4, N7 and N8 are shown and called tail transistors. Transistor N7 is controlled by WL, and it connects the cell to ground voltage. Transistor N8 is controlled by WLB, and it connects the cell to the voltage 'Vs' which is slightly higher than the ground voltage. When the cell is not in standby mode, i.e., when WL

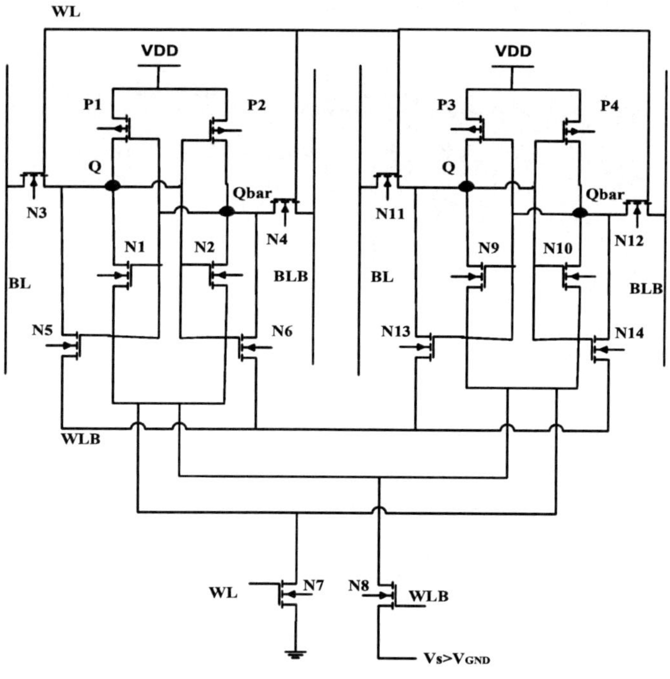

Fig. 4 Same tail transistor shared by two 10T SRAM cells from same row

is '1', transistor N7 is ON and the cell is connected to ground thereby completing the path of VDD to ground. Also, when the cell is in standby mode, i.e., when WLB is '1', transistor N8 is ON and the cell is connected to 'Vs' which is slightly higher than ground voltage thereby breaking the path of VDD to ground. These two-tail transistors are shared by more than one SRAM cell.

4 Performance Analysis

In this section, performance analysis have been done for proposed design, in terms of read access time, write access time, hold static noise margin, write static noise margin (WSNM), read static noise margin (WSNM), read-ability, write-ability, I_{read}/I_{leak} ratio, static power dissipation, dynamic power dissipation and etc.

4.1 Read Access Time (T_{RA})

The TRA (read access time) is the time difference between the point when word line (WL) rises to 50% of its final state from its initial low state and the point when BL/BLB is discharged by 0.05 V from its initial high state [15]. During the read operation, there must be a difference of at least 0.05 V between the voltage levels of BL and BLB to avoid read error. As calculated, TRA for C6T is 0.5022 ns; for PPN10T it is 1.022 ns; for TG8T it is 1.44 ns, and for P10T it is 1.2857 ns.

4.2 Write Access Time (T_{WA})

The TWA (write access time), when writing '0', is defined as the time difference between the point when WL reaches 50% of its final state (from its initial low state) and the point when 'Q' (Storing '1' initially) falls to 10% of its initial high state. Similarly, TWA, when writing '1' at 'Q' is defined as the time difference between the point when WL reaches 50% of its final state (from its initial low state) and the point when 'Q' (Storing '0' initially) rises to 90% its initial low state. TWA for C6T is 0.38 ns; for PPN10T it is 0.9775 ns; for TG8T it is 0.4511 ns, and for P10T it is 0.4955 ns.

4.3 Hold Stability

Hold static noise margin (SNM) is helpful in determining the hold stability of SRAM cell, defined as that minimum value of noise voltage (DC) which will invert the data store in the cell when it occurs (develops). The noise margins are obtained using the butterfly curves. In the smaller wing, a largest fit table square is drawn, and a side length of that square gives the value of noise margins.

The side length of that square represents the SNM (hold SNM) of the cell. Figure 5 shows the 'butterfly curve' for cells of C6T, PPN10T, TG8T. From this figure, we can see that all squares have nearly equal sides, i.e., noise margins are nearly the same. The reason behind that is the internal latch structure is more or less the same in all three cells. As observed C6T, PPN10T and TG8T cells shows 275 mV, 296 mV, and 280 mV hold SNM respectively. Figure 6, shows 'butterfly curve' of P10T. As observed, P10T cell show 145 mV hold SNM. When static noise margin value is equal to or above 25% of supply voltage, the cell is considered to have good read stability.

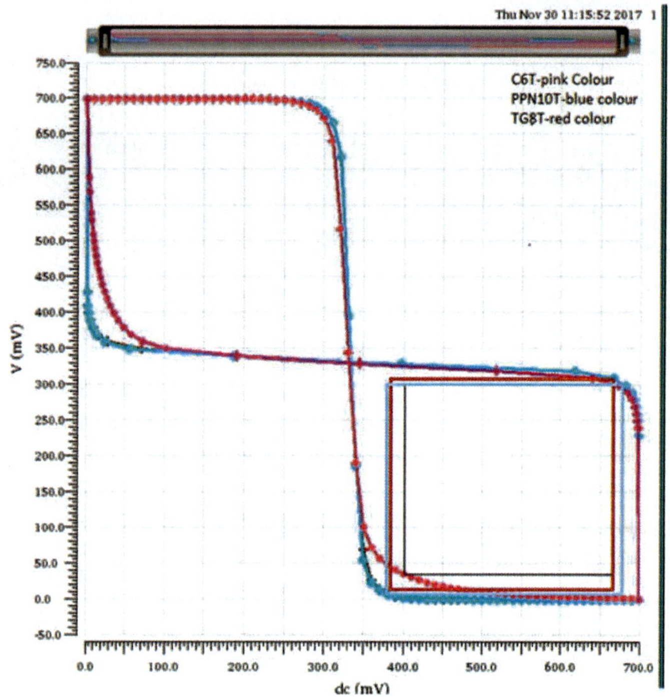

Fig. 5 Static noise margin (SNM) of C6T, PPN10T, TG8T at supply voltage of 0.7 V

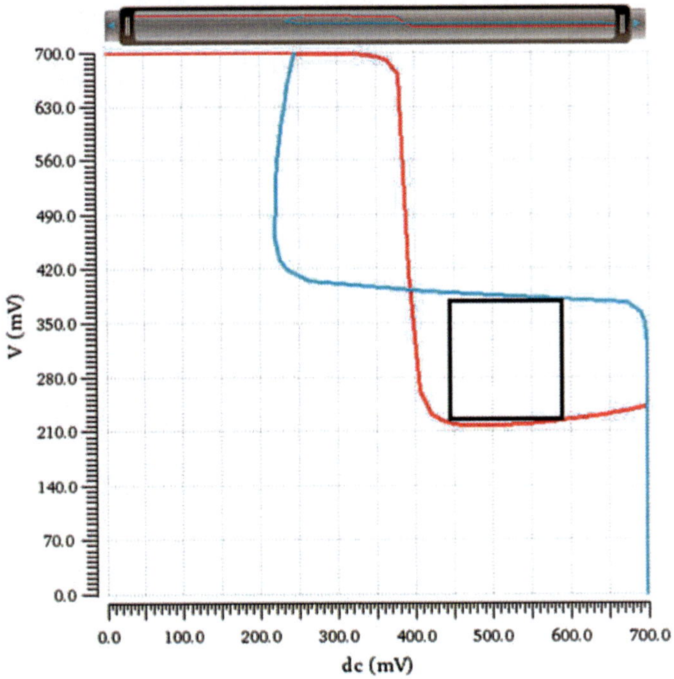

Fig. 6 Static noise margin (SNM) of P8T at supply voltage of 0.7 V

4.4 Write-Ability

Write static noise margin (WSNM) gives the write-ability of SRAM cell. Suppose a storage node is having a value ‘1’, then the ability of SRAM cell to bring down this storage node from ‘1’ to ‘0’ is determined using the write-ability of the cell. If BL is successful in bringing the voltage at ‘Q’ (storing ‘1’ initially) down to voltage level below the trip voltage of the other inverter with output ‘Qbar’ so that it content is tripped and ‘0’ is stored in ‘Q’ and ‘1’ that stored in ‘Qbar’. The ratio of strength of pull-up transistor P1 (P2) to that of access transistor N2 (N4) determines the write-ability of SRAM cell. Figure 7, shows the ‘butterfly curve’ of the cells C6T, PPN10T, TG8T for determining WSNM. As observed, in C6T, PPN10T and TG8T

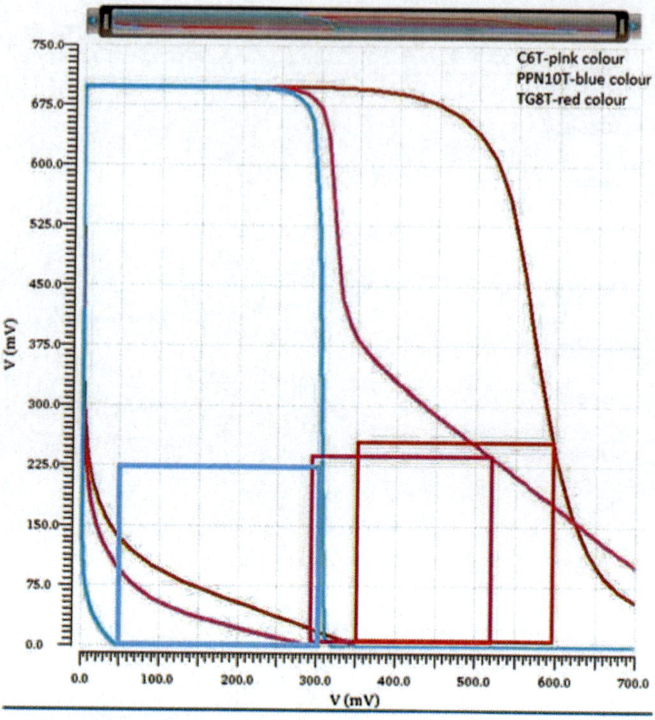

Fig. 7 Write static noise margin (WSNM) of C6T, PPN10T, TG8T at supply voltage of 0.7 V

cells show 224 mV, 168 mV, and 265 mV WSNM respectively. As observed in Fig. 8, P8T show 252 mV WSNM. The WSNM is highest for TG8T as it uses transmission gates instead of nMOS transistors for access transistors.

4.5 *Read Stability*

Read static noise margin (RSNM) is helpful in determining the read stability of an SRAM cell. Noise can easily affect the SRAM cell for the duration of the read operation because the voltage at '0' storing node rises by a small amount. The access transistor, size ratio control the rise in voltage level of storage node. If this ratio is small, then there will be high voltage at drain terminal of the pull-down transistor. Thus, there will be a high voltage difference between drain and source of pull-down transistor. This will require just a small noise voltage at the '0' storing node to invert the data stored in the cell. Therefore, read static noise margin is a critical

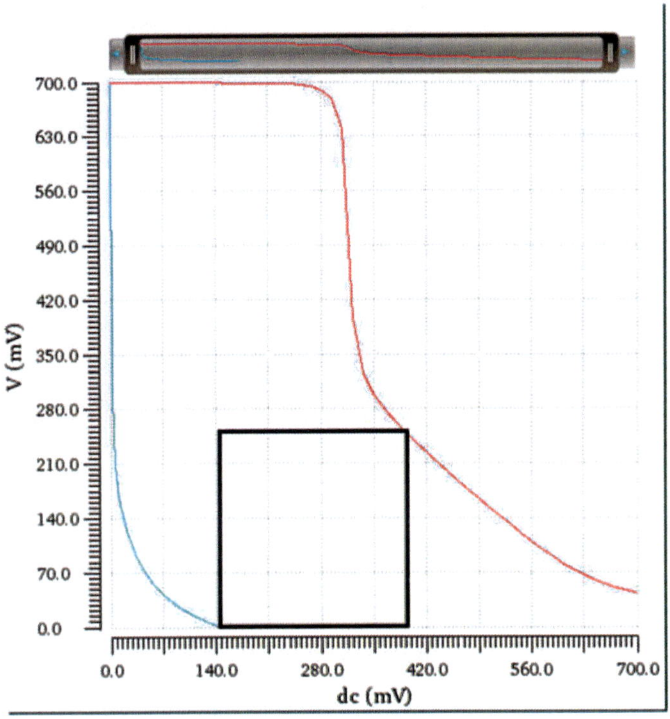

Fig. 8 Write static noise margin (WSNM) of P8T at supply voltage of 0.7 V

parameter in SRAM cell design. Figure 9 shows the 'butterfly curve' of the cells C6T, PPN10T, TG8T for determining RSNM. As observed in C6T, PPN10T and TG8T shows 105 mV, 290 mV, and 106 mV RSNM respectively. As observed in Fig. 10, P10T show 160 mV RSNM. As we can see, the RSNM is good for PPN10T and P8T cells as they use separate path for read current thereby avoiding the problem of the read disturbance.

4.6 I_{read}/I_{leak} Ratio or I_{on}/I_{off} Ratio

The on-current (I_{on}) is the current flowing into an accessed cell from the bit-line connected to it for the duration of the read operation while the off-current (I_{off}) is the current leaked to all the other an accessed cells of the same column. Usually, for reliable read operation, I_{on}/I_{off} ratio is 10 or above so that there is sufficient voltage swing between the bit-line and its complement at the time when sense amplifier starts

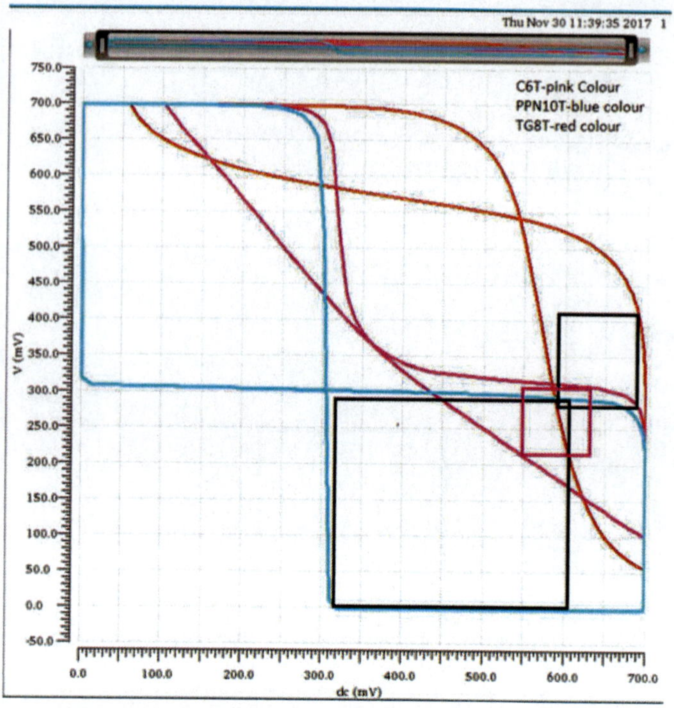

Fig. 9 Read static noise margin (RSNM) of C6T, PPN10T and TG8T at supply voltage of 0.7 V

its operation. Owing to the increased standby mode leakage current in transistors, particularly at very low voltages, this ratio has worsened. This bit-line leakage limits the bit-line length, thereby limiting the number of cells attached to a bit-line. Hence, a good I_{on}/I_{off} ratio is expected.

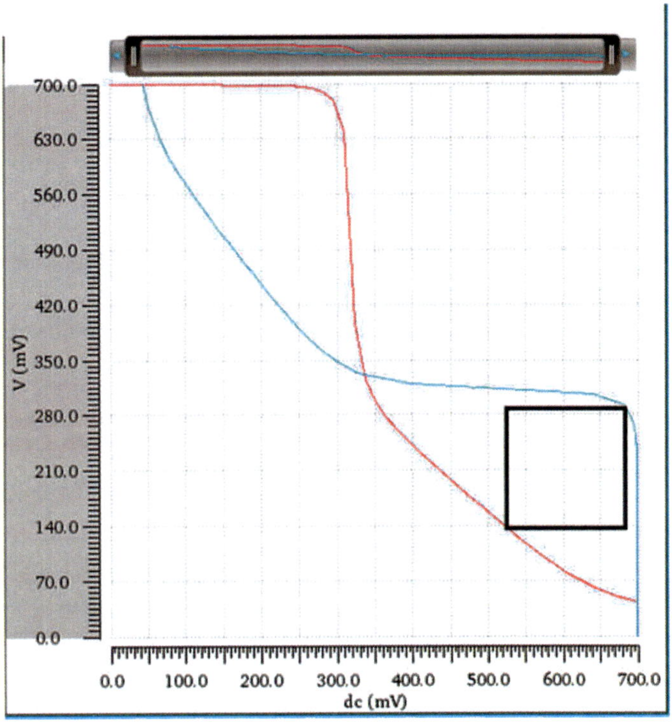

Fig. 10 Read static noise margin (RSNM) of P10T at supply voltage of 0.7 V

Table 1 gives the comparison of I_{read}/I_{leak} ratio of C6T, PPN10T, TG8T, and P8T at different supply voltages. The PPN10T and P10T show a good I_{read}/I_{leak} ratio among four circuits. Hence we can have a highest bit-line length in PPN10T and P10T, thereby allowing us to accommodate additional number of cells in a single column and increasing transistor density.

Table 2 gives the comparison of values of different performance parameters for C6T, PPN10T, TG8T, and P10T. The proposed SRAM cell has compact advantages in terms of hold, read and write SNM, delay and power dissipation over conventional 6T-SRAM, 10T-SRAM and TG8T-SRAM.

Table 1 I_{READ}/I_{LEAK} comparison

V_{dd} (V)	Conventional 6T SRAM (C6T SRAM)			PPN-based 10T SRAM cell (PPN10T SRAM)			Transmission gate-based 8T SRAM (TG8T SRAM)			Proposed 8T SRAM (P10T SRAM)		
	I_{read} (μA)	I_{leak} (nA)	I_{read}/I_{leak} (× 10^3)	I_{read} (μA)	I_{leak} (pA)	I_{read}/I_{leak}(× 10^3)	I_{read} (μA)	I_{leak} (pA)	I_{read}/I_{leak} (× 10^3)	I_{read} (μA)	I_{leak} (pA)	I_{read}/I_{leak} (× 10^3)
0.3	0.041	0.0128	3.2	0.0077	0.395	19	0.016	3.508	19	0.0071	0.3807	18.64
0.4	0.565	0.152	3.7	0.1054	0.577	182	0.169	4.356	38	0.082	0.563	145
0.5	4.132	0.239	17.28	0.797	0.797	1000	1.096	5.204	210	0.544	0.757	718
0.6	17.411	8.004	2.17	3.840	1.067	3598	5.210	6.056	860	2.7769	0.969	2865
0.7	42.814	373.02	0.11	13.226	1.401	9440	16.513	6.916	2387	9.059	1.2059	7512
0.8	86.958	0.0176	0.0049	32.297	1.815	17,790	35.070	7.792	4500	21.062	1.4735	14,293

Table 2 Comparison of four SRAM cells

Design specification	Conventional 6T SRAM (C6T SRAM)	PPN-Based 10T SRAM cell (PPN10TSRAM)	Transmission gate-based 8T SRAM (TG8T SRAM)	Proposed 10T SRAM (P10T SRAM)
Process technology (nm)	45	45	45	45
Hold SNM (@ 0.7 V) (mV)	275	296	280	145
Read SNM (@ 0.7 V) (mV)	105	290	106	160
Write SNM (@ 0.7 V) (mV)	224	168	265	252
Write access delay (T_{wa}) (ns)	0.380	0.9775	0.4511	0.4955
Read access delay (T_{ra}) (ns)	0.5022	1.0226	1.440	1.2857
Hold power dissipation (@ 0.7 V) (nW)	78.82	25.63	78.82	11.98
Dynamic power dissipation (@ 0.7 V) (μW)	2.669	2.314	1.766	0.9024

5 Conclusion

The conventional 6T SRAM cell faced many problems due to ever-increasing process variation and technology scaling. New cells with more transistors as discussed and proposed in order to overcome these problems. In this work, we have studied the overview of some previous techniques and proposed 10T SRAM cell and analyzed the circuits for different parameters like write access time (TWA), read access time(TRA), hold static noise margin, read SNM, write SNM, I_{read}/I_{leak} ratio, hold power dissipation. The performance analysis shows that in PPN10T to avoid the read disturbance problem and separate read path was used in PPN10T and hence proposed cell shows good RSNM that is necessary to make read operation successful. The projected cell gives better WSNM necessary to make write operation successful. Hold power dissipation and dynamic power dissipation is lowest in a proposed cell. The I_{read}/I_{leak} ratio of the suggested circuit is very good so that we can have a large number of SRAM cells in the single column of a memory array. Also, when we reduce leakage current to increase I_{read}/I_{leak} ratio, it gives overhead of increased read access time as seen in the case of all cells except conventional an SRAM cell. Moreover, due to the dynamic voltage scaling technique, the proposed circuit gives the lowest but enough hold SNM among four cells. Overall the proposed SRAM cell has many advantages over conventional SRAM cell, which overweight its few disadvantages.

References

1. Goll Bernhard, Zimmermann Horst (2009) A comparator with reduced delay time in 65-nm CMOS for supply voltages down to 0.65 V. IEEE Trans Circuits Syst II Express Briefs 56(11):810–814
2. Nikoozadeh Amin, Murmann Boris (2006) An analysis of latch comparator offset due to load capacitor mismatch. IEEE Trans Circuits Syst II Express Briefs 53(12):1398–1402
3. Nath Mandal D et al (2014) Analysis and design of low voltage low power dynamic comparator with reduced delay and power. Int J Eng Res Gen Sci 2
4. Babayan-Mashhadi S, Lotfi R (2014) Analysis and design of a low-voltage low-power double-tail comparator. IEEE Tran Very Large Scale Integr (VLSI) Syst 22(2):343–352
5. Ay Suat U (2011) A sub-1 Volt 10-bit supply boosted SAR ADC design in standard CMOS. Analog Integr Circ Sig Process 66(2):213–221
6. Mesgarani A et al (2010) Supply boosting technique for designing very low-voltage mixed-signal circuits in standard CMOS. In: 2010 53rd IEEE international midwest symposium on circuits and systems. IEEE
7. Maymandi-Nejad Mohammad, Sachdev Manoj (2003) 1-bit quantiser with rail to rail input range for sub-1 V $\Delta\Sigma$ modulators. Electron Lett 39(12):894–895
8. Sharifkhani M, Sachdev M (2007) Segmented virtual ground architecture for low-power embedded SRAM. IEEE Trans Very Large Scale Integr (VLSI) Syst 2:196–205
9. Kulkarni JP, Roy K (2012) Ultralow-voltage process-variation-tolerant Schmitt-trigger-based SRAM design. IEEE Trans VLSI Syst 20(2):319–332
10. Sanvale P et al (2019) An improved read-assist energy efficient single ended PPN based 10T SRAM cell for wireless sensor network. Microelectron J 104611
11. Pal S, Islam A (2015) Variation tolerant differential 8T SRAM cell for ultralow power applications. IEEE Trans Comput Aided Des Integr Circuits Syst 35(4):549–558
12. Chang IJ et al (2009) A 32 kb 10T sub-threshold SRAM array with bit-interleaving and differential read scheme in 90 nm CMOS. IEEE J Solid-State Circuits 44(2):650–658
13. Lo C-H, Huang S-Y (2011) PPN based 10T SRAM cell for low-leakage and resilient subthreshold operation. IEEE J Solid-State Circuits 46(3):695–704
14. Islam A, Hasan M (2012) A technique to mitigate impact of process, voltage and temperature variations on design metrics of SRAM Cell. Microelectron Reliab 52(2):405–411
15. Kulkarni JP, Kim K, Roy K (2007) A 160 mV robust schmitt trigger based subthreshold SRAM. IEEE J Solid-State Circuits 42:2303–2313

A Unified Approach for Calculating Outage Performance of Multi-hop Regenerative Relay Network in Nakagami-*m* Fading Channel

Himanshu Katiyar, P. K. Verma, Arun Kumar Singh, and Saurabh Dixit

Abstract Cooperation among relay nodes enhances network capacity, transmission reliability and expands the coverage in wireless networks. This is a newly emerged area which have a lot of potential to support high data rate communication. Such wireless network operates in hostile environment where signal quality is effected due to the presence of scatters in between communication link, and they are responsible for the phenomena of shadowing and path loss. Besides it, quality of signal is also effected from multi-path fading phenomena,and here, constructive and destructive combining of signals occur at the receiver. It is quite challenging for the receiver to recover the original message correctly in the presence of additive and multiplicative noise. Hence, for such type of network, a unified approach is required to attract researchers and resolve various unaddressed issues till now. In this paper, we will study multi-hop regenerative serial relay network and establish a closed form expression of outage probability for Nakagami-*m* fading channel.

Keywords Nakagami-*m* fading channel · Regenerative mode · Signal to noise power ratio · Probability density function · Outage probability

This work is supported by Collaborative Research and Innovation Program (CRIP) funding through TEQIP-III scheme of Dr. A.P.J. Abdul Kalam Technical University (AKTU), Lucknow. Authors are highly obliged for their support and gratefully acknowledge.

H. Katiyar (✉) · P. K. Verma
Department of Electronics Engineering, Rajkiya Engineering College, Sonbhadra, India
e-mail: katiyarhimanshu@gmail.com

P. K. Verma
e-mail: vermapk87@gmail.com

A. K. Singh
Department of Electronics Engineering, Rajkiya Engineering College, Kannauj, India
e-mail: aksingh.uptu@gmail.com

S. Dixit
Department Electronics and Communication Engineering, Sagar Institute of Technology, Barabanki, UP, India
e-mail: saurabh2911@ieee.org

D. Harvey et al. (eds.), *Advances in VLSI, Communication, and Signal Processing*, Lecture Notes in Electrical Engineering 683,
https://doi.org/10.1007/978-981-15-6840-4_5

1 Introduction

High data rate, spectrally efficient and highly reliable communication link are major thrust of research around the world [3]. Power, bandwidth restrictions and operation over a harsh multi-path fading channel are major bottleneck for high data rate wireless communication services. A new type of user cooperation between user terminals has been discussed in [19], and various issues of its implementation in real world are analyzed in [20]. Various low complexity protocols for cooperative communication are discussed in [15]. Initially, space-time coding for distributed relay network is analyzed in [16] by Laneman et al. Outage performance for dissimilar Rayleigh fading is analyzed in [1]. Approximate outage performance for cooperative relay network with best relay selection criteria and arbitrary channel distribution is analyzed in [2]. Outage analysis of relay-based system in which end-to-end link is established with the help of multiple relays is discussed in [23]. For the case when multi-antenna relay performs SC of signals and destination performs MRC of signals, error and outage performance is analyzed in [7] and [8], respectively. In [10], expressions of outage probability and bit error rate is derived, when signals received at multi-antenna relay and destination are coherently combined. Expressions in closed form for the average received SNR, channel capacity, average error rate and outage probability is established in [9, 14]; here, multi-antenna base station is receiving data from single antenna source and relay. In [12], outage is analyzed for the case when direct link between source and destination is absent and communication is assisted by two infrastructure-based multi-antenna relays. For multi-antenna regenerative relay network, average capacity and average SNR are derived in [11]. In [13], outage and BER expressions in closed form are established for a scenario in which communication is assisted by a infrastructure-based multi-antenna relay node. For improving coverage and capacity in remote areas, deployment of satellite–terrestrial cooperative network is beneficial which is explored in [5]. A Survey on simultaneous wireless information and power transfer with cooperative relay is done [6]. Vehicle-to-vehicle communication is useful for extending the coverage in highways with the help of cooperative communication technique explored in [24]. An expression for the average capacity of relay-based system is derived in [17] for Rayleigh channel. Regenerative relaying for the application of Internet of Things is presented in [18]. This paper investigates the outage probability of multi-hop relay system. The direct path between transmitter and receiver is completely obstructed; hence, communication is possible only through relay link. Nakagami-m fading channel is assumed in this work because we can model more versatile and realistic scenario. The contributions of this work are summarized below:

- Established a mathematical model of multi-hop regenerative serial relay network operating in Nakagami-m fading channel.
- Development of simulation model for Nakagami-m fading channel.
- Derived the expressions for outage probability.

Rest of the paper is organized as follows: Sect. 2 briefly discusses the channel modeling of Nakagami-m. Closed form expressions of outage probability for multi-hop regenerative serial relay network have been derived in Sect. 3. Numerical results are discussed in Sect. 4. In Sect. 5, tracks for future work have been discussed. Finally, conclusions are drawn in Sect. 6.

2 Channel Modeling

A N-hop serial relay system is shown in Fig. 1. Here, communication link between source (s) and destination (d) is established with the help of $N-1$ supporting relays ($r_1, r_2, \cdots, r_{N-1}$). In this system model, s transmits signal to r_1 (i.e., $s \rightarrow r_1$) in first time slot. Received signal is firstly decoded at r_1 and forwarded to r_2 (i.e., $r_1 \rightarrow r_2$) in second time slot and so on. This process will continue till signal reaches at d in $N-1$ time slot. At receivers $r_1, r_2, r_{N-1}, \cdots, d$, short-term signal variation occurs due to mobility of various scatters in environment. At any instant of time, signals transmitted from $s, r_1, r_2, \cdots, r_{N-1}$ may combine constructively or destructively at $r_1, r_2, r_{N-1}, \cdots, d$, respectively. Such fading phenomena is relatively fast and is therefore responsible for the short-term signal variations. Random variation in signal strength can be statistically modeled by various fading models. The Nakagami-m fading model is versatile because it is suitable for indoor short-distance communication, outdoor land-mobile communication and scintillating ionospheric radio links. For a special case, we can easily model one-sided Gaussian distribution by taking $m = 1/2$ and the Rayleigh distribution by taking $m = 1$. The Nakagami-m fading channel converges to a non-fading AWGN channel by taking limit as $m \rightarrow +\infty$. The PDF of Nakagami-m fading amplitude α_{ij} can be written as [21, Eq. (2.20)]:

$$f_{\alpha_{ij}}(\alpha) = \frac{2m_{ij}^{m_{ij}}\alpha^{2m_{ij}-1}}{\Omega_{ij}^{m_{ij}}\Gamma(m_{ij})}\exp\left(-\frac{m_{ij}\alpha^2}{\Omega_{ij}}\right). \quad (1)$$

where $\Gamma(\cdot)$ is gamma function [4, Eq. (8.310.1)] and m_{ij} is the Nakagami-m fading parameter, which ranges from 1/2 to ∞. α_{ij} is a random variable (RV) which models the instantaneous fading envelop between node i and j, $i \in \{s, r_1, r_2, \cdots, r_{N-1}\}$,

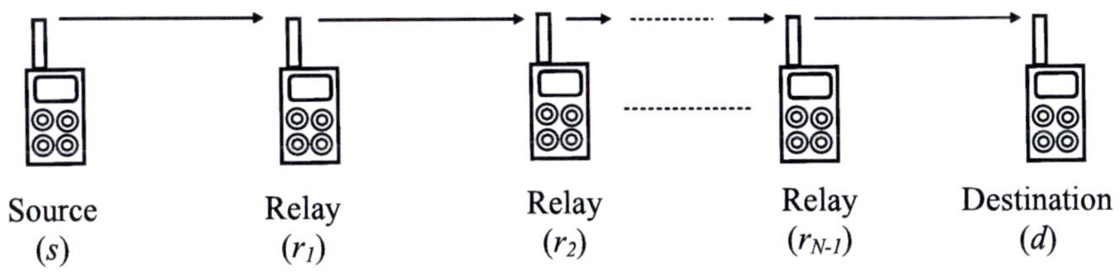

Fig. 1 Multi-hop relay network operating in Nakagami-m fading channel

$j \in \{r_1, r_2, r_{N-1}, \cdots, d\}$, $\Omega_{ij} = E\left[\alpha_{ij}^2\right]$ and $E[\cdot]$ is expected value. The PDF of received SNR (γ_{ij}) of Nakagami-m fading model can be written as [22]:

$$f_{\gamma_{ij}}(\gamma) = \frac{m_{ij}^{m_{ij}} \gamma^{m_{ij}-1}}{\bar{\gamma}_{ij}^{m_{ij}} \Gamma(m_{ij})} \exp\left(-\frac{m_{ij}\gamma}{\bar{\gamma}_{ij}}\right). \tag{2}$$

Here, $\gamma_{ij} = \alpha_{ij}^2 E_s / N_0$ is the instantaneous signal to noise power ratio per symbol, $\bar{\gamma}_{ij} = \Omega_{ij} E_s / N_0$ is the average signal-to-noise power ratio per symbol between node i and j, E_s is the energy per symbol and N_0 is one-sided power spectral density.

3 Modeling of Multi-hop Relay System

For the multi-hop regenerative serial relay system which is shown in Fig. 1, the weakest link between $s \rightarrow r_1$, $r_1 \rightarrow r_2$, $\cdots$, $r_{N-1} \rightarrow d$, will be effective. Hence, received SNR at d can be modeled by RV γ_{srd}:

$$\gamma_{srd} = \min\left(\gamma_{sr_1}, \gamma_{r_1r_2}, \cdots, \gamma_{r_{N-1}d}\right) \tag{3}$$

The cumulative distribution function (CDF) for $RV \gamma_{srd}$ can be written as:

$$\begin{aligned} F_{\gamma_{srd}}(\gamma) &= 1 - \Pr\left\{\gamma_{sr_1} > \gamma, \gamma_{r_1r_2} > \gamma, \cdots, \gamma_{r_{N-1}d} > \gamma\right\} \\ &= 1 - \Pr\left\{\gamma_{sr_1} > \gamma\right\} \Pr\left\{\gamma_{r_1r_2} > \gamma\right\} \cdots \Pr\left\{\gamma_{r_{N-1}d} > \gamma\right\} \\ &= 1 - \left[1 - \Pr\left\{\gamma_{sr_1} \le \gamma\right\}\right]\left[1 - \Pr\left\{\gamma_{r_1r_2} \le \gamma\right\}\right] \cdots \left[1 - \Pr\left\{\gamma_{r_{N-1}d} \le \gamma\right\}\right] \\ &= 1 - \prod_{\substack{i \in \{s, r_1, r_2, \cdots, r_{N-1}\} \\ j \in \{r_1, r_2, r_{N-1}, \cdots, d\}}} \left[1 - \Pr\left\{\gamma_{ij} \le \gamma\right\}\right]. \end{aligned}$$

Here, $\Pr\{\cdot\}$ is the probability of occurrence of an event. In (4), $\min\left(\gamma_{sr_1}, \gamma_{r_1r_2}, \cdots, \gamma_{r_{N-1}d}\right) > \gamma$ is true if and only if $\gamma_{sr_1} > \gamma$, $\gamma_{r_1r_2} > \gamma$, $\cdots$ $\gamma_{r_{N-1}d} > \gamma$ are true. So, (4) can be written as:

$$\begin{aligned} F_{\gamma_{srd}}(\gamma) &= 1 - \Pr\left\{\gamma_{sr_1} > \gamma, \gamma_{r_1r_2} > \gamma, \cdots, \gamma_{r_{N-1}d} > \gamma\right\} \\ &= 1 - \Pr\left\{\gamma_{sr_1} > \gamma\right\} \Pr\left\{\gamma_{r_1r_2} > \gamma\right\} \cdots \Pr\left\{\gamma_{r_{N-1}d} > \gamma\right\} \\ &= 1 - \left[1 - \Pr\left\{\gamma_{sr_1} \le \gamma\right\}\right]\left[1 - \Pr\left\{\gamma_{r_1r_2} \le \gamma\right\}\right] \cdots \left[1 - \Pr\left\{\gamma_{r_{N-1}d} \le \gamma\right\}\right] \end{aligned} \tag{4}$$

In (4), $\gamma_{sr_1}, \gamma_{r_1r_2}, \cdots, \gamma_{r_{N-1}d}$ are independent RVs because $s \rightarrow r_1$, $r_1 \rightarrow r_2$, $\cdots$, $r_{N-1} \rightarrow d$ links observe independent fading phenomena. The CDF for RV γ_{srd} can be written as:

$$F_{\gamma_{srd}}(\gamma) = 1 - \left[\left\{1 - F_{\gamma_{sr_1}}(\gamma)\right\}\left\{1 - F_{\gamma_{r_1 r_2}}(\gamma)\right\}\cdots\left\{1 - F_{\gamma_{r_{N-1}d}}(\gamma)\right\}\right]$$
$$= 1 - \prod_{\substack{i\in\{s,r_1,r_2,\cdots,r_{N-1}\}\\ j\in\{r_1,r_2,r_{N-1},\cdots,d\}}} \left[1 - F_{\gamma_{ij}}(\gamma)\right]. \quad (5)$$

Here, $F_{\gamma_{sr_1}}(\gamma) = \int_0^\gamma f_{\gamma_{sr_1}}(\gamma)d\gamma$, $F_{\gamma_{r_1r_2}}(\gamma) = \int_0^\gamma f_{\gamma_{r_1r_2}}(\gamma)d\gamma$, $F_{\gamma_{r_{N-1}d}}(\gamma) = \int_0^\gamma f_{\gamma_{r_{N-1}d}}(\gamma)d\gamma$ is CDF of RVs γ_{sr_1}, $\gamma_{r_1r_2}$, $\gamma_{r_{N-1}d}$, respectively. Similarly, CDF of remaining RV can be calculated. Communication link in outage when strength of received SNR drops below a certain value. For this case, outage probability can be calculated for a given threshold χ [21, Eq. (1.4)]:

$$P_{\text{out}} = \Pr\{\gamma \le \chi\} = F_{\gamma_{srd}}(\gamma)\big|_{\gamma=\chi} = F_{\gamma_{srd}}(\chi)$$
$$= 1 - \left[1 - \frac{\Gamma\left(m_{sr_1}, \frac{m_{sr_1}\chi}{\bar{\gamma}_{sr_1}}\right)}{\Gamma\left(m_{sr_1}\right)}\right]\left[1 - \frac{\Gamma\left(m_{r_1r_2}, \frac{m_{r_1r_2}\chi}{\bar{\gamma}_{r_1r_2}}\right)}{\Gamma\left(m_{r_1r_2}\right)}\right]\cdots\left[1 - \frac{\Gamma\left(m_{r_{N-1}d}, \frac{m_{r_{N-1}d}\chi}{\bar{\gamma}_{r_{N-1}d}}\right)}{\Gamma\left(m_{r_{N-1}d}\right)}\right]$$
$$= 1 - \prod_{\substack{i\in\{s,r_1,r_2,\cdots,r_{N-1}\}\\ j\in\{r_1,r_2,r_{N-1},\cdots,d\}}} \left[1 - \frac{\Gamma\left(m_{ij}, \frac{m_{ij}\chi}{\bar{\gamma}_{ij}}\right)}{\Gamma\left(m_{ij}\right)}\right]. \quad (6)$$

Here, $\Gamma(\cdot,\cdot)$ is upper incomplete gamma function [4, Eq. (8.350.1)].

4 Numerical Results

This section presents the outage performance of multi-hop regenerative serial relay network operating in independent and non-identically distributed Nakagami-m fading channel. Nakagami-m distributed RVs are generated as given in APPENDIX, which is mapped with (2) for $m = 1, 3$ and plotted in Figs. 2 and 3, respectively. Analytical expression of outage probability for such type of system has been derived in (6). For evaluating the outage performance in various SNR, we have fixed the value of outage threshold (χ) to unity. In Fig. 4, outage performance of multi-hop serial relay network has been plotted for various values of m and number of hops (i.e., N). Outage performance for various outage threshold is calculated by fixing the value of SNR to unity (i.e., 0 dB) and plotted in Fig. 5 for various values of m and N.

5 Tracks for Future Work

In this work, we have only analyzed the outage performance of multi-hop regenerative serial relay network operating in Nakagami-m fading channel. Other performance parameters like average fade duration, amount of fading, average error probability,

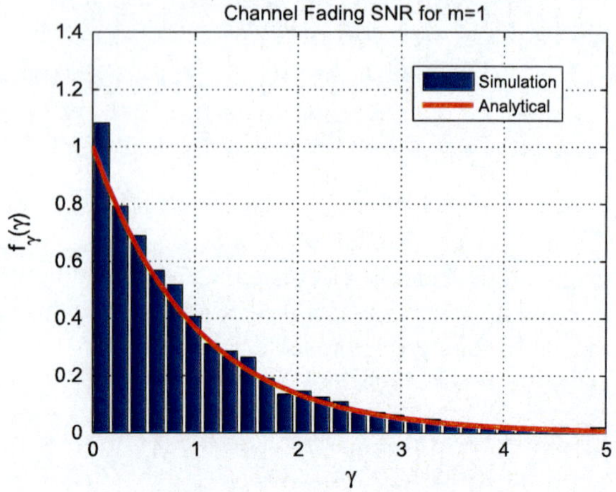

Fig. 2 Mapping of Nakagami-m distributed RV for $m = 1$

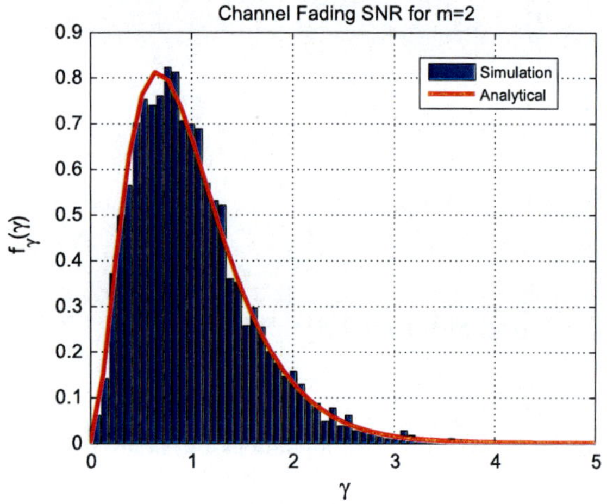

Fig. 3 Mapping of Nakagami-m distributed RV for $m = 3$

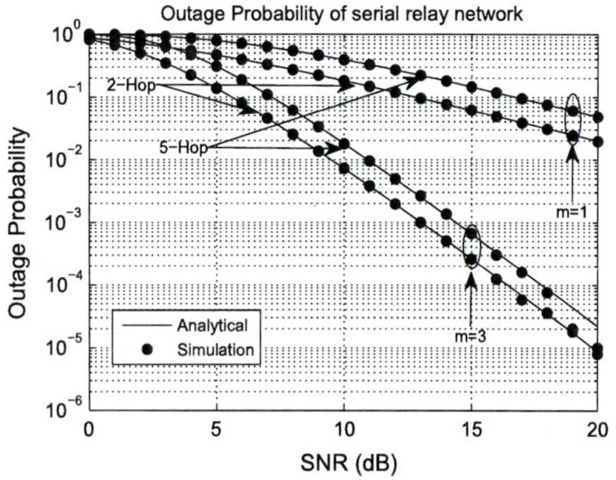

Fig. 4 Outage probability of multi-hop regenerative relay network at various SNR

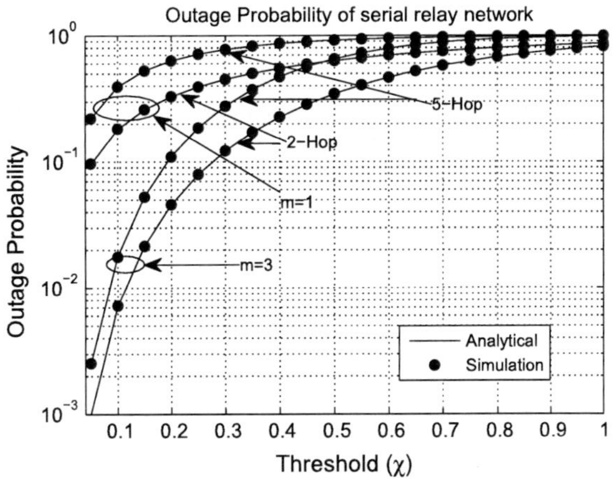

Fig. 5 Outage probability of multi-hop regenerative relay network at various outage threshold

average link capacity, end-to-end average SNR, etc., are yet to be analyzed. This work can also be further extended to other fading channel such as Ricean, Hoyt, Rice, $\kappa - \mu, \alpha - \mu, \eta - \mu$.

6 Conclusion

This paper investigates the outage performance of multi-hop regenerative relay network in Nakagami-m fading channel. Validity of analytical results can be cross-checked with the help of Monte Carlo simulation (running simulator freely with 10^6 samples), and it is found that such results are perfectly matched with the simulation results. We found that outage probability decreases with increment of SNR and m which is fading parameter of Nakagami-m channel (i.e., line of site component increases). However, outage probability increases with increment of outage threshold and number of hops. Hence, we found that in case of multi-hop communication, network coverage can be enhanced at cost of outage performance.

Appendix

Algorithm 1 Generation of Nakagami-m distributed random variable for integer value of m.

procedure NAKAGAMI- m DISTRIBUTED RV
    $m \leftarrow$Parameter of Nakagami-m fading channel
    $\mu \leftarrow$Mean of random variable
    $\sigma^2 \leftarrow$Variance of random variable
    $n \leftarrow$Channel length
    count$\leftarrow$zero matrix of order $1 \times n$
    for $i \leftarrow 1, m$ **do**
        count=count+$\frac{1}{\sqrt{2}} \left| \mathrm{N}\left(m, \sigma^2\right) + j\mathrm{N}\left(m, \sigma^2\right)\right|^2$
    end for
    $H \leftarrow count/m$
end procedure

References

1. Beaulieu NC, Hu J (2006) A closed-form expression for the outage probability of decode-and-forward relaying in dissimilar rayleigh fading channels. IEEE Commun Lett 10(12):813–815. https://doi.org/10.1109/LCOMM.2006.061048
2. Beres E, Adve R (2007) Outage probability of selection cooperation in the low to medium SNR regime. IEEE Commun Lett 11(7):589–597. https://doi.org/10.1109/LCOMM.2007.070097
3. Goldsmith A (2005) Wireless communication. Cambridge University Press, Cambridge University Press 40West 20th Street, NewYork, NY 10011–4211, USA, first edn. www.cambridge.org/9780521837163
4. Gradshteyn IS, Ryzhik IM (2007) Table of integrals, series and products. Academic Press Inc, 7th revised edn
5. Hajipour P, Shahzadi A, Ghazi-Maghrebi S (2019) Improved performance for a heterogeneous satellite-cooperative network with best relay node selection. China Commun 16(5):93–105 https://doi.org/10.12676/j.cc.2019.05.008
6. Hossain MA, Md Noor R, Yau KA, Ahmedy I, Anjum SS (2019) A survey on simultaneous wireless information and power transfer with cooperative relay and future challenges. IEEE Access 7:19166–19198. https://doi.org/10.1109/ACCESS.2019.2895645
7. Katiyar H, Bhattacharjee R (2009) Average error rate of multi-antenna decode and forward cooperative relay network. In: Proceeding of IEEE INDICON 2009, DAIICT Ahmedabad, India. https://doi.org/10.1109/INDCON.2009.5409489
8. Katiyar H, Bhattacharjee R (2009) Outage performance of two-hop multi-antenna co-operative relaying in Rayleigh fading channel. IET Electron Lett 45(17):881–883. https://doi.org/10.1049/el.2009.0379
9. Katiyar H, Bhattacharjee R (2009) Performance of regenerative relay network operating in uplink of multi-antenna base station under Rayleigh fading channel. In: Proc. of TENCON 2009, IEEE Region 10 Conference, Singapore. pp 1–5. https://doi.org/10.1109/TENCON.2009.5395929

10. Katiyar H, Bhattacharjee R (2010) Performance of MRC combining multi-antenna cooperative relay network. AEU (Elsevier) Int J Electron Commun 64(10):988–991. https://doi.org/10.1016/j.aeue.2009.07.007, http://www.sciencedirect.com/science/article/B7GWW-4X3MRFJ-1/2/032a028b69beef3ec9585bc25035a935
11. Katiyar H, Bhattacharjee R (2011) Average capacity and SNR analysis of multi-antenna regenerative cooperative relay in Rayleigh fading channel. IET Commun 5:1971–1977. https://doi.org/10.1049/iet-com.2010.0969
12. Katiyar H, Bhattacharjee R (2011) Outage performance of multi-antenna relay cooperation in the absence of direct link. IEEE Commun Lett 15(4):398–400. https://doi.org/10.1109/LCOMM.2011.020411.101863
13. Katiyar H, Bhattacharjee R (2011) Performance of two-hop infrastructure based multi-antenna regenerative relaying in Rayleigh fading channel. Phys Commun (Elsevier) 4(3):190–195. https://doi.org/10.1016/j.aeue.2011.04.007
14. Katiyar H, Bhattacharjee R (2012) On the performance of decode-and-forward relaying with multi-antenna destination. AEU (Elsevier) - Int J Electron Commun 66:1–6. https://doi.org/10.1016/j.aeue.2011.04.007
15. Laneman J, Tse D, Wornell G (2004) Cooperative diversity in wireless networks: efficient protocols and outage behavior. IEEE Trans Info Theory 50(12):3062–3080. https://doi.org/10.1109/TIT.2004.838089
16. Laneman J, Wornell G (2003) Distributed space-time-coded protocols for exploiting cooperative diversity in wireless networks. IEEE Trans Info Theory 49(10):2415–2425. https://doi.org/10.1109/TIT.2003.817829
17. Lee M, Oh S (2019) A closed-form ergodic capacity expression for a generic cooperative diversity network in rayleigh fading channels. J Commun Networks 21(3):307–318. https://doi.org/10.1109/JCN.2019.000029
18. Lyu B, Yang Z, Guo H, Tian F, Gui G (2019) Relay cooperation enhanced backscatter communication for internet-of-things. IEEE Internet Things J 6(2):2860–2871. https://doi.org/10.1109/JIOT.2018.2875719
19. Sendonaris A, Erkip E, Aazhang B (2003) User cooperation diversity. part I. system description. IEEE Trans Commun 51(11):1927–1938. https://doi.org/10.1109/TCOMM.2003.818096
20. Sendonaris A, Erkip E, Aazhang B (2003) User cooperation diversity. part II. implementation aspects and performance analysis. IEEE Trans Commun 51(11):1939–1948. https://doi.org/10.1109/TCOMM.2003.819238
21. Simon MK, Alouini MS (2005) Digital communication over fading channels. Wiley, Inc., 111 River Street, Hoboken, NJ 07030, 2nd edn. http://as.wiley.com/WileyCDA/WileyTitle/productCd-0471649538.html
22. Yang HC, Alouini MS (2005) MRC and GSC diversity combining with an output threshold. IEEE Trans Veh Technol 54(3):1081–1090. https://doi.org/10.1109/TVT.2005.844634
23. Zhao Y, Adve R, Lim TJ (2005) Outage probability at arbitrary SNR with cooperative diversity. IEEE Commun Lett 9(8):700–702. https://doi.org/10.1109/LCOMM.2005.1496587
24. Zhou J, Tian D, Wang Y, Sheng Z, Duan X, Leung VCM (2019) Reliability-optimal cooperative communication and computing in connected vehicle systems. IEEE Trans Mobile Comput pp 1–1 (2019) https://doi.org/10.1109/TMC.2019.2907491

832.5 Gb/s PM-8QAM Superchannel with 5 b/s/Hz Spectral Efficiency

Divya Sharma, Shikha Devi, and Y. K. Prajapati

Abstract The proposed work depicts a Nyquist-wavelength division multiplexing (Nyquist-WDM) superchannel, which consists of five subcarriers based on advanced modulation format, i.e., polarization multiplexed-8 quadrature amplitude modulation (PM-8QAM) format. The proposed model delivers excellent transmission capacity of 832.5 Gb/s (i.e., 5×166.5 Gb/s) at baud rate 27.75 Gb/s with 11% forward error correction (FEC) overhead. Bandwidth of each Nyquist filter at transmitter end is kept 33.33 GHz (i.e., $1.2 \times$ baud rate). At this 33.33 GHz bandwidth, all five subcarriers are aggregated for the sake of giving birth to a bandwidth-efficient superchannel with 5b/s/Hz spectral efficiency. Few key optical performance parameters are optical signal to noise ratio (OSNR), laser input power level, bit error rate (BER) and optical spectrum of the superchannel. At reference BER of value 4×10^{-3}, a performance comparison is also presented among different higher-order polarization multiplexed formats, which reflects that to achieve same BER, more OSNR is required by higher-order modulation formats. The proposed concept is applicable in terrestrial communication and flexible optical networks.

Keywords Nyquist-WDM · Superchannel · PM-8QAM

1 Introduction

An exponential growth in data traffic in the form of cloud computing, audio-video streaming, online gaming, cable television (CATV), etc., is continuously observed, which leads researcher to think over maximum utilization of the optical fiber bandwidth for the sake of satiating such a huge traffic [1]. Deployment of advance higher-order modulation format with hundreds of signal points is one of the vital solutions

D. Sharma (✉)
ECED, Thapar Institute of Engineering & Technology, Patiala, Punjab, India
e-mail: divya.fgiet@gmail.com

S. Devi · Y. K. Prajapati
ECED, Motilal Nehru National Institute of Technology Allahabad, Prayagraj, Uttar Pradesh, India

D. Harvey et al. (eds.), *Advances in VLSI, Communication, and Signal Processing*,
Lecture Notes in Electrical Engineering 683,
https://doi.org/10.1007/978-981-15-6840-4_6

to achieve terabit transmission capacity. On the other hand, hundreds of signal points results into high phase noise, hardware bandwidth problems and short reach [2]. Such a situation gave birth to superchannel concept, in which terabit transmission capacity is achieved using aggregation of lower-order subcarriers based on modulation formats containing only few signal points. In the proposed work, polarization multiplexed-8 quadrature amplitude modulation (PM-8QAM) format is preferred as subcarrier because this format does not require the as high optical signal to noise ratio (OSNR) as that of the polarization multiplexed-16 quadrature amplitude modulation (PM-16QAM) format [3]. Also, PM-8QAM format increases transmission channel capacity and spectral efficiency with a factor of 1.5 than that of the polarization multiplexed-quadrature phase shift keying (PM-QPSK) [4].

Nyquist-wavelength division multiplexing (Nyquist-WDM) is applied as a method of aggregation of multiple subcarriers in superchannel because data rate of individual subcarrier is higher than 100 Gb/s, due to which inter-carrier interference (ICI) occurs. On making Nyquist filter bandwidth approximately equal to the baud rate, decreases the chance of ICI and increases bandwidth efficiency [5].

In 2012, I. Tomkos et al. performed a successful survey over flexible optical networks and analyzed that "elasticity" is the network capability to adjust channels, modulation format, data rate, bandwidth in accordance with continuously varying real-time data traffic [5]. In year 2013, E. Palkopoulou et al. analyzed the effect of presenting elasticity through Nyquist-WDM concept in design parameters of optical networks, say laser power, modulation format, channel spacing and baud rate [6].

In this paper, we proposed the architecture of 832.5 Gb/s (i.e., 5 $\times$ 166.5 Gb/s) Nyquist-WDM superchannel using PM-8QAM subcarriers over 90 km optical fiber length. Results are presented through parameters such as laser input power optimization, bit error rate (BER), OSNR and optical spectrum through commercial simulation software tool RSoft Optsim. The whole paper is described in the following manner; Sect. 1 deals with introduction of superchannel. Section 2 presents relevant theory, while Sect. 3 describes mathematics required for superchannel transmission along with digital coherent receiver. Section 4 explains results and discussion. Section 5 gives some fruitful conclusions.

2 Architecture of the Proposed Superchannel

Here, Fig. 1 depicts the design architecture of the proposed Nyquist-WDM superchannel transmitter. Here, five subcarriers based on 166.5 Gb/s PM-8QAM format are co-transmitted to give rise to a 832.5 Gb/s superchannel. Center channel frequencies of the laser sources are as follows: 193.56, 193.53, 193.50, 193.46 and 193.43 GHz. A Gaussian bandpass filter is used as Nyquist filter which has fixed bandwidth as 33.33 GHz, which is 20% higher than the baud rate. Across the fiber channel, erbium-doped fiber amplifier (EDFA) is introduced to compensate accumulated distortion through the transmitted signal.

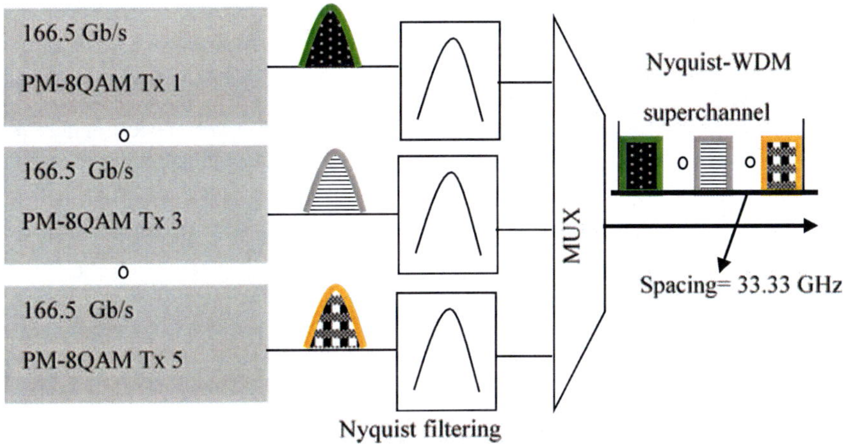

Fig. 1 Architecture of 5 × 166.5 Gb/s PM-8QAM Nyquist-WDM superchannel transmitter

In Fig. 2, digital coherent receiver is available at receiver end, which has the following key components; photodiode detector, trans-impedance amplifier and Bessel low-pass filter. At receiver, single frequency can be received at a time by setting pre received Bessel low-pass filter (LPF) cut-off frequency same as desired frequency of any of earlier mentioned fived frequencies. Five key functions of digital signal processing (DSP) unit at receiver are as follows: analog to digital conversion, electronic dispersion compensation (EDC), phase tracking, polarization de-multiplexing and digital to analog converter. Here, algorithms used across DSP for impairment compensation are time-domain chromatic dispersion compensation, least mean square (LMS) method for phase estimation and polarization de-multiplexing [7, 8]. This proposed superchannel delivers spectral efficiency of 5 b/s/Hz, as transmitted data rate is 166.5 Gb/s and the channel spacing is 33.33 GHz.

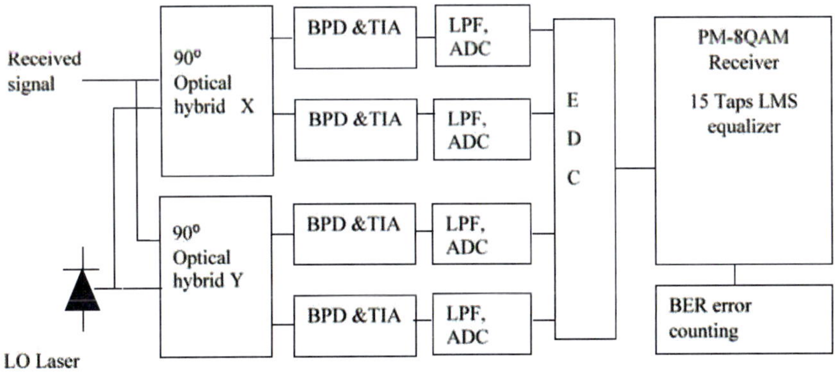

Fig. 2 Architecture of 5 × 166.5 Gb/s PM-8QAM Nyquist-WDM superchannel digital coherent receiver

3 Mathematical Modeling

Digital coherent receiver is the method of employing DSP to mitigate various linear and nonlinear channel impairments programmatically in a very faster manner. For mitigation of chromatic dispersion (linear effect), EDC algorithm is used. The impulse response of the chromatic dispersion (CD) compensation filter $I(l, t)$ is presented as [9, 10]

$$I(l,t) = \sqrt{\frac{jc}{D\lambda^2 z}} \exp\left(\frac{\pi c}{D\lambda^2 l} t^2\right) \tag{1}$$

Finite impulse response (FIR) filter is used with w weighting coefficient and tap index is

$$\alpha_k = \sqrt{\frac{jcT^2}{D\lambda^2 z}} \exp\left(-\frac{j\pi cT^2}{D\lambda^2 z} h^2\right), \quad -\frac{N-1}{2} \le w \le \frac{N-1}{2}, \tag{2}$$

Here, D is chromatic dispersion coefficient, $l = 90$ km fiber length, T is the sampling period, $N = 15$ number of taps.

Another linear effect, i.e., polarization mode dispersion (PMD) is mitigated using decision directed-least mean square (DD-LMS) filter, which has inverse channel polarization properties,

$$\begin{aligned} x_{\text{out}}(k) &= \mathbf{H_{xx}} \cdot a_{\text{in}}(k) + \mathbf{H_{xy}} \cdot b_{\text{in}}(k) \\ y_{\text{out}}(k) &= \mathbf{H_{yx}} \cdot a_{\text{in}}(k) + \mathbf{H_{yy}} \cdot b_{\text{in}}(k) \end{aligned} \tag{3}$$

Here, a_{in} and b_{in} are inputs to x and y polarization, $H_{xx}, H_{xy}, H_{yx}, H_{yy}$ are m-taps FIR filters coefficients, while $\mu = 0.0003$ is the convergence parameter.

4 Results and Discussion

Here, multiple results obtained through transmission of 5×166.5 Gb/s superchannel are analyzed and discussed. Initially, all the results are plotted for back to back transmission, i.e., in the absence of optical fiber, except Fig. 4. This back to back transmission helps us to analyze system capabilities in an ideal circumstances, when no fiber linear and nonlinear impairments are effecting fiber channel. Figure 3a depicts optical spectrum of the proposed 5×166.5 Gb/s superchannel at center frequency 193.5 THz. Centered frequency PM-8QAM subcarrier is filtered as shown in Fig. 3b to reach at receiver. Figure 3c exhibits the noisy amplified version of the centered subcarrier, due to effect of fiber channel of 90 km length.

In Fig. 4, BER is plotted against laser input power varying from −5 to 4 dBm. This inverted bell-shaped curve is helpful in finding the optimum value of transmitted

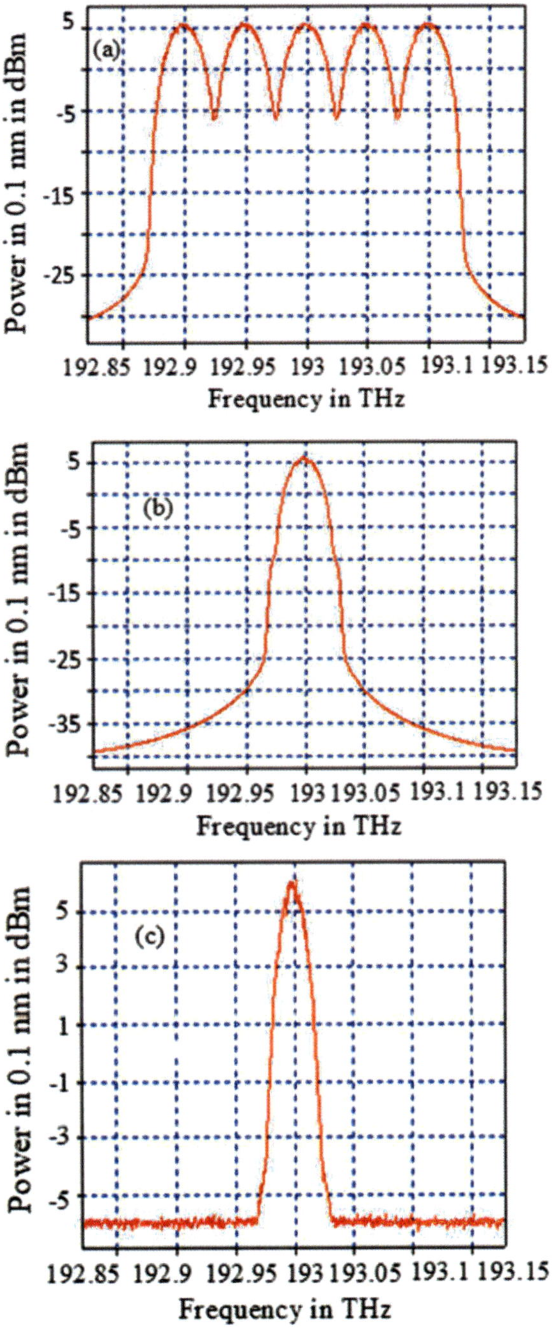

Fig. 3 **a** Optical spectrum of the proposed 5 × 222 Gb/s superchannel. **b** Optical spectrum of the centered frequency (3rd) subcarrier at 193.5 THz. **c** Optical spectrum of the amplified version of the subcarrier at 193.5 THz

Fig. 4 BER is plotted against transmitted laser input power at 20 spans of 90 km each fiber

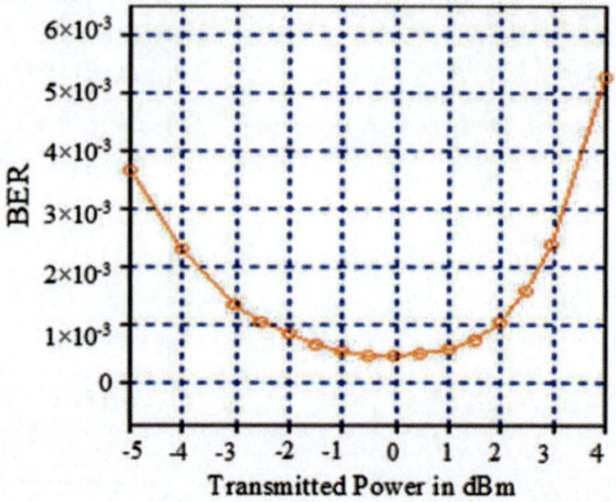

power per channel and also helps in estimating power penalty. It is analyzed that initially, on increasing laser power from −5 to 0 dBm, BER performance degrades due to nonlinearities at high power, while on increasing power from 0 to 4 dBm, OSNR increases due to which BER performance keeps on improving. Such a bell-shaped curve is obtained at 20 spans of 90 km standard single-mode fiber (SSMF) fiber type. Considered SMF parameters values are 0.22 dB/km attenuation, 16.7 ps/nm/km chromatic dispersion, 80 μm^2 effective core area and 1.26 W^{-1} km^{-1} nonlinear coefficient value.

In Fig. 5, BER is plotted against OSNR using BER analyzer in back to back configuration. Here, reference BER value, i.e., 4×10^{-3} is obtained at excellent OSNR value of 17 dB. As OSNR value from grows from 15 dB to 18.5 dB, BER performance of the proposed superchannel improves from 6×10^{-2} to 6×10^{-4}.

Fig. 5 BER is plotted against OSNR value at 0.1 nm for 5 × 166.5 Gb/s PM-8QAM superchannel

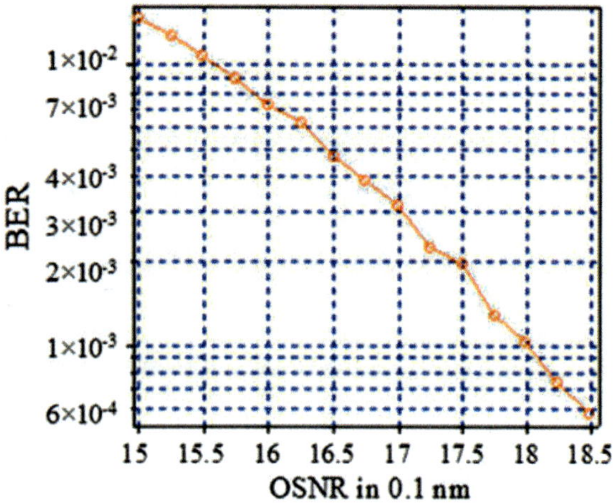

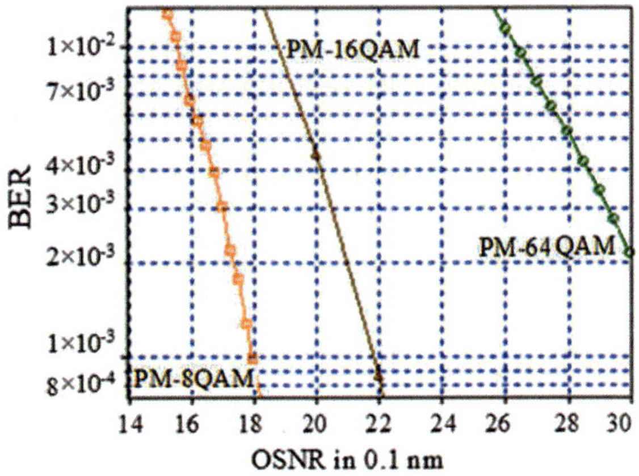

Fig. 6 BER is plotted against OSNR value at 0.1 nm for PM-8QAM, PM-16 QAM and PM-64QAM-based superchannel

In Fig. 6, a performance comparison is observed in terms of BER versus OSNR plot in back to back configuration between three different superchannels at baud rate 27.75 Gb/s, which are 5 × 166.5 Gb/s PM-8QAM, 5 × 222 Gb/s PM-16QAM and 5 × 333 Gb/s PM-64 QAM. It is realized that to achieve reference BER value of 4 × 10^{-3}, respective OSNR values are 16.6 dB, 20.1 dB and 28.4 dB. Hence, we can observe that as we move toward modulation format with higher constellation size, higher amount of OSNR is required to maintain minimum BER level [11].

5 Conclusion

In this paper, 832.5 Gb/s (5 × 166.5 Gb/s) Nyquist-WDM superchannel transmission is proposed using PM-8QAM modulation as subcarrier. An excellent spectral efficiency of 5 b/s/Hz is obtained during the work. PM-8 QAM modulation is a promising candidate to achieve nearly terabit transmission because of big constellation size. An efficient performance comparison is also presented in between PM-8QAM, PM-16 QAM and PM-64 QAM performances in terms of BER versus OSNR plot. Elastic networks are future direction of the proposed superchannel [12].

References

1. Cisco (2017) Cisco visual networking index: global mobile data traffic forecast update, 2016–2021. Cisco white paper, Document ID 1454457600805266
2. Bosco G, Curri V, Carena A, Poggiolini P, Forghieri F (2011) On the performance of Nyquist-WDM terabit superchannels based on PM-BPSK, PM-QPSK, PM-8QAM or PM-16QAM subcarriers. J Lightwave Technol 29(1):53–61
3. Sharma D, Prajapati YK (2018) Terabit Nyquist superchannel transmission using PM-QPSK subchannels. In: 2018 3rd international conference on microwave and photonics (ICMAP). IEEE, pp 1–2

4. Zhang J, Yu J, Dong Z, Jia Z, Chien HC, Cai Y, Ge C, Shi S, Chen Y, Wang H, Xia Y (2014) Transmission of 20440 Gb/s super Nyquist filtered signals over 3600 km based on single carrier 110 Gbaud PDM-QPSK with 100 GHz grid. In: Optical fiber communication, OSA, Th5B.3
5. Tomkos I, Palkopoulou E, Angelou M (2012) A survey of recent developments on flexible/elastic optical networking. ICTON, Mo.C1.1
6. Palkopoulou E, Bosco G, Carena A, Klonidis D, Poggiolini P, Tomkos I (2013) Nyquist-WDM-based flexible optical networks: exploring physical layer design parameters. J Lightwave Technol 31(14):2332–2339
7. Zhou YR, Smith K, West S, Johnston M, Weatherhead J, Weir P, Hammond J, Lord A, Chen J, Pan W, Cao C, Yang R, Zhou N, Wu S (2017) Field trial demonstration of real-time optical superchannel transport up to 5.6 Tb/s over 359 km and 2 Tb/s over a live 727 km flexible grid optical link using 64 GBaud software configurable transponders. J Lightwave Technol 35(3):499–505
8. Sharma D, Verma A, Prajapati YK, Singh V, Saini JP (2013) Forward and backward wave propagation in multilayer planar waveguide using metamaterials layer. Opt Quant Electron 45(2):105–114
9. Sharma D, Maurya JB, Prajapati YK (2015) Effect of noise on constellation diagram of 100 Gbps DP-QPSK systems under influence of different digital filters. In: International conference on microwave and photonics (ICMAP 2015), Dhanbad, India, pp 1–2
10. Sharma D, Prajapati YK, Tripathi R (2018) Success journey of coherent PM-QPSK technique with its variants: a survey. IETE Tech Rev. https://doi.org/10.1080/02564602.2018.1557569
11. Sharma D, Prajapati YK, Tripathi R (2018) Spectrally efficient 1.55 Tb/s Nyquist-WDM superchannel with mixed line rate approach using 27.75 Gbaud PM-QPSK and PM-16QAM. Opt Eng 57(7):076102
12. Xia TJ, Fevrier H, Wang T, Morioka T (2015) Introduction of spectrally and spatially flexible optical networks. IEEE Commun Mag 24–33

Design and Performance of High-Speed CMOS Double-Tail Dynamic Comparator Suitable for Mixed-Signal ICs

Avaneesh K. Dubey, Vikrant Varshney, Ankur Kumar, Pratosh K. Pal, and R. K. Nagaria

Abstract In this paper, the novel design of double-tail dynamic comparator (DTDyC) is presented using the voltage-controlled capacitive (VCC) loads and the cross-coupled dynamic inverters. The VCC loads are used in amplification stage and they are designed as the cascade of MOS capacitors with the PMOS load transistors. The combination of dynamic inverters is used in latch stage to improve its speed. The mathematical analysis of delay is presented. The proposed DTDyC is designed in CADENCE and simulated with SPECTRE using 90 nm CMOS technology to evaluate the performance. The simulation results reveal that the total delay of the proposed DTDyC is 92.77pS at 1 V of power supply. The Monte Carlo simulation is performed for 1-sigma offset voltage which is found as 2.085 mV.

Keywords Low-power dissipation · High-speed · Offset voltage · Double-tail dynamic comparator (DTDyC) · Kickback noise · MOS capacitor

1 Introduction

In today scenario, high-speed and low-power CMOS dynamic latched comparators are getting attention in the application of mixed-signal ICs such as analog-to-digital converters (ADCs) [1–3]. These ADCs are essential component to design the memory

A. K. Dubey · V. Varshney (✉) · A. Kumar · P. K. Pal · R. K. Nagaria
ECED, MNNIT Allahabad, Prayagraj, UP 211004, India
e-mail: rel1551@mnnit.ac.in

A. K. Dubey
e-mail: akd8292@gmail.com

A. Kumar
e-mail: rel1604@mnnit.ac.in

P. K. Pal
e-mail: pratoshkumar1075@gmail.com

R. K. Nagaria
e-mail: rkn@mnnit.ac.in

D. Harvey et al. (eds.), *Advances in VLSI, Communication, and Signal Processing*, Lecture Notes in Electrical Engineering 683,
https://doi.org/10.1007/978-981-15-6840-4_7

sensor amplifiers [4], medical instruments, operational trans-conductance amplifiers (OTAs) [5] and data receivers. Thus, these all applications require higher switching speed, low-power dissipation, high-input impedance and full-swing output ADCs. However, input offset in the comparator increases rapidly because of the different mismatches and process variation. This offset degrades the performance of the ADC in a large amount because single ADC has many comparators, which is the main heart of it. To accomplish the all requirement of ADCs, designed comparator should have high-speed and low-power dissipation along with less input offset voltage and less effect of kickback noise [8].

Based on the clock, comparator is classified into two categories, i.e., dynamic and static comparators. Since, static latched comparator [6] lacks because of the excess power dissipation and slow regeneration process [9], therefore, it cannot be used to design the low-power and high-speed integrated circuits (IC's). Hence, dynamic comparators [7, 8, 10] were designed to improve the speed and overcome the power dissipation. Dynamic comparators are usually used to design the high-speed ADCs because of the zero static power dissipation. Dynamic comparators consist of many stacked transistors which make it inappropriate for the low voltage operation. Thus, they require large voltage headroom for the proper functioning [14]. Moreover, dynamic comparator is not suitable for regeneration because of having only one current path and single latch for the amplification [15]. Since, power dissipation and die area become very large in order to obtain the high-speed of the comparator. Therefore, technology node is scaled down to improve the performance of the parameter. However, threshold voltage is not getting scale down in same proportion as same as the supply voltage in the technology advancement [9]. This limits the voltage headroom and input common mode range (ICMR) which degrades the offset voltage shielding range [11]. Consequently, input offset in the CMOS devices caused by the mismatches of capacitance and current factor are almost twice for each process variation at or below 90 nm CMOS technology. Therefore, this issue becomes key alarm for the designers to design a high-speed low-power comparator.

Double-tail dynamic comparators (DTDyCs) [12–22] are recently developed technique, which has two tail currents separately for amplification and for regeneration or comparison. Figure 1 demonstrates the schematic diagram of the conventional DTDyC [14] and hybrid DTDyC [15]. Here, It is noticed that by improving the speed and reducing the power dissipation in comparators, some other parameters, which have traditional trade-offs between each other, degrade the performance [20]. The offset voltage due to mismatch, PVT variation and the effect of kickback noise are such negative parameters. There are several techniques are used to design DTDyCs in recent years, such as bulk-driven PMOS load [17], charge steering technique [9], several hybrid design and low-power techniques [12, 15–18, 21, 22] and high-speed technique with kickback noise control [20]. However, these reported literatures show the trade-off between delay, power dissipation and offset voltage.

From the study, it can be observed that to design a high-speed and low-power conventional comparator is very challenging task. Therefore, a novel double-tail dynamic latch comparator is redesigned to enhance the speed and to reduce the power dissipation at optimize offset voltage. In this paper, a novel DTDyC is designed using

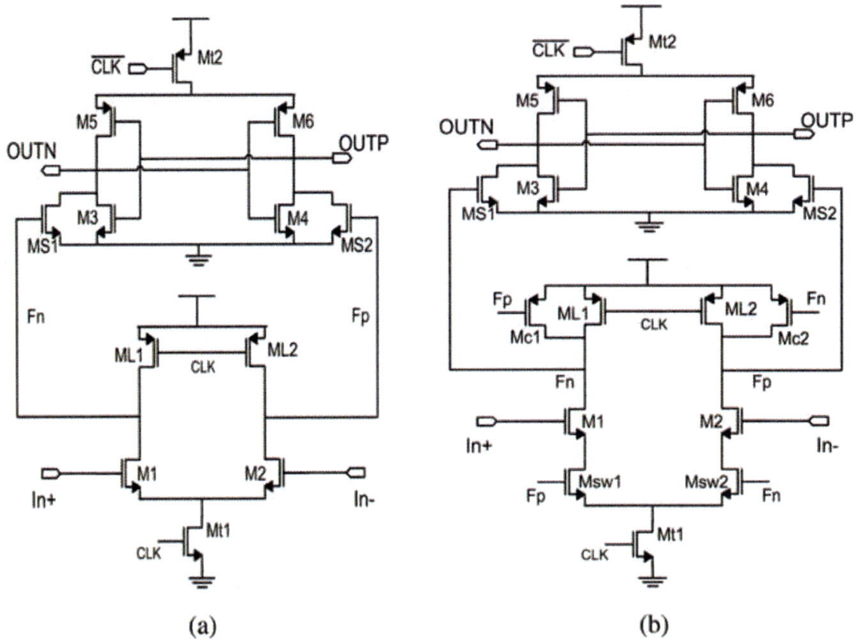

Fig. 1 Previous stated basic DTDyCs **a** conventional [14] and **b** DTDyC of [15]

the subsequent modification in amplification stage as well as in latch stage to improve the speed of latching or regeneration mode. The rest of the paper is organized as follows: Sect. 2 introduced the details of proposed DTDyC with delay analysis. The simulation results and discussions are given into Sect. 3. Finally, Sect. 4 concludes this work.

2 Proposed Double-Tail Dynamic Comparator

2.1 Circuit Implementation

The schematic diagram of the proposed DTDyC using the voltage-controlled capacitive (VCC) loads and dynamic inverter is shown in Fig. 2. In [15], transistors Mc1 and Mc2 are used to enhance the speed of the DTDyC, but it dissipates short-circuit power during the regeneration phase; hence, Msw1 and Msw2 are used to control it. In the proposed work, the VCC loads are used and to implement these loads MOS capacitors (C_{M1} and C_{M2}) are connected with the gate of M_{C1} and M_{C2}. As compare to the normal capacitors, these MOS capacitors are PMOS in nature and they will conduct only when gate voltage (V_{GSp}) is less than the threshold voltage (V_{THp}). In the proposed DTDyC, the gate voltage of load transistors will not change with the

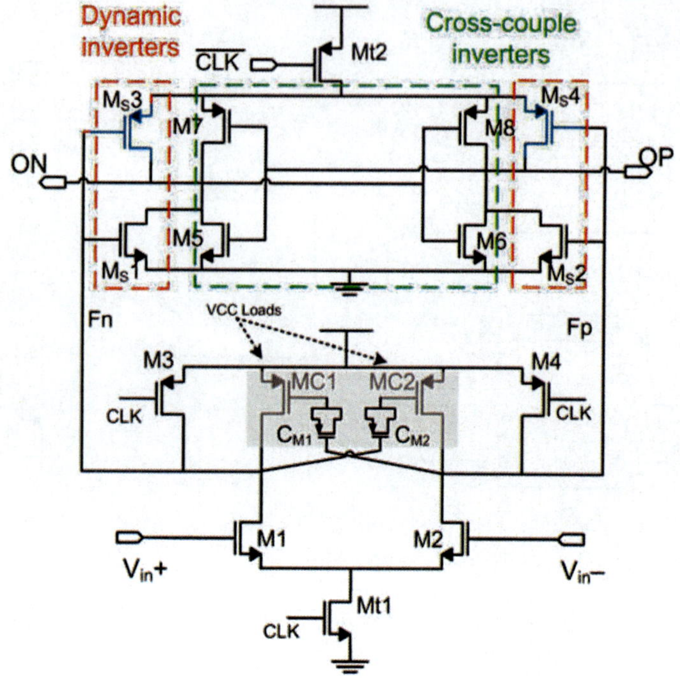

Fig. 2 Schematic diagram of the proposed DTDyC using the VCC loads (in amplification stage) and dynamic inverters (in latch stage)

change in voltage at Fn and Fp nodes. This is happened due to the application of MOS capacitors, since MC1 and MC2 are charged during the amplification mode just higher than $|V_{DD} - V_{THp}|$. They will not conduct again till voltage at Fn and Fp (V_{Fn} and V_{Fp}) nodes go below the $|V_{DD} - V_{THp}|$, i.e., near the end of the regeneration phase.

In the [14, 15], the dynamic NOR operation has been performed by the latch stage which is much slower than a dynamic NOT (inverter). Hence, in the proposed DTDyC, the latch stage is designed with dynamic inverter by the introduction of transistors M_S3 and M_S4. Now, the latch stage is designed with the four dynamic inverters, in which two inverters are cross-coupled for latching purpose. The speed of the latch stage is improved due to this implementation.

2.2 *Working of the Proposed DTDyC*

Figure 3 illustrates the transient response of the proposed DTDyC for $\Delta V_{IN} = 5$ mV, $V_{DD} = 1$ V and $V_{CM} = 0.8$ V. The amplification mode starts with CLK = logic '0'. In this mode, M3 and M4 are on, Mt1 and Mt2 are off. Initially, voltage at Fn node,

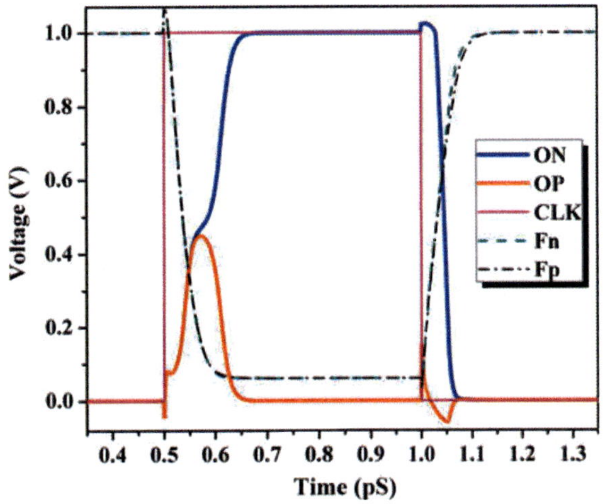

Fig. 3 Transient output of the proposed DTDyC for $\Delta V_{IN} = 5$ mV, $V_{DD} = 1$ V and $V_{CM} = 0.8$ V

V_{Fn} and voltage at Fp node V_{Fp} are at zero, due to which the voltage across C_{M1} and C_{M2} are zero and transistors MC1 and MC2 are on. Since transistors MC1, MC2, M3 and M4 are conducting, the parasitic capacitances at Fn and Fp nodes (C_{Fn} and C_{Fp}) start charging. As soon as V_{Fn} and V_{Fp} reach the threshold of Ms1 and Ms2, voltages at OP and ON nodes (V_{OP} and V_{ON}) are reset and equals to the ground. The regeneration phase starts when CLK = logic '1' (rising edge of CLK). In this phase, the Mt1 and Mt2 are on, C_{Fn} and C_{Fp} start discharging, V_{OP} and V_{ON} start increasing. Now depending on the condition of input voltage, $V_{in}+ > V_{in}-$, the discharging rate of C_{Fn} is larger than C_{Fp}, which turns on MC2 before MC1 and starts charging C_{Fp}. Similarly, the rising rate of V_{OP} is greater than the V_{ON}, which turns on M6 before M5, and hence V_{ON} starts decreasing. Finally, at the end of regeneration phase, $V_{OP} = V_{DD}$ and $V_{ON} = 0$. The proposed DTDyC will perform vice versa, for the case, when $V_{in}- > V_{in}+$.

Delay Analysis. The delay analysis of DTDyCs is presented in several styles in reported works. The delay of conventional DTDyC and hybrid DTDyC is presented in [15]. A novel analysis technique is introduced in Ref. [17, 20]. The delay equation for the proposed DTDyC is derived on the basis of these analyses. As explained in [15, 17], the total delay time (T_{Delay}) of conventional DTDyC is,

$$T_{Delay} = t_0 + T_{Latch} = 2\frac{V_{thn} \cdot C_{Lout}}{I_{t2}} + \frac{C_{Lout}}{g_{m,eff}} \cdot \ln\left(\frac{V_{DD}/2}{\Delta V_0}\right) \quad (1)$$

where C_{Lout} is the output load capacitor, t_0 is the charging time of C_{Lout} until the transistors M5 and M6 turns on. T_{Latch} is the latch stage delay, I_{t2} is the tail current used in latch stage (due to transistor Mt2), $g_{m,eff}$ is the effective transconductance of latch stage and ΔV_0 is the output voltage difference at $t = t_0$.

In the proposed DTDyC, the effective transconductance of latch stage is improved by the introduction of transistors MS3 and MS4. The parasitic capacitances at Fn/Fp node, $C_{LFn(p)}$ and MOS capacitors, $C_{M1,2}$ are in parallel during regeneration phase and during the amplification, the current will be divided into them. Hence, the ΔV_0 can be derived for the proposed DTDyC by considering the above said statements, which is given as:

$$\Delta V_0 = 4V_{thn}|V_{thp}| \cdot \frac{(g_{mS1,2} + g_{mS3,4})g_{m1,2}}{I_{t2}} \cdot \frac{C_{M1,2}}{C_{LFn(p)}} \Delta V_{IN} \cdot \exp\left(\frac{G_{m,eff1} \cdot t_0}{C_{LFn(p)} + C_{M1,2}}\right) \quad (2)$$

where $g_{mS1,2}$ and $g_{mS3,4}$ are the effective transconductance of Ms1/Ms2 and Ms3/Ms4, respectively, and $g_{m1,2}$ is the effective transconductance of driver transistors M1/M2. The ΔV_{IN} is the input voltage difference and $G_{m,eff1}$ is the total transconductance of amplification stage. Equation (2) suggests that the value of $C_{LFn(p)}$ should be small and $C_{M1,2}$ should be moderate for the effective ΔV_0 and small latch stage delay.

Similar to the derivation of T_{Latch} for DTDyCs in [15, 20], the effective transconductance of latch stage is improved because of $g_{mS3,4}$. Hence, by improving the latch stage transconductance, the total delay time in the term of ΔV_0 for the proposed DTDyC is expressed as:

$$T_{delay} = t_0 + T_{Latch} = 2\frac{V_{thn} \cdot C_{Lout}}{I_{t2}} + \frac{C_{Lout}}{g_{m,eff} + g_{mS1,2} + g_{mS3,4}} \cdot \ln\left(\frac{V_{DD}/2}{\Delta V_0}\right) \quad (3)$$

Here, the value of ΔV_0 is derived in Eq. 2 for the proposed DTDyC. From the above discussion, it may be observed that the speed of the proposed DTDyC is improved as compared to the DTDyCs of [14–16]. This improvement is seen in the terms of effective transconductance of the latch stage and output impedance of the load transistors.

3 Simulation Results and Discussion

The proposed DTDyCs are designed with CADENCE and results are simulated with SPECTRE at 90 nm CMOS technology node. Here, we have selected the values as $C_{Lout} = 2\text{fF}$, $C_{Fn} = C_{Fp} = 2\text{fF}$, $V_{DD} = 1$ V, $V_{CM} = V_{DD} - 0.2$ V, $\Delta V_{IN} = 5$ mV and $f_{CLK} = 500$ MHz. The effect of kickback noise is simulated with $R_{TH} = 8$ kΩ [8]. The energy per conversion (EPC) is defined in Ref. [12].

The variation of the performance parameters with respect to V_{DD} at different ΔV_{IN} for the proposed DTDyC (a) delay, (b) average power dissipation, (c) EPC are shown in Fig. 4. Here, the mean value of delay and average power dissipation is 130.3 pS and 2.012 μW, respectively, at $V_{DD} = 1$ V and $\Delta V_{IN} = 5$ mV. It is noted

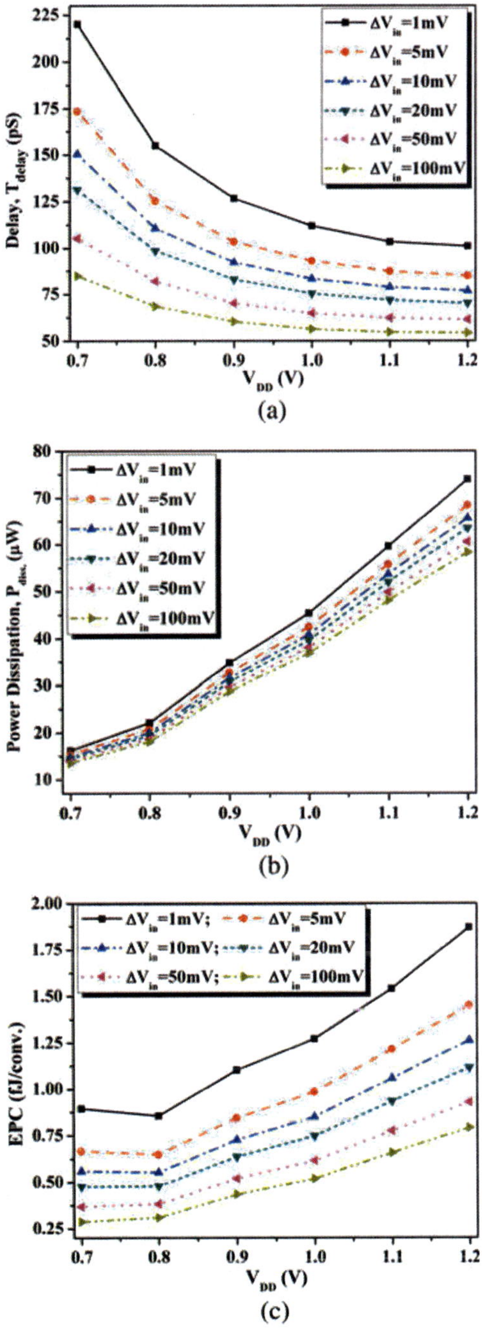

Fig. 4 Variation of the performance parameters with respect to V_{DD} at different ΔV_{IN} for the proposed DTDyC **a** delay, **b** average power dissipation, **c** EPC

that the proposed DTDyC can operate at $V_{DD} = 0.6$ V with delay = 281.48 pS, average power dissipation = 8.65 µW and EPC = 0.85 fJ/conv.

The comparison of the proposed DTDyC, DTDyCs of [14, 16] with respect to the V_{DD} at $\Delta V_{IN} = 5$ mV (a) delay, (b) average power dissipation, (c) EPC are shown in Fig. 5. Here, it is observed that the proposed DTDyC has smallest delay, consume moderate power and lowest EPC as compared to the DTDyCs of [14, 16]. The Monte Carlo simulation result for the 1-sigma offset voltage is shown in Fig. 6. Here, the 1-sigma (σ) offset voltage is 2.085 mV for $\Delta V_{IN} = 5$ mV and $V_{CM} = 0.8$ V.

The layout of the proposed DTDyC is shown in Fig. 7. The active area is 49.18 μm^2 (9.09 µm × 5.41 µm). The comparison of pre-layout and post-layout (RCX-RC) simulation of the proposed DTDyC (a) delay, (b) power dissipation is shown in Fig. 8. The variation in delay and power dissipation occurs due to the effective parasitic capacitances and resistances after RCX-RC extraction.

The performance summary and comparison of the proposed DTDyC with respect to the DTDyCs of [14–16] simulated under the same simulation environment are shown in Table 1. Here, the V_{DD}, ΔV_{IN}, V_{CM}, C_{Lout}, etc., are same for simulations. It is observed that the proposed DTDyC shows lowest delay, EPC, 1-sigma offset voltage and the effect of kickback noise as compared to the DTDyCs of [14–16].

Finally, the performance comparison of the proposed DTDyC with the recently reported DTDyCs is summarized in Table 2. The results of the reported DTDyCs are taken as it is published in the reported articles. However, this may not produce a fair comparison because of several reasons such as different simulation environment, different CMOS technology nodes, power supply and clock frequency. But it gives an idea about the performance of several DTDyCs in the terms of performance degradation parameters such as offset voltage and kickback noise. Here, it is observed that the proposed DTDyC gives low 1-sigma offset voltage and less affected by the kickback noise. Moreover, the proposed work also shows better EPC.

4 Conclusion

In this paper, a novel approach to design double-tail dynamic comparator is introduced. The proposed DTDyC is designed with a new amplification stage with VCC loads and a new latch stage with dynamic inverters. The proposed work is simulated with SPECTRE at 90 nm CMOS process with 1 V of power supply. The simulation result reveals that the proposed DTDyC has lowest delay and EPC as compared to the DTDyCs of [14, 16]. The delay of the proposed DTDyC is reduced by more than 59% and 42% as compared to the DTDyCs of [14, 16], respectively. The EPC of the proposed DTDyC is reduced to 63.69% and 50% as compared to the DTDyCs of [14, 16], respectively. Besides it, the proposed DTDyC also has small offset voltage due to mismatch and very less affected by kickback noise.

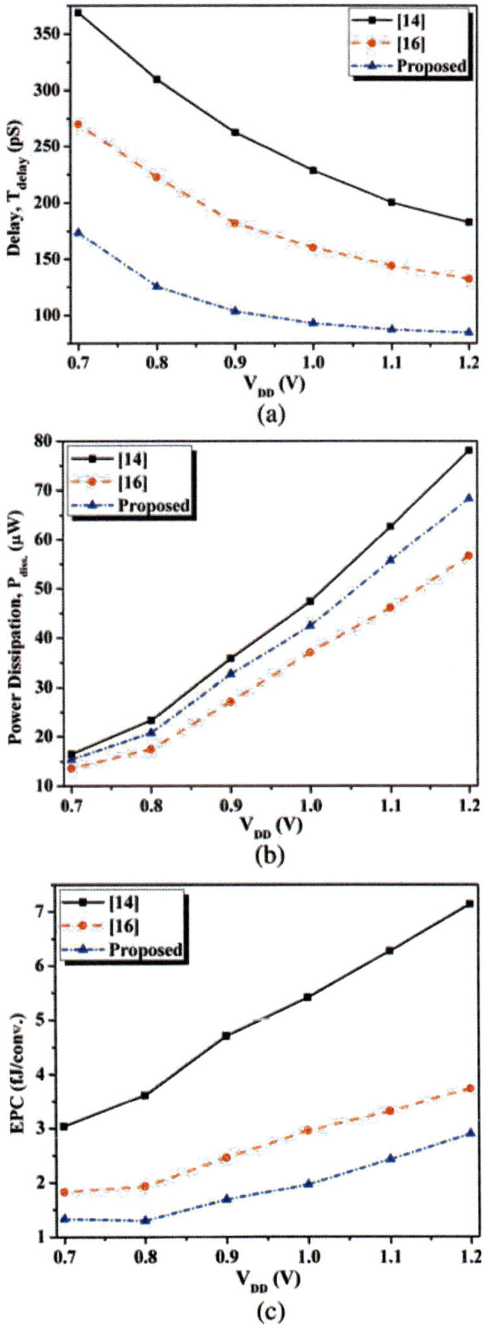

Fig. 5 Performance comparison of the proposed DTDyC, DTDyCs of [14, 16] with respect to the V_{DD} at $\Delta V_{IN} = 5$ mV **a** delay, **b** average power dissipation, **c** EPC

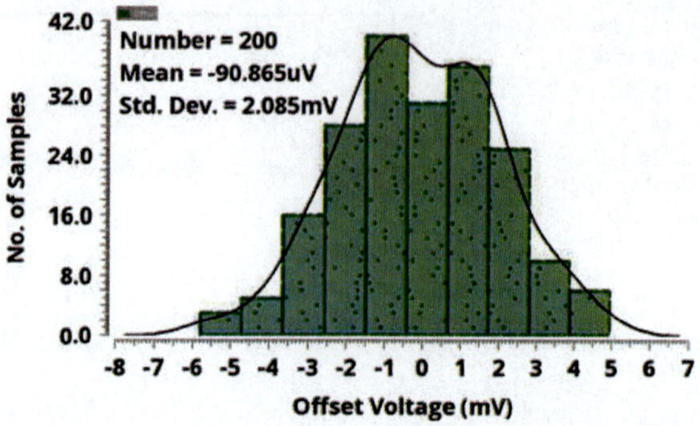

Fig. 6 Monte Carlo simulation result for the 1-sigma offset voltage

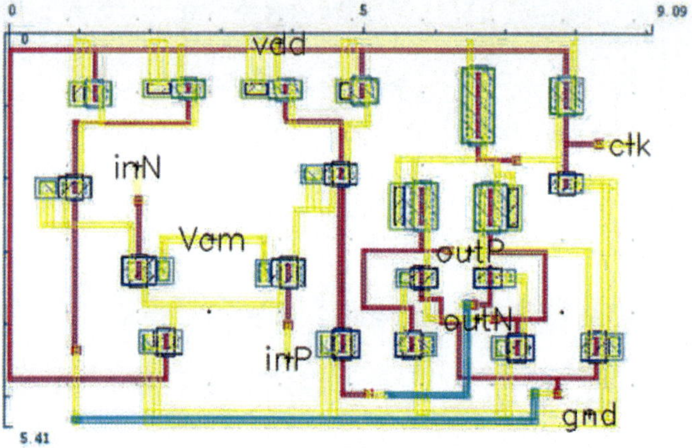

Fig. 7 Layout of the proposed DTDyC (area = 49.18 μm^2)

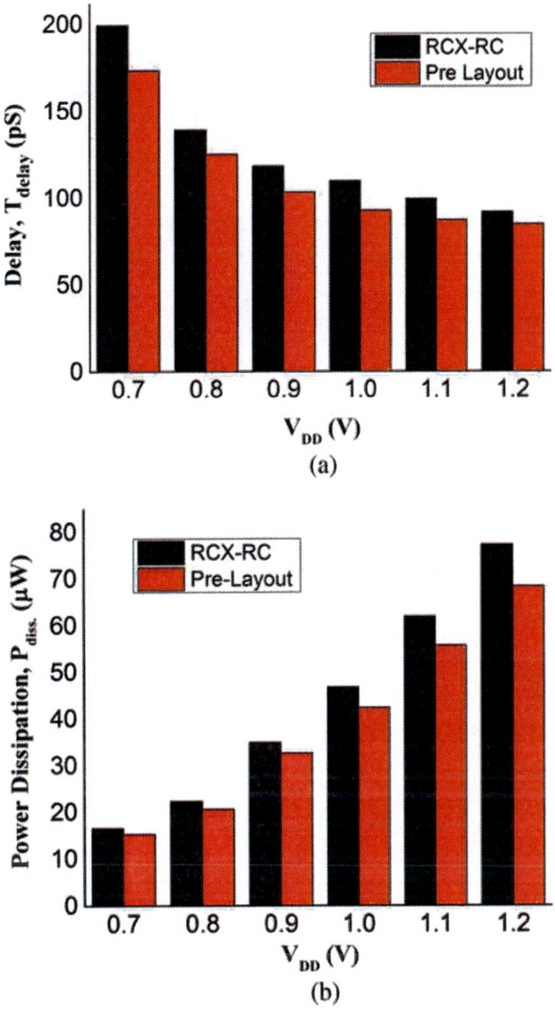

Fig. 8 Comparison of pre-layout and post-layout (RCX-RC) simulation of the proposed DTDyC **a** delay, **b** power dissipation

Table 1 Performance summary and comparison of DTDyCs simulated under the same simulation environment

Parameters	[14]	[15]	[16]	Proposed DTDyC
Maximum sampling rate (GHz)	5	8.3	7.7	12.5
Total delay time (ps)	228.49	143.67	159.84	92.77
Power dissipation (μW)	47.4	67.23	36.906	42.39
1-sigma offset voltage (mV)	3.49	5.93	3.14	2.085
EPC (fJ/Conv.)	5.415	4.829	2.949	1.966
Peak input voltage error (μV)	3.326	3.868	4.762	4.362

Table 2 Performance comparison of the proposed DTDyC with recent reported works

Parameters	[12]	[14]	[15]	[16]	[21]	Proposed DTDyC
CMOS process (nm)	90	180	180	180	90	90
Supply voltage (V)	1.1	0.8	0.8	1.8	1	1
Clock frequency (GHz)	0.25	0.5	0.5	4.54	–	0.5
Maximum sampling rate (GHz)	7	0.9	2.4	5.2	3	12.5
Total delay time (ps)	212.2	940	294	193	170	92.77
Power dissipation (μW)	8.923	–	51	420	162	42.39
1-sigma offset voltage (mV)	–	7.89	7.8	2.5	16.3	2.085
EPC (fJ/Conv.)	0.473	300	240	40.5	59.2	1.966
Peak input voltage error (V)	–	51.3 m	43 m	–	–	4.362μ

Acknowledgements The authors would like to thank Ministry of Electronics and Information Technology (MeitY), Govt. of India, New Delhi, for providing research funds.

References

1. Zhu Z, Qiu Z, Shen Y, Yang Y (2014) A 2.67 fJ/c.-s. 27.8 kS/s 0.35 V 10-bit successive approximation register analogue-to-digital converter in 65 nm complementary metal oxide semiconductor. IET Circuits Devices Syst 8(6):427–434
2. Zhu Z, Liu M, Wang Q, Yang Y (2014) A single-channel 8-bit 660MS/s asynchronous SAR ADC with pre-settling procedure in 65 nm CMOS. Microelectronics J 45(7):880–885
3. Ragab K, Chen L, Sanyal A, Sun N (2015) Digital background calibration for pipelined ADCs based on comparator decision time quantization. IEEE Trans Circuits Syst II Express Briefs 62(5):456–460
4. Yang B-D (2014) 250-mV supply subthreshold CMOS voltage reference using a low-voltage comparator and a charge-pump circuit. IEEE Trans Circuits Syst II Express Briefs 61(11):850–854
5. Yamamoto K, Carusone A (2012) A 1-1-1-1 MASH delta-sigma modulator with dynamic comparator-based OTAs. IEEE J Solid-State Circuits 47(8):1866–1883
6. Amaral P, Goes J, Paulino N, Steiger-Garção A (2002) An improved low-voltage low-power CMOS comparator to be used in high-speed pipeline ADCs. In: IEEE international symposium on circuits and systems. ISCAS 2002, vol 5, pp V-V. IEEE
7. Rahman LF, Reaz MBI, Yin CC, Marufuzzaman M, Rahman MA (2014) A high-speed and low-offset dynamic latch comparator. Sci World J
8. Figueiredo PM, Vital JC (2006) Kickback noise reduction techniques for cmos latched comparators. IEEE Trans Circuits Syst II Express Briefs 53(7):541–545
9. Dubey AK, Srivastava P, Pattanaik M (2015) Efficient technique to reduce power dissipation of Op-Amps at high speed. In: 2015 International Conference on Robotics, Automation, Control and Embedded Systems (RACE), pp 1–4. IEEE
10. Khorami A, Sharifkhani M (2016) Low-power technique for dynamic comparators. Electronics Lett 52(7):509–511

11. Dubey AK, Nagaria RK (2018) Enhanced gain low-power CMOS amplifiers: a novel design approach using bulk-driven load and introduction to GACOBA technique. J Circuits Syst Comput (JCSC) 27(13), 1850204 (17 pages)
12. Jain R, Dubey AK, Varshney V, Nagaria RK (2017) Design of low-power high-speed double-tail dynamic CMOS comparator using novel latch structure. In: 2017 4th IEEE Uttar Pradesh Section International Conference on Electrical, Computer and Electronics (UPCON), pp 217–222. IEEE
13. Huang S, Diao S, Lin F (2016) An energy-efficient high-speed CMOS hybrid comparator with reduced delay time in 40-nm CMOS process. Analog Integr Circuits Signal Process, 1–8
14. Schinkel D, Mensink E, Klumperink E, Tuijl EV, Nauta B (2007) A double-tail latch-type voltage sense amplifier with 18 ps setup + hold time. In: Solid-state circuits conference, 2007. ISSCC 2007. Digest of Technical Papers. IEEE International, pp 314–605. IEEE
15. Babayan-Mashhadi S, Lotfi R (2014) Analysis and design of a low-voltage low-power double-tail comparator. IEEE Trans Very Large Scale Integration (vlsi) Syst 22(2):343–352
16. Khorami A, Sharifkhani M (2017) Excess power elimination in high-resolution dynamic comparators. Microelectronics J 64:45–52
17. Dubey AK, Nagaria RK (2018) Optimization for offset and kickback-noise in novel CMOS double-tail dynamic comparator: a low-power, high-speed design approach using bulk-driven load. Microelectronics J 78:1–10
18. Savani V, Devashrayee NM (2018) Design and analysis of low-power high-speed shared charge reset technique based dynamic latch comparator. Microelectronics J 74:116–126
19. Gandhi PP, Devashrayee NM (2018) A novel low offset low power CMOS dynamic comparator. Analog Integr Circuits Signal Process 96(1):147–158
20. Dubey AK, Nagaria RK (2019) Design and analysis of an energy-efficient high-speed CMOS double tail dynamic comparator with reduced kickback noise effect. J Circuits Syst Comput (JCSC) 28(9), 1950157 (18 pages)
21. Jeon HJ, Kim Y-B (2010) A CMOS low-power low-offset and high-speed fully dynamic latched comparator. In: 23rd IEEE International SOC Conference, pp 285–288. IEEE
22. Dubey AK, Nagaria RK (2019) Low-power high-speed CMOS double tail dynamic comparator using self-biased amplification stage and novel latch stage. Analog Integr Circuits Signal Process. https://doi.org/10.1007/s10470-019-01518-7

Hybrid Forecasting Model Based on Nonlinear Auto-Regressive Exogenous Network, Fourier Transform, Self-organizing Map and Pattern Recognition Model for Hour Ahead Electricity Load Forecasting

Nida-e-Falak and M. M. Tripathi

Abstract Unlike other goods, electricity generated cannot be stored on an industrial scale. Adding to that, the supply and demand keeps on fluctuating in the market. If the supply and demand in the electricity market is mismatched, the change in speed of the generator causes change in system frequency. This may result in the addition or removal of either generation or connected load. The amount by which the electricity is lost in transmission and the loss due to congestion directly affects the price of electricity in the market. Therefore, it becomes vital to forecast the load of the electricity with high accuracy. A hybrid load forecasting model has been proposed in this paper that aims to reduce the overall forecasting error. Noise, the leading problem due to which load forecasting gets erroneous is addressed in this paper. The paper has also discussed the method to simplify the input waveform, for improving the performance of learning algorithm of the proposed load forecasting model.

Keywords Fourier analysis · Load forecasting · Mean absolute percentage error · Neural network · Pattern recognition · Self-organizing map

1 Introduction

Electricity load is influenced by a number of factors which includes temperature, pressure, day of the week, etc. Heuristics optimization techniques such as auto-regressive integrated moving average (ARIMA), fuzzy NNs (FNN), adaptive wavelet neural network (AWNN), hybrid PSO–ANFIS, grey model, wavelet-ARIMA–RBF, hybrid intelligent artificial neural network (ANN) and wavelet-FNN have been applied in forecasting to find optimal solution [1–12]. These techniques show good potential for load forecasting. In this paper, the effect of various factors and the weightage it

Nida-e-Falak (✉) · M. M. Tripathi
Delhi Technological University, New Delhi 110042, India
e-mail: nida_bt2k15@dtu.ac.in

D. Harvey et al. (eds.), *Advances in VLSI, Communication, and Signal Processing*, Lecture Notes in Electrical Engineering 683,
https://doi.org/10.1007/978-981-15-6840-4_8

holds in short-term forecasting of the electricity load is explored based on hybrid model. The hybrid model provides an hourly prediction of the expected load based on the model taking into account the load of the previous hours.

The core concept behind the modelling follows the Fourier analysis of the wave. It highlights the significance of the concept that any existing wave form can be constructed by superposition of the fundamental waves and their harmonics in the space-time frame. The hybrid model makes use of the curve fitting, nonlinear auto-regressive with exogenous input (NARX), self-organizing map (SOM) and pattern recognition using sigmoid hidden and softmax neuron. The curve fitting allows to add a feature which quantifies the smoothness of the curve and the degree of resemblance to the Fourier wave, obtained by superposition of sine waves. Since this work is based upon the Fourier analysis of the wave, the network is allowed to predict these waveforms once they are separated from the original condensed form using pass filters, and then, neural network is used to predict the model of the wavelet. The noise is separated out from the wave and is classified using a classifier into groups which best describes the event which caused the noise.

2 Methodology

2.1 Data Source, Analysis and Feature Vector

The electricity load varies due to a number of factors like hour of the day, day of the week, the average temperature of the day and the events being telecasted on the TV. All these factors are taken into account when computing the feature vector. Visual analysis of the system load shows periodic wave-like pattern which has been modelled using the methodology given in the paper. A nonlinear relation is observed between these variables and electricity load. However, each of the given factors has a varying correlation with the electricity load. The correlation maybe positive or negative. Each of these factors were chosen taking into consideration that they affect the electricity load in both domestic and industrial sectors of a region. The electricity consumption pattern of weekdays differs from that of weekends. Similarly, the load also differs depending on, whether it falls under office working hours or not, or the average temperature of the day. TV pickup is a term used in UK for electricity surge owning to synchronized usage of electrical appliances during commercial breaks of events/programs which attract significantly large audience. Additionally, the demand may drop during announced remembrance and energy awareness drives. Such events can be accounted with a feature vector of holidays and special days.

The dependence of these factors can be quantized using Pearson correlation coefficients. Pearson correlational coefficients, also known as bivariate correlation, measure the linear dependence of two variables. The linear dependence of variables establishes relation between variables without fitting them into a specific model. The formula to evaluate the dependence is given by Eq. 1.

$$\rho(X, Y) = \frac{1}{m-1}\sum_{i=1}^{m}\left(\frac{(X_i - \mu_A)}{\sigma_A}\right)\left(\frac{Y_i - \mu_B}{\sigma_B}\right) \tag{1}$$

where σ and μ are the standard deviation and mean, respectively. Population of data sets is represented by m.

The correlation coefficient is represented in matrix form for each pairwise variable combination as shown by Eq. 2.

$$M = [\rho(X, X)\rho(X, Y)\rho(Y, X)\rho(Y, Y)] \tag{2}$$

The value of correlational coefficients lies between −1 to 1. Data columns with linear and positive relationship are indicated by values close to 1, whereas data columns with linear and negative relationship are indicated by values close to −1. The 0 indicates no correlation between the data columns [13]. Since each column is directly related to each other, the diagonal entries are always equal to 1. Hence, correlation coefficient can be represented as Eq. 3.

$$M = [1\rho(X, Y)\rho(Y, X)1] \tag{3}$$

2.2 *Wavelet Transformation*

Wavelet analysis is used to transform the signal in time domain into frequency domain using mathematical modelling to analyse and process [14, 15]. In simple words, it is used to decompose a time series into constituent series separated by frequencies [16]. Popular choices for signal and image processing research include the Morlet and Daubechies wavelet transforms. While Daubechies wavelets exhibit sensible trade-off between parsimony and data richness, it had been observed that identical events across the considered time series are noticed in so many different fashions that it becomes impossible to acknowledge them by any prediction model [17]. Morlet wavelets, on the other hand, have a lot of consistent responses to similar events, however, have the weakness of generating more inputs than the Daubechies wavelets for the prediction models [18].

In this paper, wavelet transformation has been used for signal de-noising of the wavelets obtained after pass filtering. The maximal overlap wavelet transform also recognized as the redundant wavelet transform or stationary wavelet transform is used in this work. Using these models of wavelet transformation, several signals with different levels of resolution are obtained from a given signal by decomposition. These processes help to recover the original time domain signal without loss of information. Inverse MODWT has been used to reconstruct the waveform which is a reverse process [19].

2.3 A Subsection Sample

Fourier analysis is based on the concept that waves can be represented as sum of trigonometric functions. It follows that any wave can be mathematically be represented by the sum of a fundamental wave and its harmonics each as a multiple of a real integer as given in Eq. 4.

$$f(t) = a_0 + \sum_{m=1}^{n} [\cos(mt) + \sin(mt)] \tag{4}$$

where $f(t)$ is the wave function. In this paper, Fourier analysis for $n = 8$ has been done. Therefore, the function used for curve fitting of the wavelet changes to Eq. 5.

$$f(t) = a_0 + \sum_{m=1}^{8} [\cos(mt) + \sin(mt)] \tag{5}$$

The system load wave is divided into a number of wavelets, using pass filter, and these wavelets were fed into the neural network individually for forecasting the load. Each of these wavelets represents the most probable wave or summation of waves, which can be expressed using trigonometric functions. This has made the learning process of the model faster and has reduced the error by a significant percentage. The final wave model obtained in each part is summed together to obtain the concluding model for predicting the system load. The Fourier curve fitting has been used to calculate the goodness of the wavelet separated from the actual waveform. Once a wavelet is found satisfactory, then it is used for model prediction.

2.4 Nonlinear Auto-Regressive Exogenous Model (NARX)

NARX is a neural network algorithm that makes use of nonlinear auto-regression of exogenous input to produce a model used for prediction of variables under given input criteria. The model of such an algorithm is represented in Fig. 1. The performance of such neural network is decided by the chosen hidden layer numbers. Small number of hidden layers might not provide enough processing neurons to get the desired

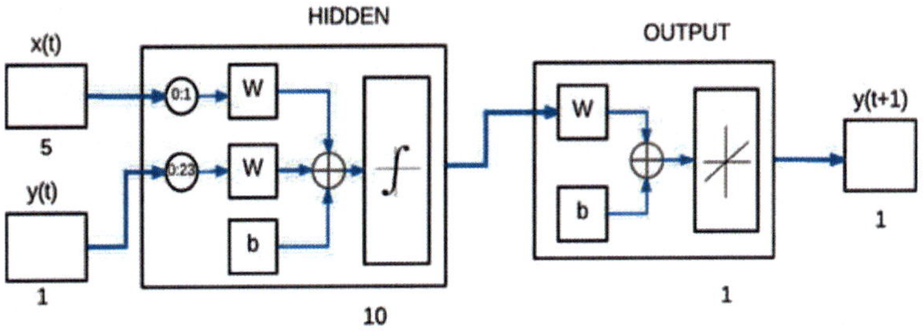

Fig. 1 Graphical representation of NARX learning model

result, ensuing under-fit results. On the other side, a large number of neurons might cause overfitting of the data in the training set and consequently, bad performance on the validation and testing data set. Modelling is started off with the default number of neurons, i.e., 10, and which turns out to be the optimized number of neurons required. The number of delays specifies the number of preceding inputs and outputs that should be taken into consideration when determining the function that forecasts the load at a particular hour. The number of input delays and feedback delays, which specifies the number of historical data to be taken into consideration while training the prediction function as given by Eq. 6, is predefined.

$$f(t) = f\big((x-1), \ldots, (x-t_i), (y-1), \ldots, \left(y-t_f\right)\big) \quad (6)$$

where
t_i = input delays, and t_f = feedback delays.

2.5 *Self-organizing Map (SOM)*

The self-organizing map introduced by T. Kohonen is known as artificial neural network with unsupervised learning [20]. A SOM transforms the income signal pattern of n dimensions to two-dimensional map of neuron. In SOM, neurons may be arranged in different structures but hexagonal lattice structure spread over a sheet where each neuron consists of weight vector is very popular [21]. The functioning of the SOM is divided into two parts such as training and mapping. Training process applies competitive learning, also called vector quantization, against error correction learning. The learning objective of SOM is to ensure that certain input patterns are responded in similar fashion by different parts of the network. This is how the human brain handles sensory information in separate parts of the cerebral cortex [22].

SOM is a very useful tool for visual analysis of the network.

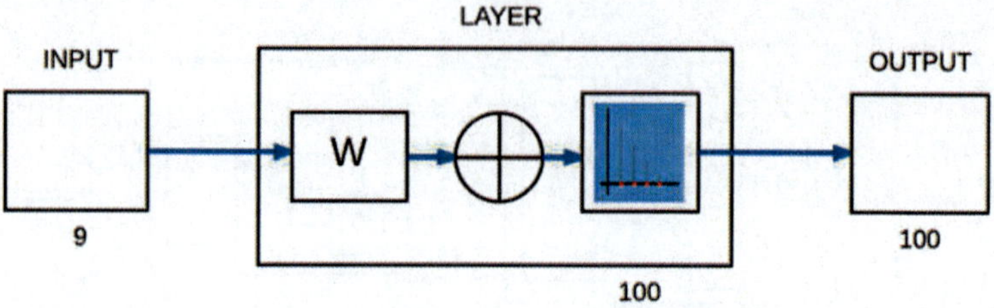

Fig. 2 Graphical representation of self-organizing map network model

1. The random values are used to initialize the weights.
2. The neurons compete against each other to become the winning neuron, provided that only one neuron will be activated in the map at any one time.
3. This neuron provides the spatial location of neighbouring neurons.
4. The excited neurons adjust their values so that the application of winning neuron improves.

The competitive process involves the use of discriminant function to measure the squared Euclidean distance between the weight vector and input vector. The discriminant function is shown in Eq. 7.

$$d_j(x) = \sum_{i=1}^{n} \left(x_i - w_{ij}\right)^2 \tag{7}$$

where x is the input vector and w_j weight vector for each neuron j.

During training, a series of calculations are done to spread the SOM throughout the input space. This process ensures that the entire set of training points is projected upon the SOM [23]. The neuron with the least Euclidean distance is declared as the winning neuron. SOM uses batch algorithm to train the data set. The final position of the neuron is mapped into a two-dimensional map for visual aid of the network functioning. The graphical representation of a SOM is shown in Fig. 2.

2.6 Pattern Recognition

The pattern recognition model is based on sigmoid neuron hidden layer and softmax function neuron output layer as shown in Fig. 3.

The SOM of the noise separated from the waveform is used as the target to train this model. Sigmoid neurons are a type of artificial neuron based on nonlinear function. The function is monotonically increasing, differentiable and bounded. These functions take multiple data as input into a single neuron or node and produce an output or activation depending on the weightage provided to each input. On the other

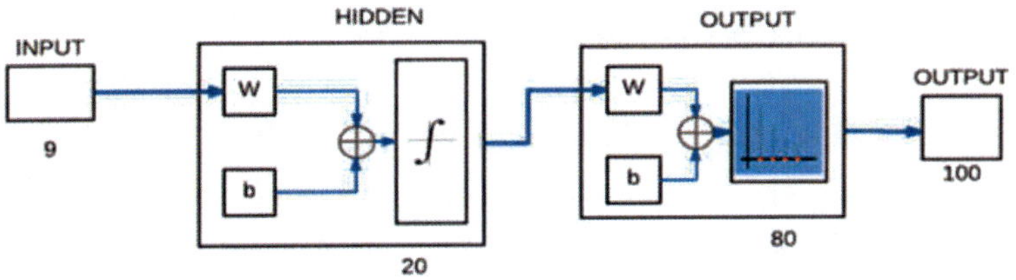

Fig. 3 Pattern recognition model with sigmoid hidden neuron and softmax output neuron

hand, the output function is based on softmax function. Softmax function is a generalized logistic function that reduces the M-dimension matrix input of real values to M-dimensional matrix of values lying between 0 and 1. Thus, the softmax function help to represent categorical distribution. The sigmoid and softmax function can be represented by the following Eqs. 8 and 9, respectively.

$$\text{Sigmoid function:}\sigma(z) = \frac{1}{1 + e^{-z}} \tag{8}$$

$$\text{Softmax function:}\delta(z)_k = \frac{e^{z_k}}{\sum_{i=1}^{M} e^{z_i}} \tag{9}$$

where $k = 1, 2\ldots$ M.

2.7 *Mean Absolute Percentage Error (MAPE)*

Mean absolute percentage error (MAPE) as shown by Eq. 10 is used as a measure of accuracy of the forecasting model. The report presents the MAPE error of each model and the final collective error of the model.

$$\text{MAPE} = \frac{100}{n} \sum_{i=1}^{n} \left| \frac{T_i - y_i}{T_i} \right| \tag{10}$$

where

n is the number of observations

T is the target/actual value of the electricity load

y is the predicted system load using the model

MAPE is scale independent, which allows to compare the forecast performance between different data sets and models.

3 Workflow of the Proposed Model

The workflow of the proposed model as shown in Fig. 4 is explained as follows.

1. Collect multi-source data: These data function as the feature vectors for the forecasting of system load. The electricity data is collected from the Pennsylvania Maryland New Jersey (PJM) electricity market website, and the weather data has been taken from the national climatic data centre (NCDC) website, USA.
2. Segregate the waveform into a number of wavelets: This is done using redundant wavelet transform. The smoothness of each wavelet so obtained is evaluated by Fourier function curve fit. The curve fit evaluation quantitates the closeness of the waveform to an eighth-degree Fourier function.
3. Filter out the noise from the waveform: Since the demand surge prediction is difficult to be forecasted with neural network only, a dedicated noise forecast network is modelled and trained separately using SOM and pattern recognition which improves the probability of forecast of the demand surge.
4. Compute the correlational covariance: It is done to compute the relation between the data and the features being used as inputs which is used to justify the use of a particular feature vector in the input matrix.
5. Training of NARX model: The multi-source data collected as the features (inputs) and the system load (without noise) as the target is used to train the NARX model. Each filtered waveform is modelled and trained into the network separately.
6. Adjustment of the input and hidden layers: The adjustment is done according to the performance required by the model. The feedback delay is set to 24 (it is the optimized number of delays as evaluated in this work, in accordance to training time and performance).
7. Training of clustering model: The noise obtained from the waveform is summed together and used to train clustering model, based on SOM algorithm, again using the multi-source data as features.
8. Evaluation of the performance of the NARX model: The MAPE is used for the performance evaluation of the SOM clustering algorithm using the SOM weight maps generated.

4 Case Study

To facilitate and validate the proficiency of the proposed model for forecasting of the load, a case study has been conducted on the data collected from PJM electricity market. The data under the case study is the system load for the month of January, in the year 2007 [24]. The weather data has been collected from various stations across the states of Illinois, Chicago and New Jersey [25].

The weather data was taken from National Climatic Data Centre (NCDC), which operates under the National Centres for Environmental Information and National

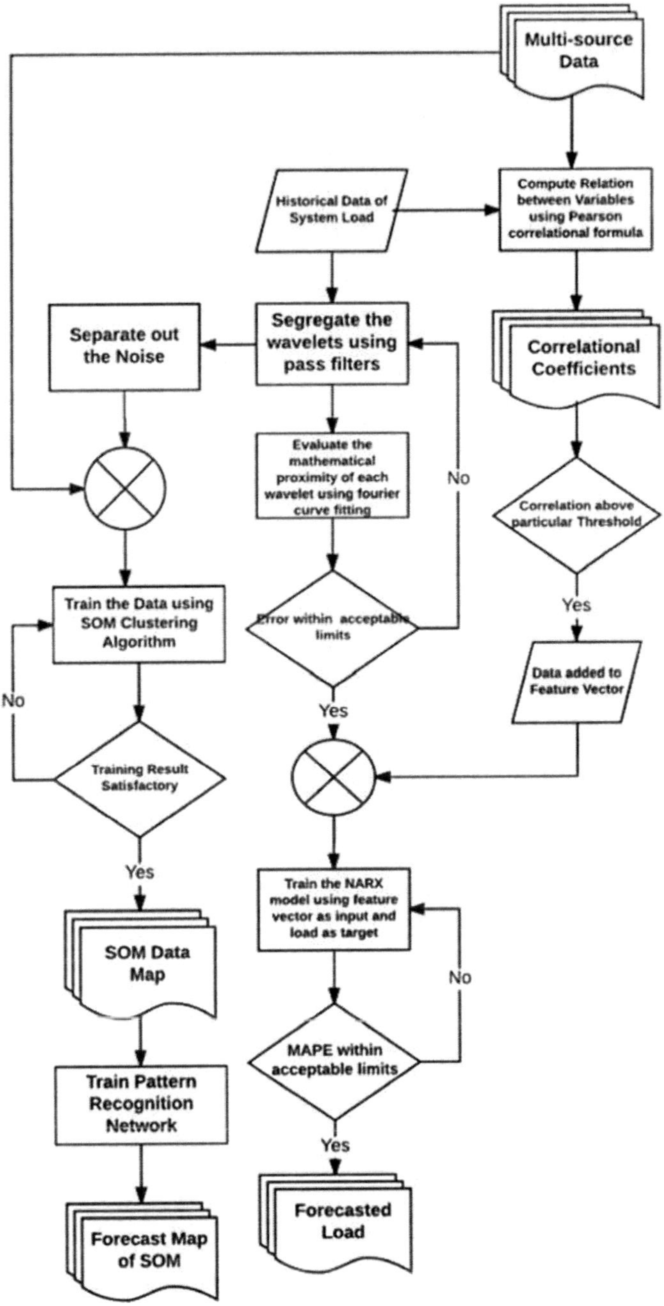

Fig. 4 Flowchart depicting the proposed methodology of forecasting loads

Table 1 Correlation

Types of correlation w.r.t system load	Correlation value (%)
Day of the week	44.88
Hour of the day	1.87
Hourly temperature	3.18
Dew point	1.08
Heat index	3.46
Wind-chill	5.48
Holiday	13.68
Periodicity of data	44.84

Oceanic and Atmospheric Administration (NOAA). The performance of the various models have been analysed using MAPE or percentage error. The forecasting model is programmed using MATLAB code and runs on an Intel Core i5-6200U CPU. Approximate runtime for a 168 period is 10 min, which means it is capable of producing real-time forecasts. The input and output information is imported through Microsoft Office Excel worksheets with a pre-established format.

4.1 Correlational Coefficient

As mentioned earlier, the correlational coefficients establishes the relational between the parameters and the system load. The correlational coefficients of each of the parameters, with respect to system load, are given in Table 1.

The evaluation of correlational coefficients shows the dependence of the system load on the given external factors. The results show that the day of the week and periodicity of data have high correlation coefficients of 44.88% and 44.84%, respectively, with system load. On holidays, the demand goes down which is indicated by good correlation of holiday with system load. Hourly temperature, heat index and wind-chill show less correlation with system load as data of three months is only considered, and variation in these parameters is not noticeable during that period. The hour of the day and dew point have the least correlation with system load as these factors do not play a significant role in predicting the fundamental wave or the subsequent filtered wave harmonics.

4.2 Curve Fitting

To examine the goodness of fit of the wavelets separated from the original waveform, the Fourier curve fitting criteria is used. One such Fourier curve fit model of the wavelets is shown in Fig. 5. Fourier analysis has been applied to the wavelets, and

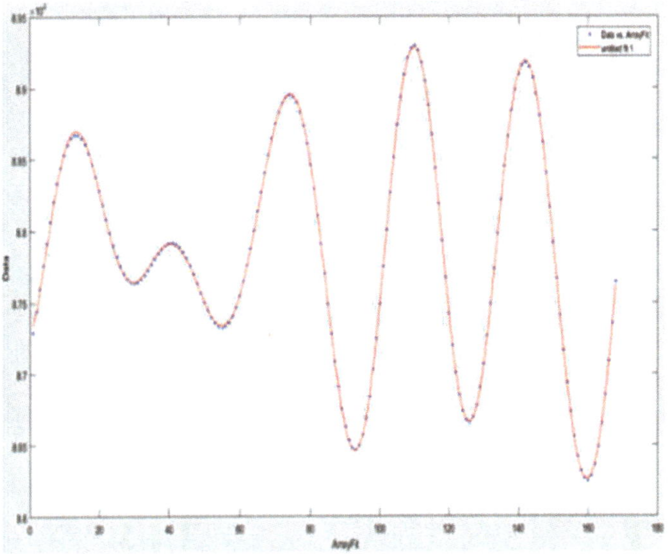

Fig. 5 Fourier curve fit model of the wavelets

error in curve fitting is measured using the root mean squared error (RMSE) value. Table 2 shows the RMSE errors obtained for various wavelets for three weeks. The maximum RMSE obtained is 49.44, minimum being 0.028 and the mode 0.028.

If the filtered wavelet does not fall under a given limit of acceptance of the Fourier curve fit error, it signifies that higher-order frequency harmonics are contributing to the given wavelet, which would obscure the prediction model performance. Therefore, the higher-order harmonics are again separated, from the wavelet, to obtain a better fit. The decisive criteria of curve fitting analysis is that the wave should consist of at most eighth-order harmonics. This enhances the performance of the model as each data set being modelled does not have more complex harmonics, with respect to the wavelet's fundamental frequency.

4.3 Model Performance

Please note that the first paragraph of a section or subsection is not indented. The first paragraphs that follows a table, figure, equation, etc., does not have an indent, either. The features presented in Table 1 were used as inputs to the proposed model of Fig. 4, and the system load (without noise) was used as target. The model was trained with 2 input delays, 24 feedback delays and 10 hidden layers. The input delay

Table 2 RMSE for Fourier curve fit of wavelets

Week waveform	Week 1	Week 2	Week 3
Lowpass band	6.8211	13.1440	13.7452
Midpass band 1	6.8211	13.1444	13.7452
Midpass band 2	0.8981	0.0827	0.5263
Midpass band 3	49.4400	8.4238	7.2738
Highpass band wavelet 1	0.3196	0.0601	0.6266
Highpass band wavelet 2	49.2729	9.8483	8.1643
Highpass band wavelet 3	0.3196	0.06014	0.6266
Highpass band wavelet 4	49.2729	9.8483	8.1643
Highpass band wavelet 5	0.3196	0.0601	0.6266
Highpass band wavelet 6	49.2729	9.8483	8.1643
Highpass band wavelet 7	0.3196	0.0601	0.6266
Highpass band wavelet 8	49.2729	9.8483	8.1643
Highpass band wavelet 9	6.821132	13.1444	13.7452
Fundamental wavelet 1	0.0027	0.0073	0.0127
Fundamental wavelet 2	0.3196	0.0601	0.6266
Fundamental wavelet 3	24.5968	1.2532	1.5297
Fundamental wavelet 4	0.8981	0.0827	0.5263
Highpass band 10	0.8981	0.0827	0.5263
Highpass band 11	49.4400	8.4238	7.2738
Highpass band 12	0.8981	0.0827	0.5263
Highpass band 13	49.4400	8.4238	7.2738
Highpass band 14	0.8981	0.0827	0.5263
Highpass band 15	49.4400	8.4238	7.2738
Fundamental wavelet 10	0.9045	0.0698	0.3390
Fundamental wavelet 11	49.2729	9.8483	8.1643
Fundamental wavelet 12	0.3196	0.06014	0.6266
Fundamental wavelet 13	49.2729	9.8483400	8.1643

ensures that the input data containing time series has finite dynamic response. From the complete data set, 70% was used for training, 15% for cross validation purpose and 15% for testing.

The performance of the model was recorded in the form of MAPE of each hour and the average MAPE of every week. The performance of the model in terms of actual and forecasted load is plotted in Fig. 6 for three consecutive weeks. The MAPE in the prediction of the load for each week has been plotted in Fig. 7. The model predicts the load (without noise) with an average MAPE of 0.0061, 0.0111 and 0.0259 for each week, respectively. As it is not possible to show the MAPE of each hour of the

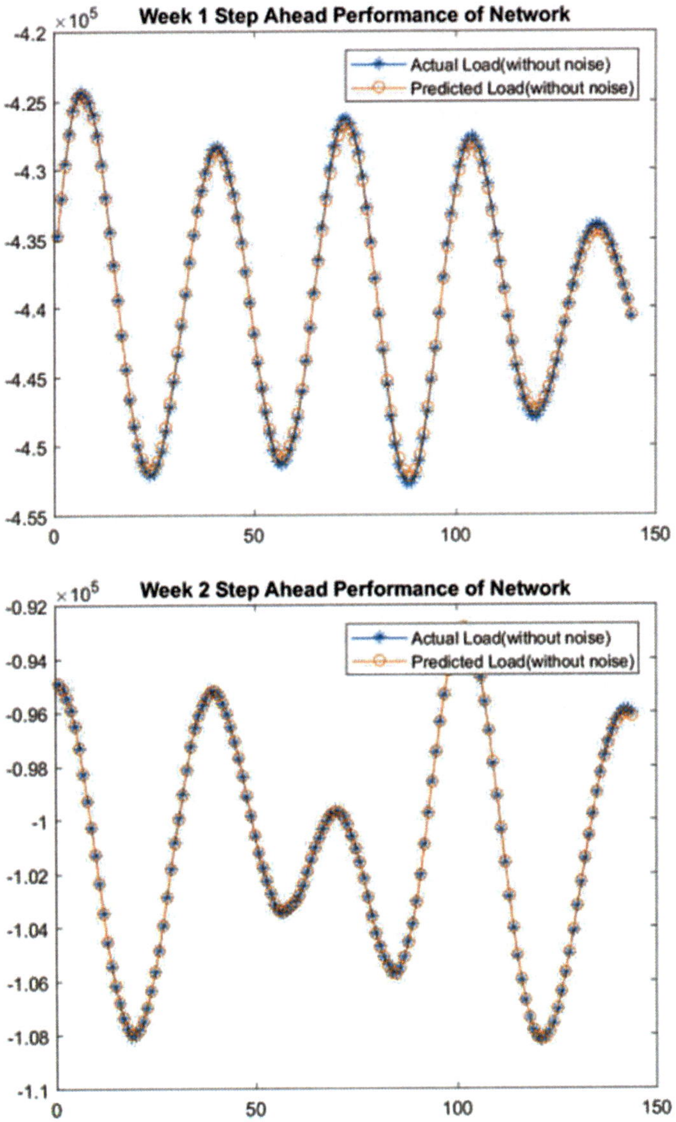

Fig. 6 Actual load versus predicted load waveform for week 1, 2 and 3

three weeks, the mode of each week has been calculated which comes to be 1.3418e-06, 1.3331e-06 and 5.4066e-07, respectively. Thus, the model displays remarkable accuracy in predicting the load. The MAPE and absolute error at each hour, incurred in the prediction load, by the model, for the three weeks have been plotted in Figs. 7 and 8, respectively.

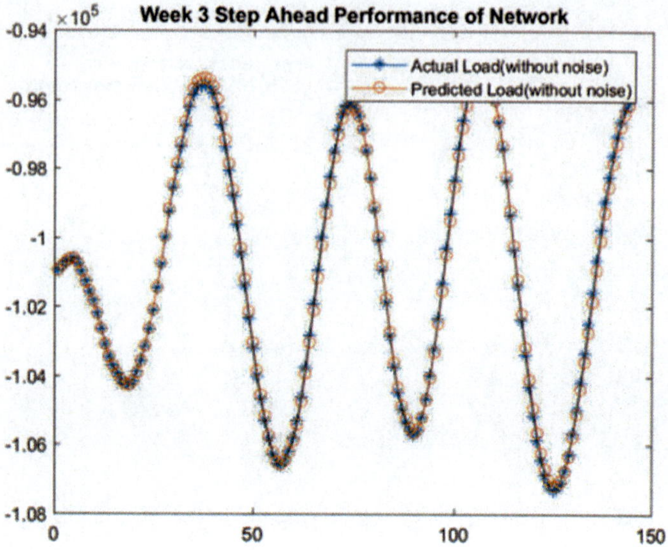

Fig. 6 (continued)

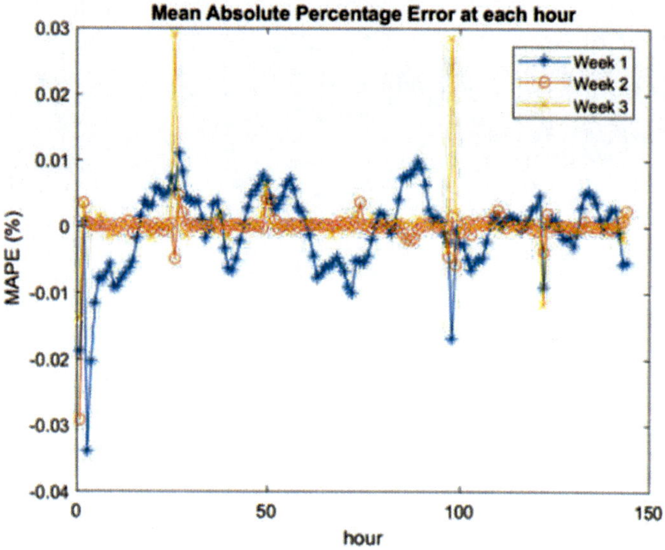

Fig. 7 MAPE of the first three weeks

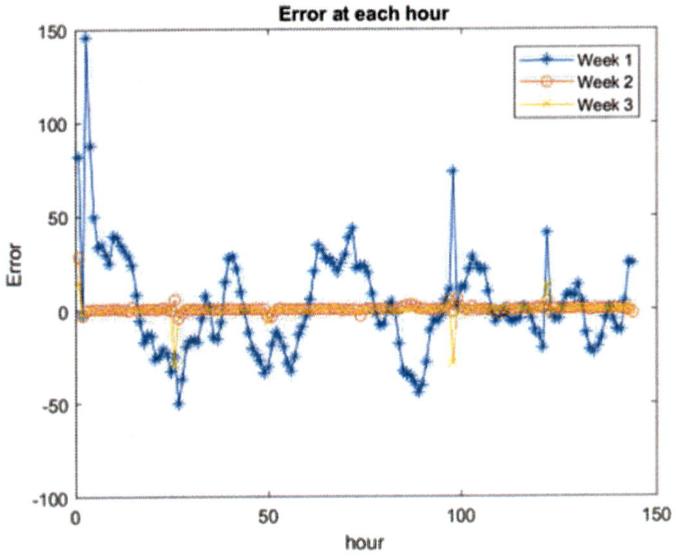

Fig. 8 Absolute error of the first three weeks

The model is applied for further three weeks. The actual load vs predicted load is plotted in Fig. 9, whereas MAPE and absolute error has been plotted in Fig. 10.

The noise obtained in the hourly load has irrational properties and is not affordable to use regression or NARX model for its learning. Therefore, clustering algorithm is used to develop a map of neurons that serves as a model for the noise. The noise in the data is due to the demand surge. The noise collected from each of the wavelets are combined with feature vectors, to form a matrix, which is then used for the training of the model. The result has been illustrated by SOM neighbourhood weight map in Fig. 11. The distance between two nodes is visually analysed by the colour code of the map. Darker the shade, farther the nodes are from each other. Each of these maps, so obtained, serve as targets for pattern recognition algorithm that develops model to learn the map and predict the expected surge in the system load in the next set of hours.

Thereafter, using the next set of hours along with its feature vectors as input and SOM map as the target output patternm recognition model learns to predict the noise. The result of the model prediction is represented by performance of the recognition model, percentage error and the MAPE as illustrated in Tables 3, 4 and 5, respectively. The error for each set of performance error has been calculated to be 0.0479, 0.0363 and 0.0438.

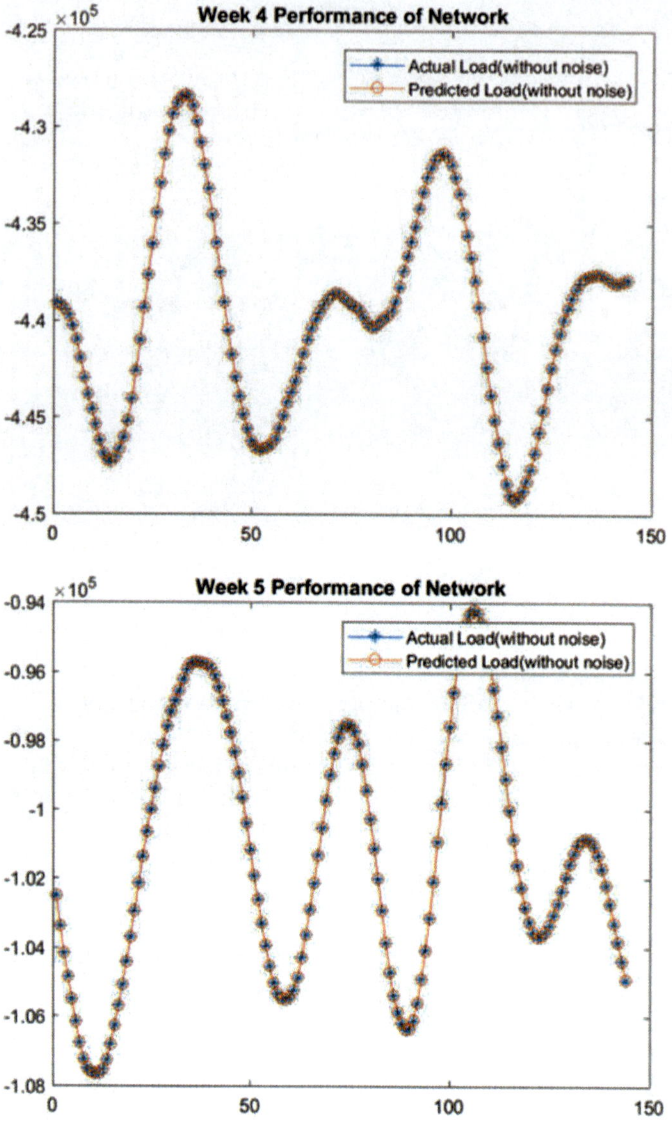

Fig. 9 Plot of actual load versus predicted load of the next three weeks

5 Conclusion

The proposed model shows considerable improvement in the accuracy of electricity load forecasting. The breakdown of the load waveform into its constituent wavelets improves the learning performance of the neural network model. Also, it is essential to treat the irregular noise in the waveform using a separate clustering model. In this

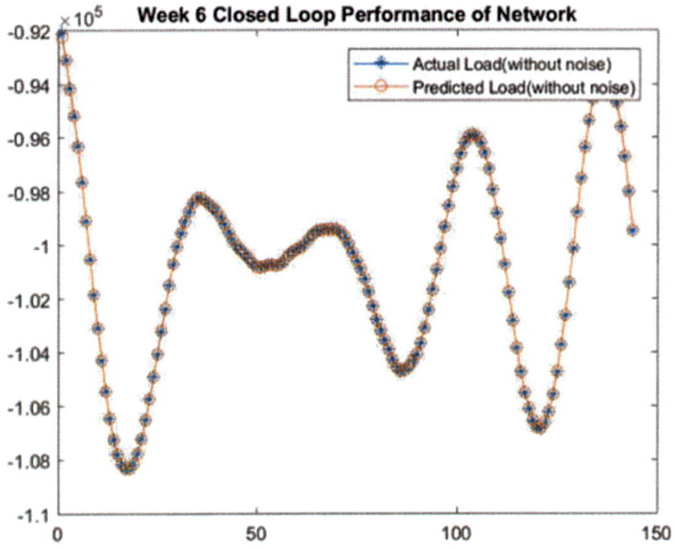

Fig. 9 (continued)

Table 3 Performance of the pattern recognition model for three weeks

Week	Performance of the noise: Filtered out of			
	Waveform	Fundamental wave	Midpass wavelet	Highpass wavelet
Week 1	0.0512	0.0391	0.0478	0.0530
Week 2	0.0354	0.0381	0.0337	0.0376
Week 3	0.0392	0.0500	0.0452	0.0405

Table 4 Percentage error of each noise set for three weeks

Week	Percentage error of noise: Filtered out of			
	Waveform	Fundamental wave	Midpass wavelet	Highpass wavelet
Week 1	0.94642	0.8511	0.9166	0.9821
Week 2	0.82142	0.8273	0.7797	0.8690
Week 3	0.85714	0.9642	0.8690	0.9107

Table 5 MAPE of each noise set for three weeks

Week	MAPE of noise: Filtered out of			
	Waveform	Fundamental wave	Midpass wavelet	Highpass wavelet
Week 1	35.6149	32.6549	45.2251	46.5078
Week 2	20.1489	35.6767	40.5609	34.8764
Week 3	26.2617	46.2516	42.2374	41.5840

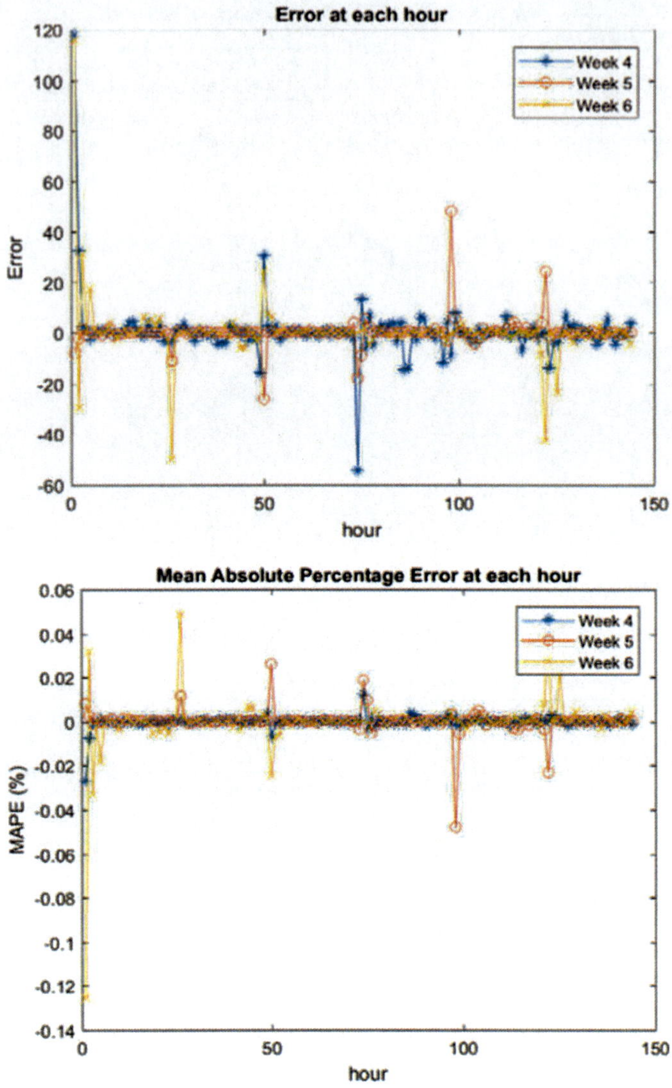

Fig. 10 Absolute error and MAPE for the consecutive three weeks

paper, the noise data is used to perform unsupervised learning on an algorithm which produces a map using clustering. The noise is then modelled by supervised learning algorithms of pattern recognition, using output of the unsupervised learning as the target. To conclude, the paper has proposed a hybrid model based on neural network algorithms, on different data sets, derived from the original system load waveform which is very accurate in forecasting the electricity load in the short term.

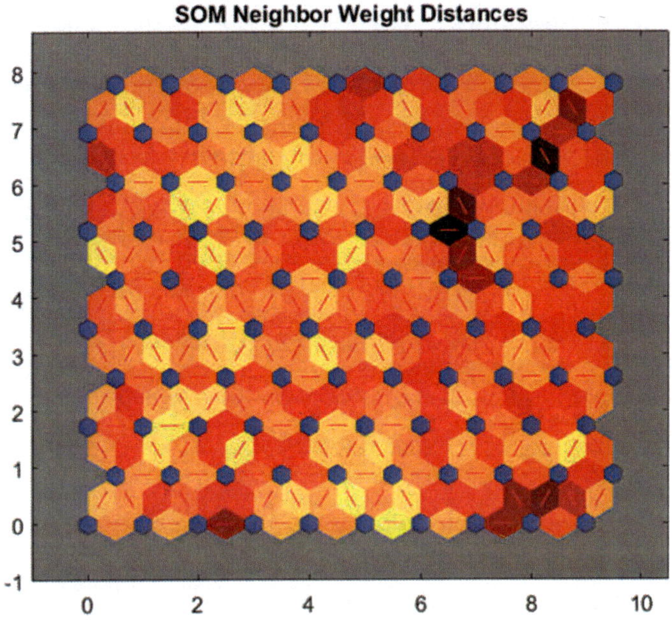

Fig. 11 SOM neighbour weight distance MAP

References

1. Contreras J, Espínola R, Nogales FJ, Conejo AJ (2003) ARIMA models to predict next day electricity prices. IEEE Trans Power Syst 18(3):1014–1020
2. Singhal D, Swarup KS (2011) Electricity price forecasting using artificial neural networks. Electron Power Energy Syst 33:550–555
3. Wu L, Shahidehpour M (2010) Hybrid model for day-ahead price forecasting. IEEE Trans Power Syst 25:1519–1530
4. Pousinho HMI, Mendes VMF, Catalo JPS (2012) Short-term electricity prices forecasting in a competitive market by a hybrid PSO–ANFIS approach. Electr Power Energy Syst 39:29–35
5. Lei M, Feng ZR (2012) A proposed grey model for short-term electricity price forecasting in competitive power markets. Electr Power Energy Syst 43:531–538
6. Shafiekhah M, Parsa Moghaddam M, Sheikh-El-Eslami MK (2011) Price forecasting of day-ahead electricity markets using a hybrid forecast method. Energy Convers Manage 52:2165–2169
7. Hooshmand RA, Amooshahi H, Parastegari M (2013) A hybrid intelligent algorithm based shortterm load forecasting approach. Electr Power Energy Syst 45:313–324
8. Amjady N (2006) Day-ahead price forecasting of electricity markets by a new fuzzy neural network. IEEE Trans Power Syst 21(2):887–896
9. Catalo JPS, Pousinho HMI, Mendes VMF (2011) Short-term electricity prices forecasting in a competitive market by a hybrid intelligent approach. Energy Convers Manage 52:1061–1065
10. Shayeghi H, Ghasemi A (2013) Day-ahead electricity prices forecasting by a modified CGSA technique and hybrid WT in LSSVM based scheme. Energy Convers Manage 74:482–491
11. Tripathi MM, Singh Sn, Upadhyay KG (2009) Price forecasting in competitive electricity markets: an analysis. In; Proceedings of international conference on energy engineering, Puducherry, India, pp 7–9

12. Yadav HK, Pal Y, Tripathi MM (2015) Photovoltaic power forecasting methods in smart power grid. In: IEEE Annual India Conference (INDICON), 2015 Annual IEEE, Delhi, India, pp 1–6
13. Correlational Coefficients. Mathworks (Online). https://in.mathworks.com/help/matlab/ref/corrcoef.htm
14. Yao SJ, Song YH, Zhang LZ, Cheng XY (2000) MODWT and networks for short-term electrical load forecasting. Energy Convers Manage 41(18):1975–1988
15. Ruey S (2005) Analysis of financial time series, 2nd edn. John Wiley & Sons, Inc
16. Aggarwal A, Tripathi MM. Performance measurement of a novel hybrid forecasting model using wavelet transform, time series time delay neural network, and error predicting algorithm for day-ahead electricity price forecasting
17. Masters T (1995) Neural novel and hybrid algorithms for time series prediction. Wiley, New York
18. Benaoudaa D, Murtaghb F, Starckc J-L, Renaud O (2006) Wavelet-based nonlinear multiscale decomposition model for electricity load forecasting. Neurocomputing 70:139–154
19. Al Wadi AS, Hamarsheh A, Alwadi H. Maximum overlapping discrete wavelet transform in forecasting banking sector
20. Kohonen T (1989) Self-organisation and associative memory, 3rd edn. Springer, Berlin
21. Vesanto J, Himberg J, Alhoniemi E, Parhankangas J (2000) SOM toolbox for Matlab5. ISBN 951-22-4951-0. www.cis.hut.fi/projects/somtoolbox/
22. Self-Organising Maps. Wikipedia (Online). https://en.wikipedia.org/wiki/Self-organizing_map
23. López M, Valero S, Senabre C, Aparicio J, Gabaldon A (2012) Application of SOM neural networks to short-term load forecasting: the Spanish electricity market case study. Electric Power Syst Res 91:18–27
24. Data Directory. PJM (Online). http://www.pjm.com/markets-and-operations/data-dictionary.aspx
25. Data Access, Land Based Station Data. National Center for Environmental Information (Online). https://www.ncdc.noaa.gov/data-access/land-based-station-data

Structural and Optical Characterization of EZO Thin Film for Application in Optical Waveguide

Lucky Agarwal, Richa Singh, and Shweta Tripathi

Abstract In the present paper, we report the growth of high-quality erbium-doped ZnO (EZO) thin films on n-silicon substrate. The deposited film may prove to be a capable material for use in optical waveguide due to high FSR and quality factor. The low-cost sol-gel spin coating method has been utilized to prepare the EZO thin film. The X-ray diffraction (XRD) result of the film exhibits single-crystalline nature of the film with (101) as the dominant orientation. In order to find the aptness of EZO thin film for application in waveguide, the ellipsometry measurement has been carried out to evaluate the optical parameters of EZO thin film. The reflectivity of EZO thin film in the visible region make it as a material of choice to use it as a transmitter in optical waveguide.

Keywords EZO · Sol-gel · Optical waveguide · n-silicon

1 Introduction

In recent years, research has been focused on various semiconductor materials. Among these materials, zinc oxide (ZnO) is a very interesting material because of its potential application in various fields such as photodetector, light-emitting diodes, gas sensor, solar cells and a transmitter in optical devices [1–4]. Zinc oxide belongs to a group of hexagonal wurtzite, 6 mm symmetry, and it is an important n-type semiconductor with a wide direct band gap 3.37 eV [5]. ZnO thin film can absorb UV radiation which renders it as a transparent conductive oxide (TCO) material for fabrication of optical devices [6]. ZnO is suitable for electronics and optical

L. Agarwal (✉)
Department of Electronics and Communication Engineering, Madanapalle Institute of Technology and Science, Madanapalle 517325, India
e-mail: rel1408@mnnit.ac.in

R. Singh · S. Tripathi
Department of Electronics and Communication Engineering, Motilal Nehru National Institute of Technology Allahabad, Prayagraj, Uttar Pradesh 211004, India
e-mail: shtri@mnnit.ac.in

D. Harvey et al. (eds.), *Advances in VLSI, Communication, and Signal Processing*, Lecture Notes in Electrical Engineering 683,
https://doi.org/10.1007/978-981-15-6840-4_9

devices because it is abundant in nature, thermally stable, amenable to wet chemically etching, bio-compatible, good thermal conductor, highly resistant to radiations and easily processable, which makes it a lower-cost material for the fabrication of high-performance optical devices [7, 8]. Among various semiconductor materials, ZnO has the highest refractive index contrast compared to that of glass materials and therefore can accomplish even smaller devices. Thus, ZnO has attracted much interest for optical devices based on microcavity since both passive and active functionalities are possible in a single material.

The properties of ZnO can be tuned by incorporating the rare earth element in its lattice. In the present paper, erbium (Er) ions are introduced in the lattice. The study of this paper reveals the influence of Er ion on the optical and structural properties of ZnO thin films for application in optical waveguide.

2 Experimental Details

Initially, the n-Si samples were cut in a size of 1×1 cm^2. Then, the cut samples were undergone for piranha cleaning. The piranha cleaning is done in two phases. In the first phase, the n-Si samples dipped in DI water were immersed in the boiling piranha solution that is concoction of H_2SO_4 and H_2O_2 in the ratio of 3:1. This phase removes the organic contamination from the Si surface. In the second phase, the DI cleaned samples were immersed in a HF solution to remove the grown silicon dioxide layer. Then, the samples were thoroughly rinsed in DI water.

The EZO thin films were deposited on the cleaned Si substrate via sol-gel spin coating method. To accomplish this, the sol was prepared by mingling the zinc acetate dehydrate ($Zn(CH_3COO)_2.2H_2O$), isopropanol and diethanolamine (DEA) to make a 0.75 M concentration solution. In the prepared solution erbium chloride hexahydrate ($ErCl_3.6H_2O$) was added as source of Er dopant atoms. The resultant solution was stirred at 70 °C and then kept for ageing. After that, EZO thin films deposited Si samples by spin coating process. The spin coater was spun with a speed of 2500 rpm for 40 s to get good adhesion of EZO film on Si substrate. The deposited thin films were prebaked at 120 °C for 10 min in oven to evaporate the solvent and organic residuals, and then the deposition was repeated several times to get the desired thickness. The deposited EZO films were annealed at 500 °C in a muffle furnace. The resultant EZO thin-film samples were then characterized to measure its structural and optical properties.

3 Result and Discussion

Figure 1 shows XRD spectra of 1 mol% EZO thin film plotted by using the extracted data obtained from X-ray diffractometer (Smart Lab by Rigaku). The deposited film shows sharp and intense peak along 100 direction. It reveals that Er doping converts

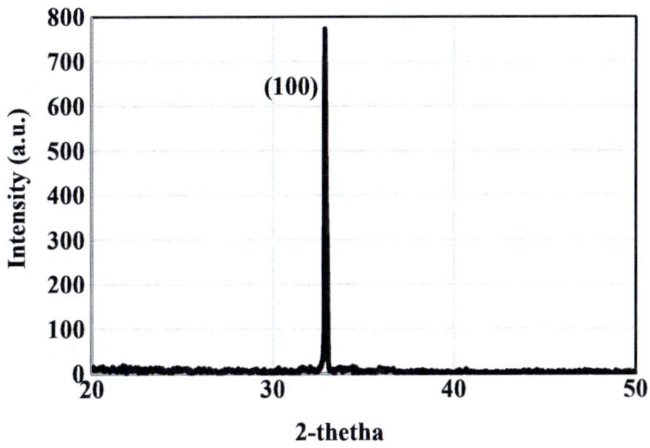

Fig. 1 XRD spectrum of EZO thin film

the polycrystalline ZnO into single-crystal structure. The interplanar distance (*d*), grain size (*D*) and FWHM are using the following relation [9].

$$\lambda = 2d \sin\theta$$

$$D = \frac{k\lambda}{\beta \cos\theta}$$

where λ is X-ray wavelength (0.154049 nm), β is FWHM of the peak, k is the constant which depends on the shape of the crystal (0.95 for spherical) and θ is Bragg's angle (in radians).

The lattice parameter *a*, *c* has been calculated using the following formula [10] and also been listed in Table 1 for all the orientations obtained by XRD result.

$$\frac{1}{d^2} = \frac{4}{3}\left(\frac{h^2 + hk + k^2}{a^2}\right) + \frac{l^2}{c^2}$$

Table 1 encompasses the structural parameter as derived from the XRD data. The XRD spectra of EZO thin films deposited on n-silicon reveal that the deposited thin film is single crystalline in nature. The XRD peak shows single peak occurred at 33.32°. The full width at half maximum of EZO thin film is 0.102°. The interplanar spacing between 100 planes is calculated as 2.68 nm.

Table 1 Structural parameters evaluated from XRD results

Type of thin film	Lattice spacing *d*	Lattice constant *a* (A°)	Lattice constant *c* (A°)	Grain size (nm)	FWHM
Undoped ZnO [11]	2.7181	3.251	5.205	12.9	0.119
1 mol% EZO	2.6812	3.0960	5.206	13.9	0.1027

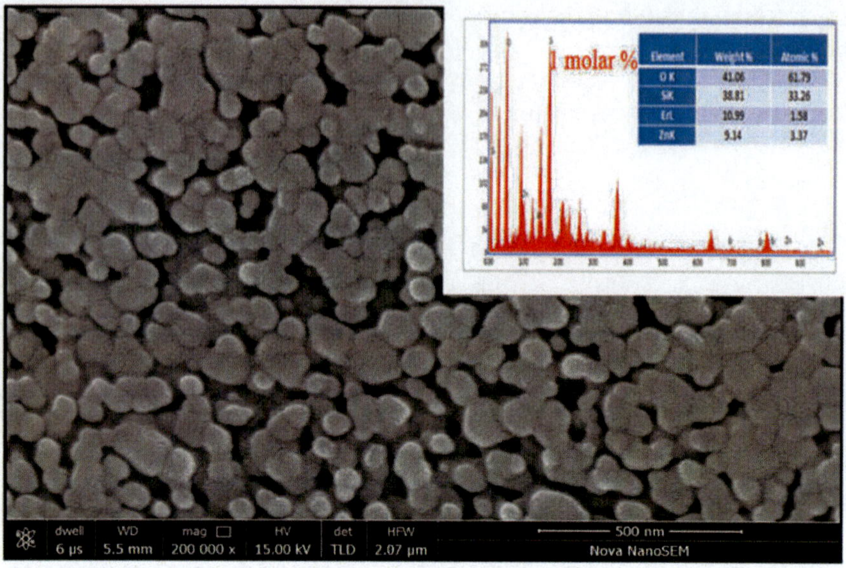

Fig. 2 FE-SEM image of EZO thin film; inset shows the EDX spectra of 1 mol% EZO thin film

The FE-SEM image of the EZO thin film is shown in Fig. 2. The SEM reveals that the deposited EZO thin film is uniformly distributed along the surface with small grains along the surface of the deposited sample. The grain sizes obtained from the FE-SEM image are around 15.2 nm which is in close agreement with the XRD data. Further, in order to confirm the presence of Er atoms in ZnO lattice, EDX analysis has been done. The EDX spectrum for 1 mol% EZO has been shown in Fig. 2. The inset of Fig. 2 reveals the atomic composition of elements present in EZO lattice. Data of Fig. 2 reveals that intended doping has been achieved in the deposited EZO thin film.

The optical reflectivity curve of 1 mol% EZO thin plot has been plotted with respect to the photon energy and has been shown in Fig. 3. It has been observed that the reflectivity of EZO thin film is more than that of undoped ZnO as reported by other authors [12]. The reflectivity is observed to be more than 90% in the visible region of the electromagnetic spectrum that ranges from 400 to 700 nm wavelength. This may be attributed due to the increase in roughness on the surface of EZO thin film due to the incorporation of Er ion in ZnO that causes more light to disperse in random direction that increases the reflectivity of EZO thin film [13]. This increased amount of reflectivity observed in EZO thin film makes it as a viable thin film to be used as a transmitting medium in any optical waveguide. Light follows the principle of total internal reflection to travel in a waveguide. This much high reflectivity of EZO film makes it an optimum choice for researchers to use EZO thin film as a transmitting medium of light in an optical waveguide.

The optical band gap energy of the EZO thin film shown in Fig. 4 has been plotted with respect to the photon energy by following Tauc's relationship between

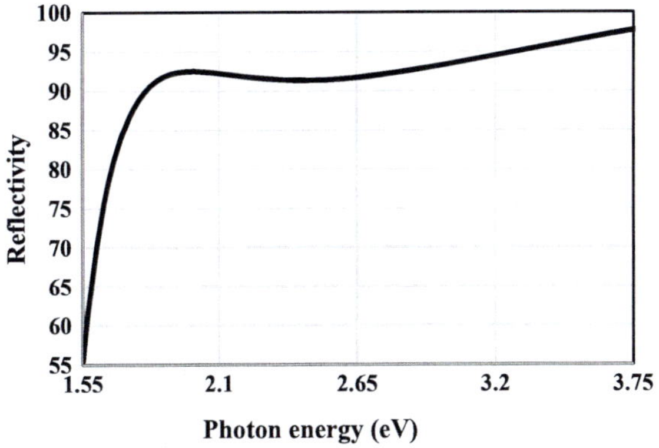

Fig. 3 Variation of reflectivity with photon energy

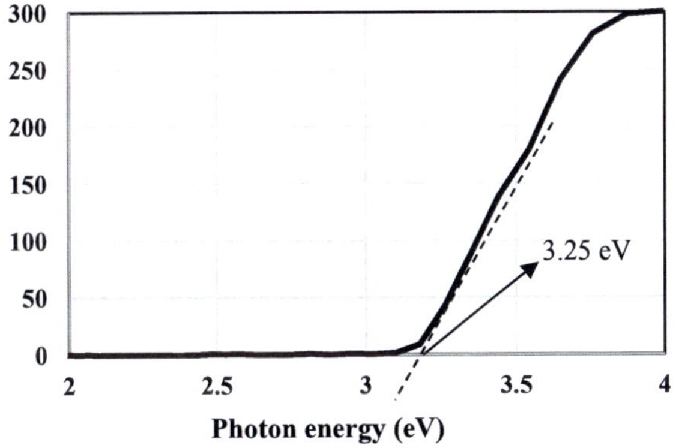

Fig. 4 Tauc's plot from reflectance spectra

absorption coefficient (α) and the photon energy ($h\nu$) and energy band gap (E_g) [14].

$$\alpha h\nu = a\left(h\nu - E_g\right)^n$$

$$\alpha = \frac{4\pi k}{\lambda}$$

where h is Planck's constant, ν is the transition frequency of incident photon, A is the constant for direct transition, k is the extinction coefficient and λ is the wavelength. The energy band gap has been evaluated by marking the extrapolation of the y-axis on the x-axis in Fig. 4. The band gap of EZO thin film is evaluated as 3.25 eV. It is observed that incorporation of Er in ZnO lattice has created stress that leads to the increase in the band gap of EZO. It may happen that Er ion creates certain surface defects which lead to the increase in the band gap of the EZO thin film.

ZnO thin film works as a transparent conducting layer in various optical applications like waveguides, so its refractive index is an important parameter. It is closely related to local field and electronic polarization of ions in the material. The refractive

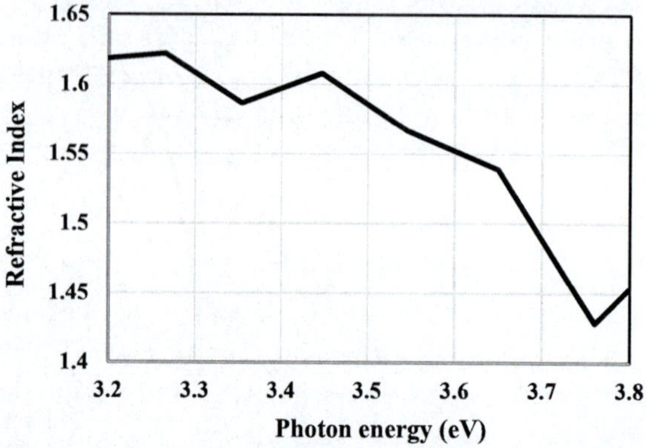

Fig. 5 Variation of refractive index of EZO thin film with photon energy

index of ZnO film [15] can be obtained using the following equations

$$n = \sqrt{\frac{4R}{(1-R)^2} - k^2} + \frac{1+R}{1-R}, \quad k = \frac{\alpha\lambda}{4\pi}$$

where R is the reflectivity of the deposited film, k is the extinction coefficient, α is absorption coefficient and λ is the wavelength. Figure 5 shows the variation refractive index with photon energy. It is observed from the drawn spectrum that the variation in refractive index with wavelength or photon energy is very less. This signifies that Fresnel's loss is very less for 1 mol% EZO thin film. Hence, 1 mol% EZO thin film can be used suitably as a good lossless transmitter in an optical waveguide. The present study of 1 mol% EZO has been concluded as the best option for use in UV optical ring resonator; therefore in the light of this, the obtained graph of Fig. 3 may supplement the present study.

In the present article, structural and optical properties of 1 mol% EZO thin films deposited via spin coating technique have been studied by XRD and ellipsometer. XRD image reveals that the EZO film has single-crystalline structure with orientation along 100 planes. FE-SEM images signify that deposited EZO film is uniformly distributed with small grain size. EDAX results confirm the presence of Er ions in ZnO lattice. The optical results show that EZO is a wide band gap semiconductor. The high reflectivity in visible region and less variation in refractive index make EZO as a cladding or transmitting material for waveguides used in optical ring resonators.

Acknowledgements The authors gratefully acknowledge Centre for Interdisciplinary Research (CIR), MNNIT Allahabad, for providing characterization facilities.

References

1. Ling Z, Zeng W (2017) Sens Actuators A Phys 267(2):242–261
2. Kumar D, Gomes T, Alves N, Kettle J (2018) IEEE Sens, 1–5
3. Bai Z, Liu Z (2017) IEEE Sens, 1–3
4. Guo Z, Zhao D, Liu Y, Shen D, Zhang J, Li B (2008) Appl Phys Lett 93(16):163501
5. Di Bartolomeo, Giubileo F, Luongo G, Iemmo L (2016) 2D Mater 29(41):015024–015032
6. Gulia S, Kakkar R (2013) Adv Mater Lett 4:876
7. Xu J, Han J, Zhang Y, Sun Y, Xie B (2008) Sens Actuators B Chem 132:334
8. Cui J, Gibson UJ (2010) J Phys Chem C 114(14):6408–6412
9. Peng W, Qu S, Cong G, Wang Z (2006) Cryst Growth Des 6(6):1518–1522
10. Ashour A, Kaid MA, El-Sayed NZ, Ibrahim AA (2006) Appl Surf Sci 252(22):7844–7848
11. Djurišić AB, Leung HY (2006) Small 2(8–9):944–961
12. Ilican S, Caglar Y, Caglar M (2008) J Optoelectr Adv Mater 10(10):2578–2583
13. Bouhssira N, Abed S, Tomasella E, Cellier J, Mosbah A, Aida MS, Jacquet M (2006) Appl Surf Sci 252(15):5594–5597
14. Zak A, Khorsand WH, Majid Abd, Abrishami ME, Yousefi R (2011) Solid State Sci 13(1):251–256
15. Kang H, Ahn SBD, Kim JH, Kim GH, Lim SH, Chang HW, Lee SY (2006) Appl Phys Lett 88(20):202108–202108

Predictive Analysis of Wind Turbine Output Power Using Support Vector Machine(SVM) Based on Genetic Algorithm(GA)

Tushar Srivastava and M. M. Tripathi

Abstract In this era, role of renewable energy is of utter most importance for the growth of power sector in any countries economy. The major types of renewable or reusable energy are: solar and wind. At present, the percentage contribution of wind energy is highest as compared to solar, in total renewable energy generation. With growth in the world population, the electrical energy demand also keeps on increasing, also with the expansion of telecom sector and electronics demand of electrical energy enhances day by day. With increase in the number of private players supplying electricity, the competition rises proportionally. So, for the existence in the market, these companies rely on the predictive future demand, which they can match up at the right time. Therefore, such players require some kind of software solution based on computing model using artificial intelligence and machine learning, so that they can predict the future demand in the most accurate way. There are some already existing predictive models, which are linear in nature, e.g. auto-regressive (AR), moving average (MA), etc. Also, there are nonlinear models, e.g. GARCH, ARCH, and neural networks. In this paper basically, we are going to discuss a novel model which is support vector machine (SVM) based upon genetic algorithm (GA) model for short-term power prediction.

Keywords Genetic algorithm (GA) · Support vector machine (SVM) · MLP · Mape · Mean square error

T. Srivastava (✉) · M. M. Tripathi
Delhi Technological University, Delhi 110042, India
e-mail: tushar300489@gmail.com

M. M. Tripathi
e-mail: mmtripathi@dce.ac.in

D. Harvey et al. (eds.), *Advances in VLSI, Communication, and Signal Processing*, Lecture Notes in Electrical Engineering 683,
https://doi.org/10.1007/978-981-15-6840-4_10

1 Introduction

Prediction or Forecasting is a phenomenon, which anticipates the upcoming or future values, by analysing or computing the past and present series of data or information. Forecasting [1] techniques can be broadly classified into following ways:

(i) Short-term forecasting (it includes time horizon of few months, weeks, days, hours, minutes, seconds)
(ii) Medium-term forecasting (it includes time horizon for one year to two years)
(iii) Long-term forecasting (it includes time horizon beyond 2 years).

Short-term predictive analysis is necessary for assuring an inexpensive surplus supply of electrical energy. Higher the accuracy of short-term forecasting, better will be the improvements in the efficiency of daily system operations. The forecasting techniques have vast applications that include business, environment, industry, economics, etc.

For the time series forecasting, the data involved there is in time series form which implies that the data or information is expressed in a particular form at equal interval of time, and the data consists of various sorts of parameters, and all parameters follow the same time series interval. In this particular paper, the parameter to be predicted is power output of wind turbine, while the input parameters are day, time and wind velocity. The data set used can be of many types depending upon a number or types of variables. It could be uni-variate or multi-variate [2]. Uni-variate data set includes only single type of data, while the multi-variate data set has more than one parameter or variable in it. Analysis of this time series data set which is also known as sequence assist in recognizing designs or patterns, trend and periods or times of cycles presented in the data or information. In the power sector or the electricity market, the prediction helps the power supply companies to easily identify the trend of demands, which may occur in future. Thus, the forecasting basically helps them to be ready for the upcoming situation. Also, the companies can become bullish or bearish with regard to electricity demands in the market, thus varying their price accordingly.

In case of time series forecasting, it essentially involves two forms of algorithms which are as follows:

(i) Linear models
(ii) Nonlinear models.

Now we know that the variety of linear models will be like: ARIMA [3], ARMA [4], AR, smooth-transition auto-regressive (STAR) [5] and its different forms. In the above models which are linear in nature, some equations which are already predefined are used to slot down or fit in a mathematical model to a uni-variate time series. The chief drawbacks of all these models are that, they do not account for hidden dynamics and nonlinearity existing within the information or data. Also, one

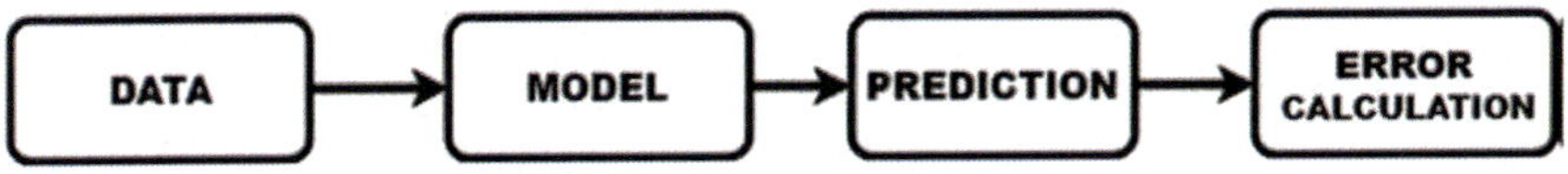

Fig. 1 Block diagram of the process

major drawback is that the model acknowledged according to one series will never be going to fit for the other type of sequence or series. The other type of models, which are nonlinear models, engross methodologies like: ARCH [6], GARCH [7], TAR [8] and deep learning algorithms [9]. Depending upon the nature of function appliance, a variety of deep learning algorithms is utilized. It includes multi-layer perceptrons (MLP) [10], recursive neural networks (RNN) [11], long short-term memory(LSTM) [12], convolutional neural network (CNN) [13], etc. These architectures or networks have been applied in varied fields like: natural language processing (NLP), image processing, software-based computational statistical analysis or time series analysis. Deep learning algorithms are able to recognize the hidden or buried designs (patterns) and underlying dynamics within the information or data provided to it, via self-training process. This paper displays a novel methodology of GA-based SVM [14] algorithm, where genetic algorithm [15, 16] is used for feature selection, and SVM is the main predictive model. Basically, we are using three input parameters, viz. day, time and wind velocity to predict the output parameter that is power output of wind turbine. Then, this hybrid model, i.e. GA-based SVM is further benchmarked by comparing with other neural network models like: RNN, linear regression and MLP/ANN. Further, with the help of performance parameters, all the algorithms are compared and the best method is suggested based on the outcomes of comparative results. Also, the scope of future work is there regarding improvement in the accuracy of hybrid model by using various other functions. In this case, data set used contains annual hourly data of wind turbine based on Kolkata region of India, having wind velocity and turbine Power output as the two parameters. Here, daily average [17] methodology is used to segregate the data for the work process.

The block diagram for the whole process can be shown in below Fig. 1.

The research paper is organized as shown: Sect. 2 describes the pre-processing of data set, Sect. 3 discusses the algorithms, Sect. 4 is on performance parameters, Sect. 5 shows results and discussion, while Sect. 6 gives conclusion and future work.

2 Pre-processing of Data set

The data of Kolkata region in India has been used for wind power forecasting. Data set, which is considered in this paper, is the hourly data of 2014 (1st January to 31st December). This data set was then normalized between 0 and 1, and further the daily average of the data (according to each day) was found out. This data set was then divided into training set and testing set. The train size and test size are 67%, 33%, respectively, of the normalized, daily average data set, which is used for the

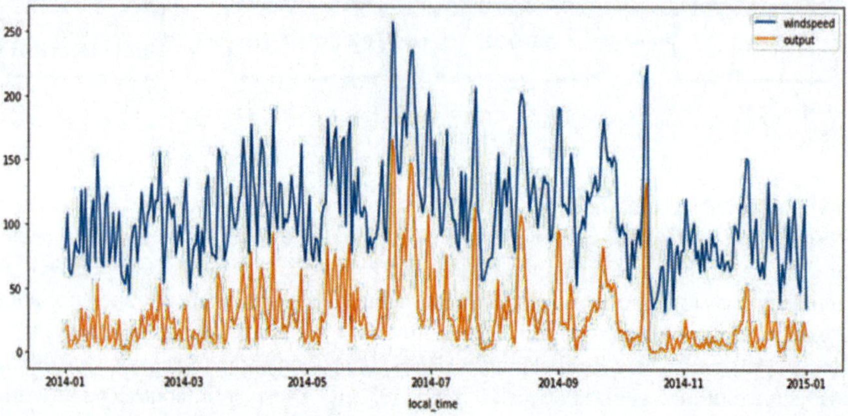

Fig. 2 Pattern of annual data set

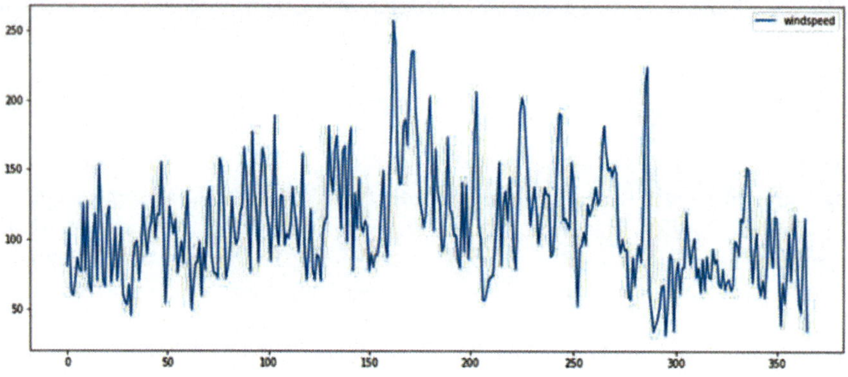

Fig. 3 Wind speed data pattern

forecasting of power output of wind turbine with the help of day, time and wind velocity.

In the above figures, viz. Figs 2, 3, and 4, the y-axis shows the normalized values of the parameters, while the x-axis shows the number of days in an year.

3 Description of Algorithm

3.1 Support Vector Machine (SVM)

It is an algorithm fallen in the category of supervised machine learning, and its main feature is that, it can be used for both condition whether of classification or regression.

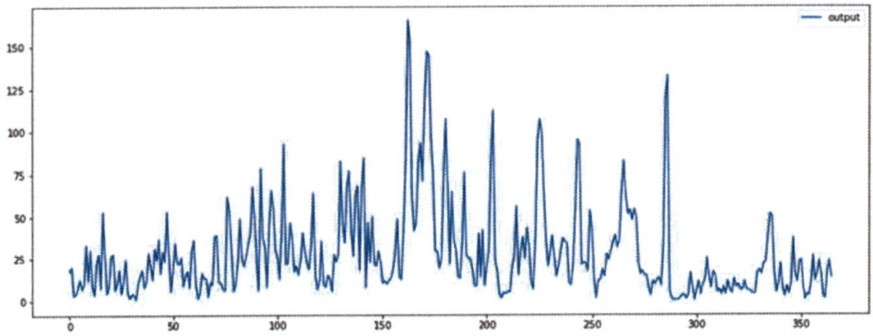

Fig. 4 Wind turbine output power data pattern

In predictive analysis, we basically apply SVM for regression analysis. SVM as a classifier involves the process of plotting up of certain points from data set in n-dimensional space (n is the number of features). Here, each and every characteristic represents return of single points or coordinate. Now thereafter, the categorization is done by deducing out the hyper-plane which discriminate between the two classes accurately (Fig. 5).

SVM as a regression analysis tool is of great use in case of forecasting like: price forecasting, energy forecasting and power forecasting. The ability of SVM in resolving out the nonlinear regression estimation problems made it quite appropriate tool for forecasting. Mathematically, for describing the SVM, let us suppose training data set as follows: $\{(X_i, y_i)\}_{i=1}^{N}$ where $X_i \in R^m$ is a vector input while y_i is its output. SVM approximating function can be shown as follows:

$$f(x, \omega) = \sum_{j=1}^{m} \omega_j \phi_j(x) + b \tag{1}$$

and the SVR will be solving a problem of reducing a given function. as shown by Eq. 2.

Fig. 5 SVM basic representation

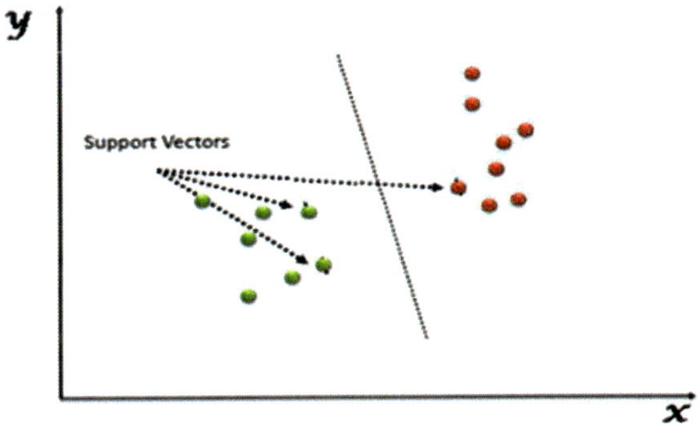

$$\min_{\omega,b,\xi,\xi^*} R(\omega,b,\xi,\xi^*) = \frac{1}{2}\omega^T\omega + C\sum_{i=1}^{N}(\xi_i + \xi_i^*) \tag{2}$$

subject to $y_i - (\omega^T\phi(X_i)+b) \leq \varepsilon + \xi_i^*$, $(\omega^T\phi(X_i)+b) - y_i) \leq \varepsilon + \xi_i$, $\xi_i, \xi_i^* \geq 0$,i = 1,2,…N.

Here, function ϕ is used to map the function X_i to a feature space of higher dimension. By doing such mapping, we basically are converting the problem of regression in nonlinearity existing in x-space into a problem of regression in linearity related to ϕ-space. ξ_i and ξ_i^* are the slack parameters, where ξ_i^* is the upper training error and ξ_i is the training error for lower bound, both subjected to the Vapnik's ε-insensitive loss function or tube $|y - (\omega^T\phi(X) + b)| \leq \varepsilon$. ω is the weight function and b is the bias. Now, here the quality of regression is guided by the researcher by selecting the legitimate values of following three parameters, viz. width of the tube ε, cost error C and function of mapping or mapping function ϕ.

Now, here one thing which we have to observe in Eq. 2 is the constraint, which implies that almost all data belonging to X_i is tried to maximum put inside the tube, $|y - (\omega^T\phi(X) + b)| \leq \varepsilon$. So, here now following condition arises, If X_i lies inside this tube, then the value of loss is zero, otherwise if X_i is not within the given tube then, an error is there which is either ξ_i or ξ_i^*, which is then minimized or reduced in one of the parameters that is cost function C. The support vector regression (SVR) bypasses the over-fitting and under-fitting of the data, which is done by reducing or minimizing the calculated error of training or training error, denoted as, $C\sum_{i=1}^{N}(\xi_i + \xi_i^*)$, also, it reduces the regularization term, represented as $(1/2)\omega^T\omega$. The above method is all followed up as per the principle of SRM. Here both regularization term and training error are reduced or minimized simultaneously at the same time. In case of traditional or simple LSR, ε is always zero and also data is not correlated, mapped or diagramed into above dimensional spaces (here LSR is least square regression). Therefore, SVR/SVM is rather a further elastic or flexible way of solving out the regression problems.

Now, as we know that ϕ might mapped the X_i in a very towering or infinite dimensional space, so this may lead to the problem of dimensionality, so in order to avoid the dimensionality problem, instead of solving ω for Eq. (2) in very high dimension, the occurrence of dual problem of Eq. (2) is resolved. So, this dual problem is organized totally in terms of training data and then, it is to minimize or reduce dual variables of Lagrangian $L_d(\alpha,\alpha^*)$. So, now the dual equation will be:

$$\min_{\alpha,\alpha^*} L_d(\alpha,\alpha^*) = \frac{1}{2}(\alpha-\alpha^*)^T G(\alpha-\alpha^*) + \varepsilon\sum_{i=1}^{N}(\alpha_i+\alpha_i^*) + \sum_{i=1}^{N} y_i(\alpha_i-\alpha_i^*) \tag{3}$$

subject to constraints,

$$\sum_{i=1}^{N}(\alpha_i - \alpha_i^*) = 0$$

$$0 \le \alpha_i \le C,$$
$$i = 1, 2, \ldots, N$$

$$0 \le \alpha_i^* \le C,$$
$$i = 1, 2, 3, \ldots, N$$

now the above Eq. 3 represents ideal or optimal optimization problem in quadratic form. Here, α_i and α_i^* are the unknown or uncharted Lagrange multipliers corresponding to ξ_i,ξ_i^* and the formulated inner multiplier or product $G_{ij} = \phi(X_i)^T\phi(X_i)$. As we know that training data cannot be available on both sides of the ε-insensitive tube therefore either α_i or α_i^* will be nonzero. Now, considering the combination of data lying inside the tube, both multiplier values will be zero. Since the $\phi(X)$ has too many elements, it will be costly and hectic to calculate inner product, i.e. G_{ij}. Hence, we will be using "Kernel Trick" to perform mapping implicitly. Few examples related to above said concept are:

polynomial kernel, $\phi(X_i)^T\phi(X_j) = (\gamma X_1^T X_2 + c_0)^p$ and radial basis function (RBF) kernel, represented as $\phi(X_i)^T\phi(X_j) = e^{-(||X_1 - X_2||^2/2\sigma^2)}$.

So, the above functions are inner multipliers or products in the uppermost dimensional space but still can be calculated in original dimensional space.

Now, learning outcomes in N Lagrange multiplier pairs are (α_i,α_i^*). After the above said learning process, the number of nonzero or free parameters α_i or α_i^* is exactly equal to the total number of SVs, i.e. Support Vectors.

Now, after the process of finding out the Lagrange multiplier vectors α and α^*, an optimal or ideal weights vector to be desired of the kernel expansion or magnification is calculated as,

$$\omega_0 = \alpha^* - \alpha \quad (4)$$

and an optimal bias b_0 as

$$b_0 = \frac{1}{N}\sum_{i=1}^{N}(y_i - g_i) \quad (5)$$

here, $g = G\,\omega_0$, and the G matrix is analogous kernel matrix $G(X_i, X_j)$.

The best hyper-function for nonlinear regression is governed by:

$$y = f(x, \omega) = G\omega + b \quad (6)$$

3.2 Genetic Algorithm (GA)

The GA is an algorithm of searching for optimization, perfectly based or relied on the mechanism of natural selection and genetics. This algorithm works on the assumption that the genuine or best solution lies in zone of space having solutions containing higher probability of genuine or most accurate solution, and these zones or regions can be found out or identified by random or robust inspection (sampling) of solution space. The ability of GA is that it can search out a very large solution space efficiently at a very low costing price, because it applies probabilistic transition rules instead of deterministic rules. Also, GA is highly applicable to those problems, where a small change in input results in huge nonlinear behaviour of solution space. The GA involves following three processes in stages as follows:

(a) Population booting/population initialization
(b) Operators
(c) Chromosome assessment/chromosome evaluation.

These are described as follows:

(a) *Population booting*:

In genetic algorithm (GA), the initial population is generated in a random procedure, but the main point is that this initial population must satisfy all precedence relations and also must comprise of valid sequences.

(b) *Operators*:

Now, operators consist of certain substages like: selection; reproduction; crossover; mutation; and migration. they are explained as follows:

(i) Selection: Selecting the selection operator is an important step in GA algorithm. Researchers have recommended various selection operators, like, e.g. "tournament selection model" given by Goldberg also similarly "expected value model" and "elitist model" given by De Jong.
(ii) Reproduction: It is another genetic operator. In reproduction, basically the images of strings are pasted into a separately new string known as "mating pool", as per their values of fitness. Therefore, it implies that the strings having higher value of fitness will be having greater chances of donating high counts of strings, as the search process progresses.
(iii) Crossover: The crossover operator main function is to examine progress of the search. Also, at the same time, it exchanges the parents string partially so as to give birth to offspring.
(iv) Mutation: The process such as crossover and reproduction creates several new chromosomes but still they failed to introduce or produce any new information at the bit level into the population. Mutation is a process in which information contained in chromosomes is occasionally and randomly oriented. Basically in mutation, genes mutate (alter or change) so as to become new genes having

lower probability. Therefore, so as to copy the evolution process, the mutation introduces bits to alter having lower probability, by inverting a bit of chromosome, which is selected randomly.

(v) Migration: It is defined as the individuals motion between the subgroups of the population. Quite often, the best performing individuals of one subpopulation replace the worst performing individual of another subgroup of population. Parameters defining the mutation are: interval, fraction and direction.

(c) *Chromosome assessment/chromosome evaluation*:

In GA finally, the chromosome quality is checked and evaluated, and the algorithm converges those genes into new offsprings, thereby resulting into new set of solution or new genes. Therefore, the algorithm terminates itself, if the algorithm fails to produce offspring, having property different from previous genes.

3.3 Genetic Algorithm (GA)-Based Support Vector Machine (SVM)

In case of SVM model, three free parameters are (C, ε, ϕ) C is cost error; ε is width of the tube; ϕ is mapping function. The performance of the SVM models is greatly affected by these free parameters, so the main task for the researcher is how to select adequate parameters value, which will lead to genuine performance. Therefore, the performance of SVM [11] models is directly dependent upon the adequate and appropriate values of parameters.

However, still there is a lack in structured methods for selecting parameters values. Also, it is unknown that at what combinational values of parameters does the SVM model performs best. So, optimizing the parameters for the SVM model is a necessary step for predicting [12] the best performance. In this research work, the optimization technique of genetic algorithm (GA) is adopted in order to find out the free parameters values, which led to better performance of SVM model.

Genetic algorithm (GA) is defined as a random search technique, which is directed in nature and it is mainly applicable to the optimization problems [13, 14], where the analytical solution is very tough to obtain, just because of large number of parameters. GA is basically utilized to find out the optimal solution [15] over a range globally.

The procedure flows as follows:

STEP-1: First of all, SVM parameters are encoded and initialized, thus establishing an initial population of chromosomes, i.e. by creating certain encoding programs, the initial population is created.

STEP-2: Now in this step, we evaluate fitness of trained SVM by taking each value of chromosome's gene as SVM parameters and then training them setwise as input and output sets, respectively. After completing, the fitness value of chromosome should be evaluated and cross-checked again.

In order to find out the fitness value of each chromosome, we use NRMSE as a evaluating parameter, here NRMSE is normalized root mean square error, as shown in Eq. 7.

$$\text{fitness function} = -\sqrt{\frac{\sum_{i=1}^{n}(a_i - f_i)^2}{\sum_{i=1}^{n} a_i^2}} \tag{7}$$

STEP-3: Selection, operation, crossover and mutation

Now out of the above-mentioned steps, elaborating them in details: Selection is done in order to select the top performing chromosomes for reproducing. On basis of the fitness function, as calculated in Eq. 7, we can predict the chromosomes yielding out much better offspring in the upcoming batch by means of roulette wheel. Crossover is randomly done so as to swap genes among the two chromosomes. The operation of mutation creeps down the crossover operation, and here it is found out that whether a chromosome can be altered or changed in next generation or not. Now, after the above four operations, the offspring generated will be replacing the old batches and thus forms the new population in the upcoming generation by the same above four operation, viz. selection, operation, crossover and mutation. This process of evolution proceeds on in same way, unless and until the termination condition are satisfied.

4 Performance Parameters

The performance of this algorithm: GA-based SVM is now compared with other neural networks like: linear regression, RNN, ANN/MLP on basis of certain parameters, mentioned below:

4.1 Mean Square Error (MSE)

The MSE is known as the second moment (about the origin) of the error and thus includes the variance of the predictor and the variance of its bias. Mathematically, MSE can be represented by Eq. 8.

$$\text{MSE} = \frac{1}{n}\sum_{i=1}^{n}(Y_i - \widehat{Y_i})^2 \tag{8}$$

Y_i = dependent variable; $\hat{Y}_i$ = predicted values.

4.2 Mean Absolute Percentage Error (MAPE)

It is an assessment of accuracy of a forecasting method. It basically implies accuracy as a percentage. MAPE is defined by Eq. 9.

$$\text{MAPE} = \frac{1}{n}\sum_{t=1}^{n}\left|\frac{A_t - F_t}{A_t}\right| \tag{9}$$

where A_t is the actual value and F_t is the forecast value.

4.3 Mean Absolute Error (MAE)

According to statistical mathematics, MAE is the determination of the difference between the two continuous variables. The mean absolute error (MAE) is given by Eq. 10.

$$\text{MAE} = \frac{\sum_{i=1}^{n}(y_i - x_i)}{n} \tag{10}$$

where y_i is the predicted value and x_i is the observed/true value. The MAE is a common way to determine forecast error in time series analysis.

4.4 Root Mean Square Error (RMSE)

It is the measure of difference between predicted values by a model and the observed values. RMSE is the measure of accuracy, commonly used to compare errors related to forecasting, for different predictive models, for the same data set. Mathematically, RMSE is expressed as Eq. 11.

$$\text{RMSE} = \sqrt{\frac{\sum_{t-1}^{T}(\hat{y}_t - y_t)^2}{T}} \tag{11}$$

$\hat{y}_t$ = predicted values; y_t = dependent variable; T = no. of data in data set.

5 Results and Discussion

In this section, the result of the simulation is presented and discussed through tabular comparative study of GA-based SVM with respect to linear regression, ANN/MLP and RNN. All the models are run for same number of epochs and also having same number of input vector size, hidden layers and output vector size. Now, Fig. 6 shows the forecast versus actual plot for linear regression. Also, Fig. 7 shows the forecast versus actual plot for ANN/MLP, while Fig. 8(i) shows the forecast versus actual plot for RNN and Fig. 8(ii) shows the tracing curve for RNN. Similarly, Fig. 9(i) shows the forecast versus actual plot for GA-based SVM and Fig. 9(ii) shows the bar plot for same, also Fig. 9(iii) shows the mape plot for GA-based SVM. The graphs or plot are made between y_true and y_pred values, where y_true is the true or actual value of the target/output while y_pred is the forecasted or predicted value of the target/output.

Now, for ANN/MLP(multi-layer perceptron), all the figures are shown below.

For, RNN all the figures are shown below:

Now, for GA-based SVM, all the graphical results are shown as follows:

In above figures right from Figs. 6, 7, 8 and 9(i), the x-axis shows the time period (which is number of days in a month, i.e. 30) while, the y-axis shows both the outcomes, viz. actual and predicted values, regarding that particular model. In Fig. 9(iii), the *y*-axis shows the time period (no. of days in a month) while the *x*-axis shows the MAPE values.

Now, on the basis of performance parameters, all the algorithms are compared to GA-based SVM, numerically and hence it is shown in the comparative Table 1.

From above comparative table, the observations are as follows:

(i) It is observed that the MAE value for GA-based SVM is good as compared to linear regression and is comparable to ANN and RNN.

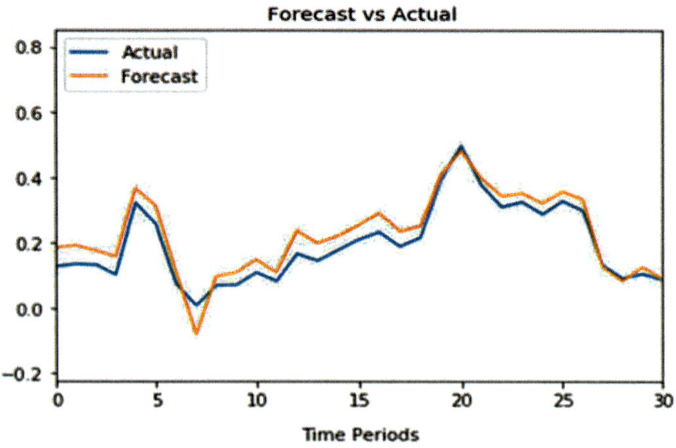

Fig. 6 Forecast versus actual plot for linear regression

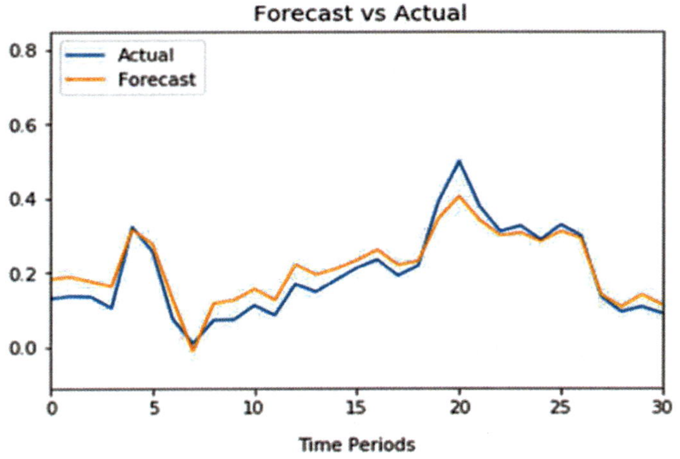

Fig. 7 Forecast versus actual plot for ANN/MLP

(ii) MAPE for GA-based SVM is better than linear regression and is comparable to that of RNN and ANN.
(iii) MSE and RMSE values are least for RNN and almost same for ANN and GA-based SVM.

Thus, it clearly proves from final outcomes that GA-based SVM outperforms linear regression and is performing equally well as ANN/MLP, however it is performing lesser as compared to RNN.

6 Conclusions

Wind power prediction is very important for the scheduling of wind power on regular basis. In this paper, GA-based SVM algorithm has been implemented to determine short-term daily average forecasting of wind power using data of Kolkata region of India. From the simulation results, it is evident that the GA-based SVM outperforms the linear regression and is almost equally well as compared to ANN or multi-layer perceptron, however GA-based SVM performs less as compared to RNN in the wind power forecasting. Further, it can be deduced that the future work could be based on the other algorithm to be used in feature selection process instead of GA. Also, other parameters such as wind direction and temperature may be taken as input to the network.

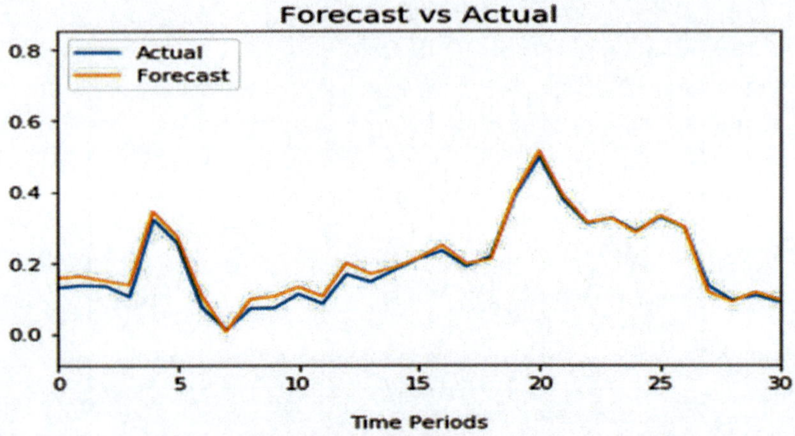

(a)

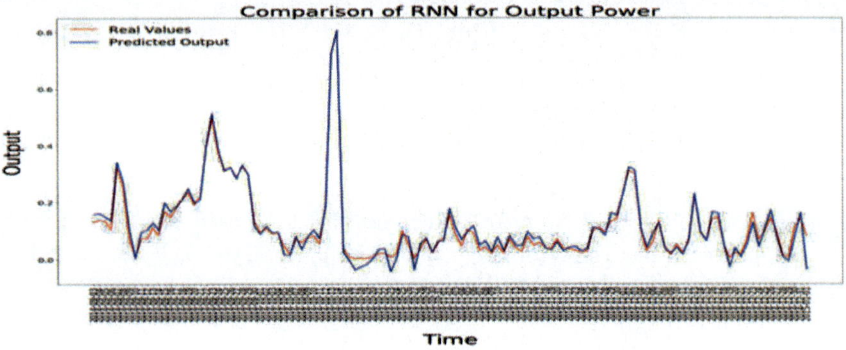

(b)

Fig. 8 **a** Forecast versus actual plot for RNN. **b** Tracing graph for RNN

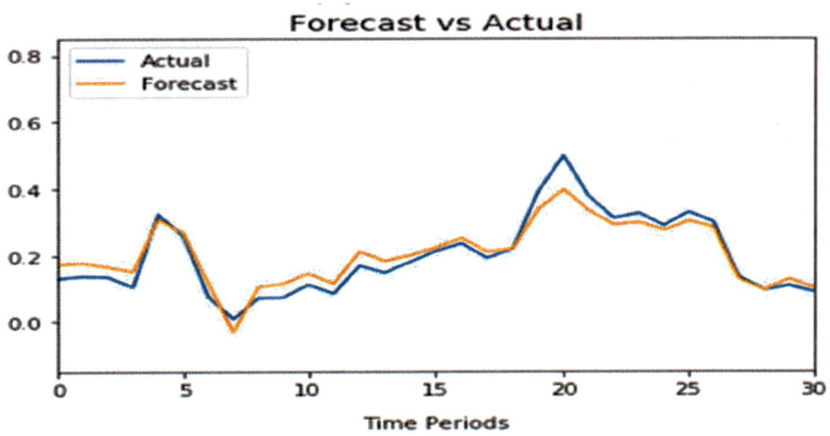

(a)

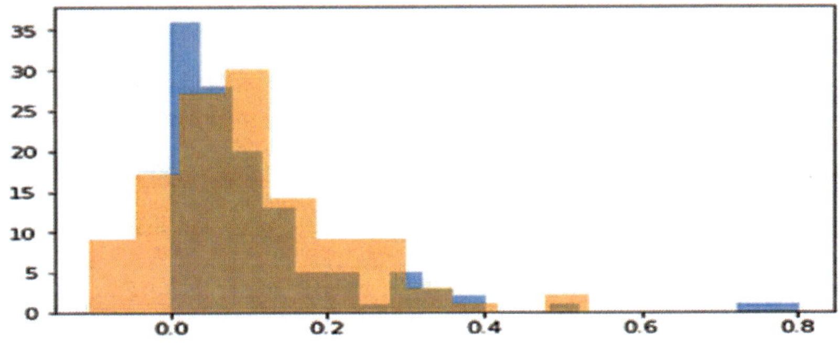

(b)

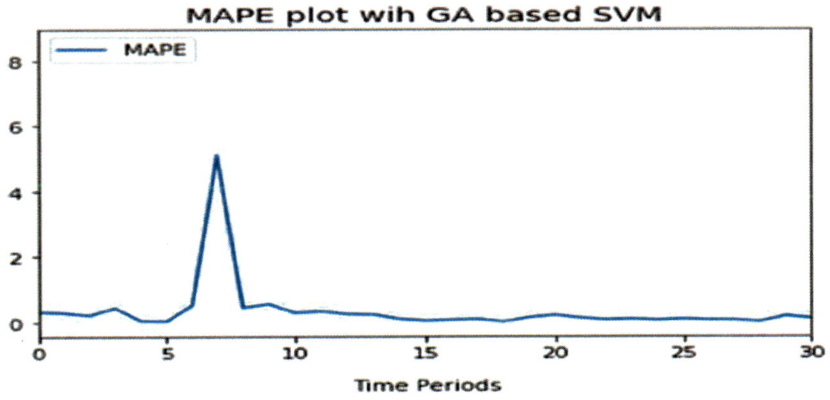

(c)

Fig. 9 **a** Forecast versus actual plot for GA-based SVM, **b** forecast versus actual bar plot for GA-based SVM. **c** MAPE plot with GA-based SVM

Table 1 Tabular analysis of GA-based SVM with other neural networks

Parameters	Linear regression	ANN/MLP	RNN	GA-based SVM
Mean absolute error (MAE)	0.047543	0.030737	0.015385	0.031417
Mean absolute percentage error (MAPE)	5.355319	1.939639	1.148229	2.972836
Mean square error (MSE)	0.003917	0.002174	0.000426	0.002427
Root mean square error (RMSE)	0.062586	0.046623	0.020640	0.049266
Variance	0.759744	0.866673	0.973868	0.851123

References

1. Devadoss AV, Ligori TAA (2013) Forecasting of stock prices using multi layer perceptron. Int J Comput Algorithm 2:440–449
2. Batres-Estrada G (2015) Deep learning for multivariate financial time series, ser. Technical Report, Stockholm
3. Shilpa GN, Sheshadri GS (2017) Short-term Load Forecasting using ARIMA Model for Karnataka State Electrical Load. Int J Eng Res Dev 13(7):75–79
4. Burlando P, Rosso R, Cadavid LG, Salas JD (1993) Forecasting of short-term rainfall using ARMA models, vol. 144, issue 1–4, April 1993, pp 193–211
5. Adebile O, Shangodoyin K, Raghunath A (2018) Forecasting performance of logistic STAR model: an alternative version to the original LSTAR models, vol 3, MASA. Model Assisted Statistics and Applications, 2018/12/16
6. Bollerslev T (2009) Glossary to ARCH (GARCH). Volatility and Time Series Econometrics: Essays in Honour of Robert F. Engle. Oxford University Press, Oxford
7. Awartani AMB, Corradi V (2005)Predicting the volatility of the S&P-500 stock index via GARCH models: the role of asymmetries. Int J Forecasting 21:167–183
8. Nieto F (2008) Forecasting with univariate TAR models. Statistical Methodol 5:263–276. https://doi.org/10.1016/j.stamet.2007.09.002
9. Ding X, Zhang Y, Liu T, Duan J (2015) Deep learning for event-driven stock prediction. In: IJCAI, pp 2327–2333
10. Victor Devadoss A, Ligori TAA (2013) Forecasting of stock prices using multi layer perceptron. Int J Comput Algorithm 2:440–449
11. Roman J, Jameel A (1996) Backpropagation and recurrent neural networks in financial analysis of multiple stock market returns. In: Proceedings of the Twenty-Ninth Hawaii international conference on system sciences, vol 2. IEEE, pp 454–460
12. Jia H. Investigation into the effectiveness of long short term memory networks for stock price prediction. arXiv preprint arXiv:1603.07893,2016
13. Selvin S, Vinayakumar R, Gopalakrishnan EA, Menon VK, Soman KP (2017) Stock price prediction using LSTM, RNN and CNN-sliding window model. 978-1-5090-6367-3/17/$31.00 ©2017 IEEE, September 2017
14. Sun W (2010) Research on GA-SVM model for short term load forecasting based on LDM-PCA technique. J Comput Inf Syst 6(10):3183–3189
15. Han X, Zhang X, Bo G (2009) Short-term wind speed prediction model of LS-SVM based on genetic algorithm. In: Wu Y (ed) International conference on WTCS 2009, AISC 116, pp 221–229

16. Saini LM, Aggarwal SK, Kumar A (2010) Parameter optimisation using genetic algorithm for support vector machine-based price forecasting model in National electricity market. IET Generation, Transmission, Distribution, vol 4, issue 1, pp 36–49
17. Dumitru C-D, Gligor A (2016) Daily average wind energy forecasting using artificial neural networks. In: 10th international conference interdisciplinarity in engineering, INTER-ENG, pp 1877-7058 © 2017 The Authors. Published by Elsevier Ltd

Dual-Band Dual-Mode Orthogonally Placed CDRA-Based MIMO Antenna for Wi-fi/WLAN Applications

Ajay Kumar Dwivedi, Anand Sharma, Pramod Narayan Tripathi, and Ashutosh Kumar Singh

Abstract A dual-port, dual-band (2 × 1) multiple-input multiple-output (MIMO) CDRA is proposed for Wi-fi and WLAN applications. A novel pentagon-shape aperture has been used to excite the CDRA that gives rise to dual-mode generation, i.e., ($HE_{11\delta}$ and $HE_{12\delta}$). The presented MIMO antenna performs well at the operable frequencies of 2.4 GHz–2.8 GHz and 5.27 GHz–6 GHz with the FW of 18.2% and 12.8%, respectively. Isolation has been compared by placing the antenna 2 orthogonally to antenna 1 and getting $|S_{12}| < -33$ dB at both the resonating frequencies. Various diversity performances of the presented MIMO antenna have been evaluated with respect to two parameters, i.e., diversity gain (DG), envelope correlation coefficient (ECC).

Keywords Multiple-input multiple-output (MIMO) · Cylindrical dielectric resonator antenna (CDRA) · Hybrid antenna · Slot · Isolation

A. K. Dwivedi (✉) · A. K. Singh
Department of Electronics & Communication Engineering, Indian Institute of Information and Technology Allahabad, Allahabad, India
e-mail: rse2018505@iiita.ac.in

A. K. Singh
e-mail: ashutosh_singh@iiita.ac.in

A. Sharma
Department of Electronics & Communication Engineering, Motilal Nehru National Institute of Technology Allahabad, Allahabad, India
e-mail: anandsharma@mnnit.ac.in

P. N. Tripathi
Department of Radio Engineering and Cybernetics, Moscow Institute of Physics and Technology, Moscow, Russia
e-mail: pntripathi@phystech.edu

D. Harvey et al. (eds.), *Advances in VLSI, Communication, and Signal Processing*, Lecture Notes in Electrical Engineering 683,
https://doi.org/10.1007/978-981-15-6840-4_11

1 Introduction

In the field of radio broadcasting system, the data transmission with elevating speed along with quality and robustness has become the need of the hour. In the fast-moving technological world, it is recommended to transmit/receive multi-data simultaneously and for this to achieve MIMO antenna plays a vital role in it [1]. In the last two decades, DRA has been able to draw the attention of the researchers because of its minimum losses, broader impedance bandwidth and availability of variable shapes and excitation methods. These attributes has proven the DRA to been inculcated in the field of MIMO applications.

Limited work has been done in the field of MIMO DRA as far as open literature is concern [2–6]. Ishimiya et al. introduced a compact DRA MIMO antenna [2]. In Ref. [3], a rectangular MIMO for 4G application was proposed. In Ref. [4], Yan et al. presented a CDRA for LTE femtocell base station. In order to transmit the data independently and efficiently, it is desirous to obtain decent isolation among the elements of the antenna and for that several methods are available in [5–8]. In [5–7], different techniques have been discussed to improvise the isolation by generating two orthogonal modes but the problem arises as the structure becomes more complex due to the use of metallic strip in the coaxial probe. Nevertheless, this technique is more suitable for single- and dual-band antennas [8].

This chapter presents 2 × 1 MIMO CDRA with microstrip feed line; pentagon-shape patch is used to stimulate the various modes ($HE_{11\delta}$ and $HE_{12\delta}$) in the DRA at 2.63 GHz and 5.7 GHz, respectively. An orthogonally placed antenna orientation is implemented to lessen the mutual coupling. The designated radiator works well over the frequency range of (2.63/5.7 GHz) covering Wi-fi and WLAN operations with good isolation of ($|S_{12}|/|S_{21}| < -35$ dB).

2 Antenna Design and Geometry

This section demonstrates the diagrammatic layout of the suggested MIMO-based DRA antenna in Fig. 1. The radiator comprises of FR4 substrate ($\varepsilon_{\text{sub}} = 4.4$ tan $\delta = 0.02$) with two alumina-based (Al_2O_3) CDRAs ($\varepsilon_{\text{Alumina}} = 9.8$ tan $\delta = 0.002$). The geometry comprises of two designs of the antenna placed orthogonally to each other so that the coupling between them can be reduced ($|S_{12}| < -35$ dB) as compared when they were placed in the same direction.

In order to design the layout, Table 1 shows the optimize value of the different parameters used.

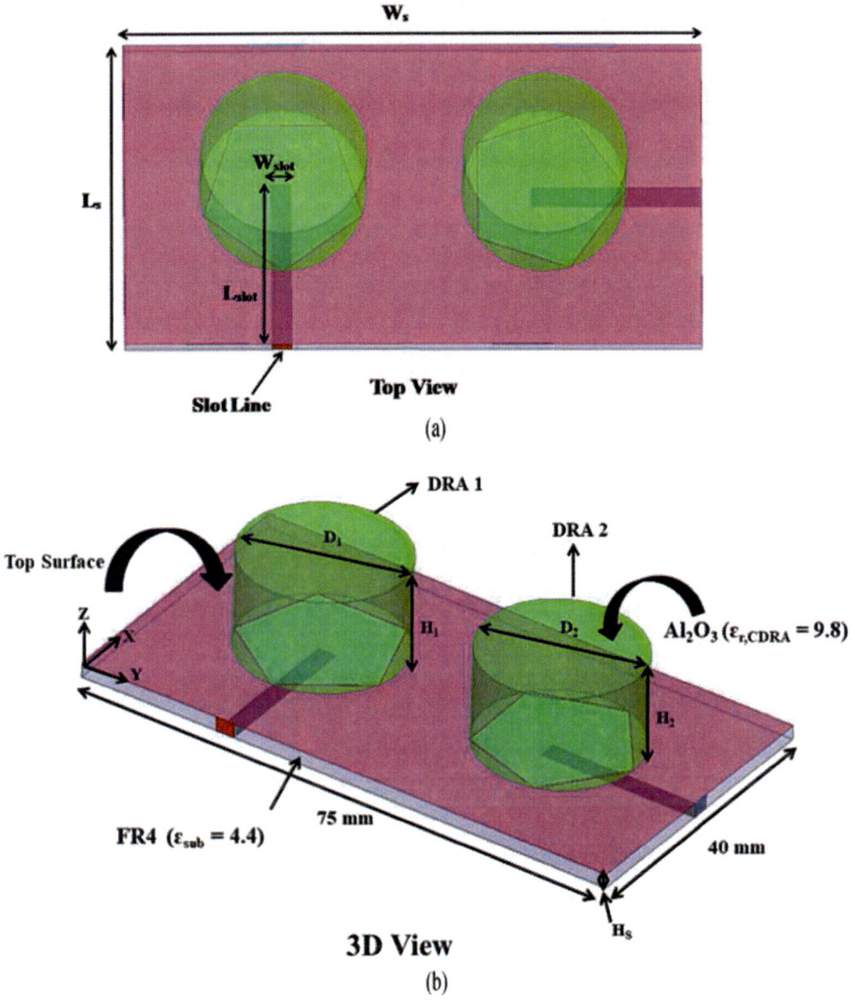

Fig. 1 Schematic layout of the MIMO-based DRA **a** top view, **b** 3D view

3 Antenna Analysis

The overall study and the mode analysis of the presented MIMO-based CDRA antenna has been done by using the EM Solver tool. Figure 2 gives the information regarding the comparison of the proposed antenna characteristics with regard to reflection coefficient, on the basis of with and without DRA. It is quite clear from the figure that lower and upper frequency is accountable due to the use of DRA which is placed on top of the pentagon shape that acts as a radiator for the modes to generate.

Figure 3 shows the variation in the reflection coefficient $|S_{11}|$ and the insertion loss $|S_{12}|$ with the different configuration of the radiator used while optimizing the

Table 1 Optimized dimensions of different parameters

Variables	Values (mm)
L_s	40
W_s	75
H_s	1.6
D_1	21.6
H_1	10.8
D_2	21.6
H_2	10.8
W_{slot}	2.5
L_{slot}	22

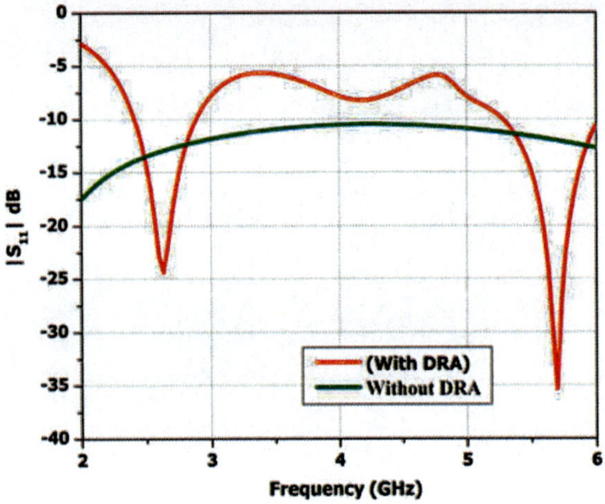

Fig. 2 $|S_{11}|$ comparative graph of proposed MIMO antenna with DR and without DRA

final structure; it is coherent from Fig. 3 that the depicted MIMO antenna works in two distinct resonating frequency bands, i.e., 2.4 GHz–2.88 GHz and 5.27 GHz–6 GHz, respectively. Figure 3 shows that the optimal values have been achieved with a pentagon-shape radiator to excite the DRA to generate $HE_{11\delta}$ and $HE_{12\delta}$ modes inside it.

4 Results and Discussion

The showcase of the presented MIMO prototype has been examined by the use of HFSS EM Solver tool, and optimized values are obtained and discussed in this section. Figure 4 displays the comparative scenario between the scattering parameter

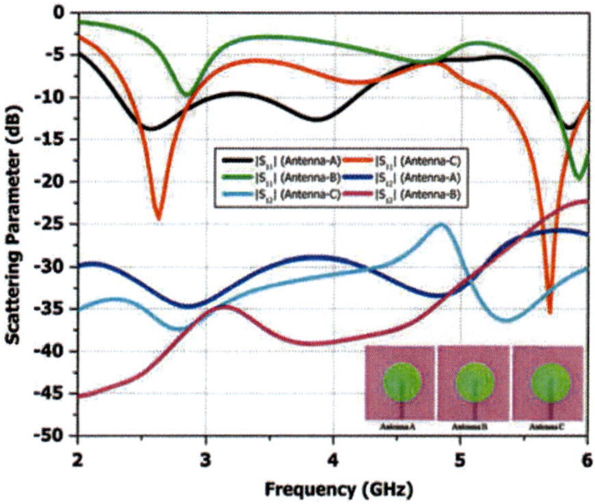

Fig. 3 Comparison between $|S_{11}|$ and $|S_{12}|$ of proposed MIMO antenna with different radiator structure

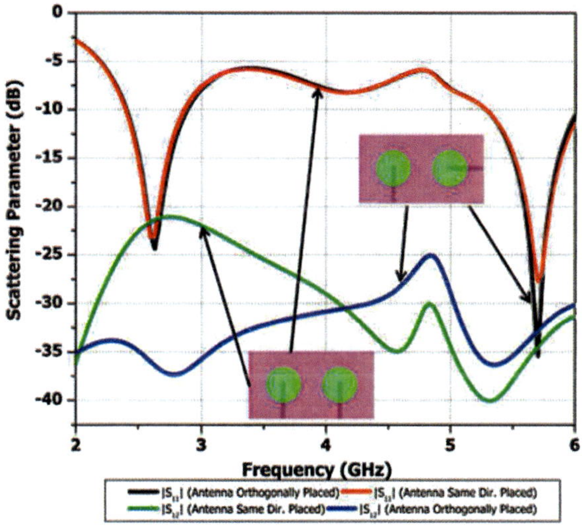

Fig. 4 Comparison of $|S_{11}|$ and $|S_{12}|$ (mutual coupling) due to the antenna orientation over the frequency band

and the isolation against the frequency range. Figure 4 gives information about the orientation of the antenna 1 and antenna 2 that positioned in the same direction and orthogonally to each other; it is clearly depicted from the graph that there is slight variation in S_{11}, whereas significant changes can be seen in $|S_{12}|$ at lower frequency band. Both the resonant peaks (2.63 GHz and 5.7 GHz) are generated because of the placement of CDRA over the pentagon-shaped radiator.

Figure 5 shows the maximum gain attained and the radiation efficiencies of the presented multiple-input multiple-output antennas; it is understandable from Fig. 5 that the positive value of the gain is achieved at both the operatable frequencies. The gain is 029 dBi and 0.24 dBi at 2.63 GHz and 5.7 GHz, respectively. Further, almost 99% radiation efficiency is achieved by the presented structure.

Figure 6 displays that the far-field radiation pattern is maximum at the operating frequency, i.e., 2.63 GHz and 5.7 GHz correspondingly. From Fig. 6, it is understandable that the acquired pattern at the operating frequencies has the extreme radiation in the broadside direction; also, it is found that the radiation patterns of other antenna are mirror image of the later one which leads to satisfy the pattern range of the proposed MIMO antenna.

Figure 7 displays the 2D far-field radiation pattern in the XZ-YZ planes at 2.63 GHz and 5.7 GHz, respectively. Figure 7 represents that the radiation from the antenna obtained at the resonating frequencies shows stable and correlated relation to the co- and cross-polarization state.

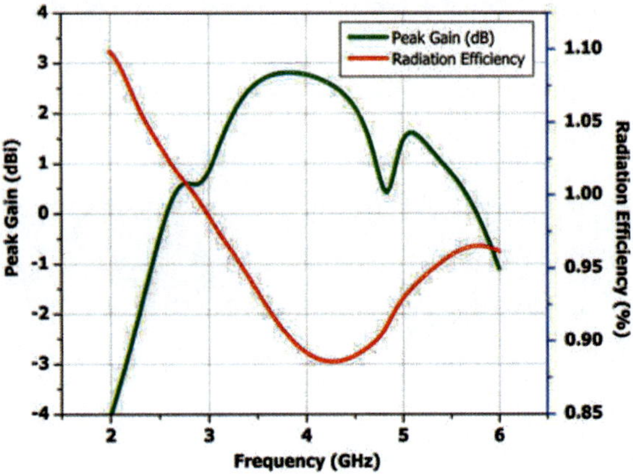

Fig. 5 Simulated peak gain and radiation efficiencies of proposed MIMO CDRA-based antenna over the frequency band

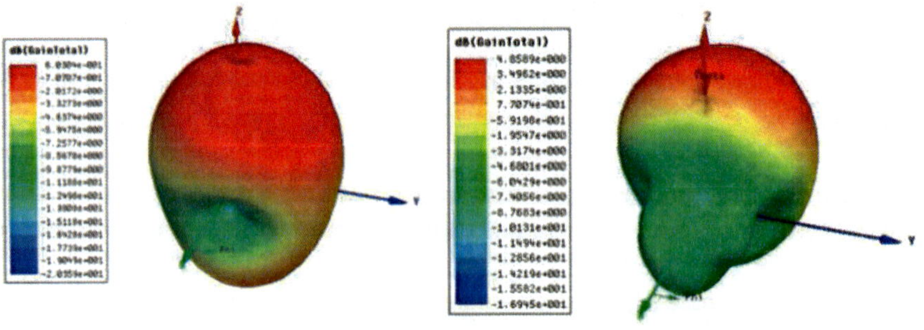

Fig. 6 3D far-field radiation patterns of proposed antenna at 2.63 GHz and 5.7 GHz

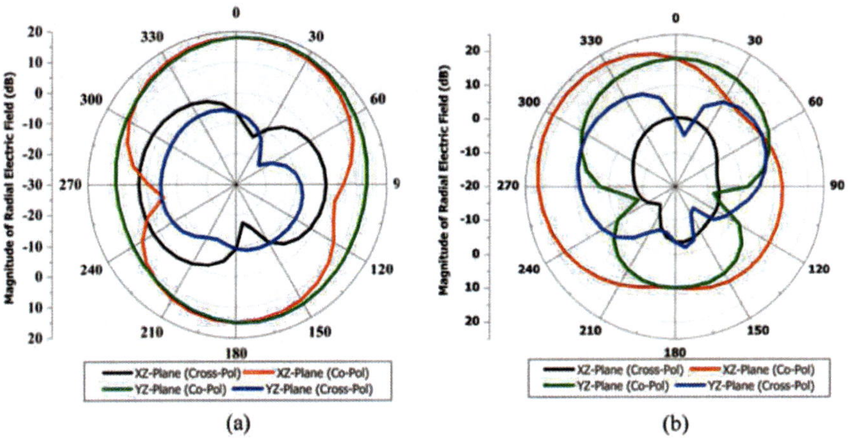

Fig. 7 2D far-field radiation pattern of XZ-YZ planes at **a** 2.63 GHz, **b** 5.7 GHz

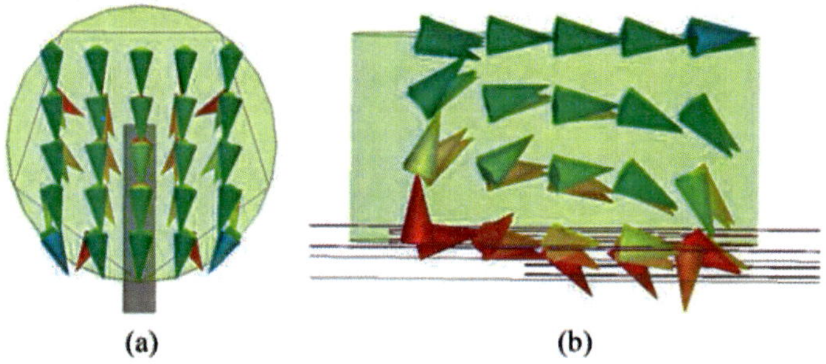

Fig. 8 E-Field distribution at 2.63 GHz **a** top view, **b** side view

Figures 8 and 9 display the electric field distribution on the top and side view of the DRA at 2.63 GHz and 5.7 GHz, respectively; it is clear from the figure that the two different modes are generated when the DRA is placed on above the pentagon-shaped aperture, $HE_{11\delta}$ mode is generated at lower frequency, and the higher mode $HE_{12\delta}$ is generated at the upper resonating frequency which can be verified from Figs. 8 and 9.

5 Diversity Performances

The various MIMO diversity parameters of the presented MIMO radiator are preconceived by examining the envelope correlation coefficient (ECC) and the diversity gain (DG). ECC and DG are the two main performance indices for the MIMO system. ECC basically gives information of the correlation between the received signals,

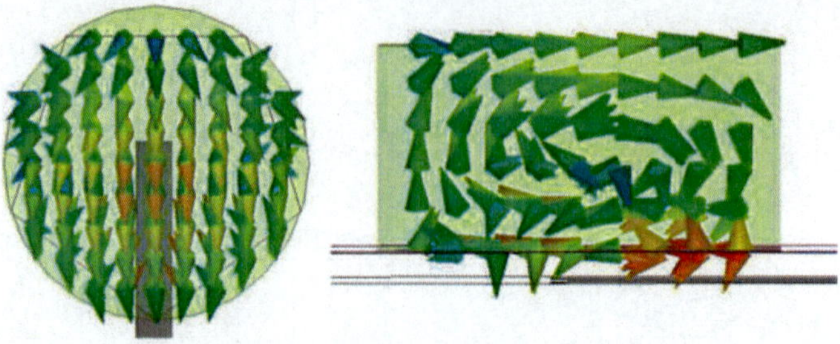

Fig. 9 E-Field distribution at 5.7 GHz **a** top view, **b** side view

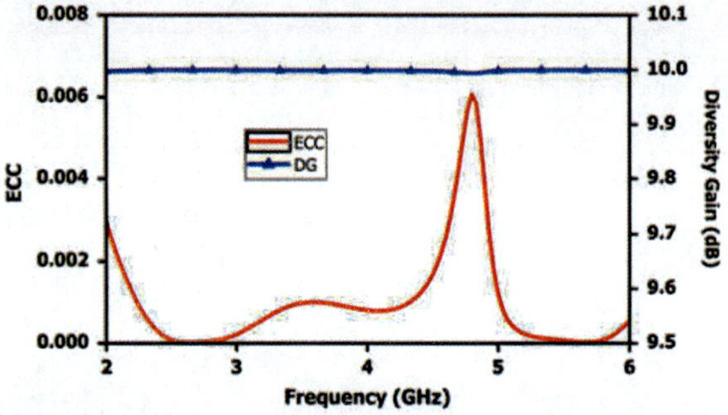

Fig. 10 ECC and DG of the proposed MIMO antenna

and it is evaluated by two methods, first by using the far-field radiation pattern and the second one by using the scattering parameters. It is observed from Fig. 10 that operable range of ECC is <0.2 which is far better than the acceptance value of 0.5.

Enhancement in the (SNR) for the multiple antenna arrangement with reference to the single antenna model is described by the diversity gain (DG). In practical applications, the operable range of DG should be close to 10 and it is well verified by Fig. 10 that the presented MIMO-based DRA antenna is having value approximately 10, depicted improved diversity performance.

6 Conclusion

In the presented chapter, a dual-band dual-mode orthogonally placed CDRA-based MIMO antenna has been inspected for various wireless applications; a novel pentagon-shape aperture has been used to excite the CDRA, thereby creating $HE_{11\delta}$ and $HE_{12\delta}$ modes. With the concept of polarization diversity, the proposed radiator is able to achieve the isolation of −33 dB between the different elements of the antenna. The presented MIMO antenna gives better results such as gain and better co- and cross-polarization (more than 12 dB); the disparate parameters examined are found within the satisfactory range.

References

1. Petosa A (2007) Dielectric resonator antenna handbook. Artech House, Norwood, MA, USA
2. Ishimiya K, Lnagbacka J, Ying Z, Takada JI (2008) A compact MIMO DRA antenna. In: Proceedings of IEEE international workshop on antenna technology: small antennas and novel metamaterials (IWAT '08), Chiba, Japan
3. Roslan SF, Kamarudin MR, Khalily M, Jamaluddin MH (2014) An MIMO rectangular dielectric resonator antenna for 4G applications. IEEE Antenna Wireless Propag Lett 13:321–324
4. Yan J-B, Bernhard JT (2012) Design of a MIMO dielectric resonator antenna for LTE femtocell base stations. IEEE Trans Antennas Propag 60(2):438–444
5. Sharawi MS, Podilchak SK, Khan MU, Antar YM (2017) Dual frequency DRA based MIMO antenna system for wireless access point. IET Microwave Antennas Propag 11(8):1174–1182
6. Das G, Sharma A, Gangwar RK (2017) Dual port aperature coupled MIMO cylindrical dielectric resonator antenna with isolation for WiMAX application. https://doi.org/10.1002/mmce.21107
7. Wang XM, Weng ZB, Jiao YC et al (2010) Dual polarized dielectric resonator antenna with isolation using hybrid feeding mechanism for WLAN applications. Prog Electromagn Res Lett 18:195–203
8. Sun J-S, Fang H-S, Lin P-Y, Chuang C-S (2016) Triple-band MIMO antenna for mobile wireless applications. IEEE Antennas Wireless Propag Lett 15:500–503

An Energy-Efficient Localization Scheme Using Beacon Nodes for Wireless Sensor Networks

Sunil Kumar, Prateek Raj Gautam, Swati Verma, and Arvind Kumar

Abstract Localization is an important scheme to find the position for the sensor nodes in the field of wireless sensor networks. In this paper, the localization is used to estimate the position of the unknown node by getting a beacon message from the nodes. The ripple localization algorithm is used, to estimate the position of unknown nodes by transmitting or receiving a beacon message from beacon nodes using multilateration process. In this algorithm, the beacon nodes have different signal transmission power. Along with this, this algorithm does not require any extra hardware for ranging and estimation of the position of unknown nodes with respect to beacon nodes. The algorithm is distributed, so the communication overhead is avoided. The transmission power is not always used by beacon nodes as it does not use maximum power always. Moreover, unknown nodes do not transmit power. Hence, energy consumption is reduced significantly.

Keywords Localization · Multilateration · Energy-efficient · Beacon node · Wireless sensor network

1 Introduction

The technological development and improvement in the field of wireless sensor network have led to low power, low cost, and multifunction sensors that are small in size. The sensor nodes perform various tasks and operations in the environment. A

S. Kumar (✉) · P. R. Gautam · S. Verma · A. Kumar
Motilal Nehru National Institute of Technology Allahabad, Allahabad, India
e-mail: rel1516@mnnit.ac.in

P. R. Gautam
e-mail: prateekrajgautam@gmail.com

S. Verma
e-mail: verma.swati.93@gmail.com

A. Kumar
e-mail: arvindk@mnnit.ac.in

D. Harvey et al. (eds.), *Advances in VLSI, Communication, and Signal Processing*, Lecture Notes in Electrical Engineering 683,
https://doi.org/10.1007/978-981-15-6840-4_12

wireless sensor field comprises of different types of sensor nodes and their battery devices so that nodes can communicate with each other. Location of sensor node is a very important key feature in a sensor network in many applications such as fire detection, monitoring forests or fields, routing protocols, topology management, etc. [1, 2]. An efficient algorithm of localization can detect the most accurate position coordinates of sensor nodes using the data available from nearby sensor nodes. The problem of determining/discovering the location of unfocalized/unknown/dumb sensor nodes is known as localization. Our main goal is to find the location of dumb/unknown node so that we can use the information present in that unknown/dumb node in any application. There are different methods exist that have already been discovered [3, 4] and described in the section below.

1.1 Classification of Localization

- Anchor free and Anchor based localization method [5]
- Centralized and Distributed localization method [6]
- Range free and Range-based localization method.

1.2 Calculation of Position

There are different methods present, to calculate the position of sensor nodes with respect to anchor node present in the sensor network field:

- Lateration (Trilateration)
- Triangulation
- Multilateration.

Lateration (Trilateration). Trilateration is a simple technique that involves the distance between unknown/dumb sensor nodes and three anchor/beacon sensor nodes to estimate the location of unknown sensor nodes as shown in Fig. 1. The location of

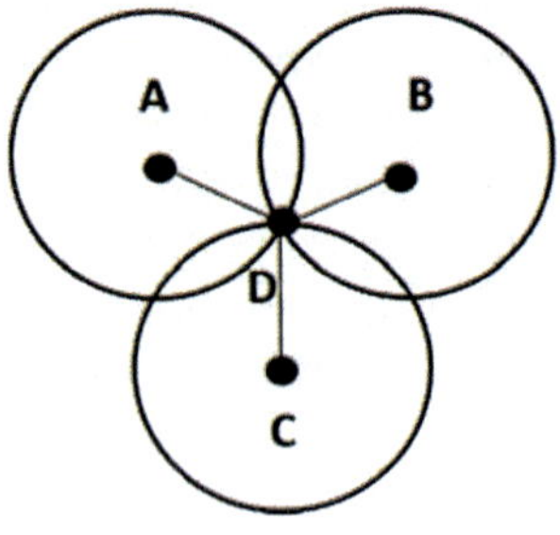

Fig. 1 Location of unknown sensor node (intersection of three circles)

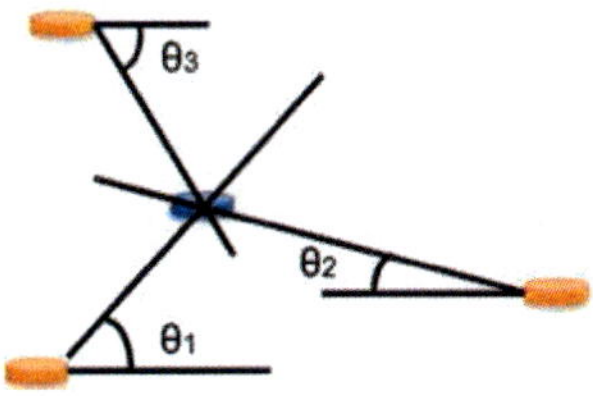

Fig. 2 Location of unknown sensor node using triangulation method

unknown sensor node is accurate when three anchor nodes are connected in 2D space [7, 8]. The point where three circles intersect each other is the coordinate location of unknown sensor nodes. If it has error than circle will not intersect at a common point.

Triangulation. Triangulation is a simple technique to find the unknown node. In this technique, angle is used instead of distance parameter, to estimate the location of the unknown node. The angle of anchor node calculates using angle of arrival method with respect to unknown node in some reference frame as shown in Fig. 2. The node position is calculated using trigonometry laws of sine or cosine [9].

Multilateration. The unknown sensor node location can also be calculated with more than three anchor nodes [10, 11]. Further, consider k anchor nodes with position coordinates (x_1, y_1), (x_2, y_2), (x_3, y_3), (x_k, y_k), and step size of anchor node is d_r.

$$\text{Estimated range } r_i = (Ri + R_{i-1})/2$$

where $R_{i-1} = R_i - d_r$

R_i = radius of outer circle of annular ring around beacon node in which unknown node lies. R_{i-1} = radius of inner circle of annular ring around beacon node in which unknown node lies.

A set of the following equation can be obtained assuming unknown sensor node coordinate to be (x, y)

$$\begin{bmatrix} (x-x_1)^2 + (y-y_1)^2 \\ (x-x_2)^2 + (y-y_2)^2 \\ \vdots \\ (x-x_k)^2 + (y-y_k)^2 \end{bmatrix} = \begin{bmatrix} r_1^2 \\ r_2^2 \\ \vdots \\ r_k^2 \end{bmatrix} \tag{1}$$

Expand square terms and rearranging them:

$$\begin{bmatrix} x^2 + y^2 - 2x_1x - 2y_1y \\ x^2 + y^2 - 2x_2x - 2y_2y \\ \vdots \\ x^2 + y^2 - 2x_kx - 2y_ky \end{bmatrix} = \begin{bmatrix} r_1^2 - x_1^2 - y_1^2 \\ r_2^2 - x_2^2 - y_2^2 \\ \vdots \\ r_k^2 - x_k^2 - y_k^2 \end{bmatrix} \tag{2}$$

Subtract last from each row above, so we obtain

$$\begin{bmatrix} 2(x_k - x_1)x + 2(y_k - y_1)y \\ 2(x_k - x_2)x + 2(y_k - y_2)y \\ \vdots \\ 2(x_k - x_{k-1})x + 2(y_k - y_{k-1})y \end{bmatrix} = \begin{bmatrix} r_1^2 - r_k^2 + x_k^2 - x_1^2 + y_k^2 - y_1^2 \\ r_2^2 - r_k^2 + x_k^2 - x_2^2 + y_k^2 - y_2^2 \\ \vdots \\ r_{k-1}^2 - r_k^2 + x_k^2 - x_{k-1}^2 + y_k^2 - y_{k-1}^2 \end{bmatrix} \quad (3)$$

Separate the unknown term (x, y) and rewrite the matrix as below:

$$\begin{bmatrix} (x_k - x_1) & (y_k - y_1) \\ x_k - x_2 & (y_k - y_2) \\ \vdots & \vdots \\ x_k - x_{k-1} & (y_k - y_{k-1}) \end{bmatrix} \begin{bmatrix} x \\ y \end{bmatrix} = \frac{1}{2} \begin{bmatrix} r_1^2 - r_k^2 + x_k^2 - x_1^2 + y_k^2 - y_1^2 \\ r_2^2 - r_k^2 + x_k^2 - x_2^2 + y_k^2 - y_2^2 \\ \vdots \\ r_{k-1}^2 - r_k^2 + x_k^2 - x_{k-1}^2 + y_k^2 - y_{k-1}^2 \end{bmatrix} \quad (4)$$

Using matrix, it can be written as

$$Az = R \quad (5)$$

where $z = \begin{bmatrix} x & y \end{bmatrix}^{\mathrm{T}}$

$$A = \begin{bmatrix} (x_k - x_1) & (y_k - y_1) \\ (x_k - x_2) & (y_k - y_2) \\ \vdots & \\ (x_k - x_{k-1}) & (y_k - y_{k-1}) \end{bmatrix} \quad (6)$$

$$R = \frac{1}{2} \begin{bmatrix} r_1^2 - r_k^2 + x_k^2 - x_1^2 + y_k^2 - y_1^2 \\ r_2^2 - r_k^2 + x_k^2 - x_2^2 + y_k^2 - y_2^2 \\ \vdots \\ r_{k-1}^2 - r_k^2 + x_k^2 - x_{k-1}^2 + y_k^2 - y_{k-1}^2 \end{bmatrix} \quad (7)$$

Using least square approximation, we get closed-form and unique solution

$$z = A^{+}R$$

where $A^{+} = (A^{\mathrm{T}} A)^{-1} A^{\mathrm{T}}$, ($A^{+}$ pseudo inverse of matrix of A).

Hence, the (x, y) coordinate computed from this algorithm is the estimated position of dumb/unknown nodes.

2 Motivation

Localization is a crucial task in WSNs as the location of nodes is required for proper interpretation of sensed and received data. The localization must be energy-efficient and must work with the help of fewer anchor nodes. The number of transmissions and receptions from nodes must be reduced so as to minimize energy consumption during the localization process. To solve these problems, this localization scheme is proposed.

3 Proposed Ripple Localization Algorithm

Ripple localization algorithm [1] is a simple technique in which each unknown node can localize its position with the help of beacon nodes. This algorithm consists of multiple power levels and annular rings around beacon nodes and further, uses multilateration method for positioning of unknown sensor nodes. Two-dimensional (2D) sensor field is considered with finite boundaries in which beacon nodes and unknown nodes are placed. As a fact, for unknown node radio range is longer compared to sensing range and for beacon node communication range is longer than unknown nodes. Hence, less beacon nodes are required to reach unknown nodes in the sensor field due to its large communication range. All nodes are equipped with omnidirectional antennas. Along with this, circular radio range is considered for easy implementation, however, due to irregularity generation in the network field, a term is introduced known as Degree of Irregularity (DOI) for irregularity in radio pattern as shown in Fig. 3 (Fig. 4).

In wireless sensor networks, the beacon nodes transmit beacon signals in circular radio patterns at regular intervals. Hence, each time unknown node receives beacon signals containing information on them that helps to estimation the location of the node. Transmitted beacon signals from beacon nodes are similar manner the ripples

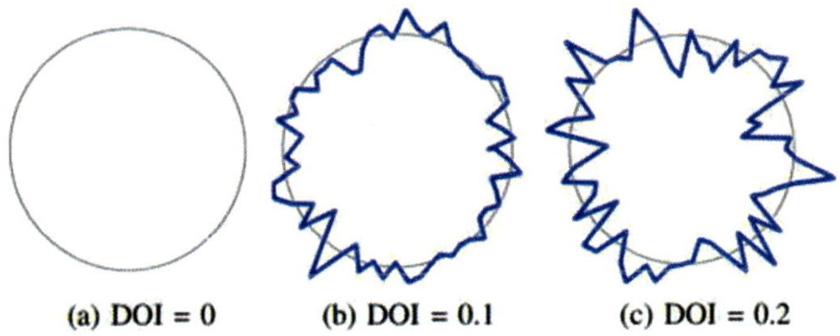

Fig. 3 Radio Pattern with DOI

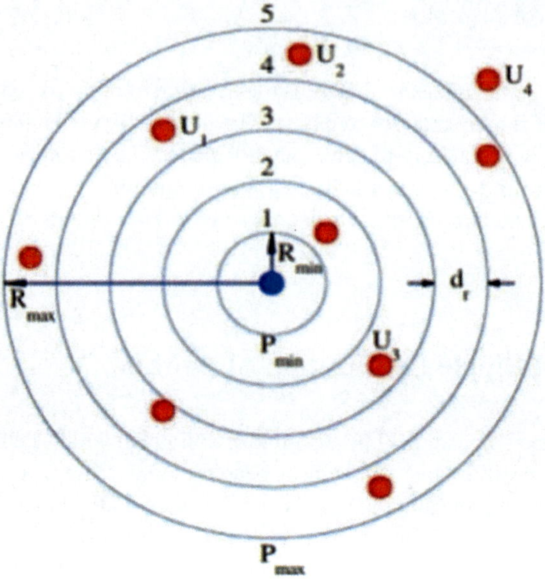

Fig. 4 Beacon signal ripples

are generated in water. The packet of a beacon signal has different sections of information as shown in Fig. 5, i.e., Time Stamp (t_0), Position Coordinates (x_b, y_b), Transmission Power (P_{ti}), Radio Range (R_i), Beacon Step Size (d_r), Minimum transmission radius ($R_{\min}$), Maximum transmission radius ($R_{\max}$).

As shown in Fig. 4, the node located at the center is a beacon node (blue color) with radio pattern ripples around it covering unknown nodes (red color). Each unknown node can estimate its inner radius and outer radius using message signal from beacon node, resulting distance is estimated between the unknown node and beacon node. Inner radius of the annular ring can be estimated by simply subtracting the step size d_r from the outer radius. Any node can estimate its location by solving multilateration equation after receiving beacons from three or more beacon nodes.

ID	t_0	(X_b, Y_b)	P_{ti}	R_i	d_r	R_{min}	R_{max}

Fig. 5 Beacon packet format

4 Simulation Results and Discussions

This section discusses the results of ripple localization algorithm using different parameters. With the help of localization algorithm, it is tried to estimate the position of unknown nodes using beacon nodes. MATLAB R2017a is used to simulate the ripple localization algorithm. In this module, the scatter graph shows the point where beacon nodes and unknown nodes are randomly deployed in two-dimensional network area. In Figs. 6 and Fig. 7 100 nodes are deployed randomly and 20 beacon nodes are selected as beacon nodes. The blue dots in Fig. 6 are beacon nodes whereas blue dots in Fig. 7 are dumb nodes.

In Fig. 8, randomly generated 20 beacon nodes and 100 dumb nodes are placed in two-dimensional network area. These beacon nodes help to estimate the actual position of dumb nodes. With the help of localization algorithm, each randomly generated dumb node is considered and finds out the position of nearby beacon nodes. Now, find out the actual coordinate of dumb node by multilateration process with the help of neighboring beacon nodes to each dumb node. In Fig. 8, blue dots are randomly generated dumb nodes whereas red dots are randomly generated beacon nodes.

In Fig. 9, shows the difference in coordinate of randomly generated dumb nodes and actual estimated coordinated of dumb nodes. The blue color dots are randomly generated dumb nodes whereas the red color nodes are actual dumb nodes position computed using multilateration process. The difference between actual and dumb nodes is reduced resulting in reduced localization error. The red spikes are focusing on the error in each dumb node between randomly generated and estimated dumb

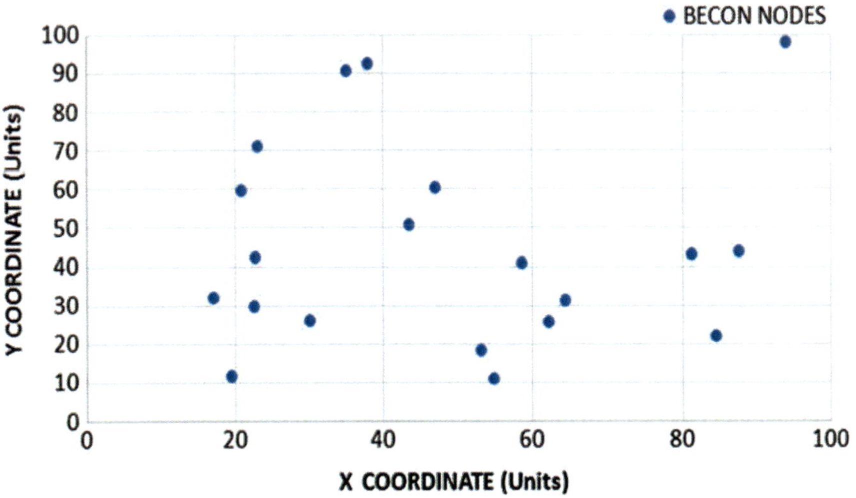

Fig. 6 Beacon deployed in 100 × 100 area

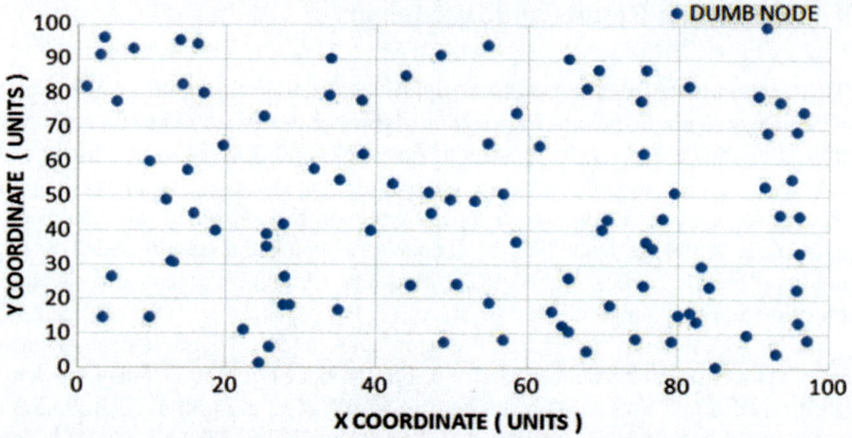

Fig. 7 Dumb nodes deployed in 100 × 100 area

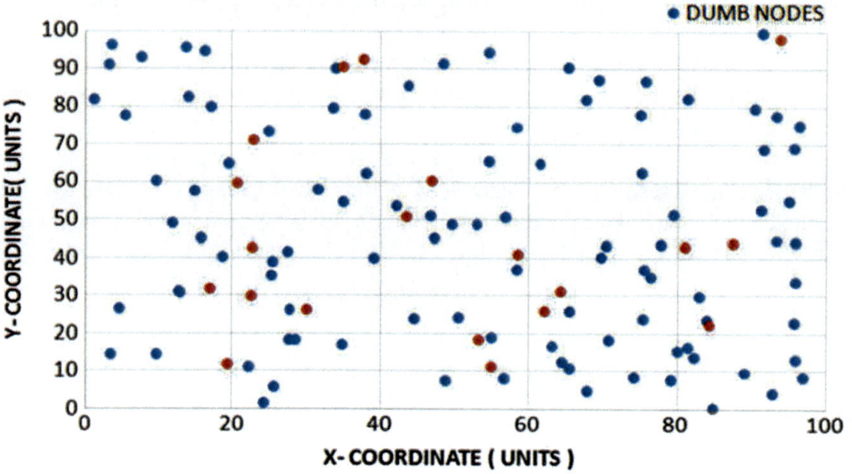

Fig. 8 Deployment of beacon nodes and dumb nodes

nodes as shown in Fig. 10. The error is calculated for 100 dumb nodes. The x-axis shows dumb node number, i.e., for each and every node error is calculated whereas on y-axis error is shown. As per the requirement, CDF is also calculated. The cumulative distribution function (CDF) of a real-valued random variable X or just distribution function of X, evaluated at x, is the probability that X will take a value less than or equal to x. Along with this, the CDF plot i.e. cumulative distributive plot for localization error is as shown in Fig. 11. This algorithm provides with actual position of dumb nodes and along with this error is also computed.

A plot is shown in Fig. 10 is of node number on x-axis and error on y-axis and hence showing the performance of localization.

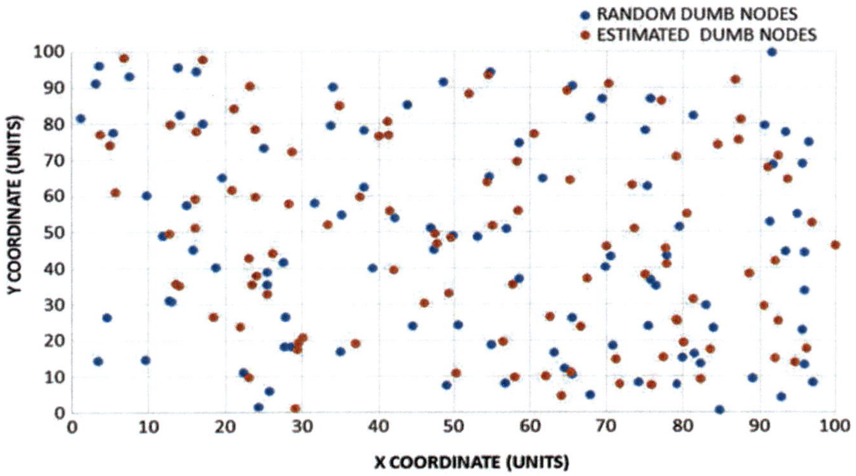

Fig. 9 Scattering graph showing difference in randomly generated dumb nodes and beacon nodes

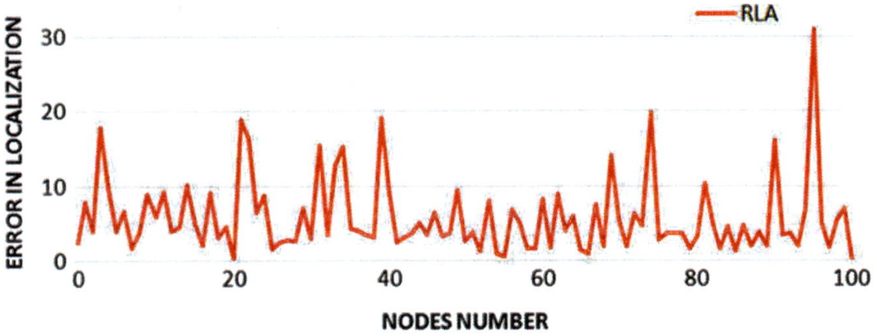

Fig. 10 Error estimation of position of unknown nodes in sensor network

Along with this, the CDF plot, i.e., cumulative distributive plot for localization error is as shown below

5 Conclusion

Research and development over the last few years have led to significant advancement in field of wireless sensor networks and more importantly in the section of localization. An easy approach for estimating the location of unknown/dumb node with help of beacon/anchor nodes in wireless sensor network using ripple localization algorithm without using any additional hardware, therefore, resulting in saving of cost, energy, and size. This simple mathematical approach is able to derive the location of

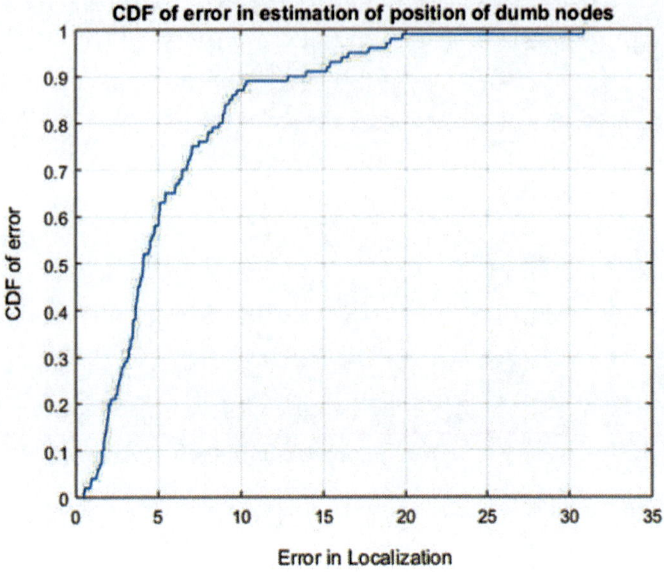

Fig. 11 CDF plot for localization error

node with accuracy to some extent. Considering the 2-dimensional area for sensor nodes, resulting in estimation of position coordinate unknown/dumb node with help of multilateration method. Along with this, other factors like distance error, localization error (mean, median, and mode), localization efficiency are also computed. The simulation results and estimated method have been implemented using the MATLAB tool.

References

1. Farooq-I-Azam M, Ni Q, Ansari EA (2016) Intelligent energy efficient localization using variable range beacons in industrial wireless sensor networks. IEEE Trans Industr Inf 12(6):2206–2216
2. Gautam PR, Kumar S, Verma A, Rashid T, Kumar A (2019) Energy-efficient localization of sensor nodes in WSNs using beacons from rotating directional antenna. IEEE Trans Ind Inf. https://doi.org/10.1109/tii.2019.2908437
3. Farrag M, Abo-Zahhad M, Doss MM, Fayez JV (2016) Different aspects of localization problem for wireless sensor networks: a review. Int J Comput Netw Commun Secur 4(5):130
4. Pal A (2010) Localization algorithms in wireless sensor networks: current approaches and future challenges. Netw Protoc Algorithms 2(1):45–73
5. Zhang J, Luo J, Luo X (2008) A robust localization algorithm for wireless sensor networks. In: 2008 4th international conference on wireless communications, networking and mobile computing, October. IEEE, pp 1–4
6. Paul A, Sato T (2017) Localization in wireless sensor networks: a survey on algorithms, measurement techniques, applications and challenges. J Sens Actuat Netw 6(4):24

7. Derr K, Manic M (2015) Wireless sensor networks—node localization for various industry problems. IEEE Trans Industr Inf 11(3):752–762
8. Wu D, Chatzigeorgiou D, Youcef-Toumi K, Ben-Mansour R (2015) Node localization in robotic sensor networks for pipeline inspection. IEEE Trans Industr Inf 12(2):809–819
9. Chraim F, Erol YB, Pister K (2015) Wireless gas leak detection and localization. IEEE Trans Industr Inf 12(2):768–779
10. Li X, Yang J, Nayak A, Stojmenovic I (2012) Localized geographic routing to a mobile sink with guaranteed delivery in sensor networks. IEEE J Sel Areas Commun 30(9):1719–1729
11. Shu L, Zhang Y, Yang LT, Wang Y, Hauswirth M, Xiong N (2010) TPGF: geographic routing in wireless multimedia sensor networks. Telecommun Syst 44(1–2):79–95

Rain Streaks Elimination Using Hybrid Median Filter and Contrast Stretching

Rais Ahmad and Som Pal Gangwar

Abstract An efficient method to enhance the rain degraded single image using Hybrid Median Filtering and contrast stretching is proposed with this paper. The rainy image consists of rain streaks and foggy appearance of the image due to rain streaks accumulation. The original image is thus having two different components—rain streaks and foggy image. There are several techniques available to enhance the rainy image using simple filters to complex algorithms. Some of the techniques are using multistage linear filtering to eliminate the rain streaks from the image and Dark channel prior to technique to enhance the foggy rain-free image. These techniques are simple but the output rain-free enhanced images are having less similarity to original image. With this paper, it is proposed that the rain streaks elimination could be augmented using a non- linear filter like Hybrid median filter before using linear filtering. Also, the contrast stretching method is implemented to improve the fog removed image obtained after dark channel prior. With our proposed technique we have shown that the similarity index of rain-free enhanced image is increased using a Hybrid Median Filtering and contrast enhancement. This technique is having lesser complexity and easy to implement. We have compared the proposed technique with Gaussian Filtering and Temporal Filtering methods. The comparison of simulation results confirms that our technique is having a higher PSNR and lower MSE. The outcome of rain-free enhanced image is having better similarity with original images.

Keywords Rain removal · Hybrid median filter · Gaussian filter · Dark channel prior · Temporal filtering · Contrast stretching

R. Ahmad (✉) · S. P. Gangwar
Kamla Nehru Institute of Technology, Sultanpur, U.P., India
e-mail: rais_hal@rediffmail.com

S. P. Gangwar
e-mail: gangwar_sp@rediffmail.com

D. Harvey et al. (eds.), *Advances in VLSI, Communication, and Signal Processing*, Lecture Notes in Electrical Engineering 683,
https://doi.org/10.1007/978-981-15-6840-4_13

1 Introduction

With the emerging recent technologies of smartphones, the need of enhancing the captured image is increased. The captured images may be degraded due to several reasons like low light, bad weather conditions, etc. Rain is one of the reasons for bad weather condition and the captured images in rainy weather are having foggy appearance and rain streaks present on them. The fogginess is due to the accumulation of rain streaks which decreases the intensity of image pixels.

The rain is a type of organized noise which causes significant degradation of original image in terms of its quality [1]. Rain affects the image in two ways—first, it adds the rain streaks on the image, and second, the image gets blurred due to the accumulation of rain streaks. The rain streaks are having characteristics of sudden change of intensity as compared to the neighboring original image pixel values. The rain streaks are thus having the high-frequency components parts while the original image is having mostly low-frequency components.

Different rainy images are shown in Fig. 1. The rain image, therefore, may be written as

$$R_i = O_i + S_r \tag{1}$$

where R_i is the rainy image, O_i is original rain-free background image and S_r is the streaks components of rain. The original image gets blurred due to rain streaks accumulation and the whole image pixel intensities are affected by a factor 'k'. Hence, Eq. 1 may become

$$R_i = kO_i + S_r \tag{2}$$

Several methods are proposed by different researchers to eliminate the rain streaks. The methods may vary from the application of simple linear filters to complex neural network algorithms. The results may differ from one method to other method. In our proposed work, we have focused on achieving a better result with a simple algorithm.

Fig. 1 Rain-affected images

The rain streaks are identified by a sudden change in intensity value than the neighboring image pixel. The same property is exhibited by the edge components of the image. So, it is necessary to separate the edge components of the image with rain streaks components. Hence, two stages of filtering along with subtraction with the original image are needed. The Hybrid median filter is having better impulsive noise reduction and can be used for preprocessing applications. For fog component enhancement, the estimation of contrast variation in the blurred image is detected [2]. Using the estimated factors, techniques like dark channel prior, de-hazing, or luminance based methods are used.

Further, this article is ordered as follows. Section 2 presents various techniques and details of the proposed scheme, Sect. 3 presents propose work and Sect. 4 presents experiment result analysis and conclusions of the study are presented in Sect. 5.

2 Techniques Used

In this paper, we have used three techniques for obtaining the rain removed image and compare the results obtained, which are described below.

2.1 Gaussian Filtering (GF) Method

The Gaussian Filtering method is an effective linear rain streak removal technique. The Gaussian Filters are used in multistage to detect and remove the high-frequency rain streaks components. The low and high-frequency components are separated and subtracted from Gaussian filtered image and the original image. While using Gaussian filter for noise reduction, the original image also gets distorted. Due to this the edge components also vanishes which leads to a reduction in the similarity index of the output image. Gaussian Filtering can also be implemented using Kernel of different sizes [3]. The Kernel accuracy depends upon floating-point multipliers which in turn consumes larger energy.

The method of multistage filtering includes the usage of Gaussian Filters in two steps to remove the rain streaks present in the rain-affected image. In the first stage, the Gaussian Filtering is applied to rain-affected image to obtain the low-frequency part of the image. The low-frequency part is then subtracted from the original rainy image to obtain the high-frequency component which consists of rain streaks and edge components. In the second stage, again Gaussian Filtering is applied to the high-frequency image to obtain the non-rain edge or texture information. This obtained result image is subtracted from the previous high-frequency resultant image to obtain the rain streaks component. Lastly, the rain streaks component image is subtracted from original rain-affected image to obtain the rain-free image [4].

Dark channel prior (DCP). In the year 2018, Z. Shi et al. proposed multi-stage filtering using Gaussian Filters for rain streaks removal and utilization of Dark channel prior to fog reduction. The fogginess present in original rainy image may be expressed as

$$O_{\text{fog}} = O_i * t + A(1 - t) \tag{3}$$

where O_i is the original rain-free background image, A is the atmospheric light and t is the scene transmission.

The DCP method adopts the calculation of statistical data of outdoor haze-free images. The dark pixels obtained from low-intensity RGB color channel gives the estimation of haziness [5]. This is a simple but powerful method of fog or haze removal. The dark channel obtained at each pixel from the least intensity value of RGB color channel [6]. Humans are more sensitive to deviations near the white color [7]. The Dark channel prior method thus gives darkness to the image by estimating the transmission and atmospheric light present in the rain-affected image. The dark channel is used to remove the fog. The continuous transmission is contrary to the increase in parameter and the quality of the image increases and noise decreases [8].

Therefore, Gaussian multistage filtering along with dark channel prior is used as our first technique for comparison with other methods.

2.2 *Temporal Filtering (TF) Method*

Temporal filtering method is used to remove the noise component present within the original image. This method is useful in neuroimaging to enhance the MRI images. The temporal filtering is an efficient tool for the design of venturing operators of diverse types [9]. As the rain contains both spatial and temporal components, temporal filtering may also be useful for reduction of noise components from the rain-affected images. It uses Fourier Transform to convert signal of intensity in time or space to convert to the frequency spectrum. The frequency range depends upon the sampling rate and hence follows the Nyquist criterion [10].

$$\Delta x \le \frac{\Delta t}{|\partial t/\partial x|} \tag{4}$$

The above equation shows the Nyquist criterion with integral operators where Δx and Δt are space and time sampling intervals, respectively. Aliasing effect occurs in the output image if the sampling criterion is not satisfied. To overcome this effect local triangle filtering is used

$$\Delta t = \Delta x |\partial t/\partial x| \tag{5}$$

This method is used as low pass filtering. The low pass filtered image is then subtracted from original image to obtain the high-frequency rain streaks components. To study the outcome of using temporal filtering method for rain image enhancement, the rain-free image obtained from temporal filtering is passed through DCP and contrast stretching. The images therefore acquired are compared with other filtering methods.

2.3 *Hybrid Median Filter (HMF) Method*

The Hybrid Median Filter belongs to non-linear windowed class of filters. This filter is widely used for impulsive noise elimination while preserving the edges components present in the image [11]. In the rain streaks detection using linear filtering, the edge components are also diminished with the rain streaks. To preserve the corners of image components, the Hybrid Median Filter is beneficial. The Hybrid Median Filter operates by taking two medians—'plus' and 'cross' centered on the pixel of interest as shown in Fig. 2. The Hybrid median filter is having adaptive nature and gives better results than the median filter of small spatial extent.

The Hybrid Median Filtering may be expressed in terms of median values as

$$F = \text{median}\left\{M_{\text{plus}},\ M_{\text{cross}},\ C\right\} \tag{6}$$

where M_{plus} is the median of horizontal and vertical pixels, M_{cross} is the median of diagonal pixels and C is the center pixel. The Hybrid median filter thus uses a similar sliding window algorithm as used in the median filter but the way of treatment of neighboring pixels is changed [12]. This filtering method is used as preprocessing of the rain corrupted image before edge subtraction to the original image to get the rain streaks free image.

Contrast Stretching (CS). Contrast stretching technique is helpful in improving the image through expanding the range of intensity values that are contained in the image. The rain-free image on passing through dark channel prior gets most of the intensity values on the darker side, due to which the output image seems to be darker than the original image. After contrast stretching the intensity values are expanded to the full frequency range.

If $r_{\max}$ and $r_{\min}$ are maximum and minimum value of pixels, respectively. The slope of the line Joining (0, 255) and ($r_{\min}$, $r_{\max}$) may be given by

$$s = 255/(r_{\max} - r_{\min}) \tag{7}$$

Fig. 2 Hybrid median filtering

Hybrid Median Filtering				
22	20	2	14	12
2	3	11	10	24
20	4	19	14	12
7	8	1	24	12
11	6	15	14	21

Then, intercept of the straight line with the axis is given by

$$c = 255 - s * r_{\max} \tag{8}$$

The transformation of image according to new slope is given by

$$i_{\text{new}} = (s * i + c)/255 \tag{9}$$

where i is the intensity value of the image. The contrast stretching is adopted for each RGB component of the image.

The method of contrast stretching is widely used in medical applications to obtain good superior quality images [13]. This method uses a linear mapping function to augment the contrast and brightness level of an image [14]. The rain degraded images are influenced by many factors, including reflection and attenuation [15].

3 Rain Streaks Elimination Using Hybrid Median Filter and Contrast Stretching

3.1 Problem Statement

1. Rain streaks detection and removal.
2. Fogginess reduction from the rain degraded image.
3. Deviation in the pixels Intensity values after application of filters and improving the similarity index.

3.2 Propose Methodology

In our proposed method, a non- linear filter is used to obtain the high-frequency components from the rain degraded images. These high-frequency components are having rain streaks and edge components information. Then the rain streaks can be separated by diminishing the edge components and subtraction from the original rain degraded image by using any simple linear or non-linear filter. The utilization of non-linear filtering at initial stage is having an additional advantage on noise removal. The rain streaks are having a close resemblance to impulse noise and salt-pepper noise. Therefore, Hybrid median Filter in our proposed method which gives better noise removal results. The outcome of Hybrid median filter image is having Low-frequency components of the rain degraded image. This image is then subtracted from original image to obtain the high-frequency components. As previously discussed, the high-frequency component is having both rain streaks and the edge components. To remove the rain streaks, the edge components are diminished using any linear or non-linear filter. We have used simple Gaussian filter for rain streaks separation. The image thus obtained is subtracted from the original image to get the rain-free image. For fog removal dark channel prior technique is used by calculating the transmission estimate and atmospheric light from the rain degraded image. The results thus obtained are having contrast accumulation mostly on the darker side. To make the result more similar to original the Contrast stretching method is used over the intensity values. The dynamic ranges of gray levels are therefore increased. Using the proposed method, the rain streaks eliminated and fog removed image is obtained from the rain degraded image

3.3 Proposed Algorithm

1. Rain affected Image is browsed from Dataset.
2. Apply Hybrid median filter (HMF) on the selected image.
3. Subtract HMF image from the selected image and obtained HF (High Frequency) image.
4. Apply Gaussian filter (GF) on this resultant image.
5. Subtract Gaussian image from HF image and obtained Intermediate image GH.
6. Subtract GH image from Gaussian image then obtained Rain-free image.
7. Then apply DCP on this rain-free image.
8. Apply Contrast stretching on DCP image.
9. Calculate parameter PSNR, MSE, and SSIM.

The rain degraded imagc is browscd from the datasets over which Hybrid Median Filtering is applied to get the filtered image. The high-frequency rain component is obtained from the subtracting HMF image from original image. A linear filter like Gaussian is applied to smoothen the HF component. The Dark Channel prior is applied to obtained rain-free image to reduce the fog component and again filtering is carried out. Thereafter Contrast Stretching is applied to obtain the rain streaks eliminated and fog component reduced image. The whole process is indicated with block diagram in Fig. 3.

4 Experimental Result Analysis

A Dataset of different rain-affected images is prepared and the picture file used is collected from that Dataset. The picked image is first converted to a similar size, data type and resolution. Figure 4 shows 15 different rain-affected images from the Dataset (I, II, and III) used in our experiment.

In our experiment work, first, browse image from Dataset of different images of rainy season and then we estimate it with Hybrid median filter. Again we apply Gaussian filter to the rainy image, so we obtain a rain-free image after that we apply

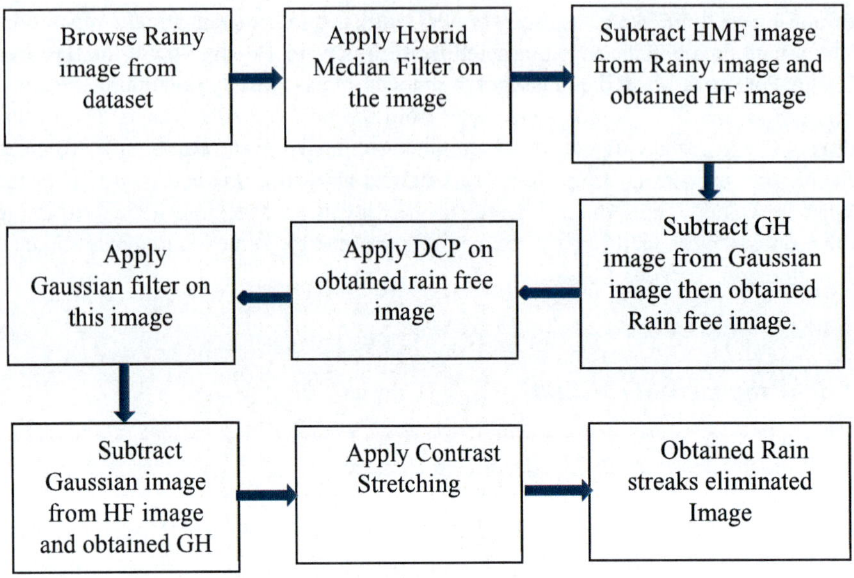

Fig. 3 Block diagram for implementation of the proposed algorithm

Fig. 4 Datasets (rain-affected Images 1–15) used in our experiment

DCP to our image thus we get our result. Figures 5, 6, 7, 8, and 9 show the outputs obtained at different stages with rain degraded Image sample using our proposed method.

Fig. 5 Rain-affected image sample

Fig. 6 Image after application of HMF

Fig. 7 Image obtained after subtraction

Fig. 8 Image after dark channel prior

The PSNR, MSE, and SSIM values of image sample using proposed algorithm are indicated in Table 1. Further comparison of different techniques of rain removal is being carried out by dividing datasets into three parts, Dataset I, II, and III.

Fig. 9 Final rain-free image as obtained after application of contrast stretching

4.1 Comparison of Results with Three Different Techniques

The Values of PSNR obtained with Dataset I (Images 1–5) using different methods shows that the higher value of Peak Signal to Noise Ratio is obtained with using Hybrid Median Filter and Contrast stretching technique as compared with Gaussian and temporal Filtering methods (refer Table 2 and Fig. 11) (Fig. 10).

The Values of MSE obtained with Dataset I (Images 1–5) using different methods show that the lower value of MSE is obtained using Hybrid Median Filter and Contrast stretching technique as compared with Gaussian and temporal Filtering methods (refer Table 3 and Fig. 12).

The values of SSIM obtained with Dataset I (Images 1–5) using different methods shows that the higher value of SSIM is obtained with using Hybrid Median Filter and contrast stretching technique as compared with Gaussian and temporal Filtering method (refer Table 4 and Fig. 13) (Fig. 14).

The Values of PSNR obtained with Dataset II (Images 6–10) using different methods shows that the higher value of Peak Signal to Noise Ratio is obtained with using Hybrid Median Filter and Contrast stretching technique as compared with Gaussian and temporal Filtering methods (refer Table 5 and Fig. 15).

The Values of MSE obtained with Dataset II (Images 6–10) using different methods show that the lower value of MSE is obtained using Hybrid Median Filter and Contrast stretching technique as compared with Gaussian and temporal Filtering methods (refer Table 6 and Fig. 16).

The values of SSIM obtained with Dataset II (Images 6–10) using different methods shows that the higher value of SSIM is obtained with using Hybrid Median

Fig. 10 Comparison of results (Dataset I, Images 1–5 from top to bottom): **a** Original rain-affected images, **b** Rain removed images using Gaussian Filtering method, **c** Rain removed images using temporal filtering method, and **d** Rain removed images using Hybrid Median Filtering and contrast stretching

Table 1 PSNR, MSE and SSIM results obtained with Image sample

Results obtained	
PSNR	41.7582
MSE	1.3287
SSIM	0.7268

Table 2 PSNR values of Dataset I (Images 1–5) obtained with different methods

Dataset I	PSNR		
	Gaussian method	Temporal filtering method	HMF & CS method
Image 1	36.2787	38.2774	41.7582
Image 2	36.1536	37.7749	41.6783
Image 3	36.1321	38.1308	41.1779
Image 4	36.3725	37.8712	41.2362
Image 5	36.2885	38.2872	41.2328

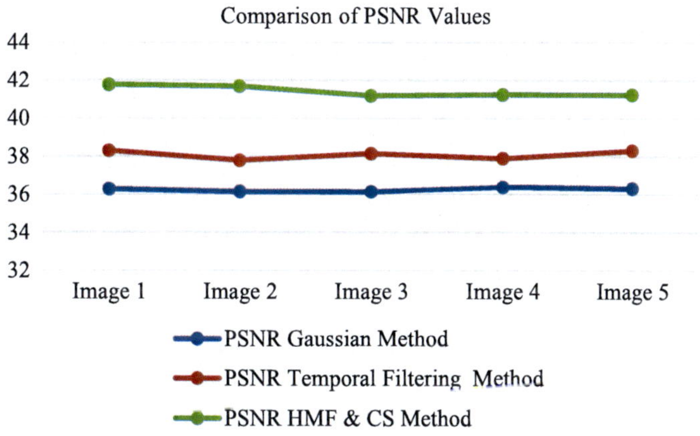

Fig. 11 Graph comparison of PSNR values of Dataset I (Images 1–5) with different methods

Table 3 MSE Values of Dataset I, (Images 1-5) obtained with different methods

Dataset I	MSE		
	Gaussian method	Temporal filtering method	HMF & CS method
Image 1	1.4107	1.3898	1.3287
Image 2	1.4138	1.4008	1.3465
Image 3	1.4139	1.3895	1.3642
Image 4	1.4129	1.3721	1.3371
Image 5	1.4124	1.3998	1.3575

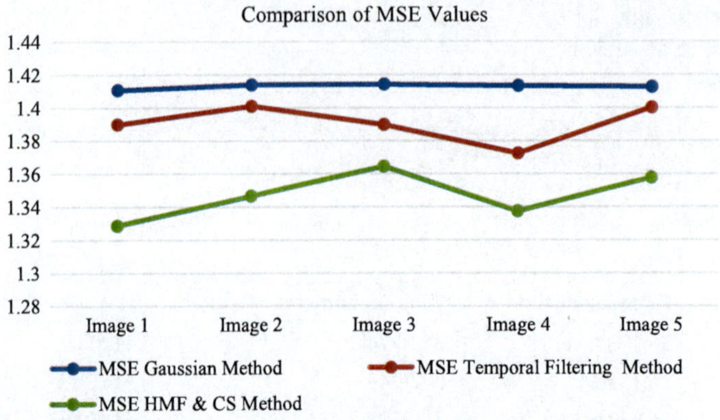

Fig. 12 Graph comparison of MSE values of Dataset I (Images 1–5) with different methods

Table 4 SSIM values of Dataset I (Images 1–5) obtained with different methods

Dataset I	SSIM		
	Gaussian method	Temporal filtering method	HMF & CS method
Image 1	0.5268	0.4356	0.7268
Image 2	0.5026	0.4229	0.7016
Image 3	0.4988	0.3948	0.6883
Image 4	0.5021	0.4368	0.6957
Image 5	0.5118	0.3998	0.7126

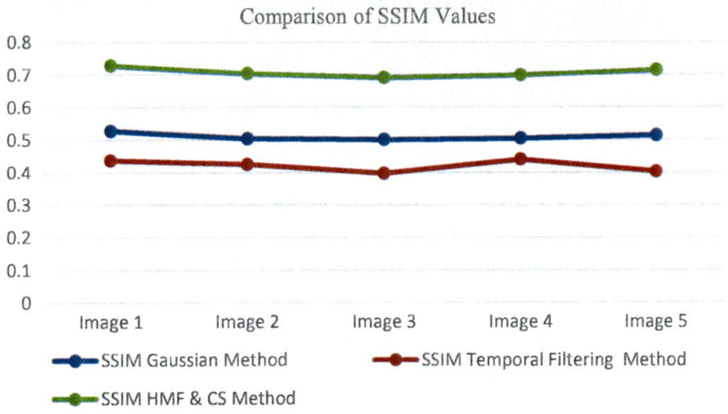

Fig. 13 Graph comparison of SSIM values of Dataset I (Images 1–5) with different methods

(a) (b) (c) (d)

Fig. 14 Comparison of results (Dataset II, Images 6–10 from top to bottom): **a** Original rain-affected images, **b** rain removal using Gaussian Filtering method, **c** rain removal using temporal filtering method, and **d** rain removal using Hybrid Median Filtering and contrast stretching

Table 5 PSNR Values of Dataset II (Images 6–10) obtained with different methods

Dataset II	PSNR		
	Gaussian method	Temporal filtering method	HMF & CS method
Image 6	36.1415	38.1462	42.1416
Image 7	36.2481	38.8108	42.7229
Image 8	36.1366	39.2021	43.3547
Image 9	36.1358	38.5795	40.7578
Image 10	36.1321	38.8476	43.3805

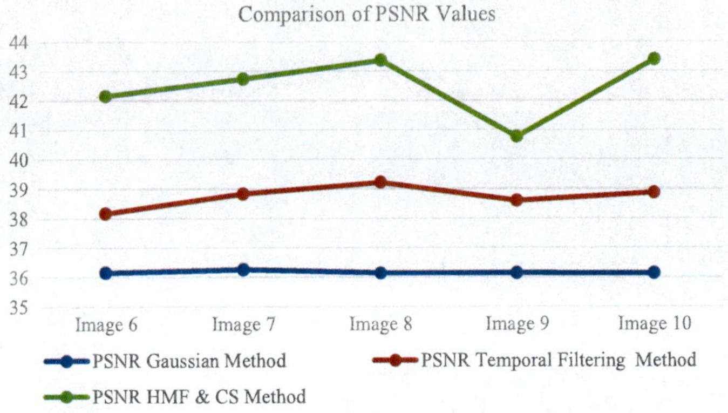

Fig. 15 Graph comparison of PSNR values of Dataset II (Images 6–10) with different methods

Table 6 MSE values of Dataset II (Images 6–10) obtained with different methods

Dataset II	MSE		
	Gaussian method	Temporal filtering method	HMF & CS method
Image 6	1.4139	1.3986	1.3709
Image 7	1.4104	1.3886	1.3641
Image 8	1.4139	1.3986	1.3524
Image 9	1.4139	1.3865	1.3394
Image 10	1.4139	1.3672	1.3333

Filter and Contrast stretching technique as compared with Gaussian and temporal Filtering methods (refer Table 7 and Fig. 17) (Fig. 18).

The Values of PSNR Obtained with Dataset III (Images 11–15) using different methods shows that the higher value of Peak Signal to Noise Ratio is obtained with using Hybrid Median Filter and Contrast stretching technique as compared with Gaussian and temporal Filtering methods (refer Table 8 and Fig. 19).

The values of MSE obtained with Dataset III (Images 11–15) using different methods show that the lower value of MSE is obtained using Hybrid Median Filter

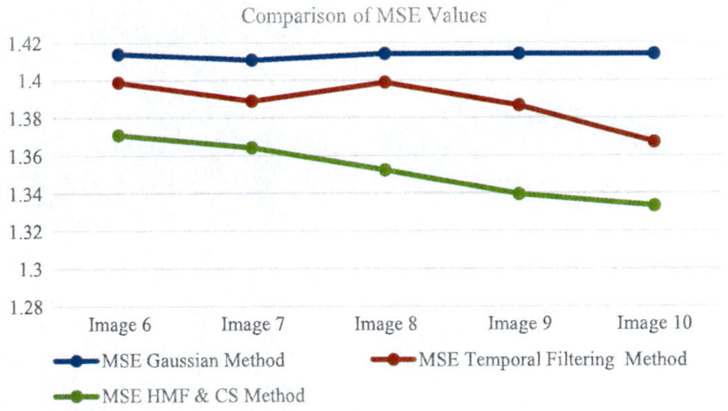

Fig. 16 Graph comparison of MSE values of Dataset II (Images 6–10) with different methods

Table 7 SSIM values of Dataset II (Images 6–10) obtained with different methods

Dataset II	SSIM		
	Gaussian method	Temporal filtering method	HMF & CS method
Image 6	0.5162	0.3006	0.6897
Image 7	0.5006	0.3854	0.6882
Image 8	0.5426	0.3986	0.7152
Image 9	0.4958	0.3994	0.6404
Image 10	0.5264	0.4126	0.7254

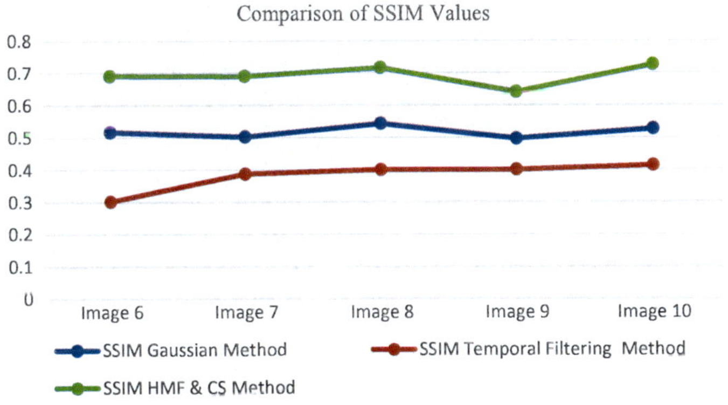

Fig. 17 Graph comparison of SSIM values of Dataset II (Images 6–10) with different methods

and Contrast stretching technique as compared with Gaussian and temporal Filtering methods (refer Table 9 and Fig. 20).

The values of SSIM obtained with Dataset III (Images 11–15) using different methods show that the higher value of SSIM is obtained using Hybrid Median Filter

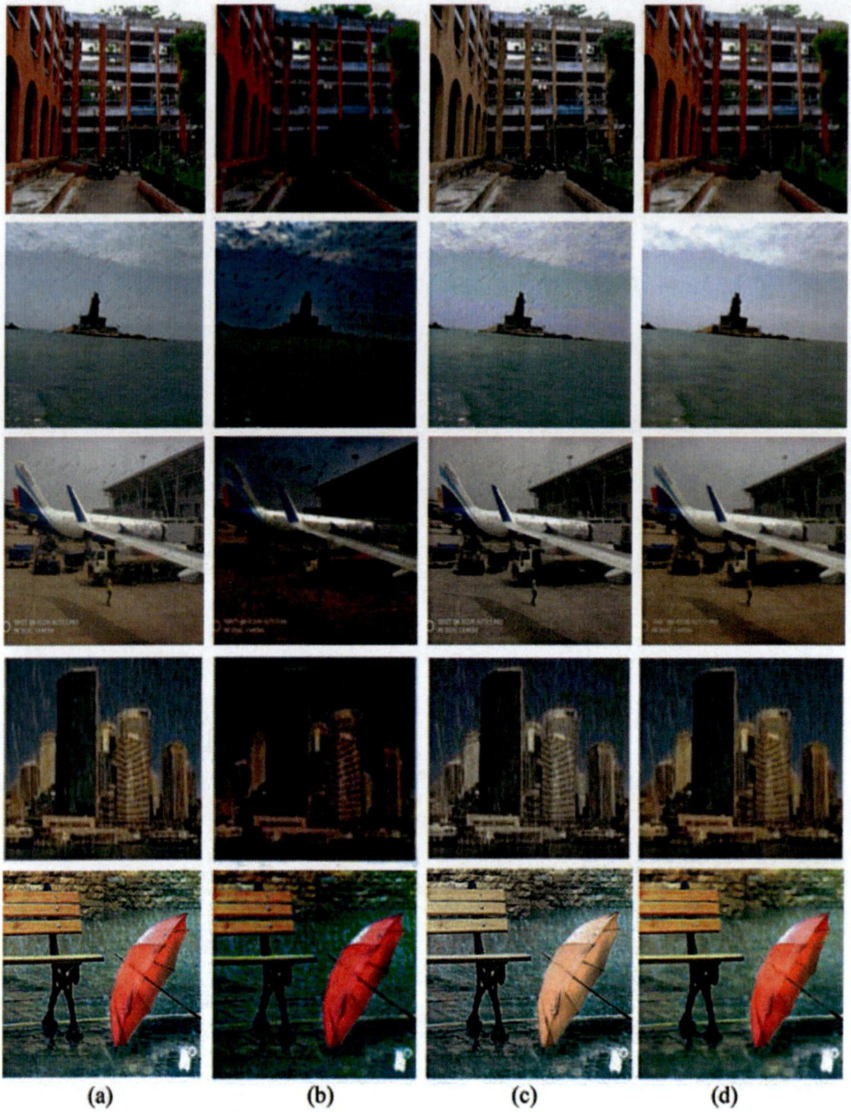

Fig. 18 Comparison of results (Dataset III, Images 11–15 from top to bottom): **a** Original rain-affected images, **b** rain removal using Gaussian Filtering method, **c** Rain removal using temporal filtering method, and **d** rain removal using Hybrid Median Filtering and contrast stretching

and Contrast stretching technique as compared with Gaussian and temporal Filtering methods (refer Table 10 and Fig. 21).

Therefore, Hybrid Median Filtering and contrast stretching is an efficient method to enhance the rain degraded single image. The rain streaks elimination could be augmented using a non-linear filter like Hybrid median filter before using linear

Table 8 PSNR values of Dataset III (Images 11–15) obtained with different methods

Dataset III	PSNR		
	Gaussian method	Temporal filtering method	HMF & CS method
Image 11	36.2764	38.9984	41.5604
Image 12	36.1334	39.2957	45.1796
Image 13	36.1321	38.8465	43.109
Image 14	36.1402	38.6526	43.2899
Image 15	36.2048	37.5849	39.7168

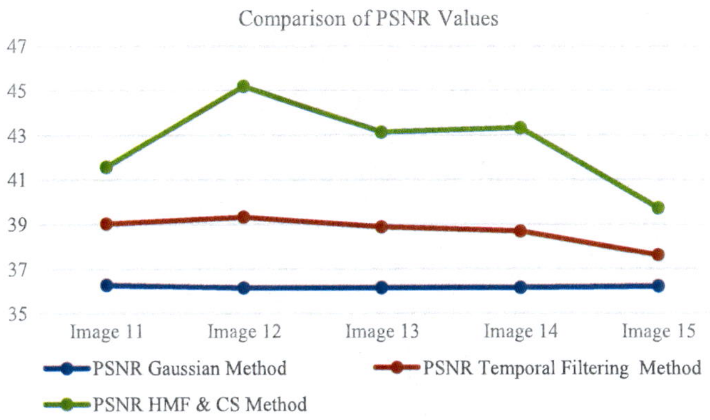

Fig. 19 Graph comparison of PSNR values of Dataset III (Images 11–15) with different methods

Table 9 MSE values of Dataset III (Images 11–15) obtained with different methods

Dataset III	MSE		
	Gaussian method	Temporal filtering method	HMF & CS method
Image 11	1.4115	1.3879	1.3577
Image 12	1.4138	1.3668	1.3243
Image 13	1.4139	1.3887	1.3473
Image 14	1.4138	1.3684	1.3365
Image 15	1.4114	1.3996	1.3452

filtering. Also, the contrast stretching method is implemented to improve the fog removed image obtained after dark channel prior. With this technique, it is shown that the similarity index of rain-free enhanced image is increased using a Hybrid Median Filtering and contrast enhancement. This technique is having lesser complexity and easy to implement. The proposed technique is compared with Gaussian Filtering and Temporal Filtering methods. The comparison of simulation results confirms that our technique is having a higher PSNR and lower MSE. The outcome of rain-free enhanced image is having better similarity with original images.

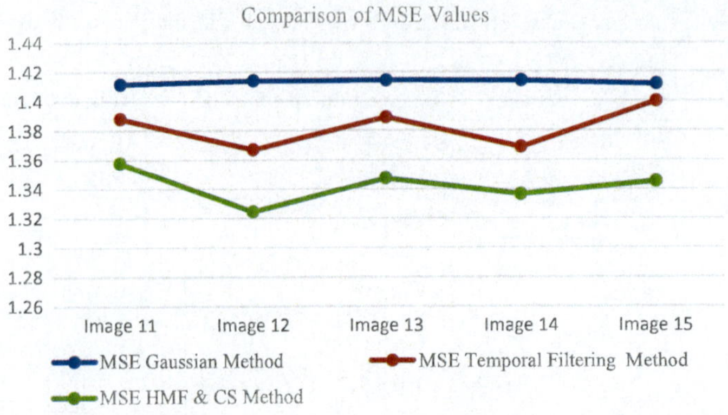

Fig. 20 Graph comparison of MSE values of Dataset III (Images 11–15) with different methods

Table 10 SSIM values of Dataset III (Images 11–15) obtained with different methods

Dataset III	SSIM		
	Gaussian method	Temporal filtering method	HMF & CS method
Image 11	0.5164	0.3648	0.7505
Image 12	0.5482	0.5116	0.6896
Image 13	0.6118	0.5421	0.7228
Image 14	0.5926	0.4882	0.6979
Image 15	0.5583	0.5118	0.7008

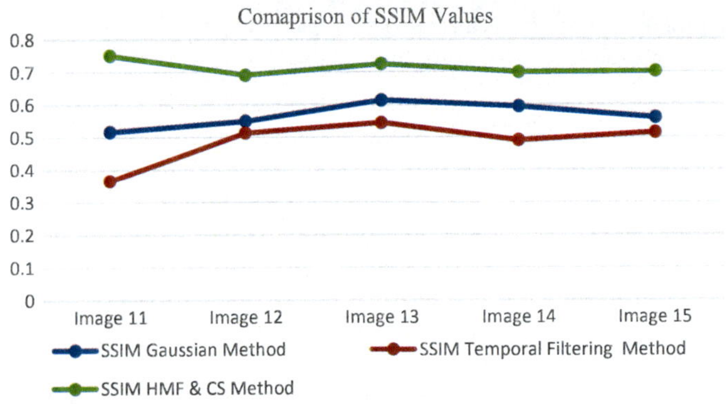

Fig. 21 Graph comparison of SSIM values of Dataset III (Images 11–15) with different methods

5 Conclusion

The comparison of results obtained for Peak Signal to Noise Ratio, Mean square error, and Structural similarity index confirms that the rainy image can be enhanced more effectively with the proposed method using simple algorithms. The method is having lesser complexity and easy to implement. The rain streaks are complex in nature as it contains both spatial and temporal characteristics. Also, the rain streaks accumulation over the original image gives foggy appearance whose removal is also a challenging job. In this age of smartphones, the single image enhancement plays a vital role, and using our proposed method the rain degraded image can be enhanced using rain streaks removal and reduction of fogginess. The use of non-linear filtering before the implementation of linear filtering followed by Dark channel prior and contrast stretching gives a better result than using a single type of recursive filtering methods.

References

1. Najiya CA, Sreera MS (2015) Single image rain removal using guided filter. Int J Adv Res Comput Sci Manag Stud 3(5)
2. Angelin Subha M, Johncy Rani T (2019) Single image de-raining using deep decomposition–composition network. Int J Recent Trends Eng Res (IJRTER) Conference on Recent Trends in Electrical and Electronics Engineering (RTEEE '19) Special Issue; March 2019. ISSN: 2455-1457. https://doi.org/10.23883/ijrter.conf.0304.006.ATCMF
3. Deng G, Cahill LW (1994) An adaptive Gaussian filter for noise reduction and edge detection. 0-7803-1487-5/94$04.008 IEEE
4. Shi Z, Li Y, Zhao M, Feng Y, He L (2018) Multi-stage filtering for single rainy image enhancement. IET Image Process 12(10):1866–1872
5. Kil TH, Lee SH, Cho NI (2013) Single image Dehazing based on reliability map of dark channel prior. 978-1-4799-2341-0/13/$31.00 ©2013 IEEE 882 ICIP
6. Sathya R, Bharathi M, Dhivyasri G (2015) Underwater image enhancement by dark channel prior. In: IEEE sponsored 2nd international conference on electronics and communication system (ICECS 2015). 978-1-4788-7225-8/15/$31.00 © IEEE 1119
7. Liu F, Yang C (2014) A fast method for single image Dehazing using dark channel prior. 978-1-4799-5274-8/14/$31.00 © IEEE
8. Neha, Aggarwal RK (2017) Effect of various model parameters on fog removal using dark channel prior. In: 2nd IEEE international conference on recent trends in electronics information & communication technology (RTEICT), 19–20 May 2017, India. 978-1-5090-3704-9/17/$31.00 © 2017 IEEE 1764
9. http://ahay.org/RSF/book/sep/antial/paper_html/node3.html. Last accessed 31 July 2019
10. Roland R et al (2013) Beyond noise: using temporal ICA to extract meaningful information from high-frequency fMRI signal fluctuations during rest. Front Hum Neurosc 7
11. Hwang Y-s, Choi S-H, Kim H-W, Lee J-M (2013) Impulse noise removal of LRF for 3D map building using a hybrid median filter. 978-1-4673-4569-9/13/$31.00 © IEEE
12. Darus MS, Noraini S, Md Tahir N, Isa NAM (2016) Modified hybrid median filter for removal of low density random-valued impulse noise in images. In: 6th IEEE international conference on control system, computing and engineering, 25–27 Nov 2016, Penang, Malaysia
13. Ramyashree BH, Vidhya R, Manu DK (2015) FPGA implementation of contrast stretching for image enhancement using system generator. Dept. of ECE KSSEM, Bangalore, India

14. Toh LB, Mashor MY, Ehkan P, Rosline H, Junoh AK, Harun NH (2016) Implementation of high dynamic range rendering on acute leukemia slide images using contrast stretching. 978-1-5090-2160-4/16/$31.00 © IEEE
15. Park K, Yu S, Jeong J (2018) A contrast restoration method for effective single image rain removal algorithm. 978-1-5386-2615-3/18/$31.00 © IEEE

Design of Efficient Ternary Subtractor

Yogesh Shrivastava and Tarun Kumar Gupta

Abstract As the time goes by binary logic is getting harder to implement on a smaller scale, so ternary logic becomes a better alternative of the same. Ternary logic has the simplicity over binary logic, and it is energy efficient also. Large number of interconnects and large chip area are the few problems of binary logic, which are reduced in ternary logic. This paper discussed a design of efficient ternary subtractor using carbon nanotube field-effect transistors (CNTFETs). The proposed design is compared with recent existing design on various performance parameters like transistor count, time delay, etc. The proposed design outmatches recent existing designs in all the parameters by 33.30%, 21.18%, and 47.32% in power consumption, delay, and PDP, respectively.

Keywords Subtractor · CNTFET · PDP · Delay

1 Introduction

It is customary to use binary logic to implement the digital circuits. The Moor's law relates with the increase of number of transistors on a chip results in higher complexity of the circuits. One of the major limitations of binary logics is area covered by 'interconnects'. The chip area dedicated to interconnect is 70%, 20% to insulation and 10% to devices. Interconnect refers to the area, which is used to connect devices; it remains there regardless of the flow of current. The area dedicated to interconnect can be used viably by transmitting more information from them; it is possible by using fuzzy logic or multiple valued logic [1].

The natural base with radix (e) is the most appropriate system to implement the digital [2], but the process is complex and difficult to use natural base. Therefore,

Y. Shrivastava (✉) · T. K. Gupta
Maulana Azad National Institute of Technology Bhopal, Link Road Number 3, Near Kali Mata Mandir, Bhopal, Madhya Pradesh 462003, India
e-mail: yogesh.21june@gmail.com

T. K. Gupta
e-mail: taruniet@rediffmail.com

D. Harvey et al. (eds.), *Advances in VLSI, Communication, and Signal Processing*, Lecture Notes in Electrical Engineering 683,
https://doi.org/10.1007/978-981-15-6840-4_14

nearer 3-valued logic found more suitable. The 3-valued logic is known as ternary logic. It is possible to implement sequential as well as combinational circuit using ternary logic [3]. The use of ternary logic limits area of interconnects and improves the availability of functions. CNTFET found more suitable to implement the ternary logic, due to its properties like variation in the threshold voltage.

In this work, ternary subtractor is designed using CNTFET and simulated on H-SPICE simulator with Stanford University model file [4].

The subtractor proposed in [3] was designed using MOSFETS gates. The different threshold voltage (V_{th}) values were obtained by using different types of transistors with resistors. In [5], they are designed using CNTFET and the same are discussed in this paper.

1.1 CNTFET (Carbon Nanotube Field-Effect Transistor)

The carbon nanotubes used in CNTFET are made of graphene sheets shown in Fig. 1. These tubes provide conducting path to charge carriers. The mobility of electron in these tubes is very high, being the very less resistive [6]. Single-walled CNT (SWCNT) consists of a single cylinder; due to that, the manufacturing process of SWCNT becomes easier. The angle at which the atoms arrange along the tube is called chirality; depending on this angle, the SWCNT can be used as a conductor or as semiconductor. The chirality vector is denoted by (n, m): Carbon nanotube will act as a conductor, if $n = m$ or $n{-}m = 3i$ (where i is an integer); otherwise, it will act as a semiconductor, where $a_0 = 0.142$ nm is the interatomic distance between each carbon and its neighbour.

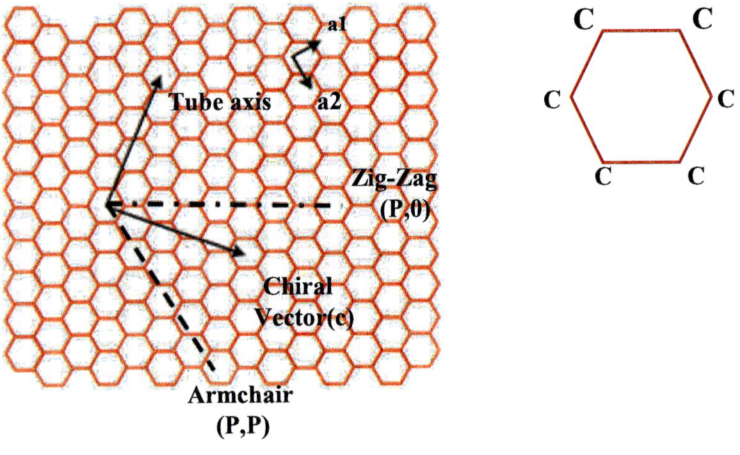

Fig. 1 Graphene sheet with chiral vector

$$D_{CNT} = \frac{\sqrt{3}a_0}{\pi}\sqrt{n^2 + m^2 + nm} \tag{1}$$

The schematics of CNTFET are shown in Fig. 2a. CNTFET has 4 terminals similar to conventional MOSFET. As shown in Fig. 2a, the semiconducting tubes are situated under the gate and in the channel, without any doping. The CNTs placed between source/drain are highly doped; these tubes provide low series resistances in the ON state [6]. As the gate voltage increases, the turned ON (N-CNTFET) or OFF(P-CNTFET) [7].

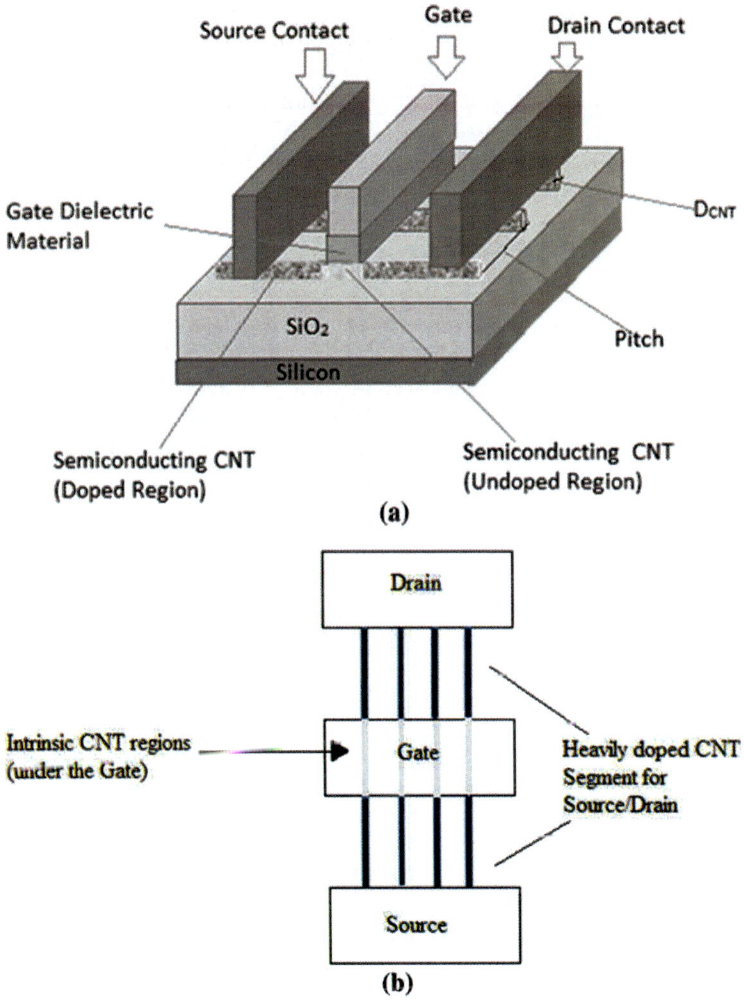

Fig. 2 Schematic diagram of a CNT transistor **a** cross-sectional view, **b** top view

$$V_{th} \approx \frac{E_g}{2e} = \frac{\sqrt{3}}{3} \frac{aV\pi}{eD_{CNT}} \tag{2}$$

where $a = 2.49$ Å is the distance between two carbon atoms, $V\pi = 3.033$ eV is the energy exists in tight bonding model of π–π carbon bond, e is the charge on unit electron, and diameter of CNT is given by D_{CNT}. The D_{CNT} of a CNT is directly proportional with chirality and inversely proportional with the V_{th}. It is confirmed by simulation results in [1]. If chirality vector m is always zero, then the ratio of the V_{th} of two CNTFETs with different chirality vector is given as

$$\frac{V_{\text{th}}1}{V_{\text{th}}2} = \frac{D_{\text{CNT2}}}{D_{\text{CNT1}}} = \frac{n_2}{n_1} \tag{3}$$

Lin et al. [7] in their work found out that the threshold voltage of a CNTFET using (10, 0), (13, 0), and (19, 0) is 0.559 V, 0.428 V, and 0.293 V, respectively.

2 Ternary Logic Operations

To satisfy Moore's law, the IC industry has to counter many problems; the area covered by interconnection is one such major problem. One of the ways to reduce interconnection is to use more number of logic levels than two levels. The one more level is introduced in the binary logic function, which results in the ternary logic function. The ternary logic makes digital design simple and energy efficient by reducing the complexity of interconnects. Because of ternary logic carries more number of information in the form of the added extra level, the transmission becomes efficient, and serial and parallel transmission becomes easier. Let 0, 1, and 2 be the ternary values to represent false, undefined, and true conditions, respectively. Any ternary function $f(x)$ of n variable $(X1, X2 \ldots Xn)$ is defined as a logic function mapping $\{0, 1, 2\}^n$ to $\{0, 1, 2\}$. The basic operations of ternary logic can be defined as:

$$Xi + Xj = \max\{Xi, Xj\} \tag{4}$$

$$Xi \cdot Xj = \min\{Xi, Xj\} \tag{5}$$

$$\overline{Xi} = 2 - Xi \tag{6}$$

$$X01 = X0 + X1 \tag{7}$$

$$X12 = X1 + X2 \tag{8}$$

$$X02 = X0 + X2 \tag{9}$$

$$X01 \cdot X12 = X1 \tag{10}$$

$$X01 \cdot X02 = X0 \tag{11}$$

$$X02 \cdot X12 = X2 \tag{12}$$

$$X0 + X1 + X2 = 2 \tag{13}$$

$$X2 = \overline{X01}\,\&\,X01 = \overline{X2} \tag{14}$$

$$X1 = \overline{X02}\,\&\,X02 = \overline{X1} \tag{15}$$

$$X2 = \overline{X01}\,\&\,X01 = \overline{X0} \tag{16}$$

$$\bar{0} = 2\,\&\,\bar{2} = 0 \tag{17}$$

3 Review of Ternary Subtractor

Subtractor is one of the most essential parts of arithmetic and logical unit (ALU); for efficient design of subtractor circuit with less number of logic gates and less number of interconnection is very important. The design proposed by Sridevi et al. [5] is pretty much similar to the Dhandhe and Ingole [3], except for negation terms. Sridharan et al. [8] proposed multidigit adder by modifying the traditional decoder. The decoder proposed in [8] gives 5 outputs. With the inspiration of the logic of [8], Sahoo et al. [9] proposed a half adder based on the same concept. Their designs can be divided into 3 main parts. First one is decoder circuit, second is the subtractor circuit, and the last one is the buffer circuit (in some research work, it is mentioned as ternary buffer circuit). Both the earlier mentioned designs are based on logic expression obtained from the K-MAP. The logic expressions of ternary subtraction and borrow are given in Eqs. 18 and 19, respectively.

$$\text{SUB} = A0B1 + A1B2 + A2B0 + 1 \cdot (A0B2 + A1B0 + A2B1) \tag{18}$$

$$\text{BORROW} = 1 \cdot (A0B1 + A0B2 + A1B2) \tag{19}$$

3.1 Decoder Circuit

Decoder outputs are high for each of the respective ternary levels, Sridevi et al. [5] and Dhandhe and Ingole [3] use the similar decoder circuit, except for, in [7] CNTFET is used and in [3] MOSFET is used. In the decoder circuit, PTI output is given into NTI and another NTI output has given to STNOR gate. The truth table of the encoder circuit is given in Table 1, and the circuit diagram is given in Fig. 3.

Ternary K-MAP

SUB

A \ *B*	0	1	2
0	0	2	1
1	1	0	2
2	2	1	0

Table 1 Truth table of ternary subtractor

A	*B*	SUB = *A* − *B*	Borrow	*A*0	*A*1	*A*2	*B*0	*B*1	*B*2
0	0	0	0	2	0	0	2	0	0
0	1	2	1	2	0	0	0	2	0
0	2	1	1	0	0	0	0	0	2
1	0	1	0	0	2	0	2	0	0
1	1	0	0	0	2	0	0	2	0
1	2	2	1	0	2	0	0	0	2
2	0	2	0	0	0	2	2	0	0
2	1	1	0	0	0	2	0	2	0
2	2	0	0	0	0	2	0	0	2

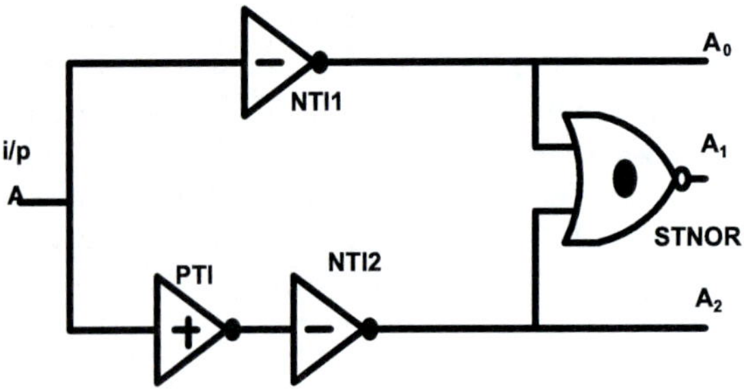

Fig. 3 Circuit diagram of ternary decoder

Table 2 Truth table for ternary decoder

Input (*A*)	*A*0	*A*1	*A*2
0	2	0	0
1	0	2	0
2	0	0	2

Borrow

A \ B	0	1	2
0	0	1	1
1	0	0	1
2	0	0	0

From truth table of ternary decoder, it is observed that the states 0, 1, and 2 with voltage levels 0, $V_{dd}/2$, and V_{dd}, respectively, are converted to the states 0 and 2 to their respective voltage levels 0 and V_{dd}; basically, decoder's main role over here is to convert the three-level logic into two-level logic (Table 2).

3.2 Intermediate Stage

The obtained signal from the decoder is applied on the intermediate state which is the combination of gates; the output of this assembly of gates is the two-level signal. To get the ternary output, the output of these gates is applied to the buffer circuit which again converts the two-level signal into three-level signals as shown in Fig. 4. The only difference between the circuit proposed in [3] to the circuit proposed in the [5] is the borrow circuit which Sridevi et al. designed with the negation of literal methods shown in Eq. 20.

$$\text{BORROW} = 1 \cdot \left(A0\overline{B0} + A1B2\right) \tag{20}$$

The circuit proposed in [5] is given in Fig. 5.

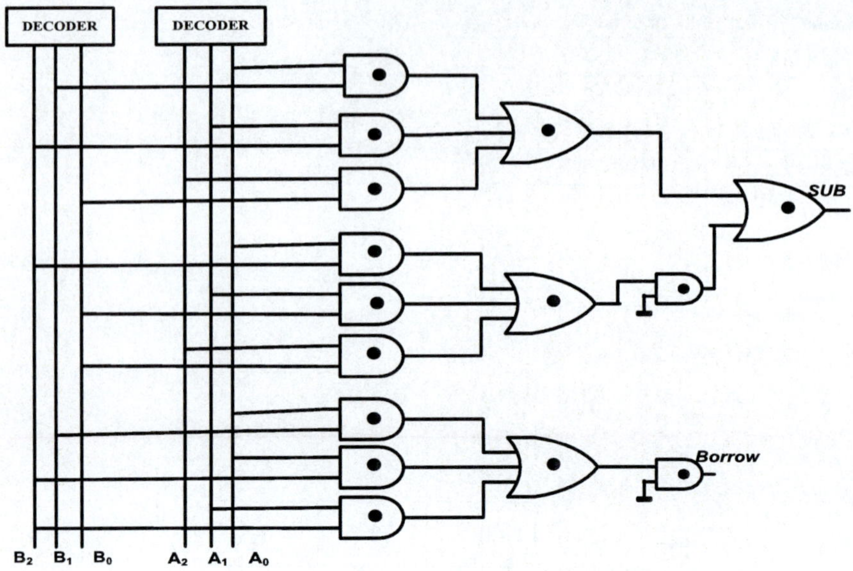

Fig. 4 Subtractor design proposed by Dhandhe and Ingole [3]

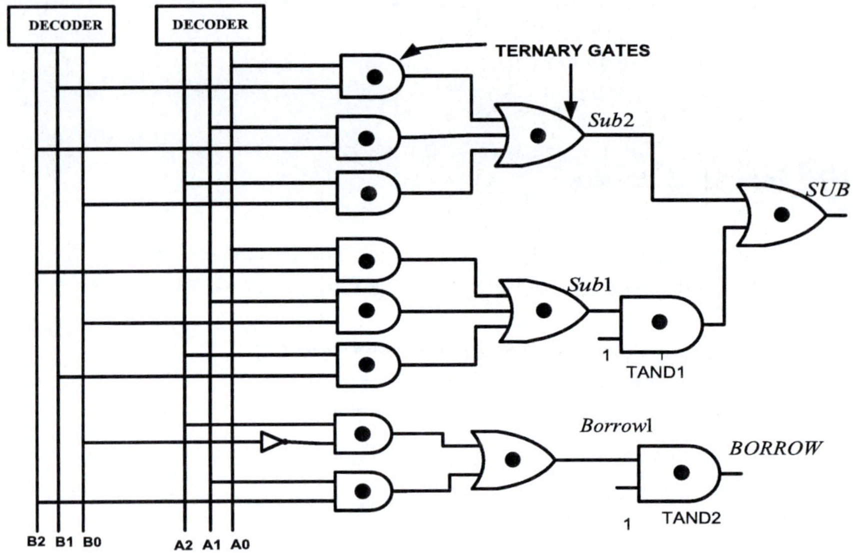

Fig. 5 Subtractor design proposed by Sridevi et al. [5]

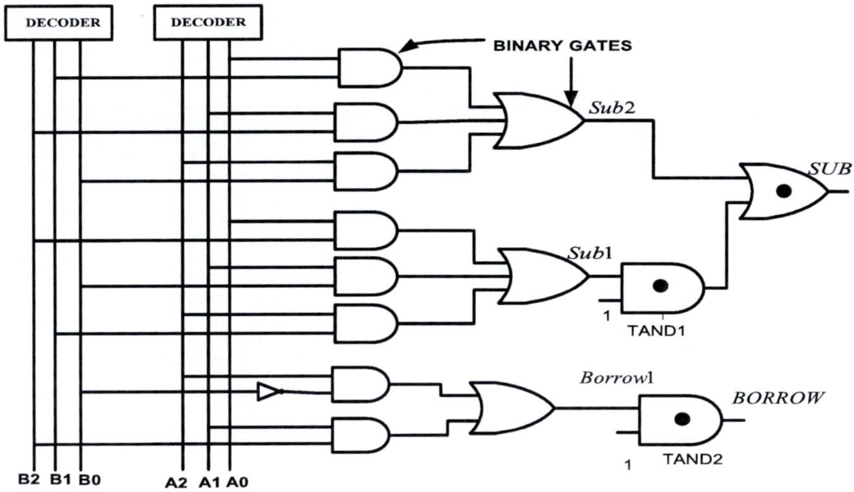

Fig. 6 Proposed subtractor design

4 Proposed Ternary Subtractor

From the circuits proposed in [7], it is clear that the ternary gates used more number of transistors than binary gates. After the conversion of the signal from the decoder, the input ternary signal is converted into two-level signal. Therefore, in place of ternary gates, binary gates can be used, as shown in Fig. 6. From this circuit, reduction in the transistor is expected from existing circuits.

5 Result and Discussion

The existing and proposed circuits are simulated on the H-Spice simulator (Hspui C-2009) with CNTFET Stanford model [4], for 10 ns at room temperature. The transient response of the proposed circuit is shown in Fig. 7. The parameters obtained from the computation of the circuit are shown in Table 3. The proposed circuit is compared with both the existing ternary subtractor circuits. The PDP (Power Dissipation Product) is obtained by multiplying power consumption with delay in the circuit [10].

After observation of Table 3, it can be seen that the proposed design shows improvement in power by 42.7%, delay by 34.09%, and PDP by 62.29% from the design proposed in [3]. From the design proposed in [5], it showed 33.30%, 21.18%, and 47.32% improvement in power consumption, delay, and PDP, respectively.

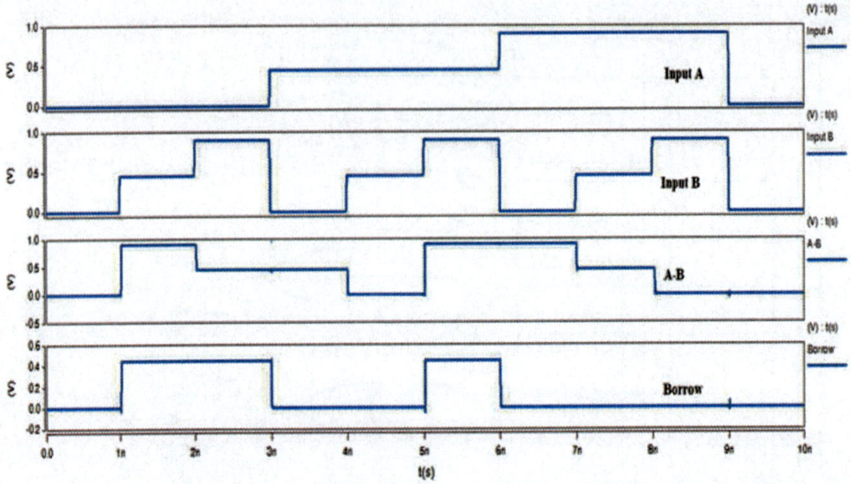

Fig. 7 Transient response of ternary subtractor

Table 3 Performance comparison

Circuit type	Power consumption (nW)	% power improvement from [3]	Total delay (ps)	% improvement in delay from [3]	PDP (aJ)	% improvement in PDP from [3]
Dhande and Ingole [3]	1400.17	–	300	–	420.05	–
Sridevi et al. [5]	1200.94	14.2	250.36	16.5	300.66	28.42
Proposed design	800.92	42.7	197.73	34.09	158.36	62.29

6 Conclusion

In this paper, design of a ternary subtractor was proposed. The proposed design was compared with existing recent designs on the parameters like power consumption, delay, and PDP. The proposed design shows improvement in all the parameters over recent existing designs by 33.30%, 21.18%, and 47.32% in power consumption, delay, and PDP, respectively.

References

1. Butler JT (1995) Multiple valued logic examining its use in ultra-high speed computation. IEEE Potentials, April/May 1995
2. Porat DI (1969) Three valued digital system. Proc IEEE 116(6):947–955
3. Dhande AP, Ingole VT (2005) Design and implementation of 2 bit ternary ALU slice. In: Proceedings of the international conference: IEEE sciences of electronic, technologies of information and telecommunications, pp 17–21
4. Stanford University (2008) Stanford University CNTFET model website. Stanford University, Stanford, CA [Online]. Available: http://nano.stanford.edu/model.php?id=23
5. Sridevi V, Jayanthy T (2014) Minimization of CNTFET ternary combinational circuits using negation of literals technique. Arab J Sci Eng 39:4875–4890
6. Appenzeller J (2008) Carbon nanotubes for high-performance electronics Progress and prospect. Proc IEEE 96(2):201–211
7. Lin S, Kim Y-B, Lombardi F (2011) Design of a CNTFET-based design of ternary logic gates and arithematic circuits. IEEE Trans Nanotechnol 10(2):217–225
8. Sridharan K, Gurindagunta S, Paudi V (2013) Efficient multiternary digit adder design in CNTFET technology. IEEE Trans Nanotechnol 12(3)
9. Sahoo SK, Member IEEE, Akhilesh G, Sahoo R, Muglikar M (2017) High-performance Ternary adder using CNTFET. IEEE Trans Nanotechnol 16(3)
10. Garg S, Gupta TK (2018) Low power domino logic circuits in deep-submicron technology using CMOS. Eng Sci Tech Int J 21(4):625–638
11. Rahman A, Guo J, Datta S, Lundstrom M (2003) Therory of ballistic nanotransistors. IEEE Trans Electron Devices 50(10):1853–1864
12. Akturk A, Pennington G, Goldsman N, Wickeden A (2007) Electron transport and velocity oscillations in a carbon nanotube. IEEE Trans Nanotechnol 6(4):469–474
13. Hashempour H, Lombardi F (2008) Devices model for ballistic CNTFETs using the first conducting band. IEEE Des Test Comput 25(2):178–186
14. Wang B, Poa P, Wei L, Li L, Yang Y, Chen Y (n,m) Selectivity of single-walled carbon nanotubes by different carbon precursors
15. Lin Y, Appenzeller J, Konch J, Avouris P (2005) High-performance carbon nanotube field effect transistor with tunable polarities. IEEE Trans Nanotechnol 4(5):481–489
16. Kleene SC (1952) Introduction to Mathematics. North-Holland, Amsterdam, The Netherlands, pp 332–340
17. Deng J, Wong H-SP (2007) A compact SPICE model for carbon- nanotube field-effect transistors including nanoidealities and its application-Part I: model of the intrinsic channel region. IEEE Trans Electron Device 54(12):3186–3194
18. Li Y, Kim W, Zhang Y, Rolandi M, Wang D (2001) Growth of single-walled carbon nanotubes from discrete catalytic nanoparticles of various sizes. J Phys Chem 105:11424–11431
19. Ohno Y, Kishimoto S, Mizutani T, Okazaki T, Shinohara H (2004) Chirality assignment of individual single-walled carbon nanotubes in carbon nanotube field effect transistors by micro-photocurrent spectroscopy. Appl Phys Lett 84(8):1368–1370
20. Bachtold A, Hadley P, Nakanishi T, Dekker C (2001) Logic circuits with carbon nanotubes transistors. Science 294(5545):1317–1320
21. Lin S, Kim Y-B, Lombardi F (2009) A novel CNTFET-based ternary logic gate design. In: Proceedings of the IEEE international Midwest symposium on circuits and systems, Aug 2009, pp 435–438
22. Lin A, Patil N, Ryu K, Badmaev A, De Arco LG, Zhou C, Mitra S, Wong HSP (2009) Threshold voltage and on-off ratio tuning for multiple-tube carbon nanotube FETs. IEEE Trans Nanotechnol 8(1):4–9
23. Mukaidono M (1986) Regular ternary logic function-Ternary logic functions suitable for treating ambiguity. IEEE Trans Comput C-35(2):179–183

24. Araki T, Tatsumi H, Mukaidono M, Ymamoto F (1998) Minimization of incompletely specified regular ternary logic functions and its application to fuzzy switching functions. In: Proceedings of the IEEE international symposium on multiple-valued logic, May 1998, pp 289–296
25. Heung A, Mouftah HT (1985) Deplition/enhancement CMOS for a lower power family of three-valued logic circuits. IEEE J Solid-State Circ 20(2):609–616
26. Raychowdhury A, Roy K (1986) Carbon–nanotube-based voltage-mode multiple-valued logic design. IEEE Trans Comput 35(2):99–106
27. Rich DA (1986) A survey of multivalued memories. IEEE Trans. Comut 35(2):99–106

A Multichannel Link-Layer Cooperation Protocol (MLCP) for Cognitive Radio Ad Hoc Network

Jahnvi Tiwari, Arun Prakash, and Rajeev Tripathi

Abstract In a cognitive radio ad hoc network (CRAHN), the secondary user node has a restricted range for sensing other users within the network. To enhance the performance of nodes (secondary and primary users) in a CRAHN, the problem of limited channel sensing range and interference should be reduced which would enable the simultaneous transmission of primary and secondary users. In this paper, a cooperative MAC (medium access control) protocol is proposed—MLCP, which exploits the concept of multichannel network, for obtaining high system throughput. The main purpose of the proposed protocol is to introduce cooperation among secondary users, using the common control channel. The presence of the primary user is detected and the information is shared among neighbouring secondary users within the network proactively. This protocol is simulated using Network Simulator 2 and analysed with respect to system throughput, packet delivery ratio, and end-to-end delay experienced by primary users. The performance of the proposed protocol is also compared with existing cooperative MAC protocol and non-cooperative MAC protocol network scenario.

Keywords Cognitive radio · Ad hoc network · Cooperative MAC · Multichannel · Throughput · Packet delivery ratio · End-to-end delay

---

J. Tiwari (✉) · A. Prakash · R. Tripathi
Motilal Nehru National Institute of Technology Allahabad, Prayagraj 211004, India
e-mail: jahnvi939@gmail.com

A. Prakash
e-mail: arun@mnnit.ac.in

R. Tripathi
e-mail: rt@mnnit.ac.in

D. Harvey et al. (eds.), *Advances in VLSI, Communication, and Signal Processing*, Lecture Notes in Electrical Engineering 683,
https://doi.org/10.1007/978-981-15-6840-4_15

1 Introduction

In the present decade, the evolution of various communication devices has resulted in spectrum scarcity. Spectrum availability has crashed in comparison to the rising number of users, all around the world. But this scarcity is nothing but an illusion as around 80% of the spectrum remains un-utilized at a given geographical location, according to FCC [1]. To mitigate this pseudo-scarcity, the concept of cognitive radio network comes into play, which is generally ad hoc and distributed in nature. The foremost objective of a CRAHN is to provide spectrum optimization by utilizing spectrum holes efficiently. The spectrum is assigned to the primary user (PU) which is the licensed owner of the spectrum. Cognitive radio nodes or the secondary users (SU) use the licensed spectrum in the absence of PU, in order to do this, SU frequently senses the spectrum to detect the re-appearance of PU. In case, the licensed PU appears in the spectrum, while SU is already using it to communicate, the SU quits the spectrum immediately to minimize interference in the network. On detecting the presence of PU, SU pauses its transmission, quits the current band and scans for a suitable band to resume its transmission and jumps to it; this process is called spectrum handoff. This scanning of the available band can be carried out in a proactive or a reactive fashion [2].

The physical and the MAC layer for wireless air interface were studied and standardized by the IEEE 802.22 Working Group [3]. Cognitive radio ad hoc networks present a challenge in the design of the MAC layer when there is mobility among nodes; the challenges faced in designing of MAC protocols rise further. The dynamic nature of the network and limited sensing range of SUs degrades throughput and packet delivery ratio of the system. A cooperation framework in link layer for throughput enhancement has grabbed the attention of researchers worldwide.

A cooperative MAC protocol provides efficient use of spectrum resources, network throughput, and reliability improvement. During communication, in a non-cooperative MAC protocol, used in a cognitive ad hoc network scenario, if the sender and destination node move out of each other's transmission range, the packet drops or the user experiences latency. If this condition arises in cooperative communications, the sender and destination node pair use the third node as a 'helper' node to achieve transmission diversity. There are many parameters which decide which node can assist in the communication; the main condition is a better channel environment and connectivity with the source as well as the destination node [4]. The mere existence of helper nodes enhances the reliability of the CRAHN even in poor channel condition and mobility. Cooperative MAC protocols also reduce the requirement of large transmission power and also increase the transmission range of nodes and coverage area of the network. The cooperative MAC protocol has another very important function in the domain of cognitive radio network which is to prevent PU-SU interference. As discussed before, the SU has a limited sensing range which results in in-efficient PU detection within CRAHN causing interference and packet drop and further results in a significant delay in the initiation of the handoff process. This problem of hidden PU can be addressed by using cooperative MAC protocol [5]. In this scenario, any node which senses PU informs all other SUs in the network about the arrival and position of the PU; helper nodes relay this information among each other and thus the SU which has to vacant the band for PU gets knowledge of the arrival of the primary user much before its actual arrival.

Here, the proposed protocol is analysed and compared with cooperative and non-cooperative protocols proposed by authors in [6, 7]. The comparison is done on the basis of system throughput, end-to-end delay faced by PU and percentage packet delivery ratio in the system. Section 2 enlists related works which have led to this protocol proposal. The system model is described in Sect. 3. The protocol is proposed in Sect. 4. Section 5 contains the simulation results of the proposed protocol and its comparison with other protocols. Conclusion and future scope are discussed in Sect. 6 of the paper.

2 Related Work

For almost a decade, the physical layer was the focal point to provide cooperative communication in a cognitive radio network. Relaying strategies and diversity gain concepts were being employed by researchers for the improvement of throughput, end-to-end delay, and network coverage area [8]. Though utilizing physical layer for applying cooperative communication is beneficial, it results in many complications in terms of extra signal overhead, increased interference, and high battery consumption. These shortcomings of cooperation through physical layer negatively affect the output parameters so much so that the benefits of cooperation effectively disappear. For example, in [9] the authors have proposed an opportunistic relay mechanism to provide cooperation among nodes in WLAN, the authors have used contention and statistical information of the network, the scheme works well in the highly dynamic network but the best possible relay is not guaranteed. In [10], the authors have designed a wireless smart grid communication system using cooperative relaying at the physical layer; they have also used IEEE 802.11 b protocol at the MAC layer. This scheme though provides a guarantee for the best relay; it is very sensitive to synchronization and channel feedback errors. Due to these issues faced in the design of cooperative protocol using the physical layer, the attention of researchers shifted to cooperative communication through the link layer. Designing cooperative MAC protocols is relatively a new area for researchers to explore [11].

The primary questions about the working of cooperative MAC protocols, i.e., When to cooperate? Who should the nodes identify as 'helpers'? and How to support concurrent transmissions? were answered by authors in [12, 13]. Most of the proposed protocols with this concept are based on IEEE 802.11 with distributed coordination function. In [14], authors have proposed a coordinated cooperative MAC protocol in WLAN for uplink transmission; this protocol reduces bottleneck congestion problem and facilitates concurrent transmissions among nodes. Most of the mentioned cooperative MAC protocols do not support the concurrent transmissions of data packets; neighbour nodes of the sender and receiver SU, communicating in a cooperative manner, are forced to pause their transmission. This issue results in hidden and exposed node problem which in turn adversely affects the network performance metrics. To mitigate this issue, authors in [15] have used node cooperation mechanisms for TDMA MAC. In this proposed protocol, cooperation is done by helper nodes which are coordinated by the access point. Also, distinct time slots are defined for cooperation and other functions which facilitate concurrent transmission.

In all the aforementioned studies, cooperation is performed on an infrastructure-based centralized network. In the case of cognitive radio, the network is distributed and dynamic in nature; in some cases, as considered in this paper, nodes can also be mobile. Therefore, cooperative MAC operations must be performed in a distributed manner for which the nodes involve cluster or slot formation for cooperative decisions. Hence, it can be established that the application of cooperative MAC protocol is not straightforward.

Cooperative MAC protocols which work effortlessly on infrastructure-less cognitive networks are relatively less in number. The authors in [16] have proposed a cooperative MAC protocol—CooperMAC which provides SUs with the capability to negotiate channels and relays. The authors have, in [17], proposed a parallel sensing and sequential channel selection scheme in MAC to help discover spectrum holes faster. In [6], the authors have proposed a multichannel and a fair MAC protocol (MC-MAC) which provides fairness in resource sharing, thus increasing the throughput of CRAHN network. The proposed protocol is developed on the base of MC-MAC and has been compared with it in the later sections of this paper.

3 System Design

The CRAHN system model is discussed in this section. Two types of nodes are considered—primary users, who possess the license of the spectrum being used, and the secondary users, who are unlicensed users of the spectrum. It is assumed that at a given time one or more nodes can move with slow speed. While the nodes are moving, any SU can come under PU's transmission range. As shown in Fig. 1, all SUs record packets which are transmitted in its vicinity for a certain time period of t sec; they also keep updating log for previous information and current packet information. It keeps on updating its table to the current time slot of t second to keep the size of table small. This information which consists of node id, origin, destination, network id, etc., is not only recorded but also analysed constantly for the presence of PU.

If SU *hears* the primary user in the network, it immediately shares this information with its neighbours. For the exchange of information, two types of channels are used: a common control channel for control signal exchange and several data channels for data transmission. The control channel operates in the ISM band which is unlicensed; the most important use of this channel is to report other SUs about the presence of PU in the network. The data channels, as the name implies, are used for data packets, and they work on licensed spectrum. With respect to nodes, the architecture is simple. Two half-duplex transmitters are used: one is used for broadcasting control signals

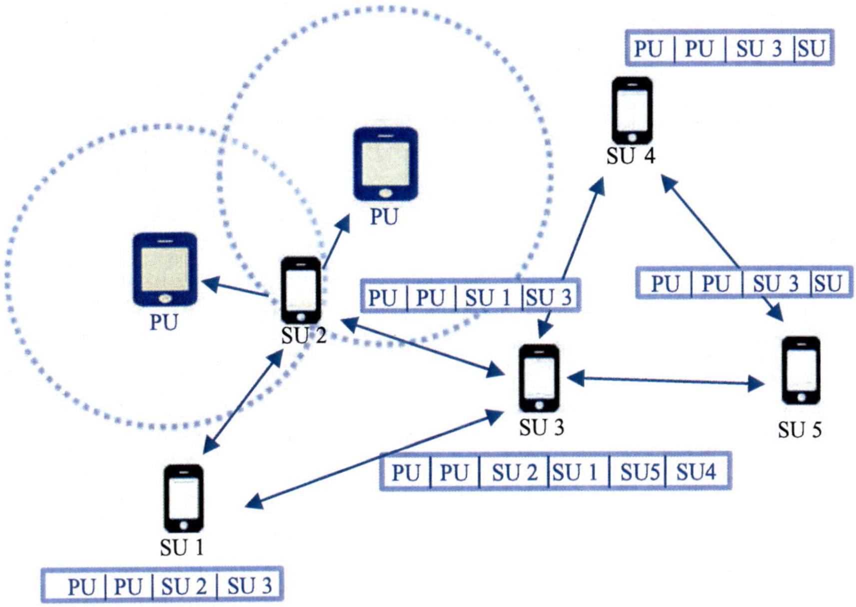

Fig. 1 System model depicting two primary users in the CRAHN for time duration '*t*'

and sensing the primary user's presence, while the other is used for data transmission and reception.

4 Proposed Protocol

In the proposed protocol, preventing a secondary user from interfering in the communication network of the primary user is the main concern. Hence, continuous and proactive PU sensing is done by SUs in the CRAHN. The proposed protocol maintains collaboration among all secondary users to detect and report the presence of primary users in the network. When a SU wants to transmit a data packet, it tries to reserve channel at the beginning of the beacon interval. It is reserved only if the channel remains idle for distributed coordination function-inter frame space (DIFS). If the channel is busy, the SU starts a back-off algorithm, where the back-off value is less than the contention window.

Later, when the channel remains vacant, the back-off value is decremented and the node transmits data when the counter reaches zero. When the counter value reaches zero, it implies that the secondary user has won contention and has the right for using the channel until a PU appears. The SU who wins the contention sends a *request to send* frame; the receiver on hearing it replies with a *clear to send* frame. Here, in this protocol, unlike non-cooperative MAC protocols, the SU again checks its table, which has a high probability of getting updated during RTS-CTS exchange by the SU's neighbours. If the presence is not recorded in the table, the SU immediately transmits data and receives an *acknowledgement* frame from the receiver. Meanwhile, the other secondary users within the network keep on updating and sensing the channels for the presence of the primary user. If a secondary user detects the arrival of the primary user, it not only updates its table, but also broadcasts this information so that its neighbours also benefit from this information. The neighbouring nodes further re-broadcast this information to their neighbours and so on until the all the SU in the vicinity of PU know about its presence. The information received by neighbouring SUs consists of the node ID of the PU, the channel it is licensed to, and the location of PU in the network; this information frame is termed as *emergency channel report* (ECR) frame. On receiving ECR, the SUs get informed about the location and range of PUs continuously which facilitate concurrent data transmission uninterrupted. The flowchart in Fig. 2 shows the working of the proposed scheme step by step. It should be kept in mind that all the signalling frames, such as RTS, CTS, ACK, and ECR, should be of small duration to prevent signalling overhead in the network. If the effective overhead duration is larger than the data packet duration, the network is deemed ineffective for practical implementation.

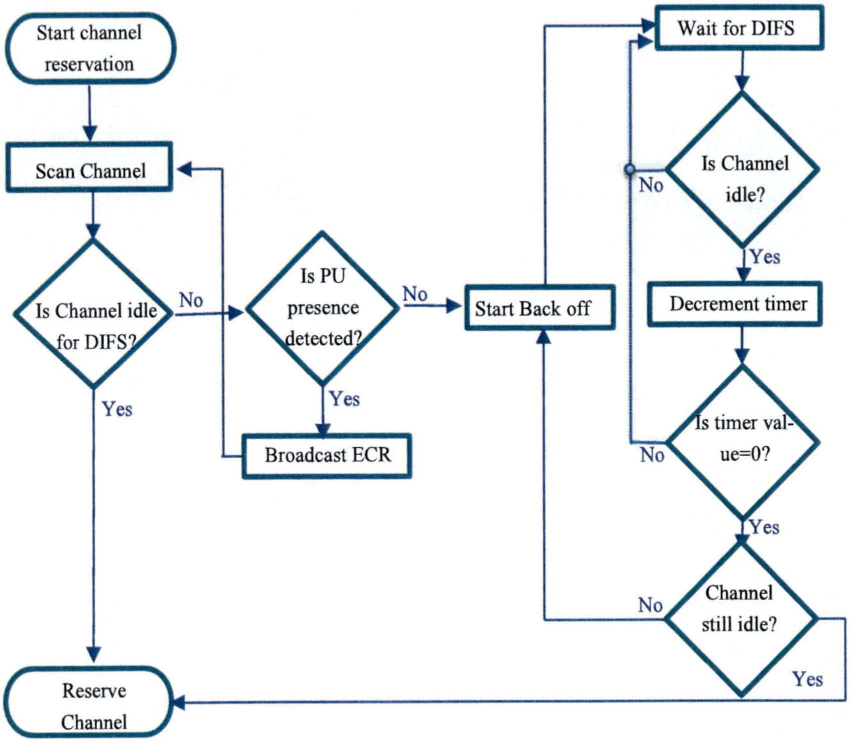

Fig. 2 Workflow for channel reservation in the proposed protocol

5 Results

In this section, the validity of the proposed protocol is verified using the network simulator-2 (NS-2) tool version ns-2.31 [18]. The performance is evaluated in terms of system throughput, packet delivery ratio (PDR), and end-to-end delay. The network area considered is 1000 m × 1000 m, and a total of 90 nodes are defined. Secondary and the primary users can move slowly for a given time interval. All the important parameters considered in this simulation are stated in Table 1. Ad hoc on-demand vector routing (AODV) protocol is used as it has minimum routing overhead and low setup delay for cognitive radio ad hoc networks [19]. It is assumed that only five secondary users are communicating at a given time and the number of primary users varies in the network, the performance of the network is analysed by increasing the number of primary users from zero to twenty while the five secondary users communicate and other secondary users may work as helper nodes for successful cooperative communication.

In Fig. 3, throughput is calculated as an indicator of successful packet delivery in the network; it is observed that the cooperative and non-cooperative MAC work similarly in the case when the number of primary users in the network is less. When their number increases, a substantial difference is observed in the system throughput. On the other hand, the proposed protocol works better than both non-cooperative and

Table 1 Network simulation parameters

Parameter	Value
Area	1000 m × 1000 m
Total no. of nodes	90
Data packet size	1000 bytes
Burst/period	0.5
No. of channels	5
Channel frequency	914 MHz
Channel bandwidth	2 Mbps
Propagation model	2- ray ground reflection
Interference range	400 m
Interference queue	50
Beacon interval duration	100 ms
Simulation time	30 s
Routing protocol	AODV
Length of RTS and CTS	352 and 304 bits
Length of ACK	304 bits
Data header	432 bits
SIFS and DIFS	8 and 16 μs
SU recording duration (t)	3 s

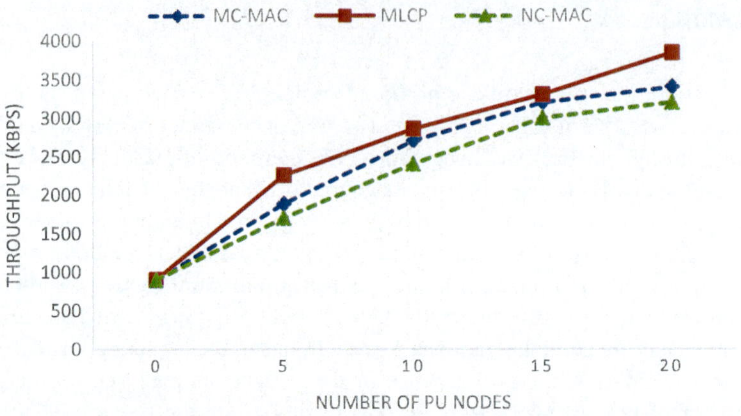

Fig. 3 Network throughput against number of primary user nodes in five channels

MC-MAC protocols even when the number of primary users within the network is less.

Similarly, packet delivery ratio (PDR) and end-to-end delay are graphically analysed in Figs. 4 and 5, respectively. The PDR% is computed as the ratio of packets that are received by the receiver node to the number of packets sent by the sender node. End-to-end delay computation considers all delays possible in the network, i.e., channel scanning, route decision, spectrum sensing, spectrum reservation, queuing, propagation, handoff, and transmission delay.

In a cognitive radio ad hoc network, the main reason for the decrease in PDR with an increase in the number of primary user nodes is interference in the SU transmission and handoff. The primary user is the legacy user of the spectrum, so when SU hears its presence, it pauses its transmission and leaves the channel to make space for PU.

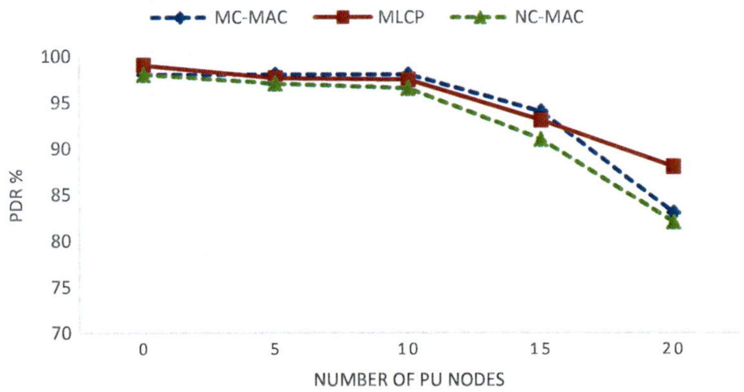

Fig. 4 Packet delivery ratio against varying number of primary user nodes

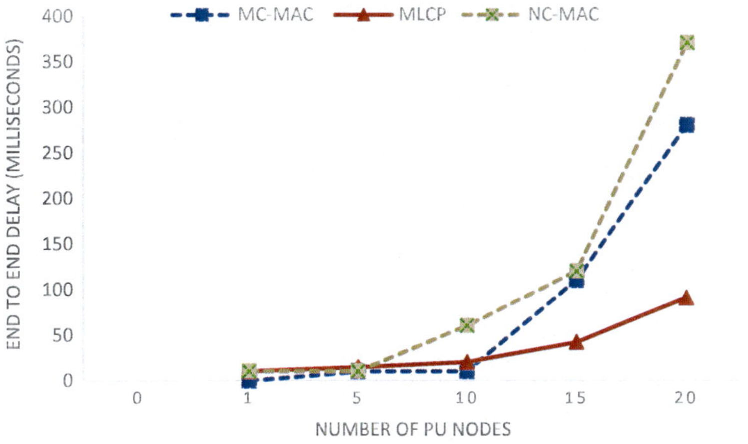

Fig. 5 End-to-end delay depicted in milliseconds against a number of primary user nodes in the network

Due to handoff, end-to-end delay increases and PDR decreases. In the case of the proposed protocol, cooperation among nodes helps in performance enhancement.

6 Conclusion

In order to prevent secondary user interference in primary user communication and seamless handoff, reliable PU sensing and collaboration among nodes is important; the protocol proposed in this paper provides both. Concept of multichannel CRAHN and proactive reporting about the PU presence is introduced and analysed graphically for important performance parameters. The simulation software used was Network Simulator 2 with CRCN patch. Moreover, the parameters, i.e., throughput, end-to-end delay, and packet delivery ratio, are compared with other MAC protocols. The proposed protocol aims to solve the hidden PU problem. From the simulation results, it can be concluded that the proposed protocol provides throughput and PDR enhancement, which is more distinct when the number of PUs increases in the network and it decreases the end-to-end delay caused by slow handoff process. Thus, it can be said that the proposed protocol effectively enhances resource utilization efficiency.

References

1. FC Commission (2002) FCC spectrum policy task force. November 15, 2002
2. Kumar K, Prakash A, Tripathi R (2016) Spectrum handoff in cognitive radio networks: a classification and comprehensive survey. J Netw Comput Appl 61:161–188
3. Shan H, Zhuang W, Wang Z (2009) Cooperation or not in mobile ad hoc networks: a MAC perspective. In: 2009 IEEE international conference on communications. IEEE, pp 1–6 (2009)
4. Sadeghi R, Barraca JP, Aguiar RL (2017) A survey on cooperative MAC protocols in IEEE 802.11 wireless networks. Wirel Pers Commun 95(2):1469–1493
5. Bharati, S., Zhuang, W.: Link-layer cooperative communication in vehicular networks. Springer (2017)
6. Sofwan A, AlQahtani SA (2017) Cooperative and fair MAC protocols for cognitive radio ad-hoc networks. Wirel Netw 23(7):2289–2306
7. So J, Vaidya NH (2004) Multi-channel mac for ad hoc networks: handling multi-channel hidden terminals using a single transceiver. In: Proceedings of the 5th ACM international symposium on mobile ad hoc networking and computing. ACM, pp 222–233
8. Dai M, Wang P, Zhang S, Chen B, Wang H, Lin X, Sun C (2014) Survey on cooperative strategies for wireless relay channels. Trans Emerg Telecommun Technol 25(9):926–942
9. Jamal T, Mendes P, Zúquete A (2011) Relayspot: a framework for opportunistic cooperative relaying. In: Proceedings of IARIA ACCESS, Luxembourg
10. Ahmed MHU, Alam MGR, Kamal R, Hong CS, Lee S (2012) Smart grid cooperative communication with smart relay. J Commun Netw 14(6):640–652
11. Gawas MA, Gudino LJ, Anupama KR (2017) AMCCR: adaptive multi-QoS cross-layer cooperative routing in ad hoc networks. J Comput Netw Commun
12. Shan H, Cheng HT, Zhuang W (2011) Cross-layer cooperative MAC protocol in distributed wireless networks. IEEE Trans Wirel Commun 10(8):2603–2615
13. Li Y, Wang P, Niyato D, Zhuang W (2011) A dynamic relay selection scheme for mobile users in wireless relay networks. In: 2011 proceedings IEEE INFOCOM. IEEE, pp 256–260
14. Hu Z, Tham CK (2008) CCMAC: coordinated cooperative MAC for wireless LANs. In: Proceedings of the 11th international symposium on modeling, analysis and simulation of wireless and mobile systems ACM, vol 54, pp 60–69. 618–630
15. Gao C, Zeng B, Lu J, Zhao G (2017) Dynamic cooperative MAC protocol for navigation carrier ad hoc networks: a DiffServ-based approach. J Sens Actuator Netw 6(3):14
16. Li Y, Zhang R, Cao B, You X, Daneshmand A (2011) A distributed cooperative MAC for cognitive radio ad-hoc networks. In: 2011 IEEE symposium on computers and communications (ISCC). IEEE, pp 1038–1043
17. Hussain R, Shakeel A, Iqbal A, Ahmed J, Alvi S, ul Hasan Q, Malik SA (2018) Efficient idle channel discovery mechanism through cooperative parallel sensing in cognitive radio network. EURASIP J Wirel Commun Netw 75(1)
18. The Network Simulator—NS 2. Available: http://www.isi.edu/nsnam/ns/
19. Royer EM, Toh CK (1999) A review of current routing protocols for ad hoc mobile wireless networks. IEEE Pers Commun 6(2):46–55

Far Field Analysis of Defected Ground Structured Wideband Antenna for RF Energy Harvesting Applications

Rashmi Pandey, A. K. Shankhwar, and Ashutosh Singh

Abstract Energy harvesting basically a green energy concept that provides RF energy from available environment sources. That RF energy is able to operate low powered devices, to recharge or replace batteries from available electronic devices, WSN, etc. Nowadays it plays an immeasurable role in the modern wireless communication system. This paper introduces a compact wideband antenna design to cover the frequency range of 1.97 GHz up to 4.744 GHz, where 2.4–2.45 GHz considered for Bluetooth/WLAN, 3.2 GHz for 3G, 2.5–3.8 GHz for LTE (Long-Term Evolution)/4G and additional 4.4–4.74 GHz is designated in the U.S. for military fixed and mobile communications, respectively. Compact octagonal microstrip patch antenna are simulated along with the defected ground structure and implemented modifications show improvement in the antenna performance in terms of return loss for low power device applications. Considered dimension of the proposed antenna is 40 mm × 45 mm × 1.635 mm and observed higher return loss −26.1, −24.94, −23.09 at 2.4, 2.45, and 2.5 GHz, respectively. For low power RF energy harvesting applications, the proposed antenna is able to cover the widest bandwidth from 1.97 to 4.744 GHz with a single radiating patch. This is the main focus of the work.

Keywords DGS · Microstrip patch antenna · RF energy harvesting · WLAN · And Wi-Fi

1 Introduction

Recently, radiofrequency energy scavenging has extract special attention in the modern wireless communication system to considered microwave energy as an available source due to their increasing applicability of RF transmitting sources. RF energy is a type of green energy that is easily available in the environment. Besides that it provides the improvement in wirelessly transferred power, charge batteries, or replace batteries, to control electronic gadgets remotely. Proposed concept provides

R. Pandey (✉) · A. K. Shankhwar · A. Singh
Department of Electronics Engineering, HBTU, Kanpur, India
e-mail: rashmi821@gmail.com

D. Harvey et al. (eds.), *Advances in VLSI, Communication, and Signal Processing*,
Lecture Notes in Electrical Engineering 683,
https://doi.org/10.1007/978-981-15-6840-4_16

Table 1 Comparative analysis of available environment sources and received output power

Available source	Available input potential	Received power	Attainability
Ambient light			
Indoor Outdoor	0.1 mW/cm^2 100 mW/cm^2	10 μW/cm^2 10 mW/cm^2	During day
Thermal energy			
Human Industrial	20 mW/cm^2 100 mW/cm^2	30 μW/cm^2 1–10 mW/cm^2	Continuous
Vibration/motion			
Human Industrial	0.5 m/s^2 @ 1 Hz 1 m/s^2 @ 50 Hz 1 m/s^2 @ 1 Hz 10 m/s^2 @ 1 kHz	4 μW/cm^2 100 μW/cm^2	Depends on the movement
Ambient RF energy	0.3 μW/cm^2	0.1 μW/cm^2 (cell phone)	Continuous

a special concentration in remote areas, where signals density and accessibility are low, the battery is hard to replace. In this manner, rather than utilizing those batteries, which have inhibited lifetime and vindictive outcomes in natural contamination, RF energy gatherers can give electrical power neatly, safely, and monetarily. Table 1 shows the comparative analysis of available power which could be obtained from environment sources [1, 2]. Here, the amount of received power totally depends on the environmental conditions.

In the modern wireless communication system antenna plays an important role to provide communication. Whether it is audio, video, or simple text message signal, etc., antenna communicates with all types of signals efficiently. Although wideband, multiband antenna is in great demand for low power devices application which depends on the ambient radio frequency (RF) energy harvesting. Wideband structure helps in energy scavenging from the surrounding environment and received RF energy depends on the frequency and power. Among the available RF energy sources, it has been developing interest between all wireless and wireline communication devices for RF energy harvesting which is able to collect surroundings microwave energy [1–7]. RF signals could be found all over the place and they emanate energy constantly. At present, RF energy harvesting showing remarkable interest to extract signals from available sources which is easily accessible such as indoor, open office, cell phones, TV transmission, Wi-Fi, Bluetooth, WLAN, GPS including itself Wi-Max is the significant powerful sources. These nearby electromagnetic energy sources are infinitely accessible and transmit Radio Frequency energy seamlessly and thus it has extraordinary capacity in the form of supplementation of other environment-friendly energy sources, for example, sun based energy. Besides, the advancement of Wireless Power Transfer advances [6, 8, 9] that permit miniaturized scale sensors [10–12], versatile electronic gadgets [13], wireless implantable collaboration [14–16] and far and near field analysis for RFID (Radiofrequency ID) [17–21] frameworks to work without batteries have activated driving force for RF energy harvesting which can give control constantly. Complete structure of the Radio Frequency energy harvesting is delineated in Fig. 1 [12]. Where EM source appeared is in charge of producing and emanating RF/microwave energy into free-space (for example environment). At the part of the output, the receiving antenna reproduces the enveloped Radio Frequency energy from the environment and transfers that energy by an applied filter at the input along with an impedance matching network to the rectifier circuit. The LPF applied at the input is conscientious to dismiss the harmonics originate by the non-linear rectifier, avoids returning this re-radiation from the radiation to the freely available space. Number of matching stages ensures that the most extreme power is taken from the available radio frequency sources to the rectifier. Without the help of the antenna matching system, the Radio Frequency energy extracted signals by output receiving antenna and that received power is reflected back on the source. Thus, there is small energy observed in the rectifier. The rectifier circuit is the main part of an energy harvesting circuit that is able to convert received Radio Frequency signals into useful DC power. Recent advances in low power devices ambient energy sources and antenna miniaturization are in great demand.

For energy harvesting available signal power is extremely weak, thus the antennas used for the system get to need better gain with omnidirectional radiation pattern from available directions. Therefore, for such a system different antennas structure performs the functions well with compact structure which is mandatory for the design of the system. Moreover, RF energy harvesting application, wideband antenna is broadly used in wireless communication technologies where multiple operating bands or broadband frequencies are preferred for the application in cell phones, multimedia appliance, etc. Wideband and multiband structure diminishes the size of antenna furthermore reduces the cost of composite systems working in multi-standard systems. Many antennas have been introduced in the literature for energy harvesting applications [22–25]. In the harvesting circuit, the antenna can be operated using an isolated frequency that can harvest power from the available isolated source of energy. Illustration, [22] presented a harvesting system with a planner uses a spiral antenna at 520 MHz and shows the various response of the proposed antenna.

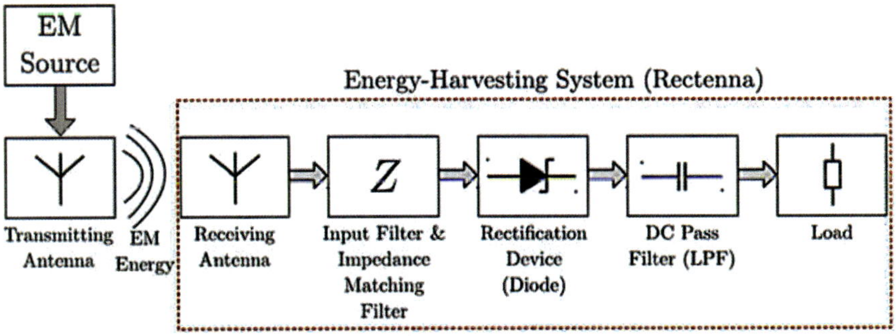

Fig. 1 Constitution of the RF energy harvesting circuit

Working on the same bands; proposed harvesting system is not capable to harvest abundant electricity available in other frequency bands.

In [24] has been introduced dual-band planar reception antenna for Radio Frequency energy harvesting. Monopole receiving antenna along with multiple microstrip lines are discussed by an author. Now, the normal impedance is coordinated in two frequency groups. Despite the fact that those two groups are used in this reception antenna, the range indicates high probabilities by collecting multiband frequencies. With the interest of expanding for the use of sustainable energy sources, the RF Energy scavenging has improved considerably. Interest is required to collect high power; these harvesting structures work on multi-frequencies. It causes such structures to scavenge energy from accessible signals of various available grades of different sources. In electromagnetic energy harvesting, many options are considered [23, 25, 26] for 350–3 GHz spectrum. This manuscript introduces the design and simulation of compact octagonal MPA with a truncated ground structure for wideband operation. Already stated antenna shows good simulated radiation and reflected result where 2.4–2.45 GHz considered for Bluetooth/WLAN, 3.2 GHz for 3G, 2.5–3.8 GHz for LTE (Long-Term Evolution)/4G and additional 4.4–4.74 GHz is designated in the U.S. for military and fixed-mobile communications, respectively, and its overall dimensions are compact in terms of RF energy harvesting applications. To accomplish broadband range, a few strategies have been proposed as of late, authors implemented Fractal Geometry [27–30], patch antenna with different slots [31–33], various shape [34], triple-port pixel antenna [35] and dielectric substrate with high permittivity have been used, different tuning stubs and few other methods could be used, like Defected Microstrip Structure (DMS), Defected Ground Structure (DGS) and electromagnetic bandgap (EBG) [36]. Different techniques of antenna miniaturization have been introduced in [37] which affect the gain of the antenna. Defected ground structure is a concept that is used in antenna for antenna miniaturization, it consists of different shapes/sizes in the ground plane, or sometimes with a complicated structure for better results. In [38] authors have been presented the concise literature relevant to various shapes and DGS structures to improve the antenna performance. In this work proposed antenna consists of rectangular shape DGS to improve antenna characteristics. Structures of the paper are designed as follows. Proposed structure of wideband antenna is introduced in Sect. 2. Analyzed results, characteristics of the structured antenna are discussed in Sect. 3 and conclusions followed in Sect. 4.

2 Antenna Design

In the modern wireless communication system antenna plays an important role to provide communication. Whether it is audio, video, or simple text message signal, etc., antenna communicates with all types of signals efficiently. Although wideband, multiband antenna is in great demand for low power devices application which depends on the ambient radio frequency (RF) energy harvesting. Wideband structure helps in energy scavenging from the surrounding environment and received RF energy depends on the frequency and power. In this work, the proposed antenna occupied an octagonal shape with geometry modification. Novelty of this work is a rectangular notch with partial ground structure effects. In the proposed antenna structure the width and length are calculated according to pre-evaluated equations mentioned in [39]. Figure 2 shows the standard extent of the antenna is 40 mm × 45 mm × 1.635 mm and the FR4 substrate is fed by a 2.60 mm × 10 mm transmission line (TL). Truncated ground structure is considered in ground plane by the removal of some parts in particular shapes such as circle, rectangular, elliptical, or cross-section shapes. With the effect of defected ground structure, the characteristics of ground are changed and after the application of DGS, it works like band stop or bandpass filter. Consequently, it generates wideband frequencies. Microstrip transmission line is used to connect the antenna feed to patch. The antenna was modeled and optimized using CST microwave studio. 50 O microstrip feed line patch antenna with a modified patch shape shows the improvement in impedance matching at wideband operational frequency. Rectangular notch is used in the proposed antenna to operate the antenna as a broadband characteristic.

The primary focus of the work is to spread a new concept of a compact and efficient antenna structure for RF energy harvesting applications rather than considering normal antenna. Proposed antenna with the defected ground is beneficial for capturing better response in any energy harvester circuit as shown in Fig. 2b. Consequently, we have observed this wideband antenna is the best candidate for the reception of wideband applications with the single radiating patch. Return loss below −10 dB for wideband frequency has been observed in Figs. 3 and 10.

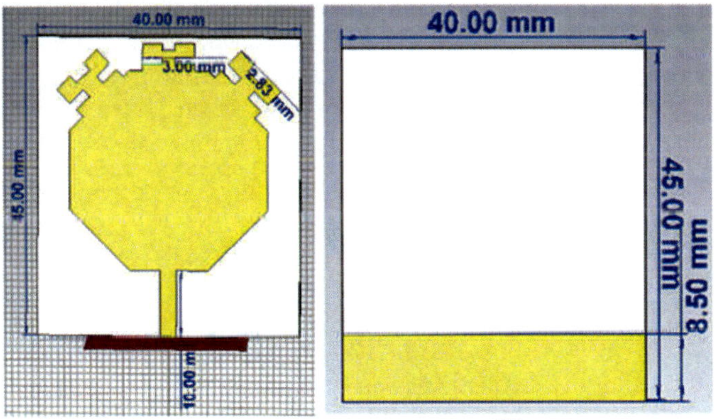

(a) (b)

Fig. 2 Structure of proposed antenna **a** Front side, **b** back side

Fig. 3 S 11 return loss plot

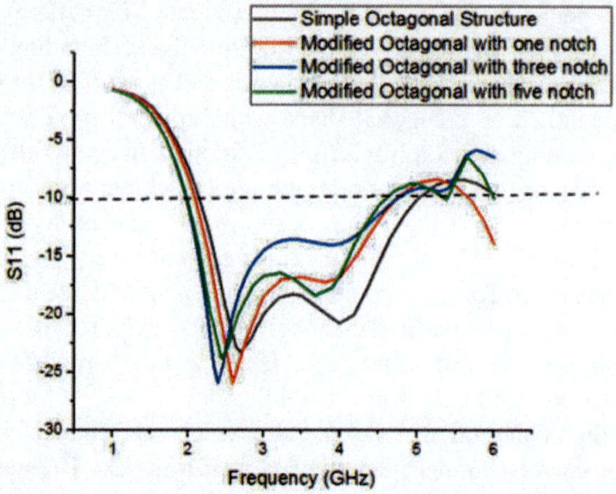

3 Simulation Results

Proposed antenna is analyzed on CST microwave studio which is a CAD tool and ability to identify the response of designed structure. Operating range of the frequency is 1.97 GHz up to 4.744 GHz, where 2.4–2.45 GHz considered for Bluetooth/WLAN, 3.2 GHz for 3G, 2.5–3.8 GHz for LTE (Long-Term Evolution)/4G and additional 4.4–4.74 GHz is designated in the U.S. for military fixed and mobile communications, respectively, in the simulation are considered. Table 2 shows the various parameter of antenna in terms of radiation pattern, return loss. The obtained return loss of the following frequency (shown in Table 2) is well below −10 dB which ensures appropriate impedance matching along with receptor antenna. Received response of proposed antenna is capable to collect energy for low powered devices.

In this paper, we have present only one structure but during the parametric analysis of proposed antenna have designed three other structure also which is a simple octagonal patch, modified octagonal with one, three and five notches. And the result

Table 2 Simulated results of directivity and gain

Operating frequency band (GHz)	Modified octagonal patch structure with three rectangular notch		
	Directivity (dB)	Gain (dBi)	Return loss (dB)
2.4	2.56	2.18	−26.1
2.45	2.60	2.22	−24.94
2.5	2.63	2.25	−23.09
3.2	3.23	2.8	−13.78
3.8	4.01	3.43	−14.01
4.73	4.79	3.86	−10.04

obtained from all these designs is shown in Figs. 3 and 10. During the comparative analysis, it has been analyzed patch structure with three notches shows good agreement between all the considered structures as shown in Fig. 10. Table 2 represents the return loss gain and directivity at various operating frequency bands of the proposed structure and identified higher gain of 3.86 dB at the 4.73 GHz of the modified octagonal patch structure with three rectangular notch.

With the help of the proposed antenna Fig. 10 shows higher return loss −26.1, −24.94, −23.09 at 2.4, 2.45, and 2.5 GHz, respectively for modified octagonal patch structure with three rectangular notch. Rest of the structure also shows the improved performance but modified octagonal patch structure with three rectangular notch has observed appropriate response. Figure 4 described the comparative return loss of all the proposed structure.

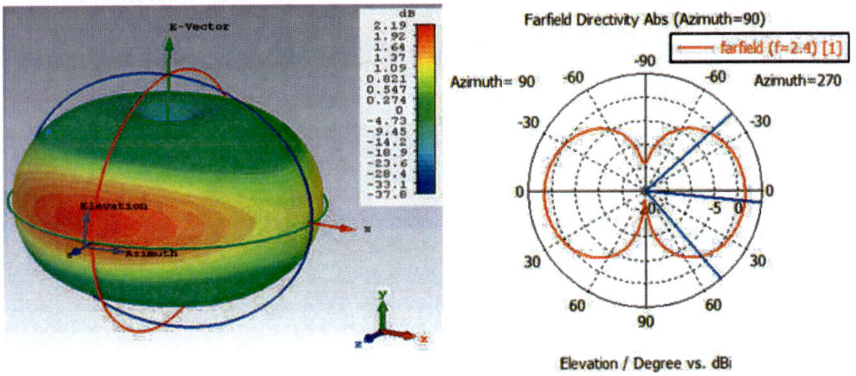

Fig. 4 2.4 GHz radiation pattern of modified octagonal patch structure with three rectangular notch

Table 3 Return loss comparison of proposed structure

Operating frequency band (GHz)	Return loss (dB)		
	Basic octagonal patch structure	Modified octagonal patch structure with one rectangular notch	Modified octagonal patch structure with three rectangular notch
2.4	−16.38	−19.53	−26.1
2.45	−17.88	−21.65	−24.94
2.5	−19.40	−23.78	−23.09
3.2	−18.59	−17.11	−13.78
3.8	−19.79	−17.23	−14.01
4.73	−12.59	−10.08	−10.08
5	−10.13	−8.72	−9.17
5.2	−9.03	−8.38	−9.21

Table 3 represents the return loss at various operating frequency band of the proposed structure and identified maximum return loss of −26.1, −24.94, −23.09 at 2.4, 2.45, and 2.5 GHz, respectively, from the proposed structure. Rest of the structure also shows improved return loss at various frequencies as also mentioned in Table 3. Figures 4, 5, 6, 7, 8, and 9 indicate simulated far field radiation pattern of the proposed antenna, here we have observed better gain 3.86 dB at 4.73 GHz. Proposed antenna performs better in a freely available spectrum, which is considered as wideband frequency (Fig. 10).

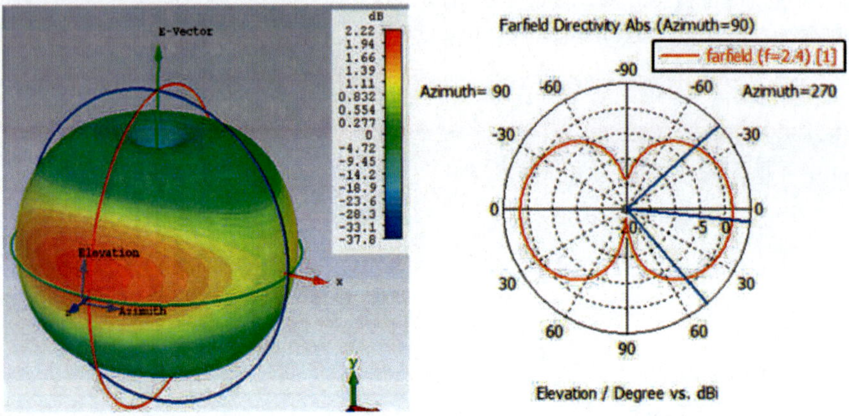

Fig. 5 2.44 GHz radiation pattern of modified octagonal patch structure with three rectangular notch

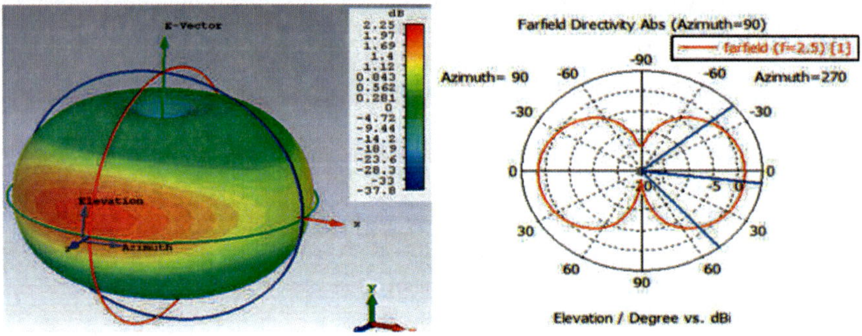

Fig. 6 2.5 GHz radiation pattern of modified octagonal patch structure with three rectangular notch

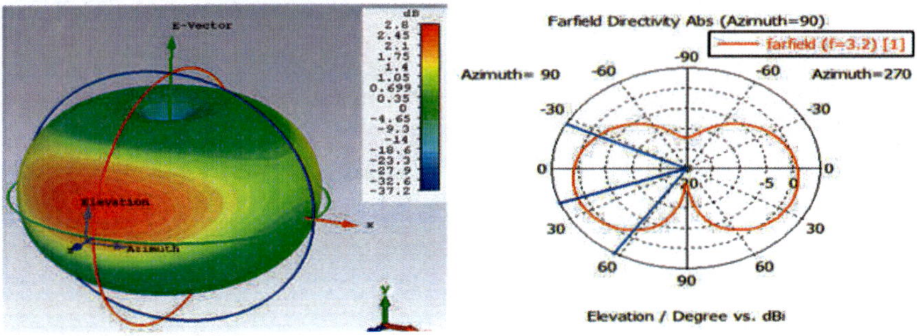

Fig. 7 3.2 GHz radiation pattern of modified octagonal patch structure with three rectangular notch

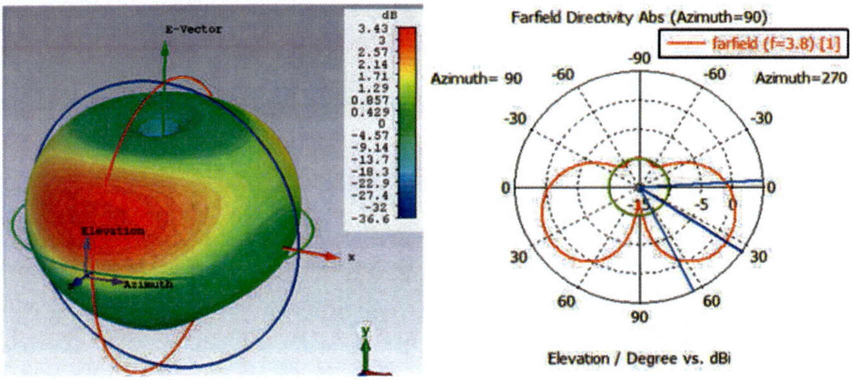

Fig. 8 3.8 GHz radiation pattern of modified octagonal patch structure with three rectangular notch

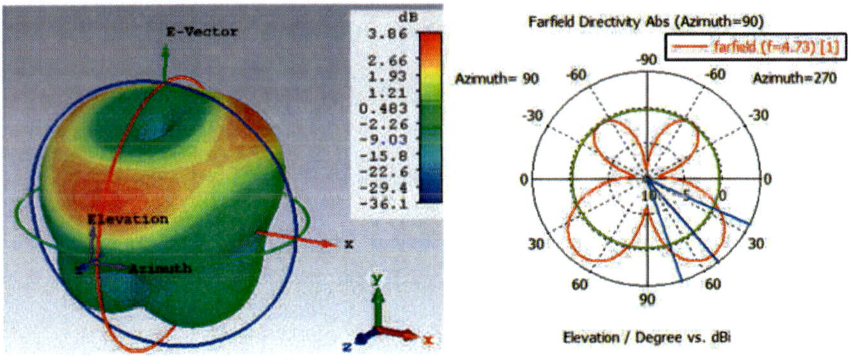

Fig. 9 4.73 GHz radiation pattern of modified octagonal patch structure with three rectangular notch

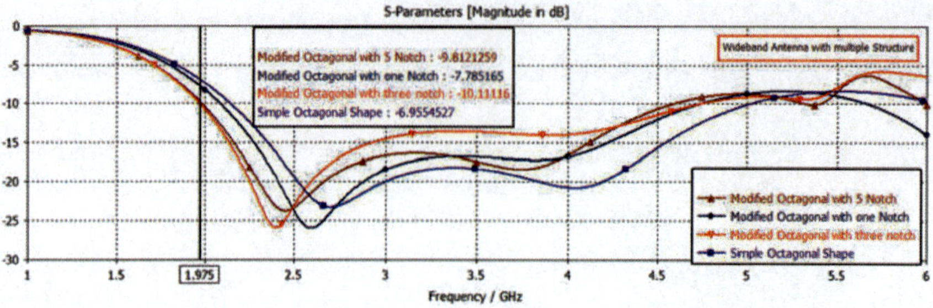

Fig. 10 S 11 return loss plot of OMPA with three rectangular notches (FR-4 substrate)

4 Conclusion

The RF energy harvesting technology extends a promising future of the appliances which consumed a low power such as wearable electronic devices and wireless sensor networks. In this work, the proposed design of the antenna provides the better return loss with modified octagonal patch 3 rectangular notch for low-frequency application and it is based on a rectangular patch antenna in order to picking up the radio frequency waves in a random manner which covers the band between 1 and 6 GHz range, the simulated design based on CST MWS. The result obtained after the simulation ended was 4.86 dB gain and shows the wideband characteristics between the bands 1.97–4.73 GHz. Proposed antenna is a good candidate to cover the wideband behavior of low-frequency energy harvesting applications. In the future, we will be present the measured results after the fabrication of the proposed antenna.

References

1. Kim S, Vyas R, Bito J, Niotaki K, Collado A, Georgiadis A, Tentzeris MM (2014) Ambient RF energy-harvesting technologies for self-sustainable standalone wireless sensor platforms. Proc IEEE 102(11):1649–1666. https://doi.org/10.1109/JPROC.2014.2357031
2. Vullers R, van Schaijk R, Doms I, Hoof CV, Mertens R (2009) Micropower energy harvesting. Solid-State Electron 53(7):684–693. https://doi.org/10.1016/j.sse.2008.12.011. Papers Selected from the 38th European solid-state device research conference—ESSDERC08
3. Brown WC (1984) The history of power transmission by radio waves. IEEE Trans Microw Theory Techn 32(9):1230–1242. https://doi.org/10.1109/tmtt.1984.1132833
4. Costanzo A, Dionigi M, Masotti D, Mongiardo M, Monti G, Tarricone L, Sorrentino R (2014) Electromagnetic energy harvesting and wireless power transmission: a unified approach. Proc IEEE 102(11):1692–1711. https://doi.org/10.1109/JPROC.2014.2355261
5. Paradiso JA, Starner T (2005) Energy scavenging for mobile and wireless electronics. IEEE Pervasive Comput 4(1):18–27. https://doi.org/10.1109/MPRV.2005.9
6. Shinohara N (2011) Power without Wires. IEEE Microw Mag 12(7):S64–S73. https://doi.org/10.1109/MMM.2011.942732
7. Brown WC (1969) Experiments involving a microwave beam to power and position a helicopter. IEEE Trans Aerosp Electron Syst AES-5(5):692–702. https://doi.org/10.1109/TAES.1969.309867

8. Borges Carvalho N, Georgiadis A, Costanzo A, Rogier H, Collado A, Garca JA, Lucyszyn S, Mezzanotte P, Kracek J, Masotti D, Boaventura AJS, de las Nieves Ruz Lavn M, Piuela M, Yates DC, Mitcheson PD, Mazanek M, Pankrac V (2014) Wireless power transmission: R D activities within Europe. IEEE Trans Microw Theory Tech 62(4):1031–1045. https://doi.org/10.1109/TMTT.2014.2303420
9. Din N, Chakrabarty C, Bin A, Devi A, Chen WY (2012) Design of RF energy harvesting system for energizing low power devices. Prog Electromagn Res 132:49–69. https://doi.org/10.2528/PIER12072002
10. Le T, Mayaram K, Fiez T (2008) Efficient far-field radio frequency energy harvesting for passively powered sensor networks. IEEE J Solid-State Circuits 43(5):1287–1302. https://doi.org/10.1109/JSSC.2008.920318
11. Paing T, Morroni J, Dolgov A, Shin J, Brannan J, Zane R, Popovic Z (2007) Wirelessly-powered wireless sensor platform. In: 2007 European conference on wireless technologies, pp 241–244. https://doi.org/10.1109/ECWT.2007.4403991
12. Visser HJ, Vullers RJM (2013) RF energy harvesting and transport for wireless sensor network applications: principles and requirements. Proc IEEE 101(6):1410–1423. https://doi.org/10.1109/JPROC.2013.2250891
13. Paulo J, Gaspar P (2010) Review and future trend of energy harvesting methods for portable medical devices. In: Lecture Notes in Engineering and Computer Science 2 (06 2010)
14. Huang F, Lee C, Chang C, Chen L, Yo T, Luo C (2011) Rectenna application of miniaturized implantable antenna design for triple-band biotelemetry communication. IEEE Trans Antennas Propag 59(7):2646–2653. https://doi.org/10.1109/TAP.2011.2152317
15. Monti G, Tarricone L (2012) Experimental investigation of a 434 MHz wireless energy link for medical applications. Prog Electromagn Res C 30:53–64. https://doi.org/10.2528/PIERC12032606
16. Walk J, Weber J, Soell C, Weigel R, Fischer G, Ussmueller T (2014) Remote powered medical implants for telemonitoring. Proc IEEE 102(11):1811–1832. https://doi.org/10.1109/JPROC.2014.2359517
17. Finkenzeller K (2010) RFID handbook: fundamentals and applications in contact-less smart cards, radio frequency identification and near field communication, 3rd edn. Wiley. https://doi.org/10.1002/9780470665121
18. Liu DS, Li FB, Zou XC, Liu Y, Hui XM, Tao XF (2011) New analysis and design of A RF rectifier for RFID and implantable devices. Sensors 11(7):6494–6508. https://doi.org/10.3390/s110706494
19. Mandal S, Sarpeshkar R (2007) Low-power CMOS rectifier design for RFID applications. IEEE Trans Circuits Syst I Regul Pap 54(6):1177–1188. https://doi.org/10.1109/TCSI.2007.895229
20. Scorcioni S, Bertacchini A, Larcher L, Ricciardi A, Dondi D, Pavan P (2012) RF to DC CMOS rectifier with high efficiency over a wide input power range for RFID applications. In: 2012 IEEE/MTT-S international microwave symposium digest, pp 1–3. https://doi.org/10.1109/MWSYM.2012.6259760
21. Shameli A, Safarian A, Rofougaran A, Rofougaran M, De Flaviis F (2007) Power harvester design for passive UHF RFID tag using a voltage boosting technique. IEEE Trans Microw Theory Tech 55(6):1089–1097. https://doi.org/10.1109/TMTT.2007.896819
22. Kang CC, Olokede SS, Mahyuddin NM, Ain MF (2014) Radio frequency energy harvesting using circular spiral inductor antenna. WAMICON 2014:1–5. https://doi.org/10.1109/WAMICON.2014.6857805
23. Nimo A, Grgi D, Reindl LM (2012) Ambient electromagnetic wireless energy harvesting using multiband planar antenna. In: International multi-conference on systems, signals devices. pp 1–6. https://doi.org/10.1109/SSD.2012.6198036
24. Shao X, Li B, Shahshahan N, Goldsman N, Salter TS, Metze GM (2011): A planar dual-band antenna design for RF energy harvesting applications. In: 2011 international semiconductor device research symposium (ISDRS), pp 1–2. https://doi.org/10.1109/ISDRS.2011.6135318

25. Tavares J, Barroca N, Saraiva HM, Borges LM, Velez FJ, Loss C, Salvado R, Pinho P, Goncalves R, Borges Carvalho N (2013) Spectrum opportunities for electromagnetic energy harvesting from 350 MHz To 3 GHz. In: 2013 7th international symposium on medical information and communication technology (ISMICT), pp 126–130. https://doi.org/10.1109/ISMICT.2013.6521714
26. Taghadosi M, Albasha L, Qaddoumi N, Ali M (2015) Miniaturized printed elliptical nested fractal multiband antenna for energy harvesting applications. IET Microw, Antennas Propag 9(10):1045–1053. https://doi.org/10.1049/iet-map.2014.0744
27. Guha D, Biswas M, Antar YMM (2005) Microstrip patch antenna with defected ground structure for cross polarization suppression. IEEE Antennas Wirel Propag Lett 4:455–458. https://doi.org/10.1109/LAWP.2005.860211
28. Raj NA, Dwivedi RP (2015) High gain antenna with DGS for wireless applications. In: 2015 2nd international conference on signal processing and integrated networks (SPIN), pp 19–24. https://doi.org/10.1109/SPIN.2015.7095317
29. Shi Y, Jing J, Fan Y, Yang L, Wang M (2018) Design of a novel compact and efficient rectenna for Wi-Fi energy harvesting. Prog Electromagn Res C 83:57–70. https://doi.org/10.2528/PIERC18012803
30. Zeng M, Andrenko AS, Liu X, Li Z, Tan H (2017) A compact fractal loop rectenna for RF energy harvesting. IEEE Antennas Wirel Propag Lett 16:2424–2427. https://doi.org/10.1109/LAWP.2017.2722460
31. Chen WS, Cheng CM, Liao BY, Chang YL, Wang HY (2018) Triple-band slot antenna array for energy harvesting for wireless sensor networks. Sens Mater 30(3(2)):587–594. https://doi.org/10.18494/SAM.2018.1827
32. Elsheakh D, Elsadek H, Abdallah E, Elhenawy H, Iskander M (2009) Miniaturized and multiband operations of inset feed microstrip patch antenna by using novel shape of defect ground structure (DGS). In: Wireless applications. PIERS proceedings, Moscow, Russia, pp 1082–1086
33. Lo TKC, Hwang Y (1997) Microstrip antennas of very high permittivity for personal communications. In: Proceedings of 1997 Asia-Pacific microwave conference, vol 1, pp 253–256. https://doi.org/10.1109/APMC.1997.659352
34. Iqbal MS, Khanzada TJ, Dahri FA, Ali A, Ali M, Khokhar AW (2019) Analysis and maximizing energy harvesting from RF signals using T-shaped microstrip patch antenna. Int J Adv Comput Sci Appl (IJACSA) 10(1):309–313. http://dx.doi.org/10.14569/IJACSA.2019.0100141
35. Shen S, Chiu C, Murch RD (2018) Multiport pixel rectenna for ambient RF energy harvesting. IEEE Trans Antennas Propag 66(2):644–656. https://doi.org/10.1109/TAP.2017.2786320
36. Yang F, Rahmat-Samii Y (2003) Microstrip antennas integrated with electromagnetic band-gap (EBG) structures: a low mutual coupling design for array applications. IEEE Trans Antennas Propag 51(10):2936–2946. https://dx.doi.org/10.1109/TAP.2003.817983
37. Divakaran SK, Krishna DD (2017) Nasimuddin: an overview of compact antennas for RF energy harvesting. In: IEEE international WIE conference on electrical and computer engineering (WIECON-ECE), pp 47–50. https://doi.org/10.1109/wiecon-ece.2017.8468923
38. Divakaran SK, Krishna DD (2019) Nasimuddin: RF energy harvesting systems: an overview and design issues. Int J RF Microw Comput-Aided Eng 29(1):1–15. https://doi.org/10.1002/mmce.21633, https://doi.org/10.1002/mmce.21633
39. Girase N, Tiwari R, Sharma A, Singh H (2014) Design and simulation of slotted rectangular microstrip patch antenna. Int J Comput Appl 103:19–23. https://doi.org/10.5120/18294-9264

New LMI Criterion to the Robust Stability of Discrete-Time Systems with Time-Varying Delays and Generalized Overflow Nonlinearities

Pushpendra Kumar Gupta and V. Krishna Rao Kandanvli

Abstract This paper handles the problem of robust stability of discrete-time systems under the influence of generalized overflow nonlinearities and time-varying state delays. The systems under consideration involve norm-bounded parameter uncertainties. By using Lyapunov–Krasovskii functional, a new delay-dependent criterion based on LMI for the stability of discrete-time systems is presented where Wirtinger-based inequalities and reciprocal convex lemma are succeedingly employed. The criterion presented in this paper employs less number of decision variables than previously presented criterion without affecting the conservativeness. Illustrative example is also given to prove the efficacy of the presented criterion.

Keywords Finite wordlength nonlinearity · Generalized overflow nonlinearity · Linear matrix inequality · Lyapunov–Krasovskii functional · Norm-bounded parameter uncertainty · Robust stability · Time-varying delay

1 Introduction

In various practical engineering systems such as networked control systems [1], Markovian jump systems [2], multipath propagation in mobile communication, congestion analysis and control in high speed networks, neural networks [3], sensor networks [4], array signal processing, time delay exists which could not be avoided and must be considered for realistic system design.

P. K. Gupta (✉)
Department of Electronics & Communication Engineering, College of Technology, Pantnagar 263145, India
e-mail: pushpendranit30@gmail.com

V. K. R. Kandanvli
Department of Electronics and Communication Engineering, Motilal Nehru National Institute of Technology Allahabad, Prayagraj 211004, India
e-mail: rao@mnnit.ac.in

D. Harvey et al. (eds.), *Advances in VLSI, Communication, and Signal Processing*, Lecture Notes in Electrical Engineering 683,
https://doi.org/10.1007/978-981-15-6840-4_17

Several stability criteria based on delay classified as delay-dependent criteria [2, 3, 5–9] and delay-independent criteria [10–15]. In general, delay-independent criteria are more conservative than delay-dependent. Free weighting matrix method [6], Jensen inequality, Wirtinger inequality [8, 16] and reciprocal convex lemma [8, 9] are the inequalities for handling the cross-terms of the difference of the Lyapunov–Krasovaskii function. In general, Wirtinger inequality-based method provides less conservative results than Jensen inequality and free weighting matrix methods [6].

Besides time delay, the presence of parameter uncertainties due to variation in system parameters, modeling errors or some ignored factors are also one of the reasons for the instability of discrete-time systems. The parameters uncertainties are assumed to be norm bounded and deterministic [6, 8, 9, 11–13, 15, 17].

While implementing stable and linear discrete-time dynamical systems using digital computer with finite wordlength processors, nonlinearities may encounter. Due to the presence of such finite wordlength nonlinearities, the system may become unstable. In this paper, generalized overflow nonlinearities [6, 10, 14, 18] are considered in the system.

In recent years, much attention has been paid on the stability properties of discrete-time systems having time-varying delays, parameter uncertainties and generalized overflow nonlinearities simultaneously [6, 10] but there is still a scope to derive improved stability results, i.e., criteria with less number of decision variables/less computational complexity without affecting the conservativeness. In the present work, a class of discrete-time systems having time-varying delay, parameter uncertainties and generalized overflow nonlinearities is considered, and a new delay-dependent stability criterion is presented which has less number of decision variables than the existing criterion [6]. A numerical example is given to show the efficacy of the presented criterion.

The organization of the remaining paper is as follows. In Sect. 2, description of the considered system and some required lemmas are given. Delay-dependent criterion for the robust stability of the discrete-time systems is presented in Sect. 3. Section 4 shows the effectiveness of the proposed criterion with a numerical example. Finally, conclusions are provided in Sect. 5.

In this paper, notations are considered as follows. $\mathbb{R}^k$ represents k-dimensional Euclidean space, $\mathbb{R}^{\alpha\times\beta}$ denotes set of $\alpha \times \beta$ real matrices, $\mathbf{0}$ is null matrix or null vector, $\boldsymbol{I}$ refers identity matrix of compatible dimension, $\boldsymbol{P}^T$ stands transpose of the $\boldsymbol{P}$, $\boldsymbol{P} > \mathbf{0}$ ($<\mathbf{0}$) shows $\boldsymbol{P}$ is positive (negative) definite real symmetric matrix, symbol $*$ stands symmetric terms in a symmetric matrix.

2 System Description

The description of the system under consideration is given by

$$\boldsymbol{x}(r+1) = \boldsymbol{f}(\boldsymbol{y}(r))$$

$$= [f_1(y_1(r))\ \ f_2(y_2(r)) \cdots\ f_n(y_n(r))]^{\mathrm{T}} \tag{1a}$$

$$\begin{aligned} \boldsymbol{y}(r) &= (\boldsymbol{A} + \Delta\boldsymbol{A})\boldsymbol{x}(r) + (\boldsymbol{A_d} + \Delta\boldsymbol{A_d})\boldsymbol{x}(r - h(r)) \\ &= [y_1(r)\ y_2(r) \cdots\ y_n(r)]^T \end{aligned} \tag{1b}$$

$$\boldsymbol{x}(r) = \boldsymbol{\varphi}(r), \forall r \in [-d_2, 0] \tag{1c}$$

$$\bar{\boldsymbol{A}} = \boldsymbol{A} + \Delta\boldsymbol{A},\ \ \bar{\boldsymbol{A}}_{\boldsymbol{d}} = \boldsymbol{A_d} + \Delta\boldsymbol{A_d} \tag{1d}$$

where $\boldsymbol{x}(r) \in \mathbb{R}^n$ is the state variable; $\boldsymbol{A},\ \ \boldsymbol{A_d} \in \mathbb{R}^{n \times n}$ are matrices (known constant); $\Delta\boldsymbol{A}$ and $\Delta\boldsymbol{A_d} \in \mathbb{R}^{n \times n}$ are the unknown matrices having uncertainties in $\boldsymbol{A}$ and $\boldsymbol{A_d}$, respectively; At time r, the initial state value is $\boldsymbol{\varphi}(r) \in \mathbb{R}^n$ and $\boldsymbol{f}(.)$ represents the generalized overflow nonlinearities.

The characteristic of generalized overflow nonlinearities $\boldsymbol{f}(.)$ is specified by

$$\left.\begin{array}{rr} L \le f_k(y_k(r)) \le 1, & y_k(r) > 1 \\ f_k(y_k(r) = y_k(r), & -1 < y_k(r) < 1 \\ -1 \le f_k(y_k(r)) \le -L, & y_k(r) < -1 \end{array}\right\},\ k = 1, 2, 3, \ldots, n \tag{2a}$$

where

$$-1 \le L \le 1. \tag{2b}$$

With applicable choice of L, (2a) covers different types of overflow arithmetics, for e.g., saturation ($L = 1$), zeroing ($L = 0$), two's complement and triangular ($L = -1$), etc. In the state matrices and the time-varying delay, $h(r)$ is considered as

$$1 \le d_1 \le h(r) \le d_2 \tag{3}$$

The parameter uncertainties are assumed as

$$\Delta\boldsymbol{A} = \boldsymbol{B}_0\, \boldsymbol{F}_0\, \boldsymbol{C}_0 \tag{4a}$$

$$\Delta\boldsymbol{A_d} = \boldsymbol{B}_1\, \boldsymbol{F}_1\, \boldsymbol{C}_1 \tag{4b}$$

where $\boldsymbol{B}_k \in \mathbb{R}^{n \times p_k}$, $\boldsymbol{C}_k \in \mathbb{R}^{q_k \times n}$ $(k = 0, 1)$ are matrices (known constant) and $\boldsymbol{F}_k \in \mathbb{R}^{p_k \times q_k}(k = 0, 1)$ is matrix (unknown) which satisfies

$$\boldsymbol{F}_k^{\mathrm{T}} \boldsymbol{F}_k \le \boldsymbol{I},\ k = 0, 1. \tag{4c}$$

The following Lemmas are useful in the proof of the main result.

Lemma 1 [6, 10, 18]. *A positive definite matrix* $\boldsymbol{G} = \boldsymbol{G}^T = [g_{kl}] \in \boldsymbol{R}^{n\times n}$ *satisfies*

$$\boldsymbol{y}^T(r)\boldsymbol{G}\boldsymbol{y}(r) - \boldsymbol{f}^T(\boldsymbol{y}(r))\boldsymbol{G}\boldsymbol{f}(\boldsymbol{y}(r)) \geq \boldsymbol{0} \tag{5}$$

if

$$(1+L)g_{kk} \geq 2\sum_{l=1, l\neq k}^{n} |g_{kl}|,\ k = 1, 2, \ldots, n \tag{6}$$

Afterward, assume the matrix $\boldsymbol{G} = [g_{kl}] \in \boldsymbol{R}^{n\times n}$

$$g_{kk} = s_k + \sum_{l=1, l\neq k}^{n} (o_{kl} + \rho_{kl}),\ k = 1, 2, \ldots, n \tag{7a}$$

$$g_{kl} = g_{lk} = \left(\frac{1+L}{2}\right)(o_{kl} - \rho_{kl}),\ k, l = 1, 2, \ldots, n\ (k \neq l) \tag{7b}$$

$$o_{kl} = o_{lk} > 0,\ \rho_{kl} = \rho_{lk} > 0\ k, l = 1, 2, \ldots, n\ (k \neq l) \tag{7c}$$

$$s_k > 0,\ k = 1, 2, \ldots, n \tag{7d}$$

where it is very easy to understand that, for $n = 1$, $\boldsymbol{G}$ corresponds to a scalar $\gamma > 0$.

Lemma 2 [8, 16]. *For* $\boldsymbol{R} > \boldsymbol{0}$ *and three non-negative integers* a_1, a_2, r*, as* $a_1 \leq a_2 \leq r$*, if*

$$\begin{aligned}\boldsymbol{\xi}(r, a_1, a_2) &= \frac{1}{a_2 - a_1}\left[\left(2\sum_{s=r-a_2}^{r-a_1-1} \boldsymbol{x}(s)\right) + \boldsymbol{x}(r-a_1) - \boldsymbol{x}(r-a_2)\right],\ a_1 < a_2 \\ &= 2\boldsymbol{x}(r-a_1),\ a_1 = a_2\end{aligned} \tag{8}$$

then

$$-(a_2 - a_1)\sum_{s=r-a_2}^{r-a_1-1} \boldsymbol{\eta}^T(s)\boldsymbol{R}\boldsymbol{\eta}(s) \leq -\begin{bmatrix}\boldsymbol{\theta}_0 \\ \boldsymbol{\theta}_1\end{bmatrix}^T \begin{bmatrix}\boldsymbol{R} & \boldsymbol{0} \\ \boldsymbol{0} & 3\boldsymbol{R}\end{bmatrix} \begin{bmatrix}\boldsymbol{\theta} \\ \boldsymbol{\theta}_1\end{bmatrix} \tag{9}$$

where

$$\boldsymbol{\theta}_0 = \boldsymbol{x}(r - a_1) - \boldsymbol{x}(r - a_2),\ \boldsymbol{\theta}_1 = \boldsymbol{x}(r - a_1) + \boldsymbol{x}(r - a_2) - \boldsymbol{\xi}(r, a_1, a_2) \quad (10)$$

$$\boldsymbol{\eta}(s) = \boldsymbol{x}(s+1) - \boldsymbol{x}(s) \quad (11)$$

Lemma 3 [8, 9]. *For any vectors $\boldsymbol{\varphi}_1$ and $\boldsymbol{\varphi}_2$, matrices $\boldsymbol{M}$, $\boldsymbol{N}$ and real numbers $c_1 \geq 0$, $c_2 \geq 0$ satisfying*

$$\begin{bmatrix} \boldsymbol{M} & \boldsymbol{N} \\ * & \boldsymbol{M} \end{bmatrix} \geq \boldsymbol{0},\ c_1 + c_2 = 1 \quad (12)$$

$$\boldsymbol{\varphi}_i = \boldsymbol{0},\ \text{if } c_i = \boldsymbol{0} (i = 1, 2) \quad (13)$$

then

$$-\frac{1}{c_1}\boldsymbol{\varphi}_1^T \boldsymbol{R}\boldsymbol{\varphi}_1 - \frac{1}{c_2}\boldsymbol{\varphi}_2^T \boldsymbol{R}\boldsymbol{\varphi}_2 \leq -\begin{bmatrix} \boldsymbol{\varphi}_1 \\ \boldsymbol{\varphi}_2 \end{bmatrix}^T \begin{bmatrix} \boldsymbol{M} & \boldsymbol{N} \\ * & \boldsymbol{M} \end{bmatrix} \begin{bmatrix} \boldsymbol{\varphi}_1 \\ \boldsymbol{\varphi}_2 \end{bmatrix} \quad (14)$$

Lemma 4 [6, 8, 9, 11–13, 15, 17]. *Let $\boldsymbol{\Xi}$, $\boldsymbol{\Theta}$, $\boldsymbol{F}$ and $\boldsymbol{U}$ be matrices (real) of suitable dimensions and $\boldsymbol{U} = \boldsymbol{U}^T$, then*

$$\boldsymbol{U} + \boldsymbol{\Xi}\boldsymbol{F}\boldsymbol{\Theta} + \boldsymbol{\Theta}^{\mathrm{T}}\boldsymbol{F}^{\mathrm{T}}\boldsymbol{\Xi}^T < \boldsymbol{0} \quad (15)$$

for all $\boldsymbol{F}^T\boldsymbol{F} \leq \boldsymbol{I}$, if there exists a scalar $\varepsilon > 0$ such that

$$\boldsymbol{U} + \varepsilon^{-1}\boldsymbol{\Xi}\boldsymbol{\Xi}^T + \varepsilon\boldsymbol{\Theta}^T\boldsymbol{\Theta} < \boldsymbol{0} \quad (16)$$

Next, we prove the main result of the paper.

3 Main Result

Theorem 1 *For given integers d_1 and d_2 ($d_2 \geq d_1 \geq 1$), the system described by (1a)–(4c) is globally asymptotically stable if there are suitable dimensioned matrices $\boldsymbol{Z} = \boldsymbol{Z}^T > \boldsymbol{0}$, $\boldsymbol{Q}_k = \boldsymbol{Q}_k^T > \boldsymbol{0} (k = 1, 2, 3)$, $\boldsymbol{R}_k = \boldsymbol{R}_k^T > \boldsymbol{0} (k = 1, 2)$, any matrix $\begin{bmatrix} \boldsymbol{X}_{11} & \boldsymbol{X}_{12} \\ \boldsymbol{X}_{21} & \boldsymbol{X}_{22} \end{bmatrix}$, positive scalars ε_0, ε_1 such that*

$$\begin{bmatrix} \boldsymbol{R}_2 & \boldsymbol{0} & \boldsymbol{X}_{11} & \boldsymbol{X}_{12} \\ * & 3\boldsymbol{R}_2 & \boldsymbol{X}_{21} & \boldsymbol{X}_{22} \\ * & * & \boldsymbol{R}_2 & \boldsymbol{0} \\ * & * & * & 3\boldsymbol{R}_2 \end{bmatrix} > \boldsymbol{0} \quad (17)$$

and satisfying (18) and (19) simultaneously.

$$\boldsymbol{\Upsilon}(h(r) = d_1) < \mathbf{0} \tag{18}$$

$$\boldsymbol{\Upsilon}(h(r) = d_2) < \mathbf{0} \tag{19}$$

where

$$\Upsilon(h(r)) = \begin{bmatrix} \boldsymbol{\Omega}_{11} + \varepsilon_0 \boldsymbol{C}_0^T \boldsymbol{C}_0 & \mathbf{0} & (\boldsymbol{Z}_3 - \boldsymbol{Z}_2)/2 - 2\boldsymbol{R}_1 & -\boldsymbol{Z}_3/2 & \boldsymbol{\Omega}_{15} + 3\boldsymbol{R}_1 & \boldsymbol{\Omega}_{16} & \boldsymbol{\Omega}_{17} & \boldsymbol{\Omega}_{18} + \boldsymbol{Z}_2^T/2 & \boldsymbol{A}^T\boldsymbol{G} & \mathbf{0} & \mathbf{0} \\ * & \boldsymbol{\Omega}_{12} + \varepsilon_1 \boldsymbol{C}_1^T \boldsymbol{C}_1 & \boldsymbol{\Omega}_{23} & \boldsymbol{\Omega}_{24} & \mathbf{0} & \boldsymbol{\Omega}_{26} & \boldsymbol{\Omega}_{27} & \mathbf{0} & \boldsymbol{A}_d^T\boldsymbol{G} & \mathbf{0} & \mathbf{0} \\ * & * & \boldsymbol{Q}_1 - 4(\boldsymbol{R}_1 + \boldsymbol{R}_2) & \boldsymbol{\Omega}_{34} & \boldsymbol{\Omega}_{35} + 3\boldsymbol{R}_1 & \boldsymbol{\Omega}_{36} + 3\boldsymbol{R}_2 & \boldsymbol{\Omega}_{37} & (\boldsymbol{Z}_3^T - \boldsymbol{Z}_2^T)/2 & \mathbf{0} & \mathbf{0} & \mathbf{0} \\ * & * & * & \boldsymbol{Q}_2 - 4\boldsymbol{R}_2 & -d_1\boldsymbol{Z}_5^T/2 & \boldsymbol{\Omega}_{46} & \boldsymbol{\Omega}_{47} & -\boldsymbol{Z}_3^T/2 & \mathbf{0} & \mathbf{0} & \mathbf{0} \\ * & * & * & * & -3\boldsymbol{R}_1 & \mathbf{0} & \mathbf{0} & d_1\boldsymbol{Z}_2^T/2 & \mathbf{0} & \mathbf{0} & \mathbf{0} \\ * & * & * & * & * & -3\boldsymbol{R}_2 & -\boldsymbol{X}_{22} & (h(r)-d_1)\boldsymbol{Z}_3^T/2 & \mathbf{0} & \mathbf{0} & \mathbf{0} \\ * & * & * & * & * & * & -3\boldsymbol{R}_2 & (d_2 - h(r))\boldsymbol{Z}_3^T/2 & \mathbf{0} & \mathbf{0} & \mathbf{0} \\ * & * & * & * & * & * & * & \boldsymbol{\Omega}_{88} & \mathbf{0} & \mathbf{0} & \mathbf{0} \\ * & * & * & * & * & * & * & * & -\boldsymbol{G} & \boldsymbol{G}\boldsymbol{B}_0 & \boldsymbol{G}\boldsymbol{B}_1 \\ * & * & * & * & * & * & * & * & * & -\varepsilon_0\boldsymbol{I} & \mathbf{0} \\ * & * & * & * & * & * & * & * & * & * & -\varepsilon_1\boldsymbol{I} \end{bmatrix} \tag{20}$$

$$\Omega_{11} = -\boldsymbol{Z}_1 + \left(\boldsymbol{Z}_2 + \boldsymbol{Z}_2^T\right)/2 - 4\boldsymbol{R}_1 + \boldsymbol{Q}_1 + \boldsymbol{Q}_2 + (d_{12} + 1)\boldsymbol{Q}_3 - \Omega_{18} \tag{21}$$

$$\Omega_{15} = d_1(\boldsymbol{Z}_4 - \boldsymbol{Z}_2)/2 \tag{22}$$

$$\Omega_{16} = (h(r) - d_1)(\boldsymbol{Z}_5 - \boldsymbol{Z}_3)/2 \tag{23}$$

$$\Omega_{17} = (d_2 - h(r))(\boldsymbol{Z}_5 - \boldsymbol{Z}_3)/2 \tag{24}$$

$$\Omega_{18} = -\left(d_1^2 \boldsymbol{R}_1 + d_{12}^2 \boldsymbol{R}_2\right) \tag{25}$$

$$\Omega_{22} = -\boldsymbol{Q}_3 - 8\boldsymbol{R}_2 + \boldsymbol{X}_{11} + \boldsymbol{X}_{11}^T + \boldsymbol{X}_{12} + \boldsymbol{X}_{12}^T - \boldsymbol{X}_{21} - \boldsymbol{X}_{21}^T - \boldsymbol{X}_{22} - \boldsymbol{X}_{22}^T \tag{26}$$

$$\Omega_{23} = -2\boldsymbol{R}_2 - \boldsymbol{X}_{11}^T - \boldsymbol{X}_{12}^T - \boldsymbol{X}_{21}^T - \boldsymbol{X}_{22}^T \tag{27}$$

$$\Omega_{24} = -2\boldsymbol{R}_2 - \boldsymbol{X}_{11} + \boldsymbol{X}_{12} + \boldsymbol{X}_{21} - \boldsymbol{X}_{22} \tag{28}$$

$$\Omega_{26} = 3\boldsymbol{R}_2 + \boldsymbol{X}_{21}^T + \boldsymbol{X}_{22}^T \tag{29}$$

$$\Omega_{27} = 3\boldsymbol{R}_2 - \boldsymbol{X}_{12} + \boldsymbol{X}_{22} \tag{30}$$

$$\Omega_{34} = \boldsymbol{X}_{11} - \boldsymbol{X}_{12} + \boldsymbol{X}_{21} - \boldsymbol{X}_{22} \tag{31}$$

$$\Omega_{35} = d_1\left(-\boldsymbol{Z}_4 + \boldsymbol{Z}_5^T\right)/2 \tag{32}$$

$$\Omega_{36} = (h(r) - d_1)(-\boldsymbol{Z}_5 + \boldsymbol{Z}_6)/2 \tag{33}$$

$$\Omega_{37} = (d_2 - h(r))(-\boldsymbol{Z}_5 + \boldsymbol{Z}_6)/2 + \boldsymbol{X}_{12} + \boldsymbol{X}_{22} \tag{34}$$

$$\Omega_{46} = -(h(r) - d_1)\boldsymbol{Z}_6/2 - \boldsymbol{X}_{21}^T + \boldsymbol{X}_{22}^T \tag{35}$$

$$\Omega_{47} = -(d_2 - h(r))\boldsymbol{Z}_6/2 + 3\boldsymbol{R}_2 \tag{36}$$

$$\Omega_{88} = \boldsymbol{Z}_1 - \Omega_{18} - \boldsymbol{G} \tag{37}$$

$$d_{12} = d_2 - d_1 \tag{38}$$

and the matrix $\boldsymbol{G}$ is given by (7a).

Proof Consider the Lyapunov–Krasovskii function [8] as

$$V(\boldsymbol{x}(r)) = \boldsymbol{\Gamma}^T(r)\boldsymbol{Z}\boldsymbol{\Gamma}(r) + \sum_{s=r-d_1}^{r-1} \boldsymbol{x}^T(s)\boldsymbol{Q}_1\boldsymbol{x}(s) + \sum_{s=r-d_2}^{r-1} \boldsymbol{x}^T(s)\boldsymbol{Q}_2\boldsymbol{x}(s)$$
$$+ \sum_{\theta=-d_2}^{-d_1} \sum_{s=r+\theta}^{r-1} \boldsymbol{x}^T(s)\boldsymbol{Q}_3\boldsymbol{x}(s)$$
$$+d_1 \sum_{\theta=-d_1+1}^{0} \sum_{s=r-1+\theta}^{r-1} \boldsymbol{\eta}^T(s)\boldsymbol{R}_1\boldsymbol{\eta}(s) + d_{12} \sum_{\theta=-d_2+1}^{-d_1} \sum_{s=r-1+\theta}^{r-1} \boldsymbol{\eta}^T(s)\boldsymbol{R}_2\boldsymbol{\eta}(s) \tag{39}$$

where

$$\boldsymbol{\eta}(r) = \boldsymbol{x}(r+1) - \boldsymbol{x}(r) = \boldsymbol{f}(\boldsymbol{y}(r)) - \boldsymbol{x}(r) \tag{40}$$

and

$$\boldsymbol{\Gamma}^T(r) = \left[\boldsymbol{x}^T(r) \;\; \sum_{s=r-d_1}^{r-1} \boldsymbol{x}^T(s) \;\; \sum_{s=r-d_2}^{r-d_1-1} \boldsymbol{x}^T(s) \right] \tag{41}$$

Defining

$$\begin{aligned} \Delta V(\boldsymbol{x}(r)) &= V(\boldsymbol{x}(r+1)) - V(\boldsymbol{x}(r)) \\ &= \boldsymbol{\chi}^T(r)\boldsymbol{\Psi}(h(r))\boldsymbol{\chi}(r) + \boldsymbol{x}^T(r)\boldsymbol{Q}_1\boldsymbol{x}(r) \\ &\quad - \boldsymbol{x}^T(r-d_1)\boldsymbol{Q}_1\boldsymbol{x}(r-d_1) + \boldsymbol{x}^T(r)\boldsymbol{Q}_2\boldsymbol{x}(r) \\ &\quad - \boldsymbol{x}^T(r-d_2)\boldsymbol{Q}_2\boldsymbol{x}(r-d_2) + \boldsymbol{x}^T(r)\boldsymbol{Q}_3\boldsymbol{x}(r) + d_{12}\boldsymbol{x}^T(r)\boldsymbol{Q}_3\boldsymbol{x}(r) \\ &\quad - \sum_{s=r-d_2}^{r-d_1} \boldsymbol{x}^T(s)\boldsymbol{Q}_3\boldsymbol{x}(s) + \boldsymbol{\eta}^T(r)\left[d_1^2\boldsymbol{R}_1 + d_{12}^2\boldsymbol{R}_2\right]\boldsymbol{\eta}(r) \\ &\quad - d_1 \sum_{s=r-d_1}^{r-1} \boldsymbol{\eta}^T(s)\boldsymbol{R}_1\boldsymbol{\eta}(s) - d_{12} \sum_{s=r-d_2}^{r-d_1-1} \boldsymbol{\eta}^T(s)\boldsymbol{R}_2\boldsymbol{\eta}(s) \end{aligned} \tag{42}$$

where

$$\begin{aligned} \boldsymbol{\chi}^T(r) = \Big[&\boldsymbol{x}^T(r) \;\; \boldsymbol{x}^T(r-h(r)) \;\; \boldsymbol{x}^T(r-d_1) \;\; \boldsymbol{x}^T(r-d_2) \;\; \boldsymbol{\xi}^T(r,0,d_1) \\ &\boldsymbol{\xi}^T(r,d_1,h(r)) \;\; \boldsymbol{\xi}^T(r,h(r),d_2) \;\; \boldsymbol{f}^T(\boldsymbol{y}(r)) \Big] \end{aligned} \tag{43}$$

$$\boldsymbol{\Psi}(h(r)) = \begin{bmatrix} -\boldsymbol{Z}_1 + (\boldsymbol{Z}_2^T + \boldsymbol{Z}_2)/2 & \boldsymbol{0} & (-\boldsymbol{Z}_2 + \boldsymbol{Z}_3)/2 & -\boldsymbol{Z}_3/2 & \Omega_{15} & \Omega_{16} & \Omega_{17} & \boldsymbol{Z}_2^T/2 \\ * & \boldsymbol{0} & \boldsymbol{0} & \boldsymbol{0} & \boldsymbol{0} & \boldsymbol{0} & \boldsymbol{0} & \boldsymbol{0} \\ * & * & \boldsymbol{0} & \boldsymbol{0} & \Omega_{35} & \Omega_{36} & (d_2 - h(r))(\boldsymbol{Z}_6 - \boldsymbol{Z}_5)/2 & (-\boldsymbol{Z}_2^T + \boldsymbol{Z}_3^T)/2 \\ * & * & * & \boldsymbol{0} & -d_1\boldsymbol{Z}_5^T/2 & -(h(r)-d_1)\boldsymbol{Z}_6/2 & -(d_2 - h(r))\boldsymbol{Z}_6/2 & -\boldsymbol{Z}_3^T/2 \\ * & * & * & * & \boldsymbol{0} & \boldsymbol{0} & \boldsymbol{0} & d_1\boldsymbol{Z}_2^T/2 \\ * & * & * & * & * & \boldsymbol{0} & \boldsymbol{0} & (h(r)-d_1)\boldsymbol{Z}_3^T/2 \\ * & * & * & * & * & * & \boldsymbol{0} & (d_2 - h(r))\boldsymbol{Z}_3^T/2 \\ * & * & * & * & * & * & * & \boldsymbol{Z}_1 \end{bmatrix} \tag{44}$$

and $\boldsymbol{\xi}(r, 0, d_1)$, $\boldsymbol{\xi}(r, d_1, h(r))$, $\boldsymbol{\xi}(r, h(r), d_2)$ are defined by (8).
Now

$$-\sum_{s=r-d_2}^{r-d_1} \boldsymbol{x}^T(s)\boldsymbol{Q}_3\boldsymbol{x}(s) \le -\boldsymbol{x}^T(r-h(r))\boldsymbol{Q}_3\boldsymbol{x}(r-h(r)) \tag{45}$$

Next, by exploiting Lemma 2, the 10th and 11th terms of $\Delta V(\boldsymbol{x}(r))$ are as follows $-d_1\sum_{s=r-d_1}^{r-1}\boldsymbol{\eta}^T(s)\boldsymbol{R}_1\eta(s)\leq$

$$-d_1\begin{bmatrix}\boldsymbol{x}(r)-\boldsymbol{x}(r-d_1)\\ \boldsymbol{x}(r)+\boldsymbol{x}(r-d_1)-\boldsymbol{\xi}(r,0,d_1)\end{bmatrix}^T\begin{bmatrix}\boldsymbol{R}_1 & \boldsymbol{0}\\ \boldsymbol{0} & 3\boldsymbol{R}_1\end{bmatrix}\begin{bmatrix}\boldsymbol{x}(r)-\boldsymbol{x}(r-d_1)\\ \boldsymbol{x}(r)+\boldsymbol{x}(r-d_1)-\boldsymbol{\xi}(r,0,d_1)\end{bmatrix} \tag{46}$$

and

$$\begin{aligned}
-\sum_{s=r-d_2}^{r-d_1-1}\boldsymbol{\eta}^T(s)\boldsymbol{R}_2\boldsymbol{\eta}(s) &= -d_{12}\sum_{s=r-d(k)}^{r-d_1-1}\boldsymbol{\eta}^T(s)\boldsymbol{R}_2\boldsymbol{\eta}(s)-\sum_{s=r-d_2}^{r-h(r)-1}\boldsymbol{\eta}^T(s)\boldsymbol{R}_2\boldsymbol{\eta}(s)\\
&\leq -\frac{(d_2-d_1)}{(h(r)-d_1)}\begin{bmatrix}\boldsymbol{x}(r-d_1)-\boldsymbol{x}(r-h(r))\\ \boldsymbol{x}(r-d_1)+\boldsymbol{x}(r-h(r))-\boldsymbol{\xi}(r,d_1,h(r))\end{bmatrix}^T\\
&\quad\times\begin{bmatrix}\boldsymbol{R}_2 & \boldsymbol{0}\\ \boldsymbol{0} & 3\boldsymbol{R}_2\end{bmatrix}\begin{bmatrix}\boldsymbol{x}(r-d_1)-\boldsymbol{x}(r-h(r))\\ \boldsymbol{x}(r-d_1)+\boldsymbol{x}(r-h(r))-\boldsymbol{\xi}(r,d_1,h(r))\end{bmatrix}\\
&\quad-\frac{(d_2-d_1)}{(d_2-h(r))}\begin{bmatrix}\boldsymbol{x}(r-h(r))-\boldsymbol{x}(r-d_2)\\ \boldsymbol{x}(r-h(r))+\boldsymbol{x}(r-d_2)-\boldsymbol{\xi}(r,h(r),d_2)\end{bmatrix}^T\\
&\quad\times\begin{bmatrix}\boldsymbol{R}_2 & \boldsymbol{0}\\ \boldsymbol{0} & 3\boldsymbol{R}_2\end{bmatrix}\begin{bmatrix}\boldsymbol{x}(r-h(r))-\boldsymbol{x}(r-d_2)\\ \boldsymbol{x}(r-h(r))+\boldsymbol{x}(r-d_2)-\boldsymbol{\xi}(r,h(r),d_2)\end{bmatrix}
\end{aligned} \tag{47}$$

Note that Lemma 3 (reciprocal convexity method) guarantees that if there exists a matrix $\begin{bmatrix}\boldsymbol{X}_{11} & \boldsymbol{X}_{12}\\ \boldsymbol{X}_{21} & \boldsymbol{X}_{22}\end{bmatrix}\in R^{2n\times 2n}$ such that (17) holds true, then the upper bound of (47) can be written as

$$\begin{aligned}
&-\sum_{s=r-d_2}^{r-d_1-1}\boldsymbol{\eta}^T(s)\boldsymbol{R}_2\boldsymbol{\eta}(s)\\
&\leq\begin{bmatrix}\boldsymbol{x}(r-d_1)-\boldsymbol{x}(r-h(r))\\ \boldsymbol{x}(r-d_1)+\boldsymbol{x}(r-h(r))-\boldsymbol{\xi}(r,d_1,h(r))\\ \boldsymbol{x}(r-h(r))-\boldsymbol{x}(r-d_2)\\ \boldsymbol{x}(r-h(r))+\boldsymbol{x}(r-d_2)-\boldsymbol{\xi}(r,h(r),d_2)\end{bmatrix}^T\begin{bmatrix}\boldsymbol{R}_2 & \boldsymbol{0} & \boldsymbol{X}_{11} & \boldsymbol{X}_{12}\\ * & 3\boldsymbol{R}_2 & \boldsymbol{X}_{21} & \boldsymbol{X}_{22}\\ * & * & \boldsymbol{R}_2 & \boldsymbol{0}\\ * & * & * & 3\boldsymbol{R}_2\end{bmatrix}\\
&\times\begin{bmatrix}\boldsymbol{x}(r-d_1)-\boldsymbol{x}(r-h(r))\\ \boldsymbol{x}(r-d_1)+\boldsymbol{x}(r-h(r))-\boldsymbol{\xi}(r,d_1,h(r))\\ \boldsymbol{x}(r-h(r))-\boldsymbol{x}(r-d_2)\\ \boldsymbol{x}(r-h(r))+\boldsymbol{x}(r-d_2)-\boldsymbol{\xi}(r,h(r),d_2)\end{bmatrix}
\end{aligned} \tag{48}$$

By considering (42)–(46) and (48), we have

$$\Delta V(\boldsymbol{x}(r))=\boldsymbol{\chi}^T(r)\boldsymbol{\phi}(h(r))\boldsymbol{\chi}(r)-\beta \tag{49}$$

where

$$\beta = \boldsymbol{y}^T(r)\boldsymbol{G}\boldsymbol{y}(r) - \boldsymbol{f}^T(\boldsymbol{y}(r)\boldsymbol{G}\boldsymbol{f}(\boldsymbol{y}(r))) \tag{50}$$

and

$$\boldsymbol{\phi}(h(r)) = \begin{bmatrix} \Omega_{11} + \bar{\boldsymbol{A}}^T\boldsymbol{G}\bar{\boldsymbol{A}} & \bar{\boldsymbol{A}}^T\boldsymbol{G}\bar{\boldsymbol{A}}_d & (\boldsymbol{Z}_3 - \boldsymbol{Z}_2)/2 - 2\boldsymbol{R} & -\boldsymbol{Z}_3/2 & \Omega_{15} + 3\boldsymbol{R}_1 & \Omega_{16} & \Omega_{17} & \Omega_{18} + \boldsymbol{Z}_2^T/2 \\ * & \Omega_{12} + \bar{\boldsymbol{A}}_d^T\boldsymbol{G}\bar{\boldsymbol{A}}_d & \Omega_{23} & \Omega_{24} & \boldsymbol{0} & \Omega_{26} & \Omega_{27} & \boldsymbol{0} \\ * & * & \boldsymbol{Q}_1 - 4(\boldsymbol{R}_1 + \boldsymbol{R}_2) & \Omega_{34} & \Omega_{35} + 3\boldsymbol{R}_1 & \Omega_{36} + 3\boldsymbol{R}_2 & \Omega_{37} & (\boldsymbol{Z}_3^T - \boldsymbol{Z}_2^T)/2 \\ * & * & * & \boldsymbol{Q}_2 - 4\boldsymbol{R}_2 & -h_1\boldsymbol{Z}_5^T/2 & \Omega_{46} & \Omega_{47} & -\boldsymbol{Z}_3^T/2 \\ * & * & * & * & -3\boldsymbol{R}_1 & \boldsymbol{0} & \boldsymbol{0} & d_1\boldsymbol{Z}_2^T/2 \\ * & * & * & * & * & -3\boldsymbol{R}_2 & -\boldsymbol{X}_{22} & (h(r) - d_1)\boldsymbol{Z}_3^T/2 \\ * & * & * & * & * & * & -3\boldsymbol{R}_2 & (d_2 - h(r))\boldsymbol{Z}_3^T/2 \\ * & * & * & * & * & * & * & \Omega_{88} \end{bmatrix} < 0 \tag{51}$$

In view of Lemma 1, the quantity β (see 50) is greater than or equal to zero. From (49), it is obvious that $\Delta V(\boldsymbol{x}(r)) < 0$ if $\boldsymbol{\phi}(h(r)) < \boldsymbol{0}$ for all $h(r) \in [d_1, \ d_2]$. Hence, $\boldsymbol{\phi}(h(r)) < \boldsymbol{0}$ together with (17) is a sufficient condition for the global asymptotic stability of the system (1a)–(4c).

Using Schur's complement, $\boldsymbol{\phi}(h(r)) < \boldsymbol{0}$ is equivalent to

$$\begin{bmatrix} \Omega_{11} & \boldsymbol{0} & (\boldsymbol{Z}_3 - \boldsymbol{Z}_2)/2 - 2\boldsymbol{R}_1 & -\boldsymbol{Z}_3/2 & \Omega_{15} + 3\boldsymbol{R}_1 & \Omega_{16} & \Omega_{17} & \Omega_{18} + \boldsymbol{Z}_2^T/2 & \bar{\boldsymbol{A}}^T\boldsymbol{G} \\ * & \Omega_{12} & \Omega_{23} & \Omega_{24} & \boldsymbol{0} & \Omega_{26} & \Omega_{27} & \boldsymbol{0} & \bar{\boldsymbol{A}}_d^T\boldsymbol{G} \\ * & * & \boldsymbol{Q}_1 - 4(\boldsymbol{R}_1 + \boldsymbol{R}_2) & \Omega_{34} & \Omega_{35} + 3\boldsymbol{R}_1 & \Omega_{36} + 3\boldsymbol{R}_2 & \Omega_{37} & (\boldsymbol{Z}_3^T - \boldsymbol{Z}_2^T)/2 & \boldsymbol{0} \\ * & * & * & \boldsymbol{Q}_2 - 4\boldsymbol{R}_2 & -h_1\boldsymbol{Z}_5^T/2 & \Omega_{46} & \Omega_{47} & -\boldsymbol{Z}_3^T/2 & \boldsymbol{0} \\ * & * & * & * & -3\boldsymbol{R}_1 & \boldsymbol{0} & \boldsymbol{0} & d_1\boldsymbol{Z}_2^T/2 & \boldsymbol{0} \\ * & * & * & * & * & -3\boldsymbol{R}_2 & -\boldsymbol{X}_{22} & (h(r) - d_1)\boldsymbol{Z}_3^T/2 & \boldsymbol{0} \\ * & * & * & * & * & * & -3\boldsymbol{R}_2 & (d_2 - h(r))\boldsymbol{Z}_3^T/2 & \boldsymbol{0} \\ * & * & * & * & * & * & * & \Omega_{88} & \boldsymbol{0} \\ * & * & * & * & * & * & * & * & -\boldsymbol{G} \end{bmatrix} < \boldsymbol{0} \tag{52}$$

From (4a), condition (52) can be expressed as

$$\boldsymbol{\phi}_0(h(r)) + \bar{\boldsymbol{B}}_0\boldsymbol{F}_0\bar{\boldsymbol{C}}_0 + \bar{\boldsymbol{C}}_0^T\boldsymbol{F}_0^T\bar{\boldsymbol{B}}_0^T < \boldsymbol{0} \tag{53}$$

$$\bar{\boldsymbol{B}}_0^T = \left[\underbrace{\boldsymbol{0}\cdots\boldsymbol{0}}_{8\text{ times}} \ \boldsymbol{B}_0^T\boldsymbol{G}\right], \quad \bar{\boldsymbol{C}}_0 = \left[\boldsymbol{C}_0 \ \underbrace{\boldsymbol{0}\cdots\boldsymbol{0}}_{8\text{ times}}\right] \tag{54}$$

$$\phi_0(h(r)) = \begin{bmatrix} \Omega_{11} & \mathbf{0} & (\boldsymbol{Z}_3-\boldsymbol{Z}_2)/2-2\boldsymbol{R}_1 & -\boldsymbol{Z}_3/2 & \Omega_{15}+3\boldsymbol{R}_1 & \Omega_{16} & \Omega_{17} & \Omega_{18}+\boldsymbol{Z}_2^T/2 & \boldsymbol{A}^T\boldsymbol{G} \\ * & \Omega_{12} & \Omega_{23} & \Omega_{24} & \mathbf{0} & \Omega_{26} & \Omega_{27} & \mathbf{0} & \bar{\boldsymbol{A}}_d^T\boldsymbol{G} \\ * & * & \boldsymbol{Q}_1-4(\boldsymbol{R}_1+\boldsymbol{R}_2) & \Omega_{34} & \Omega_{35}+3\boldsymbol{R}_1 & \Omega_{36}+3\boldsymbol{R}_2 & \Omega_{37} & (\boldsymbol{Z}_3^T-\boldsymbol{Z}_2^T)/2 & \mathbf{0} \\ * & * & * & \boldsymbol{Q}_2-4\boldsymbol{R}_2 & -h_1\boldsymbol{Z}_5^T/2 & \Omega_{46} & \Omega_{47} & -\boldsymbol{Z}_3^T/2 & \mathbf{0} \\ * & * & * & * & -3\boldsymbol{R}_1 & \mathbf{0} & \mathbf{0} & d_1\boldsymbol{Z}_2^T/2 & \mathbf{0} \\ * & * & * & * & * & -3\boldsymbol{R}_2 & -\boldsymbol{X}_{22} & (h(r)-d_1)\boldsymbol{Z}_3^T/2 & \mathbf{0} \\ * & * & * & * & * & * & -3\boldsymbol{R}_2 & (d_2-h(r))\boldsymbol{Z}_3^T/2 & \mathbf{0} \\ * & * & * & * & * & * & * & \Omega_{88} & \mathbf{0} \\ * & * & * & * & * & * & * & * & -\boldsymbol{G} \end{bmatrix} \tag{55}$$

With the help of Lemma 4, (53) can be rewritten as

$$\boldsymbol{\phi}_0(h(r)) + \varepsilon_0^{-1}\overline{\boldsymbol{B}}_0\overline{\boldsymbol{B}}_0^T + \varepsilon_0\overline{\boldsymbol{C}}_0^T\overline{\boldsymbol{C}}_0 < \mathbf{0} \tag{56}$$

where $\varepsilon_0 > 0$. Next, with the aid of Schur's complement, (56) yields

$$\phi(h(r)) = \begin{bmatrix} \Omega_{11}+\varepsilon_0\boldsymbol{C}_0^T\boldsymbol{C}_0 & \mathbf{0} & (\boldsymbol{Z}_3-\boldsymbol{Z}_2)/2-2\boldsymbol{R}_1 & -\boldsymbol{Z}_3/2 & \Omega_{15}+3\boldsymbol{R}_1 & \Omega_{16} & \Omega_{17} & \Omega_{18}+\boldsymbol{Z}_2^T/2 & \boldsymbol{A}^T\boldsymbol{G} & \mathbf{0} \\ * & \Omega_{12} & \Omega_{23} & \Omega_{24} & \mathbf{0} & \Omega_{26} & \Omega_{27} & \mathbf{0} & \bar{\boldsymbol{A}}_d^T\boldsymbol{G} & \mathbf{0} \\ * & * & \boldsymbol{Q}_1-4(\boldsymbol{R}_1+\boldsymbol{R}_2) & \Omega_{34} & \Omega_{35}+3\boldsymbol{R}_1 & \Omega_{36}+3\boldsymbol{R}_2 & \Omega_{37} & (\boldsymbol{Z}_3^T-\boldsymbol{Z}_2^T)/2 & \mathbf{0} & \mathbf{0} \\ * & * & * & \boldsymbol{Q}_2-4\boldsymbol{R}_2 & -d_1\boldsymbol{Z}_5^T/2 & \Omega_{46} & \Omega_{47} & -\boldsymbol{Z}_3^T/2 & \mathbf{0} & \mathbf{0} \\ * & * & * & * & -3\boldsymbol{R}_1 & \mathbf{0} & \mathbf{0} & d_1\boldsymbol{Z}_2^T/2 & \mathbf{0} & \mathbf{0} \\ * & * & * & * & * & -3\boldsymbol{R}_2 & -\boldsymbol{X}_{22} & (h(r)-d_1)\boldsymbol{Z}_3^T/2 & \mathbf{0} & \mathbf{0} \\ * & * & * & * & * & * & -3\boldsymbol{R}_2 & (d_2-h(r))\boldsymbol{Z}_3^T/2 & \mathbf{0} & \mathbf{0} \\ * & * & * & * & * & * & * & \Omega_{88} & \mathbf{0} & \mathbf{0} \\ * & * & * & * & * & * & * & * & -\boldsymbol{G} & \boldsymbol{G}\boldsymbol{B}_0 \\ * & * & * & * & * & * & * & * & * & -\varepsilon_0\boldsymbol{I} \end{bmatrix} < 0 \tag{57}$$

Next, by following the steps similar to (53)–(57), one can easily show that (57) is equivalent to $\boldsymbol{\Upsilon}(h(r)) < \mathbf{0}$. Under the property of matrix function affinity, the condition $\boldsymbol{\Upsilon}(h(r)) < \mathbf{0}$ if and only if (18) and (19) hold true. This concludes the proof.

Table 1 Comparison based on number of decision variables

Methods	Number of decision variables For $n = 2$ (second-order system)
Theorem 7 [6]	78
Theorem 1 (proposed)	56

Remark 1 In Table 1, based on the number of decision variables and for a case where $n = 2$, a comparison between Theorem 7 of [6] and the proposed Theorem 1 is shown. In general, the total number of decision variables for Theorem 7 [6] is given by $14n^2 + 5n + 2$ and for the proposed criterion, the number of decision variables is $12n^2 + 4n + 2$. It is clear that the requirement of decision variables for proposed criterion is less than Theorem 7 of [6]. Consequently, the proposed criterion has ensured less complexity than previous criterion [6].

Remark 2 In this paper, the proposed criterion (Theorem 1) is based on LMIs and, hence, one can easily trace the feasibility of the conditions given in Theorem 1 with the help of MATLAB and YALMIP 3.0 parser [19, 20].

Remark 3 The present work can be prolonged for the robust stability and $\boldsymbol{H}_{\infty}$ performance of discrete-time systems having external disturbances. This work may also be extended for two-dimensional and multi-dimensional systems involving generalized overflow nonlinearities and time-varying delays.

4 Numerical Example

The efficacy of the proposed criterion is shown by considering the following numerical example.

Consider the system (1a)–(4c) with

$$L = -1, \ \mathbf{A} = \begin{bmatrix} 0 & 0 \\ 0.05 & 0.9 \end{bmatrix}, \ \mathbf{A}_d = \begin{bmatrix} 0.01 & 0 \\ 0 & 0.02 \end{bmatrix}, \ \mathbf{B}_0 = \mathbf{B}_1 = \begin{bmatrix} 0 \\ 0.1 \end{bmatrix}$$
$$\mathbf{C}_0 = \begin{bmatrix} 0.01 & 0 \end{bmatrix}, \ \mathbf{C}_1 = \begin{bmatrix} 0 & 0.01 \end{bmatrix} \tag{58}$$

With the help of MATLAB and YALMIP 3.0 parser [19, 20], it is found that the conditions given in Theorem 1 are feasible for a delay range $2 \leq h(r) \leq 23$. Therefore, according to Theorem 1, the system (1a)–(4c) and (58) is globally asymptotically stable. For this example, it is also checked that Theorem 7 of [6] exhibits same level of conservatism as that of Theorem 1 but with computationally more demanding.

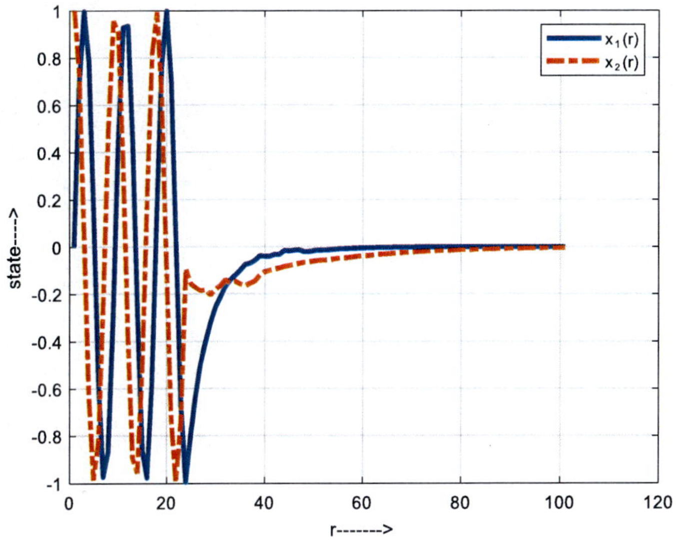

Fig. 1 State trajectories of the system in numerical example

By assuming random initial conditions, a plot (see Fig.1) for the state trajectories of the system under consideration validates the findings of Theorem 1 for the parameters taken in numerical example along with $\boldsymbol{F}_0 = \boldsymbol{F}_1 = 1$ and $2 \leq h(r) \leq 23$.

5 Conclusion

A less complex delay-dependent stability criterion for the stability of the discrete-time uncertain systems having generalized overflow nonlinearities and time-varying delays has been proposed. A comparison with the previously reported criterion with respect to number of decision variables is also given. The efficacy of the proposed criterion is numerically proved.

References

1. Wu L, Lam J, Yao X, Xiong J (2011) Robust guaranteed cost control of discrete-time networked control systems. Opt Control Appl Methods 32(1):95–112. https://doi.org/10.1002/oca.932
2. Li X, Zhang X, Wang X (2017) Stability analysis for discrete-time Markovian jump systems with time-varying delay: a homogeneous polynomial approach. IEEE Access 5:27573–27581. https://doi.org/10.1109/ACCESS.2017.2775606

3. Zhang CK, He Y, Jiang L, Wang QG, Wu M (2017) Stability analysis of discrete-time neural networks with time-varying delay via an extended reciprocally convex matrix inequality. IEEE Trans Cybern 47(10):3040–3049. https://doi.org/10.1109/TCYB.2017.2665683
4. Zhang D, Shi P, Zhang WA, Yu L (2017) Energy-efficient distributed filtering in sensor networks: a unified switched system approach. IEEE Trans Cybern 47(7):1618–1629. https://doi.org/10.1109/TCYB.2016.2553043
5. Chen WH, Guan ZH, Lu X (2003) Delay-dependent guaranteed cost control for uncertain discrete-time systems with delay. IEE Proc Control Theory Appl 150(4):412–416. https://doi.org/10.1049/ip-cta:20030572
6. Kandanvli VKR, Kar H (2012) A delay-dependent approach to stability of uncertain discrete-time state-delayed systems with generalized overflow nonlinearities. ISRN Comput Math. https://doi.org/10.5402/2012/171606
7. Mary TJ, Rangarajan P (2016) Delay-dependent stability analysis of microgrid with constant and time-varying communication delays. Electric Power Compon Syst 44(13):1441–1452. https://doi.org/10.1080/15325008.2016.1170078
8. Tadepalli SK, Kandanvli VKR, Vishwakarma A (2017) Criteria for stability of uncertain discrete-time systems with time-varying delays and finite wordlength nonlinearities. Trans Inst Meas Control 40(9):2868–2880. https://doi.org/10.1177/0142331217709067
9. Park P, Ko JW, Jeong C (2011) Reciprocally convex approach to stability of systems with time-varying delays. Automatica 47(1):235–238. https://doi.org/10.1016/j.automatica.2010.10.014
10. Kandanvli VKR, Kar H (2008) Robust stability of discrete-time state-delayed systems employing generalized overflow nonlinearities. Nonlinear Anal: Theory, Methods Appl 69(9):2780–2787. https://doi.org/10.1016/j.na.2007.08.050
11. Guan X, Lin Z, Duan G (1999) Robust guaranteed cost control for discrete-time uncertain systems with delay. IEE Proc Control Theory Appl 146(6):598–602. https://doi.org/10.1049/ip-cta:19990714
12. Bakule L, Rodellar J, Rossell JM (2006) Robust overlapping guaranteed cost control of uncertain state-delay discrete-time systems. IEEE Trans Automat Control 51(12):1943–1950. https://doi.org/10.1109/TAC.2006.886536
13. Xu S (2002) Robust H_∞ filtering for a class of discrete-time uncertain nonlinear systems with state delay. IEEE Trans Circuits Syst I 49(12):1853–1859. https://doi.org/10.1109/TCSI.2002.805736
14. Dey A, Kar H (2011) LMI-based criterion for the robust stability of 2D discrete state-delayed systems using generalized overflow nonlinearities. J Control Sci Eng p 23. https://doi.org/10.1155/2011/271515
15. Kandanvli VKR, Kar H (2009) Robust stability of discrete-time state-delayed systems with saturation nonlinearities: linear matrix inequality approach. Signal Process 89(2):161–173. https://doi.org/10.1016/j.sigpro.2008.07.020
16. Nam PT, Pathirana PN, Trinh H (2015) Discrete Wirtinger-based inequality and its application. J Franklin Inst 352(5):1893–1905. https://doi.org/10.1016/j.jfranklin.2015.02.004
17. Xu S, Lam J, Lin Z, Galkowski K (2002) Positive real control for uncertain two-dimensional systems. IEEE Trans Circuits Syst I: Fundam Theory Appl 49(11):1659–1666. https://doi.org/10.1109/TCSI.2002.804531
18. Rani P, Kokil P, Kar H (2017) l_2–$l\infty$ Suppression of limit cycles in interfered digital filters with generalized overflow nonlinearities. Circuits, Syst, Signal Process 36(7):2727–2741. https://doi.org/10.1007/s00034-016-0433-1

19. Boyd S, Ghaoui L, El Feron E, Balakrishnan V (1994) Linear matrix inequalities in system and control theory. SIAM, Philadelphia, PA
20. Lofberg J (2004) YALMIP: a toolbox for modeling and optimization in MATLAB. In: 2004 IEEE international symposium on computer aided control systems design. IEEE, New Orleans, LA, USA, pp 284–289 (2001). https://doi.org/10.1109/cacsd.2004.1393890

PAPR Reduction in OFDM for VLC System

Shilpi, M. Shukla, and Arvind Kumar

Abstract It has been reported in the literature that the orthogonal frequency division multiplexing (OFDM) scheme is an appealing modulation scheme for visible light communication (VLC) systems. OFDM systems having a lower bit error rate (BER) and advantage of multicarrier modulation while having a major disadvantage of the peak-to-average power ratio (PAPR) problem. Hence, to reduce PAPR, several PAPR reduction techniques are given in this work which are clipping and filtering scheme and μ-law mapping scheme for the optical orthogonal frequency division multiplexing (OOFDM) system. The μ-law mapping scheme gives low PAPR signals. After analysis and comparison of simulation results for both the schemes, it is found that the μ-law mapping scheme has given better result as compare to clipping and filtering scheme.

Keywords Optical orthogonal frequency division multiplexing · Visible light communication · Peak-to-average power ratio · μ-law mapping · Clipping and filtering

1 Introduction

In recent years, one of the ideas put forward for wireless optical communication is the visible light communication method. The signals with 380–780 nm (790–430TH_z) wavelength interval of the electromagnetic spectrum are the light signals that can be detected by the human eye. It is possible to achieve illumination and data transfer simultaneously by means of light-emitting diodes (LEDs) that is the

Shilpi (✉) · A. Kumar
Motilal Nehru National Institute of Technology Allahabad, Allahabad, India
e-mail: shilpi9e2@gmail.com

A. Kumar
e-mail: arvindk@mnnit.ac.in

M. Shukla
Harcourt Butler Technical University, Kanpur, India

D. Harvey et al. (eds.), *Advances in VLSI, Communication, and Signal Processing*, Lecture Notes in Electrical Engineering 683,
https://doi.org/10.1007/978-981-15-6840-4_18

prominent lighting equipment. For VLC systems, at the transmitter side, LEDs or laser diodes are used. LEDs are most preferred devices due to low cost and incoherent detection while at the receiver side photodiodes are most preferred devices for direct detection of optical power [1, 2]. The OFDM is the most suitable technique for VLC systems but it suffers from a major problem of PAPR. Due to the high value of the PAPR, system performance degrades and BER rate becomes increases. Hence, to reduce the PAPR, reduction schemes are used as clipping and filtering scheme, μ-law mapping scheme.

The remaining part of this paper is organized as follows. In Sect. 2, OOFDM modulation scheme is discussed, in Sect. 3, the PAPR problem is discussed, and in Sect. 4, PAPR reduction schemes are given. Then in Sect. 5, simulation results of the PAPR reduction schemes are given and also a comparison of these schemes is given. Finally, Sect. 6 of the paper concludes the work.

2 OOFDM Modulation Scheme

Figure 1 displays the block diagram of an OOFDM system. First, a higher rate of serial data is partitioned into N parallel data streams with lower rates. Each data stream is then mapped on a complex value using binary phase shift keen (BPSK) modulation assigned with Hermitian symmetry for the reality of the time signal [3, 4].

Performing an N-point inverse fast Fourier transform (IFFT) on X produces real-time samples $x(n)$ as:

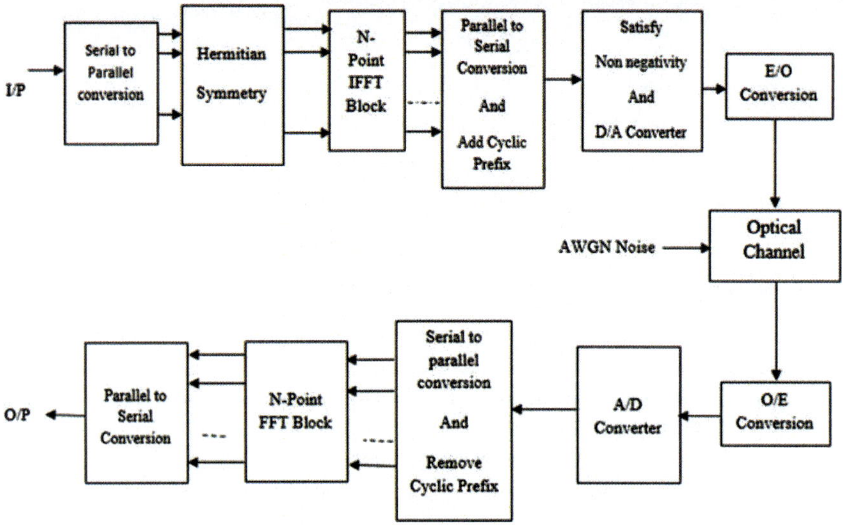

Fig. 1 Block diagram of OOFDM system

$$x(n) = \frac{1}{N}\sum_{k=0}^{N-1} X_k \exp\left(\frac{j2\pi nk}{N}\right), \quad n = 0, 1 \ldots N-1 \tag{1}$$

The resulting real-time samples are then converted back to serial, and a cyclic prefix (CP) is transmitted during the guard interval to allow linear convolution to be modeled as the circular convolution of the channel. However, the time samples are not still appropriate for transmission through the channel and the non-negativity of them must be satisfied. This is done in the block referred to as satisfy non-negativity block [5].

After satisfying the non-negativity constraint in OOFDM systems, the time samples are passed through a digital to analog converter and converted to an optical signal, $x(t)$ using an LED. The resulting optical signal is then transmitted through the optical channel. At the receiver, the demodulation and decoding of data will be performed.

3 PAPR Problem

The transmit signals in an OOFDM system can have high peak values in the time domain since many subcarrier components are added via an IFFT operation. As a result, OOFDM systems are known to have a high PAPR when compared to single carrier systems [5, 6]. In fact, the high PAPR in an OFDM system decreases the signal-to-noise ratio (SNR) of the analog-to-digital (A/D) converter and digital-to-analog (D/A) converter while degrading the efficiency of the power amplifier in the transmitter [7, 8].

The PAPR of OFDM signal $x(n)$ is expressed by the following formula [9]:

$$\text{PAPR (dB)} = \frac{P_{\text{Peak}}}{P_{\text{avg}}} = 10\log_{10}\left(\frac{\max[x(n)x^*(n)]}{E[x(n)x^*(n)]}\right) \tag{2}$$

where * corresponds to the conjugate operator.

4 PAPR Reduction Schemes

4.1 *Clipping and Filtering Scheme*

This technique of PAPR reduction is the easiest and most widely used technique. In this, clip the parts of the signals that are outside the allowed region.

Clipping operation is given [10]:

$$y = \begin{cases} -A \text{ if } x \leq -A \\ x \quad \text{if } |x| < A \\ A \quad \text{if } x \geq A \end{cases} \tag{3}$$

First of all, the maximum value of the amplitude is found, then we evaluate:

$$A = \frac{\max(x)}{\sqrt{2}} \tag{4}$$

Generally, the clipping operation is performed at the transmitter. However, the receiver needs to measure the clipping that has occurred and to compensate the received OFDM symbol accordingly. The clipping method introduces both in-band distortions and out of band radiation into OFDM signals, which degrades the system performance including BER and spectral efficiency [11].

4.2 μ-Law Mapping Scheme

Mapping techniques are used to decrease the dynamic range of the signal in order to prevent it from distortions caused by channel with limited range [12]. The mapping technique compresses the signal, making its distribution quasi-uniform, such that the maximum amplitude of the signals does not exceed the limitations of system. Thereby, no distortions will occur at the bottlenecks. At the receiver side, the original signal is obtained by reverse operation of expanding.

In the μ-law mapping, a compressor compresses the signal at the transmitter side according to the following formula [12]:

$$s_n^{'} = \frac{\max(s_n)\ln\left(1 + \frac{\mu|s_n|}{\max(s_n)}\right)}{\ln(1+\mu)}\operatorname{sgn}(s_n) \tag{5}$$

where μ is the μ-law compression parameter. At the receiver side, μ-law expander restores the original signal [12].

Figure 2 demonstrates the μ-OFDM system. The mapping procedure is an effective technique to compress more samples into the LEDs dynamic range, due to the fact that it has been ended up to be a feasible strategy to compress the dynamic range of OFDM signals.

In the above equation, $x(n)$ represents the nth time domain sample, $X(k)$ is the kth frequency domain sample while the parameter N denotes the number of subcarriers.

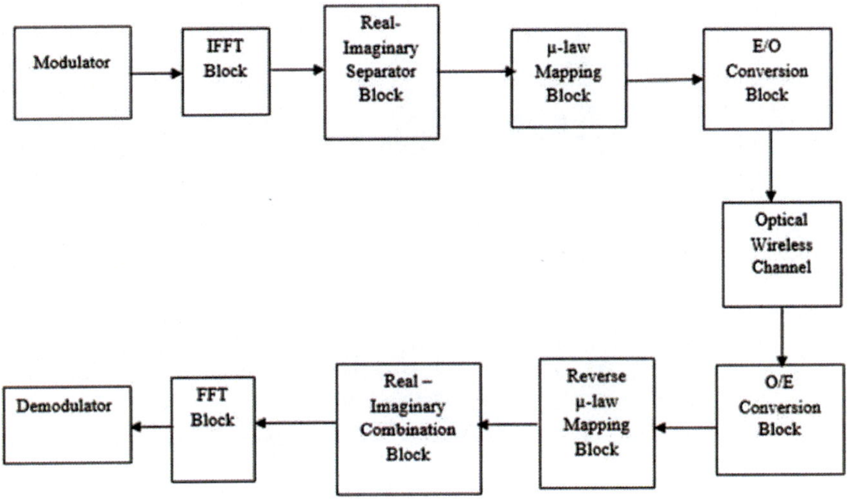

Fig. 2 Block diagram of μ-law mapping scheme [12]

5 Simulation Results

In this section, simulation results of normal OOFDM signal and also after applying PAPR reduction schemes as clipping and filtering scheme, μ-law mapping scheme is shown. In the end, a comparison of PAPR reduction schemes is represented by a bar chart.

In Fig. 3, OOFDM signal plot in terms of amplitude versus bin is shown. Here, a large variation in amplitude is observed. The maximum observed value is 0.0429, and the minimum observed value is 3.7561×10^{-4}. To tackle this large variation,

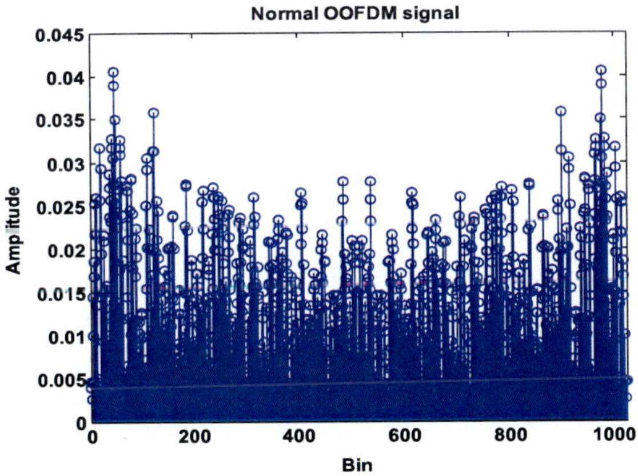

Fig. 3 OOFDM signal plot in terms of amplitude versus bin

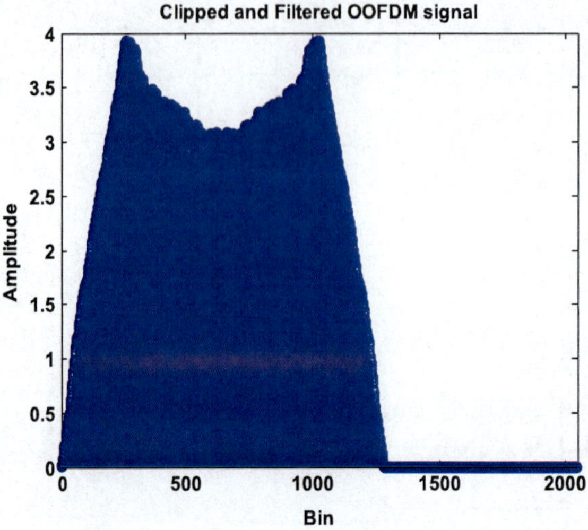

Fig. 4 Clipped and filtered scheme-based OOFDM signal plot in terms of amplitude versus bin

clipping is used. Here, if particular amplitude value is higher than 0.7 × maximum amplitude value, then it is brought down to 0.7 × maximum amplitude value, thus amplitude values above this are clipped. Now, the maximum value is 0.03.

In Fig. 4, clipped and filtered OOFDM signal plot in terms of amplitude versus bin is shown. Here, variation in amplitude has been reduced significantly. Due to the convolution gain of the filter, the amplitude has increased significantly, now the maximum value is 3.86. The PAPR observed for clipped and filtered OOFDM is 10.3900 dB for 256 subcarriers.

In Fig. 5, μ-law-based OOFDM signal plot in terms of amplitude vs. bin is shown, with a maximum amplitude as 0.0418. However, using μ-law the variation in amplitude gets reduced significantly. The PAPR value of μ-law mapping-based OOFDM is 3.1402 dB for 256 subcarriers. Thus, it can be summarized that the performance in terms of PAPR reduction is more in the μ-law mapping-based OOFDM.

In Table 1, the results of PAPR are obtained under three cases by considering different values of subcarriers:

1. PAPR of normal OOFDM (dB)
2. PAPR of clipped and filtered scheme-based OOFDM (dB)
3. PAPR of μ-law mapping scheme-based OOFDM (dB).

It is clear from Table 1 that, the μ-law mapping gives better results than clipping and filtering to a great extent.

In Fig. 6, the comparison of PAPR reduction schemes is done for different no. of subcarriers. It is clear from the figure that, in all the cases of subcarriers as for $N =$ 256,512,1024, the μ-law mapping scheme of PAPR reduction gives better results as compare to another PAPR reduction scheme.

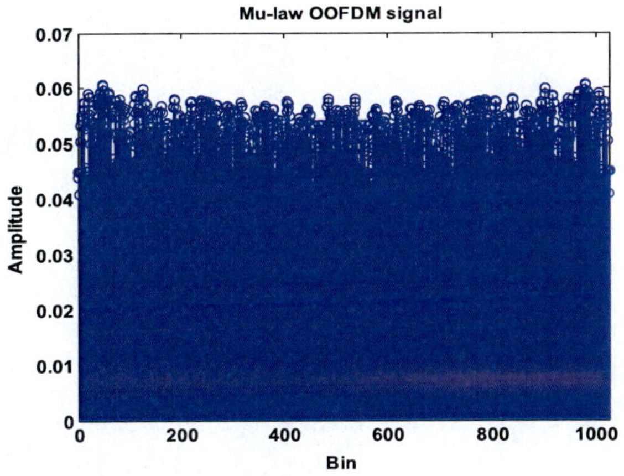

Fig. 5 μ-law mapping-based OOFDM signal plot in terms of amplitude versus bin

Table 1 Comparison of different PAPR reduction schemes for OOFDM

No. of subcarriers	PAPR of original OOFDM (dB)	PAPR of clipped and filtered OOFDM (dB)	PAPR of μ-law mapped OOFDM (dB)
256	19.0542	10.3900	3.1402
512	21.5603	9.8246	3.4424
1024	23.0988	9.2076	3.6965

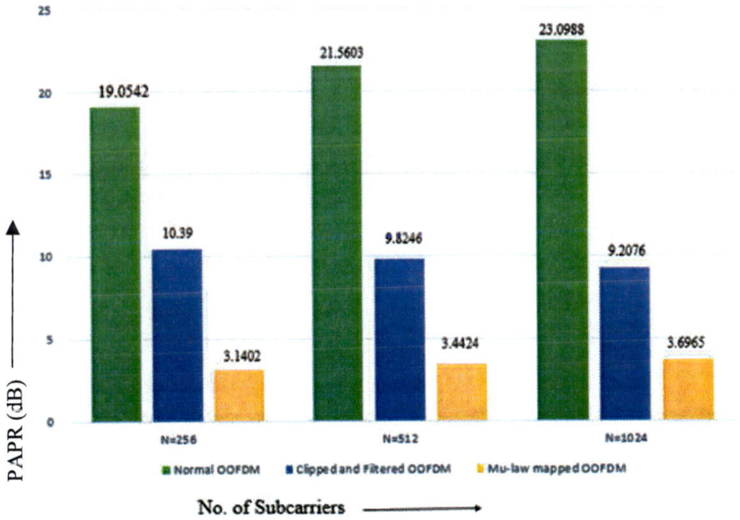

Fig. 6 Comparison of PAPR reduction schemes for different no. of subcarriers

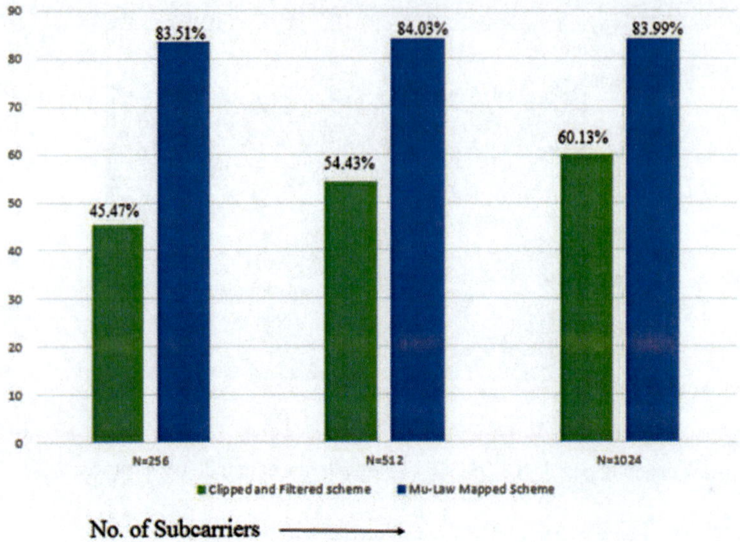

Fig. 7 Percentage variation of PAPR reduction schemes from original OOFDM

In Fig. 7, the percentage variation of PAPR reduction schemes for normal OOFDM is given. It is clear from the figure that the maximum reduction of PAPR as almost 84% is done in the case of the μ-law mapping scheme from normal OOFDM.

6 Conclusions

In OOFDM, PAPR is a problem that degrades the performance of the overall system. To combat the PAPR of the OOFDM system, various schemes such as clipping and filtering and μ-law mapping are discussed. It has been found that the performance of the μ-law mapping scheme is much superior as outperforms 7 dB in terms of PAPR in comparison to clipping and filtering scheme. A combination of both the schemes gives better results which will be presented in the future.

References

1. Medina C, Zambrano M, Navarro K (2015) LED-Based visible light communication: technology, applications and challenges—a survey. Int J Adv Eng Technol 8(4):482–495
2. Bhakte AJ, Bondre KD (2016) Review paper on reduction techniques of PAPR in VLC OFDM system. Int J Recent Trends Eng Res (IJRTER) 02(03):14–17
3. Cevik T, Yilmaz S (2015) An Overview of visible light communication systems. Int J Comput Netw Commun (IJCNC) 7(6):139–150

4. Ghassemlooy Z, Popoola W, Raj Bhandari S (2012) Optical wireless communications system and channel modelling with MATLAB. CRC Press Taylor & Francis, Boca Raton, London, New York
5. Asadzadeh K (2011) Efficient OFDM signaling schemes for visible light communication systems. Ph.D. Thesis, Sharif University Of Technology, Tehran, Iran
6. Sengar S, Pratim Bhattacharya P (2012) Performance improvement in OFDM system by PAPR reduction. Signal Image Process Int J (SIPIJ) 3(2):157–169
7. Abdulkafi AA, Alias MY, Hussein YS (2007) A novel approach for PAPR reduction in OFDM-based visible light communications. In: International conference on platform technology and Service, pp 1–4
8. Hu WW, Lee DH (2017) PAPR reduction for visible light communication systems without side information. IEEE Photon J PP(99):1–10
9. Islam MA, Ahmed N, Ahamed NU, Rahman M, Aljunid SA (2017) PAPR reduction in an OFDM system using recursive clipping and filtering technique. Wseas Trans Commun 13:291–297
10. Nakamura Y, Saito M (2013) Evaluation of clipping effect on OFDM signals for visible light communications. In: International symposium on intelligent signal processing and communication systems, pp 753–757
11. Yu Z, Baxley RJ, Zhou GT (2014) Iterative clipping for PAPR reduction in visible light OFDM communications. In: IEEE military communication conference, pp 1681–1686
12. Yang Y, Zeng Z, Fengy S, Guo C (2016) A simple OFDM scheme for VLC systems based on μ-law mapping. IEEE Photonics Technol Lett 28(6):641–644

Performance Evaluation of Energy-Efficient Adiabatic Logic Circuit-Based Multiplexer for Low Power Applications

Shivangi Jaiswal, Prashasti, Anam Khan, and Subodh Wairya

Abstract This paper presents the energy-efficient adiabatic logic (EEAL)-based logic circuits that allows VLSI designer to design and verify various complex units into simpler one. It examines low power characteristics and energy reduction benefits of adiabatic logic circuits using sinusoidal power supply. The design and evaluation of multiplexer using EEAL logic family have been illustrated. Due to the clock skew issues and high complexity in some of the adiabatic logic styles such as ECRL, CAL, PFAL, etc., they are not effective for high speed operations. So, it is focusing on high speed operations as well as energy recovery by using EEAL adiabatic logic style with efficient power clock power consumption. The power dissipation, delay and energy consumption of proposed EEAL-based MUX has been compared with the conventional CMOS-based MUX. The circuits are designed and verified by Cadence Virtuoso EDA tool using 45 nm technology at 1 V.

Keywords EEAL · Sinusoidal power clock · Multiplexer · Energy efficient

1 Introduction

With the increase of operating speed and density of CMOS chips, energy efficiency as well as power dissipation has became a matter of interest in designing of various modern digital VLSI circuits, mainly in portable devices. As the demand for low power system is increasing day by day, so the adiabatic switching style [1–4] has recently became a center of attention for the designers. Adiabatic logic circuits are the energy recovery circuit which uses adiabatic switching style to achieve energy efficiency by conducting currents and stores the energy across devices by slow charging and discharging of capacitor using a time-varying power clock. These adiabatic

S. Jaiswal (✉) · Prashasti · A. Khan · S. Wairya
Institute of Engineering and Technology, AKTU, Lucknow, Uttar Pradesh, India
e-mail: shivi.19494@gmail.com

Prashasti
e-mail: prashasti311@gmail.com

D. Harvey et al. (eds.), *Advances in VLSI, Communication, and Signal Processing*, Lecture Notes in Electrical Engineering 683,
https://doi.org/10.1007/978-981-15-6840-4_19

circuit stops operating at very high frequency but at low operating frequency, they operates efficiently and effectively [5].

In the conventional CMOS logic design, the charging and discharging of output capacitor node causes energy transition from power supply to output and from output to ground that result in power dissipation. This lost energy needs to be recovered or recycled to the supply voltage so as to decrease the power dissipation by using a technique known as adiabatic technique. Several adiabatic techniques such as efficient charge recovery logic (ECRL), clocked adiabatic logic (CAL), source coupled adiabatic logic (SCAL), pass-transistor adiabatic logic (PAL), 2N-2N2P [6–10] uses single or multiphase clocking schemes [11–15] to recover the charge that are more sensitive to clock skew and limits their high frequency performances.

EEAL is a promising design approach for ultra-low power applications which uses sinusoidal AC power supply to recover the energy of digital logic circuits. In today's era, the portable devices made up of electronic components are mainly used in every application. To run the portable devices for longer time, the power consumption must be reduced which is possible by using low power consumption techniques for designing such circuits [16]. Mostly, CMOS-based logic circuit utilizes less power but instead of that also [17], to get the ultra-low power consumption, a latest technology has been proposed known as energy-efficient adiabatic logic by Chanda et al. [18]. This newly proposed EEAL is implemented by a Differential Cascode Voltage Swing (DCVS) [19] with a single-phase power clock for correct operation. For low power and proper synchronization, these logic circuits uses single phase of sinusoidal AC power clock [20] having a voltage swing of 0–V_{dd}. It requires simple clocking schemes to achieve minimum control overheads. Due to the simpler implementation and resembled characteristics of EEAL, it decreases transistor overheads and reduces circuit complexity. It is necessary to establish various adiabatic approaches and semi-automatic tools to reduce the energy efficiency and power consumption in an easy and efficient way that makes the logic circuit friendlier to VLSI designers and which allows the designing and verification of complex circuits in a short period of time.

A parameter, adiabatic gain has been introduced to compare between the adiabatic circuits and CMOS logic circuits which gives the idea about how much energy is saved. The adiabatic gain is defined as the ratio of energy dissipation by a conventional CMOS circuit to the energy dissipation by adiabatic circuit [18].

$$\text{Adiabatic Gain} = \frac{\text{Energy dissipation by a Conventional CMOS Circuit}}{\text{Energy dissipation by Adiabatic Logic Circuit}} \tag{1}$$

2 Energy Efficiency Analysis in EEAL Inverter

For designing an inverter based on EEAL, the DCVS networks [21, 22] have been replaced by NMOS transistors. The DCVS network consists of N1, N2 that are connected with the sinusoidal power supply and N3, N4 that are connected to ground. PMOS transistors P1, P2 are cross-coupled with each other and connected in between two DCVS networks with a sinusoidal power clock connected across it as shown in Fig. 1. For the charging and discharging of capacitor at output node, the transistors N1 and P2 or N3 and P1 forms a parallel path. To control the parallel combinations of both NMOS and PMOS, both positive and negative signal polarities are needed, so dual rail input is encoded to the DCVS network. The encoded output is also dual rail, and it makes the data independent and simplifies the design circuitry.

When the sinusoidal power clock swings from V_{dd} to ground, the output node 'Outb' starts discharging through the same path and 'Out' remains at ground. To solve the floating output problem completely, a complete path is mandatory between the power clock and output nodes. When the low logic level 'logic 0' is passed through the PMOS transistor and high logic level 'logic 1' is passed through the NMOS transistor, then the threshold losses occurs. Hence, since initially logic '0' is passed through NMOS and later on logic '1' will pass through the parallel path of both NMOS and PMOS transistors, there will be small voltage drop which is negligible across the parallel path. Since the threshold loss is directly proportional to the square of small voltage drop (ΔV), so the threshold losses are also almost zero which makes the EEAL technique more robust in nature [18].

In this paper, the energy efficiency analysis of EEAL-based inverter proposed by Chanda et al. [18] has been illustrated and compared with its conventional CMOS-based circuit. The simulations are done on Cadence Virtuoso EDA tool using 45 nm technology at 1 V power clock. In the existing literature, the author had performed simulations in Cadence environment at 90 nm technology with the supply voltage of 2 V. The energy consumption and power dissipation has been reduced as compared to the existing literature. The waveforms of EEAL-based inverter and conventional CMOS-based inverter are shown in Fig. 2a, b, respectively.

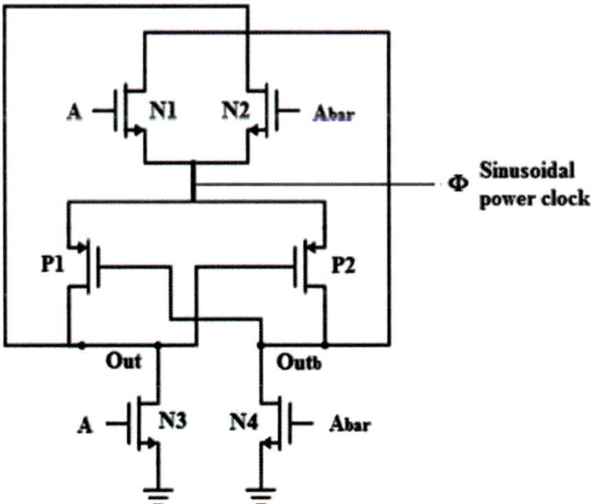

Fig. 1 EEAL inverter circuit

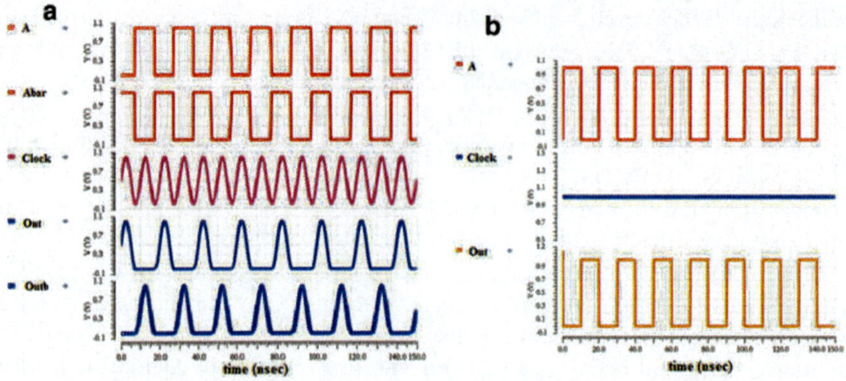

Fig. 2 **a** Waveform of EEAL inverter. **b** Waveform of CMOS inverter

Table 1 Comparison between EEAL inverter and CMOS inverter

Logic style	Power dissipation (uWatt)	Delay (ns)	Energy consumption (FJoule)
Conventional CMOS inverter	14.7	5.15	75.705
EEAL inverter	1.62	3.65	5.913

The parameters such as power, delay and energy have been calculated and compared with its conventional circuit as shown in Table 1.

From Table 1, it is clear that this EEAL technology has low energy consumption, low delay and low power dissipation as compared to the conventional CMOS circuit. The adiabatic gain for the inverter is 12.8 and the energy efficiency of EEAL-based inverter is 92.18%. It has been verified that the chosen methodology gives best performance in terms of power dissipation, energy consumption and delay as it uses AC sinusoidal power clock which act as a power supply and clock both at a time. So by using this latest adiabatic logic style, multiplexer has been proposed so as to achieve better energy efficiency and can be further used for ultra-low power applications.

3 Proposed Circuit

3.1 EEAL-Based 2:1 Multiplexer

Multiplexer is a combinational circuit that has multiple input switch and single output switch. They are the data selectors which uses many signals to share one device having single communication line. In electronics, it is a device that selects between multiple input lines and forward that signal to single output signal by using select line [23].

Multiplexers are mainly used as programmable logic devices (PLDs) and also used to transfer data in the field of digital communication. They are used to transfer huge amount of data for a certain time and certain bandwidth over a network. For n number of select lines, a multiplexer has 2^n number of inputs. The data selector is mainly used to transfer the data from input to output in a parallel to serial manner. When S (select line) is low, it selects B signal at the output and for high select line, A is selected at the output.

$$Out = \overline{S}B + SA \tag{2}$$

By using NMOS-based differential cascode voltage swing (DCVS) network, the implementation of various complex gates becomes easy. By simply replacing the DCVS network, 2:1 MUX can be implemented with EEAL topology. The proposed circuit of 2:1 MUX has two outputs 'Out' and 'Outb' which are compliment to each other. A and B are inputs, S is the select line, Φ is the power clock which is sinusoidal so the output is also sinusoidal. The circuit diagram of proposed EEAL-based 2:1 multiplexer is shown in Fig. 3.

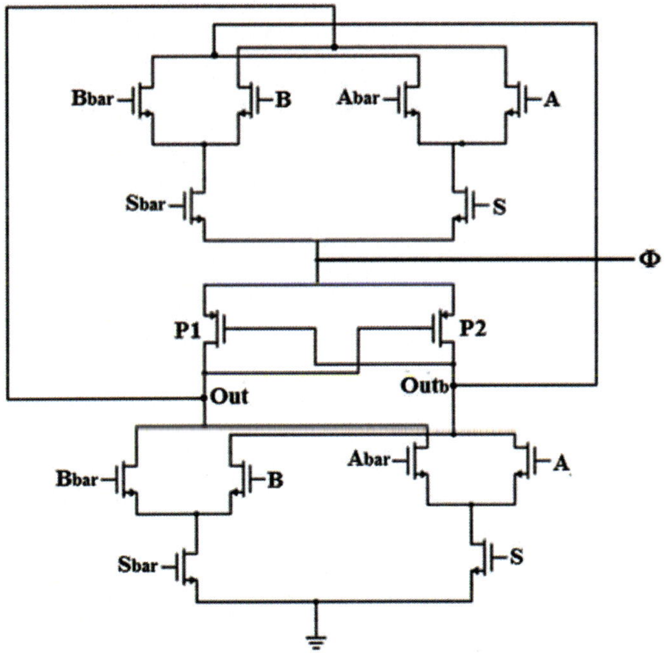

Fig. 3 Schematic diagram of proposed EEAL-based 2:1 MUX circuit

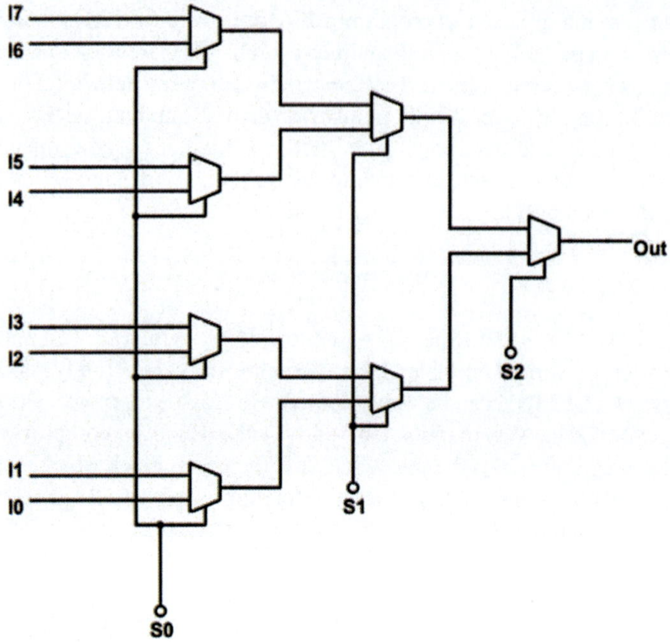

Fig. 4 Schematic diagram of proposed EEAL-based 8:1 MUX circuit

3.2 *EEAL-Based 8:1 Multiplexer*

The EEAL-based 8:1 MUX employs 2:1 MUX. Since the proposed 2:1 MUX has two outputs but 'Out' is the only required output for multiplexer which has been used for designing 8:1 MUX. So, 2:1 MUX has been designed and verified on Cadence and its symbol is created containing 2 inputs, 1 select line and 1 output pin. By using 7 such 2:1 MUX symbol, 8:1 MUX has been proposed which follow the newly proposed EEAL adiabatic logic style.

For $n = 3$ select lines, it has $2^3 = 8$ inputs, 3 select lines and single output. The output selects the particular input according to the combination of select lines and transfers the data of that input at the output. The circuit diagram of proposed EEAL-based 8:1 MUX is shown in Fig. 4.

4 Results and Graphical Analysis

The schematic of the proposed circuits and conventional CMOS circuits has been implemented and waveforms are obtained as shown in Figs. 5 and 6. All these analyses have been done on Cadence Virtuoso EDA tool using 45 nm technology with the supply voltage of 1 V at 100 MHz frequencies with a capacitive load of 20 fF.

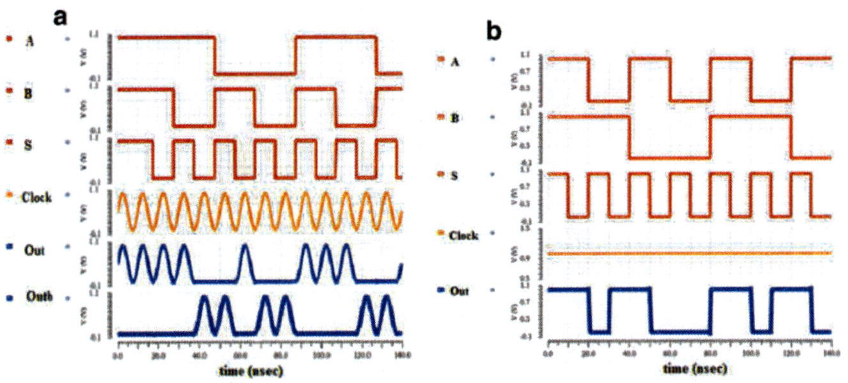

Fig. 5 **a** Proposed EEAL-based 2:1 MUX waveform. **b** CMOS-based 2:1 MUX waveform

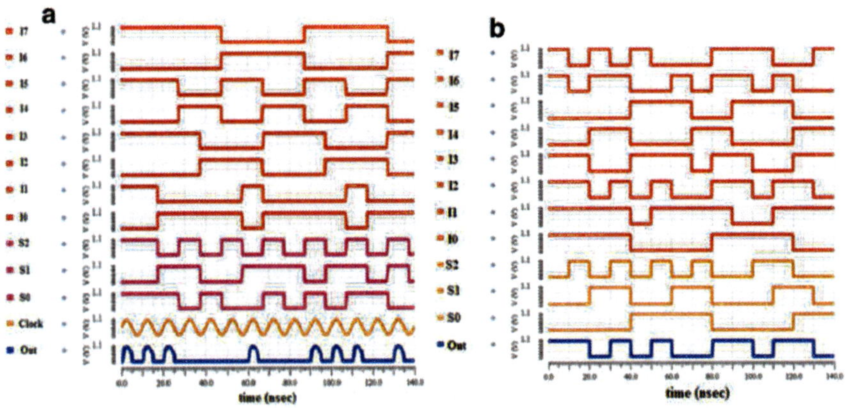

Fig. 6 **a** Proposed EEAL-based 8:1 MUX waveform. **b** CMOS-based 8:1 MUX waveform

The energy consumption, power dissipation and delay of EEAL-based 2:1 multiplexer and 8:1 multiplexer have been compared with its conventional CMOS logic circuits. The observed values are shown in Table 2. After comparison between EEAL and CMOS-based logic circuits, the adiabatic gain has been calculated and can be observed in Table 3.

Table 2 Comparative table for power, delay and energy between EEAL and CMOS-based circuits

Combinational circuits	Proposed EEAL-based circuits			Conventional CMOS-based circuits		
	Power dissipation (μWatt)	Delay (ns)	Energy consumption (FJoule)	Power dissipation (μWatt)	Delay (ns)	Energy consumption (FJoule)
2:1 MUX	7.44	4.1	30.50	20.20	7.4	149.48
8:1 MUX	13.03	4.65	60.58	48.31	7.9	381.64

Table 3 Adiabatic gain for logic circuits

EEAL-based logic circuit	Adiabatic gain
2:1 MUX	4.90
8:1 MUX	6.299

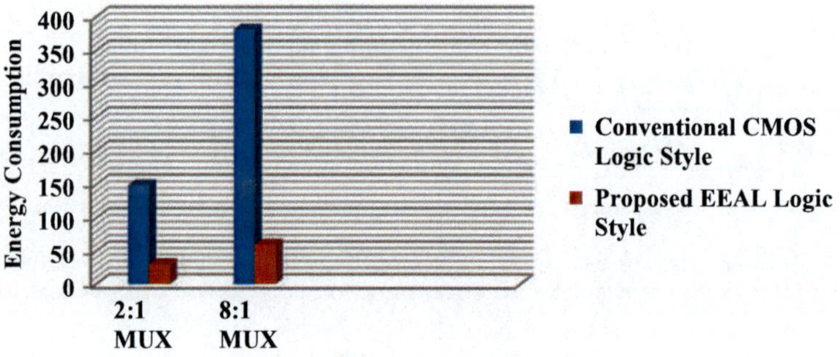

Fig. 7 Comparison of energy consumption

Table 2 shows the values of energy consumption, and by using these values, a bar graph has been created which shows the comparison between EEAL logic style and conventional CMOS logic as shown in Fig. 7. The graph shows that the proposed circuits are more energy efficient as compared to its CMOS logic circuits.

For a wide range of frequency, the energy consumption per cycle of a proposed 2:1 MUX and 8:1 MUX has been compared with its conventional circuit as shown in Fig. 8a, b, respectively.

The power consumption and delay of the proposed circuits have been calculated for different supply voltages in the range of 1-2 V and have been plotted and compared with its conventional circuit as shown in Figs. 9 and 10, respectively.

5 Conclusion

In this paper, 2:1 MUX and 8:1 MUX have been designed through CMOS logic style and EEAL-based adiabatic logic style on Cadence Virtuoso EDA tool using 45 nm technology for 1 V power clock at frequency 100 MHz. It has been found that the overall energy consumption of the proposed 2:1 MUX and 8:1 MUX has been reduced by 79.59% and 84.12%, respectively, as compared to the existing literature. It has been observed that the circuits based on EEAL are more energy efficient as compared to conventional CMOS logic style. All the results have been verified at different supply voltage in the range of 1-2 V and different frequency in the range of 10-500 MHz. Hence, EEAL logic circuits can play a significant role in ultra-low power applications where energy recovery is the main factor for high performance.

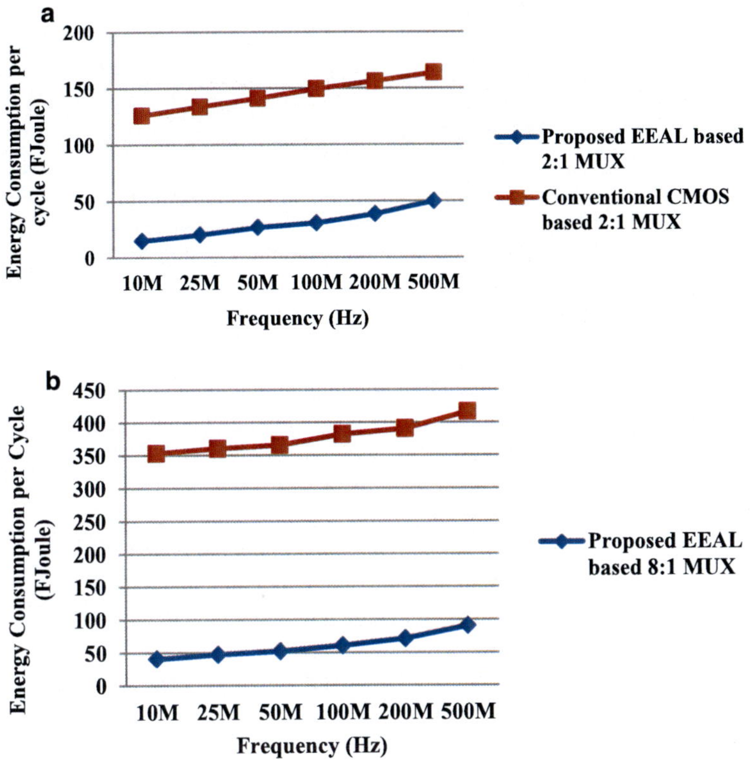

Fig. 8 **a** Energy consumption per cycle of 2:1 MUX. **b** Energy consumption per cycle of 8:1 MUX

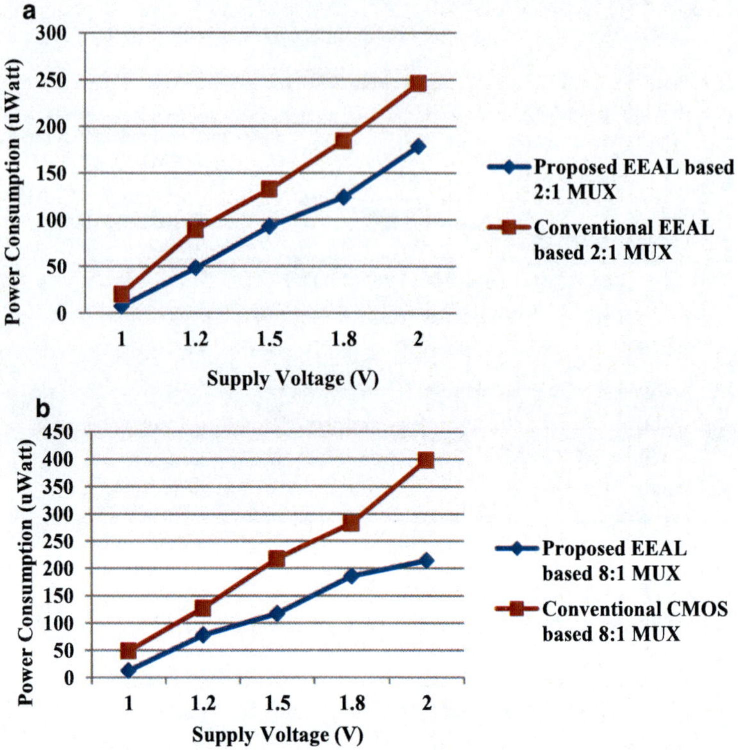

Fig. 9 **a** Power consumption of 2:1 MUX at different supply voltages. **b** Power consumption of 8:1 MUX at different supply voltages

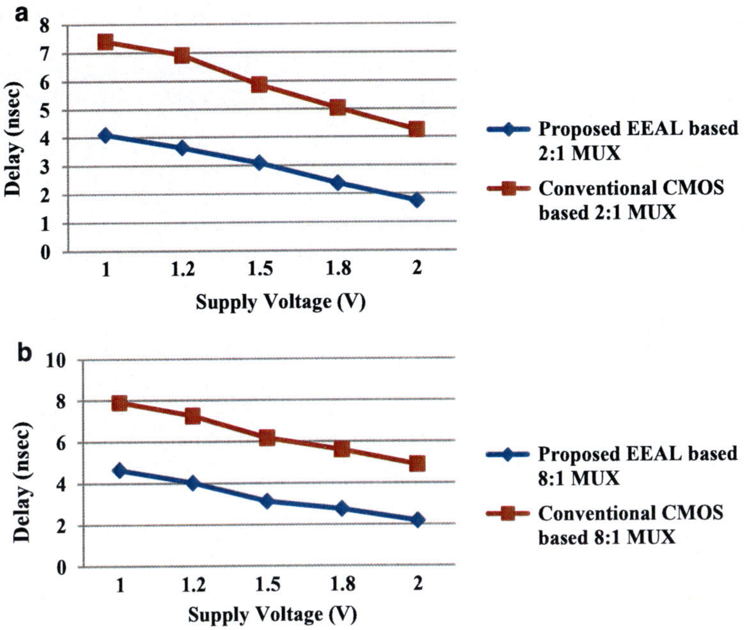

Fig. 10 **a** Delay of 2:1 MUX at different supply voltages. **b** Delay of 8:1 MUX at different supply voltages

References

1. Chanda M, Dandapat A, Rahaman H (2009) Ultra-low-power sequential circuit implementation by a quasi-static single phase adiabatic dynamic logic (SPADL). In: IEEE region 10 Conference TENCON, Singapore, pp 1–5
2. Bhaaskaran VSK, Raina JP (2010) Two-phase sinusoidal power-clocked quasi-adiabatic logic circuits. J Circuits Syst Comput 19:335–347
3. Lim J, Kim DG, Chae SI (1999) A 16-bit carry-lookahead adder using reversible energy recovery logic for ultra-low-energy systems: IEEE. J Solid-State Circuits 34:898–903
4. Hu DJ, Zhang W, Wang L (2006) Adiabatic CPL circuits for sequential logic systems. In: IEEE MWSCAS '06, pp 713–717
5. Mittal D, Niranjan A (2018) Designing of multiplexer and de-multiplexer using different adiabatic logic in 90 nm technology. In: IEEE conference (43488), 9th ICCCNT
6. Maksimovic D, Oklobdzija VG, Nikolic B, Current KW (2000) Clocked CMOS adiabatic logic with integrated single-phase power clock supply. IEEE Trans Very Large Scale Integr (VLSI) Syst 8:460–463
7. Oklobdzija VG, Maksimovic D (1997) Pass-transistor adiabatic logic using single power clock supply. IEEE Trans Circuits Syst II 44:842–846
8. Kramc A, Denkcr JS, Flower B, Moroney J (1995) 2nd order adiabatic computation with 2N-2P and 2N-2N2P logic circuits. Int Symp Low Power Des, 191–196
9. Moon Y, Jeong DK (1996) An efficient charge recovery logic circuit. IEEE J Solid-State Circuits 3:514–522
10. Kim S, Ziesler CH, Papaefthymiou MC (2003) A true single-phase energy recovery multiplier. IEEE Trans Very Large Scale Integr (VLSI) Syst 11:194–207

11. Nakata S (2004) Adiabatic charging reversible logic using a switched capacitor regenerator. IEICE Trans Electr E87-C:1837–1846
12. Ji R, Zeng X, Chen L, Zhang J (2007) The implementation and design of a low-power clock distribution microarchitecture. In: International conference on networking, architecture, and storage(NAS), Guilin, 29–31 July 2007, pp 21–30
13. Khandekar PD, Subbaraman S, Sharma A (2009) Implementation and analysis of VCO based power-clock supply generator. In: International conference on industrial and information systems (ICIIS), Sri Lanka, pp 317–320
14. Blotti A, Borghese S, Saletti R (2002) Single-inductor four-phase power-clock generator for positive feedback adiabatic logic gates. In: Proceedings of international conference on electronics, circuits systems, Dubrovnik, Croatia, pp 533–536
15. Rabaey JM, Chandrakasan A, Nikolic B (2008) Digital integrated circuits, a design perspective, 3rd edn. Peorson Education
16. Prashasti, Jaiswal S, Khan A, Wairya S (2019) High performance and low power D flip-flop using pulsed latch technique: IJAER 14(2). ISSN 0973-4562 (Special issue)
17. Heller LG, Griffin WR, Davis JW, Thoma NG (1984) Cascode voltage switch logic: a differential CMOS logic family. In: ISSCC, digest of technical papers, pp 16–17
18. Chanda M, De S, Sarkar CK (2015) Design and analysis of 32-Bit CLA using energy efficient adiabatic logic for ultra-low-power application. J Circuits Syst Comput 24(10)
19. Mohtashami H, Navi K, Mahini A, Sharifan M (2007) Pull down network minimization based on equivalent functions concept in DCVS circuit design. In: IEEE 14th international conference MIXDES
20. Voss B, Glesner M (2001) A low power sinusoidal clock. IEEE international symposium on circuits and systems (ISCAS), 6–9 May 2001, pp 108–111
21. Karoubalis T, Alexiou GP, Kanopoulos N (1995) Optimal synthesis of differential cascode voltage switch (DCVS) logic circuits using ordered binary decision diagrams (OBDDs). In: Design automation conference, pp 282–287
22. Kavehie O, Navi K, Nikoubin T, Rouholamini M (2006) A novel DCVS tree reduction algorithm. In: Integrated circuit design and technology, CICDT'06, IEEE international conference, pp 1–7
23. Sengupta R, Banerjee S, Mitra S, Barman M (2018) CMOS logic based 2X1 multiplexer. In: 2nd international conference on electronics, materials engineering and nano-technology

Design and Performance Evaluation of Highly Efficient Adders in Nanometer Technology

Prashasti, Shivangi Jaiswal, Anam Khan, and Subodh Wairya

Abstract In this paper, optimized designs of two 4-bit adders, namely ripple carry adder and look-ahead carry adder have been presented. These adder circuits are highly efficient in terms of delay, power, and PDP. A 1-bit hybrid full adder, which is the basic unit of the presented designs, is constructed using XOR and XNOR gates. Thus, energy-efficient XOR-XNOR gates are employed in the construction of 1-bit full adder and this adder when compared with conventional CMOS-based full adder, is found to have superior performance. This hybrid full adder cell is further used to implement the two 4-bit adders using Cadence Virtuoso EDA tool. The simulations, carried out at 45-nm CMOS process technology in a range of 0.6–1.2 V supply voltage, indicate that presented designs are superior in speed and power as compared to their conventional CMOS-based counterparts.

Keywords XOR-XNOR logic · Hybrid 1-bit full adder · Ripple carry adder · Look-ahead carry adder · Power-delay product

1 Introduction

Arithmetic and logical unit (ALU) is a critical element in any CPU. Efficiency of arithmetic operations in the ALU of the processors can be improved by introducing faster computational circuits [1]. Various technologies have been proposed from time to time, to improve performance of circuits in terms of low power consumption [2] and high speed. Of all the arithmetic operations in any technique, addition plays the most vital role as adders provide not only for addition but also subtraction, multiplication, division, etc. [3]. Hence, full adder designs become the center of attention for ongoing researches. Various designs have been proposed in literature,

Prashasti (✉) · S. Jaiswal · A. Khan · S. Wairya
Institute of Engineering and Technology, AKTU, Lucknow, Uttar Pradesh, India
e-mail: prashasti311@gmail.com

S. Jaiswal
e-mail: shivi.19494@gmail.com

D. Harvey et al. (eds.), *Advances in VLSI, Communication, and Signal Processing*, Lecture Notes in Electrical Engineering 683,
https://doi.org/10.1007/978-981-15-6840-4_20

so trade-offs in these existing structures must be considered before an appropriate design can be chosen to meet specific speed and power requirements of the designer. The FA cell finds application in a large number of arithmetic operations and hence offers scope to be explored explicitly in the designing of combinational circuits.

1.1 Full Adder

The FA circuits until now have been broadly classified into static and dynamic logic styles [4–7]. Each logic style has some good aspects, but at the cost of few other critical parameters. So, the hybrid-logic designs, which comprise of the best features of different logic styles, become the most suitable logic styles for realizing efficient 1-bit full adder structures [8]. The general form of a hybrid full adder is made up of three modules, which are a 2-input XOR-XNOR gate and two 2-to-1 multiplexer [9].

In a 1-bit hybrid FA cell with A, B, and C_{in} as inputs and sum, carry as outputs; the Module 1 produces XOR and XNOR which are represented by $H = A \oplus B$ and $H' = (A \oplus B)'$. And the Module 2 and 3 are 2-to-1 multiplexers which use H and H' as the select lines. The output expression for the general form 1-bit full adder is as follows.

$$\text{Sum} = A \oplus B \oplus C_{in} = H \oplus C_{in} = H.C'_{in} + H' \cdot C_{in} \tag{1}$$

$$\text{Carry} = A.H' + C_{in} \cdot H \tag{2}$$

The design form is crucial because it generates H and H' simultaneously, such that the two select lines (H and H') are generated without any delay. This typically avoids any glitches and unnecessary power consumption as the final output will be produced only when intermediate signals from Module 1 are available simultaneously. In this design technique, the XOR-XNOR gate is majorly responsible for power consumption in the FA cell [9].

1.2 XOR-XNOR Logic

In this regard, various logical circuits of XOR-XNOR gates have been explored [10]. Different logic structures, each with their own merits and demerits, were extensively reviewed to implement XOR and XNOR gates [11–15]. The XOR-XNOR gate structure, proposed by Hamed Naseri et al. [1] in the year 2018, is the latest and most efficient than others. Their designs have been shown in Fig. 1a and b. These designs offer good driving capability, full swing output, and smaller number

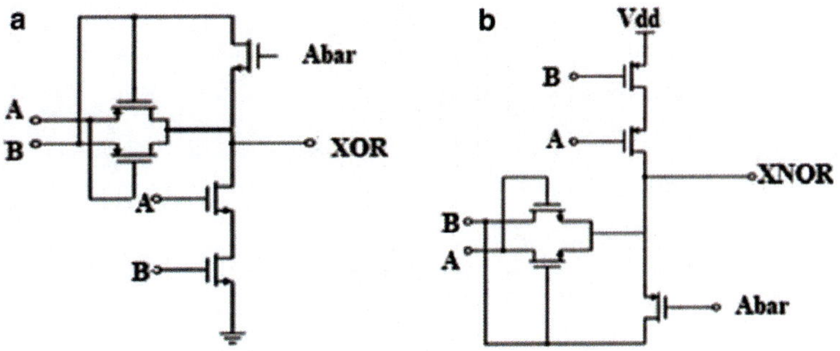

Fig. 1 **a** XOR schematic. **b** XNOR schematic

of interconnecting wires. It also has an overall reduced power consumption and delay [1].

The waveforms of the XOR and XNOR gates proposed by Naseri et al. are shown in Fig. 2a and b. These waveforms have been obtained by simulating the designs

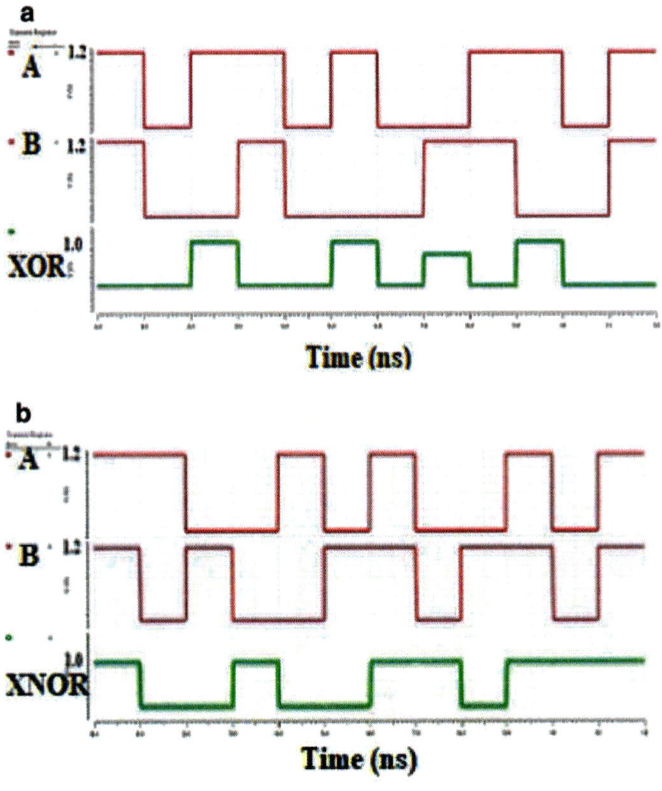

Fig. 2 **a** XOR waveform **b** XNOR waveform

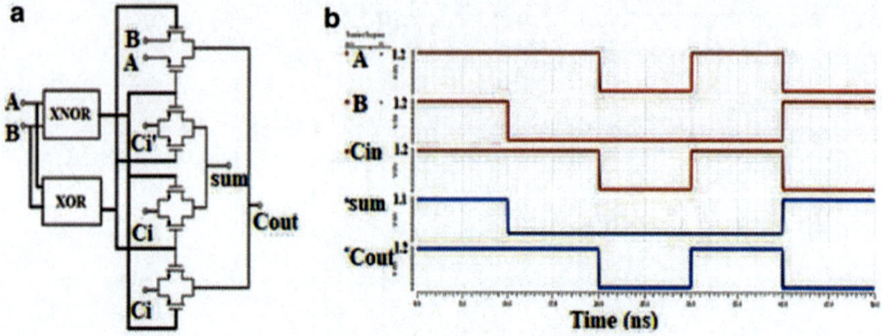

Fig. 3 **a** Schematic of HFA 22-T FA cell and **b** waveform of 1-bit FA cell (HFA-22T)

Table 1 Comparison table between HFA-22T FA and conventional CMOS-based FA

Logic style	Delay (ps)	Power (μW)	PDP (aJ)
CMOS-based 1-bit full adder	31.2	4.19	130.72
HFA-22T	5.86	3.57	20.92

using Cadence Virtuoso EDA tool at 45 nm technology. The transient analysis is done at 1.2 V supply for a simulation time of 60 ns and the rise and fall time for the input signals is 1 ps each.

Naseri et al. used their proposed designs of XOR and XNOR cells in the construction of 6 different 1-bit full adders. Simulations show that the FA cell, HFA-22T, shown in Fig. 3a, saves PDP up to 23.4% more than its best counterpart [1]. The speed and energy recovery is also found to best among all the six designs.

To analyze the efficiency of HFA-22T FA cell, simulation of the design is done in Cadence Virtuoso at 45 nm and 1.2 V and compared with its CMOS-based counterpart. The output waveform of HFA-22T obtained under these simulation conditions is shown in Fig. 3b and Table 1 presents comparison of its performance against CMOS-based FA cell.

From the above table, it is clear that the new technique has far improved performance than the conventional CMOS-based 1- bit FA as the overall power-delay product (PDP) has reduced significantly. Thus, this latest design is further used to implement fast and optimized 4- bit adders.

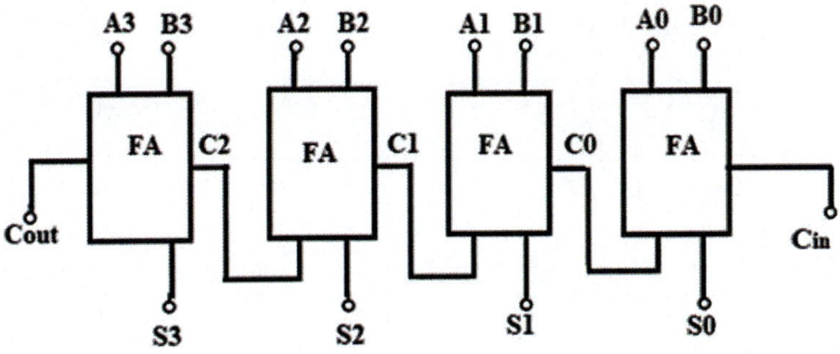

Fig. 4 Schematic of 4-bit ripple carry adder

2 Proposed Adders

2.1 Four-Bit Ripple Carry Adder

A ripple carry adder is combinational circuit where '*n*' numbers of 1-bit full adders are cascaded to obtain n-bit addition. The output carry of each FA cell is fed directly to the input carry of the next FA. For the proposed ripple carry adder, four HFA-22T FA cells have been employed to carry out the 4-bit addition as shown in Fig. 4.

In a ripple carry adder, input carry is rippled from one stage of FA cell to another and the final output is obtained only after all stages of FA have been evaluated. This increases the overall delay of the circuit as the next stage has to wait for evaluation from previous stages [16–20].

2.2 Four-Bit Look-Ahead Carry Adder

Look-ahead carry adder is proposed to obtain faster operation. This adder uses fast and power-efficient XOR structures proposed by Naseri et al. [1] to perform addition. In this type of adder, intermediate terms generate (G) and propagate (P) have been introduced to operate between the stages of FA [21] whose Boolean equations are as follows.

$$G = A \cdot B \tag{3}$$

$$P = A \oplus B \tag{4}$$

Generate (*G*) produces an output carry for a stage independent of input carry to that stage, i.e., $G = 1$ when both the inputs *A* and *B* are high. Propagate (*P*) transfers output carry of a stage as input carry to the next when either of the inputs of the

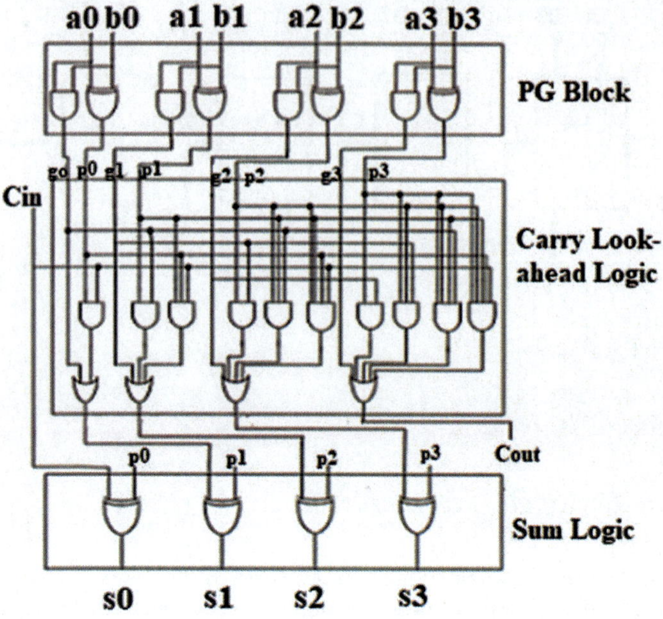

Fig. 5 Schematic of 4-bit weinberger and smith look-ahead carry adder

previous stage is high. This section of the adder helps in preprocessing whether or not to generate carry. Thus, actual addition takes place without having to wait for arrival of the carry signal from the previous stages, unlike in the ripple carry adder. Hence, the process becomes much faster and more power-efficient. The design of a 4-bit Weinberger and Smith CLA architecture [22] is shown in Fig. 5.

In this regard, the proposed look-ahead carry adder is designed by using Naseri et al. proposed XOR gate instead of normal CMOS-based XOR in the 4-bit Weinberger and Smith CLA architecture.

3 Results and Simulations

The proposed circuits are simulated using Cadence Virtuoso EDA tool at 45 nm technology with a supply voltage of 1.2 V. The transient analysis is carried out for a 60 ns time duration with the input signal rise time and fall time of 1 picosecond each. The waveforms of the proposed circuits obtained under these simulation conditions are shown in Figs. 6 and 7, that verify the addition operation in each time interval.

Delay, power, and PDP have been formulated for the proposed designs of adders. The average delay and power has been obtained over duration of 60 ns with rise and fall time of input signals being 1 ps each. The supply voltage and input signal voltage are maintained at 1.2 V for the analysis of the presented circuits. Table 2 gives

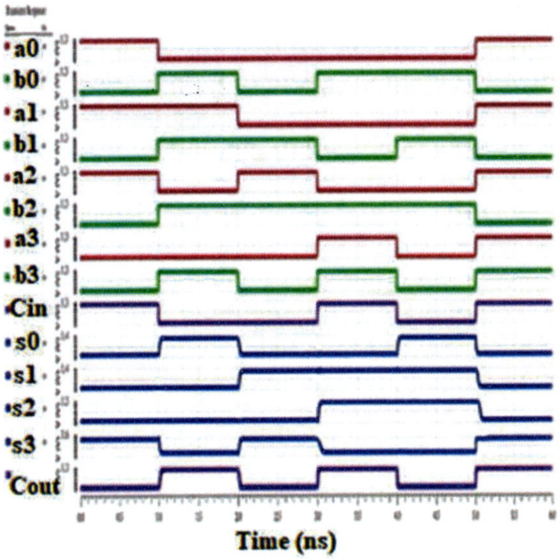

Fig. 6 Output waveform of ripple carry adder

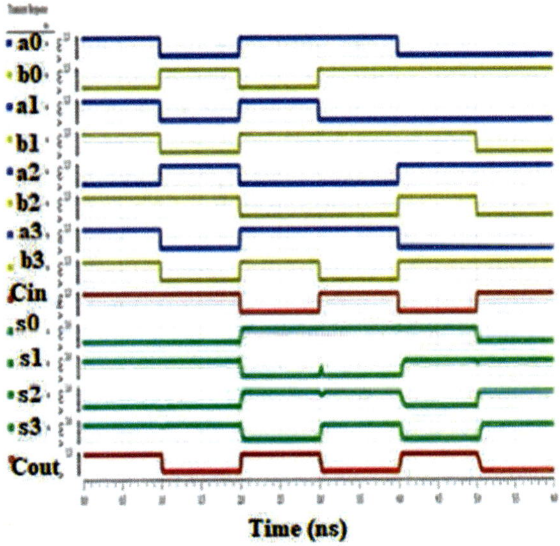

Fig. 7 Output waveform of look-ahead carry adder

a comparison of the average delay, average power, and PDP parameters between the proposed designs and the conventional CMOS-based designs.

The proposed designs show immense improvement in PDP as compared to the CMOS-based adders. Also, the performance of look-ahead carry adder is better than ripple carry adder.

Table 2 Comparison of proposed and CMOS-based designs in terms of delay, power, and PDP

Arithmetic circuits	Proposed designs			Conventional designs		
	Delay (ps)	Power (μW)	PDP (aJ)	Delay (ps)	Power (μW)	PDP (aJ)
Ripple carry adder (4-bit)	19.98	35.53	709.8	62.1	74.5	4626.4
Look-ahead carry adder (4-bit)	9.9	18.52	183.3	37.5	38.6	1447.5

4 Graphical Analysis

The effect on delay and power due to variation of V_{DD} in a range of 0.6–1.5 V on the proposed ripple carry adder and look-ahead carry adder designs against the conventional CMOS-based counterparts has been shown in Figs. 8, 9, 10, 11 and 12.

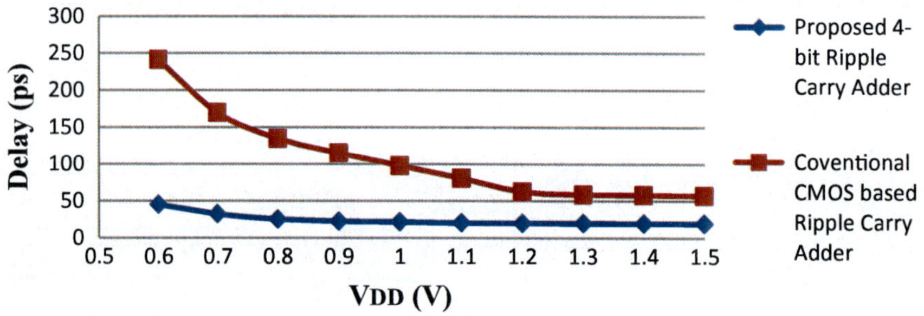

Fig. 8 Delay at different supply voltage

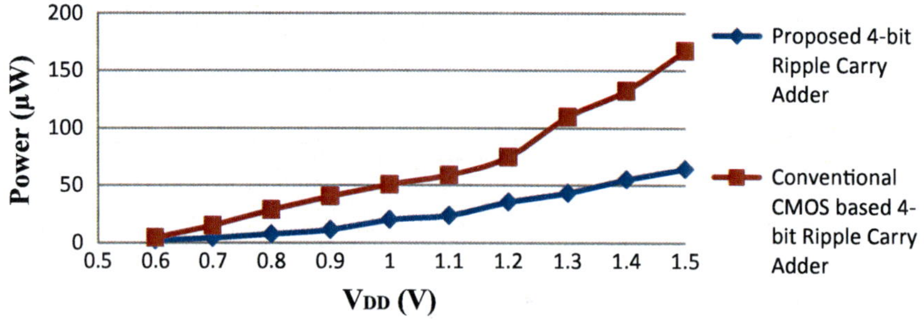

Fig. 9 Power at different supply voltage

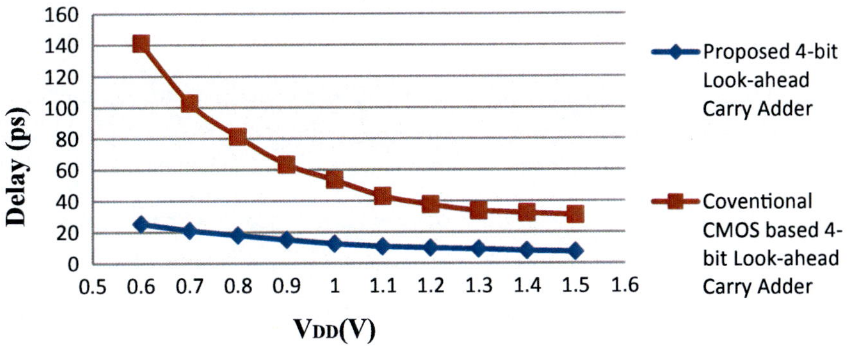

Fig. 10 Delay at different supply voltage

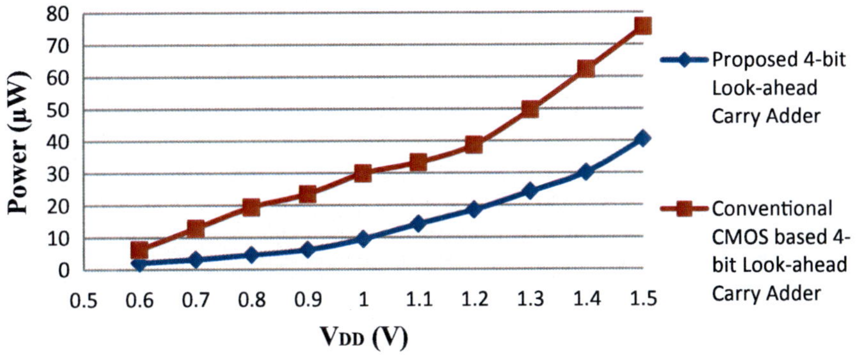

Fig. 11 Power at different supply voltage

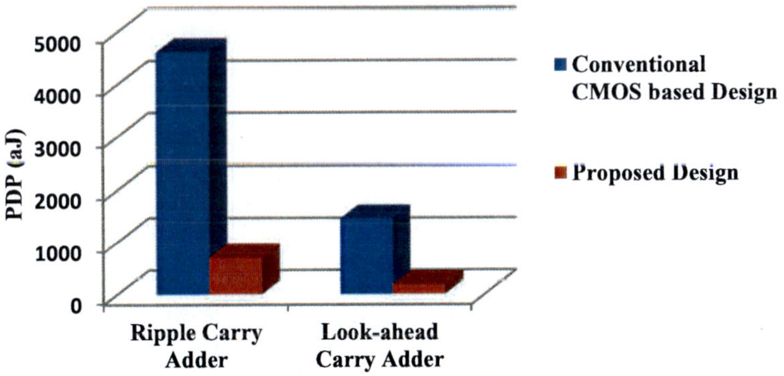

Fig. 12 Comparison of PDP

5 Conclusion

Energy-efficient adder circuits to address the ongoing demand of high-speed operations have been developed. The optimized ripple carry adder and look-ahead carry adder have 84 and 87% improved PDP than their CMOS-based counterparts. These adders show robustness against variation in supply voltage in a range of 0.6–1.5 V. Out of the two proposed designs, the look-ahead carry adder has better performance as compared to the proposed ripple carry adder as the PDP improvement is 74% more in look-ahead carry adder which makes it more suitable for use in high-speed and low-power arithmetic operations.

References

1. Naseri H, Timarchi S (2018) Low-power and fast full adder by exploring new XOR and XNOR Gates. IEEE Trans Very Large Scale Integr (VLSI) Syst
2. Prashasti SJ, Khan A, Wairya S (2019) High performance and low power D flip-flop using pulsed latch technique. IJAER, ISSN 0973-4562 14(2) (Special Issue)
3. Wang Y, Bui HT, Jiang Y (2002) Design and analysis of low-power 10-transistor full adders using novel XOR-XNOR gates. IEEE Trans Circ Syst-2: Analog and Digital Sig Proc 49(1)
4. Shalem R, John E, John LK (1999) A novel low power energy recovery full adder cell. In: Proceedings of IEEE great lakes VLSI symposium, pp 380–383
5. Weste N, Eshraghian K (1985) Principles of CMOS VLSI design. Addison-Wesley, New York, NY, USA
6. Chandrakasan AP, Sheng S, Brodersen RW (1992) Low-power CMOS digital design. IEEE J Solid-State Circuits 27:473–483
7. Bui HT, Al-Sheraidah AK, Wang Y (1999) New 4-transistor XOR and XNOR designs. Technical Report Florida Atlantic Univ, Boca Raton
8. Pedram R, Pedram M (1996) Low power design methodologies. Kluwer, Norwell, MA
9. Aranda ML, Baez R, Diaz OG (2010) Hybrid adders for high-speed arithmetic circuits: a comparison. In: 7th International Conference on Electrical Engineering Computing Science and Automatic Control (CCE), Tuxtla Gutierrez, NM, USA, pp 713–722 (2010)
10. Valashani MA, Mirzakuchaki S (2016) A novel fast, low power and high-performance XOR-XNOR cell. Proc IEEE Symp Circuits Syst (ISCAS) 1:694–697
11. Aguirre-Hernandez M, Linares-Aranda M (2011) CMOS full-adders for energy-efficient arithmetic applications. IEEE Trans Very Large Scale Integr (VLSI) Syst 19(4):718–721
12. Bhattacharyya BK, Ghosh S, Kumar V, Dandapat A (2015) Performance analysis of a low-power high-speed hybrid 1-bit full adder circuit. IEEE Trans Very Large Scale Integr (VLSI) Syst 23(10):2001–2008 (2015)
13. Wang J-M, Fang S-C, Feng W-S (1994) New efficient designs for XOR and XNOR functions on the transistor level. IEEE J Solid-State Circuits 29(7):780–786
14. Chang C-H, Gu J, Zhang M (2005) A review of 0.18-μm full adder performances for tree structured arithmetic circuits. IEEE Trans Very Large Scale Integr (VLSI) Syst 13(6):686–695
15. Goel S, Kumar A, Bayoumi M (2006) Design of robust, energy-efficient full adders for deep-submicrometer design using hybrid-CMOS logic style. IEEE Trans Very Large Scale Integr (VLSI) Syst 14(12):1309–1321
16. Wairya S, Singh G, Nagaria RK, Tiwari S Design analysis of XOR(4T) based low voltage CMOS full adder circuits. In: Proceeding IEEE Nirma University International Conference on Engineering (NUiCONE), pp 1–7 (2011)

17. Parhami B (2010) Computer arithmetic: algorithms and hardware designs, 2nd edn. Oxford University Press, New York
18. Rabaey JM, Chandrakasan A, Nikolic B (2003) Digital integrated. Circuits: a design perspective. Pearson Education
19. Koren I (1993) Computer arithmetic algorithms. Englewood Cliffs, NJ, Prentice Hall
20. Zimmermann R (1998) Binary adder architectures for cell-based VLSI and their synthesis. PhD thesis, Swiss Federal Institute of Technology (ETH) Zurich, Hartung-GorreVerlag
21. Weste N, Harris D (2004) CMOS VLSI design: a circuits and systems perspective. Pearson Education
22. Ling H (1981) High-speed binary adder. IBM J Res Dev 25(3):156–166

$\alpha - \eta - \mu$/IG Composite Fading Model for Body-Centric Communication

Prasoon Raghuwanshi and Krishan Kumar

Abstract By providing several wireless channel connectivity for communication system enables the future telecommunication. Such a solution is the upcoming frontier for telecommunication revolution and leading to the evolution of future wireless communication system. This paper is to analyse a highly efficient $\alpha - \eta - \mu/IG$ composite channel model. Here, inverse Gaussian (IG) distribution describes shadowing because it provides more accurate approximation for lognormal shadowing in additive white Gaussian noise channel. Further, two parameters named as average bit error rate and outage probability are derived in their closed form. The acquired outcome results are generic and can be straight utilized for the performance analysis of composite channel model. Some results are represented in the form of bivariate Fox-H function.

Keywords IG shadowing · $\alpha - \eta - \mu$ multipath fading · $\alpha - \eta - \mu$/IG composite fading model

1 Introduction

In today's time, the insufficiency of spectrum has become a serious issue due to the tremendous growth of communication devices [1]. Channel analysis is a powerful tool to handle this. Several channel models such as $\alpha - \mu$ [2], κ-μ [3] are used to analyse the fading effect. For the real scenarios, these models are applicable [4]. The composite fading models have been widely utilized in modeling shadowed and faded scenario. The composite fading model is jointly analysing shadowing and multipath propagation of the signal. Shadowing happens due to two different cases, line of sight (LoS) and multiplicative shadowing effect [5]. Several distribution

P. Raghuwanshi (✉) · K. Kumar
Department of Electronics & Communication Engineering, National Institute of Technology Hamirpur, Hamirpur, India
e-mail: ec15mi440@nith.ac.in

K. Kumar
e-mail: krishan_rathod@nith.ac.in

D. Harvey et al. (eds.), *Advances in VLSI, Communication, and Signal Processing*,
Lecture Notes in Electrical Engineering 683,
https://doi.org/10.1007/978-981-15-6840-4_21

channel models such as Nakagami-m and Rician are utilized for the analysis of multipath fading, whereas Gamma and lognormal distributions have been utilized for the analysis of the shadowing [6]. For the analysis of composite fading, channel models such as κ-μ/lognormal [7], Rice-Nakagami [8], and κ-μ/Nakagami [9] are mainly considered.

In body-centric communications, the received signals can be categorized with the help of path loss, large-scale shadowing, and multipath fading effect. In body-centric communications, the shadowing provoked through body. There are mainly three types of body-centric communication channels named as body-to-body communication, on body and off body. A literature related to body communication channels is available in [10]. In this work, inverse Gaussian (IG) distribution is introduced to express the shadowing effect because it results more accurate approximation for lognormal distribution. Closed form probability density function (PDF) of the $\alpha-\eta-\mu$/IG composite fading model is procured for performance analysis. Further, the propounded composite fading model perhaps utilized for all three subdivisions belonging to body-centric channels.

The rest of the work is organized into following sections. In Sect. 2, the PDF for $\alpha-\eta-\mu$/IG composite fading model is derived. The performance parameters for the composite fading model are derived in Sect. 3. The Sect. 4 persists the numerical results and outcomes of performance parameters. Finally, conclusion of the work is in Sect. 5.

2 Composite Fading Model

The envelope PDF for the generic $\alpha-\eta-\mu$multipath distribution is [11]

$$f_M(m) = \frac{2\sqrt{\pi}\alpha\varepsilon^{\mu}\mu^{\mu+\frac{1}{2}}m^{\alpha(\mu+\frac{1}{2})-1}}{\Gamma(\mu)\mathrm{E}^{\mu-\frac{1}{2}}\bar{m}^{\alpha(\mu+\frac{1}{2})}}e^{\frac{-2\mu\varepsilon m^{\alpha}}{\bar{m}^{\alpha}}}I_{\mu-\frac{1}{2}}\left(\frac{2\mu\mathrm{E}m^{\alpha}}{\bar{m}^{\alpha}}\right) \tag{1}$$

Here, $\alpha > 0$ reveals the nonlinearity belonging to broadcast channel and $\mu > 0$ reveals the aggregate of multipath bunches, its value is $\mu = \frac{E^2(M^{\alpha})}{2V(M^{\alpha})}\left(1+\left(\frac{E}{\varepsilon}\right)^2\right)$. Here, $E(.)$ used for expected value function. The term $I_{\mu-\frac{1}{2}}(.)$ is used to represent the Bessel function of order $\mu-\frac{1}{2}$. The symbol 'm' represents multipath fading and '$\bar{m}$' signifies the α-root mean value of an envelope M, and $\bar{m} = \sqrt[\alpha]{E(M^{\alpha})}$. The symbol 'E' and '$\varepsilon$' are interpreted in two forms. In form-1, $\varepsilon = (1+\eta)^2/4\eta$ and $\mathrm{E} = (1-\eta^2)/4\eta$ as long as $\eta \in [0, \infty)$. In form-2, $\varepsilon = 1/(1-\eta^2)$ and $\mathrm{E} = \eta/(1-\eta^2)$ as long as $\eta \in (-1, 1)$. Furthermore, PDF for IG distribution possesses random variable Φ be stated as [12]

$$f_{\Phi}(\phi) = \sqrt{\frac{\xi}{2\pi}}e^{\frac{\xi}{\psi}}\phi^{-\frac{3}{2}}e^{\frac{-\xi\phi}{2\psi^2}}e^{\frac{-\xi}{2\phi}} \tag{2}$$

Now, envelope PDF for $\alpha - \eta - \mu$/IG composite distribution be stated as [12].

$$f_M(m) = \int_0^{\infty} f_{M/\Phi}(m/\phi) f_\Phi(\phi) \mathrm{d}\phi \tag{3}$$

Now, $f_{\Re}(\Re) = \frac{f_M\left(\sqrt{\bar{m}^2 \frac{\Re}{\bar{\Re}}}\right)}{2\sqrt{\frac{\Re\bar{\Re}}{\bar{m}^2}}}$.

Here, '$\Re$' signifies instantaneous signal-to-noise ratio (SNR) for individual symbol and '$\bar{\Re}$' signifies average SNR for individual symbol.

$$f_{\Re}(\Re) = \sum_{q=1}^{Q} \frac{\vartheta_q \sqrt{\xi \bar{\Re}} \varepsilon^{-\frac{1}{2}-\frac{1}{\alpha}} \mu^{-\frac{1}{\alpha}} \Re^{-3/2} \zeta_q^{\mu+\frac{1}{\alpha}-\frac{1}{2}} e^{\frac{\xi}{\psi}} e^{\frac{-\xi\Re}{2\psi\bar{\Re}}\left(\frac{2\mu\varepsilon}{\zeta_q}\right)^{2/\alpha}} e^{\frac{-\xi\bar{\Re}}{2\psi\Re}\left(\frac{\zeta_q}{2\mu\varepsilon}\right)^{2/\alpha}}}{\sqrt{\theta'} 2^{\mu+\frac{1}{\alpha}} \Gamma(\mu) E^{\mu-\frac{1}{2}}} I_{\mu-\frac{1}{2}}\left(\frac{E\zeta_q}{\varepsilon}\right) \tag{4}$$

Here, $f_{\Re}(\Re)$ signifies SNR PDF for $\alpha - \eta - \mu$/IG composite fading distribution. The term $f_{\Re}(\Re)$ can be represented in standard form as

$$f_{\Re}(\Re) = \sum_{q=1}^{Q} A_q \Re^{-3/2} e^{-\Re B_q} e^{-\frac{C_q}{\Re}} \tag{5}$$

Here,

$$A_q = \frac{\vartheta_q \sqrt{\xi \bar{\Re}} \varepsilon^{-\frac{1}{2}-\frac{1}{\alpha}} \mu^{-\frac{1}{\alpha}} \zeta_q^{\mu+\frac{1}{\alpha}-\frac{1}{2}} e^{\frac{\xi}{\psi}}}{\sqrt{\psi} 2^{\mu+\frac{1}{\alpha}} \Gamma(\mu) \mathrm{E}^{\mu-\frac{1}{2}}} I_{\mu-\frac{1}{2}}\left(\frac{\mathrm{E}\zeta_q}{\varepsilon}\right) B_q = \frac{-\xi}{2\psi\bar{\Re}}\left(\frac{2\mu\varepsilon}{\zeta_q}\right)^{2/\alpha}$$

$$C_q = \frac{-\xi\bar{\Re}}{2\psi}\left(\frac{\zeta_q}{2\mu\varepsilon}\right)^{2/\alpha}$$

3 Performance Parameters of Composite Fading Models

3.1 Outage Probability

Outage probability (P_{OP}) is elucidated as the probability of '$\Re$' beneath the definite threshold '$\Re_t$' and represented as [6]

$$P_{OP} = \int_{0}^{\Re_t} f_{\Re}(\Re) d\Re \tag{6}$$

By utilizing [13], final P_{OP} be stated as

$$P_{OP} = \sum_{q=1}^{Q} A_q \sqrt{\frac{\pi}{C_q}} e^{-2\sqrt{C_q B_q}} \left(Q\left(\sqrt{\frac{2C_q}{\Re_t}} - \sqrt{2\Re_t B_q} \right) + e^{4\sqrt{C_q B_q}} Q\left(\sqrt{\frac{2C_q}{\Re_t}} + \sqrt{2\Re_t B_q} \right) \right) \tag{7}$$

3.2 Average Bit Error Rate

Average bit error rate (ABER) is elucidated as averaging the provisional BER for AWGN channels across PDF of $\Re$ [6]

$$\overline{P}(e) = \int_{0}^{\infty} P_c(e|\Re) f_{\Re}(\Re) d\Re \tag{8}$$

Here, $P_c(e|\Re) = \frac{\Gamma(\varsigma, \upsilon\Re)}{2\Gamma(\varsigma)}$ signifies provisional BER, $\Gamma(\varsigma, \upsilon\Re)$ signifies upper incomplete gamma function [6]

$$\overline{P}(e) = \frac{1}{2\Gamma(\varsigma)} \sum_{q=1}^{Q} \frac{A_q \sqrt{B_q}}{(2\pi i)^2} \int_{L_1} \int_{L_2} \frac{\Gamma(\varsigma + s)\Gamma(s)\Gamma(-t)\Gamma(-0.5 - s - t)}{\Gamma(1 + s)} \left(\frac{\upsilon}{B_q} \right)^{-s} \left(\frac{1}{B_q C_q} \right)^{-t} \mathrm{d}s\mathrm{d}t \tag{9}$$

This is calculated as [14]

$$\overline{P}(e) = \frac{1}{2\Gamma(\varsigma)} \sum_{q=1}^{Q} A_q \sqrt{B_q} H \left[\begin{matrix} \begin{pmatrix} 0\,1 \\ 1\,0 \end{pmatrix} \\ \begin{pmatrix} 0\,1 \\ 1\,0 \end{pmatrix} \\ \begin{pmatrix} 2\,0 \\ 1\,2 \end{pmatrix} \end{matrix} \middle| \begin{matrix} \left(\frac{3}{2}; 1, 1\right) \\ - \\ (1, 1) \\ - \\ (1, 1) \\ (b', 1)\ (0, 1) \end{matrix} \middle| \frac{1}{C_q B_q}, \frac{\upsilon}{B_q} \right] \tag{10}$$

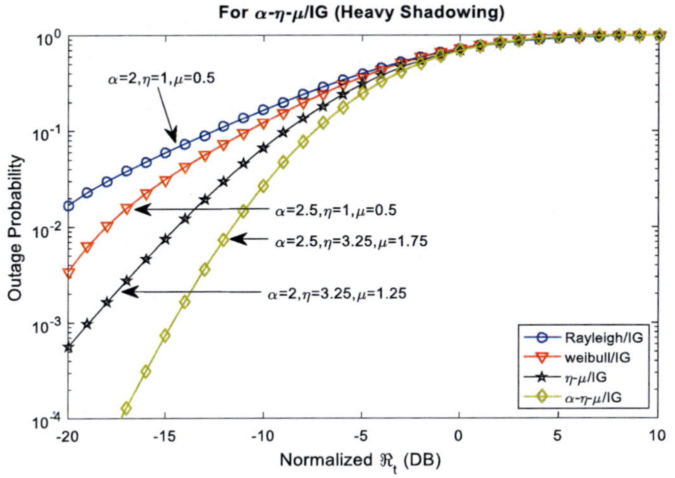

Fig. 1 Outage probability for $\alpha - \eta - \mu$/IG composite fading model

4 Performance Analysis and Discussion

In the current section, the derived closed form of equations related to PDF and ABER for $\alpha - \eta - \mu$/IG composite fading model is distinctly investigated with shadowing environment. These results are analysed and compared graphically with other existing fading models such as Rayleigh/IG, Weibull/IG, η-μ/IG, Nakagami-m/IG, κ-μ/IG. Figure 1 manifests graphically the derived equations for outage probability for $\alpha - \eta - \mu$/IG composite fading model. Figure 1 represents that the discussed $\alpha - \eta - \mu$/IG model has lower outage probability and provides better results as compare to Rayleigh/IG, Weibull/IG, η-μ/IG models. The values of channel related parameters are represented inside the results.

Figure 2 represents the plot related to average bit error rate for $\alpha - \eta - \mu/IG$ composite fading model. In this analysis, the BPSK modulation strategy is considered. From Fig. 2, it is clear that discussed $\alpha - \eta - \mu$/IG composite model has lower average bit error rate as compared to other existing channel models such as Rayleigh/IG, Weibull/IG, η-μ/IG, and Nakagami-m/IG and hence improves the performance of channel.

5 Conclusion

In this work, the fading along with shadowing is analysed by utilizing $\alpha - \eta - \mu$/IG composite fading model. This composite channel model is well suited for body area communications. The closed form expressions of the performance parameters such as average bit error rate and outage probability for $\alpha - \eta - \mu$/IG composite fading

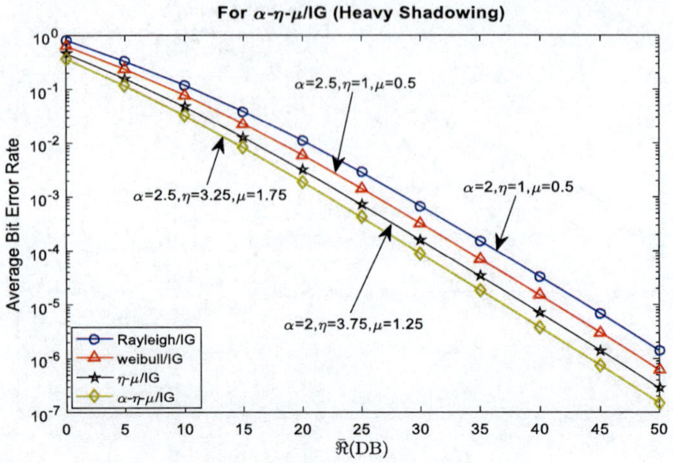

Fig. 2 Average bit error rate for $\alpha - \eta - \mu/\text{IG}$ composite fading model

model are derived in this work. Moreover, in numerical outcome section, the analysis of performance parameters for the discussed $\alpha - \eta - \mu/\text{IG}$ composite fading model is graphically represented and compared. The outage probability lies in the range of 10^{-1} to 10^{-4} and sharply decreases which leads to improvement in the wireless communication systems performance by using discussed composite fading model.

References

1. CV networking Index 2016 Forecast and methodology, 2016–2021, white paper. San Jose, CA, USA
2. Yoo SK et al (2017) The -/Inverse Gamma and -/Inverse gamma composite fading models: fundamental statistics and empirical validation. In: IEEE Transactions on Communications, pp 1–1
3. Bhatt M, Soni SK (2018) A unified performance analysis of energy detector over α-η-μ/lognormal and α-κ-μ/lognormal composite fading channels with diversity and cooperative spectrum sensing. AEU—Int J Electron Commun 94:367–376. ISSN 1434–8411
4. Al-Hmood H, Al-Raweshidy HS (2015) Performance analysis of energy detector over η-μ fading channel: PDF-based approach. Electron Lett 51(3):249–251
5. Yoo SK et al (2015) The κ-μ/Inverse gamma fading model. In: 2015 IEEE 26th annual international symposium on personal, indoor, and mobile radio communications (PIMRC), pp 425–429
6. Shankar H, Kansal A (2018) Performance Analysis of κ-μ/gamma Shadowed fading model over indoor off body communication channel. Int J Electron Commun 93:283–288
7. Cotton SL (2014) A statistical model for shadowed body-centric communications channels: theory and validation. IEEE Trans Antennas Propag 62(3):1416–1424
8. Abdi A, Lau WC, Alouini MS, Kaveh M (2003) A new simple model for land mobile satellite channels: First- and second-order statistics. IEEE Trans Wireless Commun 2(3):519–528
9. Paris JF (2014) Statistical characterization of κ–μ shadowed fading. IEEE Trans Veh Technol 63(2):518–526

10. Yoo SK et al (2016) Shadowed fading in indoor off-body communication channels: a statistical characterization using the κ-μ/gamma composite fading model. IEEE Trans Wireless Commun 15(8):5231–5244
11. Souza RAA et al (2015) On the efficient generation of α–κ–μ and α–η–μ white samples with applications. Int J Antennas Propag 1–13
12. Chauhan PS et al (2018) A unified approach to modelling of probability of detection over α-μ/IG, κ-μ/IG, and η-μ/IG composite fading channels with application to cooperative system 87:33–42
13. Rana V et al (2017) A new closed-form of ASEP and channel capacity with mrc and selection combining over inverse gaussian shadowing. Int J Electron Commun 74:107–115
14. Mittal PK, Gupta KC (1972) An integral involving generalized function of two variables. Proc Ind Acad Sci Sect A 3(3):117–123

DTMOS-Based Low-Voltage Low-Power CCII+ and Biquad Filter Using ±0.25 V Supply

Anil Kumar Gautam, Tripurari Sharan, and P. Nemthianhoi Zou

Abstract This paper presents low-voltage low-power, second-generation positive current conveyor (CCII+) comprised of 6-dynamic threshold MOSFETs (DTMOS) of pmos type and 6-nMOS operating in subthreshold region. It is designed and simulated in standard n-tub bulk-180-nm CMOS process technology. This subthreshold region biased CCII+ cell operating with dual power supply of ±0.25 V consumed total power of 2.43 μW. The circuit has been simulated using Tanner EDA 16.1 version in standard n-tub 180-nm CMOS process technology. To validation, its applications, a voltage mode multiple input single output (MISO), and a voltage mode single input multiple output (SIMO) filters as well as a current mode SIMO type biquadratic filter is designed, and their simulated responses are presented.

Keywords Second-generation current conveyor · DTMOS · Class C cascode inverter · Biquad filter

1 Introduction

The current conveyors are the basic analog design cell from the long back time when Sedra and Smith proposed its first structure in 1968 [1]. Its generation to generation development yields the first-generation, second-generation, and third-generation current conveyors [2]. The second-generation current conveyors can provide bi-directional current at its Z node. The Second-generation positive current conveyor (CCII+) has been found to be the most versatile [3] for voltage and current mode

A. K. Gautam · T. Sharan (✉) · P. Nemthianhoi Zou
Electronics and Communication Engineering, North Eastern Regional Institute of Science and Technology, Deemed to be University, Nirjuli, Itanagar, Arunachal Pradesh, India
e-mail: tsh1962mzp@gmail.com

A. K. Gautam
e-mail: anilgautam19@gmail.com

P. Nemthianhoi Zou
e-mail: nemthianhoizou@gmail.com

D. Harvey et al. (eds.), *Advances in VLSI, Communication, and Signal Processing*, Lecture Notes in Electrical Engineering 683,
https://doi.org/10.1007/978-981-15-6840-4_22

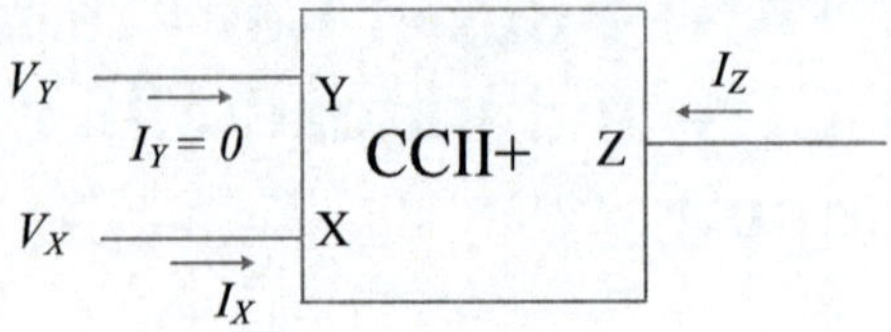

Fig. 1 Block diagram of CCII+

circuit design. The CCII+ is a 3-terminals device having Y and X as the inputs and Z as the output terminal. The block diagram of this analog cell is shown in (Fig. 1).

The input terminals Y and X ensure voltage buffering action whereas X and Z terminals provide current buffering action, i.e., $V_Y = V_X$ and $I_X = I_Z$. In CCII+ the currents I_X and I_Z flow into X and Z terminals whereas in second generation negative current conveyor (CCII-) the I_X flows into the X terminal and I_Z flows out of Z terminal. The input X terminal offers a low-impedance whereas the Y input and Z output terminals offer very high impedance. The very high impedance at Y and Z terminals makes this device suitable for voltage mode as well as current mode applications. Also, the current at X is like that of the current at Z terminal but opposite in direction of flow i.e. $I_X = - I_Z$ for CCII− [3]. The hybrid characteristic of the second-generation positive and negative current conveyor in matrix form is given in (1) which states that $I_y = 0$, $I_z = I_x$ for positive CCII+ and $I_z = - I_x$ for negative current conveyor. The $V_y = V_x$ for both types of second-generation current conveyors.

$$\begin{bmatrix} I_y \\ V_x \\ I_z \end{bmatrix} = \begin{bmatrix} 0 & 0 & 0 \\ 1 & 0 & 0 \\ 0 & \pm 1 & 0 \end{bmatrix} \begin{bmatrix} V_y \\ I_x \\ V_z \end{bmatrix} \tag{1}$$

The low-voltage and low-power CCII+ structure are used in analog circuit design for ultra-low-power and low-frequency applications. The CMOS devices biased in weak inversion region consume ultra-low-power severely limiting the maximum useable frequency of operation and are suitable for low-frequency applications. The voltage gain ratio called α is V_X/V_Y and the current gain ratio says β is I_Z/I_X. The ideal value of α and β are unity whereas they are practically very close to unity.

2 DTMOS Transistor

The DTMOS design topology suggests connecting the substrate (body) of the MOSFET to its gate [2]. Thus, it is the gate as well as bulk-driven technique. The dynamic threshold (DT) CMOS transistors DT-nMOS, DT-pMOS, and their symbols are shown in Fig. 2a, b, respectively. However, in the standard n-tub CMOS process, the bulk-terminal of the pMOS can only be kept isolated from its common substrate, so in this case, pMOS can easily be used as DTMOS transistor without the need of

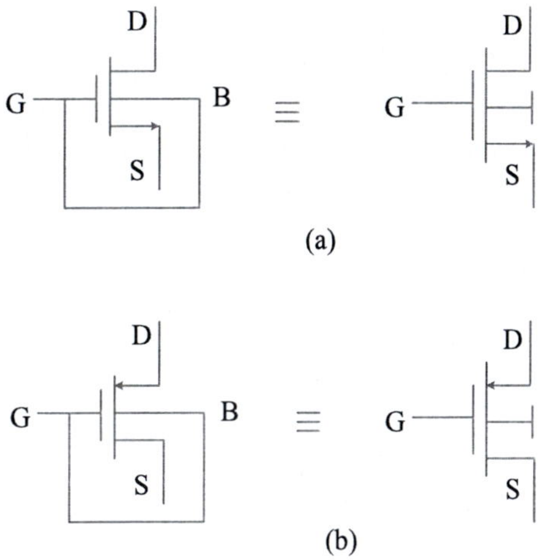

Fig. 2 DTMOS transistors and their symbols **a** nMOS, **b** pMOS

twin-tub CMOS process. So, all pMOS transistors used in the design of this CCII + are DTMOS whereas all nMOS transistors are conventional bulk transistors.

The DTMOS transistor can operate at low-voltage of 0.6 V and even below it while driving large current. This technique increases the overall transconductance of the MOS device [4]. The body-to-gate can be connected either externally or integrated during the IC fabrication [5]. The connection of gate-body leads to a body effect. The body effect can dynamically change the threshold voltage during different operation modes. The threshold voltage is given by (2).

$$V_{\mathrm{TH}} = V_{\mathrm{TH0}} + \gamma\left[\left(\sqrt{2\phi_F + V_{\mathrm{SB}}} - \sqrt{2\phi_F}\right)\right] \tag{2}$$

where V_{TH0} represents the threshold voltage when the source-body voltage is zero, γ is the bulk effect parameter, ($\gamma = 0.4\ \mathrm{V}^{1/2}$ for nMOS and $\gamma = -\ 0.5\ \mathrm{V}^{1/2}$ for pMOS and ϕ_F is a physical parameter called surface potential. In DTMOS due to the shorted body and gate structure when the input signal changes it also changes the threshold voltage of the device. So, the transconductance of the MOSFET varies dynamically which increases the current-drive ability and the transconductance.

One benefit of the DTMOS circuit is that its low-threshold voltage is not a trade-off for large leakage current. This low-threshold voltage leads to high mobility due to body charge reduction [6]. The circuit serves well if the source-body junction remains reverse biased [7]. So, this DTMOS topology is suitable for DC supply less than the cut-in voltage of silicon make diode. The subthreshold or weak inversion operation ensures circuit operation with a very low-power supply and dissipates ultra-low power. One demerit of this operation is its low-gain bandwidth product. So, it is a suitable choice for low-frequency signals, say bio-signal or sub-audio range signal processing. The DTMOS provides high ON-OFF ratio and ensures high current driveability as compared to gate or bulk-driven circuits. The power supply should be less than 0.6 V to avoid latch-up problem [7].

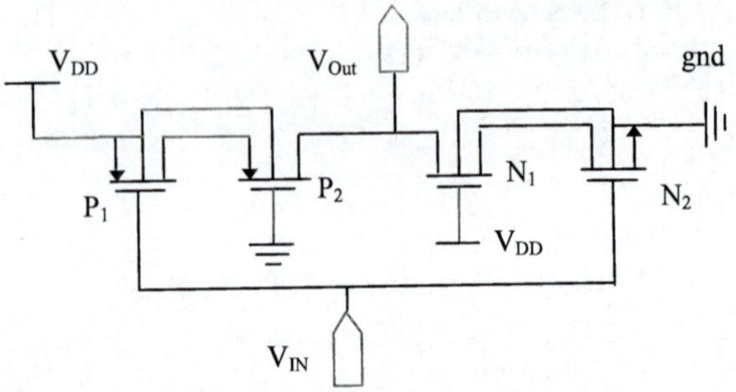

Fig. 3 Traditional class C cascode inverter comprised of four [9] CMOS devices

3 Class C Cascode Inverter

The rapid development of portable electronics market is forcing an explosive growth in the demand for micro-power IC design. A class C inverter is recently reported to replace traditional operational transconductance amplifier (OTA) [8, 9]. The input transistors of the inverter operate in a subthreshold region most of the time, thereby minimizing power dissipation. The conventional cascode class C inverter circuit is shown in Fig. 3.

The cascode class C inverter is comprised of one set of cascode pMOS and one set of cascode nMOS devices. The rail-side pMOS (P_1) and nMOS (N_2) operate in the subthreshold region of operation if the common-mode voltage is average of the rail-to-rail DC bias supply. The total rail-to-rail supply is chosen to be less than the sum of threshold voltage of nMOS and pMOS transistors, so $(V_{DD} + |V_{SS}|) < (V_{THp} + V_{THn})$ where V_{THp} and V_{THn} are the threshold voltage of pMOS and nMOS devices, respectively. This inverter works as a micro-power amplifier. The sizes of CMOS devices in cascode inverter decide the bias current and slew rate of this inverter. The DC gain of this inverter is around 42 dB. This type of CMOS inverter is sensitive to process and supply voltage variations due to its push-pull nature and subthreshold operation [8].

4 The Proposed CCII+ Circuit Schematic

The proposed circuit shown in Fig. 4 has been designed to obtain a very low-power consumption and low-power dissipation of 2.43 μW with ±0.25 V dual power supply. It consists of a pseudo-differential amplifier and a cascode class C inverter. The OTAs can be replaced with class C inverter [8]. Here all pMOS transistors used are DTMOS transistors. This CCII+ circuit is comprised of 12 CMOS devices, having 6-pMOS in DTMOS structure and six usual nMOS transistors. The transistors from P_1, P_2, N_1, and N_2 form pseudo-differential amplifier, and the remaining ones P_3–P_6 and N_3–N_6 form two-cascode class C inverters (see Fig. 4). Pseudo-amplifier does not use tail current source and save one-overdrive order of voltage. So, it operates well with

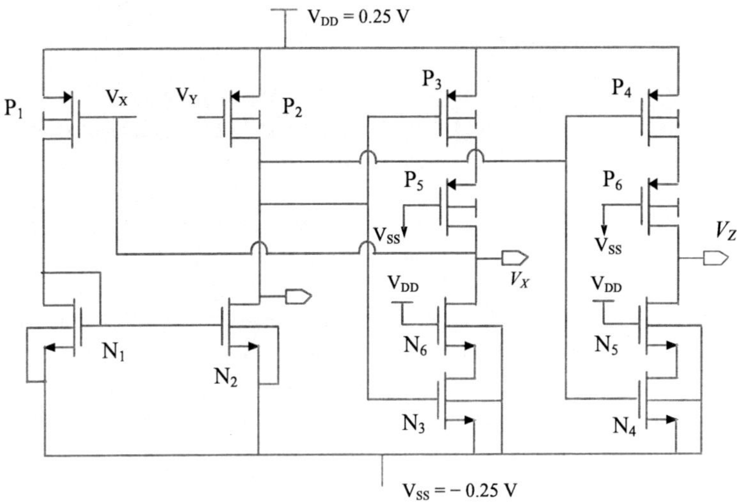

Fig. 4 The proposed CCII+ circuit

low-voltage supply, however, its CMRR is unity. The output of pseudo-input pair is applied to the input of first-class C inverter as well as to the second-class C inverter. The output of first inverter is connected to inverting input of pseudo-amplifier which forms low-impedance *X* terminal. Owing to 100% current shunt negative feedback the input impedance at *X* node is very less. The output of second-class C inverter forms the output *Z* terminal of CCII+. The non-inverting input of pseudo-amplifier forms high impedance *Y* terminal of CCII+.

The W/L ratio of the CMOS devices used in the CCII+ circuit schematic is given in Table 1. The channel length *L* of CMOS devices used in OTA sub-section of the circuit is chosen to be 1 μm to avoid channel length modulation effect in this design.

However, the CMOS devices used in inverter sections have *L* of 0.18 μm. It is chosen equal to the minimum feature size to reduce the voltage drop across these devices which in turn increases the output signal swing.

Table 1 W/L ratio of the CMOS devices

Name of CMOS	W/L in (μm/μm)
P_1, P_2	10/1
N_1, N_2	4/1
P_3, P_4	30/0.18
P_5, P_6	120/0.18
N_3, N_4	15/0.18
N_5, N_6	60/0.18

4.1 Simulation Results

The proposed model has been designed and simulated in T-spice using Tanner EDA tool of version 16.1 using standard n-tub 180 nm bulk-CMOS technology with ±0.25 dual power supply. Column 2 of Table 2 lists the various simulated parameters of this CCII+ cell. However, as suggested by reviewer-3, this circuit has also been simulated in 45 nm standard n-tub bulk-CMOS process with ±0.2 V supply and its results are listed in column 3 of Table 2. Though this subthreshold biased CCII+ when ^simulated in 45 nm CMOS process consumed less power of only 1.33 nW but its performance was limited up to 100 Hz frequency. So, its performance was found to be unsuitable even for sub-audio-frequency signal processing. Its linearity was very poor even for 1 Hz signal frequency. Owing to this inefficiency of 45 nm node CCII+ cell the further analysis of CCII+ has been performed in 180 nm CMOS process.

If the dual power supply voltage used is ±0.25 V proposed CCII+ cell dissipates total power of 2.43 μW whereas this power dissipation reduces to 720 nW for ±0.2 V dual power supply at the cost of circuit performance. The AC analysis, DC sweep analysis, and transient analysis have been performed and results are listed in Table 2.

Table 2 Performance results of the proposed CCII+

Parameters	Values	^Values
Technology length	180 nm	45 nm
Supply voltage	±0.25 V	±0.2 V
Numbers of transistors	12	12
Input impedance at *X*	7 kΩ	12 kΩ
Input impedance at *Y*	1313 MΩ	500 MΩ
Output impedance at *Z*	2 MΩ	1 MΩ
Power dissipation of single CCII+ cell	2.43 μW	1.33 nW
3-dB bandwidth at *X* and *Z*	72 kHz	15.5 kHz
Voltage gain α (V_X/V_Y) frequency range (1 Hz-10 kHz)	1.0023	1.143
Current gain β (I_Z/I_X) frequency range (1 Hz-10 kHz)	1.0000001	0.999997
Slew rates at *Z* node	[a]SR + = 100 V/ms and SR − = 134 V/ms	[b]SR + = 30 V/ms and SR − = 12 V/ms
Slew rates at *X* node	[a]SR + = 333 V/ms and SR − = 300 V/ms	[b]SR + = 50 V/ms and SR − = 33.3 V/ms

[a]Evaluated using 1 kHz pulse of 200 mV peak
[b]Evaluated using 100 Hz pulse of 100 mV peak

4.2 AC Analysis

Figure 5 shows the AC analysis result for CCII+ when VAC input has been applied to Y node and output response observed at X and Z node with load resistor of 10 kΩ at both of X and Z node. It shows a −3 dB bandwidth of 72 kHz. The V_X, V_y, and V_z node potential follow well up to 10 kHz frequencies.

The current gain ratio I_Z/I_X has been simulated in AC response mode with the condition of V_{AC} input applied at Y node and simulating the generated current frequency responses at Z and Y node. The mean of current gain β (for 1 Hz to 100 kHz range) is found to be 1.00000043 as depicted in Fig. 6 whereas the mean of this current gain β in the frequency range 1 Hz to 10 kHz is found to be 1.0000001.

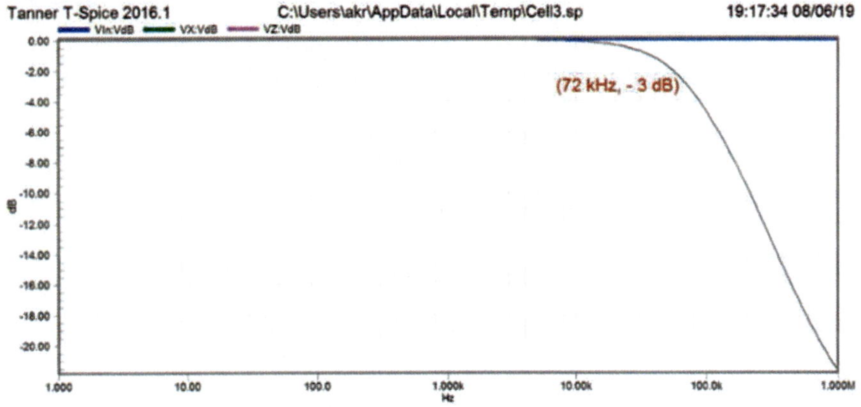

Fig. 5 AC response with input at Y terminal and output at X and Z terminals

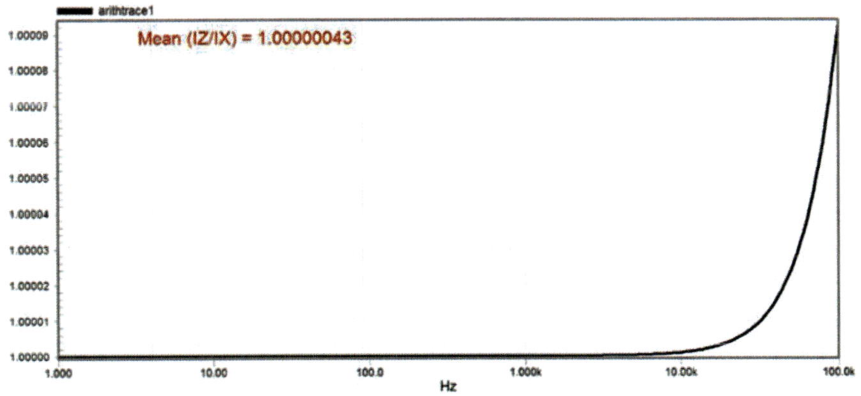

Fig. 6 AC response for β with input at Y terminal and output currents at Z and X terminals

4.3 DC Sweep Analysis

Figure 7 shows the DC sweep result with a simulation set up of DC voltage (V_1) applied at Y node and output observed at all the Y, X, Z nodes with a resistive load of 10 kΩ applied at X and Z nodes. The DC sweep result ensured that X and Z node voltages followed well for DC input range of −160 mV up to 200 mV.

Figure 8 shows the output currents flowing through load resistors at X and Z nodes when DC input is applied to Y terminal. For DC sweep rage of ±250 mV the absolute value of output current at X and Z nodes are found well within +23 μA to −21 μA.

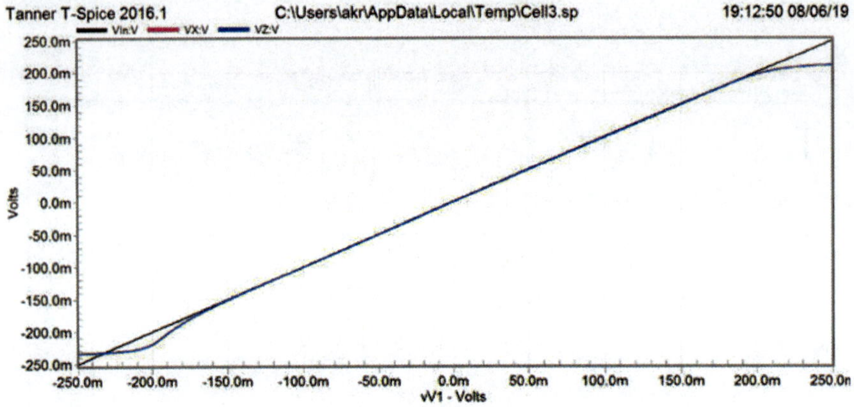

Fig. 7 DC sweep result with input DC at Y node ant output at X and Z nodes

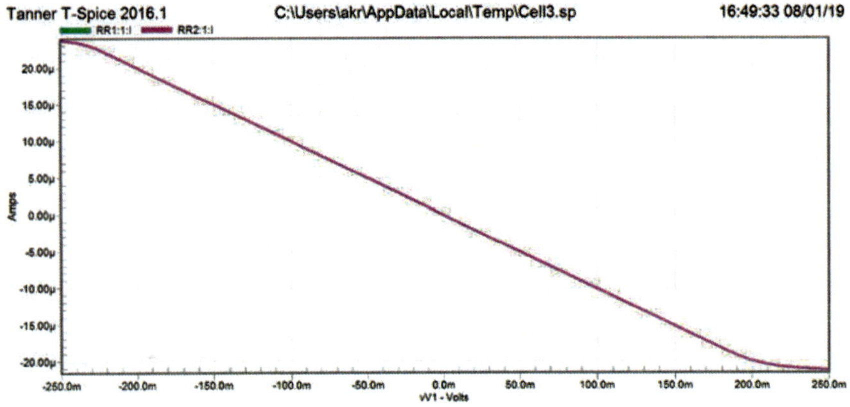

Fig. 8 DC sweep result with input DC at Y node and output currents at X and Z nodes

4.4 Transient Analysis

Figure 9 shows the transient response for CCII+ when sine wave of 100 mV peak value at 1 kHz frequency is applied to Y terminal. The simulated output indicated at X and Z nodes of the proposed CCII+ with resistive load of 10 kΩ at both the X and Z nodes.

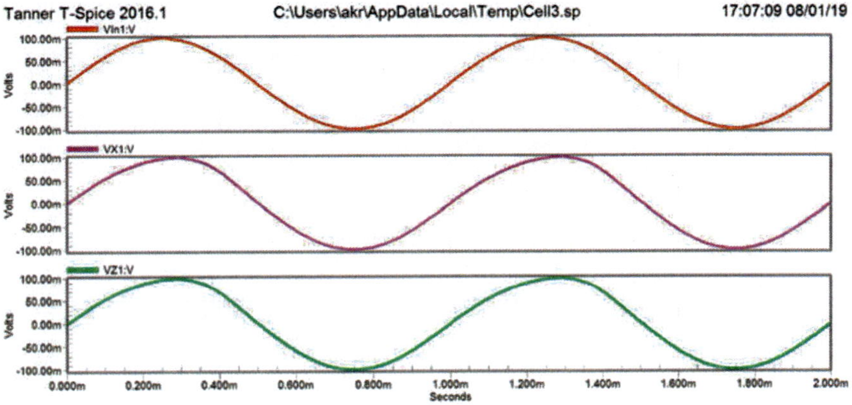

Fig. 9 Transient response of $V(Z)$, $V(Y)$, and $V(X)$ versus time (μs)

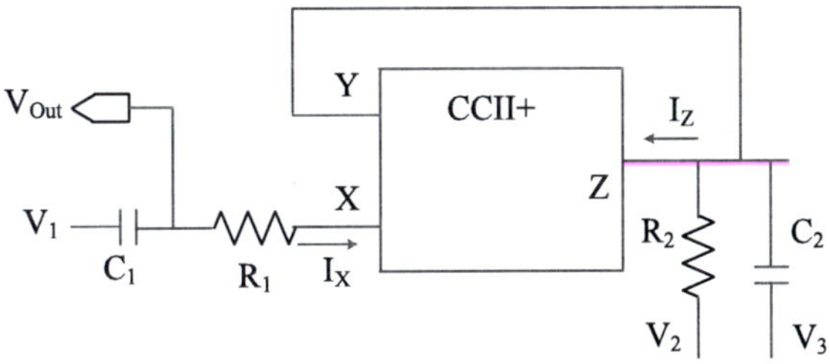

Fig. 10 Single CCII+ based MISO type biquadratic filter [10, 11]

5 Use in Design of Biquadratic Filter

5.1 *Multiple Input Single Output (MISO) Biquadratic Voltage Mode Filter*

The MISO type biquadratic filter is realized by using a single CCII+ cell and two-resistors and two-capacitors as proposed in [10] and shown in Fig. 10. Since it uses a single CCII+ cell so, it dissipates total power of 2.43 μW.

The routine analysis of the filter provides the following relation in V_o and V_1, V_2, and V_3.

$$V_O = \frac{\left(V_1 R_1 R_2 C_1 C_2 s^2 + V_3 R_2 C_2 s + V_2\right)}{\left(R_1 R_2 C_1 C_2 s^2 + (R_1 C_1 + R_2 C_2) s + 1\right)} \tag{3}$$

This filter generates all four-generic filter functions with different input conditions as listed in Table 3.

The central frequency, ω_0, and quality factor, Q of this MISO filter are given by (4–5).

$$\omega_0 = 1 \Big/ \sqrt{R_1 R_2 C_1 C_2} \tag{4}$$

and quality Factor

$$Q = \sqrt{R_1 R_2 C_1 C_2} \Big/ (R_1 C_1 + R_2 C_2) \tag{5}$$

The values of resonant frequency ω and quality factor Q are 1.8 kHz and 1, respectively. All types of filter responses have been simulated by setting all capacitors $C = 10$ nF and all resistors $R = 10$ kΩ. Figure 11 shows the low-pass (LP) function with higher cutoff frequency of 2 kHz.

In the passband of low-pass filter response a sine wave input of frequency 10 Hz and amplitude 100 mV yields a total harmonic distortion (THD) of 0.39%, equivalent to −48 dB.

Figure 12 shows the high-pass (HP) filter function response of this MISO filter which depicts the lower cutoff frequency of 1.2 kHz.

Table 3 Input conditions for various responses of MISO filter

Filter-type	Input conditions
Low-pass (H_{LP})	$V_{in} = V_2$ and $V_1 = V_3 = 0$
High-pass (H_{HP})	$V_{in} = V_1$ and $V_2 = V_3 = 0$
Band-pass (H_{BP})	$V_{in} = V_3$ and $V_1 = V_2 = 0$
Band-reject (H_{BR})	$V_1 = V_2 = V_{in}$ and $V_3 = 0$

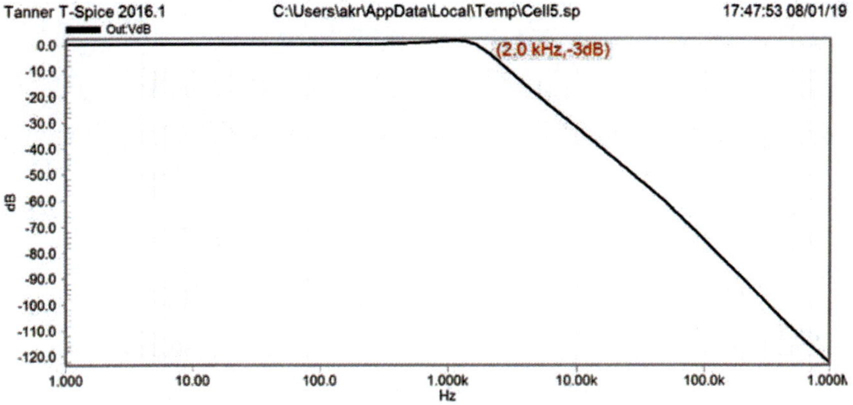

Fig. 11 Low-pass filter response of MISO biquadratic filter using single CCII+

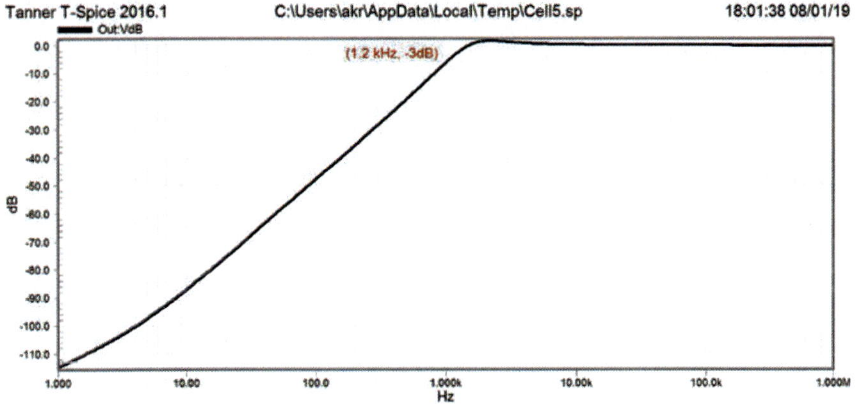

Fig. 12 High-pass Filter response of MISO biquadratic filter using single CCII+

Figure 13a shows the band-pass (BP) response of MISO biquadratic filter which showed the lower, central, and upper cutoff frequencies of 954 Hz, 1575 Hz, and 2600 Hz, respectively. The quality factor Q of this BP response is $Q = f_o/\mathrm{BW} = 1575/(2600 - 954) = 0.957$. Thus, Q is very close to unity.

Figure 13b shows the band-pass response in all device corners, typical nMOS typical pMOS (TT), fast nMOS fast pMOS (FF), fast nMOS slow pMOS (FS), slow nMOS fast pMOS (SF), and slow nMOS slow pMOS (SS) which showed a central frequency variation range of 1202–1560 Hz but its Q remained nearly same.

Figure 14a shows the band-reject (Notch) response of MISO biquadratic filter which depicted the lower, notch, and upper cutoff frequencies of 2450 Hz, 1557 Hz, and 990 Hz, respectively. The quality factor Q of this BP response is $Q = f_N/\mathrm{BW} = 1557/(2450 - 990) = 1.066$. Thus, Q is very close to unity. Figure 14b shows the

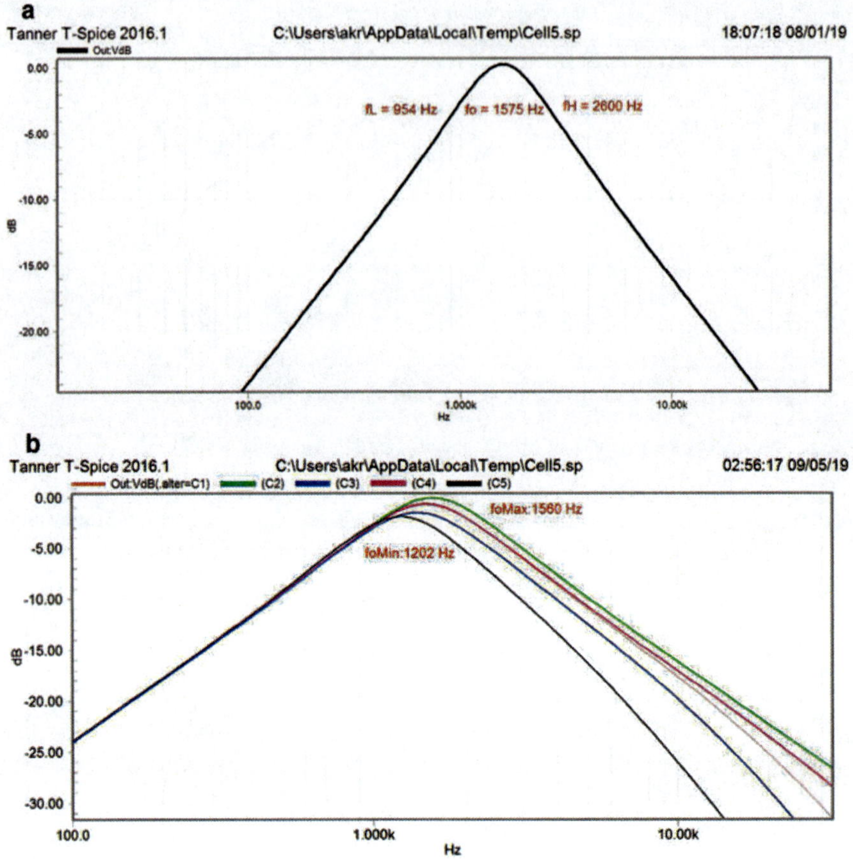

Fig. 13 **a** Band-pass Filter response of MISO biquadratic using single CCII+, **b** band-pass response in all five device corners TT, FF, FS, SF, SS

band-pass response in all device corners, TT, FF, FS, SF, and SS which showed a notch frequency variation range of 1148 to 1585 Hz, but its Q remained nearly the same.

5.2 *Single Input Multiple Output (SIMO) Voltage Mode Biquadratic Filter*

Figure 15 shows the single input multiple outputs voltage mode biquadratic filter [12] which utilized four numbers of CCII+ cells two capacitors and five resistors as proposed in [12]. The routine analysis of this SIMO type voltage mode filter provides input and output relation given by:

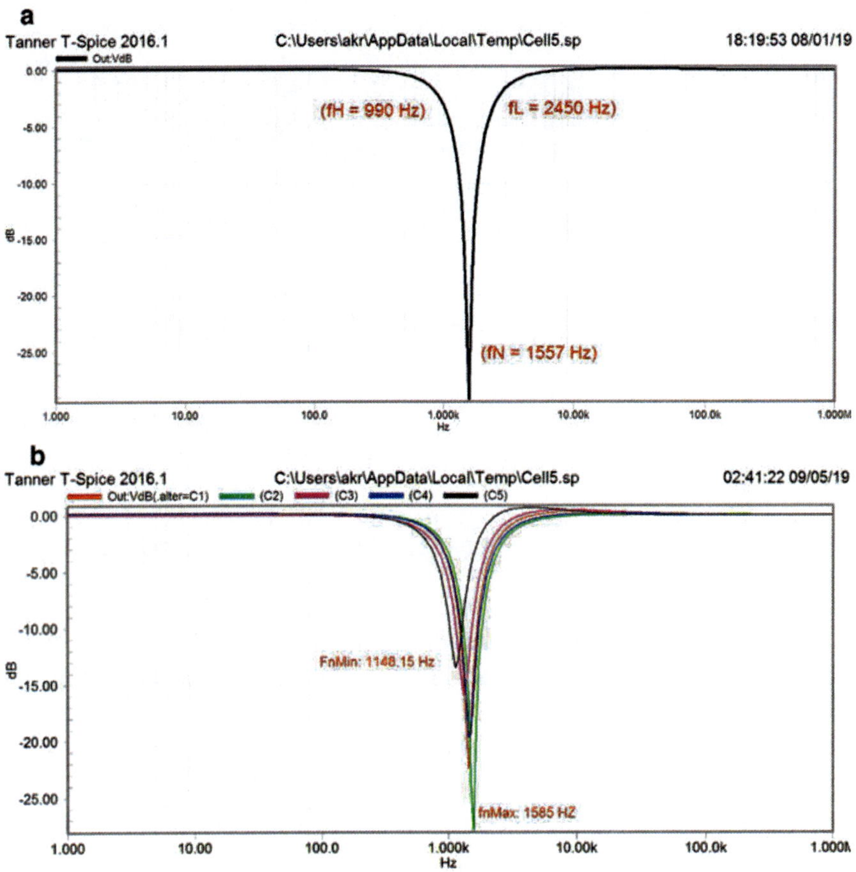

Fig. 14 **a** Band-reject Filter response of MISO Biquadratic using single CCII+, **b** band-reject (Notch) response in all five device corners, TT, FF, FS, SF, and SS

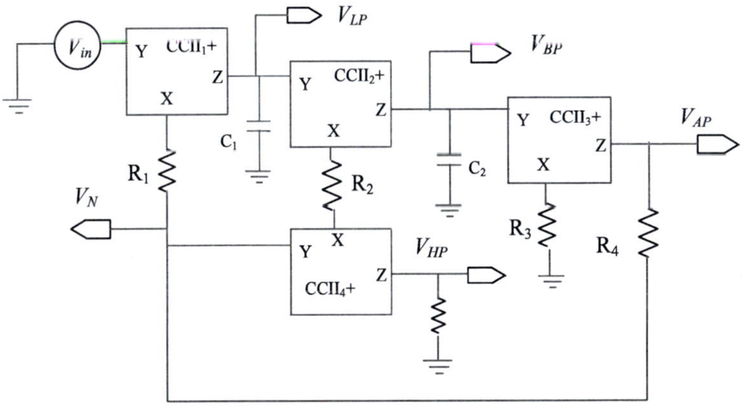

Fig. 15 Four-CCII + based voltage mode SIMO type biquadratic filter [12]

$$\frac{V_o(s)}{V_{\text{in}}(s)} = \frac{s^2\,C_1C_2G_1 - s\,C_1G_1G_2 + G_1G_2G_3}{s^2\,C_1C_2G_1 + s\,C_1G_1G_2 + G_1G_2G_3} \tag{6}$$

This filter provides LP at node V_{LP}, BP at node V_{BP}, all-pass (AP) at node V_{AP} for ($R_4 = R_1$), HP at node V_{HP} and notch responses at node V_{N} (see Fig. 15).

Its central frequency and quality factor are given by (7–8).

$$\omega_0 = 1\Big/\sqrt{C_1C_2R_2R_3} \tag{7}$$

$$\text{And } Q = \frac{\sqrt{C_2R_2R_3}}{R\sqrt{C_1}} \tag{8}$$

Figure 16a shows AC response results for low-pass, band-pass, high-pass, band-reject, and all-pass responses. It dissipates the total power of 9.7 μW. The low-pass filter response depicted a total harmonic distortion (THD) of 0.39% (−48 dB) at V_{LP} output node when a sine wave of frequency 10 Hz and amplitude 100 mV was applied to input node of this filter. The band-pass response when simulated in all device corners TT, FF, FS, SF, and SS showed a central frequency variation range of 955–1510 Hz but its Q remained nearly the same as shown in Fig. 16b. The band-reject (notch) response when simulated in four device corners FF, FS, SF, and SS showed a notch frequency variation range of 1096–1536 Hz but its Q remained nearly same as shown in Fig. 16c.

5.3 *Single Input Multiple Output (SIMO) Current Mode Biquadratic Filter*

Figure 17 shows the circuit schematic of current mode SIMO type filter which generates low-pass, band-pass, and high-pass functions [13]. The routine analysis of this filter provides the s-domain output current (I_o) and input current (I_{in}) relation given by:

$$\frac{I_o(s)}{I_{\text{in}}(s)} = \frac{s^2\,C_1C_2R_1R_2 + s\,C_1R_1 + sC_2R_1 + sC_2R_2 + 1}{s^2\,C_1C_2R_1R_2 + s\,(C_1R_1 + C_2R_2) + 1} \tag{9}$$

Central frequency ω_0 is given by:

$$\omega_0 = 1\Big/\sqrt{C_1C_2R_1R_2} \tag{10}$$

and quality factor Q is given by:

$$Q = \sqrt{R_1R_2C_1C_2}\Big/(R_1C_1 + R_2C_2) \tag{11}$$

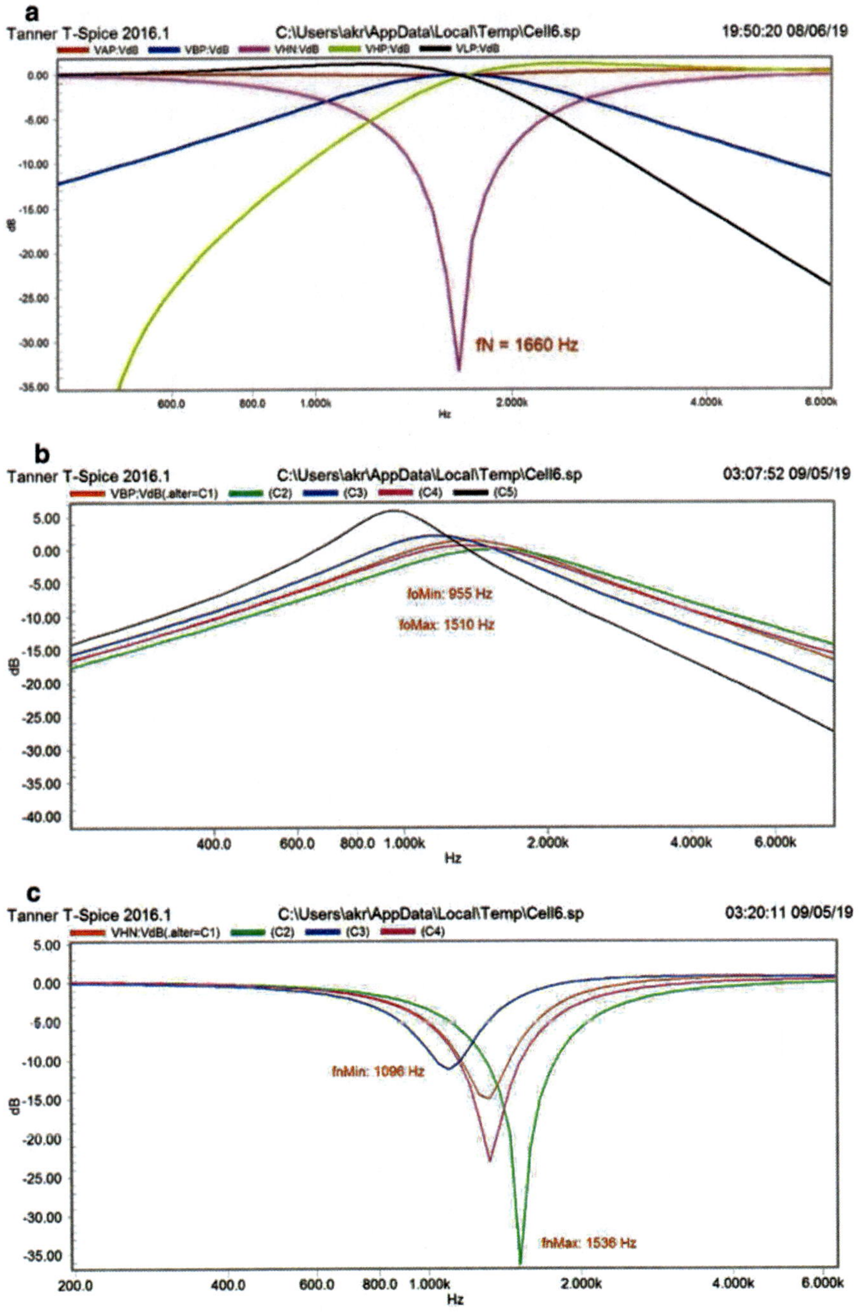

Fig. 16 **a** AC response results of SIMO type multiple outputs with $C = 10$ nF and $R = 10$ kΩ, **b** band-pass response in all device corners, **c** band-reject response in four device corners

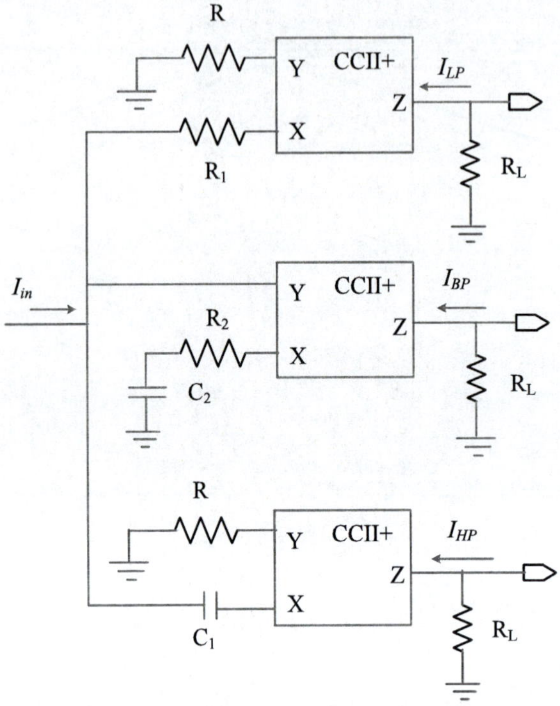

Fig. 17 The current mode SIMO type biquadratic filter realized with three CCII+ cells [13]

Figure 18 shows the AC response results of the current mode multiple outputs. It dissipates the total power of 7.3 μW. The f_H and f_L of low-pass and high-pass responses are at 1.55 kHz. The band-pass response has central frequency gain and

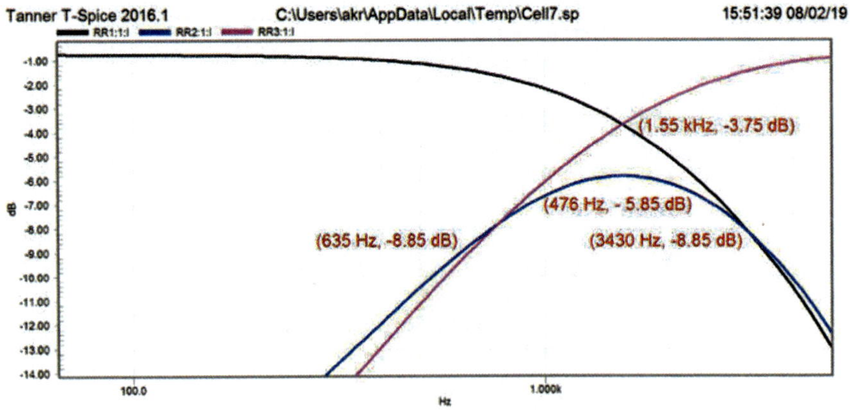

Fig. 18 AC response of SIMO type current mode biquadratic filter

Table 4 Performance comparison of CCII+ with referred work [14–18]

Parameters	This work	[14]	[15]	[16]	[17]	[18]
Power supply (V)	±0.25	0.4	±0.2	0.4	±0.2	0.5
CMOS process (nm)	180	90	180	180	90	180
BW (V_X/V_Y), kHz	72	1250	570	10	1000	15,800
BW (I_Z/I_X), kHz	72	800	570	82	1250	25,000
R_X (kΩ)	7	0.12	0.964	62.6	0.106	0.95
R_Y (MΩ)	160	–	–	2400	–	30×10^3
R_Z (MΩ)	2	0.193	0.368	41	1.0	10.9
Gain (V_X/V_Y) 1 Hz to 10 kHz	0.9998	1.004	1.00	0.998	1.004	1.00
Gain (I_Z/I_X)	1.000	1.000	1.00	0.998	1.00	0.999
Power (μW)	2.43	1.7	0.21	0.25	1.8	4.7

frequency of −5.85 dB and 1476 Hz, respectively. It has lower and upper cutoff frequencies of 635 and 3430 Hz. It has Q of 1476/(3430 − 635) = 0.528. The calculated value of Q for equal capacitor and resistor-based design from (11) is 0.5. Thus, the simulated value of Q (0.528) is very close to the calculated value of Q (0.5).

6 Performance Comparison

Table 4 presents the comparison of this proposed CCII+ with other works of low-voltage referred in [14–18]. This CCII+ has used cascode inverters at its output sides and it utilized ±0.25 V dual power supply. So, its power consumption is slightly more than the CCII+ cells biased with low-supply of 0.4 V [14–17]. However, its power dissipation is less than the CCII+ reported in [18] which also used the same DC power supply of 0.5 V.

7 Conclusions

The proposed CCII+ made of DTMOS transistors and class C cascode inverters operates at a very low-dual supply voltage of ±0.25 V. It works properly with low-power dissipation of 2.43 μW. This CCII+ has ensured very good voltage and current buffering action in between its Y, X terminals, and X, Z terminals, respectively. The second-order MISO type filter made of a single CCII+ cell with two resistors and two capacitors has proved its utility as low-voltage filter. The SIMO type biquadratic voltage mode filter which utilized four CCII+ cells, two capacitors, and five resistors have produced all five types of generic filter functions. These voltage mode MISO and SIMO filters dissipated the power of 2.43 μW and 9.7 μW, respectively whereas three CCII+ cells based current mode biquadratic filter consumed the power of 7.3 μW. These filters are suitable for low-frequency bio-signal and sub-audio range signal processing.

Acknowledgements This work has been simulated using Tanner EDA tool of version 16.1, in 0.18 μm CMOS technology, available in VLSI laboratory of ECE department of NERIST, deemed to be University, Nirjuli, Papum Pare, Arunachal Pradesh, India.

References

1. Smith KC, Sedra A (1968) The current conveyor-a new circuit building block. Proc IEEE 1368–1369
2. Uygur A, Kuntman H (2013) An ultra low-voltage, ultra low-power DTMOS-based CCII design for speech processing filters. In: 8th International Conference on Electrical and Electronics Engineering, pp 31–35
3. Bhargave G, Sheokand P, Kaur J (2015) A new low voltage and low power filter using DTMOS-CCII. In: Second international conference on advances in computing and communication engineering, Dehradun, India, pp 17–20
4. Prokop R, Musil V (2009) Current conveyor CCII as the most versatlie analog circuit building block. Ann J Electron 73–76
5. Sedra A Smith K (1970) A second generation current conveyor and its application. IEEE Trans Circuit Theory Soc Lond 17(1):132–134
6. Colinge JP (1987) An SOI-voltage controlled bipolar-MOS device. IEEE Trans Electron Devices ED-34(4):845–849
7. Assaderaghi F, Bokor J, Ko PK, Hu C (1997) Dynamic threshold-voltage (DTMOS) for ultra-low voltage VLSI. IEEE Trans Electron Devices 44(3):414–422
8. Roy K, Prasad SC Low-power CMOS VLSI design. Wiley, UK. ISBN: 978-81-256-2023-7
9. Chae Y, Han G (2009) Low, voltage, low power inverter based switched capacitor delta-sigma modulator. IEEE J Solid State Circuits 44(2):458–472
10. Chae Y, Han G (2007) A low power sigma-delta modulator using class-C inverter. In: Symposium on VLSI circuits VLSIC-digest of technical papers, pp 240–241
11. Horng JW (2010) Voltage/current-mode universal biquadratic filter using single CCII+. Indian J Pure Appl Phys 48:749–756
12. Horng JW, Chiu WY (2016) High input impedance voltage-mode biquad with one input and five outputs employing four CCII ICs. Indian J Pure Appl Phys 54:577–582
13. Naidu GA, Krishna BT (2017) Design of new universal filters with second generation current conveyor. Int J Electron Eng Res 9(6):845–858
14. Eldeeb MA, Ghallab Y, Ismail HY, Elghitani H (2016) A 0.4 V 90 nm CMOS subthreshold current conveyor. In: IEEE Canadian conference on electrical and computer engineering (CCECE), pp 1–4
15. Uygur A, Kuntman H (2014) A very compact 0.4 V DTMOS CCII employed in an audio-frequency filter. Analog Integr Circ Process 81:89–98

16. Kumngern M (2019) 0.4 V fully differential current conveyor using multiple-input bulk-driven MOST technique. Microelectron J 90:96–104
17. Eldeeb MA, Ghallab YH, Ismail Y, Elghitani H (2017) Low-voltage subthreshold CMOS current mode circuits: design and applications. AEU-Int J Electron Commun 82:251–264
18. Kumngern M, Khateb F (2014) 0.5-V bulk-driven second-generation current conveyor. In: IEEE symposium on computer applications and industrial electronics, pp 180–183

Structural and Optical Analysis of Bulk-Hetero Interface Between MoS_2: Pentacene

Anshika Srivastava and Shweta Tripathi

Abstract In recent years, the two-dimensional MoS_2 material has attained loads of attention within the entire research fraternity. Due to the nano-sheet structure formation in MoS_2, they get special attention in the area of bendable electronics and optoelectronics. 2D-organic interface devices are studied and analyzed in various forms like heterojunction and bulk heterojunction. Here, in this work for the first time, we have deposited and analyzed the MoS_2 film blended with pentacene material (MoS_2: Pentacene). The ratio of MoS_2 and pentacene in a blend was kept as 1:0.1. Then we have compared MoS_2 and MoS_2: Pentacene film via structural and optical characterization. For the structural scrutiny, we have utilized XRD and FESEM data, whereas for optical scrutiny we have used UV-Vis Spectroscopy data. The outcomes of this experimental work are reliable that gives a deeper understanding of the MoS_2 film.

Keywords MoS_2 · MoS_2: Pentacene · Blend · XRD and morphology

1 Introduction

The two-dimensional (2D) materials have delivered an enthralling podium for the investigations of essential science and technology at nano-levels. Particularly, 2D transition metal di-chalcogenides (TMDs), like Molybdenum disulfide (MoS_2), have been deeply explored owing to its exclusive optical and electronic behavior, comprising resilient spin-orbit assembly, tailorable optical band-gap, direct–indirect band-gap interaction, etc. [1–6]. As a consequence, it further allows it to be used for the forthcoming peer groups such as bendable nano-photonics and nano-electronics applications [7–9]. One of the major tasks for 2D TMD materials is the consistent and

A. Srivastava (✉) · S. Tripathi
Department of Electronics & Communication Engineering, Motilal Nehru National Institute of Technology Allahabad, Prayagraj 211004, India
e-mail: anshi230893@gmail.com

S. Tripathi
e-mail: shtri@mnnit.ac.in

D. Harvey et al. (eds.), *Advances in VLSI, Communication, and Signal Processing*, Lecture Notes in Electrical Engineering 683,
https://doi.org/10.1007/978-981-15-6840-4_23

manageable synthesis of fault-free materials. Conversely, these are in strong contrast to the deep-rooted area of organic materials. Organic materials always ensure the admirable behavior for bendable electronics [10, 11]. They can be synthesized and processed using easy and low-cost techniques. Thus, marrying 2D TMDs and organic fields will carry advantages that are not available in each material individually, facilitating even superior, versatile bendable devices [12]. The interface between organic and TMD can form either p-n heterojunction or p-n bulk heterojunction. It can be used for a number of applications including solar cells, light-emitting diodes, photodetectors, sensors, pn-diodes, etc. In applications like solar cells by interfacing, these two field materials the ultrafast charge separation of the excitons are possible, ensuing long-lasting charge transfer states.

Pentacene is an eminent organic material having tremendous charge carrier mobility among the organic materials. It has been widely used for the applications of organic electronic devices [13]. Interestingly, the pentacene interface with other materials determines the prominent properties for optoelectronic devices. Additionally in the literature, pentacene/2D-TMD heterostructures such as pentacene/MoS_2 are well premeditated. The electronic and charge transport behavior of such interfaces has been widely recognized by a number of research communities [14, 15]. By the consideration of such promising properties, several literatures represented the pentacene/MoS_2 heterostructure interface in optical devices such as photo-detector, solar cells, etc. [16]. Moreover by Shen et al. the pentacene doped MoS_2 monolayer (ML) system was engineered and well-deliberated for density functional theory (DFT) calculation and charge transfer investigation [17]. By taking all the literature, yet the interface study between these two capable materials is very few. So, in the present work, we have premeditated the structural and optical behavior of MoS_2 film blended with a small fraction of pentacene.

In the first section, a brief introduction is given for the motivation of this work. Then after in Sect. 2, the experimental process for the deposition of films is given. Subsequently, in Sect. 3 we have corroborated the results based on structural and optical characteristics. At last, the conclusion of this work is made.

2 Experimental Procedure

For the analysis of MoS_2 blended with pentacene film, we have compared two individual films deposited on indium tin oxide (ITO) substrates. Film 1 is of MoS_2 material and film 2 is of MoS_2: Pentacene material. The 99.9% pure MoS_2 and pentacene material was procured from Sigma-Aldrich, and consumed without any additional refinement. Further on one of the ITO substrates, MoS_2 was grown and on the other, MoS_2: Pentacene (1:0.1), by adapting the dispersion process. In order to form the MoS_2 solution, firstly, the fine particles of MoS_2 (0.5 g) were mixed to a 15 ml liquid solvent of 2-propanol (IPA) and stirred for 6 h. Consequently, for the MoS_2: Pentacene solution, in the same formed MoS_2 solution, we have added a pentacene solution and stirred again for 6 h. The pentacene solution was made with 0.05 g of pentacene powder in 5 ml of o-dichlorobenzene (stirred for 6 h in the separate flask). After getting a good mix of both individual solutions, we have deposited them onto the ITO substrate.

For the preparation of samples, primarily the ITO substrate was separated into small divisions of facet 2 cm × 1.2 cm. Then in the sizzling solution of distilled ionized (DI)-water and sodium hydroxide (NaOH), the sections of the substrate were sinked for thorough cleaning. After suitable cleaning, the substrates were undergone to ultrasonic bath in DI water for 1 h. In succession, the substrates were desiccated for 10 min on the hot plate, and after that, they were positioned sequentially on the spin coater sample rack. The prepared solutions were transferred onto the substrate by using 10 μl micro-pipette. Sample rack was whirled for 60 s at the speed of 800 rpm involving 5 s of acceleration time. This step was iterated for 3 times. Since the process of annealing is important to make a stable film, therefore once the films were deposited, the post-vacuum-annealing process was involved for 3 h at 60°C. At last, both the films were engaged in the characterization equipment.

3 Results and Discussions

The crystallinity and the structure of grains in the film are attained by the Field Emission Scanning Electron Microscope (FESEM) image using JEOL JEC-3000 FC equipment and X-ray diffraction spectroscopy using RIGAKU Smart-lab X-ray diffractometer (XRD) equipment having radiation Cu Kα that sustains wavelength, $\lambda = 1.54056$ Å. For the optical performance, absorption and transmittance graphs are learnt by using UV-Vis Spectroscopy equipment.

3.1 Structural Analysis

The pattern of X-ray diffractions at various angles from the crystals of MoS_2 and MoS_2: Pentacene is represented in Fig. 1. The noticeable peaks are at (002), (103), (006), (105), and (008) plans for both films that are under study. These peaks ensure the deposition of MoS_2 film having 2H stable phase. It is verified from the peaks resembling the data provided by the Joint Committee on Powder Diffraction Standards (JCPDS) card number 00-037-1492. The phase of the film is basically determined by the *d*-spacing value at a substantial *hkl* plane. So we have utilized Bragg's law (eq. 1 [18]) intended for the valuation of *d*-spacing between two successive planes.

$$k\lambda = 2d\sin\theta \quad (1)$$

where $k = 1$. The 2H stable phase with multiple peaks corresponds to the polycrystalline hexagonal structure that involves six lattice constants: α, β, γ, a, b, and c. The

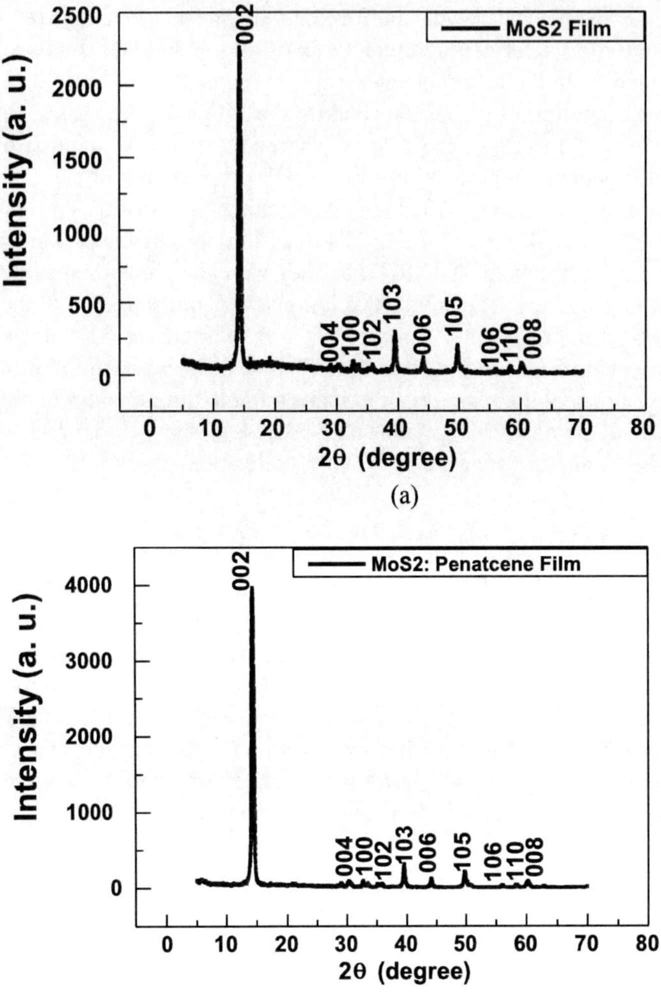

Fig. 1 XRD pattern of films on ITO substrate: (a) MoS_2 (b) MoS_2: Pentacene

parameters $\alpha = 90°$, $\beta = 90°$, and $\gamma = 120°$ are fixed for the hexagonal structure. Whereas $a = b$=3.136 Å and $c = 12.4$ Å lattice constraints are calculated from eq. 2 [19].

$$\frac{1}{d_{hkl}^2} = \frac{1}{a^2}\left[\frac{4}{3}\left(h^2 + k^2 + hk\right) + l^2\left(\frac{a}{c}\right)^2\right] \tag{2}$$

From the patterns of both films, we can detect that the growth orientation is the same for both the films. But in the MoS_2: Pentacene film the intensity is high, comparatively, i.e., the additional growth occurs in the same plane of the blended film. This exemplifies that the involvement of pentacene organic material endorses high degree of crystallinity, less strain, large crystallite size, and less dislocation density as tabulated in Table 1. The strain in the film is signified by full width at half maxima (*F*) parameter. Along with the crystallite size "*D*" is assessed from the Debye-Scherrer formula (eq. 3 [20])

$$D = \frac{M\lambda}{(F)\cos\theta} \quad (3)$$

where $M = 0.9$, "*F*" is in radians and "Θ" designates the Bragg's angle. As anticipated, the crystallite size is less for the blended film. Moreover, the dislocation density "δ" (eq. 4 [21]) have been assessed for the understanding of disorders prompted inside the film.

$$\delta = \frac{1}{D^2} \quad (4)$$

Table 1 Comparison of the properties of MoS_2 and MoS_2: Pentacene films deposited on ITO substrate at crystallography level

S.No.	Planes (hkl)	Space, d (Å)	Properties	MoS_2 film	MoS_2: Pentacene film
1.	002	6.2	*F* (radians)	0.38955	0.38703
			D (Å)	3.58	3.61
			TC	2.13	2.67
			δ ($10^{20}/m^2$)	0.078	0.076
2.	103	2.27	*F* (radians)	0.40258	0.38370
			D (Å)	3.66	3.80
			TC	0.474	0.347
			δ ($10^{20}/m^2$)	0.074	0.069
3.	006	2.04	*F*(radians)	0.41437	0.36585
			D (Å)	3.61	4.09
			TC	1.09	1.12
			δ ($10^{20}/m^2$)	0.076	0.059
4.	105	1.82	F(radians)	0.45406	0.40764
			D (Å)	3.36	3.75
			TC	0.70	0.52
			δ ($10^{20}/m^2$)	0.088	0.071
5.	008	1.53	F(radians)	0.60677	0.36886
			D (Å)	2.64	4.34
			TC	0.60	0.61
			δ ($10^{20}/m^2$)	0.143	0.053

Polycrystalline film texture expresses the distribution of crystallographic orientations. The texture coefficient (TC) quantity is subjected to the proportion of crystals taking the preferred orientation. Correspondingly, the texture of the film shows a noteworthy influence on the surface morphology of the film. Therefore, it shows a vital role in the structural study and can be deliberated by eq. 5 [22].

$$\mathrm{TC}(hkl) = \frac{I(hkl)}{\sum_{i=0}^{i=z} I(h_i k_i l_i)} \tag{5}$$

where "z" describes the entire Miller indices and the achieved atom concentration in a *hkl* plane is symbolized by "$I(hkl)$". TC is better for the MoS_2: Pentacene film illuminating better structural condition.

Figure 2a, b picture's the MoS_2 and MoS_2: Pentacene films surface morphology via FESEM image of $\times$10,000 magnification. Transparent, thin, and bulky nanosheets are witnessed in the FESEM image of both films. This transparency is the successful deposition of the bulk MoS_2 nano-particles. Figure 2c, d shows the FESEM image of magnification $\times$500. It was found that the solution-processed MoS_2 forms quite irregular film distribution at the ITO substrate, wherein, the blended film of MoS_2 with pentacene entails more ordered and even dispersion over the entire surface. In MoS_2 film, there are some soft agglomerations in the grains. This may be due to the attractive Vander Waal forces between the MoS_2 nano-particles. This agglomeration was devastated by the shear forces existing in the pentacene material. Due to the presence of pentacene in the blended film, the fragmentation of the big sheets into small sheets takes place until the stability is achieved. Thus, the blending of 2D TMD material with organic subsidizes a more stable film.

3.2 *Optical Analysis*

Figure 3 represents the variation in the absorption of as-deposited dispersed bulk films on the ITO substrate when UV-Vis light hits the surface of the sample. The absorption of MoS_2 is high in the UV region whereas it is significantly low in the visible region. It is also detected that the remarkable peaks of absorption are at 270, 455, and 685 nm. These are created from the excitonic transitions of the K point in Brillouin zone justifying the presence of excitons. In the MoS_2 film the 270, 455, and 685 nm peaks indicate the existence of C, B, and A excitons, respectively. In MoS_2, at K and K' points the transition from higher spin-orbit fragmented valence bands to the deep conduction band is associated with A and B excitons, whereas, the electron transition from the valence to conduction band leads to the occurrence of C exciton [23]. Wherein, with a suppressed and very slight (or negligible) redshift of these peaks can be observed in the MoS_2: Pentacene film. The suppression of the absorption is due to the pentacene. It can also be corroborated from the FESEM image. The surface morphology of film 1 is less ordered with some gaps resulting in high

Fig. 2 FESEM image of films on ITO substrate: with magnification ×10,000 (a) MoS_2 (b) MoS_2: Pentacene; with magnification ×500 (c) MoS_2 (d) MoS_2: Pentacene

absorption, whereas, in film 2 the more packed and the ordered film is responsible for less absorption and high reflectance.

Now, let us study the optical energy band-gap of MoS_2 and MoS_2: Pentacene film. It is an important constraint for the assessment of light energy that can be transmitted or absorbed by the film. It infers that the incident light energy will get absorbed when its energy is greater than or equal to optical band-gap. We have estimated this parameter by using the relation specified in eq. 6 [23].

$$\alpha = \frac{c(h\upsilon - E)^{1/2}}{h\upsilon} \tag{6}$$

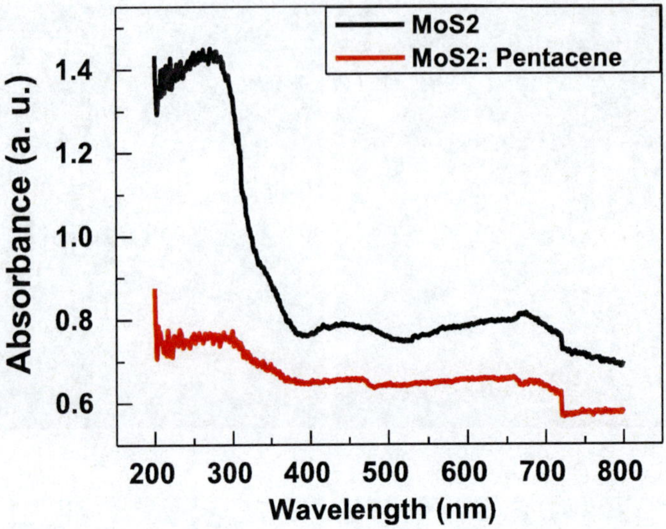

Fig. 3 Comparison of absorbance versus wavelength plot of MoS_2 and MoS_2: Pentacene films on ITO substrate

Here "E" is the optical energy band-gap, "c" is inferred as constant, "$h\upsilon$" is the energy of photon. Now for the evaluation purpose, we have plotted variation of $(\alpha h\upsilon)^{1/2}$ versus $h\upsilon$ as presented in Fig. 4. After that, a tangent was drawn along the

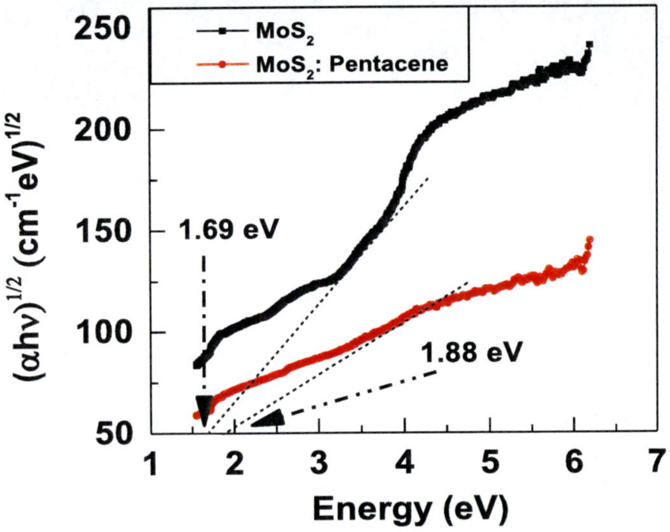

Fig. 4 Comparison of band-gap between MoS2 and MoS2: Pentacene films on ITO substrate using Tauc plot

direction of x-axis. The meeting point to the x-axis gives the energy band-gap. The probable optical band-gap for MoS_2 and MoS_2: Pentacene film was 1.69 eV and 1.88 eV, respectively. It indicates there is a noteworthy increment in optical band-gap blended film. The reason for this behavior is the high absorption coefficient edge toward longer wavelength in the visible range.

Now, in Fig. 5 the transmittance comparison of MoS_2 and MoS_2: Pentacene film on ITO substrate is presented against wavelength variation. Since the absorption is high in the UV region and low in the visible region, therefore, the transmittance is high in the visible region and low in the UV region. The blended film shows high transparency demonstrating more bulk deposition. The whole comparative study of the optical properties has been made in Table 2.

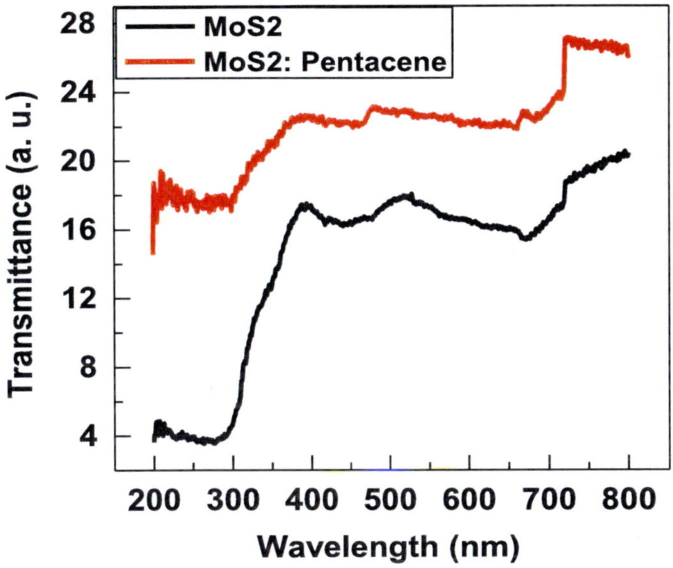

Fig. 5 Comparison of transmittance versus wavelength plot of MoS_2 and MoS_2: Pentacene films on ITO substrate

Table 2 Comparison of the properties of MoS_2 and MoS_2: Pentacene films deposited on ITO substrate at optical level

Optical properties	Wavelength range	MoS_2	MoS_2: Pentacene	Percentage change (%)
Absorbance (a.u.)	UV	1.45	0.77	88
	Visible	0.81	0.64	26.5
Transmittance (a.u.)	UV	3.47	17.55	80.22
	Visible	15.34	22.85	32.86
Band-gap (eV)		1.69	1.88	10

4 Conclusion

In this work, we have successfully investigated the structural and optical properties of the MoS_2 film blended with pentacene material (MoS_2: Pentacene). For this, we have performed the comparative study of MoS_2 and MoS_2: Pentacene film using XRD pattern and FESEM image. Moreover, for the optical scrutiny, we have performed UV-Vis Spectroscopy characterization involving absorption and transmittance plots. From the results, we have found high-intensity XRD pattern in the blended film, without any substantial alteration in the phase of MoS_2 film. Increment in the peak intensity signified better crystallinity and less strain having a larger crystallite size of the atoms of film. The FESEM image showed the more ordered grains arrangement at the surface. Additionally, the absorption curve gave three excitons A, B, and C in both films. But there is a significant suppression in the absorption and better transmission in MoS_2: Pentacene. This thought-provoking behavior allows it to be used for optical device geometry applications such as in injection and transport layer of photovoltaic cells and light-emitting diodes, optical communication system, optical fibers, etc.

References

1. Cong Q et al. (1992) Primary study of structures of r.f.-sputtered MoS_2 films. Thin Solid Films 209(1):1–8
2. Li Tianshu, Galli Giulia (2007) Electronic properties of MoS_2 nanoparticles. J Phys Chem C 111(44):16192–16196
3. Kim, Daeho, et al.: Toward the growth of an aligned single-layer MoS2 film. Langmuir 27(18), 11650–11653 (2011)
4. Wang, Shanshan, et al.: Shape evolution of monolayer MoS2 crystals grown by chemical vapor deposition. Chemistry of Materials. 26(22), 6371–6379 (2014)
5. Li, Dahai, et al.: Optical properties of thickness-controlled MoS2 thin films studied by spectroscopic ellipsometry. Applied Surface Science. 421, 884–890 (2017)
6. Theerthagiri J et al (2017) Recent advances in MoS2 nanostructured materials for energy and environmental applications–a review. J Solid State Chem 252:43–71
7. Yin, Zongyou, et al.: Single-layer MoS2 phototransistors. ACS nano. 6(1), 74–80 (2011)
8. Luo, Ai-Ping, et al. Few-layer MoS 2-deposited microfiber as highly nonlinear photonic device for pulse shaping in a fiber laser. Photonics Research. 3(2), A69-A78 (2015)
9. Mak, Kin Fai, and Jie Shan. Photonics and optoelectronics of 2D semiconductor transition metal dichalcogenides. Nature Photonics. 10(4), 216 (2016)
10. Geffroy, Bernard, Philippe Le Roy, and Christophe Prat. Organic light-emitting diode (OLED) technology: materials, devices and display technologies. Polymer international. 55(6), 572–582 (2006)
11. Srivastava, A., & Kumar, B.: Organic Light Emitting Diodes-Recent Advancements. In 2017 14th IEEE India Council International Conference (INDICON), pp. 1–6. IEEE, IIT Roorkee, (2017)
12. Huang, Yu Li, et al.: The organic–2D transition metal dichalcogenide heterointerface. Chemical Society Reviews, 47(9), 3241–3264 (2018)
13. Kim SS et al (2003) Fabrication of p-pentacene/n-Si organic photodiodes and characterization of their photoelectric properties. Appl Phys Lett 82(4):639–641
14. Kim, Jae-Keun, et al.:Trap-mediated electronic transport properties of gate-tunable pentacene/MoS 2 pn heterojunction diodes. Scientific reports 6, 36775 (2016)

15. Bettis Homan, Stephanie, et al.: Ultrafast exciton dissociation and long-lived charge separation in a photovoltaic pentacene–MoS2 van der Waals heterojunction. Nano letters. 17(1), 164–169 (2016)
16. Peng, Yingquan, et al.: High performance photodiode based on MoS2/pentacene heterojunction. Applied Surface Science. 459, 179–184 (2018)
17. Shen, Na, and Guohua Tao.: Charge Transfer and Interface Engineering of the Pentacene and MoS2 Monolayer Complex. Advanced Materials Interfaces. 4(6), 1601083 (2017)
18. Yang, Lei, et al.: Lattice strain effects on the optical properties of MoS 2 nanosheets. Scientific reports. 4, 5649 (2014)
19. Crystal Structure, Academic Resource Centre: www.iit.edu
20. Singh, Satyendra Kumar, et al.: Performance analysis of RF-sputtered ZnO/Si heterojunction UV photodetectors with high photo-responsivity. Superlattices and Microstructures 91, 62–69 (2016)
21. Akgul, Funda Aksoy, et al.: Influence of thermal annealing on microstructural, morphological, optical properties and surface electronic structure of copper oxide thin films. Materials Chemistry and Physics. 147(3), 987–995 (2014)
22. Fabrication and characterization of Au/p-ZnO Schottky contacts (2015) Singh, Brijesh Kumar, and Shweta Tripathi. Superlattices Microstruct 85:697–706
23. Singh Richa, Tripathi Shweta (2019) Structural and optical properties of few-layer MoS2 thin films grown on various substrates using RF sputtering process. J Mater Sci: Mater Electron 30(8):7665–7680

An Efficient Watermarking Process Based on Three-Level DWT and FFT Technique

Jamal Arif and Som Pal Gangwar

Abstract Digital image watermarking has become an interesting research field to face the issues created by the increasing use of Internet for distribution of digital data. For the couple of decades, authenticity of digital data has been concern for everyone. Digital image watermarking plays a vital role in protection of copy rights. In frequency domain watermarking, discrete cosine transform (DCT)-based watermarking process is most common because the majority of image compression methods are developed in DCT domain. This research presents a robust discrete wavelet transform (DWT) and fast Fourier transform (FFT) manner-based digital watermarking (DW). In the initial stage, split the original (cover) image into two horizontal and two vertical sub-bands using 3-DWT-FFT sub-band using changing their values. In this thesis, the embedding and extraction technique for watermarking is presented based on DWT & FFT transforms. In this technique, the embedding and de-embedding of the watermark is much easier than other transform techniques. Various values of PSNR's, MSE'S & NC's are analyzed for watermarked image quality and extracted watermark quality.

Keywords Digital watermarking · FFT · 3-DWT

1 Introduction

The watermarking is an old technique for hiding information. Cryptography is the oldest technique for hiding the owner information. In cryptography, the owner data is encoded such that it can be decoded only by the receiver with the help of some look up table or pre-defined rule. Next comes the steganography in which the data is hidden with the host image. Steganography is invisible and cannot be detected by

J. Arif (✉) · S. P. Gangwar
Department of Electronics Engineering, KNIT, Sultanpur, India
e-mail: ermjarif@gmail.com

S. P. Gangwar
e-mail: gangwar_sp@rediffmail.com

D. Harvey et al. (eds.), *Advances in VLSI, Communication, and Signal Processing*, Lecture Notes in Electrical Engineering 683,
https://doi.org/10.1007/978-981-15-6840-4_24

any third person except by the receiver. The third method of hiding ownership data is watermarking. The image watermarking may be visible or invisible. We shall be discussing the invisible watermarking in this paper.

In this paper, we have used frequency domain-based watermarking technique. The use of discrete wavelet transform (DWT) and fast Fourier transform (FFT) gives a better result than the DCT based image watermarking technique. 3rd level DWT & then FFT is used for embedding followed by inverse FFT (IFFT), and finally, we take inverse DWT (IDWT) that gives the watermarked image. The watermarking is a reversible operation.

There are total six sections in this paper. Section 1 gives the introduction of the digital image watermarking. Section 2 offers a few connected works. Section 3 briefs us with the techniques used in this paper. Section 4 gives the algorithm for embedding and extraction of the watermark. Section 5 includes the results of the MATLAB simulations and discussion. Section 6 is the conclusion of this paper. Finally, the references used is the paper are given.

2 Literature Review

He and Hu [1], proposed a algorithm of color image watermarking, depending upon (DWT-DCT-SVD). RGB color space is converted into YUV space, and then, embedding is done in the luminance component Y [1].

Pal et al. [2]. In this paper, a detail study of watermarking and steganography techniques are discussed. With the increasing demand of Internet, data security and authentication have been concern for every one [2].

Ginanjar et al. [3]. In this paper, a genetic algorithm based on phase shift keying is proposed for invisible watermarking. Audio watermarking has been done with this technique, and performance of the technique is evaluated [3].

Maheshwari [4]. In this paper, discrete wavelet transform and singular vale decomposition-based digital image watermarking technique has been proposed. First water mark is embedded by DWT method by breaking it into four bands LL, LH, HL, HH, then, SVD of each sub-band is taken, and second watermark is embedded. Various performance parameters are evaluated for this technique [4].

Mirzaei et al. [5]. In this paper, an invisible watermarking method using local gradient analysis has been given. Image is divided into many small-small blocks, and then, watermark is embedded into the frequency components of these blocks [5].

Muni SekharI et al. [6]. In this paper, edge-based watermarking techniques have been discussed. A new method, reference image and edge (RIE), has been introduced which can defeat the smoothing impact of earlier edge-based watermarking techniques [6].

Draganić et al. [7]. Public key cryptography signature (PKCS) is used in this technique for hiding the 32 bit serial number for the purpose of source verification and ownership rights [7].

3 Techniques Used in Research

A. Discrete Wavelet Transform (DWT)

Discrete wavelet transform is a method for converting spatial domain signal into frequency domain signal. For the purpose of image, we use two-dimensional DWT. DWT uses wavelets for decomposition of time domain signal into frequency domain. The wavelets are basically limited in time and space both, whereas the normal sine- and cosine-based transform (i.e., FFT) or the cosine only based transform (i.e., DCT) is not limited to both time and frequency domains. The most commonly used wavelet in DWT is the 'haar' wavelet. DWT may be calculated from two filters. One is low-pass filter, and another is high-pass filter. These filters can be expressed as follows (Fig. 1):

$$y_{\text{lower}}[p] = \sum_{q=-\infty}^{\infty} x[k]g[2p - q] \tag{1}$$

$$y_{\text{upper}}[p] = \sum_{q=-\infty}^{\infty} x[k]l[2p - q] \tag{2}$$

B. Fast Fourier Transform (FFT)

The FFT is a form of Fourier transform tool that is used to convert the two-dimensional picture into its SIN and COS functions [3]. The FFT is used in signal processing, image processing, audio processing, and video processing. The two-dimensional fast Fourier transform can be written as

$$G(u, v) = \frac{1}{MN} \sum_{X=0}^{M} \sum_{Y=0}^{M} g(x, y) e^{-j2\pi\left(\frac{ux}{m} + \frac{vy}{n}\right)} \tag{3}$$

The inverse transform is defined as

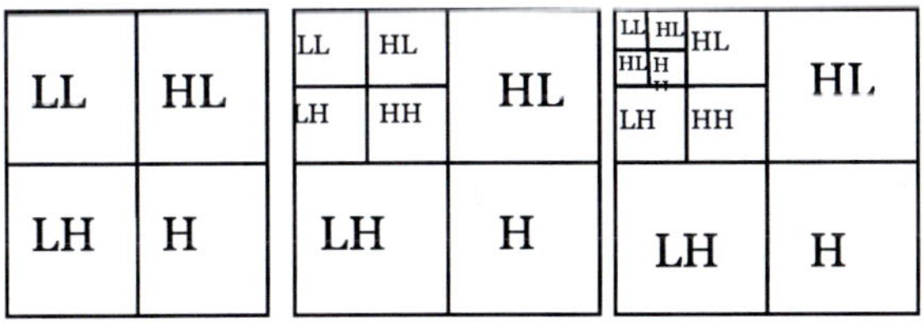

Fig. 1 **a** Level-1 DWT, **b** Level-2 DWT, and **c** Level-3 DWT

$$g(x, y) = \sum_{u=0}^{M} \sum_{v=0}^{M} G(u, v) e^{-j2\pi\left(\frac{ux}{m} + \frac{vy}{n}\right)} \quad (4)$$

4 Propose Work

A. Problem statement

DCT can be applied for an image only in block wise. Further, after DCT, the coefficients are arranged in a zig-zag way, and then, quantization is done. During quantization, the invariance property and some upper band frequency got disturbed. Also, taking block-wise DCT is very cumbersome algorithm, and it takes lot of time and hardware to implement it. Also, when we test the DCT based watermarking technique for various types of attack, we find that it is less robust to these attacks.

B. Propose Methodology

This research presents a robust discrete wavelet transform (DWT) and fast Fourier transform (FFT) manner-based digital watermarking (DW). In the initial stage, split the original (cover) image into four sub-bands using 3-DWT-FFT sub-band using changing their values. In this thesis, the embedding and extraction technique for watermarking is presented based on DWT & FFT transforms. In this technique, the insertion and extraction of the watermark are found to be simpler than other transform techniques. Various values of PSNR's, MSE'S & NC's are analyzed for watermarked image quality and extracted watermark quality.

C. Propose Algorithm

1. Third–level DWT is applied to the host image.
2. After third-level decomposition, LH band is selected.
3. Similarly, third-level DWT is taken for watermark image, and its LH band is selected.
4. The FFT coefficients of both the sub-bands are taken.
5. For invisible watermarking, suitable scaling factor is chosen.
6. Coefficients of the LH band of host image are modified by adding watermark coefficients scaled by the scaling factor.

$$I'_{\mathrm{LH}} = I_{\mathrm{LH}} + \alpha W_{\mathrm{LH}} \quad (5)$$

7. Apply IFFT and IDWT to obtain the watermarked host image.
8. Extract the invisible watermark from the LH band by using equation:

$$W'_{\mathrm{LH}} = \frac{I'_{\mathrm{LH}} - I_{\mathrm{LH}}}{\alpha} \quad (6)$$

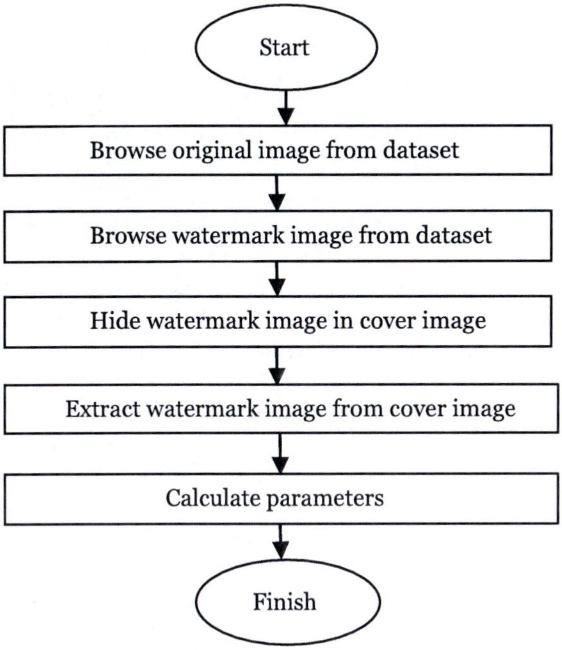

Fig. 2 Flowchart of propose methodology

9. Take the IFFT & IDWT to get the extracted watermark.
10. Compare the watermarked image with original host image and calculate the PSNR & MSE. Also calculate the normalized cross-correlation (NC) for extracted and original watermark (Fig. 2).

$$\text{MSE} = \frac{1}{M \times N} \sum_{i=1}^{M} \sum_{J=1}^{N} \left[H'(i, j) - H(i, j)\right]^2 \tag{7}$$

$$\text{PSNR} = 10\log_{10} \frac{R \times R}{\text{MSE}} \tag{8}$$

$$\text{NC} = \frac{\sum_{i=1}^{M} \sum_{j=1}^{N} W(i, j) W'(i, j)}{\sqrt{\sum_{i=1}^{M} \sum_{j=1}^{N} [W(i, j)]^2} \sqrt{\sum_{i=1}^{M} \sum_{j=1}^{N} [W'(i, j)]^2}} \tag{9}$$

5 Result

The above algorithm is tested for various data set as mentioned below. First, we calculate the PSNR and MSE values for watermarked images for various data sets. Then, we calculate the normalized cross-correlation for extracted image. Finally, we have done some attack (i.e., cropping, salt and pepper, and rotation) on the watermarked image, and the normalized cross-correlation has been calculated for various data sets and various embedding intensities (α) (Fig. 3).

If we change the scaling factor, the visible watermark becomes invisible (Fig. 4; Tables 1, 2 and 3).

Irrespective of the scaling factor (α) we get the same normalized correlation (NC) and structural similarity index (SSIM) for the extracted watermark in the absence of any noise in the system.

Now, we intentionally add some noise, i.e., salt and pepper noise, cropping, and rotation of watermarked image, and then, the robustness of the propose method is checked as follows (Table 4).

Fig. 3 Data set

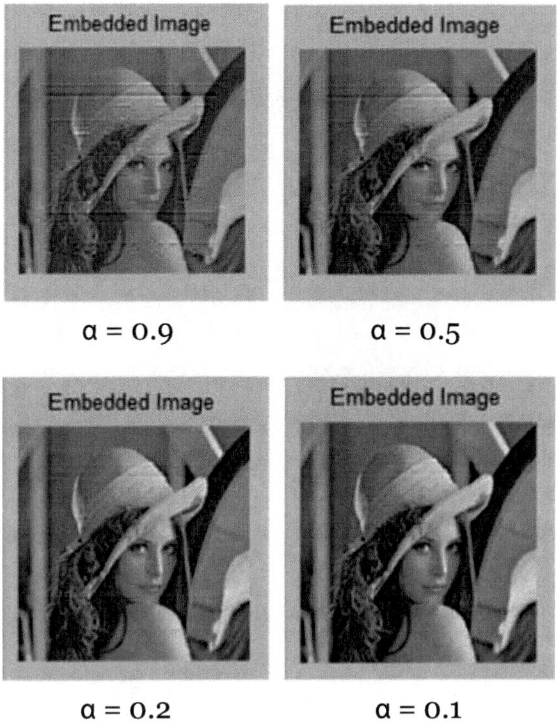

Fig. 4 Effect of scaling factor

Table 1 Comparison table of PSNR for watermarked images

Scaling factor	DCT	DWT + FFT
$\alpha = 0.9$	26.32	26.34
$\alpha = 0.5$	28.83	28.90
$\alpha = 0.2$	32.51	32.87
$\alpha = 0.1$	34.71	35.89
$\alpha = 0.01$	36.58	45.89

Table 2 Comparison table of MSE for watermarked images

Scaling factor	DCT	DWT + FFT
$\alpha = 0.9$	152.70	152.95
$\alpha = 0.5$	85.67	84.42
$\alpha = 0.2$	36.69	33.76
$\alpha = 0.1$	22.13	16.88
$\alpha = 0.01$	14.38	1.688

Table 3 Comparison table of NC & SSIM for extracted watermark

Extracted watermark quality	DCT	DWT + FFT
Normalized correlation	0.9966	1.000
SSIM	0.9963	1.000

Table 4 Table of NC for extracted watermark after various attacks/noise

Type of noise/disturbance	Watermarked image	Scaling factor	NC for DWT + FFT
Salt and pepper		$\alpha = 0.9$	0.9971
		$\alpha = 0.5$	0.9961
		$\alpha = 0.2$	0.9883
		$\alpha = 0.1$	0.9619
		$\alpha = 0.01$	0.3411
Cropping	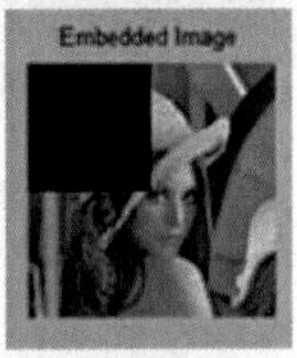	$\alpha = 0.9$	0.9993
		$\alpha = 0.5$	0.9992
		$\alpha = 0.2$	0.9983
		$\alpha = 0.1$	0.9954
		$\alpha = 0.01$	0.7506
Rotation	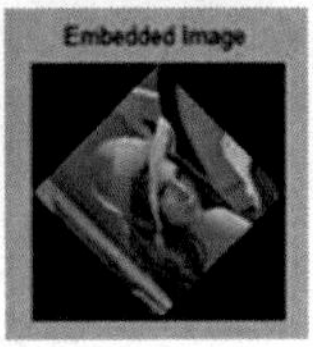	$\alpha = 0.9$	0.9963
		$\alpha = 0.5$	0.9933
		$\alpha = 0.2$	0.9717
		$\alpha = 0.1$	0.9046
		$\alpha = 0.01$	0.2097

6 Conclusions

This paper implemented watermarking for images using 3-DWT-FFT scheme which has proved a high level of robustness in opposition to DCT based watermarking. Majority of attacks including noise, blurring, other styles of IP attacks which can be tested by using getting better the watermark from any of the sub-band, which actually shows that remodel domain is extra robust than spatial domain. Generally, LL band is not changed as any kind of modifications in it could be easily perceived with the aid of human eyes. The simulation result shows that the mean square error (MSE) has decreased significantly while the peak signal-to-noise ratio (PSNR) has increased for DWT-FFT over DCT for host image. Also, the normalized cross-correlation (NC) has increased for watermark image in the DWT-FFT based technique.

The future work can be conducted by increasing PSNR and NC using principle component, scaling factor, and multiple dataset, and the second approach we can use is optimal scaling.

References

1. He Y, Hu Y (2018) A proposed digital image watermarking based on DWT-DCT-SVD. In: 2018 2nd IEEE advanced information management, communicates, electronic and automation control conference(IMCEC 2018). 978-1-5386-1803-5/18/$31.00 ©2018 IEEE
2. Pal P, Singh HV, Verma SK (2018) Study on watermarking techniques in digital images. In: Proceedings of the 2nd international conference on trends in electronics and informatics (ICOEI 2018). IEEE conference record: # 42666; IEEE Xplore ISBN: 978-1-5386-3570-4, 978-1-5386-3570-4/18/$31.00 ©2018 IEEE
3. Ginanjar RR, Kim DS, Moon CB (2018) Enhanced digital audio watermarking using genetic algorithm. In: 2018 IEEE international conference on consumer electronics-Asia (ICCE-Asia). 978-1-5386-5807-9/18/$31.00 ©2018 IEEE
4. Maheshwari DB (2018) An analysis of wavelet based dual digital image watermarking using SVD. In: 2018 International conference on advances in communication and computing technology (ICACCT), Amrutvahini College of Engineering, Sangamner, Ahmednagar, India. 8–9 Feb 2018. 978-1-5386-0926-2/18/$31.00 ©2018 IEEE
5. Mirzaei MR, Karimi M, Karimi N, Samavi S (2017) Blind image watermarking based on local analysis of gradients. In: 2017 25th Iranian conference on electrical engineering (ICEE). IEEE 20 17 © $3 1.00/978-1-5090-5963-8/17
6. Muni SekharI V, Kumar CS, RaoI KVG, Sambasiva RaoII N, Gopichand M (2017) A reversible RIE based watermarking scheme. In: 2017 IEEE 7th international advance computing conference. 978-1-5090-1560-3/17 $31.00 © 2017 IEEE. https://doi.org/10.1109/iacc.2017.179
7. Draganić A, Marić M, Orović I, Stanković S (2017) Identification of image source using serialnumber-based watermarking under compressive sensing conditions. MIPRO 2017, Opatija, Croatia, 22-26 May 2017

Blockchain-Enabled Traceable, Transparent Transportation System for Blood Bank

Nihar Ranjan Pradhan, Akhilendra Pratap Singh, and Vinay Kumar

Abstract Electronic health records (EHR) and patient health records are an Internet-based health application which is sharable to authorized stakeholders such as doctors, specialist, and patients. One of the important record is blood and sharing the blood record in secured manner is a challenging issue. Blockchain-based blood bank system can resolve the need for individuals to access, trace, manage, share their health and blood-related information, which is immutable. The blockchain-based system provides a secured environment which acts as communication hub between donors, doctors, testing laboratories, and recipients or patients. This work proposes the design of a blockchain-based solution for the blood bank system. Information related to the blood is communicated securely to the intended users from the blood bank. Additionally, the traceability feature included in the proposed system monitors detailed information of blood transfer from excess stock to the nearest place. The system model is designed, implemented and validated using solidity language platform.

Keywords Blockchain · Medical records · Secure hashing algorithm (SHA256) · Private blockchain · Elliptical curve cryptography(ECC) · RFID

N. R. Pradhan (✉) · A. P. Singh
National Institute of Technology Meghalaya, Bijini Complex, Laitumkhrah, Shillong, Meghalaya 793003, India
e-mail: niharpradhan@nitm.ac.in

A. P. Singh
e-mail: akhilendra.singh@nitm.ac.in

V. Kumar
Motilal Nehru National Institute of Technology Allahabad, Teliarganj, Prayagraj, Uttar Pradesh 211004, India
e-mail: vinay2008el03mnnit@gmail.com

D. Harvey et al. (eds.), *Advances in VLSI, Communication, and Signal Processing*, Lecture Notes in Electrical Engineering 683,
https://doi.org/10.1007/978-981-15-6840-4_25

1 Introduction

According to world health organization, in every 15 min, 901 people need blood, 202 of them are women giving birth, and 684 are over 65. Blood is a perishable raw material and mixture of various cells like plasma, nutrients, antibodies, clotting agents, proteins, salts, hormones, and waste products. It can save patient life, major precautions can be taken for Blood demand & supply mismatch, decrease in blood availability, storage cost is high, Blood is Obsolescence after 21 days, quick distribution during an emergency. Blood componentizing separates raw blood to RBC, platelets, plasma, and cryoprecipitate. Blood transfusions of RBC are required for major surgeries like liver transplant, open-heart surgery, accidents, kidney failure, sickle cell anemia, trauma due to burns and accidents, and leukemia. Transfusion of platelets is required in bleeding disorders, cancer therapy, and open-heart surgery whereas plasma transfusion is very risky [9]. Blood traceability can increase the availability system to integrate blood bank system. Coding of blood products and shipping can be created using RFID with blockchain technology. The lifetime of blood bags is 21 days. It cannot be produced like manufacturing other products because of donor willingness and need a gap of three months if donated before. Blockchain implementation has advantages like it is open (everyone can join), tokens (rewards to blood donor), transparency (anybody can see blood availability), traceability (reduce the delay in transportation, improve monitoring of blood bags), distributed (nobody owns the network), anonymization (ensures privacy), and immutable(donor/receivers can see all their previous records) [16].

2 Related Works

Data sharing, security, and privacy in healthcare is a major challenging area. In recent years, blockchain technology has provided a solution to data sharing, security and privacy. Yue et al. have proposed a layered system, cloud-based blockchain and cloud is used to store the data [18]. Private blockchain technique is used in cloud data storage. Qxia proposed a cloud-based shared blockchain system for healthcare. It uses a permissioned blockchain. The author has implemented a smart contract and accessing techniques in [17]. Kuo et al. have reviewed the modern healthcare application using blockchain system. In [8], author gives directions to researchers about blockchain challenges in the biomedical field [8]. In [2, 13, 18, 19] sevral approaches are suggested for healthcare data sharing, traceability, transparency, security, blockchain for a clinical trial etc.

The existing works have provided a diversified application of the healthcare system with blockchain. In the suggested approached blockchain technology is used for data security and sharing instead of data tracing. To provide the tracable, transparent transportion system system model has been suggested for blood bank. The proposed blockchain-based system provides blood traceability and coordination among hospitals which may reduce the delay and save precious life. It can reduce human error, improve the monitoring of blood bags, improve the safety of patients and hospital staff, and improve the efficiency of the management. Blockchain technology provide realtime monitoring in blood supply which avoids blood contamination, blood wastages, and losses due to spoilage [3, 10, 15].

2.1 Necessity of Human Blood Banks and Risks Involved in Traditional Blood Bank System

(1) Effective decision-making policy has various blood groups and components such as 8 blood groups, 4 major components and its component substitutability as depicted in Fig. 1 [14]. (2) Lack of communication between hospital and public for blood donation schedules or events. (3) Public afraid to donate blood. (4) Blood transfusion data is not secure. (5) The public is unaware of the availability of blood stocks in the hospitals. (6) No facility and rewards for donors. Blood screening result is not known to the donors. Even they do not know how many times they donor has donated blood. Physical ledgers are maintained in centralized manner. (7) Donor health condition is not shown in blood test report. (8) Different hospitals are using different software or platform which is not secure and easily modifiable [12].

Fig. 1 Blood component

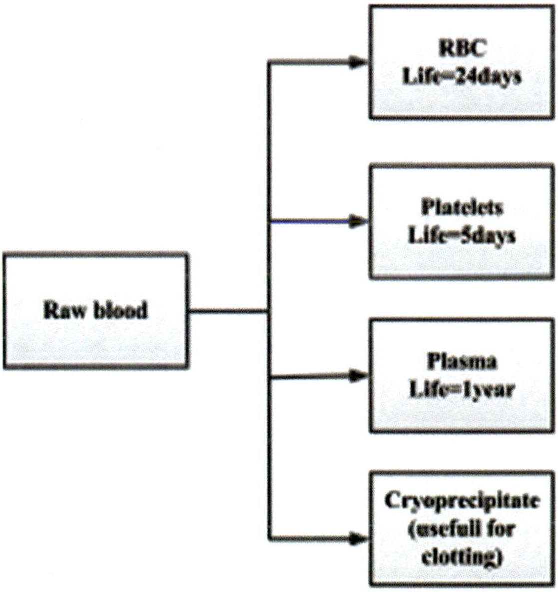

3 Proposed System—Blockchain-Based Blood Bank

Blood banks can be connected distributively and decentralized way using cloud. Information related to donors, patients, and hospitals are stored in the form of blocks. The blockchain-based blood bank can efficiently regulate the blood transporation system. The blood stock can be easily monitored and transported to the nearest blood bank having a shortage of blood. In the proposed model, private blockchain is used for storage of medical records and consortium blockchain for is used for securing indexes. Private blockchain of blood bank contains encrypted medical reports of patients [4]. It is a storage layer, where consortium blockchain provides services to end users by searching the indexes. Figure 2 describes the system architecture of blockchain-based blood bank system [11].

3.1 Implementation and Testing

The smart contract written in solidity languauge and implemented using RemixIDE: http://remix.ethereum web browser. As a case study, single-echelon which is single sender and single receiver is considered for blood bank. The proposed model considered three entities such as sender, receiver, and transport container. Each entity has Ethereum address and smart contract is designed to solve the blood supply chain tracing and transparency [6]. The proposed model considered various factors such as temperature, route, and payment terms. The smart contract generates passphrase at the sender side before shipping, and at the receiver side same passphrase is used for verifiability and delivery. This approach can be extended to multi-echelon. The smart contract calls the function at a certain time. Modifiers are used so that only the sender can call some specific function. Similarly, for transport containers and receivers, the

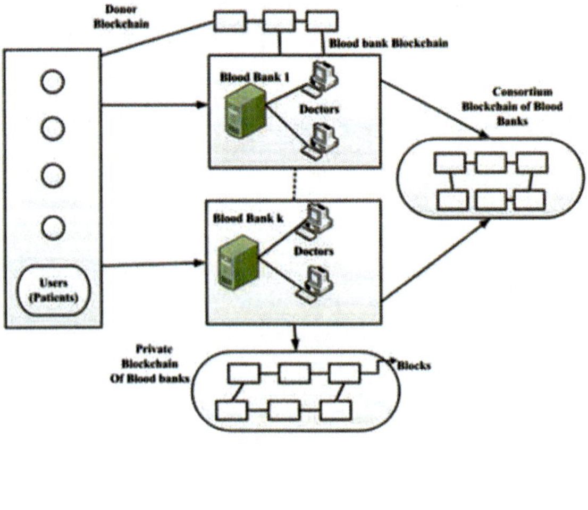

Fig. 2 Proposed system

```
PERFORMANCE_CHECK(result)
require(package_State.Package_Container_Ready_for_Self_Check)
if result = 1
    state = package_State.Ready_for_Shipment
elseif result = 1
    state = packageState.Aborted
    selfdestruct(msg.sender);
DEPOSIT_MONEY(hash)
require(package_State.Ready_for_Shipment)
state = package_State.Money_Deposited;
pass_phrase = hash;
START_SHIPMENT()
require(package_State.Money_Deposited)
state = package_state.start_shipment
startted_shipment(msg.sender)
SHIPMENT_ARRIVED()
require(package_state.money_deposited)
state = package_state.start_shipment
startted_shipment(msg.sender)
PROVIDE_PASS_PHRASE (code)
require(package_State.Waiting_for_Passphrase)
received_Code = code
if pass_phrase == keccak256
   state = package_State.Receiver_Authentiated
else state = packageState.WaitingForCorrectPasscode
PROVIDE_PASS_PHRASE_AFTER_TIME (phrase)
if block.timestamp <= startTime + daysAfter * 1 days
      PROVIDE_PASS_PHRASE(phrase)
else   state = package_State.Authentication_Failure_Aborted
      receiver.transfer(shipment_Price/2)
      Shipment_Violated_and_Refund(msg.sender)
      self_destruct(msg.sender)
UNLOCK_SHIPMENT()
require(package_state.Receiver_Authentiated)
state = package_State.Shipment_Received
Shipment_Received(msg.sender)
GET_SHIPMENT_MONEY()
require(package_State.Shipment_Received)
sender_owner.transfer(shipment_Price)
Payment_Received_by_Sender(msg.sender)
self_destruct(msg.sender)
REFUND()
require(package_State.Aborted)
if not violation = violation_Type.None
    receiver.transfer(shipment_Price)
    Shipment_Violated_and_Refund(msg.sender)
    self_destruct(msg.sender)
```

Fig. 3 Blood trace smart contract

Fig. 4 Smart contract function for violation

```
VIOLATION_OCCURED(msg, voilation,value)
require(package_State.Start_Shippment)
if violation = violation_Type.Temp
     tempertaure = value
     if tempertaure <= 1 or tempertaure >= 6
        Tempertaure_Violation( msg ,true, tempertaure)
elseif violation = violation_Type.Jerk
     jerk = value
     Sudden_Jerk(msg, true, jerk)
elseif violation = violationType.Open
     open = value
     Sudden_Container_Opening(msg, true, open)
elseif violation == violationType.Route
     on_Track = value
     Out_of_Route(msg , true, onTrack)
     Refund()
```

pseudocode is given in Figs. 3 and 4. The following are the stages involved in the blood bank process. The following are the stages involved in blood bank process.

3.2 *Donor Registration*

In this process, donor preliminary details such as name, gender, blood type, mobile no., email, city, last time of donation, age, and the weight will be filled. The figure 5 shows the process of registartion. The eligibility test of the donor's age should be between 18 and 60. Hemoglobin should be normal and weight should be more than 45 kg. Blood pressure and body temperature should be normal. Donors must not have taken medicine in the last 48 hours and should not have diseases like HIV, jaundice, syphilis, etc. The donor should not be addicted to any drugs [18] (Fig. 5).

Fig. 5 Donor registration

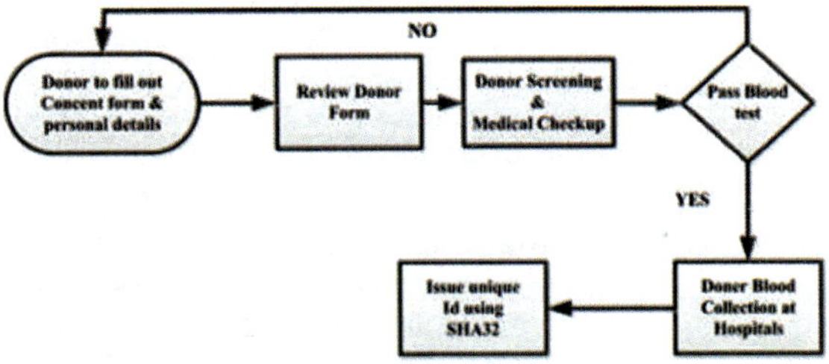

3.3 *Blood Processing*

Raw blood is batched into standard units and is labeled with RFID labels after being processed. During blood processing, the temperature is measured and monitored. Each bag contains donor number, date of collection, blood group type, component type, and date of expiry. All information is recorded in a blockchain. The blood processing steps are: (1) Testing the raw blood for infectious disease. (2) Blood component preparation. (a) This test is performed to find either plasma or serum. (b) Plasma is the liquid that contains blood and is separated from blood spinning procedure. (c) Then, it appears at the top and blood cells at the bottom. (d) The serum is like plasma, which is allowed to clot. (e) Here, blood is collected with no anticoagulant. (3) Donor patients compatible assessment. (4) Transfusion on the patients. (5) Complication control after transfusion, it evaluates the complication of transfusion [20].

3.4 *Recipient Registration*

The doctors/specialists or patients who need blood for medical operations are the recipient. Details such as hospital name, patients age, DOB, blood groups, genders, and others will be entered during registration. After registration, blood can be supplied from the hospital to admitted patient and if blood is not available in the hospital, then blood will be supplied from the nearest blood bank. The recipient's relatives or doctors can verify the donor details, laboratory test reports, blood temperature, and other details by RFID code [21].

3.5 *Application Flow*

The application flow describes process of systematic the data flow from donors, blood processing units, inventory management, blood distribution, and finally to recipients.

Fig. 6 Donor blood collection

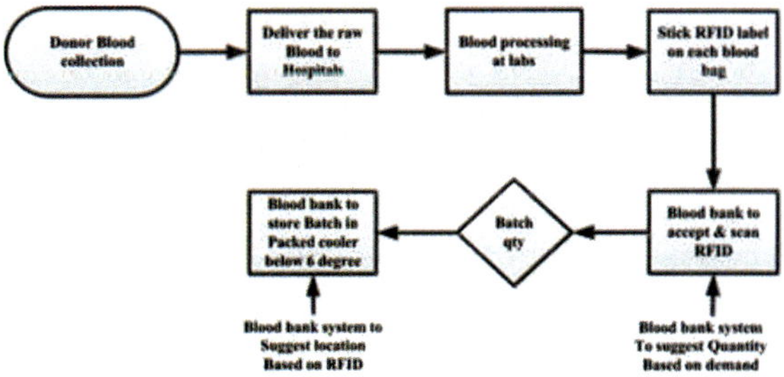

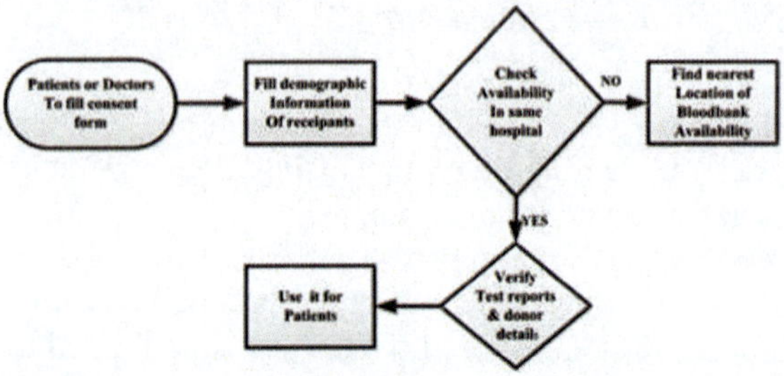

Fig. 7 Recipient registration

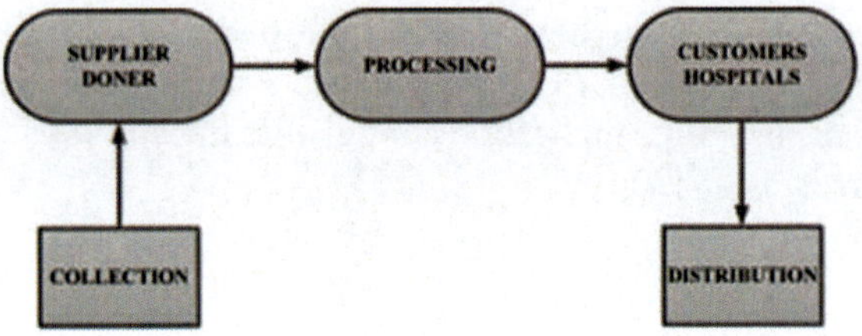

Fig. 8 Traditional blood bank structure

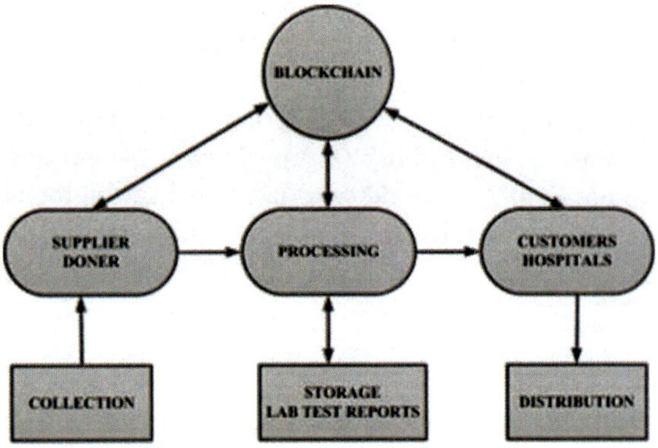

Fig. 9 Blockchain blood bank structure

Figs. 6 and Fig. 7 has shown blood collection and registration and Fig. 8 and Fig. 9 differentiates traditional and blockchain-based blood bank system.

4 Benefits of This Application

It integrates the sharing of information among all stakeholders and improves the traditional blood information system by traceability and transparency. Table 1 describes the benefits of the blockchain blood bank system [14].

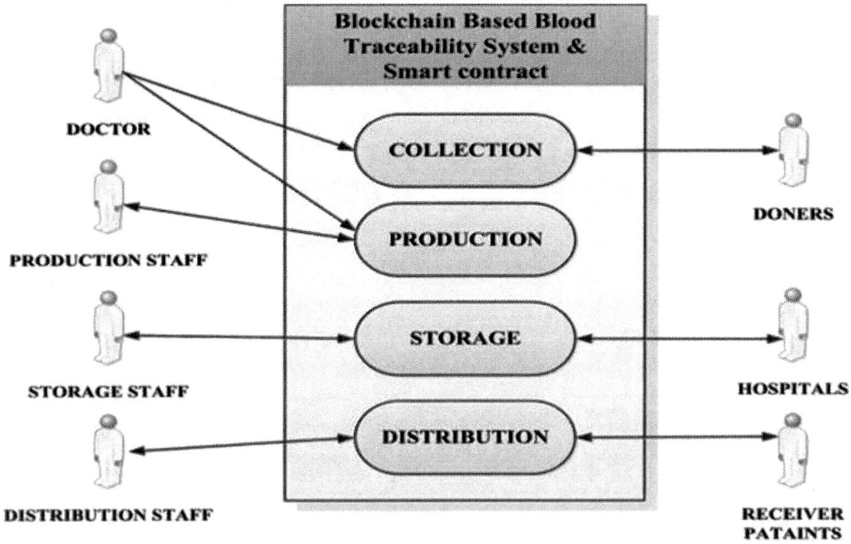

Fig. 10 Blockchain-based blood traceability system

Table 1 Benefits of blockchain-based blood bank

Benefits	Donors/recipients	Processing unit	Hospitals
Operational cost	Reduced	Reduced	Reduced cost like telephone, fax billing, etc.
Human errors	None	Reduced	Reduced
Information sharing	Understanding their health, medical, and blood record history	Increased	Increased
Traceability and monitoring	Increased	Increased records and ability of banks	Increased
Security [1, 5]	SHA is used to generate user address for privacy	NA	ECC is used to verify public and private keys.
Immutable and audit trial	unaltered patient blood record	laboratory test ledger	manage critical data
Scalability	patients increased at a exponential rate	Increased	Increased

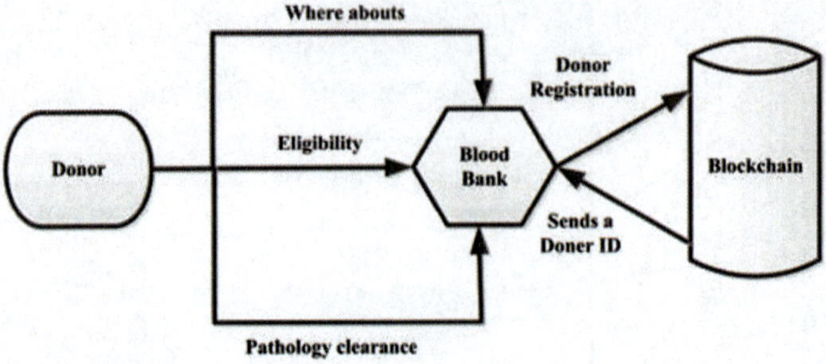

Fig. 11 Donor blockchain

5 Transaction Flow

5.1 Donor Blockchain

1. **Donor Registration**—Figure 11 shows the donor blockchain and a genesis block is created for every donor data [7]: (a) unique ID, (b) donor name, (c) KYD document. Document is stored in file, and hash of the document is stored in a block. At the time of accessing, document is verified against hash like in merkle tree (d) DOB, (e) email address, (f) contact number (Fig. 10).

5.2 Blood Bank Blockchain

Blood bank blockchain captures the data in the following steps: (a) raw blood collection, (b) batch mixing, and (c) transfusion.

5.3 Recipients Transactions Blockchain

Recipients registration—This blockchain contains the data like (a) recipients ID, (b) name, (c) identity, (d) delivery date, (e) location, and (f) email ID.

6 Conclusion

The blockchain-based blood bank system has considerd several factors such as donor test reports, transfusion and testing reports, storage of blood information, expiry reports, and brings more transparency and traceability by taking the critical data at all stages. The primary objective is the availability of blood and its component, minimizing the waste of blood, controlling the demand supply chain. The demand supply model may not be uniform for all hospitals. It is different for cities and rural areas. Rural hospitals sometimes cannot afford their blood banks. Since it deals with somebody's life, delay in transport and safety cannot be compromised. RFID code on every blood bag can be used to trace and view entirely history of the blood, which brings transparency and gives confidence in distributing to a location where there is a shortage of blood.

References

1. Abbas A, Khan SU (2014) A review on the state-of-the-art privacy-preserving approaches in the e-health clouds. IEEE J Biomed Health Inf 18(4):1431–1441
2. Ali MS, Vecchio M, Pincheira M, Dolui K, Antonelli F, Rehmani MH (2018) Applications of blockchains in the internet of things: a comprehensive survey. IEEE Communications Surveys and Tutorials
3. Buterin V (2015) On public and private blockchains. https://blog.ethereum.org/2015/08/07/on-public-and-private-blockchains/
4. Dinh TTA, Wang J, Chen G, Liu R, Ooi BC, Tan KL (2017) Blockbench: a framework for analyzing private blockchains. In: Proceedings of the 2017 ACM International Conference on Management of Data, ACM, pp 1085–1100
5. Dorri A, Steger M, Kanhere SS, Jurdak R (2017) Blockchain: a distributed solution to automotive security and privacy. IEEE Commun Manag 55(12):119–125
6. Hasan H, AlHadhrami E, AlDhaher A, Salah K, Jayaraman R (2019) Smart contract-based approach for efficient shipment management. Computers and industrial engineering
7. Hazzazi N, Wijesekera D, Hindawi S (2014) Formalizing and verifying workflows used in blood banks. Proc. Technol 16:1271–1280
8. Kuo TT, Kim HE, Ohno-Machado L (2017) Blockchain distributed ledger technologies for biomedical and health care applications. J Am Med Inf Assoc 24(6):1211–1220
9. Lowalekar H, Ravichandran N (2014) Blood bank inventory management in India. Opsearch 51(3):376–399
10. Novo O (2018) Blockchain meets IoT: an architecture for scalable access management in IoT. IEEE Internet of Things J 5(2):1184–1195
11. Saha A, Amin R, Kunal S, Vollala S, Dwivedi SK Review on blockchain technology based medical healthcare system with privacy issues. Security and Privac, p. e83
12. Selvamani K, Rai AK (2015) A novel technique for online blood bank management. Proc Comput Sci 48:568–573
13. Sethi S (2018) Healthcare blockchain leads to transform healthcare industry. Int J Adv Res Ideas Innov Technol 4(1):607–608
14. Silva Filho OS, Cezarino W, Salviano GR (2012) A decision-making tool for demand forecasting of blood components. IFAC Proc Vol 45(6):1499–1504
15. Tian F (2016) An agri-food supply chain traceability system for china based on rfid and blockchain technology. In: 2016 13th international conference on service systems and service management (ICSSSM), IEEE, pp 1–6
16. Vanany I, Maryani A, Amaliah B, Rinaldy F, Muhammad F (2015) Blood traceability system for indonesian blood supply chain. Proc Manufact 4:535–542
17. Xia Q, Sifah E, Smahi A, Amofa S, Zhang X (2017) Bbds: Blockchain-based data sharing for electronic medical records in cloud environments. Information 8(2):44
18. Yue X, Wang H, Jin D, Li M, Jiang W (2016) Healthcare data gateways: found healthcare intelligence on blockchain with novel privacy risk control. J Med Syst 40(10):218

19. Zhao H, Bai P, Peng Y, Xu R (2018) Efficient key management scheme for health blockchain. CAAI Trans Intell Technol 3(2):114–118
20. Haw MR (2018) A blockchain testbed for DoD applications. Naval Postgraduate School Monterey United States
21. Mattias S (2017) Performance and scalability of blockchain networks and smart contracts

A Compendious Analysis of Advances in HE Methods for Contrast Enhancement

D. Vijayalakshmi and Malaya Kumar Nath

Abstract Image enhancement is a crucial pre-processing technique for image processing applications such as satellite images, medical images, and aerial surveillance systems. The image enhancement aims to produce visual content of the image in more pleasing and more suitable for machine vision applications. Enhancement of image can be improved by either increasing the contrast of the image with a low dynamic range or by highlighting the prominent details of the image. In this paper, a comprehensive analysis of bi-histogram and two-dimensional histogram equalization-based contrast enhancement techniques is performed. The performance of the various algorithms has been validated through three different databases and various performance measures. From the compendious analysis, it can be interpreted that modeling an algorithm with lower absolute mean brightness error (AMBE) and higher contrast values can render a better-enhanced image. From the qualitative and quantitative analysis, edge enhancing bi-histogram equalization using guided image filter outperforms the other contrast enhancement techniques.

Keywords Contrast enhancement · Bi-histogram equalization · Two-dimensional histogram equalization · AMBE

1 Introduction

Image enhancement improves the interpretation of the information present in an image. Image enhancement may be done by increasing the contrast or by removing the noise from the particular image. The contrast enhancement improves the perception of objects in the scene by enhancing the brightness difference between objects and their backgrounds. Contrast enhancements are typically performed by contrast

D. Vijayalakshmi (✉) · M. K. Nath
National Institute of Technology Puducherry, Karaikal, India
e-mail: vijaya1183@gmail.com

M. K. Nath
e-mail: malaya.nath@nitpy.ac.in

D. Harvey et al. (eds.), *Advances in VLSI, Communication, and Signal Processing*, Lecture Notes in Electrical Engineering 683,
https://doi.org/10.1007/978-981-15-6840-4_26

stretching, where the brightness differences uniformly distributed across the dynamic range of the image. Contrast stretch can be achieved by different transformation functions like image negatives, log transformation, power law transformation, and global histogram equalization [1].

Most of the contrast enhancement methods are based on histogram calculation that provides the occurrence of each gray values and it defines the statistical distribution of the pixels in the image. Histogram equalization (Global HE) is one of the popular techniques used for contrast enhancement by transforming the distribution of pixel intensities to uniform distribution. But GHE suffers from mean brightness shifting, over enhancement, and saturation effects [2].

Image contrast enhancement introduces artifacts if the mean brightness is not preserved. In order to preserve, the mean brightness bi-histogram equalization (BHE) methods are developed. In BHE, histograms are divided by the statistical parameters of the image like mean and median values. T. Kim has proposed brightness preserving bi-histogram equalization (BBHE) [2] which divides the histogram into two parts using the mean value and equalizes the divided sub-histograms individually. By integrating the two equalized sub-histograms, an enhanced image is obtained. Dualistic sub-image histogram equalization (DSIHE) uses median value to divide the histogram [3]. Following BBHE, DSIHE recursive division of histogram-based methods has been developed to optimize the mean brightness error. Recursive mean separate histogram equalization (RMSHE) and recursive sub-image histogram equalization (RSIHE) methods divide the histograms into 2^r sub-histograms. Determination of optimal value of r is a challenge for these methods [4].

Algorithms are developed to control the enhancement rate to avoid over enhancement along with mean brightness preservation. The enhancement rate is controlled with the help of plateau limits. Plateau limits are thresholds, where the histograms are clipped beyond the threshold value. The clipped sub-histograms are equalized and integrated to get an enhanced image. Bi-histogram equalization with a plateau limit (BHEPL) [5] for digital image enhancement, segment selective dynamic histogram equalization for brightness preserving contrast enhancement (SSDHE) [6], and image enhancement via median mean-based sub-image clipped histogram equalization (MMHE) [7] are few methods that uses histogram segmentation and histogram clipping for enhancement. These methods preserve the mean brightness and information content of the image.

There are few methods that are developed based on the two-dimensional histogram (2DHE), where the distribution of pixel intensities considered along with their spatial locations for processing. T. Celik has introduced two-dimensional-based contrast enhancement. This method utilizes contextual information for equalization that transforms the difference between pixel and its neighborhood to uniform distribution. Further, spatial entropy-based contrast enhancement (SECE) and residual entropy-based contrast enhancement methods are proposed based on two-dimensional histograms for contrast enhancement. 2DHE-based methods enhance the information contained in the image [8].

In this paper, a comparative study of various contrast enhancement algorithms such as exposure-based contrast enhancement, adaptive histogram equalization, spatial entropy-based contrast enhancement, residual entropy-based contrast enhancement, contrast enhancement with feature preservation, and edge enhanced bi-histogram equalization with guided image filters are discussed and have been analyzed. Conventional contrast enhancement techniques suffer from

1. Over enhancement, if the threshold is not conditioned properly.
2. Artifacts with degradation in edge strength.
3. Negligence on information preservation with respect to gray level mapping.
4. Appropriate quantification metric for validating the effectiveness of contrast enhancement.
5. Coverage of exhaustive benchmark databases for assessment.

The above factors if conditioned will enhance the quality of images through subjective and objective measures. The significant highlights in this paper are

1. Bi-histogram equalization for benchmark database is critically analyzed.
2. Quantitative metrics validating the effectiveness of the techniques are tabled briefly.
3. Additional metrics to the references are validated for a wider scope for the effectiveness of algorithmic modeling.
4. Advances in HE techniques find a prominent application in pattern recognition and machine learning.

In Section 2, the methodology of the various contrast enhancement techniques is discussed. Section 3 discusses the performance measures of various methods for Columbia, CSIQ, and USC-SIPI databases. The conclusion of the paper is summarized in Sect. 4.

2 Methodology

In literature, various techniques have been discussed for contrast enhancement. These methods are named based on the HE process. Few techniques are discussed in detail in this section and the performance measures are computed for various databases. A conclusion is drawn for better contrast enhancement method based on certain performance metrics. In this section, the algorithm of contrast enhancement methods is given below.

2.1 Exposure-Based Sub-Image Histogram Equalization (ESHIE)

Due to the poor contrast, images are not displayed in their natural form. It can be observed from their histograms that the intensities have not been utilized for the complete dynamic range [9]. In image acquisition, exposure is the amount of light per unit area reaching the electronic imaging sensor. Based on the intensity exposition, the images are classified into underexposed, overexposed, and mixed exposed images. In under and overexposed images, the pixel intensities and their neighborhoods occupy the lower part and upper part of the entire dynamic range, respectively. So, the information is not readily seen in the image. In real time, the images are a mixture of underexposed and overexposed intensity values. To enhance these types of images, a parameter called exposure is used. This parameter helps to segment the histogram into two parts by considering underexposed values in one group and overexposed values in another group. Algorithmic steps are as follows:

1. The exposure value is obtained by using the following equation

$$\text{exposure} = \frac{1}{L}\frac{\sum_{i=1}^{L} h(i)\,i}{\sum_{i=1}^{L} h(i)} \tag{1}$$

$$I_s = (1\text{-exposure})\,L \tag{2}$$

where $h(i)$ represents the histogram of the ith intensity and L denotes number of gray levels in the image. I_s represents the intensity value which is used to segment the histogram.

2. Threshold for histogram clipping is obtained by

$$T = \frac{1}{L}\sum_{i=1}^{k} h(i) \tag{3}$$

and

$$h_m(i) = \begin{cases} h(i) & \text{if } h(i) \le T \\ T & \text{if } h(i) > T \end{cases} \tag{4}$$

3. From the modified histogram bins, cumulative distribution functions (CDF) are obtained.
4. In the end, sub-histograms are equalized using the mapping functions

$$F_L = I_s \times C_L \tag{5}$$

$$F_u = (I_s + 1) + (L - I_s + 1) \times C_U \tag{6}$$

where C_L and C_U represent cumulative distribution functions (CDF) of lower and upper histograms, respectively. F_L and F_u represent mapping functions of lower and upper histograms, respectively. Equalized sub-histograms are combined to get an enhanced image.

2.2 *Adaptive Image Enhancement Based on Bi-histogram Equalization with a Clipping Limit (AIEHE)*

Various techniques have been proposed to solve the problem of preserving mean brightness and controlling the enhancement rate. The enhancement rate is controlled to reduce the dominant effect of high-frequency histogram bins that can be achieved by histogram clipping. In this method, the histogram is divided by using the median value in order to preserve the mean brightness. Then, the sub-histograms are clipped by an adaptive plateau limit which is calculated by considering the minimum value among histogram bins, mean, and median value of histogram bins [10].

The probability of the sub-histograms are defined as

$$p_{\text{low}}(k) = \frac{H_{\text{low}}(k)}{n_{\text{low}}} \tag{7}$$

where $p_{\text{low}}(k)$ represents the probability of kth intensity in lower sub-histogram, $H_{\text{low}}(k)$ denotes the number of times the intensity k has occurred, and n_{low} denotes the total number of pixels in the lower histogram.

$$p_{\text{up}}(k) = \frac{H_{\text{up}}(k)}{n_{\text{up}}} \tag{8}$$

where $p_{\text{up}}(k)$ represents the probability of kth intensity in upper sub-histogram, $H_{\text{up}}(k)$ denotes the number of times the intensity k has occurred, and n_{up} total number of pixels in the upper histogram.

$$\text{Clipping limit} = \min(p(k), \text{mean}(p(k)), \text{median}(p(k))) \tag{9}$$

The modified sub-histograms are obtained by using adaptive histogram clipping limit. The following mapping function is used to yield an enhanced image from the clipped sub-histograms.

$$y(k) = \begin{cases} x_0 + (x_{\lambda-1} - x_0) \,.\, \sum_{k=0}^{\lambda-1} \text{new\_low}(k) \,. \\ \text{for} \quad k = 0, 1, \ldots \lambda - 1 \\ x_\lambda + (x_{L-1} - x_0) \,.\, \sum_{k=\lambda}^{L\ 1} \text{new\_up}(k) \,. \\ \text{for} \quad k = \lambda, \lambda + 1, \ldots L - 1 \end{cases} \tag{10}$$

where x_0, x_λ, and x_{L-1} represent minimum, median, and maximum intensities of the input image, respectively. The new_low (k) and new_up (k) denote the modified lower and upper histograms.

2.3 Spatial Entropy-Based Global and Local Image Contrast Enhancement (SECE)

In this method, histograms are calculated based on the distributions with respect to the spatial locations called as two-dimensional histograms and it is used for enhancement [8]. Low contrast image is divided into non-overlapping spatial grids. Then, the histograms are calculated in the spatial grids for the pixel intensities available in the input image. The two-dimensional histogram is represented by

$$h_i = \{h_i\,(m, n)\,|1 \le m \le M,\ 1 \le n \le N \tag{11}$$

where $h_i\,(m, n)$ represents the two-dimensional histogram of the ith intensity value and m, n represent the spatial grid values on the image. M and N represent the total number of grids in the input image. From the two-dimensional histogram, spatial entropy is obtained in order to calculate the mapping function.

$$S_i = -\sum_{m=1}^{M}\sum_{n=1}^{N} h_i\,(m, n)\log_2\,(h_i\,(m, n)) \tag{12}$$

where S_i represents the spatial entropy, and a discrete function f_i is derived from S_i, which is defined as

$$f_i = \frac{S_i}{\sum_{l=1,l\neq i}^{k} S_l} \tag{13}$$

f_i is further normalized to obtain cumulative distribution function F_i. From F_i, a mapping function is derived to obtain an enhanced image.

$$y_i = \lfloor F_i \times (y_u - y_d) + y_d \rfloor \tag{14}$$

where y_i represents the mapping function for the input intensity x_i, y_u and y_d represent the minimum and maximum intensities of the dynamic grayscale, respectively.

2.4 Residual Spatial Entropy-Based Image Contrast Enhancement (RESE)

To utilize contextual information around each pixel and to overcome the drawbacks of using a one-dimensional histogram, methods based on two-dimensional histograms are proposed [11]. The two-dimensional histogram and the spatial entropy are obtained by using (11) and (12). From the two-dimensional histogram, joint spatial histogram is calculated by

$$h_{k,l}(m,n) = \max(h_k(m,n)\,h_l(m,n)) \tag{15}$$

From the joint spatial histogram, joint spatial entropy is obtained by

$$S_{k,l} = -\sum_{m=1}^{M}\sum_{n=1}^{N} h_{k,l}(m,n)\log_2 h_{k,l}(m,n) \tag{16}$$

With the help of spatial and joint entropies, residual entropy R_k is calculated as follows.

$$R_k = \frac{w_k s_k - w_{k,r}\sum_{l=1,l\neq k}^{K} s_{k,l}}{w_k + w_{k,r}} \tag{17}$$

where w_k and $w_{k,r}$ are weighting functions obtained from two-dimensional histograms and joint histograms, respectively.

$$w_k = \sum_{m=1}^{M}\sum_{n=1}^{N} h_k(m,n) \tag{18}$$

$$w_{k,r} = \sum_{l=1,l\neq k}^{K}\sum_{m=1}^{M}\sum_{n=1}^{N} h_{k,l}(m,n) \tag{19}$$

Residual function R_k is used to compute the discrete function f_k and a cumulative distribution function F_k by using (13). A mapping function is derived from the cumulative distribution function to get an enhanced image with the help of (14).

2.5 *Contrast Enhancement Using Feature Preserving Bi-histogram Equalization (CEF)*

This algorithm mainly serves for mean brightness preservation, which avoids over enhancement and controls the histogram pits. Adaptive gamma transformation is performed on the low contrast input image to restrain the histogram spikes which leads to over enhancement [12]. Gamma value is calculated based on the exposure value which is obtained by the histogram of the image using (1).

$$\gamma = \begin{cases} 1 - \text{exposure}, & 0.5 \le \text{exposure} < 1 \\ \text{exposure}, & \text{exposure} < 0.5 \end{cases} \tag{20}$$

After gamma transformation, histogram is segmented based on the exposure value which is given as

$$T_S = L \times \text{exposure} \tag{21}$$

where L represents the maximum intensity of the dynamic scale. T_S splits the histogram into under and over histogram. The underexposed and overexposed histograms are equalized individually. To preserve the details due to histogram pits, equalized sub-histograms are smoothened by adding the standard deviations of under and overexposed histograms. After histogram addition, an enhanced image is obtained.

2.6 *Edge Enhancing Bi-histogram Equalization Using Guided Image Filter (EEBHE)*

Histogram of the input image is divided into two parts using mean intensity value (I_m) of the image which is calculated by

$$I_m = \text{floor}\left(\frac{\sum_{i=0}^{l-1} I_i \times n_i}{N}\right) \tag{22}$$

where I_m represents the mean intensity, n_i represents the occurrence of the ith intensity, and N represents the total number of pixels in the image. To prevent over enhancement, plateau limits are calculated with the help of entropy values of segmented histogram and histogram of the input image for histogram clipping [13].

$$T_L = \frac{H_L}{H_I} \times \frac{1}{m+1} \sum_{i=0}^{m} h_L(i) \tag{23}$$

$$T_H = \frac{H_H}{H_I} \times \frac{1}{L-m-1} \sum_{i=m+1}^{L-1} h_H(i) \tag{24}$$

where H_L and H_H denote the entropy values of lower and upper sub-histograms, respectively. H_I represents the entropy value of input image. h_L and h_H represent lower and upper histograms.

From the modified histogram, the cumulative distribution function is obtained, and linear coefficients of the guided filter are computed for every pixel in the input image. The enhanced image is generated by using the mapping function and linear coefficients of the filter.

3 Results and Discussion

In this section, the simulation results and the performance measures of the ESIHE, AIEHE, SECE, RESE, CEF, and EEHBE methods are presented. The simulation results are obtained from three databases, namely Columbia [14], USC-SIPI [15], and CSIQ [16] databases. Six algorithms are evaluated on these databases. First, the low contrast images are identified based on the distribution of pixel intensities on the dynamic grayscale. If the distribution of pixel intensities is in a narrow range, then those images are considered for enhancement. Similarly, from each database, five low contrast images are selected to which the above-discussed algorithms are applied and the performance metrics are calculated for analysis. The performance metrics used in this study are absolute mean brightness (AMBE), contrast, structural similarity index (SSIM), standard deviation (SD), contrast improvement index (CII), and difference in entropy (DE). Details about the performance measures are discussed below.

3.1 Absolute Mean Brightness Error (AMBE)

AMBE [17, 18] is used as an objective measure to examine the preservation of mean brightness in the enhanced image. It is defined as:

$$\text{AMBE} = |M(I) - M(J)|, \tag{25}$$

where $M(I)$ and $M(J)$ represent the mean values of the low contrast (I) and enhanced (J) images, respectively. Lower values of AMBE indicate that the mean brightness is well preserved.

3.2 Contrast Improvement Index (CII)

CII [19] computes the contrast improvement of the enhanced image by means of a local window in order to measure the improvement with respect to the spatial locations. It is given as:

$$\text{CII} = \frac{E\left(C_{\text{loc}}(J)\right)}{E\left(C_{loc}(I)\right)} \tag{26}$$

where

$$C_{loc} = \frac{\max - \min}{\max + \min} \tag{27}$$

where max and min represent the maximum and minimum intensity values in a 3 × 3 window, respectively. Greater CII represents better locally enhanced image.

3.3 Standard Deviation (SD)

Standard deviation [19] measures contrast globally and it is defined as:

$$\mathrm{SD} = \sqrt{\sum_{k=0}^{L-1} (J_k - M(J))^2 \times pdf(J_k)} \tag{28}$$

where J_k indicates kth intensity of the enhanced image and $M(J)$ is the mean intensity of the enhanced image and $\mathrm{pdf}(J_k)$ represents the probability density of kth intensity. Higher value of standard deviation indicates better enhancement.

3.4 Contrast

It measures the deviation of pixel intensity values from the mean intensity of the image. Through this, contrast improvement of the output image is calculated [10]. It is defined as:

$$C = \frac{1}{M \times N} \sum_{x=1}^{M} \sum_{y=1}^{N} J^2(x, y) - \left| \frac{1}{M \times N} \sum_{x=1}^{M} \sum_{y=1}^{N} J(x, y) \right|^2 \tag{29}$$

Higher value of C indicates the utilization of the entire dynamic range and it is expressed in decibels as follows:

$$C_{\text{in dB}} = 10 \log_{10} C \tag{30}$$

3.5 Structural Similarity Index (SSIM)

SSIM [20] is a perceptual quality measurement of a processed image with respect to reference image. It varies from 0 to 1, where '1' indicates the structural information of the image is prevented and '0' indicates structural information is lost during enhancement. It is calculated from the statistical parameters of input and enhanced images. It is defined as,

$$\mathrm{SSIM}(I, J) = \frac{(2\mu_I \mu_J + c_1)\left(2\sigma_{I,J} + c_2\right)}{\left(\mu_I^2 + \mu_J^2 + c_1\right)\left(\sigma_I^2 + \sigma_J^2 + c_2\right)} \tag{31}$$

where μ_I and σ_I^2 denote mean and variance of the input image, μ_J and σ_J^2 denote mean and variance of the enhanced image, respectively. $\sigma_{I,J}$ is the covariance of images I and J. c_1 and c_2 are constants. They are defined as:

$$c_1 = (k_1 L)^2, c_2 = (k_2 L)^2 \tag{32}$$

$k_1 \ll 1, k_2 \ll 1$ and L is the maximum intensity of the image.

3.6 Difference in Entropy (DE)

Image entropy defines the uncertainty in the image values. It measures average quantity of information to encode the image values [21, 22]. A higher value of entropy specifies that the image has more details. Entropy is computed by:

$$E(I) = -\sum_{l=1}^{P} p(i_l)\log_2 p(i_l) \tag{33}$$

where $p(i_l)$ is the probability of the intensity i_l. P indicates the total number of intensities. Difference in entropy is given as:

$$\mathrm{DE} = |E(I) - E(J)| \tag{34}$$

where $E(I)$and $E(J)$ represent the input and enhanced image entropy values, respectively. Lower value of DE indicates better preservation information details of the processed image.

3.7 Qualitative Analysis

Contrast enhancement is a subjective process, so qualitative assessment is needed along with quantitative assessment. The performance of the contrast enhancement algorithms can be analyzed by visual assessment. Figures 1, 2, and 3 show the qualitative results of databases Columbia, CSIQ, and USC-SIPI, respectively. Figure 1 shows the enhanced output images by various methods. The first column represents low contrast images from the Columbia database. Column two to column seven represents the enhanced images by EEBHE, CEF, RESE, SECE, AIEHE, and ESIHE method, respectively. Images obtained by EEBHE have better visual clarity compare to other methods. In this method, edges are retained due to the usage of the guided filter.

Images in fourth and fifth columns have good contrast that leads to over enhancement. Although RESE and SECE produce better results in terms of contrast, as seen

Fig. 1 Contrast enhanced images for Columbia database. First column: low contrast images(image 1,image 2 … image 5); enhanced image: second column: EEBHE; third column: CEF; fourth column: RESE; fifth column: SECE; sixth column: AIEHE, seventh column: ESHIE

from the images, in this case, the edges are not preserved and artifacts are introduced near the edges. The objective of the enhancement technique is balancing contrast increment with mean brightness and edge preservation. The balance is absent in the enhanced images of AIEHE and ESIHE methods which are shown in the sixth and seventh columns. On analysis, it can be inferred that EEBHE outperforms when it is compared with other state of art algorithms with perfect balance between the edge preservation and contrast enhancement. On visual interpretation, this trend follows for the databases: CSIQ and USC-SIPI. Enhanced images from CSIQ and USC-SIPI by different techniques are shown in Figs. 2 and 3, respectively. For these databases, good quality enhanced images are obtained by the EEBHE technique (Tables 1, 2, 3, 4, 5, 6, 7, 8, 9, 10, 11 and 12).

Table 1 AMBE values of various contrast enhancement techniques (Columbia)

I	M_1	M_2	M_3	M_4	M_5	M_6
I_1	4.34	3.04	7.34	57.13	10.33	9.48
I_2	4.69	12.23	58.97	82.57	13.77	34.81
I_3	13.07	11.51	56.92	78.46	21.71	71.64
I_4	3.34	28.12	73.49	78.33	0.14	76.70
I_5	15.48	31.21	45.16	109.14	3.36	62.96
Avg	8.18	17.22	48.38	81.13	9.86	51.12

I = Image, M_1 = EEBHE, M_2 = CEF, M_3 = RESE, M_4 = SECE, M_5 = AIEHE, M_6 = ESIHE, Avg = average value

Table 2 CII values of various contrast enhancement technique (Columbia)

I	M_1	M_2	M_3	M_4	M_5	M_6
I_1	1.13	2.80	2.04	2.60	0.92	2.05
I_2	1.14	1.34	0.93	1.38	1.48	1.25
I_3	1.78	2.02	0.97	2.11	2.08	1.16
I_4	1.16	1.24	0.90	1.28	1.42	0.92
I_5	1.34	0.99	0.95	0.90	1.05	0.98
Avg	1.31	1.68	1.16	1.65	1.39	1.27

I = Image, M_1 = EEBHE, M_2 = CEF, M_3 = RESE, M_4 = SECE, M_5 = AIEHE, M_6 = ESIHE, Avg = average value

Table 3 SD values of various contrast enhancement techniques (Columbia)

I	M_1	M_2	M_3	M_4	M_5	M_6
I_1	27.51	41.51	41.63	71.87	18.02	39.24
I_2	40.96	36.32	82.94	67.37	70.02	68.69
I_3	34.82	35.95	32.54	76.99	21.76	42.06
I_4	14.30	29.70	43.19	57.27	18.02	44.89
I_5	41.51	36.78	46.03	74.20	18.59	57.76
Avg	31.82	36.05	49.27	69.54	29.28	50.53

Table 4 Contrast values of various contrast enhancement techniques (Columbia)

I	M_1	M_2	M_3	M_4	M_5	M_6
I_1	66.29	74.51	74.57	85.49	57.83	73.39
I_2	74.25	71.85	88.36	84.20	84.97	84.59
I_3	71.00	71.64	69.652	86.87	61.60	74.78
I_4	53.21	67.82	75.31	80.95	57.83	76.08
I_5	74.52	72.09	76.58	86.13	58.46	81.12
Avg	67.85	71.58	76.89	84.73	64.14	77.99

Table 5 SSIM values of various contrast enhancement techniques (Columbia)

I	M_1	M_2	M_3	M_4	M_5	M_6
I_1	0.99	0.79	0.89	0.67	0.97	0.89
I_2	0.96	0.78	0.70	0.44	0.69	0.83
I_3	0.70	0.65	0.64	0.31	0.54	0.51
I_4	0.95	0.57	0.25	0.22	0.86	0.23
I_5	0.82	0.75	0.60	0.30	0.97	0.47
Avg	0.89	0.71	0.622	0.39	0.80	0.59

Table 6 Entropy values of various contrast enhancement techniques (Columbia)

I	M_1	M_2	M_3	M_4	M_5	M_6
I_1	0.015	0.157	0.001	0.076	0.891	0.019
I_2	0.006	0.228	0.005	0.211	0.685	0.017
I_3	0.659	0.167	0.005	0.061	1.612	0.013
I_4	0.211	0.072	0.015	0.021	0.185	0.002
I_5	0.499	0.057	0.055	0.063	0.443	0.009
Avg	0.27	0.13	0.01	0.08	0.76	0.01

Table 7 AMBE values of various contrast enhancement techniques (CSIQ)

I	M_1	M_2	M_3	M_4	M_5	M_6
I_1	4.45	15.50	18.57	132.41	10.07	6.02
I_2	3.67	33.02	16.96	19.77	10.88	27.19
I_3	2.54	74.25	10.27	16.99	8.35	56.20
I_4	6.52	1.72	24.07	69.14	15.97	16.76
I_5	3.81	3.67	9.78	19.31	16.32	18.54
Avg	4.20	25.63	15.93	51.52	12.32	24.94

Table 8 CII values of various contrast enhancement techniques (CSIQ)

I	M_1	M_2	M_3	M_4	M_5	M_6
I_1	1.43	1.69	1.51	1.24	1.44	2.29
I_2	1.36	3.92	2.02	3.72	1.46	2.90
I_3	1.44	10.26	3.41	9.95	2.23	2.32
I_4	1.54	3.58	1.85	3.70	0.97	3.16
I_5	1.32	3.29	1.86	3.41	1.10	2.55
Avg	1.42	4.55	2.13	4.14	1.44	2.64

Table 9 SD values of various contrast enhancement techniques (CSIQ)

I	M_1	M_2	M_3	M_4	M_5	M_6
I_1	36.17	32.55	48.79	44.59	35.32	53.12
I_2	25.51	53.69	42.09	76.51	29.18	36.52
I_3	11.88	38.55	30.12	58.04	19.44	27.62
I_4	34.419	34.52	46.71	63.61	21.19	59.38
I_5	40.89	63.77	54.69	71.34	32.71	42.95
Avg	29.77	44.62	44.48	62.82	27.57	43.92

Table 10 Contrast values of various contrast enhancement techniques (CSIQ)

I	M_1	M_2	M_3	M_4	M_5	M_6
I_1	71.76	69.66	77.75	75.95	71.29	79.45
I_2	64.78	79.66	74.79	86.75	67.47	71.96
I_3	49.50	73.04	68.10	81.22	59.35	66.37
I_4	70.77	70.83	76.88	83.05	61.07	81.68
I_5	74.22	83.10	80.03	85.34	69.75	75.20
Avg	66.20	75.26	75.51	82.46	65.79	74.93

Table 11 SSIM values of various contrast enhancement techniques (CSIQ)

I	M_1	M_2	M_3	M_4	M_5	M_6
I_1	0.915	0.850	0.853	0.411	0.857	0.678
I_2	0.972	0.657	0.877	0.623	0.895	0.835
I_3	0.959	0.358	0.650	0.332	0.720	0.644
I_4	0.962	0.881	0.894	0.626	0.923	0.780
I_5	0.965	0.676	0.890	0.593	0.948	0.856
Avg	0.95	0.68	0.83	0.51	0.86	0.75

Table 12 DE values of various contrast enhancement techniques (CSIQ)

I	M_1	M_2	M_3	M_4	M_5	M_6
I_1	0.310	0.177	0.007	0.208	0.377	0.012
I_2	0.182	0.087	0.012	0.090	0.458	0.006
I_3	0.226	0.065	0.015	0.052	0.986	0.006
I_4	0.237	0.069	0.003	0.028	1.444	0.010
I_5	0.215	0.118	0.011	0.182	0.531	0.025
Avg	0.23	0.10	0.01	0.11	0.75	0.01

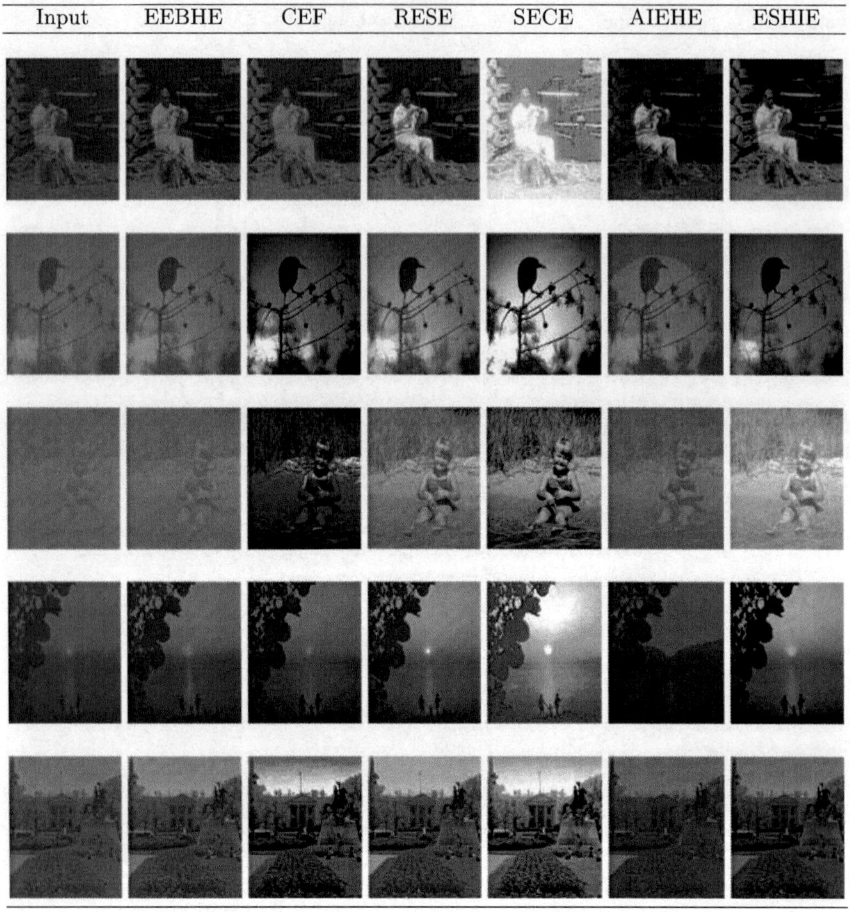

Fig. 2 Contrast enhanced images for CSIQ database. First column: low contrast images (image 1, image 2 … image 5); enhanced image: second column: EEBHE; third column: CEF; fourth column: RESE; fifth column: SECE; sixth column: AIEHE, seventh column: ESHIE

3.8 Quantitative Analysis

The combination of various performance metrics gives the quantitative assessment of different algorithms. The evaluation results of six metrics for each image from three databases are given in Tables where I_1, $I_2 \ldots I_{10}$ represent image 1, image 2 … image 10.

Tables 1, 7, and 13 show the AMBE value of different methods for Columbia, CSIQ, and USC-SIPI databases, respectively. The lower value of AMBE shows better the preservation of mean brightness. It is observed that from Table 1 EEBHE method has a low average AMBE value with the comparison of other methods. A

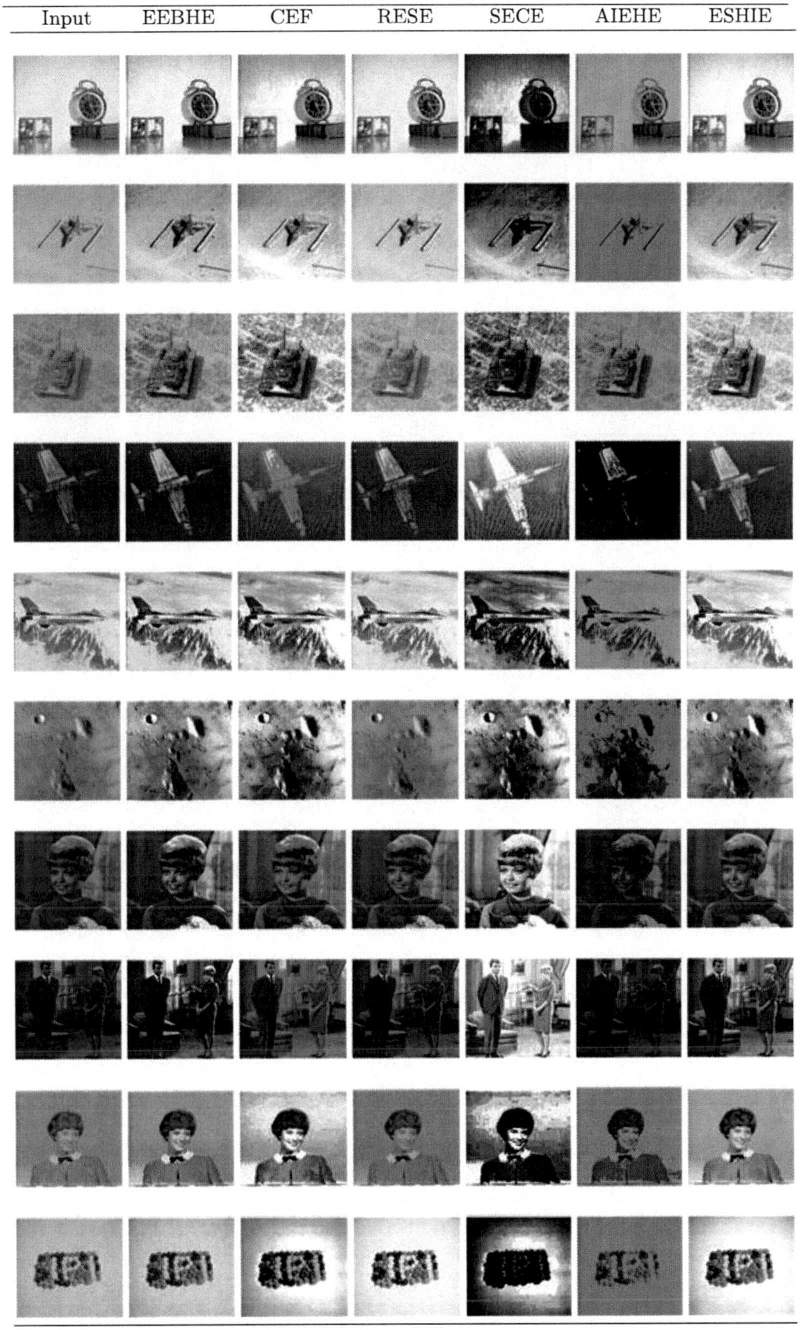

Fig. 3 Contrast enhanced images for USC-SIPI database. First column: low contrast images (image 1, image 2 ... image 10); enhanced image: second column: EEBHE; third column: CEF; fourth column: RESE; fifth column: SECE; sixth column: AIEHE, seventh column: ESHIE

Table 13 AMBE values of various contrast enhancement techniques (SIPI)

I	M_1	M_2	M_3	M_4	M_5	M_6
I_1	0.85	15.49	0.395	83.71	43.36	1.76
I_2	4.01	14.48	24.00	58.41	54.80	1.15
I_3	7.44	23.36	28.70	23.29	7.55	42.55
I_4	2.70	40.10	11.30	103.42	22.45	32.14
I_5	3.62	11.58	10.80	66.57	47.37	14.02
I_6	8.59	9.19	3.54	3.19	41.40	7.70
I_7	11.81	15.66	8.45	78.02	17.04	19.69
I_8	18.63	33.17	7.84	113.89	4.41	39.24
I_9	6.61	30.87	8.68	12.30	22.46	24.91
I_{10}	5.07	1.14	32.81	75.08	50.16	26.42
Avg	6.93	19.51	13.65	61.79	31.10	20.96

similar trend is followed in Tables 7 and 13. From the observation, it is interpreted that if histogram segmentation uses the mean value as the separation point, then the mean brightness of the image is preserved well after enhancement.

The CII value of various methods for different databases is shown in Tables 2, 3, 8, 9, 14, and 15 show the performance metrics SD of the three databases. Contrast value is shown in Tables 4, 10, and 16 for Columbia, CSIQ, and USC-SIPI database, respectively. These three performance metrics are correlated with each metric attribute toward enhancing the contrast. This helps to understand the quantitative validation of contrast enhancement. It is inferred from the tables that the SECE method gives the highest value of these metrics when compared to the other methods. This is due to the fact that spatial distributions of pixel intensities help to utilize the entire dynamic range of the grayscale results in a high contrast image (Tables 13, 14, 15 and 16).

Tables 5, 11, and 17 show the quantitative parameter value of SSIM. In general, It is observed that the higher the SSIM closer to 1.0, the better the edges are preserved in the image. From the tables, the SSIM is closer to 1.0 for the methods EEBHE, AIEHE, and CEF are observed. The variation shown in the tables is due to inherent image characteristics and algorithmic modeling. Irrespective of the database, EEBHE performs better for all the images when compared to other techniques. Due to the usage of the guided filter, the EEBHE method retains the structural information.

Tables 6, 12, and 18 give the performance measure of the difference in entropy (DE). The lower value of DE shows the better preservation of entropy. Merging of intensities during enhancement causes decrement in the entropy value of the processed image. From the tabulated value, it is observed that the RESE method preserves the entropy value the same as input when compared to the other methods. This is due to the fact that joint spatial histogram helps to enhance image while preserving the entropy value.

Table 14 CII values of various contrast enhancement techniques (SIPI)

I	M_1	M_2	M_3	M_4	M_5	M_6
I_1	1.44	1.49	1.17	2.25	1.07	1.31
I_2	2.44	2.28	1.11	7.14	0.84	2.36
I_3	2.02	2.66	1.05	3.96	1.53	1.87
I_4	1.83	2.06	0.96	2.12	1.51	1.51
I_5	1.84	1.92	1.08	2.98	1.33	1.24
I_6	2.18	2.81	0.99	3.19	2.45	2.06
I_7	1.19	1.16	0.96	1.14	1.12	1.00
I_8	1.01	0.70	0.90	0.62	0.88	0.88
I_9	1.6	2.15	1.18	3.41	1.36	1.58
I_{10}	1.17	1.87	1.31	2.61	0.93	1.45
Avg	1.67	1.91	1.07	2.94	1.30	1.52

Table 15 SD values of various contrast enhancement techniques (SIPI)

I	M_1	M_2	M_3	M_4	M_5	M_6
I_1	67.55	63.86	62.93	78.98	42.42	64.99
I_2	41.45	44.54	27.47	70.99	17.12	41.58
I_3	47.59	69.25	34.38	68.41	39.39	58.23
I_4	41.85	41.37	33.32	74.11	30.34	45.06
I_5	64.57	66.64	52.12	75.98	43.05	55.86
I_6	57.55	76.92	28.22	76.92	50.14	56.87
I_7	61.09	47.40	44.63	77.03	36.18	50.78
I_8	62.11	39.63	37.81	71.01	28.32	58.41
I_9	43.94	64.63	32.19	89.27	38.16	44.31
I_{10}	45.90	63.63	53.34	80.15	23.41	55.49
Avg	53.36	57.79	40.64	76.28	34.85	53.16

In summary, RESE is suited for the preservation of information content. SECE gives good contrast enhancement in terms of CII, Contrast, and SD. Though RESE and SECE methods have the capability to preserve the entropy while increasing contrast but they fail to preserve the mean brightness and structural information, which leads to undesired artifacts in the enhanced image. EEBHE has good performance in terms of AMBE and SSIM. It produces better results validating CII, Contrast, and SD performance metrics. From the observations, it is suggested that EEBHE provides good enhancement output in many aspects (Table 17 and 18).

Table 16 Contrast values of various contrast enhancement techniques (SIPI)

I	M_1	M_2	M_3	M_4	M_5	M_6
I_1	84.25	83.13	82.84	87.38	74.95	83.48
I_2	74.49	75.92	66.26	85.25	56.81	74.55
I_3	77.25	84.75	70.75	84.51	73.47	81.29
I_4	74.68	74.45	70.12	86.11	68.25	76.16
I_5	83.35	83.98	79.07	86.61	75.25	80.45
I_6	81.05	86.85	66.80	86.85	78.29	80.81
I_7	82.24	77.17	75.96	86.88	71.77	78.55
I_8	82.57	73.59	72.65	85.25	66.87	81.35
I_9	75.66	83.37	69.43	89.83	72.83	75.82
I_{10}	76.53	83.06	79.53	87.67	63.06	80.32
Avg	79.21	80.63	73.34	86.63	70.15	79.28

Table 17 SSIM values of various contrast enhancement techniques (SIPI)

I	M_1	M_2	M_3	M_4	M_5	M_6
I_1	0.974	0.844	0.994	0.379	0.918	0.977
I_2	0.834	0.716	0.985	0.351	0.842	0.749
I_3	0.824	0.562	0.962	0.411	0.907	0.754
I_4	0.892	0.364	0.939	0.144	0.235	0.591
I_5	0.935	0.805	0.994	0.464	0.842	0.968
I_6	0.723	0.525	0.999	0.492	0.444	0.728
I_7	0.926	0.885	0.985	0.552	0.847	0.931
I_8	0.833	0.608	0.957	0.249	0.958	0.610
I_9	0.923	0.655	0.994	0.403	0.836	0.921
I_{10}	0.986	0.84	0.969	0.409	0.907	0.949
Avg	0.88	0.68	0.97	0.38	0.77	0.81

4 Conclusion

In this paper, a compendious analysis of advances in histogram equalization for contrast enhancement has been discussed. The various methods are tested on publicly available Columbia, CSIQ, and USC-SIPI databases. The quality of obtained enhanced images by different techniques is evaluated by subjective and objective measures. From the analysis, it is observed that bi-histogram-based methods produce low AMBE value, indicates the preservation of mean brightness. By controlling the enhancement rate, the performance of bi-histogram equalization methods can be improved. Spatial locations of intensities in the low contrast image for calculating the histograms are used for improving the contrast. Hence, SECE and RESE methods produce high contrast images compared to the other methods and they preserve the

Table 18 DE values of various contrast enhancement techniques (SIPI)

I	M_1	M_2	M_3	M_4	M_5	M_6
I_1	0.10	0.27	0.0	0.67	1.13	0.07
I_2	0.17	0.12	0.01	0.27	3.25	0.04
I_3	0.37	0.07	0.04	0.20	0.16	0.03
I_4	0.21	0.29	0.0	0.31	2.40	0.05
I_5	0.17	0.26	0.02	0.47	1.27	0.04
I_6	0.65	0.20	0.01	0.30	2.65	0.09
I_7	0.15	0.28	0.00	0.26	0.74	0.07
I_8	0.24	0.23	0.01	0.27	0.45	0.06
I_9	0.10	0.27	0.01	0.63	2.14	0.08
I_{10}	0.10	0.19	0.00	0.57	1.90	0.03
Avg	0.22	0.22	0.01	0.39	1.6	0.06

information content after enhancement. But they fail to preserve the mean brightness. EEBHE method gives better contrast enhancement along with mean brightness preservation and preservation of structural information. It is concluded that the EEBHE method performs well compared to other algorithms. This was verified by the low average AMBE value and high average SSIM value. The limitations of various algorithms may be overcome by considering the strength and potential of different algorithms. This combination may be utilized for machine vision applications.

References

1. Gonzalez CR (2012) Digital inage processing. Pearson Prentice Hall Publisher
2. Kim YT (1997) Contrast enhancement using brightness preserving bi-histogram equalization. IEEE Trans Consum Electron 43(1):1–8. https://doi.org/10.1109/30.580378
3. Wang Y, Chen Q, Zhang B (1999) Image enhancement based on equal area dualistic sub-image histogram equalization method. IEEE Trans Consum Electron 45(1):68–75. https://doi.org/10.1109/30.754419
4. Sim K, Tso C, Tan Y (2007) Recursive sub-image histogram equalization applied to gray scale images. Pattern Recogn Lett 28(10):1209–1221. https://doi.org/10.1016/j.patrec.2007.02.003
5. Ooi CH, Pik Kong NS, Ibrahim H (2009) Bi-histogram equalization with a plateau limit for digital image enhancement. IEEE Trans Consum Electron 55(4):2072–2080. https://doi.org/10.1109/TCE.2009.5373771
6. Khan MF, Khan E, Abbasi Z (2014) Segment selective dynamic histogram equalization for brightness preserving contrast enhancement of images. Optik 125(3):1385–1389. https://doi.org/10.1016/j.ijleo.2013.08.005
7. Chen S-D, Ramli AR (2003) Minimum mean brightness error bi-histogram equalization in contrast enhancement. IEEE Trans Consum Electron 49(4):1310–1319. https://doi.org/10.1109/TCE.2003.1261234
8. Celik T (2014) Spatial entropy-based global and local image contrast enhancement. IEEE Trans Image Proc 23(12):5298–5308. https://doi.org/10.1109/TIP.2014.2364537

9. Singh K, Kapoor R (2014) Image enhancement using exposure based sub image histogram equalization. Pattern Recogn Lett 36:10–14. https://doi.org/10.1016/j.patrec.2013.08.024
10. Tang JR, Isa NAM (2014) Adaptive image enhancement based on bi-histogram equalization with a clipping limit. Comput Electr Eng 40(8):86–103. https://doi.org/10.1016/j.compeleceng.2014.05.017
11. Celik T, Li HC (2016) Residual spatial entropy-based image contrast enhancement and gradient-based relative contrast measurement. J Modern Opt 63(16):1600–1617. https://doi.org/10.1080/09500340.2016.1163427
12. Wang X, Chen L (2018) Contrast enhancement using feature-preserving bi-histogram equalization. Sign Image and Video Proc 12(4):685–692. https://doi.org/10.1007/s11760-017-1208-2
13. Mun J, Jang Y, Nam Y, Kim J (2019) Edge-enhancing bi-histogram equalisation using guided image filter. J Visual Commun Image Represent 58:688–700. https://doi.org/10.1016/j.jvcir.2018.12.037
14. Ng TT, Chang SF, Hsu J, Pepeljugoski M (2004) Columbia photographic images and photorealistic computer graphics dataset. Tech Rep 205-2004-5, ADVENT, Columbia University
15. USC-SIPI Database: http://sipi.usc.edu/database/
16. Larson EC, Chandler MC (2010) Most apparent distortion: full-reference image quality assessment and the role of strategy. J Electron Imag 19(1):1–21. https://doi.org/10.1117/1.3267105
17. Ooi CH, Mat Isa NA (2010) Adaptive contrast enhancement methods with brightness preserving. IEEE Trans Consum Electron 56(4):2543–2551
18. Kar M, Giritharan R, Elangovan P, Kumar M (2019) Analysis of diagnostic features from fundus image using multiscale wavelet decomposition. ICIC Express Lett, Part B: Appl 10:175–184 https://doi.org/10.24507/icicelb.10.02.175
19. Liang K, Ma Y, Xie Y, Zhou B, Wang R (2012) A new adaptive contrast enhancement algorithm for infrared images based on double plateaus histogram equalization. Infrared Phys Technol 55(4):309–315
20. Li C, Bovik AC (2010) Content-partitioned structural similarity index for image quality assessment. Sign Proc: Image Commun 25(7):517–526
21. Shannon CE (1948) A mathematical theory of communication. Bell Syst Techn J 27(3):379–423
22. Nath MK, Dandapat S (2012) Differential entropy in wavelet sub-band for assessment of glaucoma. Int J Imag Syst Technol 22:161–165

A Survey of Semantic Segmentation on Biomedical Images Using Deep Learning

Yellamelli Prajna and Malaya Kumar Nath

Abstract This survey presents an overview of recent attempts on semantic segmentation on biomedical images using deep learning techniques. Deep learning is a revolutionary technology and recent advances in it have made it possible to improve the performance of semantic segmentation methods, which are common tasks in medical imaging. Recently, semantic segmentation is a breakthrough in biomedical image segmentation. Semantic segmentation is the process of classifying each and every pixel of an image into a class. This paper explicitly deals with various semantic segmentation methods applied to different biomedical images. Further, diverse architectures of deep neural networks for semantic segmentation of images such as U-Net, fully convolution network (FCN), and SegNet are examined. In brief, datasets namely Dristhi-GS, ALL-IDB1, H & E stained images, Flair-MRI Brats 2015, and Japanese Society of Radiological Technology (JSRT) are studied. In this paper, a complete taxonomy of various methods of semantic segmentation for medical images is given.

Keywords Semantic segmentation · Deep learning · Convolution network · U-Net

1 Introduction

Modern image segmentation techniques are powered by deep learning technology. Due to advances in technology, deep learning [1] is penetrating deep into the field of medical image segmentation, especially semantic segmentation. The aim of semantic image segmentation is to label each and every pixel of an image with an equivalent class [2]. As prediction is on every pixel in the image, it is generally referred to as dense prediction [2]. Semantic segmentation, which plays a key role in image processing and computer vision domain has its significance in multiple domains such as medicine, robotics, and navigation. This paper deals with the segmentation

Y. Prajna (✉) · M. K. Nath
National Institute of Technology Puducherry, Karaikal, India
e-mail: prajnayellamelli@gmail.com

M. K. Nath
e-mail: malaya.nath@nitpy.ac.in

D. Harvey et al. (eds.), *Advances in VLSI, Communication, and Signal Processing*, Lecture Notes in Electrical Engineering 683,
https://doi.org/10.1007/978-981-15-6840-4_27

performed on biomedical images. Some are (i) segmenting nuclei boundary and detecting geometric center of nuclei [3], (ii) optic disk and cup segmentation from fundus image [4], (iii) blood cells segmentation from blood smear image [5], (iv) brain tumor segmentation [6], and (v) lung segmentation in chest radiographs [7].

Especially semantic segmentation is instrumental in the field of medicine for diagnosis of a disease, as segmenting manually is very time consuming and intensive process for pathologists. The accuracy and precision are more effectively achieved with computer-based methods rather than human intervention.

Researchers have used different architectures like U-Net [8], SegNet [9], FCN [10] for designing the segmentation models. Convolutional neural network [11] is the basic building block for the above-mentioned nets. It is a branch of neural networks, which consists of a stack of layers each performing special operation, e.g., convolution, pooling, loss calculation, etc. FCN [10] is obtained by replacing the last fully connected layer of CNN with the fully convolutional layer. The U-Net model [8] is built upon the elegant architecture of FCN, which integrates downsampling and upsampling on the input image in the spatial direction between successive convolutional layers making it symmetric. SegNet [9] architecture uses pairs of encoders and decoders for feature mapping, the encoder has a convolutional layer, batch normalization, and ReLU activation function, after that max pooling is performed to reduce the dimensionality. Decoder stage includes upsampling and last step of the decoder is connected to softmax classifier to classify each pixel in an image.

Examining various literatures it is revealed that researchers have employed prevailing datasets such as Drishti-GS [12] for disk and cup segmentation of fundus image, H & E stained images [13] for cell nuclei detection & segmentation, ALL-IDB1 [14] for blood cells segmentation, Flair-MRI Brats 2015 [15, 16] for brain tumor segmentation, and JSRT [17, 18] for chest radiographs lung segmentation for their research work.

The rest of the paper is organized as follows: Sect. 2 provides details on various methods of segmentation performed on various biomedical images. Section 3 briefs about results, database description, and the performance measure. Finally, the paper ends with a conclusion in Sect. 4.

2 Taxonomy for Semantic Segmentation of Medical Images

This section deals with the various semantic segmentation methods applied on various biomedical images using existing deep learning architectures and publicly available databases. Some of the methods and applications are for boundary segmentation of nuclei (& also finding its geometric center), cup & optic disk segmentation, red blood cells & white blood cells separation, segmentation of brain tumor, and chest radiographs of the lung. Taxonomy of segmentation, architectures, and their applications are shown in Table 1.

Table 1 Taxonomy of segmentation

Segmentation	Architectures used	Applications
Cell nuclei (2019) [3]	U-Net	Cancer diagnosis
Blood cells (2018) [5]	SegNet	Anemia, leukemia, blood disorders diagnosis
Brain tumor (2018) [6]	U-Net-based CNN along with MLP	Glioblastoma diagnosis
Lung (2015) [7]	Deep belief network (DBN)	Pulmonary pathology identification mainly tuberculosis

2.1 *Segmentation of Nuclei Boundary and Detection of Geometric Center of Nuclei [3]*

Cell nuclei segmentation is mostly essential for acquiring the features and functionality of cell to aid in pathology-based diagnosis, especially cancer, which is a perilous disease. Previous literature work on nuclei boundary segmentation and detection proposed several methods based on threshold segmentation with Otsu algorithm [19], morphological segmentation with watershed algorithm [20], and also some of the machine or deep learning technologies, widely embraced for computational pathology. However, segmentation and detection of nuclei pose a challenge because nuclei exhibit different morphologies and exist close to each other [21], which makes it difficult for their separation.

Chen et al. [3] have overcome the above challenges. They have proposed an approach, which initially progresses with the utilization of deep learning to estimate each pixel probability to be a cell nuclei followed by boundary smoothening and detection of individual nuclei using image processing techniques.

The author partitioned the sample of color tissue into overlapping patches and fed this to the U-net model [8]. The model is a multi-layer convolutional neural network (CNN) [11]. It uses both ground truth mask and edge map in the loss function. Due to this, during training, it enhances the separation between nuclei. The output of the model is a gray-scaled mask patches. A mask is constructed by averaging the overlapping patches and eventually boundary smoothening by the erosion operation on the mask. At last, each ellipse is isolated and depicted as individual nuclei as shown in Fig. 2. The complete process is shown in Fig. 1, which consists of a sampling of color tissue, image partition into overlapping patches, U-net model, gray-scaled mask patches, mask, boundary segmentation & erosion, nuclei detection, and center calculation.

To enhance the edge separation, a loss function is adopted in training as a combination of mask and edge difference represented in Eq. (1).

$$\text{Loss} = w.\sum_{i \epsilon K}(\hat{y}_i - y_i) + (1 - w).\sum_{j \epsilon S}(\hat{e}_j - e_j) \qquad (1)$$

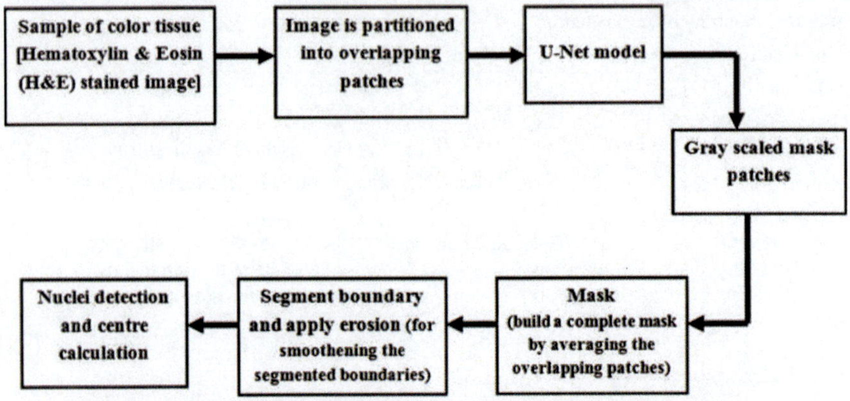

Fig. 1 Cell nuclei segmentation and detection process flow [3]

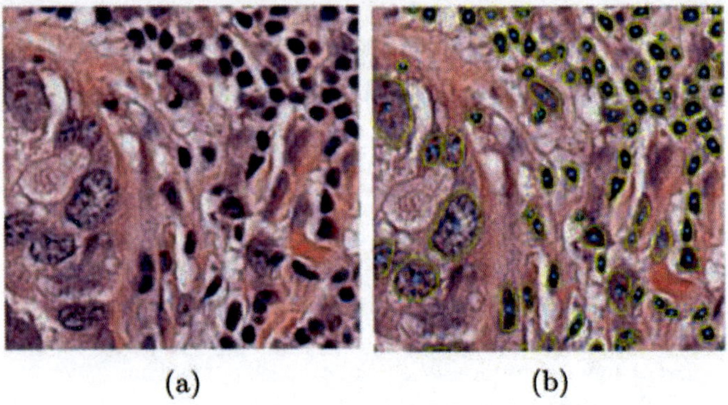

Fig. 2 **a** Tissue sample and **b** edge detection and center calculation [3]

where

$\hat{y}_i$ is ground truth mask,
$y_i (i \epsilon K)$ is output of the model,
$\hat{e}_j (j \epsilon S)$ is the edge map of $\hat{y}_i$,
$e_j (j \epsilon S)$ is the edge map of y_i ground truth.

In this, an input RGB image size of 128 × 128 × 3 is reduced to an output image size of 128 × 128 × 1 by using a number of filters having various sizes. An average precision of 79%, recall of 95%, F-score of 86%, and intersection over union (IoU) of 83% has been achieved by the above method on H & E dataset [13].

The method has been extended to color tissues like bladder, colon, liver, stomach, prostate, kidney, and breast tissues with H & E stained images dataset [13].

2.2 Optic Disk (OD) and Cup Segmentation of Fundus Image [4]

Automated segmentation of optic disk and cup of fundus image is crucial for achieving clinical parameters such as cup to disk ratio (CDR), to aid glaucoma diagnosis. Prior works are carried out using traditional and deep learning methods. Some of the traditional methods include estimating the depth of discontinuity of cup boundary by comparing the acquired images from motion model; correlation of color, texture, shading of depth, and discontinuity from optical coherence tomography (OCT). Deep learning methods include extraction of features from CNN [11] (pre-trained on ImageNet [27]) is inclusive of specialized convolutional layers to anticipate blood vessels, training of two U-Nets for cup and disk segmentation separately, etc.

Edupuganti et al. [4] have proposed a single-shot segmentation pipeline instead of separately segmenting cup and disk. They have presented a system using FCN8s architecture using one deep neural network for segmenting cup and disk in one shot. Then, prioritized specific regions to retrieve boundaries using special procedures and subsequently developed pre-processing techniques are used to reduce false positives.

Predominantly, in the fundus image, the cup and OD occupy a relatively small area compared to the background. In order to account for that experiments are modified as mentioned below. It consists (i) A FCN is trained using color fundus image (CFI) and ground truth is generated, (ii) to account for imbalance between no. of background pixels and cup/OD regions, a weighted loss function (which assigns a 10× higher loss to pixels belonging either to OD or cup regions) is used, (iii) a weighted loss function (which assigns 10× higher loss to pixels belonging only to OD and cup boundaries) similar to (ii) is initiated to find CDR accuracy as it is directly related to boundary quality [4]. Among these three, the first performs best.

The experiments were conducted on a fully convolution network (FCN) [10] with a stride of 8 pixels on Drishti-GS [12] dataset. Due to less number of training images, the FCN [10] encoder is initialized with weights from a model trained on ImageNet [27].

At the end, a filter module is used to obtain clean cup and disk boundaries. This helps in reducing false positives. The cup and disk boundaries of a healthy eye and glaucomatous eye are shown in Fig. 3 [4].

Correlating the previously used traditional methods and deep learning methods, the latter provides the best performance in terms of F-score. The achieved F-score for the first experiment is 89% for cup and 96% for disk on Drishti-GS dataset [12].

2.3 Blood Cells Image Segmentation from Blood Smear [5]

Red blood cells (RBC) carry oxygen throughout our body, white blood cells (WBC) help to fight against infections and platelets are responsible for blood clotting. RBC, WBC, and platelets are the main focus in blood smear that help doctors to diagnose

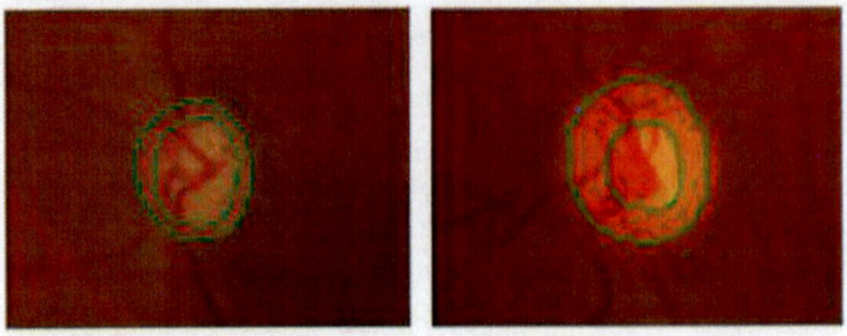

Fig. 3 Cup and disk boundaries **a** glaucomatous eye and **b** healthy eye [4]

for blood disorders. So a complete blood count (CBC), i.e., the measure of number and quality of these components in blood cell images aids in identifying various disorders.

Recent works on blood cells segmentation include Gram-Schmidt orthogonalization and morphological enhancement in segmenting WBC including nucleus & cytoplasm; shifting of invariant complex wavelet transform into watershed segmentation for WBC segmentation; k-means, fuzzy c-means, and moving k-means clustering for leukocyte segmentation; image processing techniques followed by histogram equalization [5].

Tran et al. [5] adopted the method, which reduces the complex image processing steps for image enhancement. Initially, the steps are as discussed below: label the RBC and WBC with different colors, to identify each pixel to its corresponding class (i.e., semantic segmentation) using SegNet [9] architecture and is applied to ALL-TIDB1 [14] database, which results in leukocytes and erythrocytes separation. The segmented result is shown in Fig. 4 [5].

An input RGB image of blood cells is segmented using SegNet [9] where its encoder stack contains convolutional layer, batch normalization, and ReLu activation function followed by max pooling. Max pooling reduces the feature map size that leads to blurring of object boundaries in images. This is resolved by storing the boundary information in encoder feature maps before sub-sampling. The output image and the input image should have the same size, so upsampling is performed in the decoder stack of SegNet [9] using the stored max-pooling indices in the encoder. The last stage of the decoder is connected to softmax classifier to classify each pixel in an image.

The current approach presented an experimental outcome of 89.45% accuracy as compared to the previously available methods on ALL-IDB1 database [14]. Segmentation of WBC, RBC, and the background of blood smear image reached 94%, 91%, and 87% accuracy, respectively.

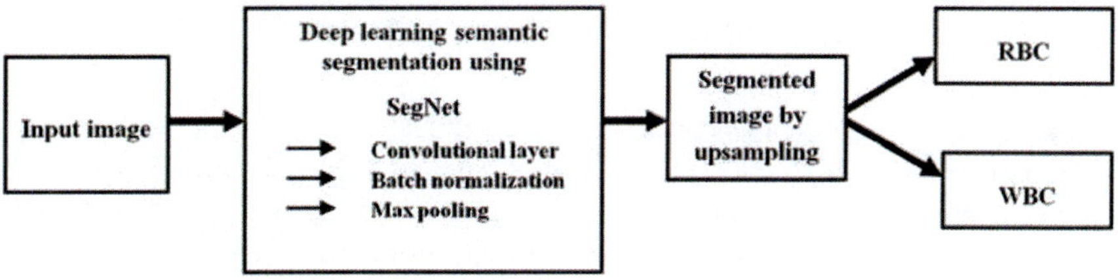

Fig. 4 Process flow of blood cells segmentation [5]

2.4 *Brain Tumor Segmentation by Optimization of a Variational Model [6]*

Premature detection of brain tumor is helpful in early prevention of disease and increases the probability of survival of a patient. Several models have come into limelight for the segmentation of the brain tumor.

Ramirez et al. [6] have acquainted a variational model for brain tumor segmentation. This model emerges by integrating the saliency term to the existing total variation-based restoration function. In computer vision, a saliency map is an image that shows each pixel's unique quality. The purpose of saliency detection in image processing is mainly to distinguish between relevant and non-relevant pixels in a given image.

Brain MRI (Magnetic Resonance Imaging) [16, 22] of patients suffering from glioblastoma is chosen as a test dataset for the above-mentioned framework, which achieved 85.7% dice similarity coefficient.

2.5 *Lung Segmentation in Chest Radiographs [7]*

Lung segmentation is a major step toward the diagnosis of lung health by analyzing the parameters such as lung volume and lung shape. Segmentation mainly assists in screening of pulmonary pathologies, specifically tuberculosis [23]. Several hybrid methodologies are already proposed based on the integration of different techniques like pixel classification and deformable models for lung segmentation. However, automated lung segmentation faces challenges because of the ribcage and clavicle edges, the inconsistent appearance of clavicle bone at the lung apex and discrepant lung shape of individuals [7].

Ngo et al. [7] proposed a model for lung segmentation using deep belief network (DBN) [24]. Model is based on a combination of distance regularized level set [25] and deep structured inference. It was tested on Japanese society of radiological technology (JSRT) dataset [17, 18] and average accuracy vary from 94 to 98% depending on the initialization.

3 Results, Datasets, and Performance Measures

Researches have used various datasets and performance measures for semantic segmentation. Some of them are: (i) H & E stained images [13] for cell nuclei detection and segmentation, (ii) Drishti-GS [12] for disk and cup segmentation of fundus image, (iii) ALL-IDB1 [14] for blood cells segmentation, (iv) Flair-MRI Brats 2015 [15] for brain tumor segmentation, and (v) JSRT [17] for chest radiographs lung segmentation. The description of different datasets and performance measures for various applications are described below and shown in Tables 2 and 3.

Table 2 Description of Datasets for semantic segmentation in medical images

Organ for segmentation	Dataset	Size of dataset	Dimension	Modality	Performance measure	Image
Cell nuclei	H & E stained images [13]	30 images	2D (1000 × 1000)	Microscopy images	Recall, F-score, Precision, IoU	Colored
Retina (fundus image)	Drishti-GS (2013) [12]	101 images (50 training + 51 testing)	2D (2047 × 1760)	Color fundus imaging (CFI)	F-score	Colored
Blood smear	ALL-IDB1 (2005) [14]	108 images	2D (2592 × 1944)	Optical laboratory microscope	Accuracy	Colored (24-bit depth)
Brain	Flair-MRI Brats (2015) [15]	220 flair-MRI volume with 155 slices per volume	3D	MRI	Precision, Recall	Gray scale
Lung	JSRT (1998) [17]	154 nodule & 93 non-nodule images	2D (2048 × 2048)	Chest radiographs	Dice coefficient, Jaccard similarity coefficient	Gray scale (12-bit)

Table 3 Performance measure used in literature

Performance measure	Formula
Precision	$\frac{TP}{TP+FP}$
Recall	$\frac{TP}{TP+FN}$
F-score	$2 * \frac{\text{precmision} * \text{recall}}{\text{precision} + \text{recall}}$
Intersection over Union (IoU)	$\frac{TP}{TP+FP+FN}$
Dice score	$\frac{2*\text{no. of TP}}{2*\text{ no. of TP}+\text{ no. of FP}+\text{ no. of FN}}$

TP True positive, *FP* False positive, *FN* False negative

3.1 H & E Stained Images

Chen et al. [3, 13] have used H & E stained image dataset [13] for segmentation of nuclei to diagnose the various stages of cancer. They have used 30 H & E stained color images of size 1000 × 1000. The performance of this method was measured by the recall (95.5%), precision (79.9%), F-score (87%), and mask IoU (83.5%).

3.2 Drishti-GS

Edupuganti et al. [4] have used [12] Dristhi-GS [12] dataset for segmenting OD and cup of fundus image. Ground truth for OD and cup segmentation is available for segmentation purpose. They have taken 40 training images and augmented them to 160 training images using image flips. They have chosen a validation set of 10 images and testing set of 51 images. Each experiment is trained for 100 epochs using an Adam optimizer. FCN8s network with 134M parameters is initialized with ImageNet weights. The performance of this method was measured by F-Score (89% for cup and 96% for disk).

3.3 ALL-IDB1

Tran et al. [5] have [14] used ALL-IDB1 dataset [14] for blood cells segmentation from the blood smear images. The chosen data lacks the corresponding segmentation ground truth images. So, GNU Image Manipulation Program (GIMP 2.10.2) is used to label red blood cells and white blood cells in images for the purpose of segmentation. They collected 42 images of different size and their corresponding ground truth images. The size of the images and corresponding pixel label images in dataset are reduced to 360 × 480 × 3 to decrease the memory usage and training time. Out of total, they divided 70% as the training set (i.e., 29 images and 29-pixel label images) and 30% as testing set (i.e., 13 images and 13-pixel label images). The performance of this method is measured by accuracy (94% for WBC, 91% for RBC, and 87% for background).

3.4 Flair-MRI Brats 2015 [15]

Ramirez et al. [6] have used Flair-MRI Brats 2015 dataset [15] for brain tumor segmentation. From the available slices, they utilized 16114 slices out of which 80% are for training and 20% are for validation. The CNN [26] considered in this model has 64 features in each layer with a kernel size of 3 × 3. The performance of this method is measured by precision (73% for TVS, 74% for U-net CNN, and 84% for CNN+TVS) and recall (63% for TVS, 87% for U-net CNN, and 88% for CNN+TVS).

3.5 Japanese Society of Radiological Technology [17]

Ngo et al. [7] have used publicly available JSRT [17] (Japanese Society of Radiological Technology) dataset, which contains manual segmentations of lung fields, heart, and clavicles for lung segmentation in chest radiographs. The images are 12-bit gray-scaled images. This dataset is randomly split into 84 training images, 40 validation images, and 123 test images by them. The performance of this method is measured by the Dice coefficient (99.2%) and jaccard similarity coefficient (98.5%).

4 Conclusion

In this paper, a taxonomy for various semantic segmentation methods and their challenges has been discussed and compared. We had a walk-through of several methods for diagnosis of diseases of several organs like eye, lung, and brain, etc., using diverse architectures of deep neural networks. Each and every method performance measure differs based on their architecture and the dataset used. Every method has its own importance and can be further developed for better disease diagnosis in the field of medicine.

References

1. LeCun B, Hinton G (2015) Deep learning. Nature 521:436–444
2. Towards Data Science. http://www.towardsdatascience.com
3. Chen K, Zhang N, Powers L, Roveda J (2018) Cell nuclei detection and segmentation for computational pathology using deep learning. J Diagn Tech Biomed Anal 7. https://doi.org/10.4172/2469-5653-C3-018
4. Edupuganti VG, Chawla A, Kale A (2018) Automatic optic disk and cup segmentation of fundus images using deep learning. In: 25th IEEE international conference on image processing, pp 2227–2231. https://doi.org/10.1109/ICIP.2018.8451753
5. Tran T, Kwon O-H, Kwon K-R, Lee S-H, Kang K-W (2018) Blood cell images segmentation using deep learning semantic segmentation. In: IEEE international conference on electronics and communication engineering, pp 13–16. https://doi.org/10.1109/ICECOME.2018.8644754
6. Ramírez I, Martín A, Schiavi E (2018) Optimization of a variational model using deep learning: an application to brain tumor segmentation. In: 15th IEEE international symposium on biomedical imaging, pp 631–634. https://doi.org/10.1109/ISBI.2018.8363654
7. Ngo TA, Carneiro G (2015) Lung segmentation in chest radiographs using distance regularized level set and deep-structured learning and inference. In: IEEE international conference on image processing, pp 2140–2143. https://doi.org/10.1109/ICIP.2015.7351179

8. Ronneberger O, Fischer P, Brox T (2015) U-Net: convolutional networks for biomedical image segmentation. In: International conference on medical image computing and computer-assisted intervention, Spring, Cham. LNCS, vol 9351, pp 234–241. https://doi.org/10.1007/978-3-319-24574-4_28
9. Badrinarayanan V, Kendall A, Cipolla R (2016) SegNet: a deep convolutional encoder-decoder architecture for image segmentation. IEEE Trans Pattern Analy Mach Intell 39:2481–2495. https://doi.org/10.1109/TPAMI.2016.2644615
10. Shelhamer E, Long J, Darrell T (2017) Fully convolutional networks for semantic segmentation. IEEE Trans Pattern Anal Mach Intell 39:6401–651. https://doi.org/10.1109/TPAMI.2016.2572683
11. Stutz D, Beyer L (2017) Understanding convolutional neural networks
12. Sivaswamy J, Krishnadas S, Joshi GD, Jain M, Tabish AUS (2014) Drishti-gs: retinal image dataset for optic nerve head (onh) segmentation. In: 2014 IEEE 11th international symposium on biomedical imaging (ISBI). IEEE, pp 53–56. https://doi.org/10.1109/ISBI.2014.6867807
13. Kumar N, Verma R, Sharma S, Bhargava S, Vahadane A, Sethi A (2017) A Dataset and a technique for generalized nuclear segmentation for computational pathology. IEEE Trans Med Imag 36(7). https://doi.org/10.1109/TMI.2017.2677499
14. Acute Lymphoblastic Leukemia Image Database for Image Processing. https://homes.di.unimi.it/scotti/all/
15. Korfiatis P, Kline TL, Erickson BJ (2016) Drishti-gs: automated segmentation of hyperintense regions in FLAIR MRI using deep learning. J Tomogr 2(4):334–340. https://doi.org/10.18383/j.tom.2016.00166
16. Menze B et al (2014) The multimodal brain tumor image segmentation benchmark (BRATS). IEEE Trans Med Imag 34:1993–2024. https://doi.org/10.1109/TMI.2014.2377694
17. Japanese Society of Radiological Technology, Digital Image Database. http://db.jsrt.or.jp/eng.php
18. Shiraishi J et al (2000) Development of a digital image database for chest radiographs with and without a lung nodule: receiver operating characteristic analysis of radiologists, detection of pulmonary nodules. Am J Roentgenol 174(1):71–74. https://doi.org/10.2214/ajr.174.1.1740071
19. Otsu N (1979) A threshold selection method from Gray-Level histograms. IEEE Trans Syst Man, and Cybern 9:62–66. https://doi.org/10.1109/TSMC.1979.4310076
20. Kowal M, Zejmo M, Skobel M, Korbicz J, Monczak R (2019) Cell nuclei segmentation in cytological images using convolutional neural network and seeded watershed algorithm. J Digital Imag 1–12. https://doi.org/10.1007/s10278-019-00200-8
21. Irshad H, Veillard A, Roux L, Racoceanu D (2013) Methods for nuclei detection, segmentation,and classification in digital histopathology: a review—current status and future potential. IEEE Rev Biomed Eng 7:97–114. https://doi.org/10.1109/RBME.2013.2295804
22. Gordillo N, Montseny E, Sobrevilla P (2013) State of the art survey on MRI brain tumor segmentation. Magn Reson Imag 31:1426–1438. https://doi.org/10.1016/j.mri.2013.05.002
23. Daley CL, Gotway M, Jasmer R (2006) Radiographic manifestation of tuberculosis: a primer for clinicians. In: Curry International TB Center
24. Hinton G, Salakhutdinov R (2006) Reducing the dimensionality of data with neural networks. J Sci 313:504–507. https://doi.org/10.1126/science.1127647
25. Li C et al (2010) Distance regularized level set evolution and its application to image segmentation. IEEE Trans Image Proc 19:3243–3254. https://doi.org/10.1109/TIP.2010.2069690
26. Pereira S, Pinto A, Alves V, Silva CA (2016) Brain tumor segmentation using convolutional neural networks on MRI images. IEEE Trans Med Imag 35:1240–1251. https://doi.org/10.1109/TMI.2016.2538465
27. Krizhevsky A, Sutskever I, Hinton GE (2012) Imagenet classification with deep convolutional neural networks. In: 25th international conference on neural information processing systems, vol 1, pp 1097–1105. https://doi.org/10.1145/3065386

HT-IWT-DCT-Based Hybrid Technique of Robust Image Watermarking

Priyank Khare and Vinay Kumar Srivastava

Abstract Nowadays, several malpractices are performed for violating integrity of digital multimedia. In this paper, a hybrid technique of image watermarking is proposed which finds its potential capability to overcome this issue. The proposed scheme efficaciously uses homomorphic transform (HT), integer wavelet transform (IWT), discrete cosine transform (DCT) along with singular value decomposition (SVD). Selection of reflectance component of the host image for watermark embedding ensures that robustness and imperceptibility are achieved concurrently. Embedding of watermark is done by modifying it with singular values (SV) of LH subband of the host image. The proposed scheme is highly robust and imperceptible as numerous performance metrics attain high values under attacks. Performance of the proposed algorithm is compared with other formerly reported techniques and it is found that the developed scheme is superior over them.

Keywords Robust image watermarking · Perceptual invisibility · Intellectual rights · HT · IWT

1 Introduction

The extensive growth of Internet technologies has provided a suitable way for transmission of multimedia data. But due to this transmission, a threat of infringement of sole proprietary of ownership rights is also enhanced rapidly. So, there must be technique that should be developed which can easily sort out this shortcoming. Digital image watermarking approach provides solution to this problem [1]. In this methodology, important information is obscured inside cover image such that it becomes invisible. By means of this technique, meaningful information can be transmitted

P. Khare (✉) · V. K. Srivastava
ECED, MNNIT Allahabad, Allahabad, India
e-mail: rel1606@mnnit.ac.in

V. K. Srivastava
e-mail: vinay@mnnit.ac.in

D. Harvey et al. (eds.), *Advances in VLSI, Communication, and Signal Processing*, Lecture Notes in Electrical Engineering 683,
https://doi.org/10.1007/978-981-15-6840-4_28

safely. A watermarking technique should be resilient against attacks which highlights the robustness of the scheme, whereas there should exist visual similarity among original and watermarked images for scheme to be imperceptible [2]. Security is another key feature which makes the scheme secure against data manipulation and tampering. Usually, watermarking is performed in two domains (i) spatial domain [3] and (ii) transform domain. In spatial domain approach, only pixel values are modified, whereas in transform domain, several transforms like discrete wavelet transform [4, 5] and IWT [6] are used. Transform domain approach offers better robustness than spatial domain technique.

In [7], DWT along with shuffled SVD (SSVD) is used for watermarking. Original host image is decomposed using DWT and blockwise SVD is performed, whereas watermark is shuffled with SSVD. To remove false positive (FP) problem, principal components of watermark are embedded into largest SV of each block. This approach tries to develop watermarking method free from FP. Göker et al. [8] suggested a technique which uses quadtree algorithm with DWT. Medical image is considered as host image and watermark is embedded into region of non-interest of the host image without affecting region of interest. Several optimization algorithms like bacterial foraging optimization (BFO) and particle swarm optimization (PSO) are used for medical image watermarking in combination with DWT and DCT transforms in [9]. Gupta and Parmar [10] presented a comparative study between IWT-SVD and DWT-SVD techniques for same set of attacks. It appears from the results that IWT-SVD technique performs better than DWT-SVD.

IWT-SVD transforms using watermarking technique is developed by Makbol and Khoo [11], in which direct insertion of watermark is done into IWT subbands of the host image. This technique is reliable as they used digital signature for embedding and extraction processes. It seems that developed scheme achieves sufficient robustness and imperceptibility. Khare and Srivastava [12] investigated a HT-DWT watermarking technique where multiresolution analysis of reflectance component of the host image is done with DWT. LL subband is chosen for embedding of watermark. The developed technique demonstrates fair robustness against attacks. In [13], secure blind watermarking approach is stated using HT–Arnold transform (AT), where AT of watermark is performed so that key information confidentiality is sustained. HT of host image is computed and further reflectance component is directly added with watermark SV. This scheme proves to be blind and secure simultaneously. Abdallah et al. [14] suggested HT–Karhunen–Loeve transform (KLT)-based technique in which KLT is applied to the reflectance component. Watermark image is altered with transformed reflectance component. They also suggested a second method in which partition of reflectance component is done into blocks using spiral scanning and watermark is individually added to KLT transformed blocks. This approach exhibits better robustness as compared to former developed method. Haar wavelet transform (HWT) is used with fusion of DCT in [15]. LL subband of host image is partitioned in 8 × 8 blocks and block processing DCT is implemented. Watermark is altered with each block DC coefficients for obtaining the watermarked image. Developed methodology achieves good robustness.

Rest of the paper is ordered as: Sect. 2 explores significant attributes of the proposed technique. Section 3 established the proposed algorithm while experimental results are demonstrated in Sect. 4. Proposed work is concluded in Sect. 5.

2 Significant Attributes of the Proposed Approach

Proposed watermarking scheme employs HT [16], IWT [17], DCT [18, 19] and SVD [20] transforms. In this developed work, original cover image is processed with HT which yields illumination and reflectance components. Reflectance component is preferred over illumination because it shows better robustness and imperceptibility. Embedding in high frequency component makes the watermark invisible which enhances imperceptibility. IWT decomposes reflectance component in distinct frequency subbands, and DCT processes LH subband of the reflectance component. LH subband is suitable choice for embedding as robustness and perceptual invisibility both are balanced together. SV of modified host image is computed using SVD that provides robustness to the scheme as they remain invariant under effect of attacks. These SV are altered with watermark directly. So, with use of these hybrid transforms, an efficient watermarking technique is developed.

3 Proposed Technique

Watermark embedding and extraction techniques are explored in this section. Figure 1a, b depicts these processes diagrammatically while steps for these processes are listed in Sects. 3.1 and 3.2.

3.1 Embedding Technique

- Illumination 'G^{I}' and reflectance 'G^{R}' components of host image 'G' are obtained using HT.
- Frequency domain transformation of 'G^{R}' is attained with first level IWT.

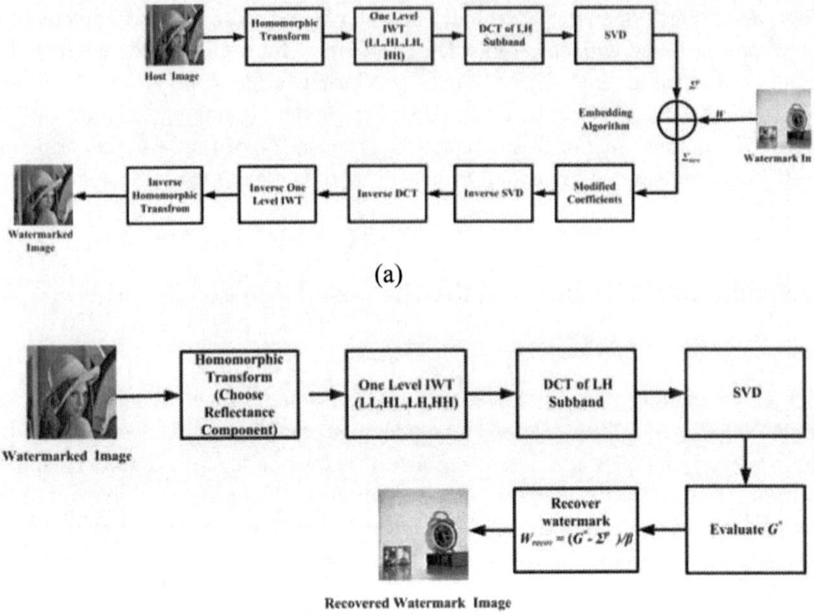

Fig. 1 Proposed watermarking technique **a** embedding and **b** extraction

$$G^{R}\underrightarrow{\text{IWT}}\left[G^{RLL}, G^{RHL}, G^{RLH}, G^{RHH}\right] \tag{1}$$

- 'G^{RLH}' subband is chosen for DCT transformation and further SV are obtained as '$\sum^{p}$' using SVD.
- Embedding of watermark takes place according to Eq. (2):

$$\sum_{\text{new}} = \sum^{p} + \beta W \tag{2}$$

here 'β' symbolizes scaling factor which is chosen as 0.02.

- Again apply SVD to '$\sum_{\text{new}}$' for obtaining modified coefficients.

$$\sum_{\text{new}} = U_{wi} \sum_{wi} V_{wi}^{T} \tag{3}$$

- Successively inverse SVD, DCT, IWT and HT are applied to the modified coefficients for obtaining 'G_{W}' as watermarked image.

3.2 Extraction Technique

- 'G_W^I' and 'G_W^R' are obtained as illumination and reflectance components of 'G_W' image through HT.
- First level IWT of 'G_W^R' is performed and LH subband of it is transformed with DCT resulting in 'G_{DCT}^R'.

$$G_W^R \underrightarrow{\text{IWT}} [G_W^{RLL}, G_W^{RHL}, G_W^{RLH}, G_W^{RHH}] \tag{4}$$

$$G_W^{RLH} \underrightarrow{\text{DCT}} G_{DCT}^R \tag{5}$$

- SV are again evaluated for 'G_{DCT}^R' matrix as per Eq. (6):

$$G_{DCT}^R = U^{*p} \sum^{*p} V^{*pT} \tag{6}$$

- 'G^*' matrix is computed as:

$$G^* = U_{wi} \sum^{*p} V_{wi}^T \tag{7}$$

- Watermark is recovered in accordance with Eq. (8):

$$W_{recov} = (G^* - \sum^{p})/\beta \tag{8}$$

4 Experimental Results

Experimental results are discussed in given present section. The proposed algorithm is examined for five gray scale images as 'Zelda', 'Lena', 'Splash', 'Aerial' and 'Mandrill' of 512 × 512 size. 'Clock' image of 256 × 256 is taken as watermark image. All experimental results are simulated in MATLAB. Robustness and perceptual invisibility are thoroughly investigated under attacks by computing peak signal to noise ratio (PSNR) [21], normalized correlation (NC) [21] and structural similarity index (SSIM) [22] parameters. Figure 2 illustrates various images used in the proposed work, whereas Fig. 3 depicts watermarked images and recovered watermark for developed algorithm.

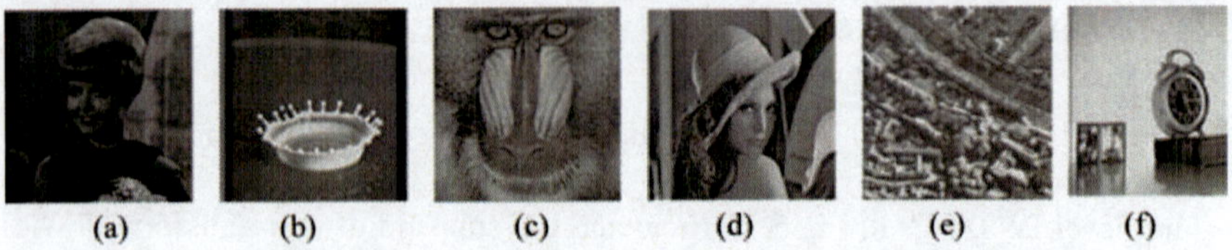

Fig. 2 Cover images **a** Zelda, **b** Splash, **c** Mandrill, **d** Lena, **e** Aerial, **f** Watermark 'Clock'

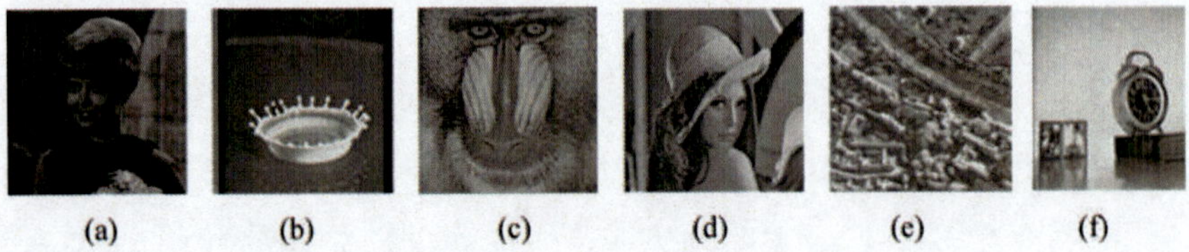

Fig. 3 Watermarked images **a** Zelda, **b** Splash, **c** Mandrill, **d** Lena, **e** Aerial, **f** Recovered watermark

Table 1 Metrics values under no attacks

Metrics	Lena	Mandrill	Zelda	Splash	Aerial
SSIM	0.9963	0.9973	0.9962	0.9952	0.9976
PSNR (dB)	49.3884	44.3323	49.8551	47.4567	48.4435
NC	1.00	1.00	1.00	1.00	1.00

Table 1 tabulates different metrics values for the proposed technique without attacks. Average PSNR value is 47 dB, whereas SSIM metric is near to unity. Hence, the proposed algorithm displays good perceptual invisibility while NC metric values are unity. Table 2 pictorial presents several watermarks recovered from attacks. Resilience of the proposed technique toward various attacks such as sharpening; scaling is presented in Table 3 by computing NC metric values. As these values are close to unity which clearly makes emphasis for high robustness of the proposed scheme.

SSIM and PSNR values under attacks are graphically demonstrated in Fig. 4a, b. 0.9992 is observed as highest SSIM value for Aerial image under scaling attack as shown in Fig. 4a, whereas 52.7988 dB is maximum PSNR value achieved for Zelda image under scaling attack as shown in Fig. 4b.

Proposed method is superior to other developed schemes like [11] and [15] as illustrated in Table 4, whereas Fig. 5 represents graphical comparison among method in [14] and developed scheme. Thus, the proposed method is clearly distinguished from other investigated methods.

Table 2 Robustness analysis under attacks

Attacks	Watermarked image	Recovered watermark image
Gaussian noise (0, 0.01)	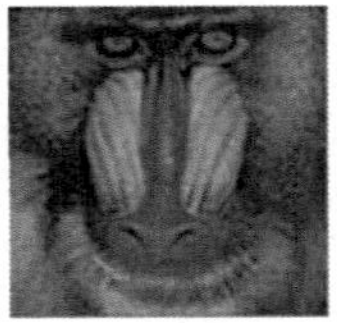	
Sharpening	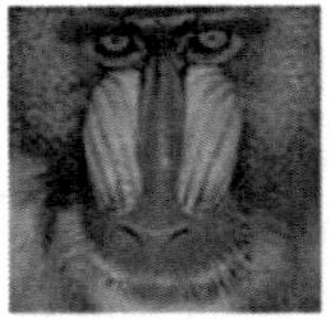	
Scaling	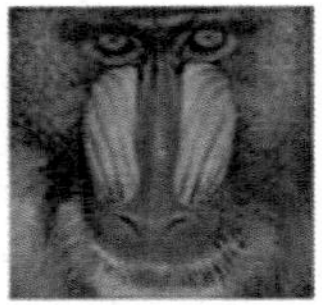	
Histogram equalization	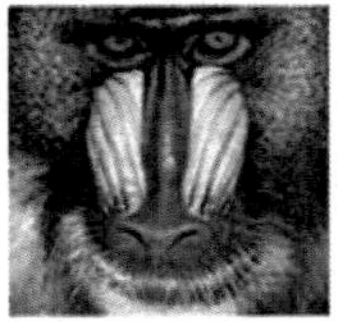	

(continued)

Table 2 (continued)

Attacks	Watermarked image	Recovered watermark image
JPEG (60)	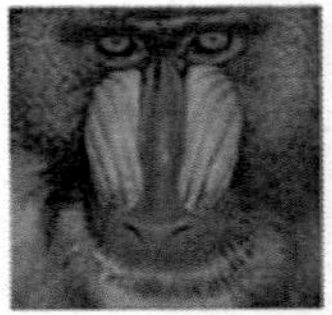	
Gamma correction (0.9)	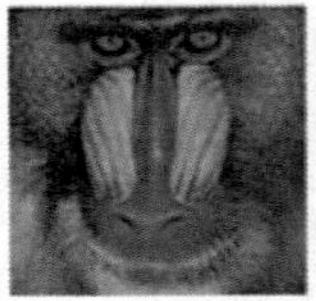	
Rotation (10°)	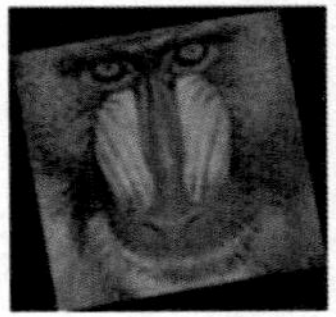	

Table 3 NC metric values for the proposed scheme

Attacks	Lena	Mandrill	Zelda	Splash	Aerial
Salt and pepper noise (0.02)	0.9998	1.00	0.9998	0.9997	0.9998
Gaussian noise (0, 0.01)	0.9997	0.9998	0.9996	0.9997	0.9998
Histogram equalization	0.9978	0.9969	0.9985	0.9970	0.9963
Sharpening	0.9994	0.9976	0.9997	0.9996	0.9989
Gaussian filter (3 × 3)	1.00	0.9999	1.00	0.9999	0.9996
Median filter (3 × 3)	0.9999	0.9869	0.9999	0.9976	0.9967
Wiener filter (3 × 3)	1.00	0.9975	1.00	0.9988	0.9965
JPEG compression (60)	0.9999	0.9999	1.00	0.9986	0.9971
Rotation (10°)	0.9916	0.9863	0.9923	0.9878	0.9971
Gamma correction (0.9)	1.00	1.00	1.00	1.00	1.00
Scaling	1.00	0.9997	1.00	0.9998	0.9991
Cropping	0.9939	0.9959	0.9967	0.9983	0.9554

5 Conclusion

Robust and imperceptible technique of hybrid image watermarking is proposed in this paper. Imperceptibility and robustness are achieved by selecting reflectance component for embedding. IWT removes rounding errors by direct mapping of integers as compared to DWT. DCT and SVD further enhance robustness of the proposed technique. Hence, a proficient watermarking approach is proposed with these transforms. Experimental results outline that developed scheme sustain attacks well as performance metrics are high. Proposed work is compared with [11, 14, 15] schemes and found to better in robustness. Further, in future work, the proposed technique can also be extended for various multimedia watermarking techniques.

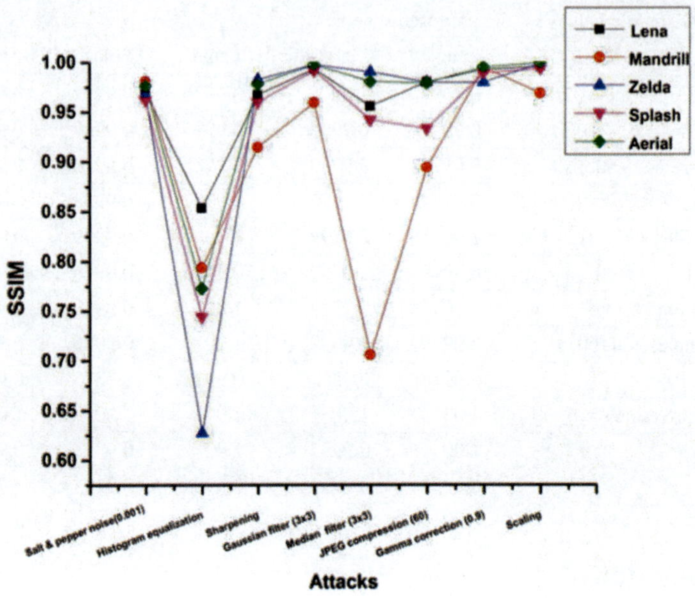

(a)

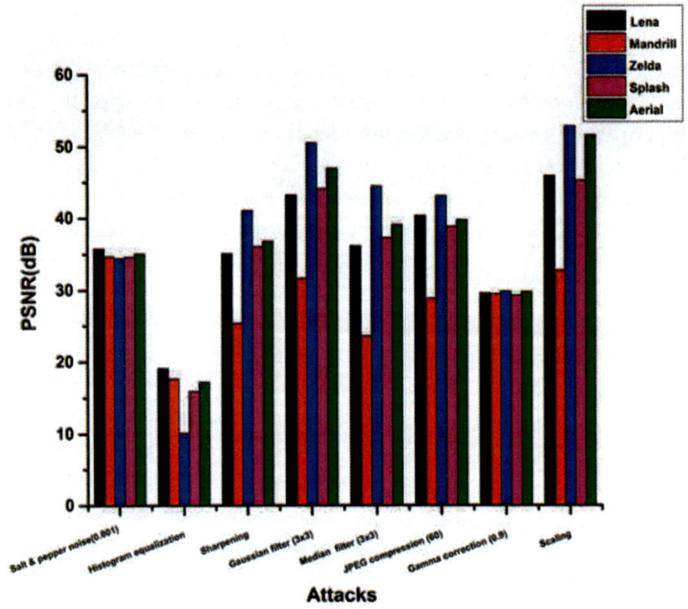

(b)

Fig. 4 Perceptual invisibility analysis under attacks for developed technique **a** SSIM and **b** PSNR

Table 4 Proposed method comparison with schemes in [11, 15]

Attacks	[11]	[15]	Proposed method
Median filter	0.9758	0.9785	0.9999
Gaussian filter	0.9324	0.9229	1.00
Wiener filter	0.9772	–	1.00
Salt and pepper noise (0.005)	0.9286	0.8545	1.00
Gaussian noise (0, 0.01)	0.8998	0.7565	0.9997
JPEG compression	–	0.9477	0.9999
Scaling	0.9854	–	1.00

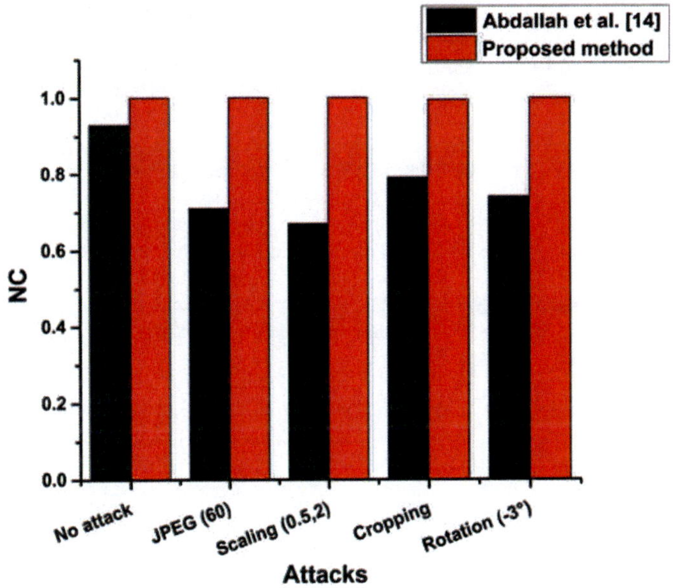

Fig. 5 NC comparison of the proposed scheme with [14] scheme

References

1. Kumar C, Singh AK, Kumar P (2018) A recent survey on image watermarking techniques and its application in e-governance. Multimedia Tools Appl 77(3):3597–3622
2. Kadhim IJ, Premaratne P, Vial PJ, Halloran B (2019) Comprehensive survey of image steganography: techniques, evaluations, and trends in future research. Neurocomputing 335:299–326
3. Su Q, Chen B (2018) Robust color image watermarking technique in the spatial domain. Soft Comput 22(1):91–106
4. Pham TD, Tran D, Ma W (2015) A proposed blind DWT-SVD watermarking scheme for EEG data. In: International conference on neural information processing. Springer, Cham, pp 69–76

5. Singh D, Singh SK (2017) DWT-SVD and DCT based robust and blind watermarking scheme for copyright protection. Multimed Tools Appl 76(11):13001–13024
6. Makbol NM, Khoo BE (2013) A hybrid robust image watermarking scheme using integer wavelet transform, singular value decomposition and Arnold transform. In: International visual informatics conference. Springer, Cham, pp 36–47
7. Bhuyan T, Srivastava VK, Thakkar F (2016) Shuffled SVD based robust and secure digital image watermarking. In: 2016 International conference on electrical, electronics, and optimization techniques (ICEEOT). IEEE, pp 1229–1233
8. Göker O, Nazli N, Erol MM, Choupani R, Dogdu E (2018) A robust watermarking scheme over quadrant medical image in discrete wavelet transform domain. In: 2018 5th International conference on control, decision and information technologies (CoDIT). IEEE, pp 277–282
9. Bharati S, Rahman MA, Mandal S, Podder P (2018) Analysis of DWT, DCT, BFO & PBFO algorithm for the purpose of medical image watermarking. In: 2018 International conference on innovation in engineering and technology (ICIET). IEEE, pp 1–6
10. Gupta P, Parmar G (2017) Image watermarking using IWT-SVD and its comparative analysis with DWT-SVD. In: 2017 International conference on computer, communications and electronics (Comptelix). IEEE, pp 527–553
11. Makbol NM, Khoo BE (2014) A new robust and secure digital image watermarking scheme based on the integer wavelet transform and singular value decomposition. Digit Signal Proc 33:134–147
12. Khare P, Srivastava VK (2018) Image watermarking scheme using homomorphic transform in wavelet domain. In: 2018 5th IEEE Uttar Pradesh section international conference on electrical, electronics and computer engineering (UPCON). IEEE, pp 1–6
13. Khare P, Srivastava VK (2019) Secure image watermarking approach using HT-AT. In: 2019 6th International conference on signal processing and integrated networks (SPIN). IEEE, pp 547–551
14. Abdallah HA, Faragallah OS, Elsayed HS, Shaalan AA, El-samie FEA (2016) Robust image watermarking method using homomorphic block-based KLT. Optik Int J Light Electron Opt 127(4):2374–2381
15. Susanto A, Sari CA, Rachmawanto EH (2017) Hybrid method using HWT-DCT for image watermarking. In: 2017 5th International conference on cyber and IT service management (CITSM). IEEE, pp 1–5
16. Abdallah HA, Ghazy RA, Kasban H, Faragallah OS, Shaalan AA, Hadhoud MM, El-Samie FEA (2014) Homomorphic image watermarking with a singular value decomposition algorithm. Inf Process Manage 50(6):909–923
17. Naik K, Trivedy S, Pal AK (2018) An IWT based blind and robust image watermarking scheme using secret key matrix. Multimedia Tools Appl 77(11):13721–13752
18. Sinhal R, Ansari IA (2018) A blind and secured image watermarking method using DWT-DCT techniques. In: 2018 Conference on information and communication technology (CICT). IEEE, pp 1–6
19. Singh N, Joshi S, Birla S (2019) Suitability of singular value decomposition for image watermarking. In: 2019 6th International conference on signal processing and integrated networks (SPIN). IEEE, pp 983–986
20. Ansari IA, Pant M, Ahn CW (2016) Robust and false positive free watermarking in IWT domain using SVD and ABC. Eng Appl Artif Intell 49:114–125
21. Khare P, Srivastava VK (2018) Robust digital image watermarking scheme based on RDWT-DCT-SVD. In: 2018 5th International conference on signal processing and integrated networks (SPIN). IEEE, pp 88–93
22. Wang Z, Bovik AC, Sheikh HR, Simoncelli EP (2004) Image quality assessment: from error visibility to structural similarity. IEEE Trans Image Process 13(4):600–612

Reliability and Circuit Timing Analysis with HCI and NBTI

S. Udaya Shankar and P. Kalpana

Abstract One of the key challenges in the electronic industries is the need to detect aged or recycled integrated circuits (ICs) before they enter into their production line. The use of aged or recycled ICs may lead to degradation in the quality, performance and reliability of electronic products. The performance of these ICs is adversely affected by aging induced by reliability issues such as bias temperature instability (BTI) and hot-carrier injection (HCI) effects. Both BTI and HCI affect transistor electrical parameters depending on operating environment and time of usage of ICs. In this paper, the frequency shift in ring oscillator and delay variation in adder output due to aging is analyzed with BTI and HCI using Cadence RelXpert tool.

Keywords Reliability · Stress · Aging · Hot-carrier injection · Bias temperature instability

1 Introduction

Technology scaling has created a serious threat to the performance of the integrated circuit (IC) in terms of reliability. Maintaining the desired functionality of system throughout the lifetime is difficult. Reliability effects need to be checked in the early stages of design in order to have sufficient lifetime margins to give a better performance. In the era of nanotechnology, most of the manufacture defects are responsible for the degradation of the device lifetime. Most reliability effects are concerned with the degradation of device performance due to its aging. There are various reliability effects like transistor degradation due to HCI, BTI, transistor abrupt failure due to field oxide breakdown and interconnect degradation due to electro-migration and self-heating. The process scaling of IC results in increasing vertical electric field. In addition to this, the power density is increasing, which in turn increases the operating temperature of the device.

S. Udaya Shankar (✉) · P. Kalpana
Electronics and Communication Engineering, PSG College of Technology, Coimbatore, India
e-mail: udayashankar01@gmail.com

D. Harvey et al. (eds.), *Advances in VLSI, Communication, and Signal Processing*,
Lecture Notes in Electrical Engineering 683,
https://doi.org/10.1007/978-981-15-6840-4_29

Aging analysis of the device is carried out with the effects of HCI, positive BTI (PBTI) and negative BTI (NBTI) [1]. HCI degrades the performance of PMOS and NMOS transistors, PBTI degrades only NMOS transistors and NBTI degrades only PMOS transistors. The lifetime of the device is reduced mainly due to the leakage current generated in the device. HCI affects the lifetime of the CMOS devices by increase in the sub-threshold voltage and decrease in the carrier mobility (μ_c) in the NMOS transistors. When the device is biased in strong inversion with large value of V_{DS} (drain-source voltage), HCI will be a main dependent factor. When the voltage supply is given to the gate terminal of the PMOS devices, it increases the threshold voltage leading to the NBTI effect. Increase in the temperature of the device speed up the increase in the value of the drain current (I_D) and trans-current. PBTI is also another reliability issue, which has a negligible effect in the device performance. The reliability parameters affecting the performance are noise margin, leakage current and delay between input-output responses.

The degradation in the circuit occurs with respect to time and this time-dependent degradation of the circuit results in the aging effect of the device. The shift in the device parameters will degrade the performance of the devices and also causes the circuit failure over a time period as shown in Fig. 1. The aging effects result in threshold voltage shift (ΔV_{th}) of transistors and lead to a rise in the gate delay value. Hot-carrier (HC) effects in MOSFET transistors are caused by the acceleration of carriers like electrons or holes under lateral electric fields in the channel, to the point where they gain high energy and momentum to break the barriers of the surrounding dielectric like gate and sidewall oxides [2]. The presence of hot carriers triggers sequences of physical processes that affect the characteristics of the circuit under normal operating conditions. These effects that are build up over a long period of time, causes aging of the integrated circuit with time, resulting in the performance degradation. This may further eventually lead to the complete failure of the circuit.

Over the past few years, several techniques have been proposed to improve circuit lifetime in the presence of transistor aging [3–6]. Among them, adding timing margins is the most common technique [5]. The degradation in MOSFET parameters like V_{th} (threshold voltage), G_m (trans-conductance), etc., is due to interface states created by electrons (HCI) with high energy [7]. The bathtub curve model for reliability defines the failure rate in operating time of the device [8]. The frequency degradation and impact of switching due to HCI and NBTI for a ring oscillator are modeled in [9,

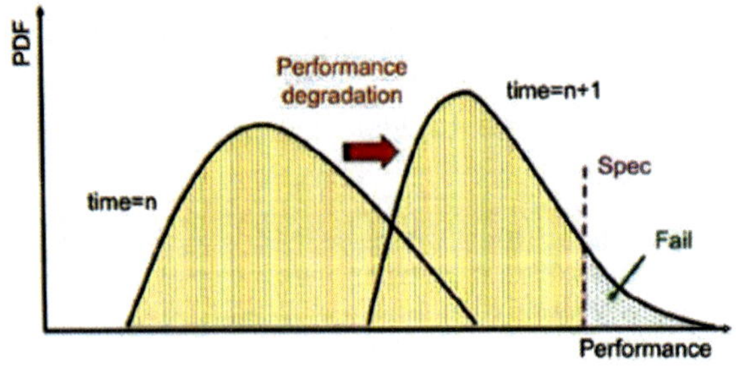

Fig. 1 Yield degradation caused by the aging effects [2]

10]. The aging and stress analysis of eight transistor hybrid adder is analyzed in [11]. The shift in threshold voltage due to NBTI is discussed.

The reliability concept in terms of tool flow has been discussed in [12]. The issues related to the effect HCI and BTI on devices are explained with a simulation. The reliability analysis under stress conditions at various levels of designs like transistor, gate and layout is also explained. The dynamic simulation is needed to analyze the device parameters like substrate current and gate current after simulating the fresh and aged circuit to identify the lifetime of the circuit. Static rule-based technique is conservative because it performs reliability analysis only under worst-case DC and AC stress conditions at short channel device.

The challenges related to the modeling of aging models with design-reliability flow are discussed in [13]. The complete reliability aging model is embedded within the tool to perform aging simulation. This allows determining the most degraded device due to reliability effects like HCI, BTI and TDDB. The degraded device parameters like drive current, linear current, trans-conductance and threshold voltage can also be identified. Aging of active devices has only been concentrated to determine the lifetime of the device. But the passive devices like resistor and capacitor becomes a critical issue in degrading the performance of the device due to aging. Hence, electro-thermal simulation is used to determine the degradation of passive components using the electrical and thermal netlist.

A complementary lookup table (LUT) is analyzed for aging under HCI, NBTI and PBTI in [14]. The leakage current and noise margin are determined to differentiate the degradation due to aging in different technology nodes.

The reliability and aging analysis for comparator under the effect of HCI, PBTI and NBTI are described in [15]. The voltage output and time increases due to NBTI and HCI. The lifetime of the MOSFET transistors is minimized with increase in threshold voltage and degrades the other parameters like substrate current and gate current.

Aging analysis is performed in Artix 7 family of FPGAs as described in [16]. Ring oscillator (RO)-based sensor is used to identify the aged FPGA based on the RO frequency. The tests were carried out in 17 FPGAs for different circuits under test placed in FPGA operating at different temperatures under DC and AC stress conditions. As temperature increases the degradation increases with HCI and frequency is degraded mainly due to the effect of NBTI.

An incremental aging process is discussed along with a sampling method in [17]. The repeated iterative simulation process reduces the simulation time and maintaining the accuracy. The parameter change during aging analysis is updated and reused to reduce the simulation with same amount of accuracy. The time for performing the aging analysis is reduced by the delta model. The aging sampling technique skips some of the simulations steps. This further reduces the simulation time with a small amount of loss in accuracy.

The reliability analysis of analog-to-digital sigma-delta modulators (SDMs) is discussed in [18]. Statistical analysis is also performed to analyze the process parameter variation due to reliability. Lifetime is evaluated to predict the aging performance of the circuit. MOSFET model for 28 nm circuit is designed and simulated for aging under electrical stress to predict its lifetime is shown in [19]. The MOSFET is modeled based on bias, temperature and geometry. Two types of MOSFET models are considered in [19]. Parameterized model degrades the threshold voltage and table-based model degrades current and conductance.

The paper is organized as follows. Section 2 describes impact of HCI, NTBI and PTBI on reliability analysis, and Sect. 3 describes simulation results of the reliability analysis of ring oscillator and the adder circuit.

2 Impact of HCI, NBTI and PBTI on Reliability

Device aging gradually shifts the value of the threshold voltage (V_{th}) of the transistors and results in the degradation of the delay of the different gates present in the circuit. Eventually, when the total time delay of the integrated circuit (IC) exceeds the limits in terms of timing, the circuit starts to fail. Transistors are aged due to the effects like NBTI, PBTI and HCI. NBTI affects only the P-MOSFET transistors, while PBTI affects only N-MOSFET transistors. Both NBTI and PBTI effects cause the threshold voltage (V_{th}) of the transistors to increase over a period of time. Injection of hot carriers occurs when there is a transition in the gate-source terminal of the transistor and results in increase in the value of the threshold voltage. The various aging factors and their models are described in detail in the following sections.

2.1 Hot-Carrier Injection (HCI)

Carriers moving with high energy in the MOSFET channel in the form of a drain current (I_{d}) can create a new electron-hole pairs upon the impact of ionization with the atoms in the channel. The newly generated minority carriers are attracted toward the gate electrode and trapped inside the gate oxide layer. This leads to the alteration of the threshold voltage of transistor.

The large voltage drop in saturation region results in the high electric field close to the drain region. When this carries pass through the electric field they attain energy to generate electron-hole pairs by ion implantation. Once gate-source voltage is up to the drain-source voltage, the injection of carrier happens.

$$V_{\mathrm{gs}} = V_{\mathrm{ds}} \tag{1}$$

Due to this, changes will occur in the V_{th} by the occupied traps within the gate oxide. The growth rate of charge from [11] is expressed in the equation below.

$$\frac{\mathrm{d}N_{\mathrm{it}}}{\mathrm{d}t}[1 + AN_{\mathrm{it}}] = KI_{\mathrm{bb}} \tag{2}$$

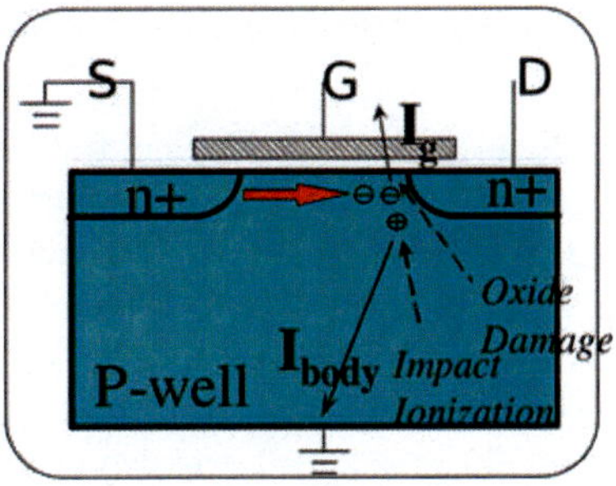

Fig. 2 Degradation of transistor due to HCI

where derivative of the N is the growth rate of the traps in the oxide layer, I_{bb} is the bond breaking current. The bond breaking current is the current that the particles acquire the maximum kinetic energy and I_{ds} is the value of the drain-source current. V_{th} is the threshold voltage growth rate and time is denoted by t.

$$I_{bb} = \frac{C_1}{W} I_{ds} \tag{3}$$

The hot-carrier injection into the layer of gate oxide is the important problem in the reliability of modern integrated circuits. The degradation process is slow during normal operation of the circuit. In order to characterize the hot carrier response, the degradation process is more bias than the normal circuit operation. By this, the stress analysis of the device can be performed.

In the HCI, the resistance of the channel increases due to the charge carriers. In the NMOS transistor, the damage occurs when the kinetic energy of the electrons is higher than the energy of the band gaps leading to the breaking of the bond as shown in Fig. 2.

In HCI, the aging parameter expressed in Eq. (4) can be measured using drain-source current (I_{ds}), substrate current (I_b) and the amount of stress applied to the device [7, 8].

$$\mathrm{Age}(t) = \frac{I_{ds}}{WH}\left(\frac{I_b}{I_{ds}}\right)^m * t \tag{4}$$

Here, W is the width of the MOSFET channel, H is a constant parameter and m is the value of an acceleration factor equivalent to −3.0. The amount of degradation is calculated using the expression shown below in Eq. (5).

$$\Delta t_f(t) = f(\mathrm{Age}) \tag{5}$$

2.2 *Temperature Bias Instability (TBI)*

If the bias voltage is applied for the MOSFET, then the gate channel shifts in the threshold voltage at the particular temperature. In BTI, degraded mobile charge carriers are measured. By the device level measurement, the lifetime of the device can be calculated [9].

NBTI and PBTI are observed in temperature instability. The property is observed on applying a stress voltage. The degradation of the device is recovered after removing the applied stress voltage.

$$\Delta V_{\mathrm{th}} \propto \exp(\alpha_1 V_{\mathrm{GS}})t^{np} + V_{\mathrm{GS}}^{\alpha 2}(C_{\mathrm{R}} + n_{\mathrm{R}} \log_{10}(t)) \exp\left(-\frac{E_{\mathrm{a}}}{KT}\right) \tag{6}$$

Positive Bias Temperature Instability (PBTI) PTBI occurs in NMOS transistors [10] when a bias voltage is supplied. If voltage is applied to the gate oxide layer of the NMOS transistor, outputs are positive and it is not stable due to the degradation of the trapped electrons. Due to this, there is a change in the voltage of the oxide layer of the channel. The effect of the PBTI is negligible in comparison with the effect of NBTI.

Negative Bias Temperature Instability (NBTI) When the negative bias voltage is applied, NBTI occurs on PMOS transistors and the traps are formed. When the voltage is provided to the gate terminal of PMOS then the output will be negative [6]. Due to this, threshold voltage gets altered and reduction in the I_{d} (drain current) occurs.

$$\Delta V_t = q(\Delta N_{\mathrm{it}} + \Delta N_{\mathrm{f}})/C_{\mathrm{ox}} \tag{7}$$

Here, q is the charge, C_{ox} is the oxide capacitance, N_{it} is the density of charge interface states and N_{f} is the density of traps. NBTI follows an influence law and the power exponent is not modified along with the stress voltage.

$$\Delta V_{\mathrm{th}} = g V_{\mathrm{gs}} t^m t^n \tag{8}$$

ΔV_{th} is the shift of the threshold voltage, g is generation constant and t is the stress time. The I_{sub} and the V_{dsat} equations are

$$I_{\mathrm{sub}} = \frac{A_i}{B_i}(V_{\mathrm{ds}} - V_{\mathrm{dsat}}) I_{\mathrm{d}} \exp\left(\frac{B_i L_{\mathrm{C}}}{V_{\mathrm{ds}} - V_{\mathrm{dsat}}}\right) \tag{9}$$

$$V_{\mathrm{dsat}} = \frac{E_{\mathrm{crit}} L_{\mathrm{eff}}(V_{\mathrm{gs}} - V_t)}{E_{\mathrm{crit}} L_{\mathrm{eff}} + V_{\mathrm{gs}} - V_t} \tag{10}$$

When V_{gs} (gate to source voltage) is negative, the holes from the source are trapped at the gate terminal. If V_{gs} is positive, then it shifts the threshold voltage

(V_{th}) and supports to the I_d (drain current). The damage happened at the positive bias is called NBTI relaxation [7]. The V_{th} and mobility of the mobile charge carriers are decreased due to the holes trapped in the gate terminal.

The aging and stress reliability analysis is performed for the circuit ring oscillator and a combinational circuit full adder. The degradation due to HCI and BTI for both circuits is analyzed.

3 Simulation Results

The ring oscillator and full adder circuits are designed. The reliability simulation is performed on these circuits using the Cadence RelXpert simulator on a computer with 4 GB RAM. The degradation in parameters of these above circuits due to the effect of HCI and NBTI are analyzed.

3.1 Reliability Analysis of Ring Oscillator

Ring oscillator is designed in a circular chain composed of an odd number of inverter as shown in Figs. 3 and 4. The final output is asserted a finite amount of time after the first input is asserted and the feedback of the last output to the input causes oscillation. Ring oscillator is exposed to the stress condition in the particular temperature and the voltage to study the effect of HCI and NBTI. Then, the time-dependent degradation

Fig. 3 Inverter used in oscillator

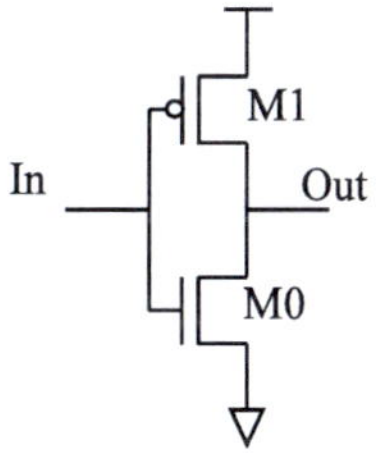

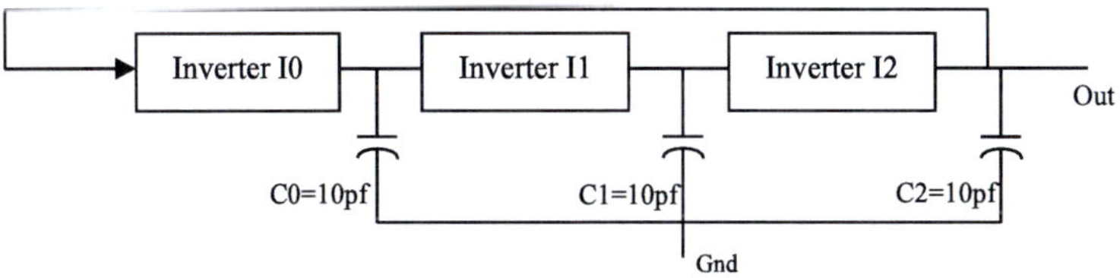

Fig. 4 Three-stage ring oscillator

Table 1 Lifetime and degradation due to NBTI

Transistor	max V_g (V)	max V_d (V)	Degradation	Lifetime (years)	V_{th} shift (V)
I0.M1	1.10e+00	1.25e+00	2.33e−05	5.83e−85	2.6273e−02
I1.M1	1.25e+00	1.10e+00	2.35e−05	6.99e−85	2.6615e−02
I2.M1	1.10e+00	1.10e+00	2.39e−05	9.29e−85	2.5534e−02

appeared in the device parameters are determined due to these effects. The inverter used to design the ring oscillator is shown in Fig. 3.

Degradation Due to NBTI Due to the applied stress in ring oscillator, each and every transistor in the circuit gets degraded. The PMOS transistors are degraded due to NBTI effects. Ring oscillator is simulated under the stress for a period of 10 years. The threshold voltage (V_{th}) of the PMOS device will shift due to the change on the maximum absolute gate-source voltage (max V_g) and maximum absolute drain-source voltages (max V_d) leading to the degradation. The shift in threshold voltage (V_{th} Shift) and lifetime (age) of the transistor is calculated and tabulated in Table 1. The threshold voltage of the transistor M1 of instance I0 (PMOS of first inverter) is shifted to 2.6273e−02, due to the degradation in the max V_g of 1.10e+00 and max V_d of 1.25e+00. Degradation rate of the transistors M1 (in instance I0) is 2.33e−05 for the lifetime of 5.83e−85 years. In M1 (in instance I1) and M1 (in instance I2), the threshold voltage change happens and the degradation rates are 2.35e−05 and 2.39e−05 for the lifetime of 6.99e−85 and 9.29e−85 years, respectively.

Degradation Due to HCI and NBTI Both HCI and NBTI affect PMOS and causes degradation in the substrate current and gate current. This leads to the degradation of the PMOS device. max I_b and avg I_b are the maximum and average substrate current. max I_g and avg I_g are the maximum and average gate current. The degradation of gate and substrate currents due to the HCI and NBTI effects along with lifetime are shown in Table 2.

The effect of HCI and NBTI varies with different PMOS transistors. The maximum substrate current and maximum gate current of PMOS transistor M1 (from instance I0) are 2.09e−15 and 9.67e−16. The degradation value is 4.73e−05 for a lifetime of 1.00e−14 years.

Table 2 Degradation values with lifetime due to HCI and NBTI

Transistor	max I_b (A)	avg I_b (A)	max I_g (A)	avg I_g (A)	Degradation	Lifetime (years)
I0.M1	2.09e−15	6.82e−17	9.67e−16	5.01e−17	4.73e−05	1.00e−14
I1.M1	6.03e−15	1.33e−16	9.57e−16	5.65e−17	5.42e−05	2.10e−14
I2.M1	1.68e−15	4.99e−17	9.52e−16	3.95e−17	4.55e−05	7.24e−15

The age degradation of NMOS devices with the HCI and NBTI in ring oscillator is observed by applying a stress of 10 years.

Aging Analysis in Ring Oscillator

Transistor aging due to HCI. Due to the applied stress for 10 years, the NMOS transistors will degrade due to HCI. This affects the lifetime of the device.

The aging parameters of NMOS transistor's values are shown in Table 3. The degraded value of the NMOS transistor M0 (of instance I0) after 1 year is 1.2980e−03, 2.8 years is 3.6344e−03, 4.6 years is 5.9708e−03, 6.4 years is 8.3072e−03, 8.2 years is 1.0644e−02 and 10 years is 1.2980e−02, respectively.

Transistor aging due to NBTI. In NBTI, the PMOS transistors are degraded and this affects the age of the PMOS transistors. The degradation of the age of the PMOS transistors occurs due to NBTI is tabulated in Table 4. The age of the PMOS transistor M1 (of instance I0) is degraded by 1.1291e−08 after 1 year, 3.1616e−08 after 2.8 years, 5.1941e−08 after 4.6 years, 7.2265e−08 after 6.4 years, 9.2590e−08 after 8.2 years and 1.1291e−07 after 10 years.

Output Response Analysis in Ring Oscillator Due to the effects of HCI and NBTI on the PMOS and NMOS transistors, the output frequency of ring oscillator is changed. The influence of reliability on threshold voltage, gate current, substrate current, gate-source voltage and drain-source voltage has delayed the output response over a period of time. The change in the output response of the ring oscillator for a lifetime of 10 years is shown in Fig. 5.

The delay between the new and the aged response due to aging of the devices is calculated and tabulated in Table 5.

The oscillation frequency of the ring oscillator is calculated. Due to the aging along with the stress in the devices, the decrease in the value of the oscillator frequency is observed. Over a period of 10 years, the frequency has been reduced from 15.76 to 10.48 GHz as shown in Table 6 and Fig. 6.

3.2 *Reliability Analysis of Full Adder*

In digital design circuits, most of the logical operations are performed by the adders. The adder is used as the main function element in the processers. Adder is implemented by using the basic gates and simulated using Cadence Virtuoso tool. The full adder has three inputs A, B and C with two outputs Sum and C_{out} as shown in Fig. 7.

Degradation Due to NBTI The degradation of the PMOS in full adder with NBTI due to the applied stress is shown in Table 7. The maximum degradation value is 5.04e−02 for a lifetime of 3.82e−07. The maximum threshold voltage shift (V_{th} shift) is 3.2058e−02.

Table 3 Age degradation for transistors due to HCI

Transistor	Age (0 years)	Age (1 years)	Age (2.8 years)	Age (4.6 years)	Age (6.4 years)	Age (8.2 years)	Age (10 years)	Age degradation (%)
I0.M0	2.4603e−04	1.2980e−03	3.6344e−03	5.9708e−03	8.3072e−03	1.0644e−02	1.2980e−02	74.35
I1.M0	9.0551e−04	1.6445e−06	4.6046e−06	7.5647e−06	1.0525e−05	1.3485e−05	1.6445e−05	93.32
I2.M0	9.0856e−04	1.4806e−06	4.1456e−06	6.8106e−06	9.4756e−06	1.2141e−05	1.4806e−05	94

Table 4 Age degradations of the PMOS in NBTI

Transistor	Age (0 years)	Age (1 years)	Age (2.8 years)	Age (4.6 years)	Age (6.4 years)	Age (8.2 years)	Age (10 years)	Age degradation (%)
I0.M1	9.9999e−04	1.1291e−08	3.1616e−08	5.1941e−08	7.2265e−08	9.2590e−08	1.1291e−07	99.43
I1.M1	9.9999e−04	1.1900e−08	3.3320e−08	5.4739e−08	7.6159e−08	9.7579e−08	1.1900e−07	99.40
I2.M1	9.9999e−04	1.2306e−08	3.4457e−08	5.6608e−08	7.8759e−08	1.0091e−07	1.2306e−07	99.38

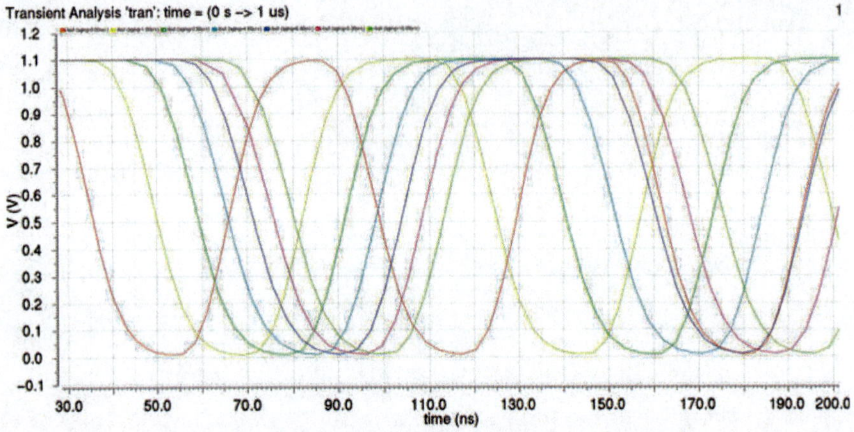

Fig. 5 Delayed output response of ring oscillator

Table 5 Oscillator output variation between new and aged response

Age in years	Delay (s)
1	14.16e−9
2.8	23.04e−9
4.6	29.64e−9
6.4	35.00e−9
8.2	38.91e−9
10	43.77e−9

Table 6 Frequency degradation due to aging

Frequency (GHz)	Age in years
15.76	0
13.32	1
12.26	2.8
11.67	4.6
11.18	6.4
10.81	8.2
10.48	10

Degradation Due to HCI and NBTI The degradation of the PMOS in full adder with HCI and NBTI due to the applied stress is shown in Table 8.

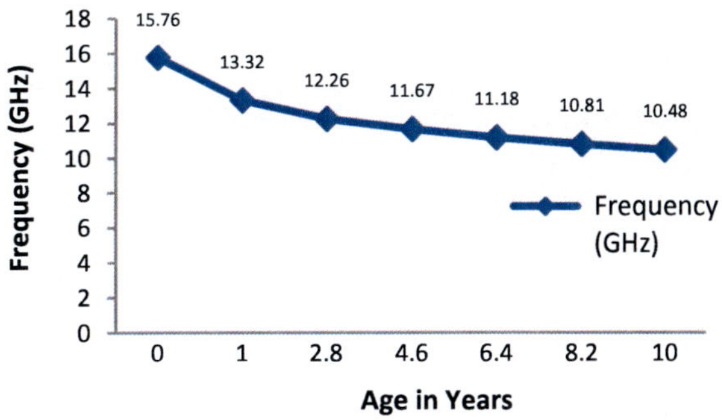

Fig. 6 Degradation in the frequency of the ring oscillator

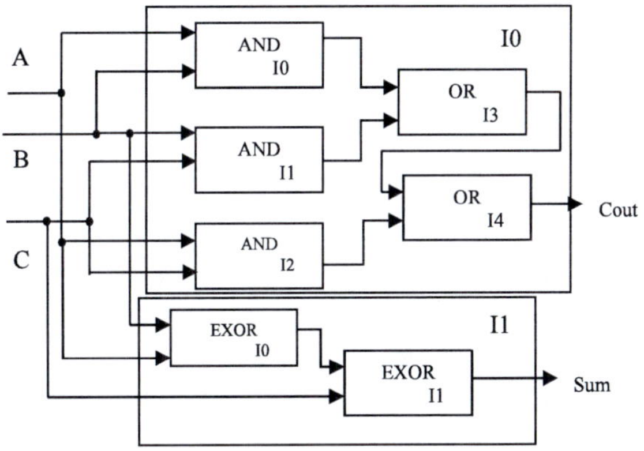

Fig. 7 Full adder circuit

The degradation after 10 years for maximum substrate current is 8.42e−23, average substrate current is 1.81e−25, maximum gate current is 2.49e−16 and average gate current is 1.06e−18. The maximum degradation rate is 6.84e−02.

Aging Analysis in Full Adder

Transistor aging due to HCI. The age of the degraded NMOS device due to HCI is shown in Table 9. By using the aging analysis with a stress of 10 years, the age of different transistors at a different time period is calculated. The maximum degraded age of a NMOS transistor in the circuit after 10 years is 5.0296e−18.

Transistor aging due to NBTI. NTBI occurs in PMOS causing degradation in the device. Using aging analysis, the aged PMOS transistors of determined with a stress. The degradation value of age of PMOS device for 10 years is shown in Table 10. The maximum age degradation value of PMOS after 10 years is 6.0777e−08.

Output Response Analysis in Full Adder The degraded transistors affect the output voltages of the circuit as shown in Figs. 8 and 9. The outputs Sum and C_{out} of the full adder are delayed due to the HCI and NBTI effects in the transistors. For a gradual period of 10 years, the outputs are analyzed.

Table 7 Lifetime and degradation due to NBTI

Transistor	max V_g (V)	max V_d (V)	Degradation	Lifetime (years)	V_{th} shift (V)
I0.I0.I0.M1	1.10e+00	1.14e+00	4.60e−02	2.46e−07	2.9024e−02
I0.I0.M0	1.00e+00	1.00e+00	3.99e−02	1.21e−07	2.0191e−02
I0.I0.M1	1.00e+00	1.00e+00	3.97e−02	1.18e−07	2.0087e−02
I0.I1.I0.M1	1.20e+00	1.19e+00	4.57e−02	2.36e−07	2.9065e−02
I0.I1.M0	1.00e+00	1.10e+00	3.97e−02	1.19e−07	2.5068e−02
I0.I1.M1	1.00e+00	1.10e+00	3.94e−02	1.13e−07	2.4855e−02
I0.I2.I0.M1	1.13e+00	1.20e+00	4.56e−02	2.34e−07	2.9249e−02
I0.I2.M0	1.00e+00	1.03e+00	3.99e−02	1.21e−07	2.0206e−02
I0.I2.M1	1.00e+00	1.03e+00	3.93e−02	1.13e−07	2.2771e−02
I0.I3.I0.M1	1.19e+00	1.16e+00	4.79e−02	3.00e−07	3.0952e−02
I0.I3.M0	1.08e+00	1.05e+00	4.19e−02	1.55e−07	2.4186e−02
I0.I3.M1	1.04e+00	6.24e−01	4.35e−02	1.85e−07	2.2271e−02
I0.I4.I0.M1	1.18e+00	1.20e+00	4.98e−02	3.62e−07	3.1449e−02
I0.I4.M0	1.09e+00	1.05e+00	4.01e−02	1.23e−07	2.3039e−02
I0.I4.M1	1.06e+00	6.86e−01	4.20e−02	1.56e−07	2.1522e−02
I1.I0.I2.M1	1.10e+00	1.11e+00	4.68e−02	2.65e−07	2.4121e−02
I1.I0.I3.M1	1.10e+00	1.13e+00	4.66e−02	2.60e−07	2.7710e−02
I1.I0.I4.M1	1.15e+00	1.13e+00	4.64e−02	2.54e−07	2.8517e−02
I1.I0.M1	1.00e+00	6.37e−01	3.99e−02	1.21e−07	2.0202e−02
I1.I0.M14	1.00e+00	1.06e+00	3.47e−02	6.08e−08	1.9941e−02
I1.I0.M15	1.03e+00	7.23e−01	4.28e−02	1.72e−07	2.2037e−02
I1.I0.M16	1.00e+00	1.14e+00	3.92e−02	1.11e−07	2.2966e−02
I1.I1.I2.M1	1.13e+00	1.15e+00	5.04e−02	3.82e−07	3.1986e−02
I1.I1.I3.M1	1.10e+00	1.14e+00	4.61e−02	2.48e−07	2.9665e−02
I1.I1.I4.M1	1.14e+00	1.16e+00	4.93e−02	3.44e−07	3.2058e−02
I1.I1.M1	1.03e+00	6.02e−01	4.29e−02	1.73e−07	2.1923e−02
I1.I1.M14	1.00e+00	1.02e+00	3.32e−02	4.90e−08	1.8268e−02
I1.I1.M15	1.04e+00	6.88e−01	4.25e−02	1.66e−07	2.1779e−02
I1.I1.M16	1.01e+00	1.12e+00	3.32e−02	4.89e−08	1.6515e−02

Due to the degradation and aging affect in the circuit, the delay occurs in both Sum and C_{out} which are shown in Tables 11 and 12, respectively. The delay value between new and aged output response increases up to 8.911e−12 for Sum and 12.23e−12 for C_{out} after a period of 10 years.

Table 8 Degradation values with lifetime due to HCI and NBTI

Transistor	max I_b (A)	avg I_b (A)	max I_g (A)	avg I_g (A)	Degradation	Lifetime (years)
I0.I0.I0.M1	3.79e−15	3.79e−18	9.24e−15	2.34e−17	6.20e−02	6.68e−13
I0.I1.I0.M1	3.95e−15	5.22e−18	9.00e−15	3.73e−17	6.23e−02	8.87e−13
I0.I1.M0	2.06e−15	2.58e−18	6.55e−15	1.49e−17	5.48e−02	4.48e−13
I0.I1.M1	2.06e−15	2.58e−18	6.55e−15	1.49e−17	5.45e−02	4.48e−13
I0.I2.I0.M1	4.97e−15	6.30e−18	9.08e−15	3.84e−17	6.29e−02	1.12e−12
I0.I2.M1	1.96e−16	1.62e−19	4.38e−15	9.41e−18	4.88e−02	1.89e−14
I0.I3.I0.M1	1.01e−14	8.07e−18	1.26e−14	4.12e−17	6.62e−02	1.62e−12
I0.I3.M0	1.96e−16	1.33e−19	4.40e−15	6.38e−18	5.12e−02	1.58e−14
I0.I4.I0.M1	4.32e−15	5.53e−18	1.43e−14	3.75e−17	6.66e−02	9.46e−13
I0.I4.M0	1.64e−16	1.19e−19	4.25e−15	6.93e−18	4.91e−02	1.39e−14
I0.I4.M1	5.60e−24	5.27e−27	7.34e−17	7.32e−20	4.25e−02	5.06e−23
I1.I0.I3.M1	5.23e−16	6.37e−19	9.32e−15	1.62e−17	5.83e−02	7.98e−14
I1.I0.I9.M1	1.76e−15	1.92e−18	9.62e−15	2.69e−17	6.06e−02	2.95e−13
I1.I0.M14	1.02e−16	1.26e−19	3.24e−15	6.08e−18	4.38e−02	1.40e−14
I1.I0.M15	1.32e−22	4.42e−26	1.51e−16	7.07e−20	4.36e−02	6.69e−22
I1.I0.M16	5.91e−16	2.33e−19	5.05e−15	4.55e−18	4.94e−02	3.29e−14
I1.I1.I2.M1	7.75e−15	6.03e−18	9.46e−15	1.98e−17	6.78e−02	1.20e−12
I1.I1.I3.M1	5.40e−15	7.16e−18	9.92e−15	5.86e−17	6.37e−02	1.24e−12
I1.I1.I9.M1	4.92e−15	1.24e−17	8.74e−15	6.35e−17	6.84e−02	2.24e−12
I1.I1.M1	7.46e−31	3.89e−34	3.53e−18	1.84e−21	4.29e−02	3.53e−31
I1.I1.M14	2.92e−17	2.10e−20	2.97e−15	4.84e−18	3.99e−02	1.80e−15
I1.I1.M15	1.04e−24	8.23e−28	6.54e−17	7.50e−20	4.29e−02	5.68e−24
I1.I1.M16	8.42e−23	1.81e−25	2.49e−16	1.06e−18	3.41e−02	2.27e−21

4 Conclusion

BTI and HCI reliability issues play a major role in CMOS circuits. The degradation in device parameters and performance due to these effects on the three-stage ring oscillator and full adder circuits are analyzed through aging and stress reliability simulations. The results show that after a period of 10 years, the age of the full adder is degraded from 91.8 to 99.99% for NMOS and from 95.14 to 99.44% for PMOS. The age of the ring oscillator is degraded from 74.35 to 94% for NMOS and from 99.38 to 99.43% after a period of 10 years. The frequency of the ring oscillator shifts by 25.95% after 4.6 years and 33.5% after 10 years. The influence of HCI and NTBI effects on transistors leads to change in the frequency of ring oscillator and delay variation in CMOS full adder output. The age of the transistor at which the rate of degradation happens is also analyzed. Analysis of the change in device parameters due to the aging helps to detect the aged or recycled devices.

Table 9 Age degradation of the NMOS in HCI

Transistor	Age (0 years)	Age (1 years)	Age (2.8 years)	Age (4.6 years)	Age (6.4 years)	Age (8.2 years)	Age (10 years)	Age degradation (%)
I1.I1.I3.M0	9.0630e−04	1.6010e−06	4.4827e−06	7.3645e−06	1.0246e−05	1.3128e−05	1.6010e−05	93.5
I0.I4.I0.M0	9.1363e−04	1.2333e−06	3.4533e−06	5.6733e−06	7.8932e−06	1.0113e−05	1.2333e−05	95.03
I1.I1.I9.M0	9.1909e−04	1.0002e−06	2.8006e−06	4.6009e−06	6.4013e−06	8.2016e−06	1.0002e−05	95.99
I1.I0.I2.M0	9.3018e−04	6.2331e−07	1.7453e−06	2.8672e−06	3.9892e−06	5.1111e−06	6.2331e−06	90.93
I1.I0.I3.M0	9.3232e−04	5.6431e−07	1.5801e−06	2.5958e−06	3.6115e−06	4.6273e−06	5.6431e−06	91.80
I0.I1.I0.M0	9.3428e−04	5.1333e−07	1.4373e−06	2.3613e−06	3.2853e−06	4.2093e−06	5.1333e−06	92.56
I0.I2.I0.M0	9.3555e−04	4.8231e−07	1.3505e−06	2.2186e−06	3.0868e−06	3.9550e−06	4.8231e−06	93.02
I0.I3.I0.M0	9.3832e−04	4.1885e−07	1.1728e−06	1.9267e−06	2.6806e−06	3.4346e−06	4.1885e−06	93.95
I1.I1.I2.M0	9.4206e−04	3.4268e−07	9.5950e−07	1.5763e−06	2.1931e−06	2.8100e−06	3.4268e−06	95.07
I0.I4.M2	9.5014e−04	2.1173e−07	5.9284e−07	9.7395e−07	1.3551e−06	1.7362e−06	2.1173e−06	96.98
I1.I0.I9.M0	9.5277e−04	1.7796e−07	4.9829e−07	8.1862e−07	1.1389e−06	1.4593e−06	1.7796e−06	97.47
I0.I3.M2	9.6496e−04	6.8358e−08	1.9140e−07	3.1445e−07	4.3749e−07	5.6053e−07	6.8358e−07	96.47
I0.I0.M2	9.7923e−04	1.2774e−08	3.5766e−08	5.8759e−08	8.1751e−08	1.0474e−07	1.2774e−07	99.35
I0.I2.M2	9.7927e−04	1.2698e−08	3.5554e−08	5.8411e−08	8.1267e−08	1.0412e−07	1.2698e−07	99.35
I0.I4.M3	9.7998e−04	1.1353e−08	3.1789e−08	5.2225e−08	7.2661e−08	9.3097e−08	1.1353e−07	99.42
I0.I3.M3	9.8073e−04	1.0037e−08	2.8105e−08	4.6172e−08	6.4239e−08	8.2307e−08	1.0037e−07	99.49
I0.I1.M2	9.8248e−04	7.3980e−09	2.0714e−08	3.4031e−08	4.7347e−08	6.0663e−08	7.3980e−08	98.62
I1.I1.M17	9.9684e−04	3.0492e−11	8.5378e−11	1.4026e−10	1.9515e−10	2.5003e−10	3.0492e−10	99.92
I1.I1.M18	9.9893e−04	9.3620e−13	2.6213e−12	4.3065e−12	5.9916e−12	7.6768e−12	9.3620e−12	99.96
I1.I0.M17	9.9898e−04	8.0829e−13	2.2632e−12	3.7182e−12	5.1731e−12	6.6280e−12	8.0829e−12	99.97
I1.I1.M13	9.9964e−04	2.7847e−14	7.7972e−14	1.2810e−13	1.7822e−13	2.2835e−13	2.7847e−13	99.99

(continued)

Table 9 (continued)

Transistor	Age (0 years)	Age (1 years)	Age (2.8 years)	Age (4.6 years)	Age (6.4 years)	Age (8.2 years)	Age (10 years)	Age degradation (%)
I1.I0.M13	9.9969e−04	1.7157e−14	4.8041e−14	7.8924e−14	1.0981e−13	1.4069e−13	1.7157e−13	99.99
I1.I1.M19	9.9981e−04	3.7878e−15	1.0606e−14	1.7424e−14	2.4242e−14	3.1060e−14	3.7878e−14	99.99
I0.I2.M3	9.9999e−04	1.3004e−19	3.6413e−19	5.9821e−19	8.3229e−19	1.0664e−18	1.3004e−18	99.99
I0.I1.M3	9.9996e−04	3.0748e−17	8.6094e−17	1.4144e−16	1.9679e−16	2.5213e−16	3.0748e−16	99.99
I0.I0.M3	9.9999e−04	5.0296e−19	1.4083e−18	2.3136e−18	3.2189e−18	4.1243e−18	5.0296e−18	99.99

Table 10 Age degradation of the PMOS in NBTI

Transistor	Age (0 years)	Age (1 years)	Age (2.8 years)	Age (4.6 years)	Age (6.4 years)	Age (8.2 years)	Age (10 years)	Age degradation (%)
I1.I1.I9.M1	9.9999e−04	3.4446e−08	9.6450e−08	1.5845e−07	2.2046e−07	2.8246e−07	3.4446e−07	98.28
I1.I1.I2.M1	9.9999e−04	3.8240e−08	1.0707e−07	1.7591e−07	2.4474e−07	3.1357e−07	3.8240e−07	98.09
I0.I4.I0.M1	9.9999e−04	3.6209e−08	1.0138e−07	1.6656e−07	2.3173e−07	2.9691e−07	3.6209e−07	98.19
I0.I3.I0.M1	9.9999e−04	2.9969e−08	8.3914e−08	1.3786e−07	1.9180e−07	2.4575e−07	2.9969e−07	98.50
I1.I1.I3.M1	9.9999e−04	2.4804e−08	6.9451e−08	1.1410e−07	1.5874e−07	2.0339e−07	2.4804e−07	98.76
I0.I2.I0.M1	9.9999e−04	2.3366e−08	6.5425e−08	1.0748e−07	1.4954e−07	1.9160e−07	2.3366e−07	98.83
I0.I1.I0.M1	9.9999e−04	2.3553e−08	6.5948e−08	1.0834e−07	1.5074e−07	1.9313e−07	2.3553e−07	98.83
I0.I0.I0.M1	9.9999e−04	2.4563e−08	6.8776e−08	1.1299e−07	1.5720e−07	2.0141e−07	2.4563e−07	98.77
I1.I0.I9.M1	9.9999e−04	2.5417e−08	7.1167e−08	1.1692e−07	1.6267e−07	2.0842e−07	2.5417e−07	98.73
I1.I0.I3.M1	9.9999e−04	2.5965e−08	7.2703e−08	1.1944e−07	1.6618e−07	2.1292e−07	2.5965e−07	98.7
I0.I1.M0	9.9999e−04	1.1863e−08	3.3217e−08	5.4572e−08	7.5925e−08	9.7280e−08	1.1863e−07	99.4
I0.I1.M1	9.9999e−04	1.1331e−08	3.1726e−08	5.2121e−08	7.2516e−08	9.2911e−08	1.1331e−07	99.43
I0.I3.M0	1.0000e−03	1.5480e−08	4.3344e−08	7.1208e−08	9.9072e−08	1.2694e−07	1.5480e−07	97.09
I1.I0.I2.M1	1.0000e−03	2.6519e−08	7.4252e−08	1.2199e−07	1.6972e−07	2.1745e−07	2.6519e−07	95.14
I0.I4.M0	1.0000e−03	1.2337e−08	3.4543e−08	5.6749e−08	7.8955e−08	1.0116e−07	1.2337e−07	97.74
I1.I0.M16	9.9999e−04	1.1056e−08	3.0956e−08	5.0857e−08	7.0757e−08	9.0657e−08	1.1056e−07	99.44
I0.I2.M1	1.0000e−03	1.1278e−08	3.1579e−08	5.1879e−08	7.2179e−08	9.2480e−08	1.1278e−07	97.93
I0.I3.M1	1.0000e−03	1.8548e−08	5.1936e−08	8.5323e−08	1.1871e−07	1.5210e−07	1.8548e−07	96.60
I1.I0.M15	1.0000e−03	1.7153e−08	4.8029e−08	7.8906e−08	1.0978e−07	1.4066e−07	1.7153e−07	96.86
I1.I1.M1	1.0000e−03	1.7288e−08	4.8406e−08	7.9525e−08	1.1064e−07	1.4176e−07	1.7288e−07	96.83
I1.I1.M15	1.0000e−03	1.6571e−08	4.6398e−08	7.6226e−08	1.0605e−07	1.3588e−07	1.6571e−07	96.96

(continued)

Table 10 (continued)

Transistor	Age (0 years)	Age (1 years)	Age (2.8 years)	Age (4.6 years)	Age (6.4 years)	Age (8.2 years)	Age (10 years)	Age degradation (%)
I0.I4.M1	1.0000e−03	1.5623e−08	4.3746e−08	7.1868e−08	9.9990e−08	1.2811e−07	1.5623e−07	97.14
I1.I0.M14	1.0000e−03	6.0777e−09	3.3928e−08	5.5740e−08	7.7550e−08	9.9362e−08	1.2117e−07	97.78
I0.I2.M0	1.0000e−03	1.2117e−08	3.3899e−08	5.5691e−08	7.7482e−08	9.9274e−08	1.2107e−07	97.78
I1.I0.M1	1.0000e−03	1.2107e−08	3.3819e−08	5.5561e−08	7.7301e−08	9.9043e−08	1.2078e−07	97.78
I0.I0.M0	1.0000e−03	1.2078e−08	3.3071e−08	5.4331e−08	7.5591e−08	9.6851e−08	1.1811e−07	97.84
I0.I0.M1	1.0000e−03	1.1811e−08	1.7017e−08	2.7957e−08	3.8897e−08	4.9837e−08	6.0777e−08	95.90
I1.I1.M14	1.0000e−03	4.8979e−09	1.3714e−08	2.2530e−08	3.1347e−08	4.0163e−08	4.8979e−08	96.69
I1.I1.M16	1.0000e−03	4.8935e−09	1.3702e−08	2.2510e−08	3.1318e−08	4.0127e−08	4.8935e−08	96.70

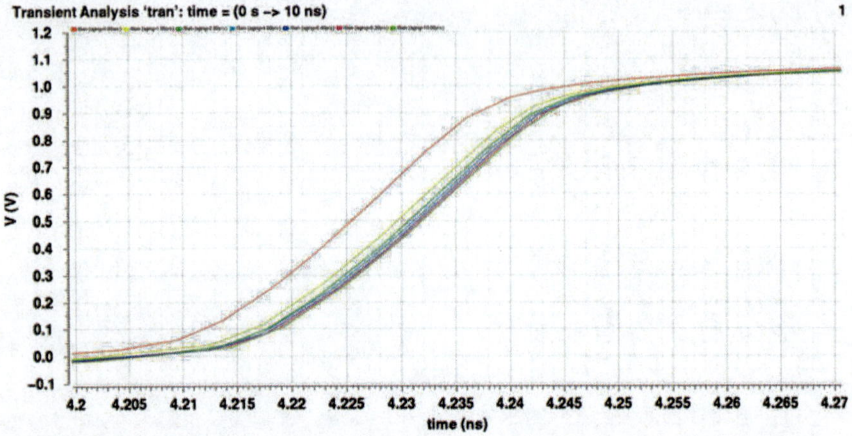

Fig. 8 Gradual output delay of sum

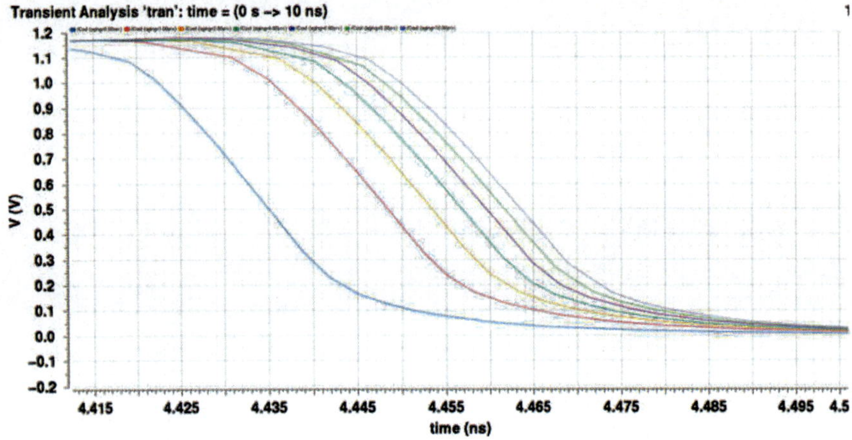

Fig. 9 Gradual output delay of C_{out}

Table 11 Delay variation in sum output between new and aged response

Age in years	Delay variation (s)
1	4.241e−12
2.8	5.855e−12
4.6	6.866e−12
6.4	7.672e−12
8.2	8.349e−12
10	8.911e−12

Table 12 Delay variation in CARRY output (C_{out}) between new and aged response

Age in years	Delay variation (s)
1	6.636e−12
2.8	8.986e−12
4.6	9.904e−12
6.4	10.79e−12
8.2	11.77e−12
10	12.23e−12

References

1. Bernstein K, Frank D, Gattiker A, Haensch W, Ji B, Nassif S, Nowak E, Pearson D, Rohrer N (2006) High-performance CMOS variability in the 65-nm regime and beyond. IBM J Res Dev 50(4.5):433–449
2. Takeda E, Yang CY, Miura-Hamada A (1995) Hot-carrier effects in MOS devices, 1st edn. Academic Press, New York, USA
3. Zhang L, Dick R (2009) Scheduled voltage scaling for increasing lifetime in the presence of NTBI. In: Proceedings of the ASP-DAC. IEEE, Japan, pp 492–497
4. Wang Y, Chen X, Wang W, Balakrishnan V, Cao Y, Xie Y, Yang H (2009) On the efficacy of input vector control to mitigate NBTI effects and leakage power. In: Proceedings of the ISQED. IEEE, pp 19–26
5. Gupta M, Rivers J, Bose P, Wei G, Brooks D (2009) Tribeca: design for PVT variations with local recovery and fine-grained adaptation. In: Proceedings of the 42nd annual IEEE/ACM international symposium on microarchitecture (MICRO). IEEE, New York, USA, pp 435–446
6. Fu X, Li T, Fortes J (2008) NBTI tolerant microarchitecture design in the presence of process variation. In: Proceedings of the IEEE/ACM international symposium on microarchitecture. IEEE, Italy, pp 399–410
7. Hu C, Simon C Tam, Fu-Chieh H, Ping-Keung K, Tung-Yi C, Terrill KW (1985) Hot-electron induced MOSFET degradation-model, monitor and improvement. IEEE Trans Electron Dev 20(1):295–305
8. Feinberg A (2017) How aging laws influence parametric and catastrophic reliability distributions. In: Proceedings of the annual reliability and maintainability symposium. IEEE, Orland, FL, USA
9. Nigam T, Parameshwaran B, Krause G (2009) Accurate product lifetime predictions based on device-level measurements. In: Proceedings of the IEEE international reliability physics symposium. IEEE, Canada, pp 634–639
10. Ahmed F, Milor L (2010) Ring oscillator based embedded structure for decoupling PMOS/NMOS degradation with switching activity replication. In: Proceeding of the IEEE international conference on microelectronic test structures (ICMTS). IEEE, Japan
11. Prem Kumar ASB, Shyam Akashe B (2017) Aging analysis of low power hybrid adder based on HCI and NBTI. IEEE VLSI Circ Syst Lett 3(3):17–26
12. Liu Z, McGaughy BW, Ma JZ (2006) Design tools for reliability analysis. In: Proceedings of the design automation conference, pp 182–187
13. Selim M, Jeandeau E, Descleves C (2016) Design–reliability flow and advanced models address IC reliability issues. In: Proceeding of the workshop on early reliability modeling for aging and variability in silicon systems (2016)
14. Prabhu Deva Kumar SVSV, Akashe S, Kim H-J (2019) Reliability analysis & performance of 3 input CMOS look up table (LUT) in various nano technologies. IEEE VLSI Circ Syst Lett 5(1)

15. Prabhu Deva Kumar SVSV, Akashe S (2019) Reliability analysis of comparator: NBTI, PBTI, HCI, AGEING. In: Verma S, Tomar R, Chaurasia B, Singh V, Abawajy J (eds) Communication, networks and computing, CNC 2018. Communications in computer and information science, vol 839. Springer, Singapore
16. Slimani M, Benkalaia K, Naviner L (2017) Analysis of ageing effects on ARTIX7 XILINX FPGA. Microelectron Reliab 76–77:168–173
17. He S-R, Qui NC, Kuo Y-H, Liu C-NJ (2017) An incremental aging analysis method based on delta circuit simulation technique. In: Proceedings of the Asia test symposium, pp 60–65
18. Cai H (2012) Reliability of sigma-delta modulator: reliability. Research Report. Telecom Paris-tech
19. Tudor B, Wang J, Chen Z, Tan R, Liu W, Lee F (2012) An accurate MOSFET aging model for 28 nm integrated circuit simulation. Microelectron Reliab 52(8):1565–1570

Realization of a Low Profile, Wideband Omni-directional Antenna for Ku-band Airborne Applications

Pramendra Kumar Verma and Raj Kumar

Abstract In this paper, the design, analysis, and development of a low profile, wideband omni-directional antenna with wide beam coverage for Ku-band line of sight (LOS) data link communication system has been discussed. The antenna mainly consists of a shaped quarter wave radiating element, a metallic ground plate, and a dielectric material-based radome. A tapered conical section combined with a cylinder has been used to achieve wideband of operation as well as wide elevation beam coverage. For simulation and optimization of antenna, ANSYS's HFSS 19.3 EM simulation tool has been used. The effect of the configurations of the radome on the electrical performance of the antenna also has been carried out. The antenna has minimum reflection of RF signal at the input port over the frequency band 14.0–16.0 GHz. The gain of the antenna is better than 0dBi over the elevation beam coverage of 60° from horizon and omni beam coverage (i.e., 360°) in azimuth plane. The antenna shows stable radiation pattern over the desired frequency band.

Keywords Airborne platform · LOS antenna · Omni-directional · Radiation pattern · VSWR

1 Introduction

The airborne platform demands all direction radiation from its antenna to have communication with the ground terminal during its motion. The antenna on the airborne platform needs to establish a communication link from near as well as far distances during its operational time.

For airborne applications, the antenna should be low profile, light in weight, and rugged structure. The antenna should provide sufficient angular beam coverage and a stable radiation pattern over the operating frequency band. Presently, C-band omni-directional antenna is being used for LOS data link which is having the limitation of

P. K. Verma (✉) · R. Kumar
Antenna System Group, Defence Electronics Applications Laboratory, Raipur Road, Dehradun, Uttrakhand, India
e-mail: pramendra@deal.drdo.in

D. Harvey et al. (eds.), *Advances in VLSI, Communication, and Signal Processing*, Lecture Notes in Electrical Engineering 683,
https://doi.org/10.1007/978-981-15-6840-4_30

low data rate, and the size of the ground terminal antenna system is comparatively larger. Also, the congestion at C-band frequency needs to push the technologies toward the Ku-band frequency. Ku-band data link system provide wider bandwidth, high data rate, and smaller antenna size compared to C-band [1, 2].

A low profile, wide beam Ku-band airborne omni-directional antenna is required for transmission and reception of intelligence contained in electromagnetic waves from airborne terminal to ground terminal with larger data rate. Conventional monopoles are narrowband and provide low elevation coverage, hence are not suitable for near and far range application. Considering the data link parameter (data rate, range, path loss, etc.) along with radiation pattern requirements (bandwidth, beam coverage, polarization, gain, etc.), a wideband, wide beam linearly polarized omni-directional antenna is proposed for (Ku-band) LOS application. During operation period, one antenna may be masked by body of aircraft. So, two antennas one at top and another at the bottom would be placed on the airborne. A Teflon-based radome is integrated with the antenna to protect it from outer environmental conditions.

The paper discusses the design, analysis, and realization of Ku-band omni-directional antenna along with the effects of different configurations of the radome on the electrical performance of the antenna.

2 Design and Simulation

Several variants of monopole antennas suitable for airborne applications have been reported in the literature [3]. Generally, dipole, monopole, monocone, discone, planar trapezoidal monopole, valcano antenna, etc., are used to obtain omni-directional radiation pattern [4–9]. Dipole and monopole are narrow band antenna, and its bandwidth is only 1–2% which can be further improved by increasing the diameter of the radiating element of antenna, but it is limited to only 5%. As the diameter of the antenna increases, its effective capacitance increases, resulting in reduction of quality factor of antenna and hence increases its operational bandwidth. More increase in antenna diameter results in increase of conductor losses which cause the reduction in antenna radiation efficiency. Further, bandwidth can also be enhanced by flaring the antenna in the shape of discone, monocone, bicone, etc. These antennas have larger volume and weight.

To mitigate the above disadvantages, a tapered conical section combined with a cylinder has been selected as a radiating element. The conical structure of monocone is responsible for the bandwidth of operation of the antenna while the cylindrical section shapes the radiation pattern. The relation between electrical volume, gain, efficiency, and bandwidth of the antenna is decided through Eq. 1 [10].

$$\frac{\text{Electrical Volume}}{\text{Gain} * \text{Efficiency} * \text{B.W.}} = \text{Constant} \tag{1}$$

Hence for the constant value of gain and efficiency, the volume of the antenna will be more for large bandwidth of operation.

The overall height of the antenna is decided by operating frequency. The distance of shaped conical tapered element with ground plate and the lower radius of conical section improve the impedance matching of the antenna. The radiated beam of the antenna can also be shaped according to requirements, using different sizes of the ground plane as well as by modifying cylinder height. The diffraction of electromagnetic wave from the corner of circular ground plate produces the radiation in the bottom half space to lowest with a null at back lobe.

The shape, size, material electrical parameters (loss tangent and dielectric constant), and distance from radiating element of radome also affect the performance of the antenna. Lossy and thicker dielectric material of radome reduces the gain of antenna, and also, squint in the beam may occur. Hence, the loss tangent and thickness of radome should be as low as possible. The present antenna uses a Teflon-based radome (dielectric constant = 2.1 and loss tangent = 0.001) with a thickness of 1.5 mm to protect it from outer atmosphere.

After designing the antenna, it has been modeled on full wave HFSS 19.3 Electromagnetics Suite. Simulation and optimization of antenna have been carried out after applying proper boundary conditions. The three configurations of antenna, i.e., antenna without radome and antenna with top screw and with side screw radome, have been considered here. The three-dimensional CAD model of the antenna without radome and with radome is shown in Fig. 1.

The antenna has been simulated over the frequency band 14.0–16.0 GHz. The antenna has been analyzed for three configurations, i.e., antenna without radome, the antenna with the top screwed radome, and antenna with side screwed radome. The dimensions of the radiating element and ground plane have been optimized over the frequency band of 14–16 GHz for 60° from horizon plane. Figure 2 shows the sketch of antenna and radome with its dimensions (in mm). The simulated voltage standing wave ratio (VSWR), three-dimensional radiation pattern of antenna, elevation, and azimuth radiation pattern are shown in Figs. 3 and 4.

In HFSS software, a quasi-Newton (gradient) technique has been used for optimization of antenna. Parametric analysis of antenna also has been carried out to see the effects of gap (between radiating element and ground plane) and cone upper radius (ur_cone) on VSWR, which are shown in Fig. 5. The value of VSWR is improved by increasing the gap as well as cone upper radius. Variation of input port impedance over the frequency band is shown in Fig. 6. The radiation efficiency and peak gain of antenna over the frequency band are shown in Fig. 7. The simulated elevation (EL) pattern of the antenna over the frequency band is shown in Fig. 8.

The simulated VSWR plot shows that for antenna without radome the value of VSWR is less than 1.15 from 14.0 to 16.0 GHz. After covering the antenna with radome, value of VSWR increased to 1.48. VSWR for top screw-based design degrades more as compared to side screw-based and the optimized VSWR for antenna with side screw radome is 1.4:1 over 14–16.0 GHz. The ripples in omni-pattern are observed more in the case of top screw-based design. Hence, the radome with side screw has been finalized in the present case. The radiation efficiency and port

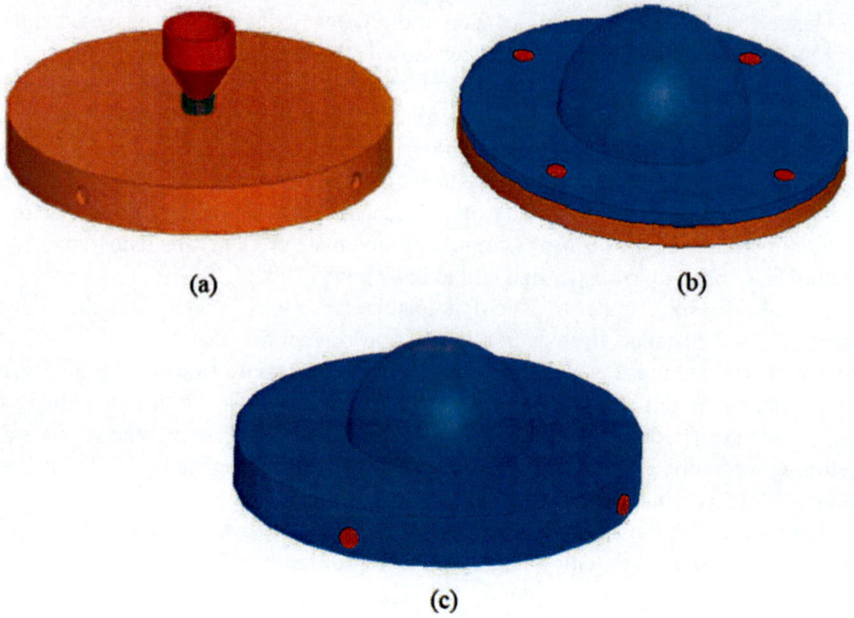

Fig. 1 CAD model of antenna **a** without radome, **b** radome with top screw, **c** radome with side screw

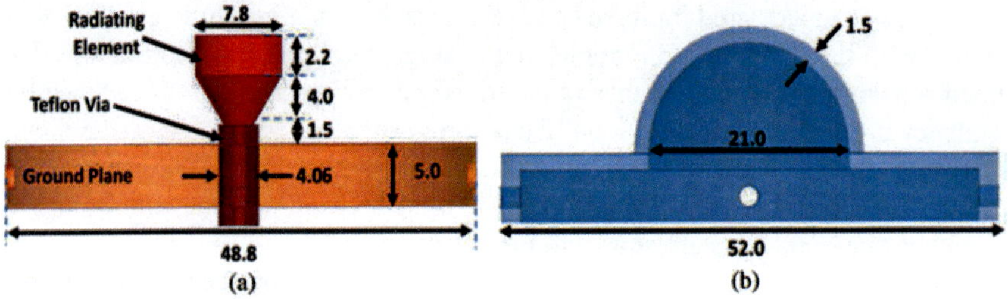

Fig. 2 Optimized dimensions of **a** antenna , **b** radome

impedance of antenna are better than 97.5% and 48 ohms, respectively. Also, the peak gain of antenna is better than 3.6 dBi over the desired frequency band.

3 Fabrication and Measurement

The photograph of the developed antenna without radome and with radome is shown in Fig. 9. Brass and aluminum alloys have been used for fabrication of radiating element and ground of antenna. SMA female connector has been used to feed the

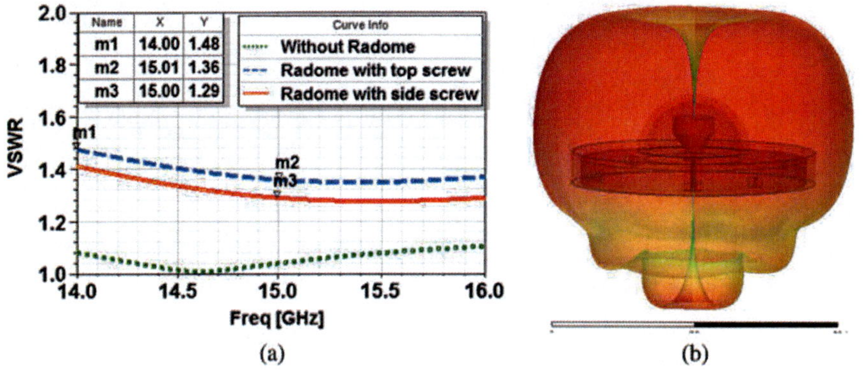

Fig. 3 Simulated **a** VSWR of antenna, **b** three-dimensional radiation pattern at 15 GHz

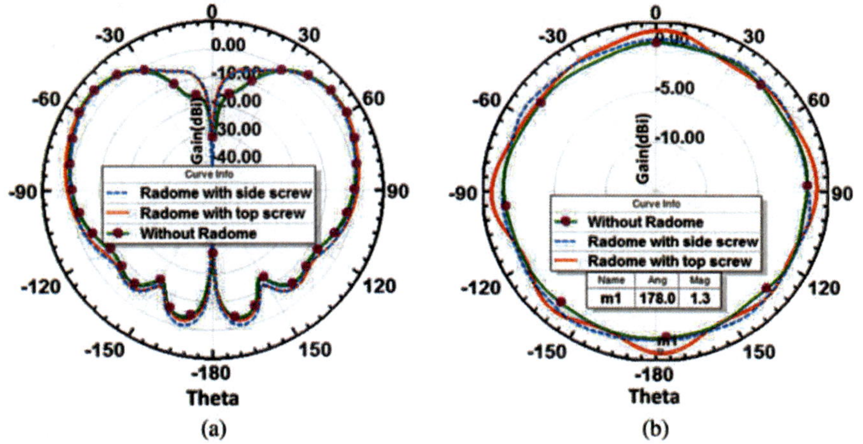

Fig. 4 Simulated radiation pattern at 15 GHz **a** elevation, **b** azimuth

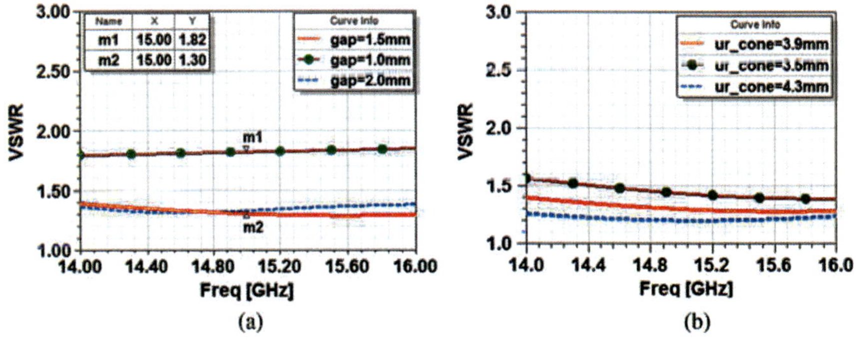

Fig. 5 Effects of **a** gap **b** cone upper radius on VSWR

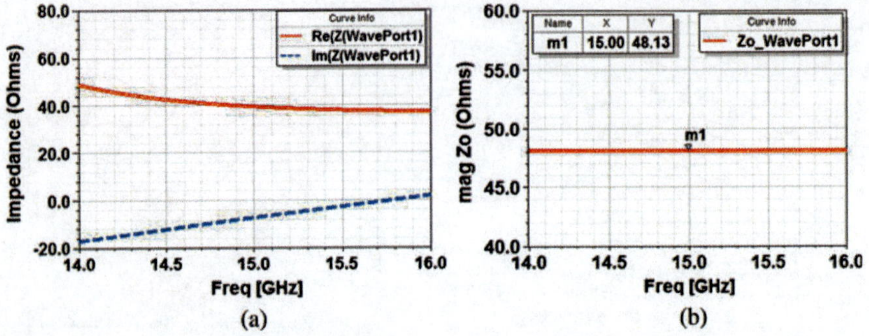

Fig. 6 Impedance plot **a** real and imaginary **b** magnitude

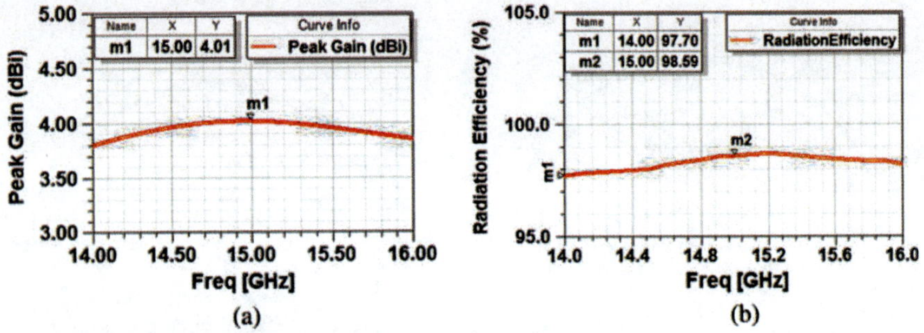

Fig. 7 **a** Peak gain, **b** radiation efficiency over the frequency band

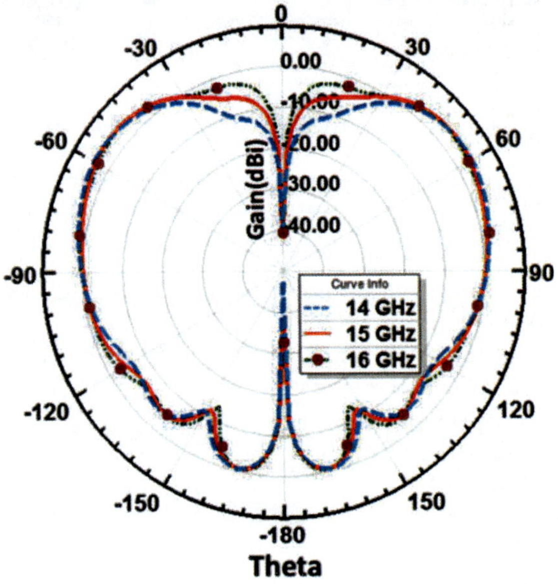

Fig. 8 Simulated elevation radiation pattern at 14, 15, and 16 GHz

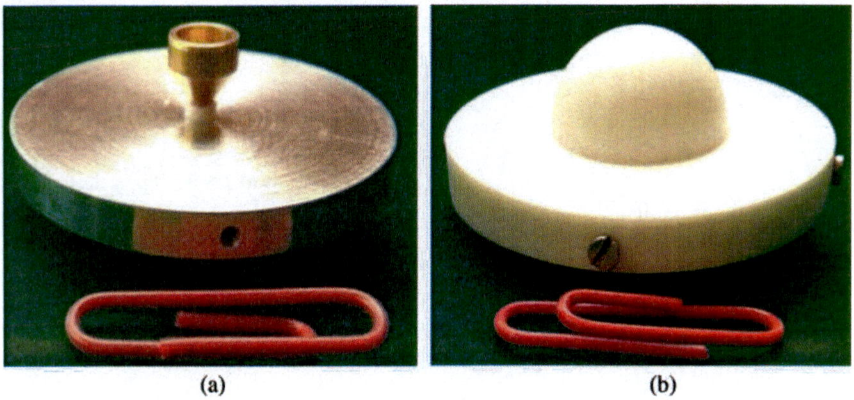

Fig. 9 Developed antenna **a** without radome, **b** with radome

antenna. A Teflon-based hemispherical radome has been developed. The total weight of the antenna is near to 35 grams. The size of realized omni-directional antenna with radome is 20.5 mm (height) × 52 mm (dia.).

The VSWR measurement of the developed antenna has been carried out on Agilent vector network analyzer. The antenna has been measured at antenna test range facility. The comparison of simulated and measured VSWR and radiation pattern at 15 GHz are shown in Figs. 10 and 11. The measured VSWR is less than 1.54 over the complete frequency band. The measured peak gain of the antenna is 3.8 dBi at 15 GHz, and elevation coverage is better than 60°. The antenna shows a less variation in radiation pattern (i.e. ±0.6 dB) over the desired beam coverage.

Fig. 10 Measured VSWR of antenna

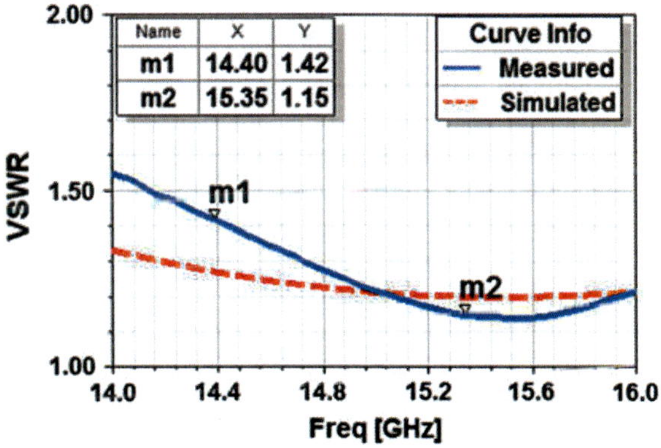

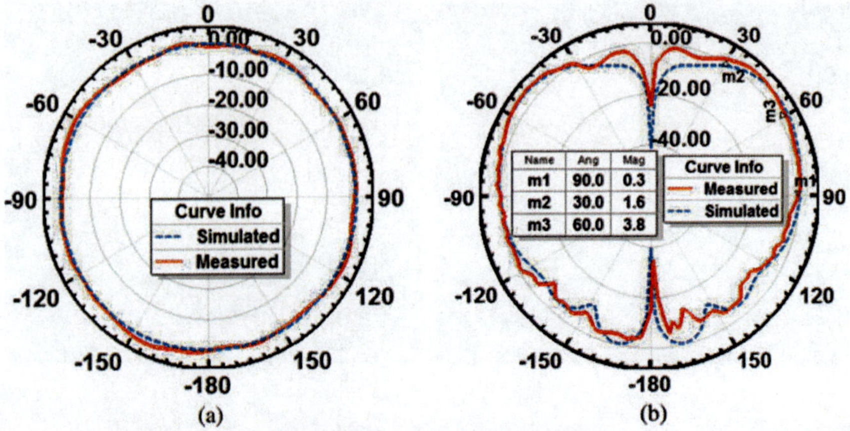

Fig. 11 Measured radiation pattern of antenna **a** azimuth, **b** elevation

4 Conclusion

Design, analysis, and realization of a low profile, light weight, linearly polarized, Ku-band omni-directional antenna for airborne LOS data link has been carried out. Measured results show very good resemblance with simulated values. The measured VSWR of the antenna with side screwed radome is less than 1.54 over the frequency band 14–16 GHz. Elevation coverage better than 60° (from horizon) has been achieved over the operating frequency band of the antenna. The antenna has a stable radiation pattern over the frequency band. The antenna is most suitable for LOS data link as well as for fast-moving airborne platforms.

Acknowledgements The authors would like to thank Director DEAL for giving permission to publish this paper.

References

1. Verma PK, Kumar R, Singh M (2014) C-band Omni-directional antenna for Stealth application. In: Proceeding of international symposium on antenna & propagation APSYM2014, Coachi University of Science & Technology, India, pp 255–258, 17–19 Dec 2014
2. Verma PK, Kumar R, Singh M (2016) Wide band waveguide fed linearly polarized MM Wave omnidirectional antenna. In: International conference on emerging trends in communication technologies, Dehradun, India, 18–19 Nov 2016
3. Jonson RC (1993) Antenna engineering handbook, 3rd edn. Mc Graw-Hill, New York, pp 37(7)–37(11)
4. Poljak D, Doric V (2006) "Wire antenna model for transient analysis of simple grounding systems. Part I: The vertical grounding electrode. Prog Electromagnet Res 64:149–166
5. Zhong L, Qiu J, Zhang N (2008) Study of a circular disc monopole ultrawide-band miniature antenna. PIERS Online 4(3):326–330

6. Ghosh D, Sarkar TK, Mokole EL (2009) a spherically-capped discone antenna for ultra-wideband operation. Prog Electromagnet Res B 16:229–245
7. Liu Y-T, Su C-W (2008) Wideband omnidirectional operation monopole antenna. Prog Electromagnet Res B 1:255–261
8. Taniguchi T, Maeda A, Kobayashi T (2006) "Development of an omnidirectional and low-VSWR ultra wideband antenna. Int. J Wirel Opt Comms 3(2):145–157
9. Evans J-A, Ammann M-J (1999) Planar trapezoidal and pentagonal monopoles with impedance bandwidths in excess of 10:1. In: Proceedings of the IEEE Antennas and Propagation Society international symposium, Orlando, FL, pp 1558–1561, July 1999
10. Schantz H (2005) Ultra wide band antennas. Artech House, Boston/London

Effect of Structural Metal on Metamaterial-Based Absorber Performance

Sajal Agarwal and Y. K. Prajapati

Abstract This study is done to analyze the effect of structural metal of metamaterial on the performance of the absorber. Absorbers are important part of various applications because of the increased demand of radiation absorption. This study is aimed to study the effect of various metals and compared based on their absorption performance. It is observed that bismuth provides the best absorption characteristics among all the considered metals.

Keywords Metamaterial · Absorber · Bismuth · Helix

1 Introduction

In last decade, optical devices are very well used in many interdisciplinary applications like sensors, photography, microscope, etc., because of their outstanding performance. Evolution of nano-optics is done due to increasing demand of smaller, faster, sensitive, and highly integrated optical and electronic devices for all applications [1]. The prime concern of researchers is to improve the performance of the optical devices for better efficiency and performance with small size and area.

Absorber is very crucial part of most of the devices, as this optical device is used to absorb all the radiation according to its operating region. Absorbers are widely used in solar cells to improve the total captured spectrum and improves the efficiency [2]. Therefore, it is highly required to design an electromagnetic (EM) absorber which can be used to absorb different frequencies without disturbing the useful frequencies [3].

It is researched that for the absorber modeling, a number of materials can be used; however, as the device gets thinner (nano regime), it becomes a huge problem

S. Agarwal (✉)
Jaypee Institute of Information Technology, Noida, India
e-mail: sajal.engg@gmail.com

Y. K. Prajapati
Motilal Nehru National Institute of Technology Allahabad, Prayagraj, Uttar Pradesh, India
e-mail: yogendrapra@mnnit.ac.in

D. Harvey et al. (eds.), *Advances in VLSI, Communication, and Signal Processing*,
Lecture Notes in Electrical Engineering 683,
https://doi.org/10.1007/978-981-15-6840-4_31

to get good enough absorption through the same material. Metamaterial is one of advanced materials attracts great attention because of its extraordinary properties like high absorption, negative refractive index [4, 5]. First metamaterial-based absorber is proposed by N. I. Landy et al. in 2008 [6]. This simulation as well as experimental study used finite difference time domain (FDTD)-based modeling for analysis. Split ring resonator geometry is used with multiple layers of elements. After a number of studies, it is observed that constituent material greatly affects the metamaterial absorber performance [7–10]. Then, in 2009, J. K. Gansel et al. [11] proposed helix-based metamaterial structure which showed polarization insensitive nature. In 2015, S. Agarwal et al. [12] proposed a helix-based absorber for the absorbing application having aluminum (Al) as the constituent metal and displayed that helix metamaterial has wide operating range but very complex geometry. Although, it is observed that there are fluctuations in absorption spectrum which reduce the overall absorption. Recently, A. Ghobadi et al. [13] proposed a bismuth-based metamaterial absorber which displayed very promising results but the operating range is limited. After that, a number of studies have been published based on the bismuth-based metamaterial for various applications [14, 15]. Thus, based on the previously done research, it is observed that the efficiency of the metamaterial-based absorbers can be improved with proper design and material. Thus, in this study, single helix metamaterial is designed with different metals as the comparative study.

2 Design and Simulation

Proposed study consists of helix-based metamaterial over glass substrate. For the modeling, 20 nm diameter helix wire is used. Separation between the opposite-handed helices is 150 nm. Radius of helices is 50 nm. Since, FDTD method is used for the modeling of metamaterial absorber, Lumerical FDTD software is used. Metals are chosen to be modeled using Lorentz--Drude model. *p*-polarized plane wave light source is used with the wavelength ranging from 200 to 600 nm. Figure 1 has the modeled structure having two opposite-handed helices to ensure the polarization independent nature [12]. For the analysis, various metals have been used and their performance is compared in terms of absorption, peak absorption. Systematic study is done by first observing the absorption of the considered metals. Following section has the observed results.

3 Results and Discussion

From the initial study of thin-film absorption spectrum of different metals, it is observed that various metals displayed different operating regions and absorbance. Table 1 has the peak absorbance value and the wavelength for different materials. From the above data, it is clear that the absorbance of the thin film varies with

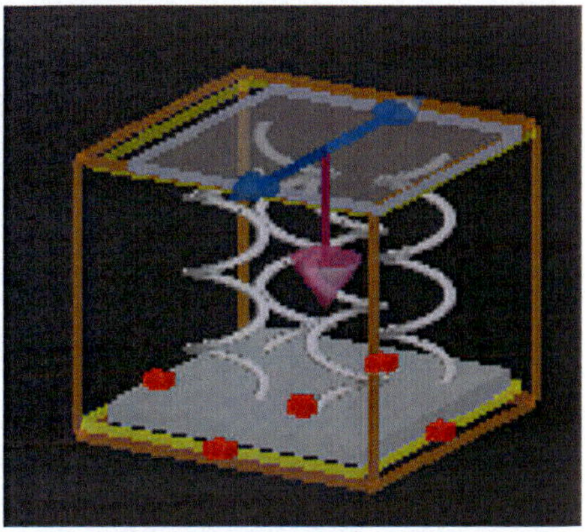

Fig. 1 Modeled helical metamaterial absorber

Table 1 Peak absorbance and wavelength

Metal	Peak absorbance (a.u.)	Wavelength (nm)
Gold (Au)	0.92	420
Tungsten (W)	0.98	380
Bismuth (Bi)	0.98	450
Aluminum (Al)	0.90	378
Tin (Sn)	0.91	514

the material. From all the considered metals, W and Bi displayed the best peak absorbance near unity. However, for the detailed study of the constituent metal effect, all the metals are considered for the helix formation in metamaterial structure. Figure 2 has the absorbance graph for different metals of helix. It is observed from the above study that for different constituent metals of helix, there is no large difference in the wavelength of peak absorption. This is because of the helical geometry of the metamaterial. As the wavelength depends on the diameter of helix and the distance between the helices.

Since, in this study, parameters of the metamaterial are not varied, peak absorbance is near by for all the metals. However, it is seen that because of the high imaginary part of the refractive index of Al, W and Bi, these metals display high peak absorbance of 0.78, 0.84, and 0.99 a.u., respectively. This is the initial study to explore the effect of Bi-based helix metamaterial absorber, and this absorbance can further be optimized through intertwining of the helices.

4 Conclusion

It is observed that the thin film of different metals provides different peak absorbance at different visible wavelengths; however, metamaterial absorber gives similar type

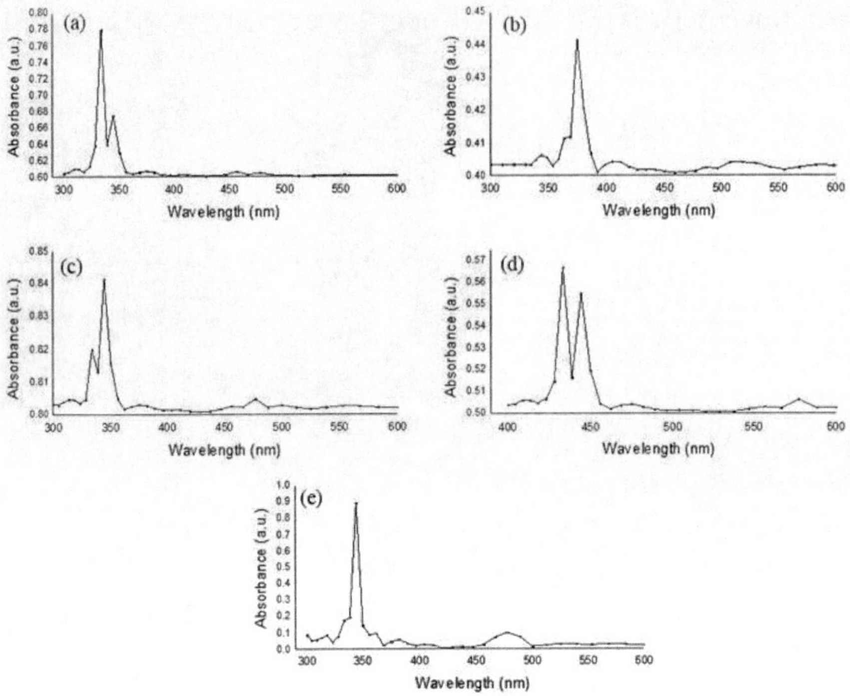

Fig. 2 Absorbance curve for different helix metals, **a** Al, **b** Au, **c** W, **d** Sn, **e** Bi

of nature of absorbance curve. It is also observed from this study that the constituent metal greatly affects the absorber performance in term of peak absorbance and Bi provides the best absorbance for the narrow wavelength region ranging from 338 to 350 nm.

References

1. Laszlo BK (2002) End of moores law: thermal (noise) death of integration in micro and nano electronics. Phys Lett A 305(3):144149
2. Chen HT, Padilla WJ, Zide JM, Gossard AC, Taylor AJ, Averitt RD (2006) Active terahertz metamaterial devices. Nature 444(7119):597–600
3. Agarwal S, Prajapati YK (2017) Analysis of metamaterial-based absorber for thermo-photovoltaic cell applications. IET Optoelectron 11(5):208–212
4. Yang J, Gong C, Sun L, Chen P, Lin L, Liu W (2016) Tunable reflecting terahertz filter based on chirped metamaterial structure. Sci Rep 6
5. Cai W, Shalaev V (2009) Optical metamaterials: fundamentals and applications. Springer Science & Business Media
6. Landy NI, Sajuyigbe S, Mock JJ, Smith DR, Padilla WJ (2008) Perfect metamaterial absorber. Phys Rev Lett 100(20):207402

7. Wu C, Neuner B III, John J, Milder A, Zollars B, Savoy S, Shvets G (2012) Metamaterial-based integrated plasmonic absorber/emitter for solar thermo-photovoltaic systems. J Opt 14(2):024005
8. Shchegolkov DY, Azad AK, Ohara JF, Simakov EI (2010) Perfect subwavelength fishnetlike metamaterial-based film terahertz absorbers. Phys Rev B 82(20):205117
9. Andryieuski A, Lavrinenko AV (2013) Graphene metamaterials based tunable terahertz absorber: effective surface conductivity approach. Opt Express 21(7):9144–9155
10. Slobozhanyuk AP, Lapine M, Powell DA, Shadrivov IV, Kivshar YS, McPhedran RC, Belov PA (2013) Flexible helices for nonlinear metamaterials. Adv Mater 25(25):3409–3412
11. Gansel JK, Thiel M, Rill MS, Decker M, Bade K, Saile V, Freymann G, Linden S, Wegener M (2009) Gold helix photonic metamaterial as broadband circular polarizer. Science 325(5947):1513–1515
12. Agarwal S, Prajapati YK, Singh V, Saini JP (2015) Polarization independent broadband metamaterial absorber based on tapered helical structure. Opt Commun 356:565–570
13. Ghobadi A, Hajian H, Gokbayrak M, Butun B, Ozbay E (2019) Bismuth-based metamaterials: from narrowband reflective color filter to extremely broadband near perfect absorber. Nanophotonics 8(5):823–832
14. Sadecka K, Toudert J, Surma HB, Pawlak DA (2015) Temperature and atmosphere tunability of the nanoplasmonic resonance of a volumetric eutectic-based Bi_2O_3-Ag metamaterial. Opt Express 23(15):19098–19111
15. Cheng XS, Ahmad H, Harun SW (2010) Broadband ASE source using bismuth-based erbium doped fibers in double-pass setup. Microw Opt Technol Lett 52(7):1636–1638

Effect of Various Parameter Variations on Electrical Characteristics of Rectangular Gate All Around Junctionless Transistor

Manish Kumar Rai, Sanjeev Rai, and Abhinav Gupta

Abstract Junctionless transistors in the present day scenario are the trend which the engineers are focusing onto. These devices are not only small scale but also devoid of the effect of junctions. This research paper presents a detailed analysis of the effect of various parameters like fin width, fin height, gate work function, and oxide thickness. The analysis done in this research paper is quite useful to control the electrical behavior or characteristics of the device by varying the device parameters. Increase in channel width increases the I_{on}, and on the other hand, high work function of gate material provides proper gate control over the channel and reduces leakage.

Keywords Rectangular Gate All Around (ReGAA) · Junctionless transistor (JLT) · Fin · Short-channel effect (SCE)

1 Introduction

Multigate transistors without junctions are used for the device scaling beyond 20 nm node since they provide better control of the channel charges by the gate. Due to its junction-free nature, it has advantages such as low thermal budget, improved short-channel effect, and comparatively simple fabrication process [1]. Junctionless transistors are basically accumulation-mode devices with a very small silicon thickness (~5 to 10 nm). The requirement of thin silicon layer in the channel region is to have complete depletion of carriers to turn off the device. Therefore, JLT offers good subthreshold characteristics. However, a typical accumulation-mode device

M. K. Rai (✉) · S. Rai
Motilal Nehru National Institute of Technology, Allahabad, Prayagraj, India
e-mail: mrai@mnnit.ac.in

S. Rai
e-mail: srai@mnnit.ac.in

A. Gupta
Rajkiya Engineering College, Sonbhadra, Churk, India
e-mail: abhinavkit87@gmail.com

D. Harvey et al. (eds.), *Advances in VLSI, Communication, and Signal Processing*,
Lecture Notes in Electrical Engineering 683,
https://doi.org/10.1007/978-981-15-6840-4_32

is made of relatively thick silicon films (typically higher than 20 nm or so) and hence, it exhibits worst short-channel performances. However, one advantage of the accumulation-mode transistor is that the drain current varies less with channel doping concentration, because in JLT, gradient of the doping concentration between source and channel or drain and channel is zero and no diffusion can take place. The other major difference of JLT with accumulation-mode transistor is that in the former, accumulation of carriers takes place at relatively high threshold voltage than later. The junctionless transistor which is also called "gated resistor" or "nanowire pinch-off FET" is highly doped (typically ~8 × 10^{18} cm^{-3} to 8 × 10^{19} cm^{-3}) to have an acceptable drain current. High doping concentration is required in JLT to increase the ON current for its satisfactory operation [2]. High doping concentration in channel region results in variation in threshold voltage [3]. Uniform doping in channel region is technologically difficult to obtain because after thermal annealing, doping profile becomes nonuniform. This non-uniform doping profile reduces ON current [4]. To further improve the ON current, negative capacitance technique using ferroelectric material is used in latest junctionless transistors [5]. Leakage current can be reduced by reducing the cross-sectional area of the channel region [6]. JLTs have reduced effect of drain-induced barrier lowering (DIBL) and better subthreshold slope (SS) resulting in better scalability, less sensitivity to doping fluctuations, greatly simplified process flow, and low thermal budget [7]. JLTs have flexibility in the choice of materials for gate metal and dielectric which make it suitable for wide range of applications. Because of uniform and homogeneous doping in the channel region, a JLT device eliminates the millisecond annealing process and thus can be fabricated with reduced channel lengths [3]. In addition, JLTs relatively consume lesser standby power whereas gate-induced drain leakage (GIDL) is a negligible component. JLTs have lesser fabrication steps which reduce process cost significantly as compared to junction-based devices with similar dimensions. JLTs exhibit lesser 1/f noise and random telegraph-noise. The vertical electric field in a JLT is much lower as compared to junction-based MOSFETs and accumulation-mode devices. Therefore, mobility in a JLT is not reduced much because of vertical electric field [8]. Due to reverse-biased drain junction, conventional junction-based device is normally an OFF device; thus, it prevents current flow through the device. To turn the device on, an inverted channel is created by applying a gate bias. However, JLT is normally an ON-device, where the work function difference between the gate electrode and silicon nanowire (~1.1 eV) shifts the flatband voltage and turns the threshold voltage into a positive value [9–11]. In the ON-state, the device is in flatband condition. Therefore, there is zero vertical electric field. GAA transistors are very useful to fully deplete the transistor and provide proper control to the gate [12]. Section 2 of this paper explains the device structure and simulation tool. Simulation results are presented in Sect. 3 followed by conclusion in last section.

2 Device Structure and Simulation Setup

Rectangular gate all around junctionless transistor is used for the simulation and analysis in presented work. This device provides better electrostatic control over channel resulting in reduced short-channel effects. The dimension of the device is shown in Table 1. Fin thickness is kept small for better electrostatic control over the channel (Fig. 1).

Device is simulated using SILVACO ATLAS software by including drift-diffusion, CVT, and SRH models for parameter extraction. Simulation model MOS PRINT is used which enables the CVT, SRH, and FERMIDIRAC models. Numerical techniques GUMMEL and NEWTON are included to solve the iterations.

Table 1 Dimension of ReGAA JLT

Parameter	Value
Channel length	20 nm
Fin width	15 nm
Fin thickness	7 nm
Device doping conc.	1×10^{19} atoms cm^{-3}
Oxide thickness	2 nm
Source/drain work function	4.2 eV
Gate electrode work function	4.8 eV

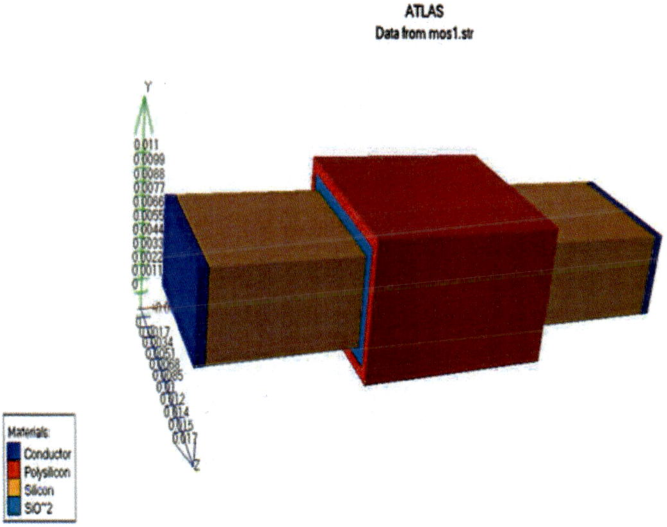

Fig. 1 ReGAA Junctionless transistor

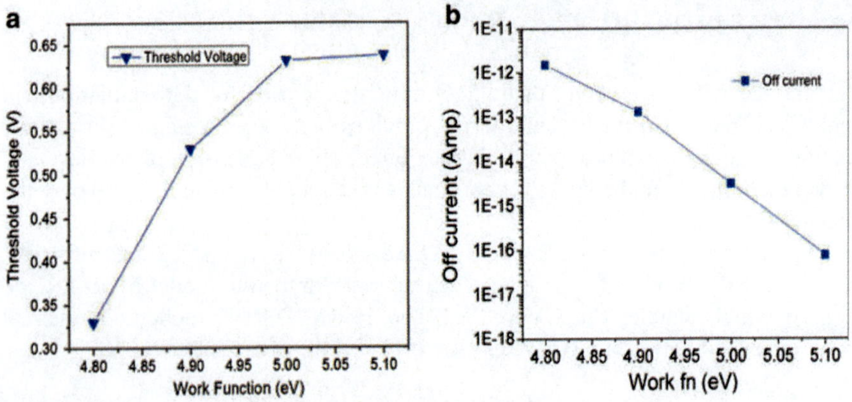

Fig. 2 Variation of **a** threshold voltage and **b** OFF current with gate work function

3 Effect of Parameter Variation

This section presents the effect of variation in gate material work function, device fin width, fin height, and oxide thickness on the performance of junctionless transistor.

3.1 Variation in Gate Work Function

Junctionless transistor works due to work function difference in the off state. As we increase the work function of gate material, the charge carriers are depleted from the channel and results in reduction of the OFF current and increase in threshold voltage. From simulation results, we find that workfunction of gate can be varied from 4.2 to 5.1 eV (Fig. 2).

3.2 Variation in Fin Width

Fin width of the device has direct effect on the ON current because increase in width of the device reduces the channel resistance and increases the carriers due to increased area. This reduces the control of gate due to increased area and hence reduces the device threshold voltage. Along with increase in device ON current (I_{on}), OFF current also increases due to reduced gate control over channel (Figs. 3 and 4).

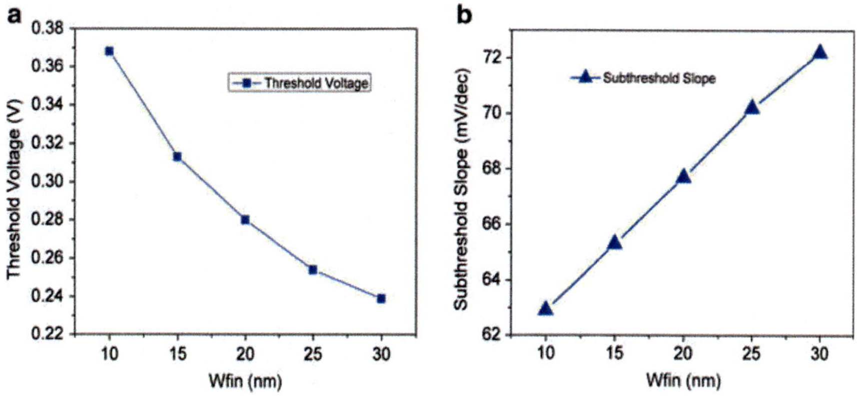

Fig. 3 Variation of **a** threshold voltage and **b** subthreshold slope with fin width

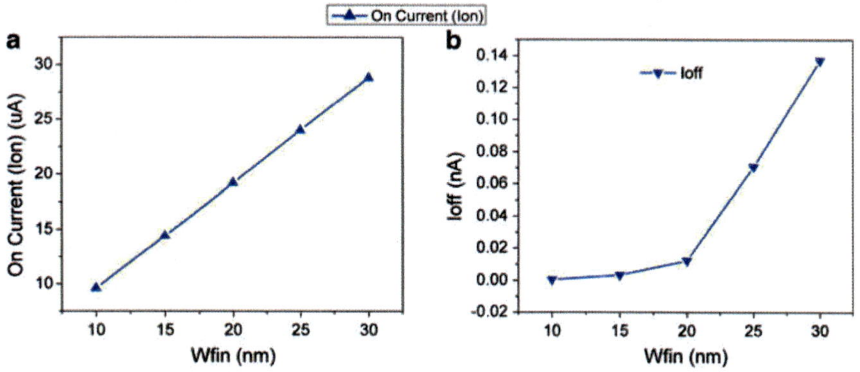

Fig. 4 Variation of **a** ON current and **b** OFF current with fin width of the JLT

3.3 Fin Height Variation

In junctionless transistor, there is similar type of doping in the source, drain, and channel region; therefore, this device operates in accumulation region. In this condition for $V_{gs} = 0$, charge carriers are depleted from the channel by difference in work function between Si and gate material. The effect of work function difference removes charge carriers effectively only when Si thickness is small. By increase in fin height, this thickness increases and gate control reduces which results in reduced threshold voltage and increased subthreshold swing and OFF current (Figs. 5 and 6).

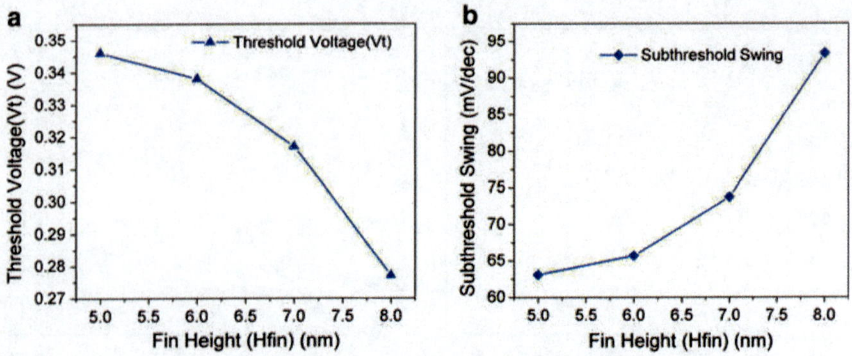

Fig. 5 Variation of **a** threshold voltage and **b** subthreshold swing with fin height

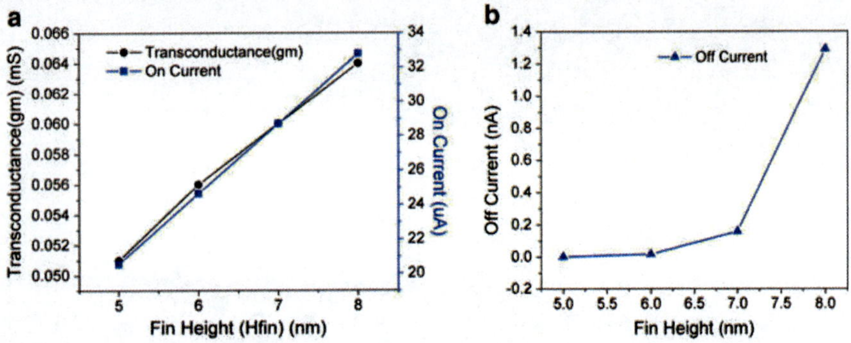

Fig. 6 Variation of **a** transconductance, ON current and **b** OFF current with fin height

3.4 Variation in Thickness of Gate Oxide

The oxide thickness has significant effect on the device behavior. Thin oxide increases the gate control over channel but it also increases the leakage current through tunneling. With increase in oxide thickness, the control of gate over channel reduces and this results in reduction in threshold voltage and increase in subthreshold slope. With increase in oxide thickness, although the leakage current through tunneling of gate oxide reduces, the leakage current between source and drain increases. Leakage through gate oxide tunneling is negligible in comparison to the leakage between source and drain resulting in overall increase in OFF current (Fig. 7).

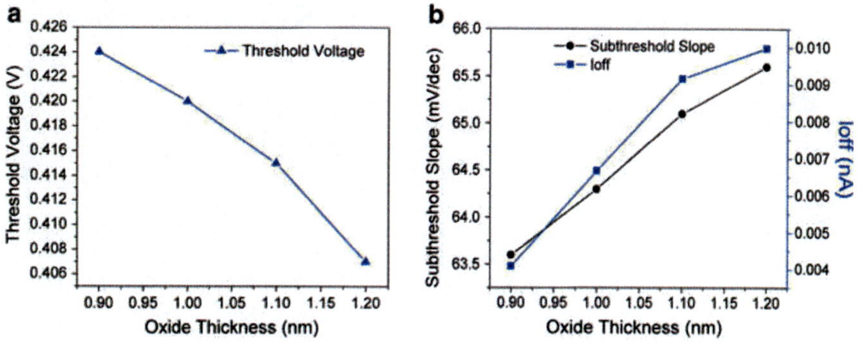

Fig. 7 Variation of **a** threshold voltage, **b** subthreshold slope and OFF current with oxide thickness

4 Conclusion

The research work presented here has a detailed analysis of various effects on GAA obtained as a variation in device parameter. These results are very helpful in analyzing a new device to improve the performance. To minimize various short-channel effects, device parameters are adjusted to their optimized values. By analyzing the simulation results obtained, we find that work function of the gate material between 4.6 and 5.2 eV gives increased threshold voltage and reduced leakage current. Fin height between 5 and 7 nm and oxide thickness between 1 and 2 nm results in increased I_{on} and reduced I_{off}. Simulation results obtained in this paper can be used to set the dimensions of the device and other parameters to improve the performance and increase reliability of semiconductor devices.

References

1. Kranti A, Lee CW, Ferain I, Yu R, Akhavan ND, Razavi P, Colinge JP (2010) Junctionless nanowire transistor: properties and design guide lines. In: Proceedings of IEEE 34th European solid-state device research conference, pp 357–360
2. Colinge JP, Lee CW, Afzalian A, Akhavan ND, Yan R, Ferain I, Razavi P, Neill B, Blake A, White M, Kelleher AM, McCarthy B, Murphy R (2010) Nanowire transistors without junctions. Nat Nanotechnol 5(3):225–229
3. Sahu C, Singh J (2014) Charge-plasma based process variation immune junctionless transistor. IEEE Electron Device Lett 35(3)
4. Mondal P, Ghosh B, Bal P, Akram MW, Salimath A (2015) Effects of non-uniform doping on junctionless transistor. Appl Phys A 119:127–132
5. Jiang C, Liang R, Wang J, Xu J (2016) Simulation-based study of negative capacitance double-gate junctionless transistors with ferroelectric gate dielectric. Solid-State Electron 126:130–135
6. Sahay S, Kumar MJ (2017) Diameter dependence of leakage current in nanowire junctionless field effect transistors. IEEE Trans Electron Devices 64(3)
7. Ehteshamuddin M, Loan SA, Member, IEEE, Rafat M (2018) Planar Junctionless silicon-on-insulator transistor with buried metal layer. IEEE Electron Device Lett 39(6)

8. Gundapaneni S, Ganguly S, Kottantharayil A (2011) Bulk planar junctionless transistor (BPJLT): an attractive device alternative for scaling. IEEE Trans Electron Devices 32(3):261–263
9. Koukab A, Jazaeri F, Sallese JM (2013) On performance scaling and speed of junctionless transistors. Solid-State Electron 79:18–21
10. Zakir Hossain M, Khosru QDM (2013) Threshold voltage roll-off due to channel length reduction for a nanoscale nchannel FinFET. Int J Emerg Technol Comput Appl Sci (IJETCAS) 13–125
11. Choi SJ, Moon DI, Kim S, Duarte J, Choi YK (2011) Sensitivity of threshold voltage to nanowire width variation in junctionless transistors. IEEE Electron Device Lett 32(2):125–127
12. Su CJ, Tsai TI, Liou YL, Lin ZM, Lin HC, Chao TS (2011) Gate-all-around junctionless transistors with heavily doped polysilicon nanowire channels. Electron Device Lett 32(4):521–523

Slotted I-Patch with Capacitive Probe Fed Microstrip Antenna for Wideband Applications

Santosh Kumar Gupta and Sangaraju Varun

Abstract The design and analysis of a capacitive fed slotted I-patch microstrip antenna are presented and compared with reported coplanar capacitive fed different slotted patch microstrip antennas hanging above ground plane. It has been shown that the I-patch antenna provides good impedance bandwidth and gain for various microwave bands. The proposed I-slot is fed with a co-axial probe using capacitive feed strip which is not attached to the radiating patch directly. In the present work, a rectangular patch from truncated microstrip transmission line has been used for capacitive feeding. The I-patch antenna configuration can be used in applications where large bandwidth is required such as wireless local area network (WLAN).

Keywords Slotted patch · Microstrip antenna · Capacitive fed · Wideband

1 Introduction

The microstrip antennas best suit for modern broadband applications due to their desirable characteristics possessed [1–3]. Microstrip antenna in its fundamental form shows limited bandwidth (BW) which can be improved by incorporating certain modifications. Also, several researches have shown that BW of the microstrip antenna may be improved by changing the feed type or utilizing impedance matching techniques [2]. Most of the techniques used for improving the bandwidth are by using a stacked metal, dielectric layers [4], and modified probe shapes like T-, L-, meander shaped, etc., which avoid the primary advantage of microstrip antenna such as easy fabrication and assembly [5].

A capacitive fed microstrip antenna using single-layer coplanar structure has been reported by Mayhew-Ridgers et al. [3]. The antenna reported in [3] has a bandwidth of 28%. This can be enhanced by elaborating feed strip dimensions and its placement w.r.t. the radiator patch [5]. Radiation pattern of such an antenna can be improved up

S. K. Gupta (✉) · S. Varun
ECED, Motilal Nehru National Institute of Technology Allahabad, Prayagraj, Uttar Pradesh 211004, India
e-mail: skg@mnnit.ac.in

D. Harvey et al. (eds.), *Advances in VLSI, Communication, and Signal Processing*, Lecture Notes in Electrical Engineering 683,
https://doi.org/10.1007/978-981-15-6840-4_33

to 50% by modifying the radiator patch edge closer to feed patch [6]. The maximum bandwidth of this antenna can be achieved by placing the radiator patch $0.16\lambda_0$ [5] over the ground plane. This antenna provides a unidirectional radiation pattern in the above geometries due to ground plane. The capacitive feed strip placement alongside the radiating edges of patch counteracts for probe inductance [7]. The design, optimization of the coplanar capacitive fed wideband microstrip antenna is presented in this paper.

2 Antenna Architecture

Figure 1 represents the architecture of proposed I-slot antenna. The normal-, cross-, diagonally cross-, I-slotted microstrip patches are presented and compared for compact antenna. This consists of a hanging microstrip antenna of which both radiating patch and feed strip are putted over substrate of '*h*' thickness. A longer pin SMA connector connects the feed strip capacitively coupling energy to radiating patch. The exhaustive parametric study is carried out to get optimized geometry of the antenna. The salient design parameters that affect antenna performance include

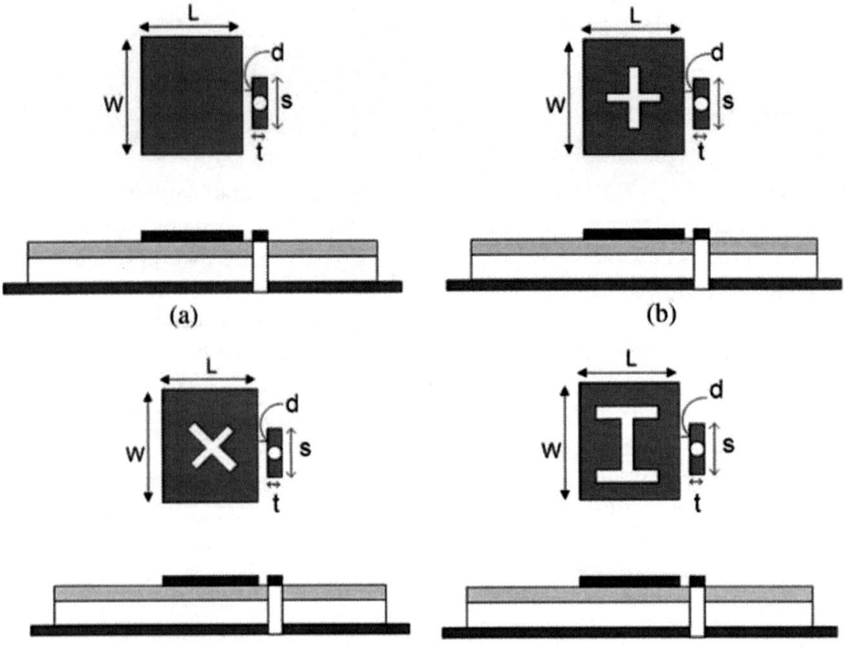

(a) (b)

(c) (d)

Fig. 1 Different Slotted Patch Microstrip Antennas: **a** normal patch, **b** cross-slotted patch, **c** diagonal cross-slotted patch, **d** I-slotted patch

Table 1 Antenna parameters for 5.9 GHz

Parameter	Value
Radiator patch length (L)	15.5 mm
Radiator patch width (W)	16.4 mm
Feed strip length (s)	3.7 mm
Feed strip width (t)	1.2 mm
Feed strip to patch distance (d)	0.5 mm
Air gap between substrates (g)	6 mm
Relative dielectric constant (ε_r)	3.0
Substrate thickness (h)	1.56 mm

the parting between radiator patch and feed strip, air gap, length and width of the feed strip. The RO3003 is used for substrate of antenna fabricated having dielectric constant = 3, thickness $h = 1.56$ mm, and loss tangent = 0.0013. The proposed antenna is simulated using HFSS.

The fundamental design of microstrip antenna is executed by choosing the center frequency of operating band. Here, 5.9 GHz is decided as the center frequency. The dimensions of the radiator patch are computed from standard design expressions after due corrections for the hanging $(g + h)$ dielectric, where g is height of the substrate above ground [1, 8]. Above corrections include the total height over ground and effective dielectric constant of hanging microstrip [9]. The impedance BW of the antenna can be maximized by the design expression

$$g \cong 0.16\lambda_0 - h\sqrt{\varepsilon r} \quad (1)$$

where ε_r is the dielectric constant of the substrate. However, this equation predicts only the initial value whereas optimum value may be within ±10% [5]. The dominating dielectric property remains largely unaffected for small variations in dielectric constant of hanging substrate and hence the bandwidth. But the change in the feed strip reactance and thickness of dielectric layer (h) could result in the shifting of center frequency (Table 1).

3 Comparison of Slotted Patches

The reported antennas along with proposed one have been implemented using HFSS Software with a 1.4 mm probe diameter. The resonant frequency corresponds to the peak negative value in the S_{11} characteristics. The S_{11} parameter plots of the normal patch, cross-shaped, diagonally cross-shaped, and I-shaped slotted microstrip patch antennas are shown in Fig. 2 and VSWR plots are represented by Fig. 3. The −10 dB bandwidth obtained for the normal patch, cross-shaped, diagonally cross-shaped, I-shaped slotted microstrip antennas is 2.6868, 2.8232, 2.8073, and 2.6949 GHz. From

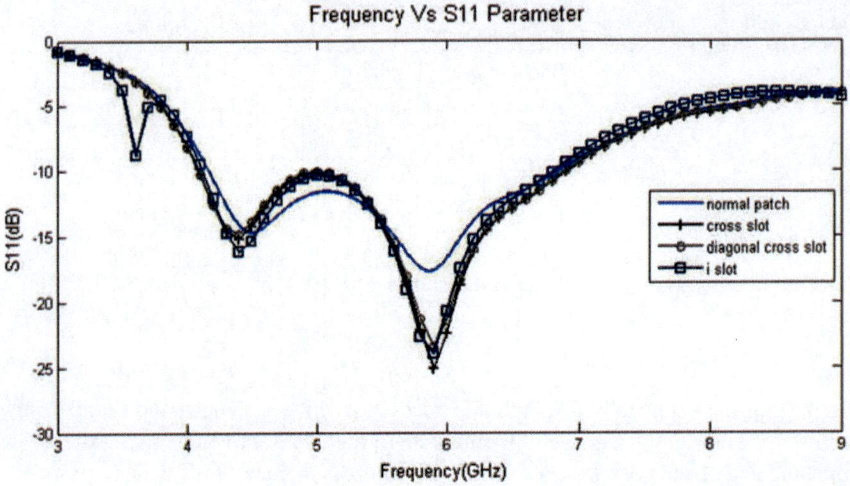

Fig. 2 S_{11} characteristics of different slotted patch microstrip antennas

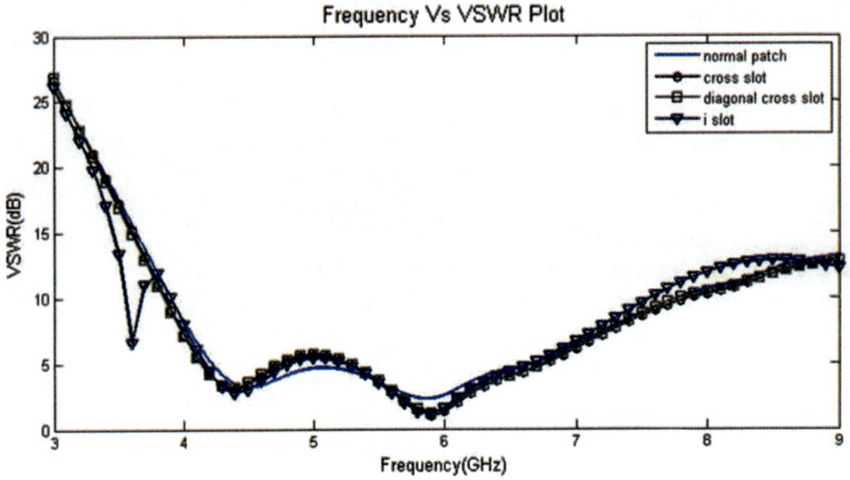

Fig. 3 VSWR plot of different slotted patch antennas

Fig. 2, the minimum return loss from the S_{11} characteristics is obtained at 5.9 GHz which is the required center frequency. The E-field co- and cross-polarization of the reported antennas is shown by Figs. 4 and 5, respectively.

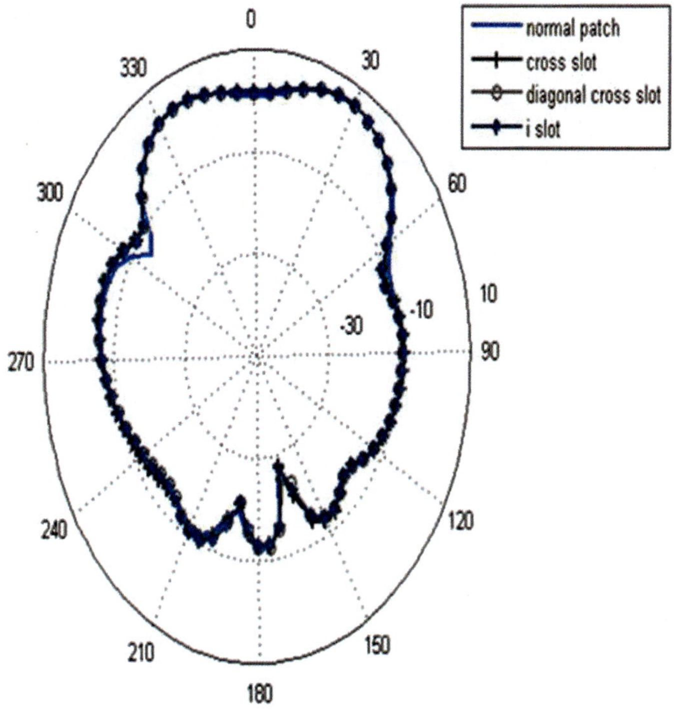

Fig. 4 E-field co-polarization of different antenna shapes

4 Parametric Study of I-Shaped Slot

The above results show that the I-slot provides the minimum cross-polarization levels. Since the I-slot area is more and the attachment area of the slot with the patch is more, the I-slot antenna has minimum cross-polarization levels. Here, the I-slot is enhanced by carrying parametric analysis in which length and width of slot are varied to get enhanced value of cross-polarization.

4.1 Analysis of Slot Length Variation

The impact of slot length variation on I-slot performance is carried by maintaining width to be fixed. Having fixed width of 4 mm, the slot length is varied from 8 to 12 mm with a step of 0.5 mm. The S_{11} parameter plot of frequency and the return loss for the change in slot length is shown in Fig. 6 and the VSWR plot is shown in Fig. 8. From Fig. 6, the bandwidth of the I-slot antenna is more when the length of the slot is equal to 10 mm. The bandwidth is increased from $l = 8$ mm to 10 mm and decreased from $l = 10$ to 12 mm. The maximum bandwidth achieved at $l = 10$ mm

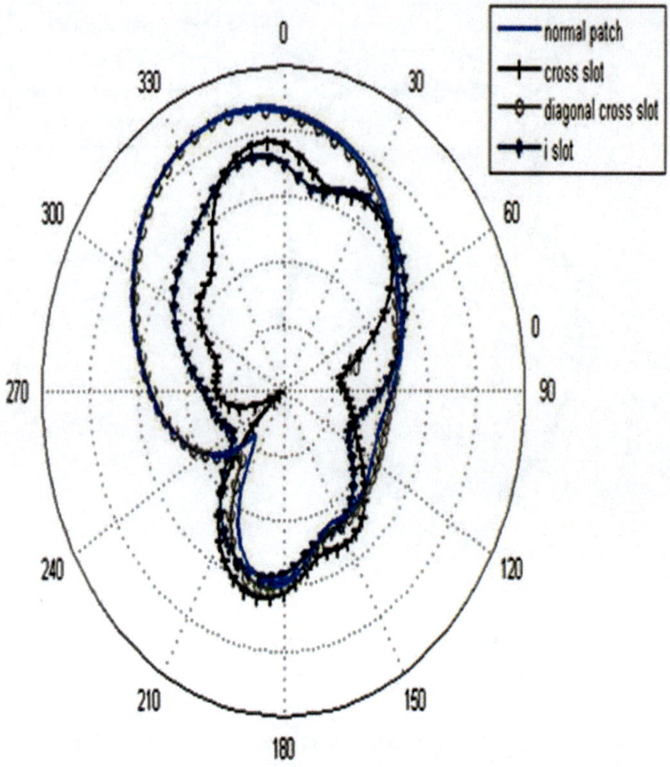

Fig. 5 E-field cross-polarization of different antenna shapes

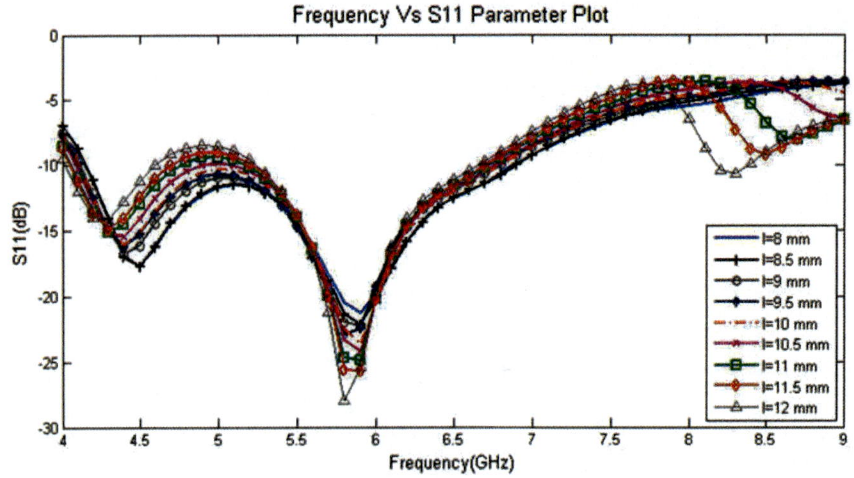

Fig. 6 S_{11} parameter characteristics with the variation in the slot length from $l = 8$ to 12 mm

is 2.6938 GHz. The variation of bandwidth with respect to slot length is presented by Fig. 10 (Figs. 7 and 9).

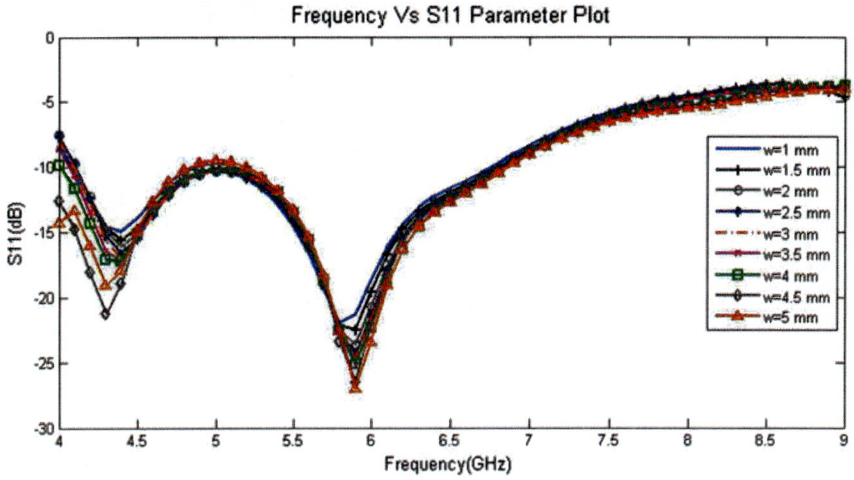

Fig. 7 S_{11} parameter characteristics with the variation in the slot width from $w = 1$ to 5 mm

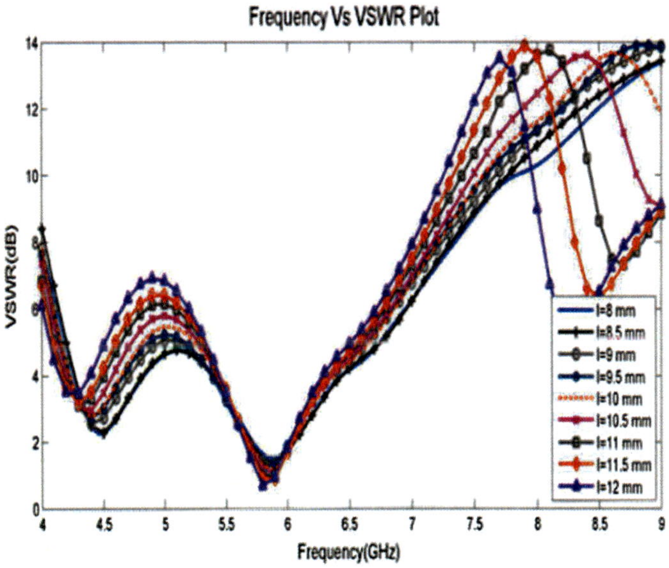

Fig. 8 VSWR plot with the variation in the slot length from $l = 8$ to 12 mm

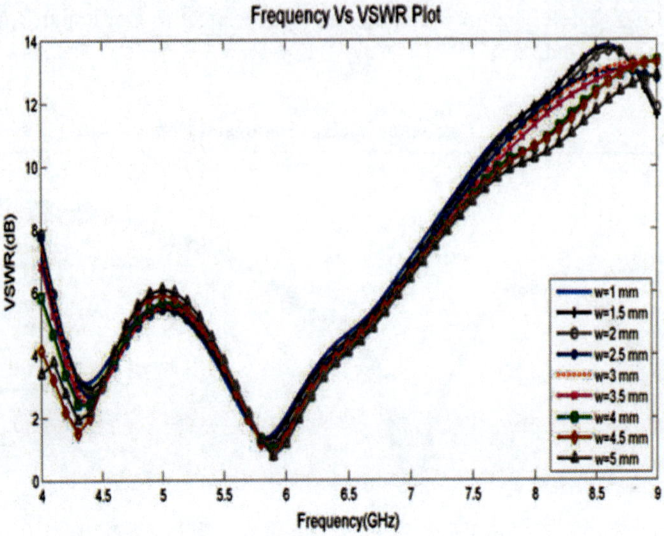

Fig. 9 VSWR plot with the variation in the slot width from $w = 1$ to 5 mm

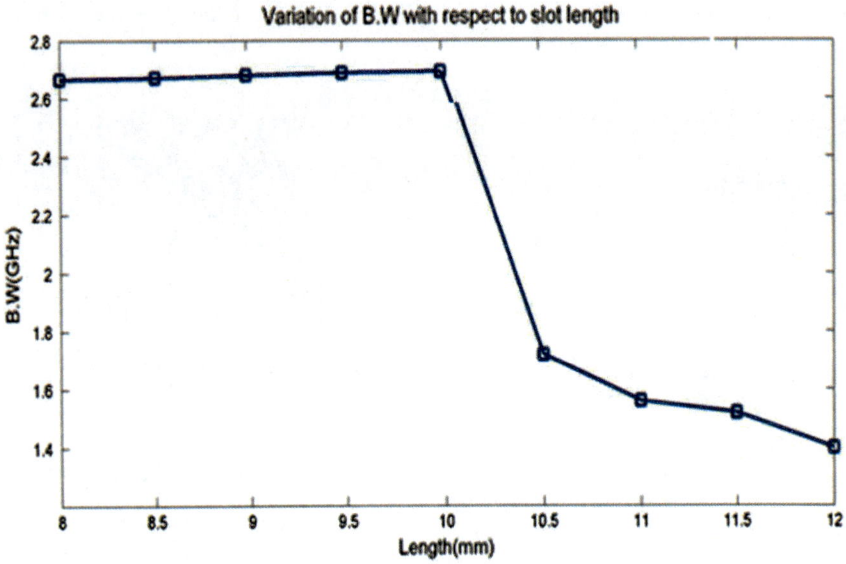

Fig. 10 B.W. versus slot length

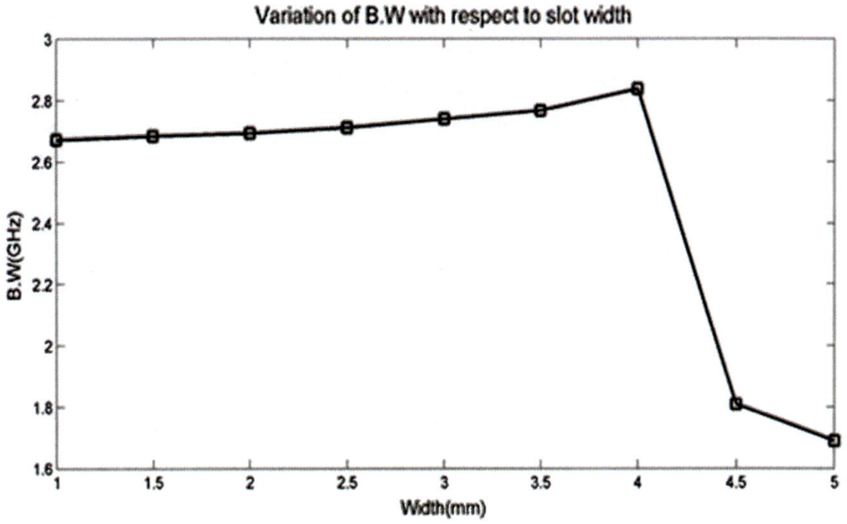

Fig. 11 B.W. versus slot width

4.2 Analysis of Slot Width Variation

The impact of slot width variation on I-slot performance is carried by maintaining the length to be fixed. With fixed slot length of 10 mm, the slot width is varied from 2 to 6 mm with a step of 0.5 mm. The S_{11} parameter plot of frequency and the return loss for the change in slot width is shown in Fig. 7 and the VSWR plot is shown in Fig. 9. From Fig. 7, the bandwidth of the I-slot antenna is more when the width of the slot is equal to 4 mm. The bandwidth is increased from w = 1 mm to 4 mm and decreased from $w = 4$ to 5 mm. The maximum bandwidth achieved at $w = 4$ mm is 2.8383 GHz. The variation in B.W with respect to slot width is shown in Fig. 11.

4.3 Optimization of I-Slot Dimensions

From the parametric study of the I-slot dimensions above and the results obtained, the bandwidth is more when the length and width of slot are 10 and 4 mm, respectively. From Fig. 12, the cross-polarization level is -13.0327 dB. The co- and cross-polarization levels at slot length and width of 10 mm and 4 mm, respectively, are shown in Fig. 12.

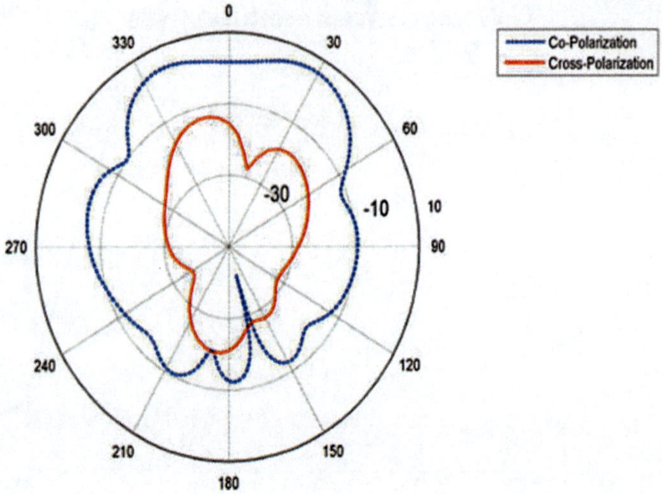

Fig. 12 E-field co- and cross-polarization of I-slot for $l = 10$ mm and $w = 4$ mm

5 Conclusions

A compact I-shaped slotted coplanar capacitive-coupled microstrip antenna for wideband application has been presented and compared with three different shapes: normal patch, cross-slot patch, and diagonal cross-slot patch. The proposed I-slot microstrip antenna gives the minimum cross-polarization levels of -13.0327 dB with a return loss of -24.5903 dB. It also provides the minimum return loss at 5.9 GHz center frequency.

References

1. Garg R, Bhartia P, Bahl I, Ittipiboon A (2001) Microstrip antenna design handbook. Artech House, Norwood
2. Kumar G, Ray KP (2003) Broadband microstrip antennas. Artech House, Norwood
3. Mayhew-Ridgers G, Odondaal JW, Joubert J (2003) Single-layer capacitive feed for wideband probe-fed microstrip antenna elements. IEEE Trans Antennas Propag 51(6):1405–1407
4. Kokotoff DM, Aberle JT, Waterhouse RB (1999) Rigorous analysis of probe fed printed annular ring antennas. IEEE Trans Antennas Propag 47(2):384–388
5. Kasabegoudar VG, Upadhyay DS, Vinoy KJ (2007) Design studies of ultra wideband microstrip antennas with a small capacitive feed. Int J Antennas Propag 2007:1–8
6. Kasabegoudar VG, Vinoy KJ (2008) A wideband microstrip antenna with symmetric radiation patterns. Microw Opt Technol Lett 50(8):1991–1995

7. Hall PS (1987) Probe compensation in thick microstrip patches. Electron Lett 23(11):606–607
8. Bahl IJ, Bhartia P (1980) Microstrip antennas. Artech House, Boston
9. Schellenberg JM (1995) CAD models for suspended and inverted microstrip. IEEE Trans Microw Theory Tech 43(6):1247–1252

Development of Cloud-Based Multi-Modal m-Cardiac Management System

Naveen Kumar, Arun Kumar, Prabal Gupta, Rohini Srivastava, R. P. Tewari, Nitin Sahai, and Basant Kumar

Abstract This paper presents a cloud-based remote cardiac health monitoring system using 10-lead ECG, oxygen saturation and heart rate sensors. In this paper, we propose an IoT-enabled m-cardiac care management system to monitor the patient's vital cardiac parameters, like- ECG, heart rate, and oxygen saturation through the medical sensor network. The proposed IoT-enabled m-cardiology system provides inter-facing between the patient and the medical expert via cloud server [1]. An android application has been developed to transmit the sensor data to the IoT-based cloud server. Traditionally, ECG signals are examined for the detection of any abnormalities in the functioning of the heart. However, in many medical conditions, ECG may not be fully conclusive. In cardiac patients, if the oxygen level is not sufficient, their heart will not be able to pump properly. Therefore, this paper proposes a multi-modal m-cardiac management platform involving acquisition and transmission of oxygen saturation and heart rate parameters in addition to 10-lead ECG data. The

N. Kumar · A. Kumar · P. Gupta
ECED, Indian Institute of Information Technology, Una, Una, Himachal Pradesh, India
e-mail: naveenkumar209202@gmail.com

A. Kumar
e-mail: arungupta7221@gmail.com

P. Gupta
e-mail: guptaprabal444@gmail.com

R. Srivastava (✉) · B. Kumar
ECED, Motilal Nehru National Institute of Technology, Allahabad, Prayagraj, Uttar Pradesh, India
e-mail: rohini@mnnit.ac.in

B. Kumar
e-mail: singhbasant@mnnit.ac.in

R. P. Tewari
AMD, Motilal Nehru National Institute of Technology, Allahabad, Prayagraj, Uttar Pradesh, India
e-mail: rptewari@mnnit.ac.in

N. Sahai
BMED, North East Hill University, Shillong, Meghalaya, India
e-mail: nsahai@nehu.ac.in

D. Harvey et al. (eds.), *Advances in VLSI, Communication, and Signal Processing*, Lecture Notes in Electrical Engineering 683,
https://doi.org/10.1007/978-981-15-6840-4_34

main purpose of this paper is to propose a multi-modal system to acquire data from different sensors and transmit it to the cloud server using the developed android app. The proposed cardiac management system can facilitate the remote cardiac patients in getting emergency healthcare services, while doctors may get benefited from the proposed system by observing their patients remotely, without visiting in person. Pulse oximeter sensor (MAX30100) and 10-lead ECG sensor (ADS1298) are used to monitor the patient's cardiac health. After successful implementation and prototyping of the proposed remote cardiac management system, its performance has been analyzed. It is concluded that the proposed multi-modal cardiac care management system is cost-effective and has the potential to provide improved diagnostic assessments with regard to remote cardiac care.

Keywords Cloud server · ECG · Oxygen saturation · MAX310100 · ADS1298

1 Introduction

In India, the death rate due to cardiovascular diseases has increased by 34% being the leading cause of death, according to a new international study [2]. Early diagnosis and prevention of heart diseases are very important in our daily life. Therefore, additional development and education of the research of heart diseases will be required. Cardiac activities of the patient can be easily monitored by Electrocardiogram (ECG) [3]. ECG can be measured using single-channel and multi-channel system. Multi-channel and long-time ECG monitoring is much better than the single-channel ECG system as it diagnoses the heart disease more accurately [4, 5]. This paper presents a low cost and long-time portable 10-lead ECG acquisition system using Raspberry PI. This system is composed of a commercial Analog Front End board (ADS1298) for ECG acquisition and an easily available Raspberry PI board. For proper measurement of cardiac activity, only ECG sensors will not be sufficient. There are some other parameters that may affect the proper functioning of the heart. One of the parameters is oxygen saturation, which can be measured using pulse oximeter and heart rate sensor (MAX30100). If the oxygen saturation level of the patient is not sufficient, the heart muscles will not be able to pump properly. Therefore, measurement of oxygen saturation is an inevitable process for the proper cardiac management system development. MAX310100 is a pulse oximeter and heart rate sensor and having the advantage of compatibility with Raspberry PI board [6]. Therefore, these oxygen saturation data can also be sent to the cloud server. Some earlier work has also been reported, regarding cloud-based healthcare system. Vidhaydhar et al. [7] have presented a healthcare system using two microcontrollers with temperature, pulse rate, and ECG sensor. Use of two microcontrollers increases its complexity. Chhabra et al. [8] have presented ECG acquisition and transmission to the cloud server using Raspberry PI. They have used 3-lead ECG sensor, which is very sensitive to noise and may lead to improper acquisition. To overcome the limitations of the reported work, we have proposed a multi-modal m-cardiac care management system. The rest of the paper is organized as follows: Sect. 2 proposes the complete methodology for m-cardiac care management system. Section 3 presents the proposed methodology for the acquisition of ECG and SpO2 sensor data. Section 4 presents the database management system for IoT board data along with the server data. Development of android application for multi-modal m-Cardiac management system is discussed in Sect. 5. Results and conclusions are discussed in Sects. 6 and 7, respectively.

2 Proposed Methodology: m-Cardiac Care Management System

In this section, the complete block diagram is mentioned for multi-modal cardiac care system. The data is obtained from the 10-lead ECG sensor and oxygen saturation sensor, then the data is sent to the cloud, using Raspberry PI. The Raspberry PI has the ability to easily upload the data to the cloud server. From the patient's end, the data is transmitted to the cloud server. For receiving the data at the doctor's end, an android application has been developed. The complete frame work of the proposed m-cardiology system is shown in Fig. 1.

The figure shows that the proposed system architecture in this paper leads to an idea of developing an accurate, portable, robust, compact, and low cost device for monitoring the complete cardiac activity.

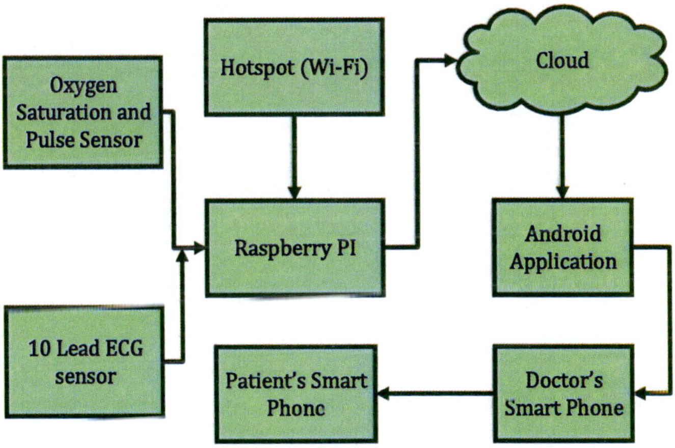

Fig. 1 Block diagram of IoT-enabled cloud-based m-cardiac management system

3 Acquisition of 10-Lead ECG Data and Pulse Oximeter Data

The proposed system is composed of the two low cost commercial boards, Analog Front End (AFE) and Raspberry PI. Analog Front End is a 10 lead, 8 channel ECG acquisition board for acquisition of an accurate ECG and Raspberry PI is a single board computer used for data processing and transmission on the cloud server. AFE board is a low-power, multi-channel, 24-bit delta-sigma analog-to-digital converters (ADCs) with integrated programmable gain amplifiers (PGAs) [9]. This device incorporates various ECG-specific functions that make it well-suited for scalable electrocardiogram (ECG) applications. This device is also used in high performance, multichannel data acquisition systems by powering down the ECG-specific circuitry. ECG data is acquired on the computer by using AFE board and the ECG data is converted into image format. ECG in image form is transmitted to the cloud using microcontrollers and transceiver modules. Oxygen saturation sensor is a compact system solution for measuring pulse oxygen saturation and heart rate. The oxygen saturation is measured using ambient light cancellation (ALC), 16-bit analog-to-digital converter, and discrete time filter. This discrete time filter is used in the sensor to reject power interference and low-frequency residual ambient noise. Two LEDs are used in this device; one emits red light and another emits infrared light. When the heart is contracting, the volume of oxygenated blood increases, whereas when the heart is relaxing, volume of oxygenated blood reduces. The pulse rate is determined by knowing the time between the increment and decrement in oxygenated blood volumes. The main function of oxygen saturation sensor, infrared light is absorbed by the oxygenated blood and red light is absorbed by deoxygenated blood. The absorption level for both the LEDs is stored in a buffer by this sensor. This buffer can be read on the Raspberry PI [10].

4 Cloud-Based Health Monitoring: m-Cardiac Care Management System

In the proposed work, data transmission to the cloud server is performed in two stages: In first stage, data is sent from the microcontroller to the cloud server and in second stage, from computer to the cloud server. Oxygen saturation data has been uploaded from Raspberry PI to the Google Cloud using Google Sheet Application Program Interface (API). For uploading the data, we have to create a service account which will create a Java script file for authentication. After the authentication, the Raspberry PI data is uploaded to the cloud. For ECG data, we have to transmit the data to the cloud server from computer by which AFE board is accessed. Figure 2 shows the experimental setup for IoT-enabled e-healthcare system. AFE is connected to laptop via USB and oxygen saturation sensor is connected to Raspberry PI.

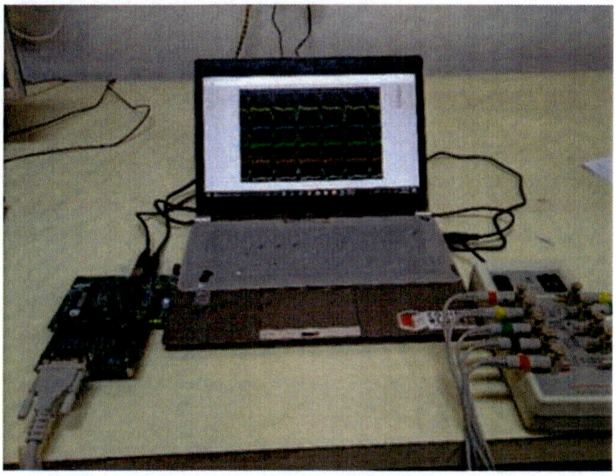

Fig. 2 IoT-enabled m-cardiac management system

This transmission is performed using Google Drive API [11]. When both the data is successfully uploaded to the cloud, an android application is designed to fetch the data from the cloud server to the smart phone of the patient as well as of the medical expert [12]. For ECG and heart rate (oxygen saturation), two android applications has been designed: Heart Rate Page and ECG Page. At the Heart Rate Page of the application, data is acquired using Google API key data of SpO_2 sensor and heart rate is fetched from Google Sheet. Similarly, at the ECG Plot Page, data is acquired using Google API key data of ECG plot fetched from Google Drive. The data is uploaded to cloud using Google API and then data are fetched to the android phone of the patient and the medical expert.

5 Development of Android Application for Multi-Modal m-Cardiac Management System

In this section, android application development for multi-modal m-Cardiac care management system is discussed. Sensors data: oxygen saturation sensor and ECG sensor, is uploaded on cloud server and an application for android mobile phone is designed using AppsGeyser: free App creator for Android. The developed app has three screens: Start Page, Heart Rate Page, and ECG Plot Page. The Heart Rate Page of the android application is accessed with cloud using Google Sheet API and the ECG Page is accessed with cloud using Google drive API. When the doctor receives data, the required diagnostic advice is sent to the patient. Figure 3 presents the working algorithm of the developed android application for multi-modal m-cardiac care management system.

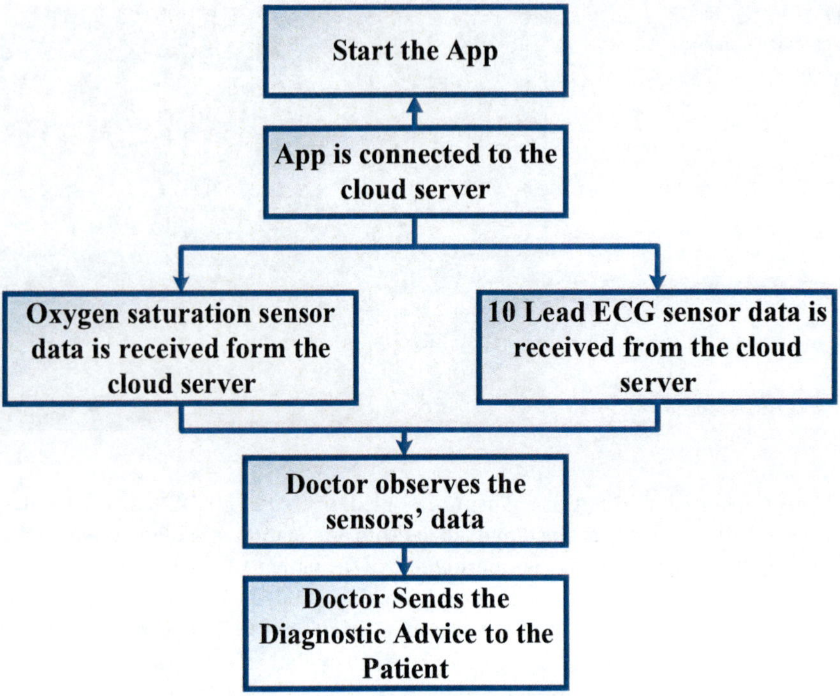

Fig. 3 Algorithm for developed android application of m-cardiac care management system

6 Results

In the proposed work, experiments were performed by acquiring 10-lead ECG data from ECG simulator and oxygen saturation & heart rate data is obtained by human test subjects. Analog Front End board gives the ECG data in an image form and oxygen saturation sensor provides oxygen saturation data along with the heart rate. Figure 4 shows images of Android Application Screens. ECG output from 10-lead ECG sensor is shown in Fig. 5.

Oxygen saturation sensor provides the saturation and heart rate data in numeric form at distinct time instants. The sensor's data is presented in Table 1. For different instants of time, oxygen saturation data along with heart rate is obtained and it is graphical represented is shown in Fig. 6.

7 Conclusion

This paper presented development of a multi-modal management system for cardiac care management. In the proposed system, AFE board and oxygen saturation sensor

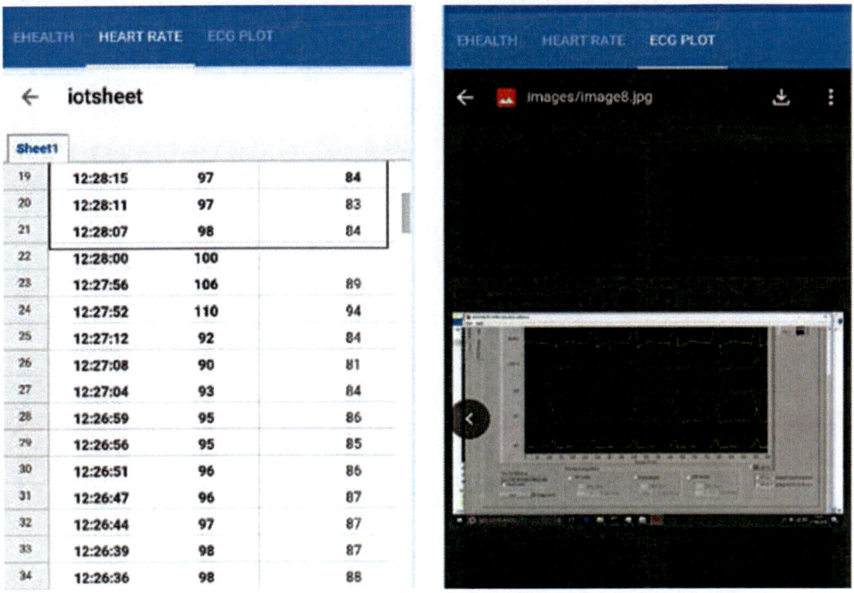

Fig. 4 Android application images

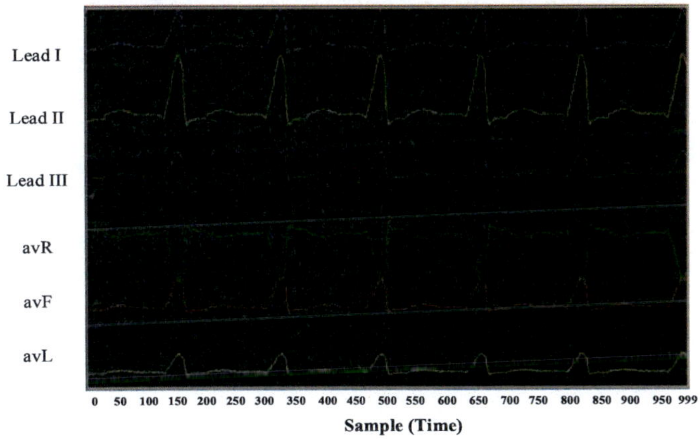

Fig. 5 ECG output from ADS1298 sensor

were used for multi-channel ECG monitoring and acquisition of oxygen saturation, respectively. For home patient monitoring, an android application was developed as the mobile platform for m-cardiology. Data transmission on the cloud server was performed using Google Application Program Interface. The proposed system not only saves the time for the patients but it also reduces the work pressure of

Table 1 Oxygen saturation and heart rate at different time instants

S. No.	Time	Oxygen saturation	Heart rate
1	12:26:36	98	88
2	12:26:39	98	87
3	12:26:44	97	87
4	12:29:47	96	87
5	12:26:51	96	86
6	12:26:56	95	85
7	12:26:59	95	86
8	12:27:04	93	84
9	12:27:08	90	81
10	12:27:12	92	84
11	12:28:07	98	84
12	12:28:11	97	83
13	12:28:17	97	84
14	12:28:26	94	80

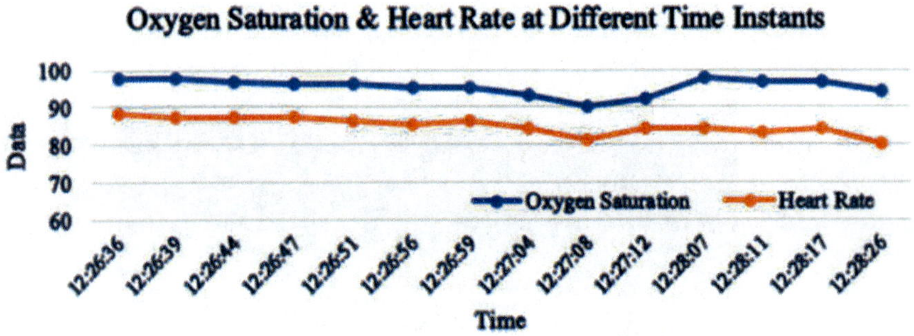

Fig. 6 Graphical representation of oxygen saturation and heart rate data

the hospitals. Therefore, the system can be used as a helpful cardiac management platform for the remote home patients as well as the medical experts.

Acknowledgements The work presented in the paper is a part of a project, "Design of Neural Network based Pacemaker" funded by Department of Biotechnology, Govt. of. India. Therefore, the authors are grateful to the funding agency.

References

1. Thilakanathan D, Chen S, Nepal S, Calvo R, Alem L (2014) A platform for secure monitoring and sharing of generic health data in the Cloud. Future Gener Comput Syst 35:102–113
2. Onubeze A (2016) Augustine Onubeze developing a wireless heart rate monitor with MAX30100 and nRF51822, Oct 2016
3. Engineering I (2014) A review paper on analysis of electrocardiograph (ECG) signal for the detection of arrhythmia abnormalities, pp 12466–12475
4. Sneha S, Madhav T, Nitin T, Shaikh PJH (2018) A review paper on smart health monitoring system. 6(02):1113–1116
5. Singh M et al (2014) Application of handheld Tele-ECG for health care delivery in Rural India. Int J Telemed Appl 2014:1–6
6. Strogonovs R, Led IR, Leds IR (2017) Implementing pulse oximeter using, pp 1–20
7. Sonawane SS, Aski VJ, Soni U (2019) IoT enabled ubiquitous healthcare data acquisition and monitoring system for personal and medical usage powered by cloud application: an architectural overview, vol 740. Springer, Singapore
8. Chhabra M, Kalsi M (2017) Real time ECG monitoring system based on Internet of Things (IoT). Int J Sci Res Publ 7(8):547
9. Information D (2015) ADS129x low-power, 8-channel, 24-bit analog front-end for biopotential measurements
10. Hema LK, Mohana Priya R, Indumathi R (2018) Design and development of IoT based pulse oximeter. 119(16):1863–1868
11. Patil S, Pardeshi S (2018) Health monitoring system using IoT. Irjet, pp 1678–1682
12. De Lucena SE, Sampaio DJBS, Mall B, Meyer M, Burkart MA, Keller FV (2015) ECG monitoring using Android mobile phone and Bluetooth. In: IEEE international instrumentation & measurement technology conference, vol 2015–July, pp 1976–1980

PDP Analysis of CNTFET Full Adders for Single and Multiple Threshold Voltages

M. Elangovan, R. Ranjith, and S. Devika

Abstract Adder is a basic building block of the arithmetic logic unit (ALU). Designing of optimized adder circuit inherently makes a pavement for obtaining optimized ALU design. The implementation of metal–oxide–semiconductor field-effect transistor (MOSFET)-based very large-scale integration (VLSI) circuits in the nanoscale range is reached saturation condition. This is due to the MOSFET that meets significant issues like producing more leakage current and highly dependent on PVT variation during nanoscale fabrication. The carbon nanotube field-effect transistor (CNTFET) can overcome the demerits of MOSFET, and it supports low-power, delay-optimized VLSI circuit design. In this paper, different types of full adders are implemented using CNTFET and their power delay product (PDP) is analysed for single and multiple threshold voltages of CNTFET. From the simulation, the low and high PDP of full adders are identified. The PDP of full adders is optimized by varying the threshold voltage of CNTFET. The simulation is carried out using the HSPICE simulation tool. The Stanford University 32-nm-CNTFET model is used for the simulation.

Keywords CNTFET · Full adder · PDP

M. Elangovan · R. Ranjith (✉) · S. Devika
Department of Electronics and Communication Engineering, Government College of Engineering, Bargur, Tamil Nadu, India
e-mail: r.ranjith@gcebargur.ac.in

M. Elangovan
e-mail: m.elangovan@gcebargur.ac.in

S. Devika
e-mail: devikasiva26@gmail.com

D. Harvey et al. (eds.), *Advances in VLSI, Communication, and Signal Processing*, Lecture Notes in Electrical Engineering 683,
https://doi.org/10.1007/978-981-15-6840-4_35

1 Introduction

1.1 CNTFET

The MOSFET bulk semiconductor channel is replaced by cylindrically shaped carbon nanotube which forms the structure of CNTFET. The CNTFET has its excellent electrical properties as compared to MOSFET. The performance of CNTFET is excellent in all aspects. The speed of CNTFET is 5–8 times larger than MOSFET. The power consumption of CNTFET is 2–7 times lesser than the MOSFET, and the power delay product is 15–20 times lower than MOSFET [1]. The chiral vectors (*m*, *n*) determine the diameter of the CNT (D_{CNT}), threshold voltage of CNTFET (V_{th}) and metallic or semiconductor behaviour of the carbon nanotubes. When $n = m$ or|*n*-*m*|=3*i,* the CNT behaves as conducting medium (where *i* is an integer); otherwise, the CNT acts as a semiconducting medium. When $m = 0$, the carbon nanotubes form a *zigzag* structure; when $n = m$, the carbon nanotubes give an *armchair* structure; for all other cases, the carbon nanotubes form *chiral* structure [2]. The structural view of CNTFET and sheet of graphene roll are shown in Fig. 1. In CNTFET, the threshold voltage (V_{th}) is a function of the diameter of carbon nanotube (D_{CNT}). This depends on the chirality vector as given below.

$$D_{\mathrm{CNT}} = \frac{a\sqrt{n^2 + nm + m^2}}{\pi} \tag{1}$$

$$V_{\mathrm{th}} = \frac{E_g}{2e} = \frac{\sqrt{3}}{3}\frac{aV_\pi}{eD_{\mathrm{CNT}}} \tag{2}$$

The dependency of chiral vectors with threshold voltage of CNTFET is analysed. It is found that the chiral vectors are negatively proportional to the threshold voltage of the CNTFET. Hence, increasing of chiral vectors decreases the threshold voltage of the CNTFET. The variation of threshold voltage with respect to chiral vector is shown in Fig. 2.

1.2 Full Adder

Adders play a vital role in any digital systems. Arithmetic logic unit (ALU) is the heart of a computer processor. Adder is a basic building block of ALU, and it is used to perform the fundamental arithmetic operation of binary addition. The full adder consists of three binary inputs A, B, C and two outputs sum (S) and carry (Cout) [1].

where

$$S = A \text{ XOR } B \text{ XOR } C \tag{3}$$

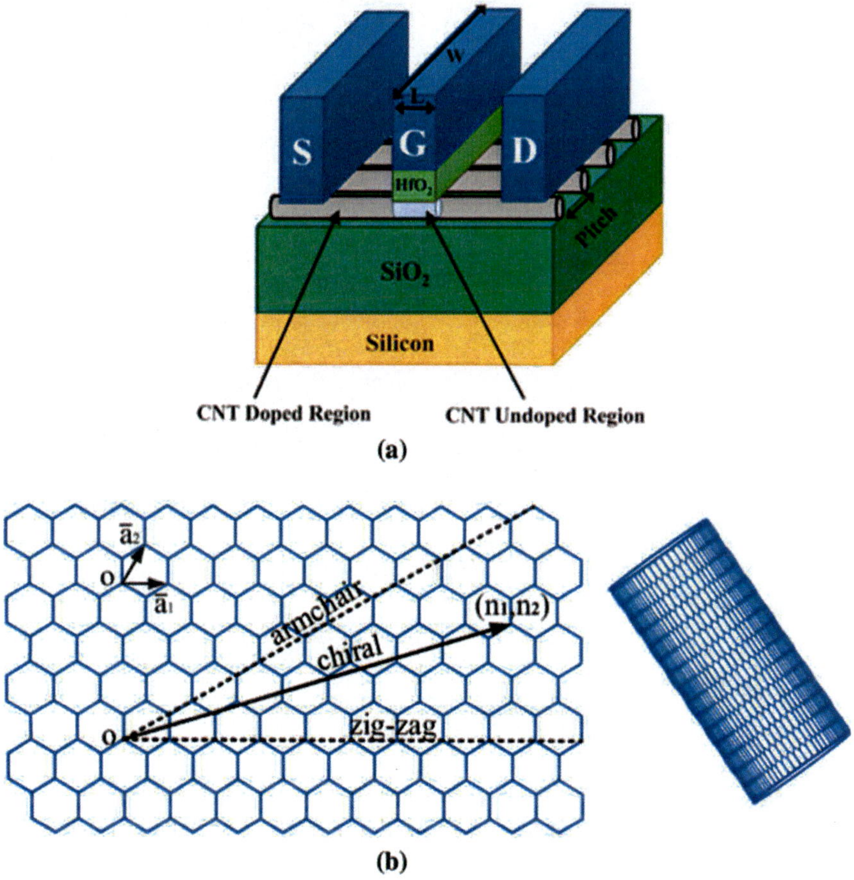

Fig. 1 **a** Structural view of CNTFET. **b** A sheet of graphene roll in different structures

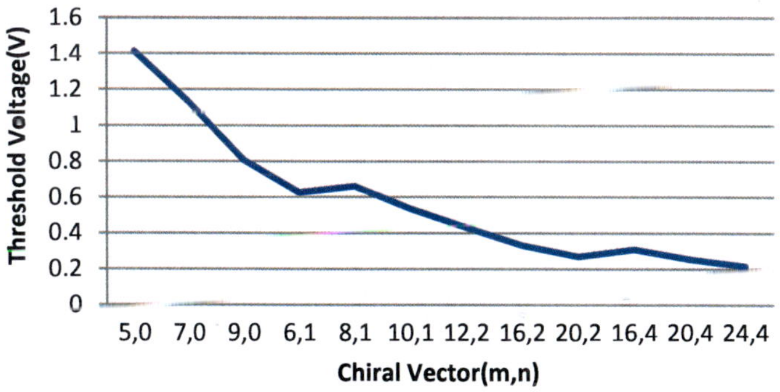

Fig. 2 Variation of threshold voltage with respect to chiral vector

$$C_{out} = A \text{ AND } B \text{ OR } C\ (A \text{ XOR } B) \quad (4)$$

2 Proposed Work

Here different types of full adder circuits like GDI full adder cell, SERF adder cell, proposed full adder using XOR module, proposed full adder using XNOR module [3], proposed CNTFET implementation of one-bit full adder [4], conventional full adder cell, proposed full adder [5], NEW-PT-FA, NEW-FL-FA, NEW-DD-FA, NEW-SD-FA, NEW-RSD-FA, NEW-ND-FA [6] are implemented using CNTFET. The power, delay and power delay product of the above adders for single threshold voltage (all CNTFET with $m = 19$, $n = 0$) and multiple threshold voltages (N-CNTFET with $m = 19$, $n = 0$ and P-CNTFET $m = 16$, $n = 0$) are analysed and compared. For single threshold voltage concept, both N-CNTFET and P-CNTFET having chiral values of $m = 19$ and $n = 0$. Thus, the threshold voltage of the CNTFET is 0.289 V. In the multiple threshold case, N-CNTFETs and P-CNTFET threshold voltages are, respectively, 0.289 V and −0.343 V [7].

3 Results and Discussions

The power, delay and power delay product of different full adders for single and multiple threshold voltages are calculated and summarized in Table 1. The simulation results show that for both single and multiple threshold voltage SERF adder cell consume the least power, respectively, 1.53E−10 W and 1.25E−10 W. The NEW-SD-FA cell dissipates the highest power of 2.29E−05 W and 2.12E−05 W, respectively, for single and multiple threshold voltages. The least delay offered by NEW-ND-FA of 3.48E−13S for a single threshold voltage, and proposed full adder [5] gives 3.91E−13S for multiple threshold voltage. The SERF adder cell gives the least PDP 7.27E−23 J and 5.86E−23 J, respectively, for single and multiple threshold voltages. The NEW-SD-FA provides the worst PDP value of 1.03E−16 for single threshold voltage, and the NEW-RSD-FA gives the highest PDP of 4.81E−17 for multiple threshold voltage. Figures 3, 4 5 show the power, delay and PDP comparison of full adders for single and multiple threshold voltages.

4 Conclusion

The simulation results show that the SERF adder cell offers the lowest PDP of 7.27e−23 J, and NEW-SD-FA gives the highest PDP of 1.03e−16 J for a single

Table 1 Power, delay PDP comparison of different CNTFET full adders

S. No.	Adders	Power (W)		Delay (S)		PDP (J)	
		Single threshold voltage	Multiple threshold voltage	Single threshold voltage	Multiple threshold voltage	Single threshold voltage	Multiple Threshold Voltage
1	GDI Full adder cell [3]	2.02E−09	1.22E−09	2.74E−12	3.85E−12	5.54E−21	4.72E−21
2	SERF adder cell [3]	1.53E−10	1.25E−10	4.73E−13	4.68E−13	7.27E−23	5.86E−23
3	Proposed full adder using XOR module [3]	1.21E−06	2.50E−07	1.70E−12	7.19E−13	2.06E−18	1.80E−19
4	Proposed full adder using XNOR module [3]	7.05E−07	1.66E−07	1.02E−12	5.26E−13	7.22E−19	8.78E−20
5	Proposed CNTFET implementation of one-bit full adder [4]	3.59E−09	2.50E−09	1.60E−12	1.63E−12	5.77E−21	4.09E−21
6	conventional full adder cell [3]	2.65E−09	2.71E−09	3.85E−12	3.09E−12	6.15E−21	8.38E−21
7	Proposed full adder [5]	1.85E−05	1.33E−05	4.15E−13	3.91E−13	7.71E−18	7.87E−18
8	NEW-PT-FA [6]	2.12E−08	3.19E−08	4.81E−12	7.02E−13	1.02E−19	2.24E−19
9	NEW -FL-FA [6]	6.54E−09	1.13E−08	3.16E−12	5.85E−13	2.07E−20	6.67E−20
10	NEW-DD-FA [6]	7.37E−09	2.60E−09	4.53E−12	7.40E−12	3.34E−20	1.92E−20
11	NEW-SD-FA [6]	2.29E−05	2.12E−05	4.49E−12	2.20E−12	1.03E−16	4.67E−17
12	NEW-RSD-FA [6]	8.78E−06	8.81E−06	5.89E−12	4.87E−12	5.18E−17	4.81E−17
13	NEW-ND-FA [6]	3.89E−09	2.68E−09	3.48E−13	4.90E−13	1.35E−21	1.31E−21

threshold voltage. The SERF adder cell offers the lowest PDP of 5.86e−23 J, and NEW-RSD-FA gives the highest PDP of 4.81e−17 J for multiple threshold voltages. The comparative study shows that the multiple threshold voltages-based full adder circuits PDP is lower than that of single threshold-based full adder circuit. In multiple threshold voltages-based circuits, the P-CNTFETs have chiral vectors $m = 16$ and $n = 0$. This in turn increases the threshold voltage of P-CNTFETs. Hence, the power

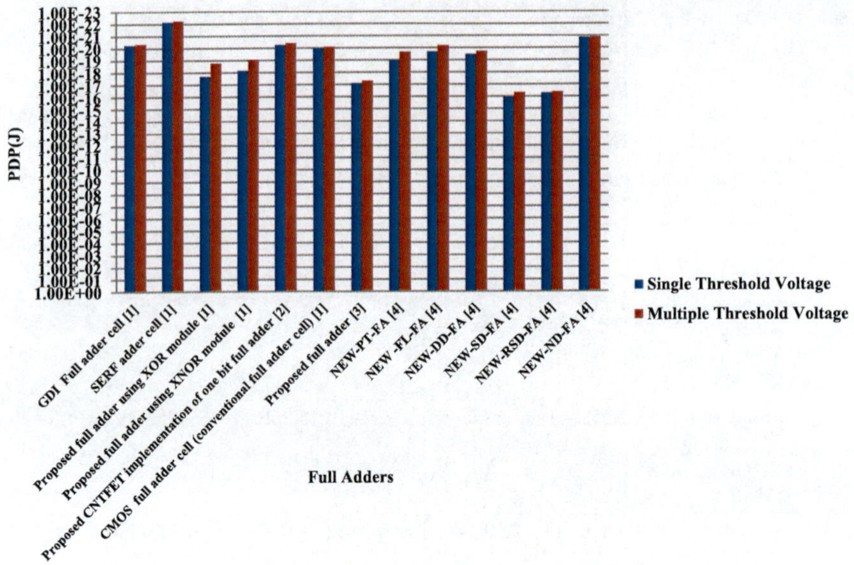

Fig. 3 PDP comparison of full adders for single and multiple threshold voltages

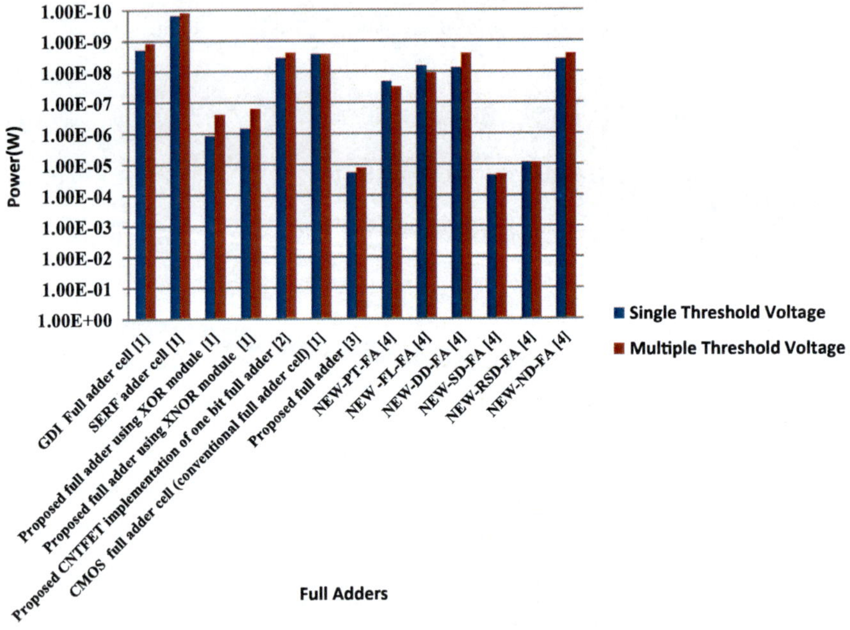

Fig. 4 Power comparison of full adders for single and multiple threshold voltages

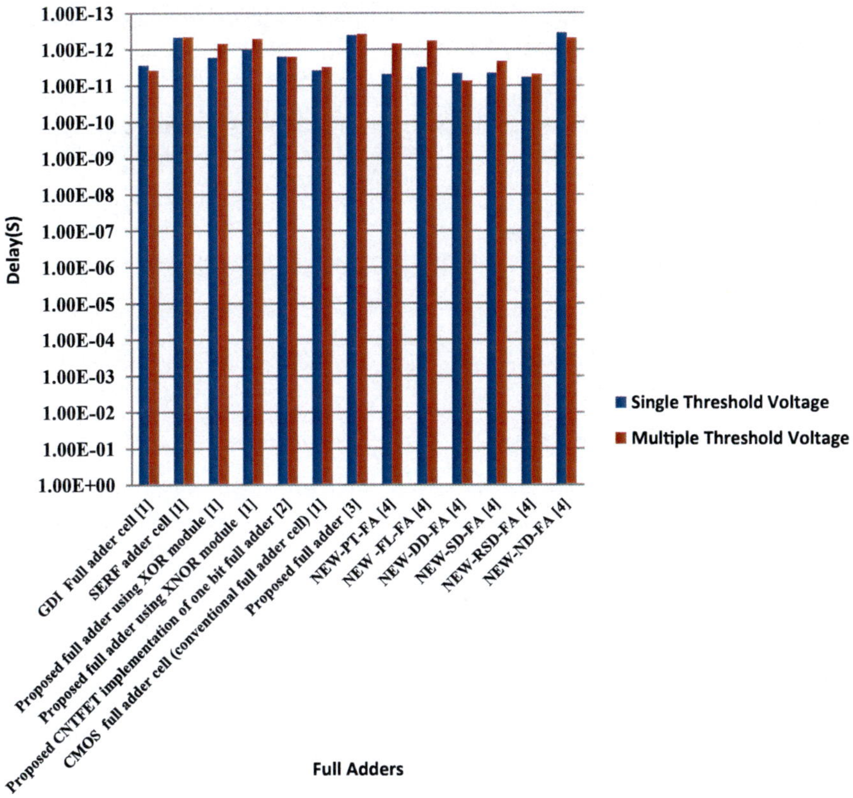

Fig. 5 Delay comparison of full adders for single and multiple threshold voltages

consumption of multiple threshold full adder circuits becomes lesser than that of single threshold voltage circuits. In multiple threshold case, N-CNTFETs have $m = 19$, $n = 0$. This in turn increases the speed of the full adder circuits.

References

1. Mehrabani YS, Eshghi M (2016) Noise and process variation tolerant, low-power, high-speed and low energy full adders in CNTFET technology. IEEE Transactions on Very Large Scale Integration (VLSI) Systems
2. Bhaskar A, Reddy D, Sarananan S, Naidu J (2016) A low power and high speed 10 transistor full adder using multi threshold technique. In: 11th international conference on industrial and information systems, pp 371–374
3. Agrawal K, Shilpa R, Kottilingel R, Sonukusare M, Rathod SS (2014) Performance characteristics of a single walled carbon nanotube field effect transistor (SWCNT-FET). In: International Conference on circuits Systems, Communication and Information Technology Applications(CSCITA)

4. Gautam H, Bindra P (2015) Structure and a detailed analysis of various simulation results of CNTFET: a review. Int J Sci Eng Technol 4(2)
5. Yadav AK, Bhavana PS, Dadoriya AK (2017) Low power high speed 1-bit full adder circuit design at 45 nm CMOS technology. In: Proceeding international conference on recent innovations is signal processing and embedded systems (RISE-2017), pp 427–432
6. Jitendra KS, Srinivasulu A, Prasad Singh B (2017) A new low power full adder cell for low voltage using CNTFETs. ECAI 2017—international conference-9th edn
7. Lin S, Kim Y-B, Lombardi F (2010) Design of a CNTFET-based SRAM cell by dual-chirality selection. IEEE Trans Nanotechnol 9(1)
8. Kumar M, Baghel RK (2017) Ultra low power high-speed single-bit hybrid full adder circuit. In: IEEE-40222, 8th ICCCNT 2017
9. Niranjan NK, Bahadur Singh R, Navaid Rizvi Z (2016) Parametric analysis of a hybrid 1-bit full adder in UDSM and CNTFET technology. Int Conf Electr Electron Optim Tech 4267–4272
10. Harish B, Sivani K, Rukmini MSS (2017) Performance comparison of various CMOS. In: International conference on energy communication, Data analysis and soft computing pp 3789–3792
11. Chaitali P, Sharma M (2017) Area-improved high-speed hybrid 1bit full adder circuit using 3T-XNOR Gate. In: 3rd international conference on computing, communication, control and automation
12. Ravi Kumar K (2017) A novel approach to design area optimized, energy efficient and high speed wallace-tree multiplier using GDI based full adder. In: IEEE international conference on computational intelligence and computing research
13. Madhu JGP, Dhiman G (2017) An Architecture for energy-efficient hybrid full adder and its CMOS Implementation. In: Conference on information and communication technology
14. Naseri H, Timarchi S (2018) Low-power and fast full adder by exploring new XOR and XNOR gates. IEEE transactions on very large scale integration (VLSI) systems
15. Phanindra LS, Rajath MN, Rakesh K, Patel KV (2016) A novel design and implementation of multi-valued logic arithmetic full adder circuit using CNTFET. In: IEEE international conference on recent trends in electronics information communication technology, pp 563–568

A Linear OTA Using Series-Connected Source-Degenerated Bulk-Driven Floating Gate Differential Pairs

Tanmay Dubey and Vijaya Bhadauria

Abstract In this paper, a source-degenerated OTA utilizing bulk-driven floating gate input stage is proposed. The linearity of the OTA is enhanced by combining bulk-driven and floating gate techniques along with source degeneration. The series connection of two differential pairs at input stage of OTA lessens the nonlinearity and also reduces the overall transconductance. The proposed circuit shows 32.9 μA/V of transconductance and −78 dB of HD_3 when subjected to 0.2 V_{pp} input differential voltage at the frequency of 1 MHz. The power consumption of the proposed OTA is 218 μW.

Keywords Bulk-driven · Floating gate · Linearity · OTA · Source degeneration

1 Introduction

An operational transconductance amplifier (OTA) is considered as one of the popular internal blocks of various analog and mixed signal circuits like *Gm-C* filters, tunable signal generator and variable gain amplifiers [1–4]. The principle operation of an OTA is to convert the input voltage into the output current; hence, it is also referred as voltage-to-current (V-to-I) converter. This voltage-to-current conversion through OTA is affected by the nonlinear behaviour of its basic element, i.e. MOSFET. This creates a challenge for analog circuit designers to overcome this nonlinearity and come with a structure of OTA which performs V-to-I conversion linearly over a significant range of input voltage signal depending upon the applications. Various linearization techniques have been reported in the literature like bulk-driven [5, 6], floating gate [7, 8], nonlinear term cancellation [9, 10] and source degeneration [11, 12]. Sometimes, more than one linearization techniques are combined to achieve the

T. Dubey (✉) · V. Bhadauria
Electronics and Communication Engineering Department, Motilal Nehru National Institute of Technology Allahabad, Prayagraj 211004, India
e-mail: dubey.tanmay@gmail.com

V. Bhadauria
e-mail: vijaya@mnnit.ac.in

D. Harvey et al. (eds.), *Advances in VLSI, Communication, and Signal Processing*, Lecture Notes in Electrical Engineering 683,
https://doi.org/10.1007/978-981-15-6840-4_36

desired level of linearity and the most popular technique to combine with any other technique is source degeneration due to its simple application. The combination of source degeneration and nonlinearity cancellation is used in [13, 14] to achieve the better linearity while in [15, 16] and [7], source degeneration is used with bulk-driven and floating gate techniques, respectively. Recently, a linear OTA is proposed in [2] using bulk-driven floating gate transistor with source degeneration.

In this paper, a linear OTA suitable for low *Gm* applications is proposed. The input stage of this OTA comprises of series connection of two differential pairs which effectively reduces the overall transconductance and the nonlinearity term present in the output current expression of OTA [17]. The input transistors of the proposed OTA are bulk-driven floating gate transistors which improve the linearity. Along with this, for biasing, flipped voltage follower (FVF) is used at the input stage of OTA to reduce the power consumption.

The structure of this paper is as follows: Sect. 2 gives the introduction of bulk-driven floating gate MOSFET, and the proposed OTA is presented in Sect. 3. The simulation results are presented and discussed in Sect. 4 of the paper, and as a final point, Sect. 5 concludes the paper.

2 Bulk-Driven Floating Gate MOSFET

As discussed in the previous section, both bulk-driven and floating gate techniques provide a method to linearize the OTA. In bulk-driven technique, the input is provided at the bulk of the MOSFET while gate is supplied with a proper biasing voltage to create the channel. This eliminates the threshold voltage from the signal path and hence making it suitable for low voltage operations. Another important feature of bulk-driven MOSFET is its lower transconductance value which is 3–5 times lesser than its gate counterpart. This property of bulk-driven MOSFET favours its use in low-frequency application like biomedical and neural network. On the other hand, in the case of the floating gate MOSFET, the input is provided through the capacitive network form at the floating gate terminal. This capacitive network also reduces the effective threshold voltage of the floating gate MOSFET.

In bulk-driven floating gate MOSFET, the bulk terminal is also utilized by connecting it to one of the inputs of the floating gate MOSFET as shown in Fig. 1. The voltage at the floating gate of the bulk-driven floating gate MOSFET is [2]

$$V_{\mathrm{FG}} = \frac{(C_1 + C_{\mathrm{GB}})V_1}{C_{\mathrm{T}}} + \frac{C_2}{C_{\mathrm{T}}} V_{\mathrm{Bais}} + \frac{C_{\mathrm{GD}}}{C_{\mathrm{T}}} V_D + \frac{C_{\mathrm{GS}}}{C_{\mathrm{T}}} V_{\mathrm{S}} \tag{1}$$

where C_1 and C_2 are the capacitors connected to the gates of the bulk-driven floating gate MOSFET; C_{GB}, C_{GD} and C_{GS} are the parasitic capacitance connected between floating gate and respected terminals, i.e. bulk, drain and source, respectively. C_{T} is the total capacitance associated with the floating gate that can be defined as

Fig. 1 Bulk-driven floating gate MOSFET

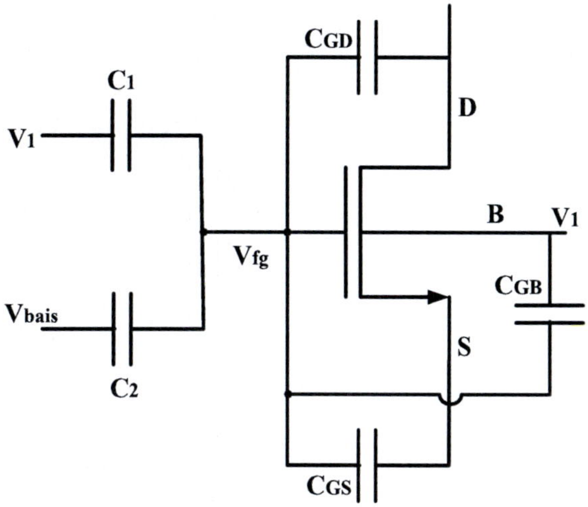

$$C_{\mathrm{T}} = \sum_{i=1}^{N} C_i + C_{\mathrm{GB}} + C_{\mathrm{GD}} + C_{\mathrm{GS}} \tag{2}$$

The overall transconductance of the bulk-driven floating gate MOSFET ($g_{\mathrm{m,bdfg}}$) is larger than both bulk-driven transconductance (g_{mb}) and floating gate transconductance ($g_{\mathrm{m,fg}}$) individually as in (3).

$$g_{\mathrm{m,bdfg}} = g_{\mathrm{m,fg}} + g_{\mathrm{mb}} \tag{3}$$

3 Proposed OTA

A linear OTA based on bulk-driven floating gate technique has been recently reported [2] named as BDFG-FVF OTA (shown in Fig. 2). The proposed OTA is the modified version of BDFG-FVF OTA. In proposed OTA, instead of using single differential pair, a series combination of two differential pairs is used to further improve the linearity. The block diagram representation and the complete schematic of the proposed OTA are shown in Figs. 3 and 4, respectively. The series-connected differential pairs are formed using bulk-driven floating gate transistors M_1, M_3 and M_2, M_4, respectively. Both the differential pairs are source-degenerated with the help of active source degeneration resistance formed by the parallel connection of M_5, M_6 and M_7, M_8, respectively. This series connection of differential pair decreases the nonlinearity present in the output current expression and also reduces the overall transconductance [17]. To decrease the power consumption, FVF configuration is used which eliminates the requirement of dedicated tail current source and hence reduces the

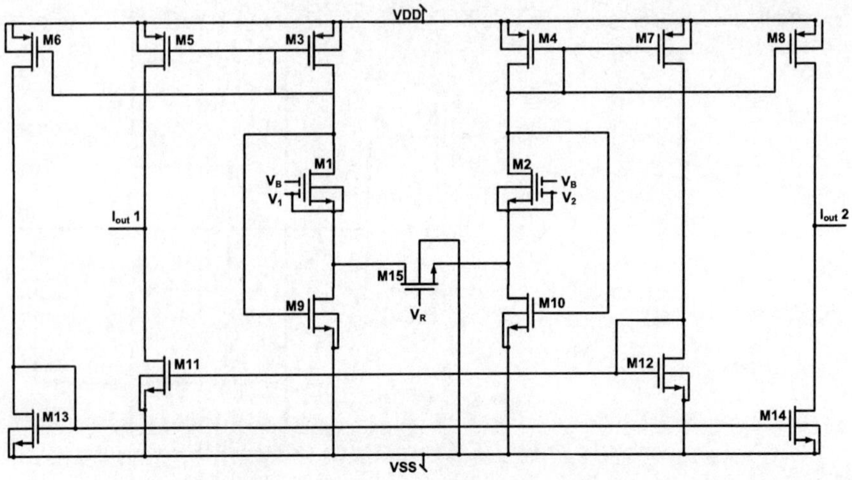

Fig. 2 BDFG-FVF OTA [2]

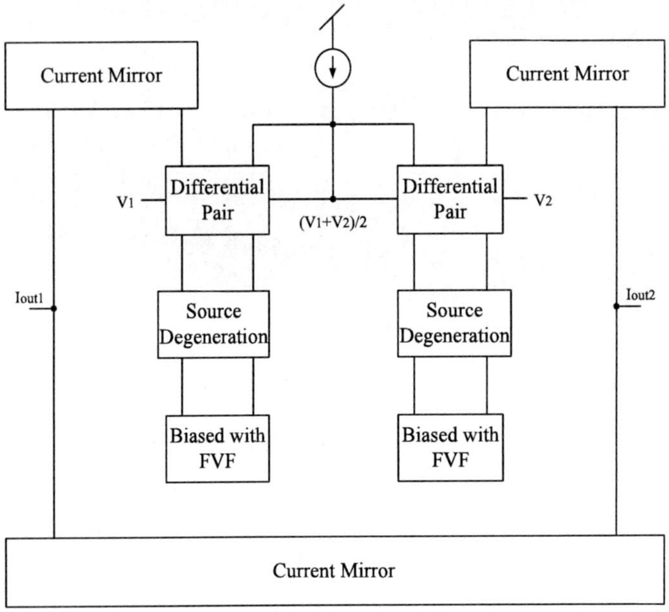

Fig. 3 Block diagram of proposed OTA

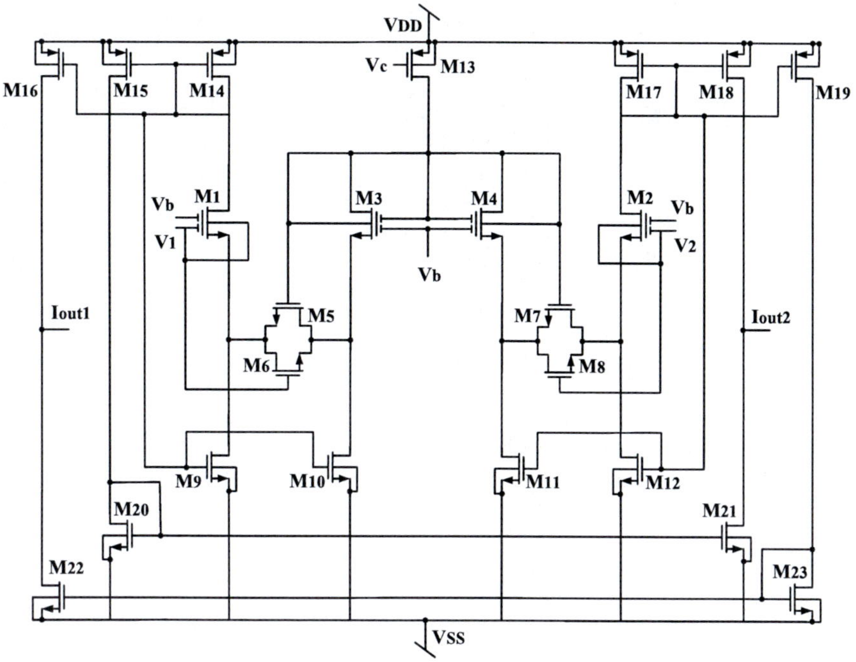

Fig. 4 Schematic of proposed OTA

overall power consumption. The negative feedback loop created by the transistors M_1 and M_9 (M_2 and M_{12}) in FVF configuration lessens the source resistance of M_1 (M_2) making FVF a superior choice over the conventional source follower. Finally, the drain currents of M_1 and M_2 are conveyed to the output nodes with the help of simple current mirrors.

The drain current equation of bulk-driven floating gate MOSFETs M_1–M_4 is

$$I_{\mathrm{Di}} = k_{\mathrm{i}}(V_{\mathrm{FGSi}} - V_{\mathrm{Ti}})^2 \tag{4}$$

where V_{FGSi} is the floating gate voltage with respect to the source terminal of the ith transistor. The output current I_{out1} can be defined as

$$I_{\mathrm{out1}} = I_{\mathrm{D1}} - I_{\mathrm{D2}} \tag{5}$$

Combining (4) and (5) gives

$$I_{\mathrm{out1}} = k(V_{\mathrm{FGS1}} - V_{\mathrm{Th}})^2 - k(V_{\mathrm{FGS2}} - V_{\mathrm{Th}})^2 \tag{6}$$

By simplifying (6)

$$I_{\mathrm{out1}} = k[(V_{\mathrm{FGS1}} - V_{\mathrm{Th}}) + (V_{\mathrm{FGS2}} - V_{\mathrm{Th}})](V_{\mathrm{FGS1}} - V_{\mathrm{FGS2}}) \tag{7}$$

where V_{FGS1} and V_{FGS2} can be expressed as

$$V_{\text{FGS1}} = \frac{(C_1 + C_{\text{GB}})}{C_{\text{T}}}(V_1 - V_{\text{S1}}) + \frac{C_2}{C_{\text{T}}}(V_{\text{b}} - V_{\text{S1}}) + \frac{C_{\text{GD}}}{C_{\text{T}}}(V_{\text{D1}} - V_{\text{S1}}) \tag{8a}$$

$$V_{\text{FGS2}} = \frac{(C_1 + C_{\text{GB}})}{C_{\text{T}}}(V_2 - V_{\text{S2}}) + \frac{C_2}{C_{\text{T}}}(V_{\text{b}} - V_{\text{S2}}) + \frac{C_{\text{GD}}}{C_{\text{T}}}(V_{\text{D2}} - V_{\text{S2}}) \tag{8b}$$

by using (8a, b) in (7), I_{out1} can be simplified as

$$I_{\text{out1}} = k\left[\sqrt{\frac{I_B + \frac{I_{out1}}{2}}{k}} + \sqrt{\frac{I_B - \frac{I_{out1}}{2}}{k}}\right]\left[\left\{\frac{C_1 + C_{\text{GB}}}{C_{\text{T}}}\right\}V_{\text{in}} - 2I_{\text{out1}}R\right] \tag{9}$$

where I_B is the drain current of the input transistors M_1 and M_2 when there is no input signal is applied and R is the effective source degeneration resistance of parallel combination of M_5 and M_6 (M_7 and M_8). The output current I_{out1} can be obtained by further simplifying (9).

$$I_{\text{out1}} = k\left[\left\{\frac{C_1 + C_{\text{GB}}}{C_{\text{T}}}\right\}V_{\text{in}} - 2I_{\text{out1}}R\right]\sqrt{\frac{4I_B}{k} - \left[\left\{\frac{C_1 + C_{\text{GB}}}{C_{\text{T}}}\right\}V_{\text{in}} - 2I_{\text{out1}}R\right]^2} \tag{10}$$

4 Simulation Result and Discussion

The proposed OTA as well as BDFG-FVF OTA is designed and simulated in Cadence Virtuoso tool with 180-nm-CMOS process technology library. A ±0.5 V dual power supply is used for both the OTAs. The bias voltage V_{b} is also set at 0.5 V. The variation of the output current with respect to the input differential voltage is shown in Fig. 5. From this plot, it is clear that output current is varying linearly for the entire range of input differential voltage. The comparison of transconductance variation with respect to input differential voltage for both the OTAs is shown in Fig. 6. The transconductance of suggested OTA is reduced to half of the transconductance of BDFG-FVF OTA due to the series connection of two differential pairs. The transconductance G_{m} of proposed OTA is found to be 33 μA/V while the transconductance of BDFG-FVF OTA is 66 μA/V.

The frequency response of the proposed OTA is presented in Fig. 7. This figure depicts the voltage gain and the phase plot of the OTA. The DC voltage gain of the OTA is found out to be 17 dB while the unity gain bandwidth is 9.18 MHz. The phase margin of the OTA is 93° which validates the stability of the OTA. The comparison of HD_3 variation of both the OTAs with respect to input voltage is displayed in Fig. 8. This plot is obtained for an input differential voltage of 0.2 V_{pp} at the frequency of

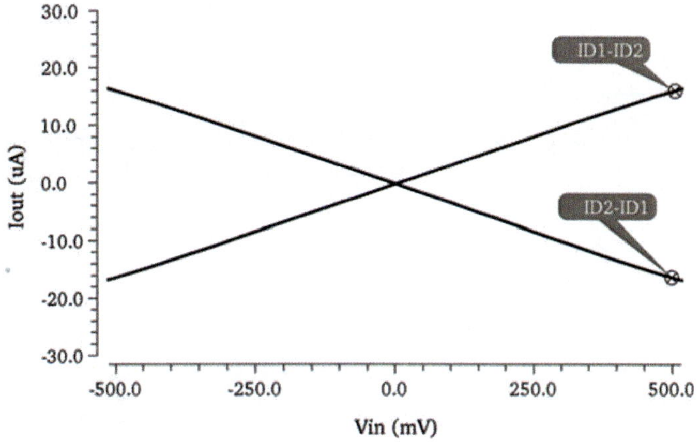

Fig. 5 Output current variation with respect to input voltage

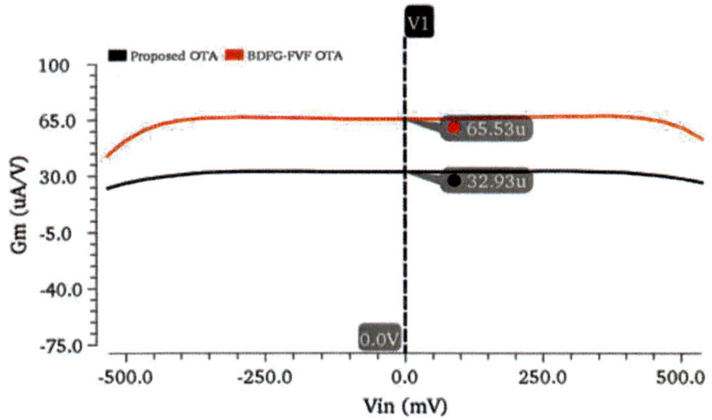

Fig. 6 Comparison of transconductance variation of proposed OTA and BDFG-FVF OTA

1 MHz. The proposed OTA shows HD_3 of − 78 dB which is 7 dB better than that of BDFG-FVF OTA. The HD_3 of both the OTAs is also examined up to 0.4 V_{pp} input signal, and it is found that the HD_3 of the proposed OTA remains better by more than 6 dB that of BDFG-FVF OTA. The CMRR of both the OTAs is compared in Fig. 9. The transient behaviour in unity gain configuration of the proposed OTA is shown in Fig. 10. From this response, it is clear that the output voltage swing of the OTA is nearly following the 1 V_{PP} input differential voltage. The comparison summary of various performance parameters of proposed OTA and BDFG-FVF OTA is given in Table 1.

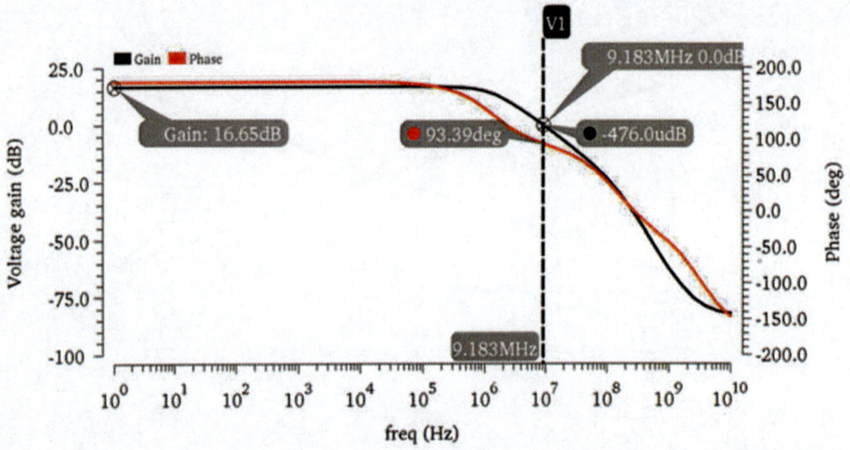

Fig. 7 Frequency response of proposed OTA

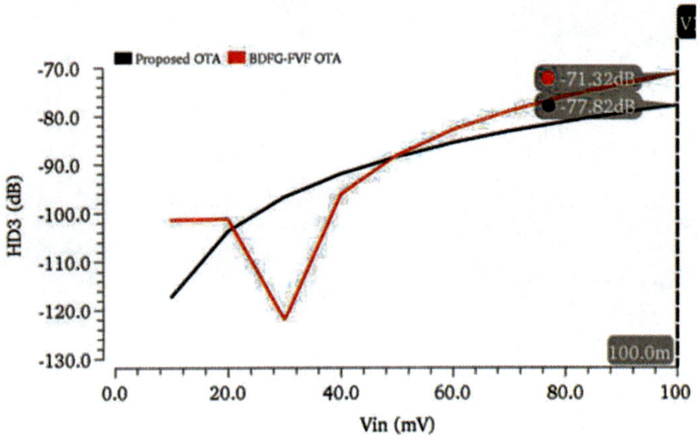

Fig. 8 Comparison of HD_3 variation of proposed OTA and BDFG-FVF OTA

5 Conclusion

Considering the linearity of the OTA is one of its important parameters, an OTA named as BDFG-FVF OTA and linearized by combining more than one linearization techniques as bulk-driven, floating gate and source degeneration is presented in [2]. In this paper, a modified version of this BDFG-FVF OTA using series connection of the two source-degenerated differential pairs which further enhances the linearity is presented; however, this reduces the overall transconductance which is useful for biomedical application. The circuit shows –78 dB and −64 dB of HD_3 for 0.2 V_{pp} and 0.4 V_{pp} input signal at 1 MHz, respectively. The OTA is designed by using UMC

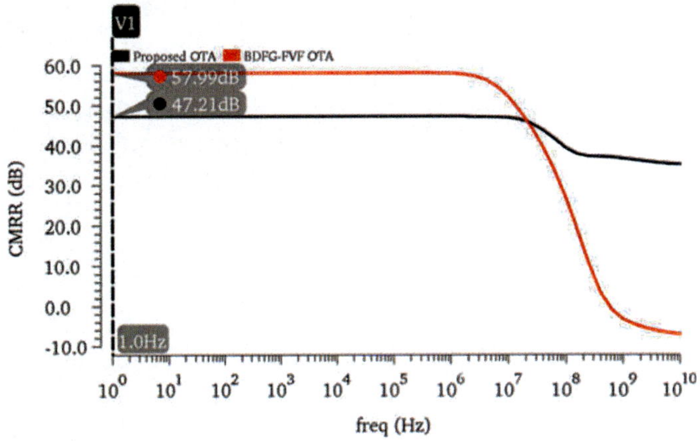

Fig. 9 Comparison of CMRR of proposed OTA and BDFG-FVF OTA

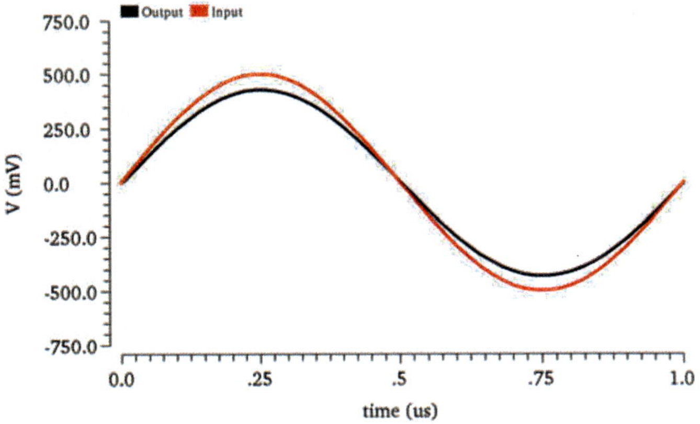

Fig. 10 Transient analysis of proposed OTA

180 nm CMOS process technology. The power supply given to the circuit is ±0.5 V, and the power consumed by it is 218 μW.

Table 1 Comparison of BDFG-FVF OTA and proposed OTA

Parameter	[2]	Proposed work
Technology (μm)	0.18	0.18
Power supply (V)	±0.5	±0.5
G_m (μA/V)	65.5	32.9
HD_3 (dB)	– 71	– 78
for V_{PP} (V)	0.2	0.2
HD_3 (dB)	– 58	– 64
for V_{PP} (V)	0.4	0.4
@ Freq. (MHz)	1	1
Voltage gain (dB)	20.67	16.65
GBW (MHz)	18.20	9.18
CMRR (dB)	58	47.21
Phase margin	87.26°	93.39°
P_{Diss} (μW)	158	218

Acknowledgements This work has been executed utilizing the resources of VLSI Laboratory under Special Manpower Development Programme Chip to System Design (SMDP-C2SD) project funded by the Ministry of Electronics and Information Technology (MeitY), Government of India.

References

1. Ohbuchi T, Matsumoto F (2013) A new design of a linear local-feedback MOS transconductor for low frequency applications. Analog Integr Circuits Signal Process 75:257–266. https://doi.org/10.1007/s10470-012-0006-6
2. Dubey T, Bhadauria V (2019) A low-voltage highly linear OTA using bulk-driven floating gate MOSFETs. AEU - Int J Electron Commun 98:29–37
3. Kar SK, Sen S (2011) Tunable square-wave generator for integrated sensor applications. IEEE Trans Instrum Meas 60:3369–3375
4. Calvo B, Celma S, Sanz MT et al (2008) Low-voltage linearly tunable CMOS transconductor with common-mode feedforward. In: IEEE transactions on circuits and systems I: regular papers, pp 715–721
5. Khateb F, Kulej T, Vlassis S (2017) Extremely low-voltage bulk-driven tunable transconductor. Circ Syst Sig Process 36:511–524. https://doi.org/10.1007/s00034-016-0329-0
6. Sharan T, Bhadauria V (2017) Fully differential, bulk-driven, class AB, sub-threshold OTA with enhanced slew rates and gain. J Circ Syst Comput 26:1750001. https://doi.org/10.1142/S0218126617500013
7. Dubey T, Pandey R (2018) Low-voltage highly linear floating gate MOSFET based source degenerated OTA and its applications. Inf MIDEM 48:19–28
8. Sánchez-Rodríguez T, Muñoz F, Galán J et al (2015) Low voltage linear tunable transconductor for high speed filters. Analog Integr Circ Sig Process 82:329–333. https://doi.org/10.1007/s10470-014-0435-5
9. Khumsat P (2018) Linearisation technique for low-voltage tuneable Nauta's transconductor in Gm − C filter design. IET Circ Dev Syst 12:347–361. https://doi.org/10.1049/iet-cds.2017.0177

10. Shen D-L, Chu Y-J, Chen H-W (2012) A linearized technique in an All-MOS transconductance amplifier. Microelectronics J 43:1023–1028. https://doi.org/10.1016/j.mejo.2012.07.017
11. Elamien MB, Mahmoud SA (2018) On the design of highly linear CMOS digitally programmable operational transconductance amplifiers for low and high-frequency applications. Analog Integr Circ Sig Process 2. https://doi.org/10.1007/s10470-018-1128-2
12. Rezaei F (2017) Linearity enhancement in the entire tuning range of CMOS OTA using a new tune compensated source degeneration technique. Microelectronics J 66:128–135. https://doi.org/10.1016/j.mejo.2017.06.008
13. Lewinski A, Silva-Martinez J (2006) A high-frequency transconductor using a robust nonlinearity cancellation. IEEE Trans Circ Syst II Express Briefs 53:896–900. https://doi.org/10.1109/TCSII.2006.880025
14. Kuo KC, Leuciuc A (2001) A linear MOS transconductor using source degeneration and adaptive biasing. IEEE Trans Circ Syst II Analog Digit Sig Process 48:937–943. https://doi.org/10.1109/82.974782
15. Yodtean A, Thanachayanont A (2013) Sub 1-V highly-linear low-power class-AB bulk-driven tunable CMOS transconductor. Analog Integr Circ Sig Process 75:383–397. https://doi.org/10.1007/s10470-013-0044-8
16. Veeravalli A, Member S, Member S (2002) Transconductance amplifier structures with very small transconductances: a comparative design approach. IEEE J Solid-State Circ 37:770–775
17. Torrance R, Viswanathan T, Hanson J (2004) CMOS voltage to current transducers. IEEE Trans Circ Syst 32:1097–1104. https://doi.org/10.1109/tcs.1985.1085644

Compressive Sensing-Based Continuous EEG Monitoring: Seizure Detection Performance Comparison of Different Classifiers

Meenu Rani, S. B. Dhok, and R. B. Deshmukh

Abstract Compressive sensing (CS) is a newer sensing modality, which samples the signals at a rate much below the Nyquist rate and still allows the faithful reconstruction from fewer samples. Acquisition of EEG signals using Nyquist sampling generates too many samples, which invokes the need for compression before storage and transmission of these samples. In this scenario, CS has been proved to be better candidate, suppressing the need of compression by generating the fewer samples, which can be stored or transmitted directly. In this paper, at the transmitter side, the acquisition of EEG signal is done using CS, and then, at the receiver side, the reconstruction is performed using orthogonal matching pursuit (OMP) algorithm of CS. After reconstructing the signal for different undersampling factors, the features are extracted from these signals. Several classifiers are trained and tested on these features to detect the epileptic seizure. Performance comparison of these classifiers shows that even at higher undersampling factors like 64, a high seizure detection accuracy of 98.9% by these classifiers. All the simulations are done on the EEG signal taken from CHB-MIT database using MATLAB 2017a.

Keywords Compressive sensing · EEG-monitoring · Random demodulator · OMP · Classifiers

1 Introduction

Requirement of continuous EEG monitoring for early detection of disease poses certain challenges on traditional signal acquisition techniques. For faithful reconstruction, sampling such signals by conventional methods results in huge amount

M. Rani (✉) · S. B. Dhok · R. B. Deshmukh
Visvesvaraya National Institute of Technology, Nagpur 440010, MH, India
e-mail: meenubanait@gmail.com

S. B. Dhok
e-mail: sanjaydhok@gmail.com

R. B. Deshmukh
e-mail: mona1810@yahoo.com

D. Harvey et al. (eds.), *Advances in VLSI, Communication, and Signal Processing*, Lecture Notes in Electrical Engineering 683,
https://doi.org/10.1007/978-981-15-6840-4_37

of samples, and hence, compression is mandatory before storage and transmission. Recently, a rapid growth has been reported in the area of remote health care, in which health is monitored remotely using battery operated devices. Longer battery life is major requirement of these systems for continuous monitoring. But, the difficulty is the requirement of high sampling rate for error-free operation consumes a considerable amount of power, and as a consequence of it, the huge amount samples generated has to be compressed, which requires extra power for further processing. In such situation, compressive sensing (CS) proves itself to be a better candidate to overcome the disadvantages of conventional methods. The acquisition scheme offered by compressive sensing, samples a signal at much lower rate, generating far fewer samples and avoids the need of compression, thereby lowering the power consumption. In literature, use of CS for EEG monitoring has been reported, addressing the applicability of CS for remote heath care [1–7].

Compressive sensing (CS) is an emerging signal processing technique, introduced in 2004 by Donoho, Candès, Romberg and Tao [8, 9]. They have developed its mathematical foundation and demonstrated its applicability for sparse or compressible signals. A signal is said to be sparse if it has fewer significant components either in original domain or in some transformed domain, compared to the total number of samples generated when sampled at Nyquist rate. Similarly, compressible signals are those whose sorted components decay according to the power law, (Fig. 1). In compressive sensing paradigm, the sampling rate of a signal is sparsity dependent and is independent of the frequency of the highest frequency component present in the signal. Compressive sensing utilizes random sampling mechanism for sampling signals generating far fewer random samples which can be easily stored or transmitted. This mechanism gives an impression of compression at the time of sampling/sensing and hence the name compressive sensing. At the receiver end, the original signal can be reconstructed by the nonlinear techniques, which are more complex than the conventional techniques. Hence, it can be said that CS shifts the complexity of signal processing from acquisition end towards reconstruction end [10–12].

CS is advantageous over conventional techniques specifically where the number of measurements that can be taken is limited may be because of factors like cost, speed and power etc. Examples of such prominent areas where CS finds its application are

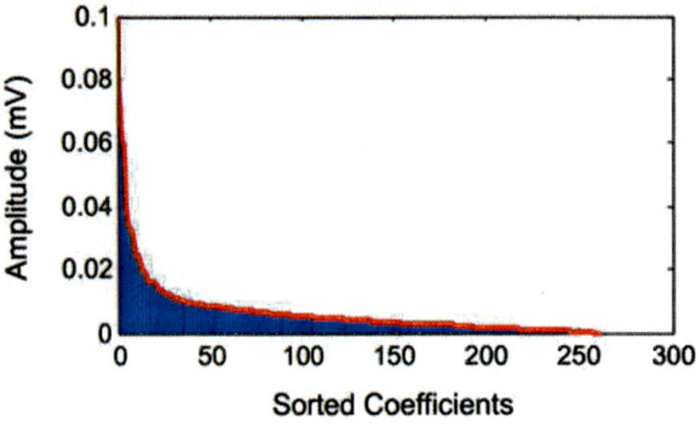

Fig. 1 Sorted components of a signal depicting power law

medical imaging, ultra-wideband communication, seismology and telemonitoring, etc. In this regard, the early prototype hardware implementation by Duarte et al., namely the single-pixel camera, demonstrates image acquisition using compressive sensing [13]. Other implementations reported in the literature for signal acquisition using compressive sensing are random demodulator which is a CS ADC for wide-band signals [14], modulated wideband converter for ultra-wideband signals [15] and compressive multiplexer [16], etc. In this paper, we have used random demodulator (RD) for EEG signal acquisition. The organization of this paper is as: Sect. 2 presents the preliminary introduction of compressive sensing. Section 3 describes the EEG signal acquisition using random demodulator technique and the reconstruction methods used in this paper for recovering the original signal back from compressive random measurements. Section 4 presents the method for feature extraction and seizure detection from the original signal as well as from reconstructed signals. In Sect. 5, the seizure detection performance comparison of different classifiers has been presented for different undersampling factors.

2 Preliminaries and Background of CS

2.1 Acquisition Model

Compressive sensing is different from traditional methods in the way it samples signals. CS works by taking fewer random measurements. The measurements are non-adaptive, i.e., not learning from previous measurements. Further, to reduce the number of measurements required for perfect reconstruction, the signal sparsity domain and acquisition domain should be incoherent from each other. For example, time and frequency are incoherent domains in the sense that a signal having sparse representation in frequency domain spreads out in time domain. The CS acquisition method can be described mathematically by (1) and is shown in Fig. 2a.

$$y = \varphi x, \tag{1}$$

where $x \in \mathbb{R}^n$ is input signal of length n, $\varphi \in \mathbb{R}^{m \times n}$ or $\mathbb{C}^{m \times n}$ is an $m \times n$ random measurement matrix and $y \in \mathbb{R}^m$ or $\mathbb{C}^m$ is the measurement vector of length m, here number of measurements $m \ll n$ (the length of input signal) [10–12].

2.2 Reconstruction Model

The input arguments to the CS reconstruction algorithm are y and Θ (Fig. 2b), where matrix $\Theta = \varphi \times \psi \in \mathbb{R}^{m \times n}$ or $\mathbb{C}^{m \times n}$ is the reconstruction matrix and ψ is the sparsifying basis of the signal x. The sparsifying basis is the basis in which signal has sparse representation and signal can be written as a linear combination of its basis vectors (columns of ψ) (2).

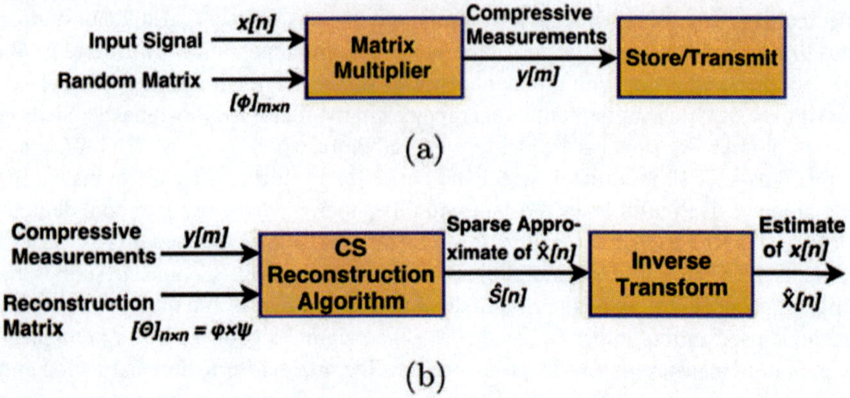

Fig. 2 CS Model: **a** Acquisition model, **b** Reconstruction model

$$x = \sum_{i=1}^{n} s_i \psi_i = \psi s, \tag{2}$$

where $s \in \mathbb{R}^n$ is the sparse coefficient vector of length n, having fewer significant/nonzero entries. The original signal can be recovered back from compressive measurements by solving (1), which is an underdetermined system of linear equations having infinite number of solutions. However, it has been shown in the literature that (1) can be solved by posing it as a convex optimization problem and finding a solution having minimum ℓ_1-norm (3). This is considered as a feasible solution because solvers available from linear programming can be used for solving ℓ_1-minimization problems [17–19].

$$\hat{s} = \arg\min_{s} \|s\|_1 \quad \text{sub. to} \quad \Theta s = y, \tag{3}$$

where $\hat{s}$ is the estimate of s and $\|s\|_1$ denotes the ℓ_1-norm of s. The output of CS reconstruction algorithm is the sparse vector s, from which x can be obtained by taking inverse transform.

2.3 *Necessary and Sufficient Conditions for Perfect Recovery*

2.3.1 Restricted Isometry Property (RIP)

Let k be the sparsity of vector s, then the necessary condition for recovering s from measurements y is that the reconstruction matrix Θ must obey RIP of order k, (4).

$$1 - \delta \leq \frac{\| \Theta v \|_2}{\| v \|_2} \leq 1 + \delta, \tag{4}$$

where v is a vector having the same k-nonzero entries as s and $\delta > 0$ is called restricted isometry constant. This inequality states that matrix Θ must preserve the distance between two k-sparse vectors. However, a sufficient condition for a stable solution for both k-sparse and compressible signals is that Θ satisfies (4) for an arbitrary 3 k-sparse vector v. As calculating δ is itself a very tough task, another condition on measurement and reconstruction matrices, which guarantees stable solution is incoherence [11, 12].

2.3.2 Incoherence

This condition says that the measurement basis φ and sparse basis ψ must be incoherent. This enables each measurement to capture some part of information present in signal. The coherence between two matrices is calculated by the largest correlation between them (5).

$$\mu(\varphi, \psi) = \sqrt{n}. \max_{1 \leq i, j \leq n} \mid \langle \varphi_i, \psi_j \rangle \mid, \tag{5}$$

where $\mid \langle \cdot, \cdot \rangle \mid$ represents the inner product operator. The range of coherence is $\mu(\varphi, \psi) \in [1, n]$. Lower value of coherence is desired which in turn lowers the number of measurements required for CS reconstruction. For random matrices, the relation between number of measurements required for faithful reconstruction and coherence is given by (6), where c is a constant.

$$m \geq \mu c k \log n. \tag{6}$$

The general measurement matrices used in CS are the matrices drawn from Gaussian and Bernoulli distributions. Partial Fourier matrices obtained by randomly selecting the rows from Fourier matrices are also considered as good CS measurement matrix. It has been shown in the literature that these three matrices are incoherent with any sparsifying basis and also satisfy restricted isometry property [12, 20]. Recently, toeplitz and circulant matrices have been proposed as a good option for CS measurement matrices. These matrices are shown to have performance comparable to random matrices. The advantages of these matrices over random matrices are that they can be easily stored and reproduced at reconstruction end and hence lowers the transmission overhead [21].

3 EEG Signal Acquisition and Reconstruction Using CS

3.1 Acquisition

The segment of EEG signal data taken from the CHB-MIT database is shown in Fig. 3a [22]. Now, this signal is sampled using CS ADC also known as random demodulator to obtain the compressive measurements Fig. 3b. First stage of RD is a

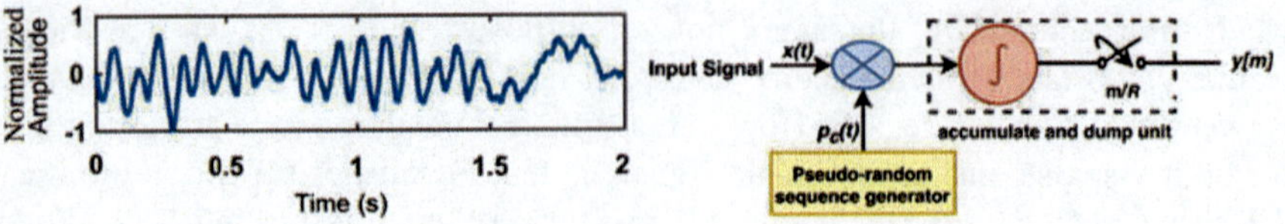

(a) (b)

Fig. 3 EEG signal and its acquisition method **a** Segment of EEG signal taken from CHB-MIT database, **b** EEG signal acquisition via random demodulator

multiplier, which multiplies the input signal by a pseudorandom chipping sequence of ± 1s. This introduces randomness in the signal and smear its frequency contents over whole of the frequency band. This smearing generates to a unique frequency signature corresponding to each frequency component present in signal. This signal is then passed through a low-pass filter to retain the lower frequencies. Now, this signal is sampled at Nyquist rate which is much lower compared to sampling rate required without random demodulator [14].

The matrix form of RD is governed by (7) and (8). The experimental settings done for EEG signal acquisition using CS are $n = 512$ and $m = 256$, which implies input signal x is a vector of size 512×1, measurement matrix $\varphi = H \times P$ is of size 256×512 and y is of 256×1, where P is an 512×512 diagonal matrix of chipping sequence having diagonal elements as pseudorandom sequence $(p_c(t))$ of ± 1s and H is 256×512 accumulate and dump matrix serving as low pass filter. The number of terms to be accumulated for one measurement generally equals the ratio $R = (n/m)$.

$$P = \begin{bmatrix} p_1 & & \\ & \ddots & \\ & & p_n \end{bmatrix}; \quad H = \begin{bmatrix} 111\cdots & & \\ & 111\cdots & \\ & & 111\cdots \end{bmatrix} \tag{7}$$

$$\left.\begin{aligned} \tilde{x} &= Px \\ y &= H\tilde{x} = \varphi x \\ \varphi &= HP \end{aligned}\right\} \tag{8}$$

3.2 *Reconstruction*

As described in Fig. 2b, inputs to the reconstruction algorithm are measurement vector y and reconstruction matrix Θ. At this stage, we already have generated the measurement vector y, now is the turn to form a reconstruction matrix that will help us to reconstruct the original signal from y. As we know, $\Theta = \varphi \times \psi$, the structure of φ has already been discussed, now next is the sparsifying basis ψ. Here, Fourier basis has been used as sparsifying basis ψ of EEG signal as per [23].

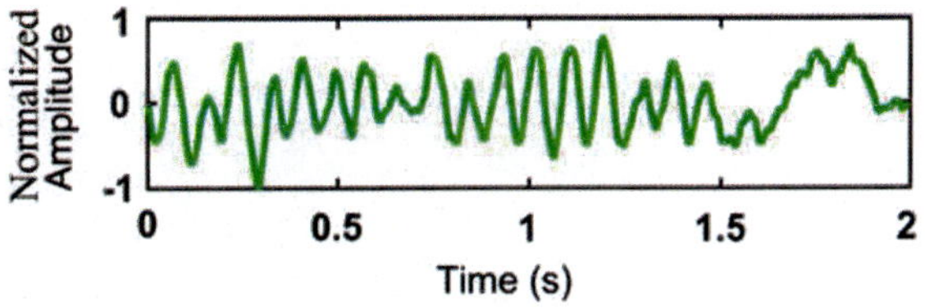

Fig. 4 EEG signal reconstructed using OMP

However, there are various CS reconstruction algorithms available in the literature [24] In this paper, we have used OMP for recovering the EEG signal from CS measurements, as per the performance comparison given in [23]. The OMP algorithm belongs to the category of greedy algorithms, which are relatively simpler and faster [25]. According to the experimental settings of n, m and φ, the size of reconstruction matrix Θ should be 256×512 for an undersampling factor of 2. The other undersampling factors considered here are 4, 8, 16, 32 and 64, and accordingly is the size of φ and Θ. Now, $\Theta = \varphi\psi$, so the size of Fourier matrix (ψ) should be 512×512. The resulted matrix Θ and compressive measurements y are given as inputs to the reconstruction algorithms. The output of reconstruction algorithms is the sparse vector $\hat{s}$, i.e. the frequency spectrum of reconstructed signal in our case, from which the time domain signal is obtained by taking the inverse Fourier transform. The signal reconstructed by OMP is shown in Fig. 4. For implementing OMP algorithm, the MATLAB function which is available at [26] has been used.

4 Feature Extraction and Seizure Detection

After reconstruction, this signal is further processed segment-wise to detect the presence of seizure, as shown in Fig. 5. The features are extracted from the original and reconstructed signals with the help of filtering and norm-2 computations segment-wise. Here, norm-2 is used to extract the energy from filtered signal, which is then used as a feature for further processing. The total signal length processed is 921,600 samples in the segment of size 512. The length of the feature vector so obtained is 1800. The processes of feature extraction and classification are then repeated for different CS undersampling factors. Feature extraction actually reduces the dimensionality and increases the class separability of signal under consideration, which eases the detection process. The features so extracted are then used to train and test a classifier. The training and testing is done using tenfold cross-validation. The classification problem here is a two-class classification problem.

5 Seizure Detection Performance Comparison

Several classifiers like simple tree, linear SVM, Quadratic SVM and medium Gaussian SVM have been tried to classify the features obtained in the previous step. A comparison of the classification accuracy obtained by these classifiers for original

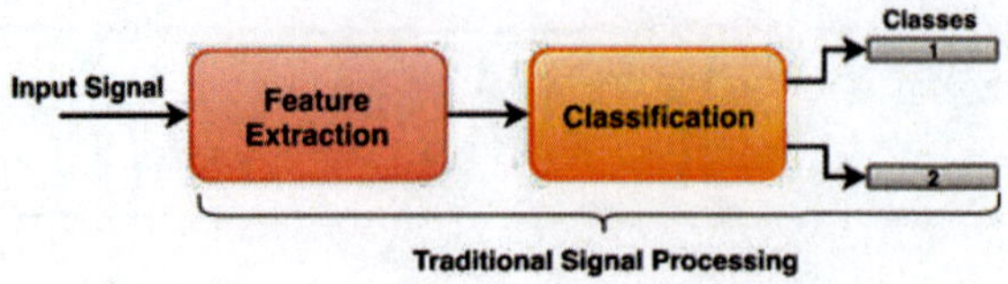

Fig. 5 Signal processing steps for detecting seizure from original and reconstructed EEG signals

Table 1 Performance comparison of different classifiers in detecting seizure from CS reconstructed signal for different undersampling factors and from original EEG signal

Classifier	Classification accuracy						
	Original signal	CS reconstructed signal					
		$u = 2$	$u = 4$	$u = 8$	$u = 16$	$u = 32$	$u = 64$
Simple tree	99.2	99.2	99.2	99.1	98.9	99.1	98.9
Linear SVM	99.1	99.1	99.2	99.1	99.2	99.1	98.9
Quadratic SVM	99.2	99.1	99.3	99.1	99.2	99.0	98.9
Medium gaussian SVM	98.9	98.9	98.9	98.9	98.8	98.9	98.9

EEG signal and for CS reconstructed signal for different undersampling factors is shown in Table 1. The classification results of only those classifiers are presented, which have nearly same classification accuracy. It has been found that CS performance is quite satisfactory in detecting the seizure from EEG signal, even at higher undersampling factors like 32 and 64 because the seizure and non-seizure classes are highly separable.

6 Conclusion

Compressive sensing is an attractive signal processing paradigm that performs compression at the time of sensing. CS works by random sampling at a rate proportional to the sparsity of the underlying signal. There are wide variety of signal processing areas where CS has been used. In this paper, signal acquisition via random demodulator has been done. For reconstructing original signal back, OMP has been used because of simpler, faster and efficient nature. The features have been extracted from the reconstructed signal for different undersampling factors, as well as from original EEG signal for comparison purpose. The performance comparison of several classifiers have been done in detecting seizure from the extracted features. It has been found that even at higher undersampling factors like 64, a high seizure detection accuracy of 98.9% has been achieved.

References

1. Aviyente S (2007) Compressed sensing framework for EEG compression. IEEE/SP 14th workshop on Stat Sig Proc Madison, WI, USA, pp 181–184. https://doi.org/10.1109/SSP.2007.4301243
2. Zhang Z et al (2013) Compressed sensing of EEG for wireless telemonitoring with low energy consumption and inexpensive hardware. IEEE Trans Biomed Engg 60(1):221–224. https://doi.org/10.1109/TBME.2012.2217959
3. Abdulghani AM et al (2012) Compressive sensing scalp EEG signals: implementations and practical performance. E Med Biol Eng Comp 50:1137–1145. https://doi.org/10.1007/s11517-011-0832-1
4. Lay-Ekuakille A et al (2014) Entropy index in quantitative EEG measurement for diagnosis accuracy. IEEE Trans Inst Meas 63:1440–1450
5. Lay-Ekuakille A et al (2013) Multidimensional analysis of EEG features using advanced spectral estimates for diagnosis accuracy. IEEE Int Symp Med Meas App (MeMeA), Gatineau, QC, pp 237–240. https://doi.org/10.1109/MeMeA.2013.6549743
6. Vergallo P et al (2014) Identification of visual evoked potentials in EEG detection by emprical mode decomposition. IEEE IEEE 11th international multi-conference on systems, signals & devices (SSD14), Barcelona, p 1–5. https://doi.org/10.1109/SSD.2014.6808848
7. Vergallo P et al (2014) Spatial filtering to detect brain sources from EEG measurements. In: International symposium on medical measurements and applications (MeMeA), Lisboa, pp 1–5
8. Candès EJ et al (2006) Robust uncertainty principles: exact signal reconstruction from highly incomplete frequency information. IEEE Trans Inf Theory 52(2):489–509. https://doi.org/10.1109/TIT.2005.862083
9. Donoho DL (2006) Compressed sensing. IEEE Trans Inf Theory 52(4):1289–1306. https://doi.org/10.1109/TIT.2006.871582
10. Baraniuk RG (2007) Compressive sensing [Lecture Notes]. IEEE Sign Process Mag 24(4):118–121. https://doi.org/10.1109/MSP.2007.4286571
11. Candès EJ, Wakin MB (2008) An introduction to compressive sampling. IEEE Sig Process Mag 25(2):21–30. https://doi.org/10.1109/MSP.2007.914731
12. Baraniuk R et al (2011) An introduction to compressive sensing. OpenStax-CNX http://legacy.cnx.org/content/col11133/1.5/
13. Duarte MF et al (2008) Single-pixel imaging via compressive sampling. IEEE Signal Process Mag 25(2):83–91
14. Tropp JA et al (2010) Beyond nyquist: efficient sampling of sparse bandlimited signals. IEEE Trans Inf Theory 56(1):520–544. https://doi.org/10.1109/TIT.2009.2034811
15. Mishali M, Eldar YC (2010) From theory to practice: sub-nyquist sampling of sparse wideband analog signals. IEEE J Sel Top Signal Process 4(2):375–391
16. Slavinsky JP, Laska JN, Davenport MA, Baraniuk RG (2011) The compressive multiplexer for multi-channel compressive sensing. In: IEEE international conference on acoustics speech and signal process (ICASSP), Prague, Czech Republic, pp 3980–3983 (2011)
17. Donoho DL (2006) For most large underdetermined systems of linear equations the minimal L1-norm solution is also the sparsest solution. Commun Pure and Appl Math 59(6):797–829
18. Candès E, Tao T (2005) Decoding by linear programming. IEEE Trans Inform Theory 51(12):4203–4215

19. Chen S, Donoho DL, Saunders MA (1999) Atomic decomposition by basis pursuit. SIAM J Sci Comp 20(1):33–61
20. Candès EJ, Romberg J (2007) Sparsity and incoherence in compressive sampling. Inverse Prob 23(3):969–985
21. Yin W, Morgan S, Yang J, Zhang Y (2010) Practical compressive sensing with Toeplitz and circulant matrices. Visual Comm Image Process
22. Goldberger AL et al (2000) PhysioBank, PhysioToolkit, and PhysioNet: components of a new research resource for complex physiologic signals. Circulation 101(23):e215–e220 [Circulation Electronic Pages; http://circ.ahajournals.org/cgi/content/full/101/23/e215]
23. Rani M, Dhok SB, Deshmukh RB (2019) EEG monitoring: performance comparison of compressive sensing reconstruction algorithms. Info Syst Des Intell Appl
24. Rani M et al (2018) A systematic review of compressive sensing: concepts. Implementations Appl IEEE Access 6:4875–4894. https://doi.org/10.1109/ACCESS.2018.2793851
25. Pati YC, Rezaiifar R, Krishnaprasad PS (1993) Orthogonal matching pursuit: recursive function approximation with applications to wavelet decomposition. In: Proceeding of 27th asilomar conference on signals, systems, and computers, Pacific Grove, CA, vol 1, pp 40–44
26. Becker S (2016) A matlab function: CoSaMP and OMP for sparse recovery, version 1.7. Available online at:https://in.mathworks.com/matlabcentral/fileexchange/32402-cosamp-and-omp-for-sparse-recovery

A Low Harmonic High Gain Subthreshold Flipped Voltage Follower-Based Bulk-Driven OTA Suitable for Low-Frequency Applications

Sougata Ghosh, Saumya Tripathi, and Vijaya Bhadauria

Abstract This paper addresses the efficient configuration of a bulk-driven flipped voltage follower-based single-stage operational transconductance amplifier (OTA) for audio frequency applications. Partial positive feedback (PPF) introduced in diode-connected MOS transistor of the flipped voltage follower differential pair is capable of producing full ($-V_{SS}$ to V_{DD}) input as well as output dynamic range. This subthreshold operated OTA utilizing PPF techniques makes an improvement of voltage gain from its typical corner value of 55.89 dB to 71.34 dB and reduces the input-referred noise (IRN) from 2.166 μV/sqrt(Hz) to 779.43 nV/sqrt(Hz) at 1 kHz frequency. The simulation is performed by Cadence Virtuoso environment using UMC 180 nm CMOS process. The proposed topology using source degenerative resistor of 1 MΩ achieves a 71.35 dB of open-loop gain, unity gain frequency (UGF) of 1.073 kHz, phase margin of 89.07° while driving a high capacitive load of 50 pF. The performance comparison among all the reported OTA designs available in the literature intensifies that proposed OTA contributes an enhancement of 1.5 times in large-signal performances and maintains small-signal performances. The proposed OTA consumes only 140 nW of power and suitable for low-frequency biomedical and wireless sensor networks. The OTA is analyzed for five corners (TT, SS, FF, SNFP, and FNSP) with different temperatures of 0 °C–100 °C.

Keywords Operational transconductance amplifier (OTA) · Flipped voltage follower (FVF) · Unity gain frequency (UGF) · Partial positive feedback (PPF) · Source degeneration (SD)

S. Ghosh (✉) · V. Bhadauria
Department of Electronics and Communication Engineering, Motilal Nehru National Institute of Technology, Allahabad, Uttar Pradesh 211004, India
e-mail: sougata.vlsi@gmail.com

V. Bhadauria
e-mail: vijaya@mnnit.ac.in

S. Tripathi
Department of Electronics and Communication Engineering, Dr B R Ambedkar National Institute of Technology, Jalandhar, Punjab 144011, India
e-mail: saumya_96@hotmail.com

D. Harvey et al. (eds.), *Advances in VLSI, Communication, and Signal Processing*, Lecture Notes in Electrical Engineering 683,
https://doi.org/10.1007/978-981-15-6840-4_38

1 Introduction

In the recent past, the massive development of ultra-scaled CMOS technologies has forced the modern consumer electronics industry to design the portable handheld gadgets such as notebook and microcomputer, wireless telephone, personal digital assistants, distributed autonomous sensors, and biomedical implantable devices [1] in our daily life. This demand is steered by the endless integration of complex analog and digital modules in a single chip. Due to the expanded density of integrated circuits, low power consumption and silicon area are only criteria for portability and longer battery life. Analog designers are still relentlessly working to meet the above requirements through the appropriation of low voltage design techniques [2–14] in analog/mixed signal circuit. The demand of energy saving VLSI circuits is becoming more as result of downscaled CMOS processes with specific supply voltages. Integration of more components eventually increases chip power density. To reduce the power consumption, scaling of supply voltage and smaller components are essentially required. In present ultra-scaled VLSI processes, threshold voltage is one of the most important concerns for analog designers since downscaling of supply voltages and the threshold voltages of MOS devices are not proportional. Hence, it is difficult to maintain reliable performance under this stringent condition as the headroom of the analog operation gets reduced. So for $V_{DD} \leq 0.7$ V, bulk-driven amplifiers operating in subthreshold region with low bias currents (few nA) are often used to overcome the threshold voltage barrier problem from the signal path as well as for low power consumption. Therefore, sub-μW power-based biomedical applications within a few kHz, amplifiers are designed using subthreshold transistors due to drain-to-source voltage (V_{DS}) of around 78 mV as compared to 250 mV in strong inversion [15]. Also, sub-nm channel length devices associated with second-order effects reduce its inherent gain. Single-stage amplifiers with capacitive loads are most power-efficient but single-stage cascode amplifier (e.g., combination of common source with common gate device) is no longer used since this approach leads to limit output voltage swing due to the increased number of transistors vertically stacked together and also very difficult to achieve a large gain. In order to maintain a sufficient gain and wide output swing, two-stage amplifiers exploiting the cascoding techniques in the first stage are generally preferred but it requires a frequency compensation technique to stabilize the amplifier. If appropriate compensation is not used then almost 30–50% power efficiency might be lost. To increase the gain, a self-cascode [16] transistor is adopted for having high resistance value with single V_{DS} compared to cascade transistor.

OTA is the most fundamental building block used in the design and employment of medical equipment like pacemakers, human implants, bio-instruments and intra-body communication transceivers. As bio-potential signals are attributed to a bandwidth range of few kHz and amplitude in milli-volts, the OTA is required to have high gain with $-V_{SS}$ to V_{DD} input/output range, low noise and low harmonics. Since OTA comprised of simple MOS differential pair is not able to achieve linear transconductance (G_m) for large input differential voltages, various design techniques have been reported to improve the linearity [17–20]. Among these source degeneration techniques is widely followed, which can be used along with other linearization techniques to enhance the performances [21, 22]. Linearity of a transconductor is considered as a major issue concerning the design of low-frequency G_m-C filters. In last few years, several microelectronic circuits using bulk-driven transistors such as differential amplifiers, current mirrors, and voltage buffers have been designed to increase the ICMR range as well as output swing.

In this work, efforts have been given to design a high gain single-stage low power 0.7 V operated bulk-driven OTA using SD technique to improve the harmonics. Improvement of gain is achieved by increasing the output impedance using FVF current mirror load and decreasing the conductance by introducing partial positive feedback in FVF differential pair.

This paper has been organized in five sections. Section 2 deals with the analysis of LV and LP FVF OTA. The detailed analysis with required expressions are derived in Sect. 3, followed by the simulated results in Sect. 4. Comparison of state-of-the-art OTA designed with 180 nm technologies is shown in Sect. 5. Finally, the paper is ended in Sect. 6.

2 Circuit Description of Low Voltage and Low Power FVF OTA

Basically, OTA is a voltage to current converter which converts input differential voltage ($V_{in+} - V_{in-}$) into equivalent current depending on the transconductance factor at output node. V_{in+} and V_{in-} are referred to as non-inverting and inverting input voltages; similarly, V_{out} is single-ended output terminal of the OTA. The schematic diagram of the proposed FVF-based OTA is shown in Fig. 1. The proposed OTA is designed by flipped voltage follower-based differential pair (M_1, M_2). In this proposed circuit, identical transistors are related by $M_i = M_{i+1}$ where $i = 1, 2\ldots,$ 19. The biasing current sources of the FVF differential pair are replaced by diode-connected MOSFETs M_5 and M_6 respectively. Three current mirrors are used to copy the small-signal currents i_1 and i_2 generated by the applying differential input voltage, respectively. Out of three current mirrors two are high-performance low voltage FVF current mirrors and other one is self-cascode current mirror. The current transfer ratio of the mirrors is fixed to 1. The configuration has been designed in such a way that differential small current $\pm(i_1 - i_2)$ is allowed to flow into the corresponding load (C_L) connected to output terminal. The low voltage FVF current mirror load increases the output impedance and eventually increases the overall gain without any additional power. The expression for input and output impedances is given as:

$$R_{in} = \frac{1}{g_{m13} g_{m15} r_{out15}} \tag{1}$$

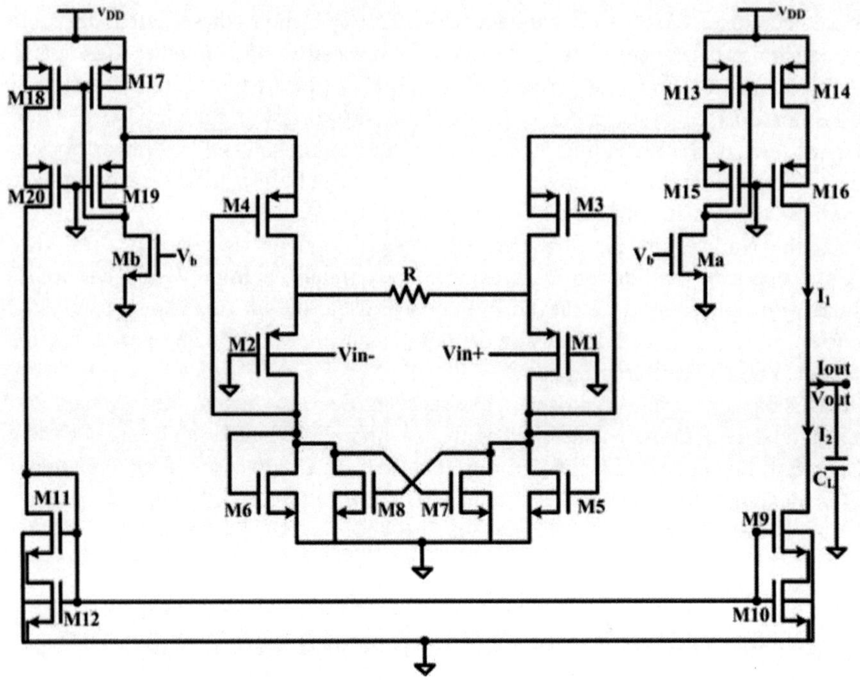

Fig. 1 Schematic of the proposed OTA

$$R_{\text{out}} = r_{\text{out15}} r_{\text{out16}} g_{\text{m16}} \tag{2}$$

The load of the FVF differential pair is taken as passive source degeneration resistor R. Tuning the values of R the overall transconductance of the circuit can be changed. Two additional MOSFETs M_7 and M_8 have been connected in cross-coupled manner between the two diode-connected MOS transistors M_5 and M_6 in FVF differential pair with 1:1 ratio. This connection separates the different paths for AC and DC currents. As a result, generated partial positive feedback reduces the conductance and increases the gain significantly. All the MOSFETs are operated in subthreshold region to reduce the power consumption of the proposed OTA. The source degeneration technique and bulk driving design perspective improves linearity, input common-mode range (ICMR) and output dynamic range. Almost rail-to-rail ICMR and voltage swing is observable from the simulation results.

One important trade-off lies between passive degeneration resistor and input-referred noise. For high values of R, input-referred noise increases but eventually decreases the overall gain. For $R = 1$, MΩ input-referred noise at 1 kHz is found to be 779.33 nV/sqrt(Hz).

The schematic of the proposed bulk-driven FVF-based OTA is shown in Fig. 1.

3 Design Analysis

This section correlates to the behavior of the proposed OTA mathematically using small-signal equivalent circuit.

3.1 Input Core Transconductance ($G_{m,I}$)

Input core flipped voltage follower-based differential pair is shown in Fig. 2. The gate terminal of the two p-MOS driving transistors is grounded which creates an inversion layer and generates a bias current I_{BIAS} through each transistor of the differential pair under no ac signal. The potential of source terminal of transistors M_1 and M_2 are denoted as V_{D1} and $V_{D2,}$ respectively. If a small-signal differential voltage is applied across the bulk of the M_1 and $M_{2,}$ then small-signal drain currents i_1 and i_2 is allowed to flow through the differential pair. Also, current across the source degenerative load resistor R is given by

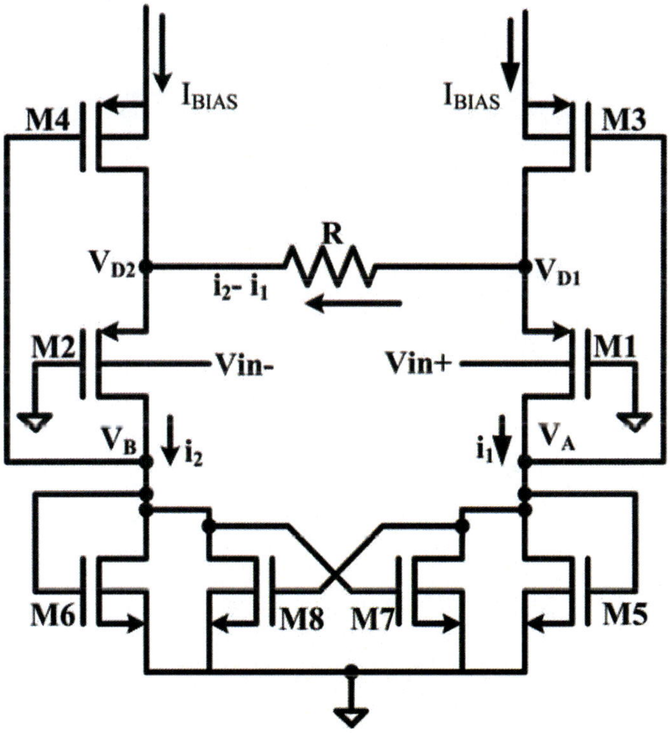

Fig. 2 Input core FVF differential pair (sub-circuit of proposed OTA)

$$i_{\mathrm{R}} = \frac{V_{\mathrm{D1}} - V_{\mathrm{D2}}}{R} \tag{3}$$

The differential input current, i_{id}, flows across the resistor R under small-signal condition but no currents flow under no signal condition. This differential input current is allowed to flow across output node through the capacitive load. Applying KCL across node at drain of M_1 and M_2, the following expressions for i_1 and i_2 are given by

$$i_1 = -g_{\mathrm{mb1}} V_{\mathrm{in+}} + g_{\mathrm{mb1}} V_{\mathrm{D1}} + g_{\mathrm{m1}} V_{\mathrm{D1}} \tag{4}$$

$$i_2 = -g_{\mathrm{mb1}} V_{\mathrm{in-}} + g_{\mathrm{mb1}} V_{\mathrm{D2}} + g_{\mathrm{m1}} V_{\mathrm{D2}} \tag{5}$$

Subtracting (4) from (5), the small-signal differential input current is given by

$$(i_2 - i_1) = g_{\mathrm{mb1}}(V_{\mathrm{in+}} - V_{\mathrm{in-}}) - g_{\mathrm{mb1}}.R.(i_2 - i_1) - g_{\mathrm{m1}}.R.(i_2 - i_1) \tag{6}$$

Therefore, input core transconductance can be written as

$$G_{m,i} = \frac{(i_2 - i_1)}{(V_{in+} - V_{in-})} = \frac{g_{mb1}}{1 + (g_{m1} + g_{mb1})R} = \frac{(n_p - 1)g_{m1}}{[g_{m1} + (n_p - 1)g_{m1}]R} \tag{7}$$

where n_p is subthreshold slope factor, and it is technology-dependent parameter ($1 < n_p < 2$). For 180 nm technology, $n_{\mathrm{p}} = \frac{C_{\mathrm{ox}} + C_{\mathrm{depl}}}{C_{\mathrm{ox}}} = 1.5$

3.2 Differential Gain

Differential voltage gain is expressed by

$$A_{\mathrm{V,diff}} = G_{\mathrm{m,i}} R_{\mathrm{out}} \tag{8}$$

where r_{out} is the overall output impedance of FVF current mirror which is given by

$$R_{\mathrm{out}} = r_{\mathrm{out15}} r_{\mathrm{out16}} g_{\mathrm{m16}} \tag{9}$$

The unity gain frequency of this OTA is found by

$$\text{Unity gain frequency (UGF)} = \frac{G_{\mathrm{m,i}}}{2\pi C_{\mathrm{L}}} \tag{10}$$

4 Simulation Results

The main circuit depicted in Fig. 1 is implemented and simulated using Cadence Virtuoso environment with UMC 180 nm standard CMOS technology. A supply voltage of 0.7 V is used to bias the circuits. In this design, the aspect ratio of all transistors has been selected by trading off all the design parameters in order to achieve the desired gain, low input-referred noise as well as low harmonics.

The deviation of input core currents I_{D1} and I_{D2} using different values of passive source degeneration resistor (R) is observed by DC transfer characteristics of the proposed OTA displayed in Fig. 3, which is executed by sweeping the input differential voltage from −0.7 V to +0.7 V. It is observed that the differential input core current ($I_{D1} - I_{D2}$) of OTA with passive SD resistor of 1 MΩ varies linearly throughout the entire range of input voltage.

Simulation of input transconductance $G_{m,i}$ with all passive SD resistor values is displayed in Fig. 4. $G_{m,I}$ has a wide dynamic range in the interval [−0.7 V, +0.7 V] for SD resistor of 1 MΩ and delivers a maximum value of 337.332 nA/V which is almost same as expected by (7), shown in (11).

For high value of $R = 1$ MΩ and $n_p = 1.5$, the overall input core transconductance $G_{m,i}$ is

$$G_{m,i} \simeq \frac{g_{mb1}}{(g_{m1} + g_{mb1})R} \simeq \frac{(n_p - 1)g_{m1}}{[g_{m1} + (n_p - 1)g_{m1}]R} \simeq \frac{1}{3R} \simeq 333\frac{\text{nA}}{\text{V}} \quad (11)$$

To perform the open-loop magnitude and phase response of the proposed OTA, differential input AC signal of 1 V magnitude is applied to the input with C_L of 50 pF. The response is shown in Fig. 5. A differential voltage gain of 71.35 dB, an UGF of

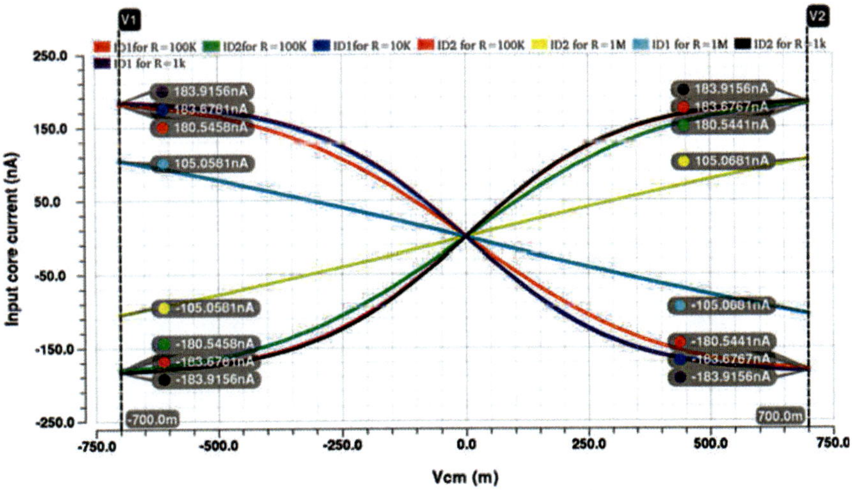

Fig. 3 DC transfer characteristics of proposed OTA

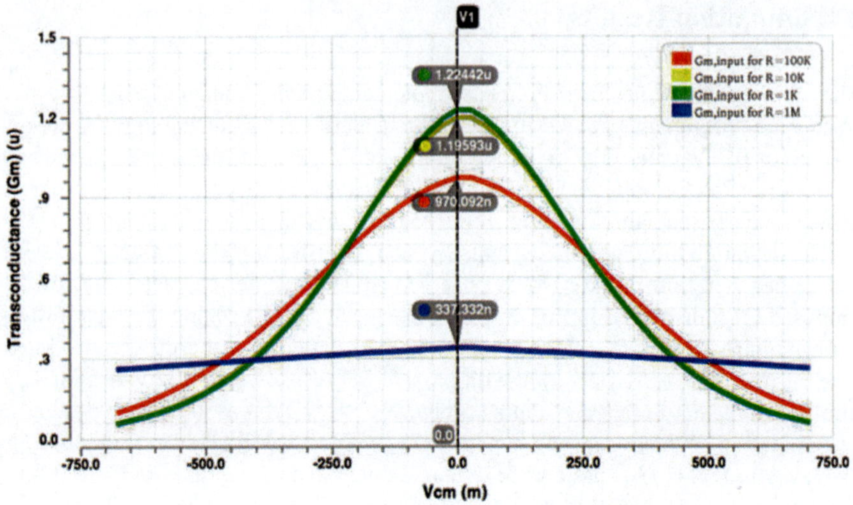

Fig. 4 Simulated transconductance curve for different SD resistor

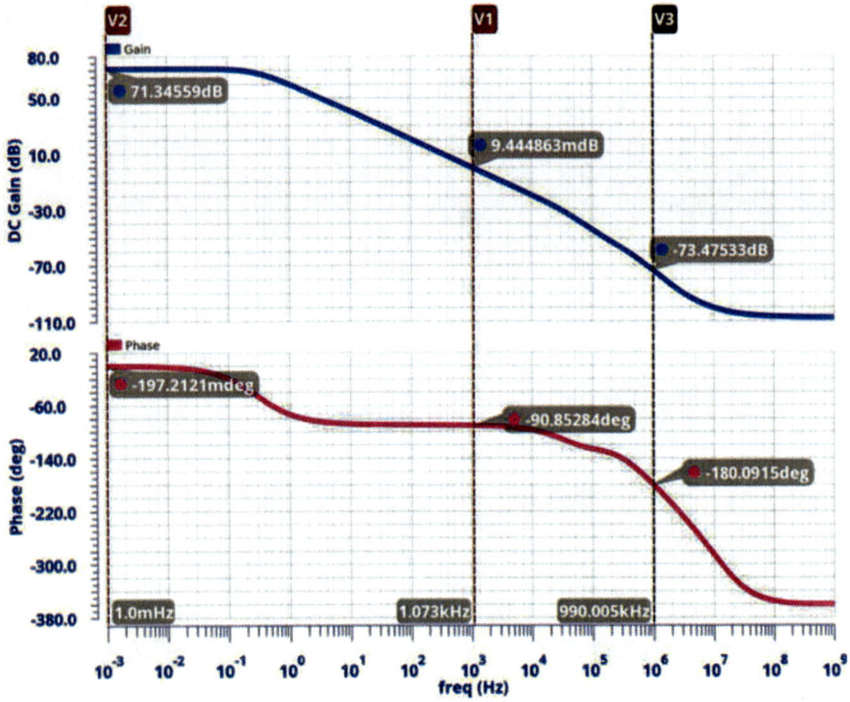

Fig. 5 Open-loop frequency response for gain and phase of the proposed OTA

1.073 kHz with phase margin of 89.07° is observed with SD resistor of 1 MΩ. The UGF appears to be low since the proposed OTA is driven by high value of capacitive load which is 50 pF.

UGF is found to be 1.073 kHz from open-loop gain and phase response shown in Fig. 5 which validates Eq. (10) for $C_L = 50$ pF and $G_{m,i}$ value in Eq. (11).

Figure 6 indicates the CMRR and PSRR ± response of OTA. The differential and common-mode voltage gain is observed to be 71.35 dB and −67.18 dB in Fig. 6a, which results a CMRR to be very high of 138.53 dB. Similarly, positive power supply gain of −5.74 dB and negative power supply gain of −6.34 dB from Fig. 6b and c result PSRR+ and PSRR− to be 77.08 dB and 77.68 dB, respectively, for V_{CM} of 0.35 V.

The voltage characteristic of IRN of the proposed OTA is presented in Fig. 7. Due to the large passive SD resistor of 1 MΩ, the noise is seen to be slightly high, i.e., 779.43 nV/sqrtHz as compared to 582.8 nV/sqrtHz using 100 KΩ.

In order to inspect the robustness of gain and phase response against the process and mismatch variations in this work, a Monte Carlo simulation for 1000 samples has been analyzed which is shown in Fig. 8. From the simulation results, the obtained sensitivity σ/μ (%) does not exceed more than 8% so the proposed OTA is processed insensitive due to process and mismatch variation. The Monte Carlo iterations of different performance parameters for 1000 samples have been shown in Table 1.

To observe the variation of different performance parameters under five process corners (nominal, FF, FNSP, SS, and SNFP) for room temperatures, corner analysis for gain, CMRR, and PSRR± has been shown in Fig. 9. The gain varies from minimum 63.71 dB in FF corner to maximum 78.94 dB in SS corner. It closely follows each other in TT, SNFP, FNSP corner and values are 71.34 dB, 71.20 dB, 71.03 dB, respectively. CMRR and PSRR± values in SS corner at room temperature are maximum, and values are 146.51 dB, 85.01 dB, and 84.95, respectively. Also, all parameters are analyzed to observe variations on different corners as well as for temperatures 0 °C, 50 °C, 100 °C. The variations of all parameters for all the five corners at temperatures of 0 °C, 50 °C, 100 °C are summarized in Table 2.

Figure 10 illustrates the simulation result of input common mode range in unity gain closed-loop configuration and output voltage swing of proposed OTA using DC sweep analysis. The obtained ICMR and output voltage swing range are found to be rail-to-rail.

To calculate slew rates (±), large-signal pulse response of the proposed OTA has been done in unity gain feedback configuration using step input of 0.7 V amplitude and 100 Hz frequency having 50% duty cycle with a C_L of 50 pF. The slew rate at

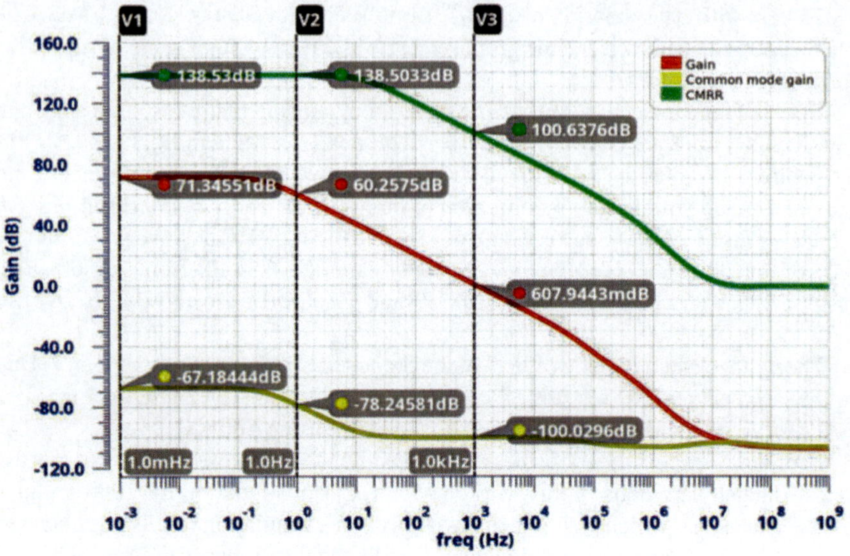

(a)

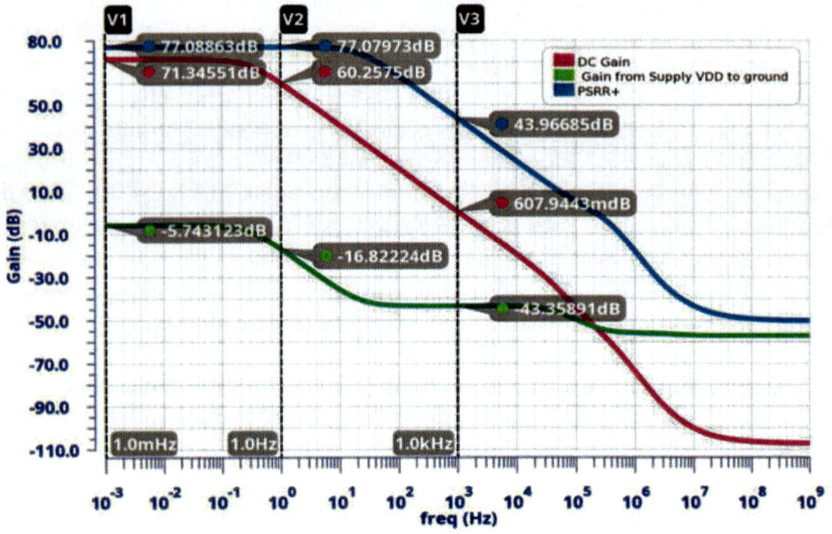

(b)

Fig. 6 CMRR (**a**), PSRR+ (**b**) and PSRR− (**c**) response at V_{CM} of 0.35 V

rising and falling edges from simulation results have been observed to be 1.579 V/ms and plot is shown in Fig. 11.

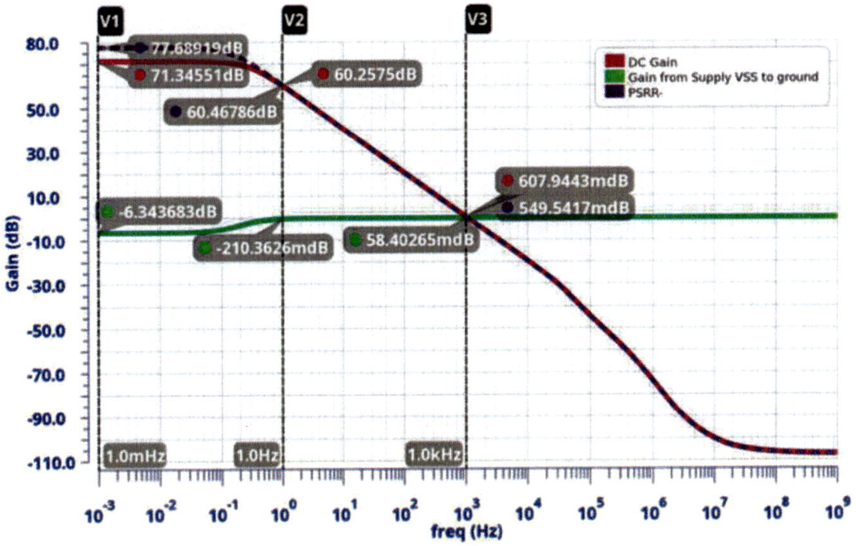

(c)

Fig. 6 (continued)

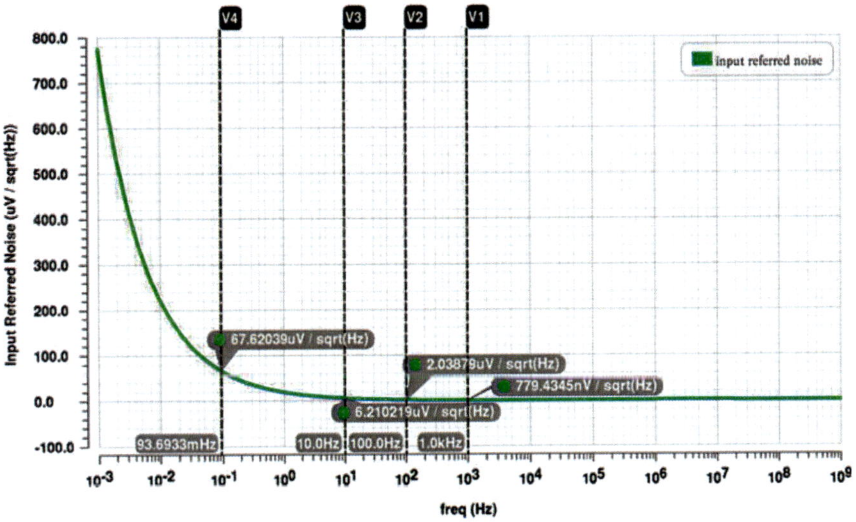

Fig. 7 Input-referred noise at single-end output (V_{out}) of the proposed OTA

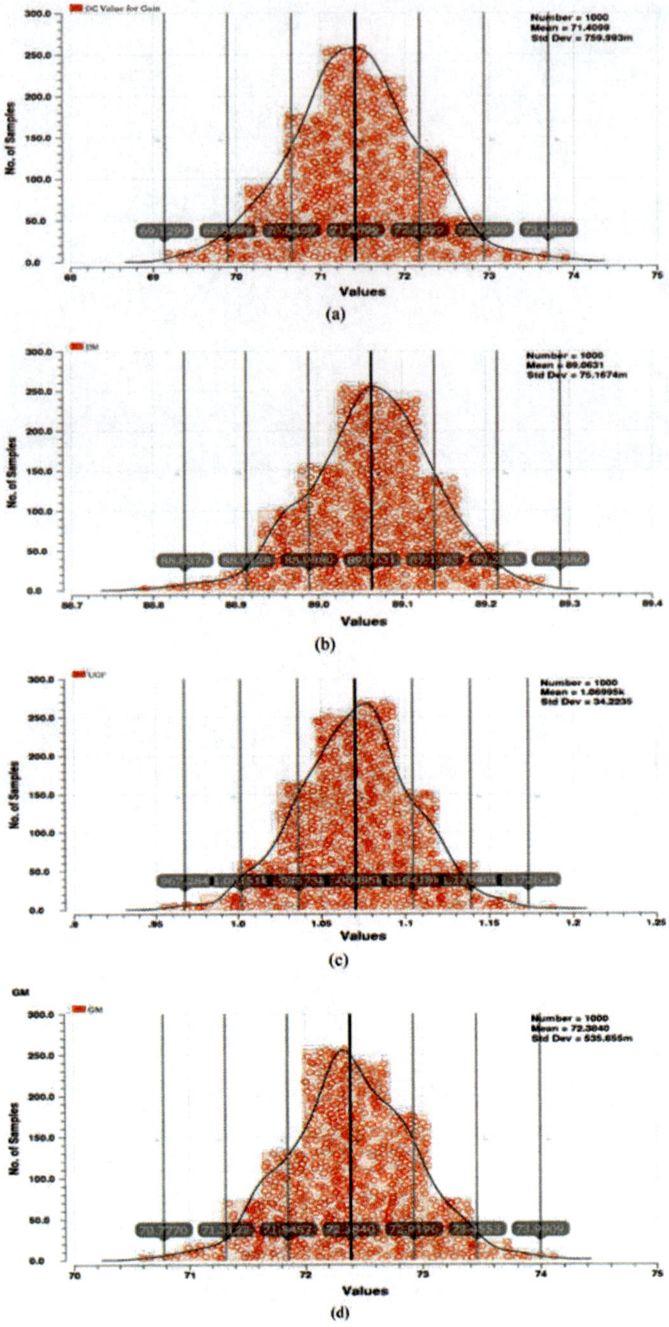

Fig. 8 Monte Carlo simulation results for 1000 iterations: gain (**a**), phase margin (**b**), GBW (**c**), and gain margin (**d**)

Table 1 Monte Carlo iteration results for 1000 samples

Parameter	Mean value (μ)	Standard deviation (σ)	σ/μ (%)
Open-loop DC gain (dB)	71.4099	759.993 m	1.06
UGF (kHz)	1.06995	34.2235 m	3.19
Gain margin (dB)	72.3840	535.655 m	0.74
Phase margin (°)	89.0631	75.1674 m	0.08
Slew rate (±) (V/ms)	1.57566 at 50 pF load	51.8473 m	3.29
Input-referred noise (nV/sqrt(Hz))	778.981	14.1690 n	1.81
Common-mode Gain ($A_{v,cm}$) (dB)	−67.1903	43.0743 m	0.06
Power supply gain (A_{vdd}) (dB)	−5.74592	35.7474 m	0.62
CMRR (dB) at 1 MHz	138.60	802.176 m	0.57
PSRR+ (dB) at 1 MHz	77.1558	795.572 m	1.03
Rise time (μs)	333.114	2.18660 u	0.65
Fall time (μs)	333.114	2.18660 u	0.65
Power dissipation (nW)	139.0669	11.0151 n	7.92
Total stand by current consumption (nA)	198.667	15.7358 n	7.92

5 Comparison

The proposed OTA contributes second-largest DC gain compared to OTAs powered by 0.7 V supply and exhibits a highest CMRR and phase margin among all studied works shown in Table 3. Although the power consumption is second-lowest among all, it shows little bit high input-referred noise than other works. As the proposed OTA drives a high capacitive load of 50 pF, it appears to be specified by lower UGF and slew rate values than others which consume a remarkably greater amount of power. Therefore, two figures of merit ($\mathbf{FOM_{UGF}}$ and $\mathbf{FOM_{SR}}$) are embraced to compare relevantly the effective performances of the OTAs under examination.

This work shows a significant improvement in $\mathbf{FOM_{SR}}$ which grants more appropriate compensation between speed, power, and load to be compared with other reported OTAs. The large-signal driving capability of proposed design provides 8.21% more than [16], 1.46% more than [24], 1.62% more than [25]. Though small-signal driving ability is 1.33% and 1.18% more than [16, 23], 3.51% and 1.60% lower than [24, 25].

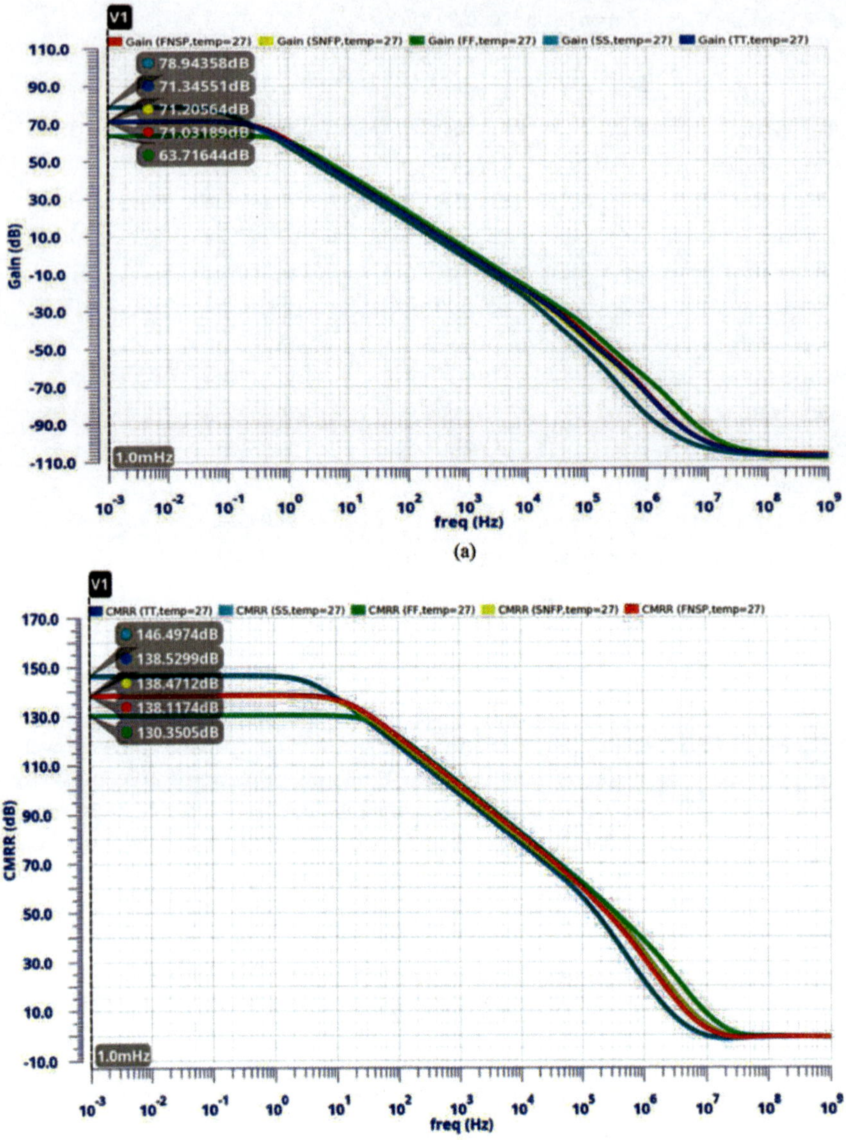

(a)

(b)

Fig. 9 Effect of process corners on (**a**) Gain, (**b**) CMRR, (**c**) PSRR+, (**d**) PSRR−, at 27 °C

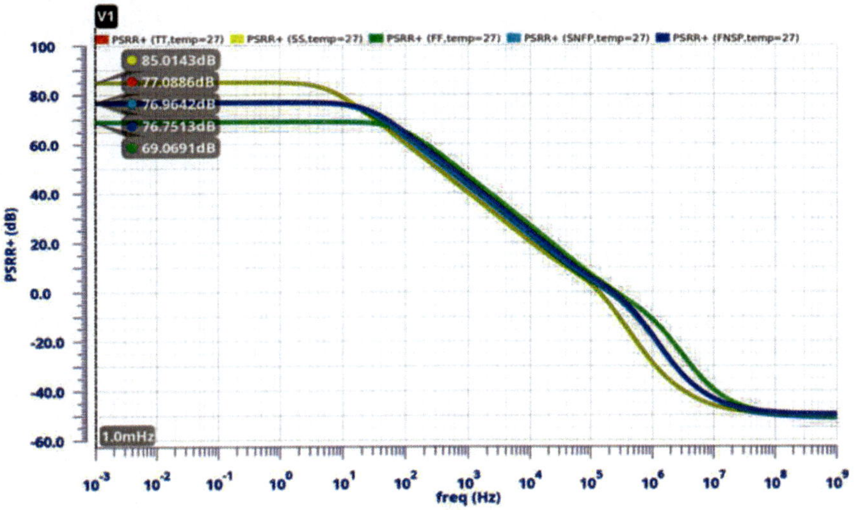

(c)

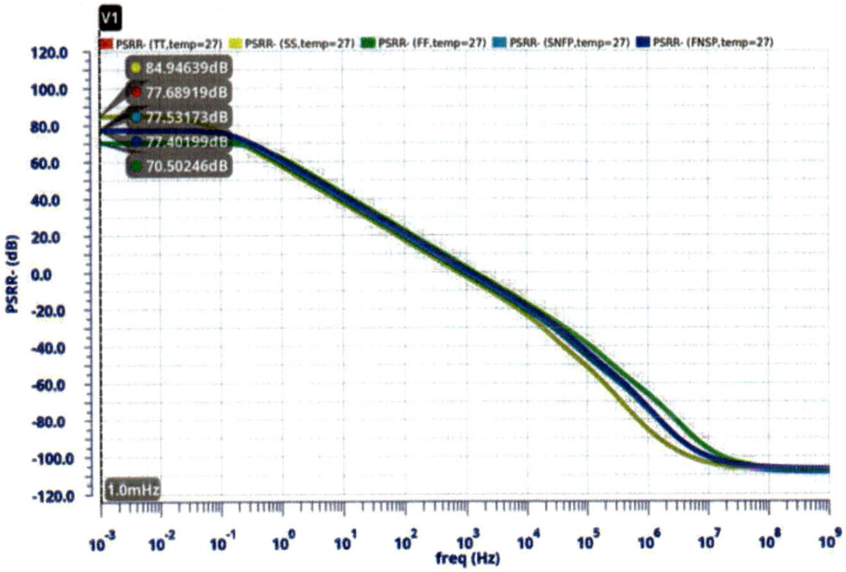

(d)

Fig. 9 (continued)

6 Conclusion

This work addresses the techniques of LV, LP OTA based on FVF-based differential pair along with low voltage FVF-based current mirror. The proposed OTA exhibits

Table 2 Effect of process and temperature variation on different parameters

Performance parameter	Corner for temperature = 0 °C				
	SS	TT	FF	SNFP	FNSP
DC Gain	83.448	75.45	67.23	75.40	75.06
UGF	0.637 k	0.983 k	1.32 k	0.854 k	1.12 k
Phase margin	86.26	88.38	89.33	88.53	88.27
GM	61.62	66.94	73.99	69.16	66.94
CMRR	151. 71	143.40	134.83	143.36	142.99
PSRR+	89.57	81.24	72.65	81.22	80.82
PSRR−	89.39	81.74	73.93	81.66	81.38
I_T (nA)	44.22	106.705	254.942	112.62	104.335
Power (nW)	31.155	75.06	179.095	79.215	73.395
Performance parameter	Corner for temperature = 50 °C				
	SS	TT	FF	SNFP	FNSP
DC gain	76.07	69.04	61.79	68.65	68.86
UGF	0.802	1.12	1.47	1.00	1.26
Phase margin	87.81	89.03	89.60	89.12	88.97
GM	69.07	73.77	77.95	74.97	72.77
CMRR	147.55	137.06	127.90	136.10	137.79
PSRR+	82.14	74.70	67.01	74.33	74.50
PSRR−	82.07	75.48	68.75	75.07	75.33
I_T (nA)	205.61	401.63	775	418.81	394.04
Power (nW)	143.925	281.14	545	293.165	275.825
Performance parameter	Corner for temperature = 100 °C				
	SS	TT	FF	SNFP	FNSP
DC gain	68.69	62.62	56.35	61.90	62.67
UGF	0.967	1.27	1.62	1.14	1.40
Phase margin	89.37	89.68	89.88	89.71	89.67
GM	76.52	79.62	81.91	80.77	78.60
CMRR	143.40	130.71	120.98	128.83	132.60
PSRR+	74.72	68.16	61.37	67.45	68.19
PSRR−	74.75	69.21	63.57	68.47	69.29
I_T (nA)	366.995	695	1300	725	680
Power (nW)	256.895	487.585	910	505	478.62

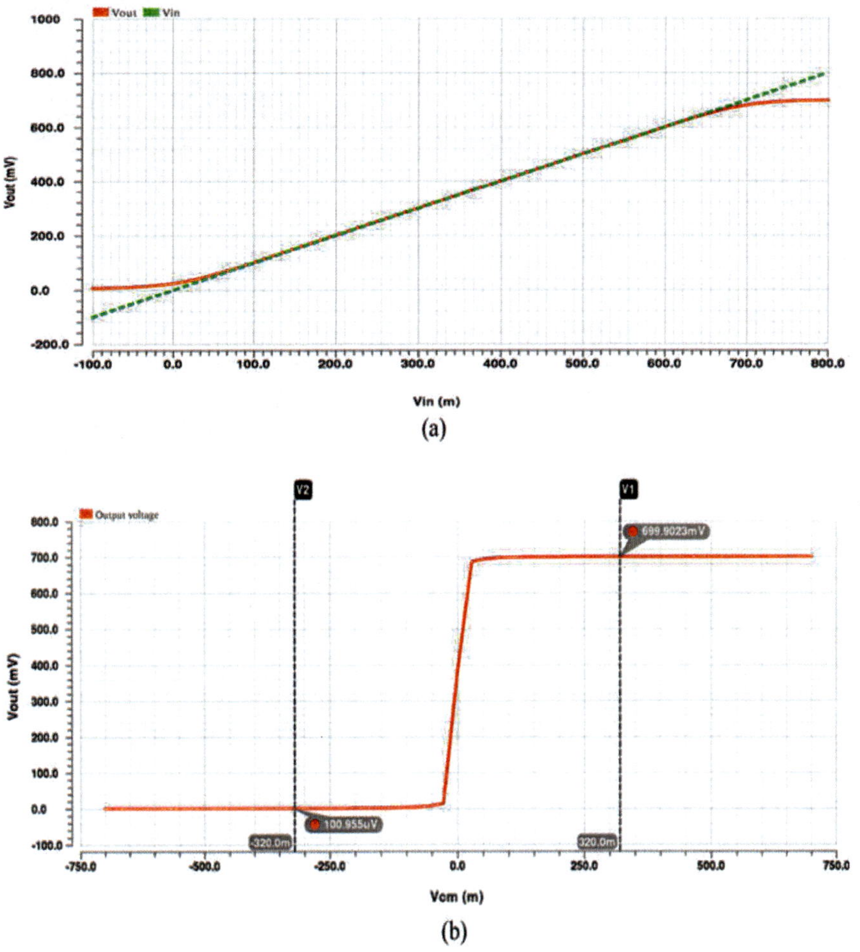

Fig. 10 ICMR response (**a**), output voltage swing (**b**), of proposed OTA

full range ($-V_{SS}$ to V_{DD}) ICMR and output voltage swings. Linearity improvement is obtained at the cost of DC gain. A open-loop gain of 85.12 dB is observed for source degeneration resistance of 100 KΩ or lower; however, power consumption remains the same. Compared to report published works of OTA, the proposed low harmonic design shows enriched performances with LV, LP environment. The bulk-input subthreshold operated FVF differential pair consumes only 140 nW power at 0.7 V supply; therefore, it is suitable for LP data acquisition system for bio-implantable devices.

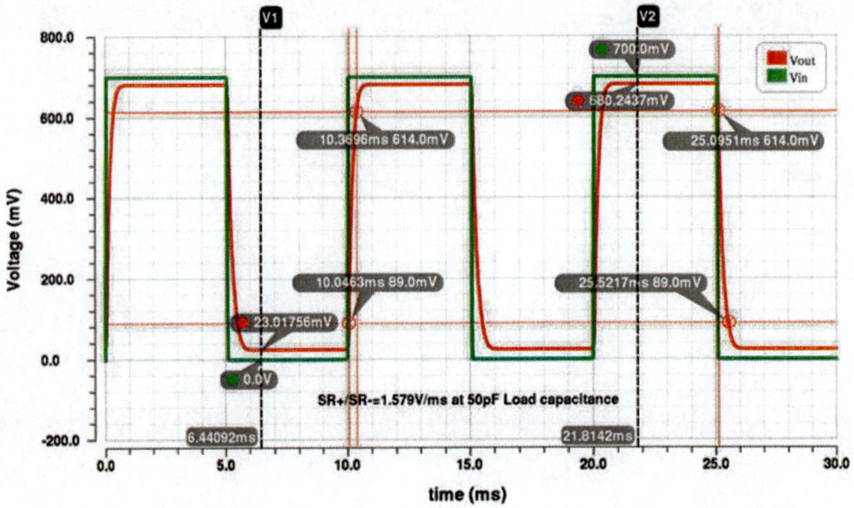

Fig. 11 Large-signal pulse response in unity gain feedback configuration

Table 3 Performance comparison of proposed OTA with other OTAs designed in 180 nm technology

Parameters	[16]	[23]	[24]	[25]	[This work]
Power supply (V)	0.5	0.5	0.6	0.5	0.7
CMOS process node (μm)	0.18	0.18	0.18	0.18	0.18
Output type	FD	FD	SE	SE	SE
Configuration	BD	BD	BD	BD	BD
DC gain	65	38.5	82	67.8	71.35
UGF (MHz)	0.55	0.027	0.019	0.00326	0.00107
Phase margin	50	69	60	68.9	89.07
CMRR at 1 Hz	86 at 5 kHz	53.15	130.2 at 100 Hz	NA	138.50
PSRR+	76	NA	NA	NA	77.08
PSRR−	NA	NA	NA	NA	60.23
IRN ($\mu V/\sqrt{Hz}$)	0.432 at 1 kHz	2 at 0.1 kHz	0.16 at 1 kHz	0.56 at 1 kHz	0.779 at 1 kHz
Slew rate± (V/ms)	230	NA	12	0.84/0.59	1.579/1.579
Load capacitance (pF)	20	5	15	15	50

(continued)

Table 3 (continued)

Parameters	[16]	[23]	[24]	[25]	[This work]
I_T (μA)	56	0.6	0.667	0.052	0.2
Power dissipation (μW)	28	0.3	0.4	0.026	0.14
FOM_S	20	22.5	43	94	26.75
FOM_L	8.21	NR	27	24.23	39.47

References

1. Chatterjee S, Tsividis Y, Kinget P (2005) 0.5-V analog circuit techniques and their applications in OTA and filter design. IEEE J Solid-State Circ 40(12):2372–2387
2. Calhoun BH, Daly DC, Verma N, Finchelstein DF, Wentzloff DD, Wang A, Cho SH, Chandrakasan AP (2005) Design considerations for ultra-low energy wireless microsensor nodes. IEEE Trans Comput 54(6):727–740
3. Tabesh A, Fréchette LG (2009) A low-power stand-alone adaptive circuit for harvesting energy from a piezoelectric micropower generator. IEEE Trans Industr Electron 57(3):840–849
4. Romani A, Filippi M, Tartagni M (2013) Micropower design of a fully autonomous energy harvesting circuit for arrays of piezoelectric transducers. IEEE Trans Power Electron 29(2):729–739
5. Harrison RR, Charles C (2003) A low-power low-noise CMOS amplifier for neural recording applications. IEEE J Solid-State Circuits 38(6):958–965
6. Chandrakasan AP, Verma N, Daly DC (2008) Ultralow-power electronics for biomedical applications. Annu Rev Biomed Eng 10:247–274
7. Ghosh S, Bhadauria V (2018) 0.5 V two-stage subthreshold fully differential miller compensated OTA using voltage combiners. In: Advances in VLSI, communication, and signal processing: select proceedings of VCAS 2018, 587, 463
8. Abdelfattah O, Roberts GW, Shih I, Shih YC (2015) An ultra-low-voltage CMOS process-insensitive self-biased OTA with rail-to-rail input range. IEEE Trans Circ Syst I Regul Pap 62(10):2380–2390
9. Kulej T, Khateb F (2015) 0.4-V bulk-driven differential-difference amplifier. Microelectron J 46(5):362–369
10. Ferreira LH, Pimenta TC, Moreno RL (2007) An ultra-low-voltage ultra-low-power CMOS miller OTA with rail-to-rail input/output swing. IEEE Trans Circ Syst II Express Briefs 54(10):843–847
11. Veldandi H, Shaik RA (2018) A 0.3-V Pseudo-differential bulk-input OTA for low-frequency applications. Circ Syst Sig Proc 37(12):5199–5221
12. Akbari M, Nazari M, Sharifi L, Hashemipour O (2015) Improving power efficiency of a two-stage operational amplifier for biomedical applications. Analog Integr Circ Sig Process 84(2):173–183
13. Ferreira LH, Sonkusale SR (2014) A 60-dB gain OTA operating at 0.25-V power supply in 130-nm digital CMOS process. IEEE Trans Circ Syst I Regul Pap 61(6):1609–1617
14. Sharan T, Chetri P, Bhadauria V (2018) Ultra-low-power bulk-driven fully differential subthreshold OTAs with partial positive feedback for G_m-C filters. Analog Integr Circ Sig Process 94(3):427–447
15. Sharan T, Bhadauria V (2017) Fully differential, bulk driven, class AB, sub-threshold OTA with enhanced slew rates and gain. J Circ Syst Comput 26(1):1750001
16. Trakimas M, Sonkusale S (2009) A 0.5 V bulk-input OTA with improved common-mode feedback for low-frequency filtering applications. Analog Integr Circ Sig Process 59(1):83–89

17. Chen J, Sanchez-Sinencio E, Silva-Martinez J (2006) Frequency-dependent harmonic-distortion analysis of a linearized cross-coupled CMOS OTA and its application to OTA-C filters. IEEE Trans Circ Syst I Regul Pap 53(3):499–510
18. Sengupta S (2005) Adaptively biased linear transconductor. IEEE Trans Circ Syst I Regul Pap 52(11):2369–2375
19. Kimura K (1997) U.S. Patent No. 5,602,509. U.S. Patent and Trademark Office, Washington, DC
20. Baruqui FAP, Petraglia A (2006) Linearly tunable CMOS OTA with constant dynamic range using source-degenerated current mirrors. IEEE Trans Circ Syst II Express Briefs 53(9):797–801
21. Ohbuchi T, Matsumoto F (2013) A new design of a linear local-feedback MOS transconductor for low frequency applications. Analog Integr Circ Sig Process 75(2):257–266
22. Rezaei F, Azhari SJ (2010) A highly linear Operational transconductance amplifier (OTA) with high common mode rejection ratio. In: 2010 international conference on signal acquisition and processing. IEEE, pp 18–22
23. Suda N, Nishanth PV, Basak D, Sharma D, Paily RP (2014) A 0.5-V low power analog front-end for heart-rate detector. Analog Integr Circ Sig Process 81(2):417–430
24. Akbari M, Hashemipour O (2016) A 0.6-V, 0.4-μW bulk-driven operational amplifier with rail-to-rail input/output swing. Analog Integr Circ Sig Process 86(2):341–351
25. Zhao X, Fang H, Ling T, Xu J (2015) Transconductance improvement method for low-voltage bulk-driven input stage. Integration 49:98–103

Performance Analysis of MoS_2FET for Electronic and Spintronic Application

Kavindra Kumar Kavi, R. A. Mishra, and Shweta Tripathi

Abstract A new class of 2D materials that is transition metal dichalcogenide (TMDCs) provides better control over channel the thickness and bandgap. Among the various TMDCs, the molybdenum disulfide (MoS_2) is one of the most promising candidates for electronic and optoelectronic switches, including large on/off current ratio, small sub-threshold swing, and high field-effect mobility and future digital applications. In this paper, the analysis of molybdenum disulfide field-effect transistors (MoS_2FETs) is done in terms of the variation of drain current with drain voltage, top gate and back gate voltage, and variation of mobility with the thickness of MoS_2 layered materials. A commercially available 2D device simulation software ATLAS TM has been used to obtain the simulation results.

Keywords Two-dimensional (2D) materials · Transition metal dichalcogenide (TMDCs) · Molybdenum disulfide (MoS_2) · Molybdenum disulfide field-effect transistors (MoS_2FETs) · Subthreshold swing (SS)

1 Introduction

In general, there are many applications of semiconductor materials in the field of electronics and optics. Among them, one of the most important applications of semiconductor is a transistor. Since the last couple of decade study, scaling and fabrication of metal-oxide-semiconductor field-effect transistors (MOSFETs) have been performed. However, the saturation of Moore's law is being observed because successive scaling reached its limit caused by quantum effects and difficulty in heat dissipation [1]. Such circumstance encourages us to search for new device designs or materials. Fortunately, 2D materials have been found as potentials candidate to overcome the short-channel effects that can limit device performance significantly.

Generally, to become an ideal field-effect transistor, the following favorable characteristics must be there such as a high on/off ratio for switching applications, low

K. K. Kavi (✉) · R. A. Mishra · S. Tripathi
Motilal Nehru National Institute of Technology Allahabad, Prayagraj, Uttar Pradesh 211004, India
e-mail: kavipsu@gmail.com

D. Harvey et al. (eds.), *Advances in VLSI, Communication, and Signal Processing*, Lecture Notes in Electrical Engineering 683,
https://doi.org/10.1007/978-981-15-6840-4_39

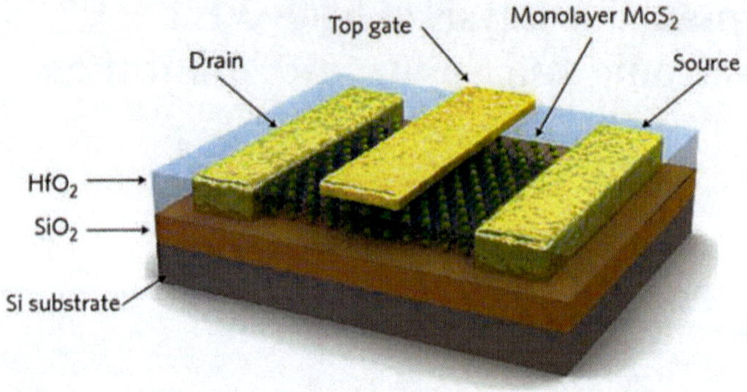

Fig. 1 MoS_2FET [11]

off-state conductance for low power applications, and high carrier mobility for high-speed operation [2]. Source, drain, and channel materials are doped, to increase the charge density, which, however, results in scattering and mobility degradation [3]. An on/off ratio of 10^4–10^7 is required for digital logic switching applications. Even though graphene is promising for RF applications because of its extremely high carrier mobility and excellent transconductance [4–7], it is not suitable for logic applications because of the absence of bandgap resulting in large leakage current. Therefore, new ultra-thin material is required with a reasonable bandgap and large mobility so that a large on/off ratio and a large on-current can be achieved. Among the many possible materials, molybdenum disulfide (MoS_2) is one of the most suitable candidates for future digital applications. It shows promising characteristics for electronic and optoelectronic switches, including a large on/off current ratio, a small subthreshold swing, and high field-effect mobility [8]. MoS_2 is two-dimensional layered materials which have a tunable bandgap [9]. The Bulk MoS_2 has an indirect bandgap of 1.3 eV, whereas monolayer MoS_2 has a direct bandgap of 1.8 eV [10].

The first n-type MoS_2 transistor was introduced by Kis, et al. [11], as shown schematically in Fig. 1. The top-gated structure with single-layer MoS_2 showed significant on/off current ratio (~10^8), large room temperature mobility (200 cm^2 V^{-1} s^{-1}), and 74 mV/decade of subthreshold swing (SS). In order to increase the mobility and dielectric engineering, there are lots of theoretical studies have been done. The layer concern research works for single layer as well as multilayer channel using different high-*k* dielectrics are exploring by many research group globally. A single-layer MoS_2 transistor having 0.65 nm channel thickness is fabricated by Radisavljevic et al. In this process, the technique used in order to synthesize the MoS_2 layer was the scotch tape method. 30 nm thick oxide layer of HfO_2 used as gate dielectric which is deposited by the atomic layer deposition technique, and for the contacts, electron beam lithography technique is applied. Due to the presence of high-*k* dielectric high mobility of 217 $cm^2/(V\ s)$ and large on/off current ratio of 10^8 achieved.

In this paper, the MoS_2FET has been proposed and analyzed to obtain the electrical parameters including drain currents variation for drain voltages and gate voltage and mobility variation with the thickness of channel materials. A commercially available 2D device simulation software ATLAS ^{TM} has been used to get the simulation results.

2 Devices Structure

Silicon is used as substrate material having a thickness of 230 μm, which also works as back gate contact. Figure 2 shows the schematic device for a single-layer MoS2FET

A single layer of MoS_2 (thickness, 6.5 Å) is used as a channel and the length of the channel is 20 nm placed over 150 nm thin silicon dioxide SiO_2 layer. For charge trapping, the SiO_2 layer has been used. In order to improve the mobility and the performance of device, there are few terms or parameters which having the proper use of (i) source/drain contact engineering, (ii) the high-*k* gate dielectric material, and (iii) doping techniques. HfO_2 is used as high-k gate dielectric in this device for boosting the mobility with dielectric engineering having a thickness of 30 nm. The device structure of MoS_2FET obtained during simulation using commercially available 2D device simulation software ATLAS^{TM} is shown in Fig. 3.

Choosing an appropriate metal contact to improve the device performance by achieving low contact resistance for MoS_2-based FET is still challenging. The contact materials gold is used as conductor electrodes for source and drain where drain electrode is used for supply voltage and other source electrode is grounded. Chromium or gold made electrode used for gate terminal and applied an appropriate voltage to turn on the devices and created the channel in order to give a proper motion of carriers. The dimensions of device parameters like channel length and thickness, oxides thickness, and electrodes are given in Table 1.

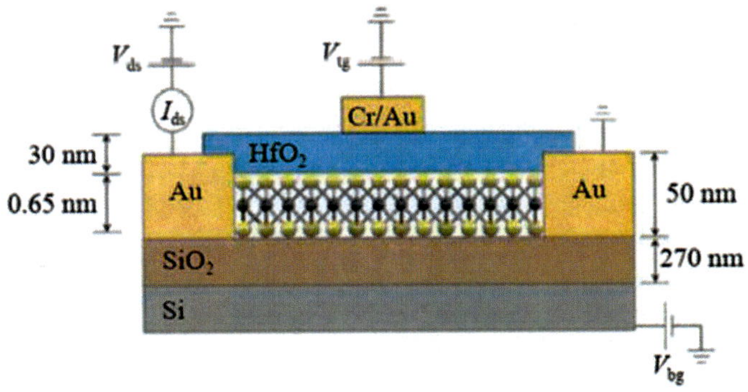

Fig. 2 Schematic device structure for a single-layer MoS2FET

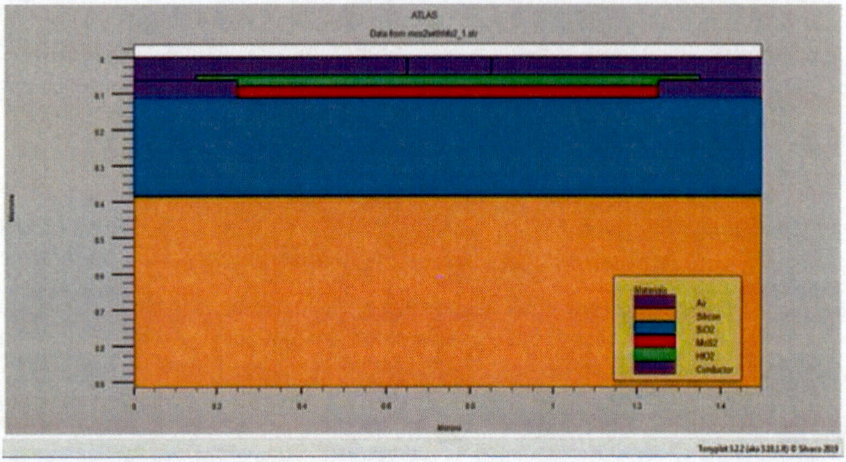

Fig. 3 ATLAS devices structure of MoS_2 FET

Table 1 MoS_2FET devise parameters

S. No.	Parameters	Values (nm)
1	Thickness of MoS_2 (t_{MoS2})	0.65
2	Oxide thickness (t_{ox}) SiO_2	270
3	Channel length (L)	20
4	Substrate thickness (t_s)	230
5	Oxide thickness (t_{ox}) HfO_2	30

3 Results and Discussions

MoS_2-based FET has been demonstrated to exhibit high on/off current ratio exceeding 10^8, high mobilities and low sub-threshold swing at room temperature, indicating its potential employment in future electronic devices [13]. Figure 4a the drain-source current (I_{ds}) versus drain-source voltage (V_{ds}) curves at a different gate-source voltage (V_{gs}). Since highly p-type doped silicon has been used as substrate so when a small V_{gs} is applied there is no flow of electron but as the gate voltage is increased and fixed at a certain level a huge number of electron accumulate in the MoS_2 channel. Keeping gate voltage constant, the applied drain voltage is varied from 0 to 1 V with the step of 0.2 V. It can be observed that for small drain voltage, there is no flow of drain current up to the threshold voltage (V_t) but beyond V_t the flow of electron is observed in form of current from drain to. Similarly, high I_{ds} is observed at the different larger value of V_{gs}.

In Fig. 4b, the variation of I_{ds} with respects to top gate voltage V_{tg} is shown for the high and low value of V_{ds}. It can be observed that almost 20 μA and 10 μA drain current is found for the high and lower value of $V_{ds,}$ respectively. In conventional MOSFETs, the threshold voltage depends on the ratio of gate oxide capacitance to

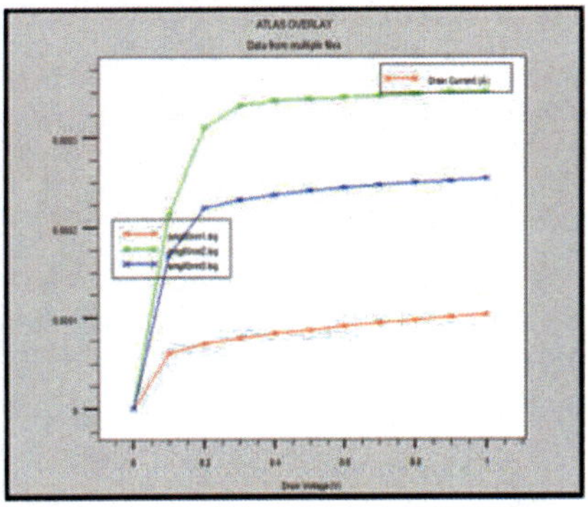

(a) Plot of I_d Vs V_{ds}

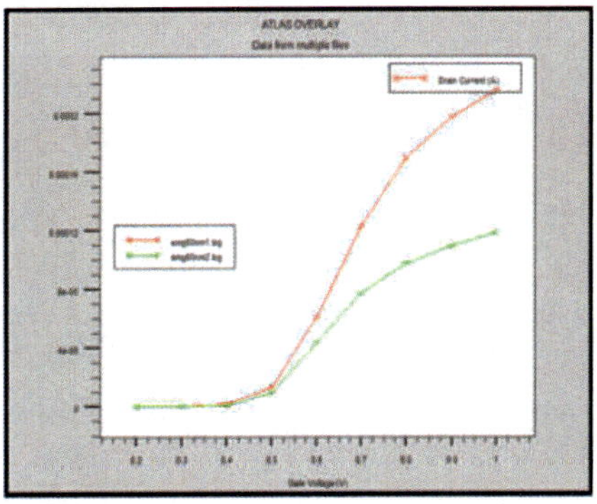

(b) Plot of I_d Vs V_{tg}

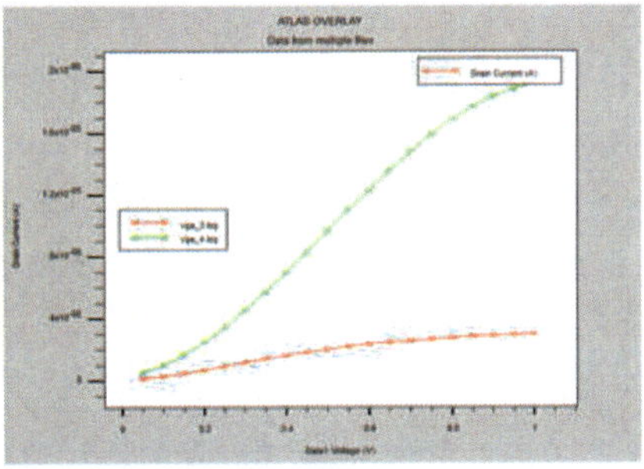

(c) Plot of I_d Vs V_{bg}

Fig. 4 **a** Plot of I_d versus V_{ds} **b** Plot of I_d versus V_{tg} **c** Plot of I_d versus V_{bg}

the other parasitic capacitance such as channel capacitance and interface trap-state capacitance and has a theoretical limit of about 0.7 V at room temperature but due to layered MoS_2, the present structure has the capacity of accumulating the huge number of electron at comparatively low voltage and the threshold voltage is found as 0.5 V.

Figure 4c illustrates the variation of I_{ds} with respect to back gate voltage V_{bg} for the high and low value of V_{ds}. The substrate for the MoS_2-based FET has used highly doped silicon so the huge number of electron accumulates at top of the oxide layer and in the channel after applying the V_{bg}. Here, the silicon substrate itself treated as the back gate contact terminal. For the high and low value of $V_{ds,}$, the drain current is found as 0.2 μA after applying the back gate voltage beyond the threshold voltage.

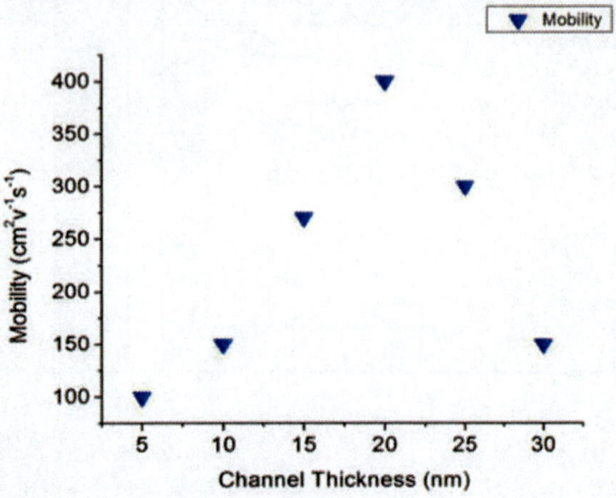

Fig. 5 Variation of mobility versus channel thickness

4 Channel Thickness Versus Mobility Curve

In order to evaluate the layer thickness-dependent electronic properties, we have calibrated the back gate configurations of MoS_2FET. In general, a thinner MoS_2 can usually offer a lower off-state current, but a thicker one has a larger on-state current. Due to the partial screening of electric field by the bottom layers, interlayer resistance and/or Coulomb scattering effect the mobility varies according to the thickness [12]. Above the 20nm thickness of MoS_2, back gate get very hard to turn on and off due to the increasing screening effect. A plot of mobility versus thickness shows that a MoS_2 with a thickness between 5 and 30 nm exhibits both the high mobility and high on--off ratio (Fig. 5). Therefore, we have mainly focused on the MoS_2 flakes with a thickness between 5 nm and 30 nm for the simulations of high-performance MoS_2FETs in this study.

5 Conclusions

In this paper, we have analyzed the drain current characteristics by variations of different electrical parameters and changes in mobility for different channel thickness in MoS_2FET by numerical simulation approach. A commercially available 2D device simulation software ATLAS ^{TM} has been used to get the simulation results. The optimizations of drain current are done by applying the high-k dielectric materials and channel engineering. To achieve high-performance MoS_2FET, the ferromagnetic materials can be used as contact for layered materials.

References

1. ITRS (2013) The international technology roadmap for semiconductors
2. Lu W, Lieber CM (2007) Nanoelectronics from the bottom up. Nat Mater 6(11):841–850
3. Yoon Y, Ganapathi K, Salahuddin S (2011) How good can monolayer MoS_2 transistors be? Nano Lett 11(9):3768–3773
4. Schwierz F (2010) Graphene transistors. Nat Nanotechnology 5(7):487–496
5. Wu Y, Jenkins KA, Valdes-Garcia A, Farmer DB, Zhu Y, Bol AA, Dimitrakopoulos C, Zhu W, Xia F, Avouris P, Lin Y-M (2012) State-of-the-art graphene high frequency electronics. Nano Lett 12(6):3062–3067
6. Wu Y, Lin Y, Bol AA, Jenkins KA, Xia F, Farmer DB, Zhu Y, Avouris P (2011) High frequency, scaled graphene transistors on diamond-like carbon. Nature 472(7341):74–78
7. Lin Y, Valdes-Garcia A, Han S-J, Farmer DB, Meric I, Sun Y, Wu Y, Dimitrakopoulos C, Grill A, Avouris P, Jenkins KA (2011) Wafer-scale graphene integrated circuit. Science 332(6035):1294–1297
8. Yoon Y, Ganapathi K, Salahuddin S (2011) How good can monolayer MoS_2 transistors be? Nano Lett 11:3768–3773
9. Wang QH, Kalantar-Zadeh K, Kis A, Coleman JN, Strano MS (2012) Electronics and optoelectronics of two-dimensional transition metal dichalcogenides. Nature Nanotechnol 7:699–712
10. Liu L, Kumar SB, Ouyang Y, Guo J (2011) Performance limits of monolayer transition metal dichalcogenide transistors. IEEE Trans Electron Devices 58:3042–3047
11. Radisavljevic B, Radenovic A, Brivio J, Giacometti V, Kis A (2011) Single-layer MoS_2 transistors. Nat Nanotechnol 6(3):147–150
12. Jena D, Konar A (2007) Enhancement of carrier mobility in semiconductor nanostructures by dielectric engineering. Phys Rev Lett 98(13):136805
13. Tong X, Ashalley E, Lin F, Li, H, Wang ZM (2015) Advances in MoS_2-based field effect transistors (FETs). Nano-Micro Lett 7(3):203–218

Split Behavior of Supervised Machine Learning Algorithms for Phishing URL Detection

Madhurendra Purbay and Divya Kumar

Abstract *Phishing* is a fraudulent technique that is an effort by an individual or a group of the individual over the Internet to defraud users to steal their personal information such as username, password, bank account, and ATM card information. Personal information may be used by 'phishers' for their gain, and it is the primary objective of the phishing. It costs the online company and the numerous stakeholders hundreds of million dollars per year. This paper deals with methods to detect phishing URLs by examining different components of URLs using machine learning and deep learning techniques. We have discussed different supervised learning methods used for phishing URL detection based on lexical feature, WHOIS properties, PageRank, Traffic Rank details and page importance properties. Further, we have analyzed how different volumes of training data affect the classification accuracy. The algorithms being analyzed are support vector machine (SVM), K-NN, decision tree classification (DTC), random forest classification (RFC), and artificial neural network (ANN).

Keywords Phishing URL · Features · Classification · Confusion matrix · Lexical analysis · SVM · K-NN · DTC · RFC · ANN

1 Introduction

Phishing is a type of social engineering attack, which is an attempt by an individual or a group to extract secret data such as user id passwords, ATM card data from unsuspecting victims for fraud, gain, and completely different deceitful activities [1, 2]. In the new scenario, when the required user wants to access various websites for

M. Purbay (✉) · D. Kumar
Motilal Nehru National Institute of Technology Allahabad, Prayagraj, Uttar Pradesh 211004, India
e-mail: madhurendra.mnnit@gmail.com

D. Kumar
e-mail: divyak@mnnit.ac.in

D. Harvey et al. (eds.), *Advances in VLSI, Communication, and Signal Processing*, Lecture Notes in Electrical Engineering 683,
https://doi.org/10.1007/978-981-15-6840-4_40

services like fund transfer, hostel booking, online shopping, etc., then the user has to enter credentials like user id, passwords, ATM No/Pins, date of birth, etc., on the login page. But, this data, when transferred over the Internet, is prone to be captured by hackers or phishers [2].

On the Internet phishing attack, phishing portals or websites are created by the hacker (phisher) that looks identical to some of the popular website to scam the online user to get their sensitive monetary and private data. Phishing starts with sending some spoofed email to target victims with some URL or attacker put that link with some hot offer in some other popular website. The email comes from phisher are spoofed, these emails, however, look identical to that of an email generated by an authentic source. For example, if the manager of some organization "ABC" has email address mgr@abcbank.co.in, then phisher spoofs that email address using a similar address like mger@abcbank.com and sends some hot offer like a loan with very minimal interest and put phishing website link which is similar to bank website, and user believes that email comes from the bank. When the user clicks the Web link given in the spoofed email or link provided on another popular website with some hot offer, control is transferred to the misleading website created by the phisher. As this misleading website interface seems to be quite similar to the original website, a new user often fails to identify and judge this fake website. The user then enters the requested information, and the result is a successful phishing attempt. Some of the examples of these kinds of websites are given in Table 1, taken from [3].

Phishing has been in the scene since the year 1996 and stealing the <user name>, <password> pair is a very common activity. In most cases, a successful phishing attack is made through advertisement/luring emails and website spoofing [4].

In the past, various mechanisms have been employed by the researchers to identify a phishing Uniform Resource Locator (URL) based on their features. The prominent features are: using IP Addresses instead of names, URL length, domains of registration, DNS record, page rank, and Web traffic [5]. A lot of classification algorithms [6] have also been deployed to classify a URL as phishing or authentic. Further, these algorithms can be categorized into two classes: supervised or unsupervised learning algorithms [7]. The performance of these algorithms relies on the volume and accuracy of training dataset [8]. The performance of any classification algorithm is directly proportional to the accuracy; however, the volume of the training dataset may result in an underfit or overfit models in the case of less or more training data, respectively. The paper aims to find the accuracy of the several classification algorithms

Table 1 Some authentic and phishing website from [3]

S. no.	Authentic website	Phishing website
1	https://www.bestchange.com/	http://bestxchange.net.ru/
2	https://www.bestchange.com/	http://bestxhange.pp.ru/
3	http://inkasator.hr/	http://inkasator.org.ru/
4	https://24paycoin.top.cutestat.com/	http://www.24paycoin.pro/
5	https://www.paypal.com/in/signin	http://paypal.secure-gb.org

based on the volume of training data. In the pursuit of this objective, this manuscript is designed as discussed below. In section two, some of the well-known classification algorithms have been briefed. Further, the applications of these algorithms in phishing URL detection have been surveyed. In section three, testing methodology has been described with results and observations on different combinations of the volume of training and testing data. Finally, the conclusions are presented.

2 Phishing Website Detection

2.1 Phishing Website

For stealing the user's identity and authentication credentials, the phisher sometimes create a dummy website that looks similar to the initial website. Later, the phisher sends a dummy email, containing some luring hyperlinks, to sufferers so as to reprehensibly perform dishonorable money transactions on behalf of the users [9]. The phisher continuously sends emails to several other Internet users together with hyperlinks to the fake website in as conceive to fool Internet users. Most of the Internet users who are not aware of the security issues and potential Internet dangers follow the given URL contained in the email. The website linked to the URL prompts the users to enter the user name and passwords. Thus, the phisher gets the secure information by fooling the innocent people.

2.2 Heuristic-Based Technique for Phishing Website Detection

Heurictic-based techniques start by collecting some features of websites, and later on, the evaluation is done on the basis of these features values. These features play a vital role in detecting phishing websites. Some of the features are:

IP:	IP address used in place of domain name	SU:	URLs containing sub-domains
LU:	Long URLs, containing more than 50 characters	HU:	Existence of protocols other than HTTP
GI:	Google Index	PU:	Use of non-standard ports.
RU:	Redirection URLs, using //	PR:	Page rank of the website
DU:	URLs containing a dash (–)	AG:	Age of domain

There are several other features also. The heuristic algorithms are supplied with the previously categorized data, and a trained model is prepared according to it. Afterward, this model is used to classify a URL whose category is unknown. From the existing literature, some of the most referred and cited heuristic algorithms that have been used for phishing URL detection is as given in Table 2.

3 Testing Methodology and Results

This work comprises of lexical feature extraction of collected URLs and investigation. The primary step is the collecting of phishing and legitimate URLs. The lexically based feature extraction is used to shape a database of feature values. In database, feature values are in integer form (−1, 0, 1) and in final column show website characteristics +1 = legitimate URL, 0 = suspicious URL, −1 = phishing URL. Figure 1 shows the flow diagram of supervised classification algorithms.

The phishing URL dataset is taken from the UCI Machine Learning Repository [25] to guide the supervised machine learning classifiers utilized in phishing sites identification. This dataset contains 4898 phishing URLs and 6157 authenticated URLs. These URLs are used for training and testing of the accuracy of classifiers dealing with phishing URL identification. Each URL is described using 30 attributes.

Table 2 Well-known classification algorithms

S. no.	Algorithm	References
1	Support vector machine (SVM)	[2, 10–12]
2	K-nearest neighbor (K-NN)	[13, 14]
3	Decision tree classification (DTC)	[14–16]
4	Random forest classification (RF)	[17–20]
5	Artificial neural network (ANN)	[21–24]

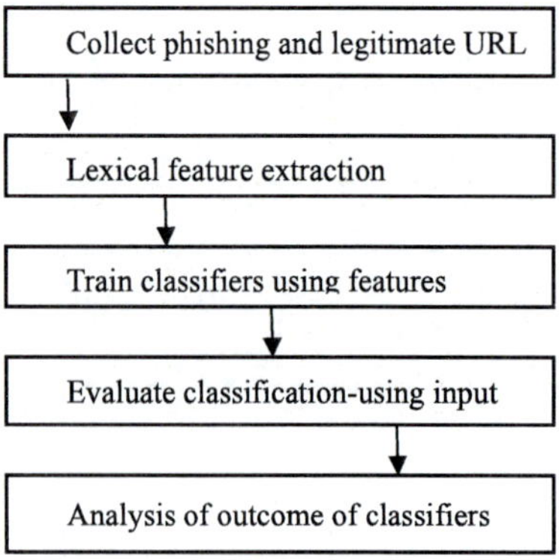

Fig. 1 Flow diagram of the proposed work

For each site, a site design vector was extricated and shaped to be utilized as a case in the preparation of training data set that has 30 imperative highlights for that site. The site design vector relating to the real site is allocated to a class with mark +1 and the phishing site is allocated to a class with mark −1.

We have examined the prepared URL feature dataset using support vector machine (SVM), K-NN, decision tree classification, random forest classification, and artificial neural network (ANN). The performance is evaluated on the basis of the confusion matrix. The confusion matrix for each classifier shows detection accuracy, true positive rate, and false positive rate.

3.1 Confusion Matrix

A confusion matrix is a tabular structure that is used to explaining the performance of a classification model (or "classifier") on a set of test data for which the result is known [6]. A confusion matrix of size "m × m" related to a classifier shows the expected and actual classification, where m is the number of various categories. Table 3 shows a confusion matrix for "m" = two, whose entries have the following meanings.

a: number of correct predictions for negative instance.
b: number of incorrect predictions for negative instance.
c: number of wrong predictions for positive instance.
d: number of correct predictions for positive instance.

Accuracy prediction and classification error prediction can be obtained from the confusion matrix as follows, for more information on error evaluation and confusion matrix, we redirect the readers to [26, 27]:

$$\text{Accuracy} = (a + d) / (a + b + c + d) \tag{1}$$

$$\text{Error} = (b + c) / (a + b + c + d) \tag{2}$$

$$\text{Total URL tested} = (a + b + c + d) \tag{3}$$

Table 3 Confusion matrix for "m" = two

	Predicted negative	Predicted positive
Actual negative	a	b
Actual positive	c	d

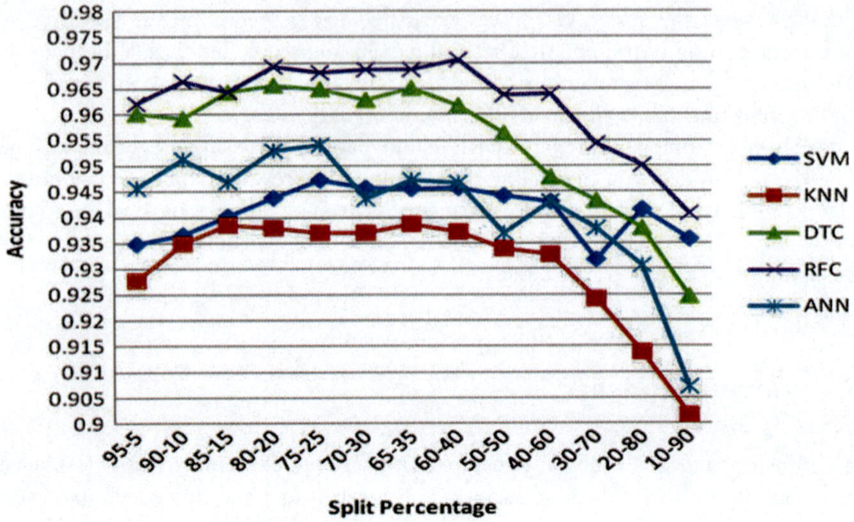

Fig. 2 Accuracy versus split percentage of various algorithms

3.2 *The* x–y *Split Case*

Here, percentage split of data set is *x*–*y*, i.e., *x* percentage of the data set is taken as coaching data with which classifier is able to understand a classification model, and once the training phase is complete with given x percentage of data, the classifier is given unknown URLs as input here, *y* percentage as test data, and a predicted class is returned as output. Various confusion matrices for different algorithms on different splits are given below in Table 4. Afterward, the accuracy trends of the classification algorithms are plotted in Fig. 2. The code is also available at github repository.[1]

4 Conclusion and Future Work

In this paper, we compared several features of the Web URL using various machine learning and data mining algorithms. The results indicate the efficiency that can be obtained using the lexical features of URLs. Random forest classification provided better efficiency as compared to other algorithms such as SVM, K-NN, decision tree classification, and ANN. It can also be observed that for most of the algorithms, an optimum split percentage range is between 80 and 20 and 50 and 50. To secure end-users from visiting phishing sites, we can try to recognize phishing URLs by

[1] https://github.com/manucp123/Phishing-Url-detection-project-.

Table 4 Confusion matrix result of different classification algorithm

Algorithm	Split *x*–*y*	a	b	c	d	Accuracy
SVM	95–5	236	25	11	281	0.9349
	90–10	466	47	23	570	0.9367
	85–15	682	66	33	878	0.9403
	80–20	936	78	46	1151	0.9439
	75–25	1156	93	53	1462	**0.9472**
	70–30	1387	111	69	1750	0.9457
	65–35	1622	127	84	2037	0.9455
K-NN	95–5	239	22	18	274	0.9277
	90–10	470	43	29	564	0.9349
	85–15	686	62	40	871	**0.9385**
	80–20	932	82	55	1142	0.938
	75–25	1149	100	74	1441	0.937
	70–30	1378	120	89	1730	0.937
	65–35	1608	141	96	2025	0.9388
DTC	95–5	245	16	6	286	0.9602
	90–10	480	33	12	581	0.9593
	85–15	707	41	18	893	0.9644
	80–20	961	53	23	1174	**0.9656**
	75–25	1184	65	32	1483	0.9649
	70–30	1417	81	42	1777	0.9629
	65–35	1669	80	55	2066	0.9651
RFC	95–5	245	16	5	287	0.962
	90–10	484	29	8	585	0.9665
	85–15	704	44	15	896	0.9644
	80–20	966	48	20	1177	**0.9692**
	75–25	1187	62	26	1489	0.9682
	70–30	1424	74	30	1789	0.9686
	65–35	1669	80	41	2080	0.9687
ANN	95–5	249	12	18	274	0.9458
	90–10	494	19	35	558	0.9512
	85–15	720	28	60	851	0.947
	80–20	985	29	75	1122	0.953
	75–25	1210	39	88	1427	**0.9541**
	70–30	1469	29	158	1661	0.9436
	65–35	1709	40	164	1957	0.9473

The bold point indicates the highest accuracy for the respective algorithm scanning their lexical and host-based features. As a part of future work, we shall try to improvise the random forest classification by mixing it with evolutionary optimization algorithms.

References

1. Sahoo D, Liu C, Hoi SCH (2017) Malicious URL detection using machine learning: a survey. arXiv preprint arXiv:1701.07179
2. Agrawal, P, Mangal D (2015) A novel approach for phishing URLs detection. Int J Sci Res (IJSR)
3. Online Phishing Repository by Open DNS: https://www.phishtank.com/
4. Varshney G, Misra M, Atrey PK (2016) A survey and classification of web phishing detection schemes. Secur Commun Networks 9(18):6266–6284
5. James J, Sandhya L, Thomas C (2013) Detection of phishing URLs using machine learning techniques. In: 2013 International conference on control communication and computing (ICCC). IEEE (2013)
6. Hlavac V (2016) Classifier performance evaluation. Czech Technical University. http://people.ciirc.cvut.cz/~hlavac/TeachPresEn/31PattRecog/13ClassifierPerformance.pdf
7. Kotsiantis SB, Zaharakis I, Pintelas P (2007) Supervised machine learning: a review of classification techniques. Emerg Artif Intell Appl Comput Eng 160:3–24
8. Yadav DP et al (2017) A novel ensemble based identification of phishing e-mails. In: Proceedings of the 9th international conference on machine learning and computing. ACM
9. Ali W (2017) Phishing website detection based on supervised machine learning with wrapper features selection. Int J Adv Comput Sci Appl 8(9):72–78
10. Furey TS et al (2000) Support vector machine classification and validation of cancer tissue samples using microarray expression data. Bioinformatics 16(10):906–914
11. Tong S, Koller D (2001) Support vector machine active learning with applications to text classification. J Mach Learn Res 2:45–66
12. Basnet R, Mukkamala S, Sung AH (2008) Detection of phishing attacks: a machine learning approach. Soft computing applications in industry. Springer, Berlin, Heidelberg, pp 373–383
13. Keller JM, Gray MR, Givens JA (1985) A fuzzy k-nearest neighbor algorithm. IEEE Trans Syst Man, and Cybern 4:580–585
14. Toolan F, Carthy J (2009) Phishing detection using classifier ensembles. eCrime Researchers Summit, 2009. eCRIME'09. IEEE
15. Safavian SR, Landgrebe D (1991) A survey of decision tree classifier methodology. IEEE Trans Syst Man, Cyber 21(3):660–674
16. Aggarwal A, Rajadesingan A, Kumaraguru P (2012) PhishAri: automatic realtime phishing detection on twitter. 2012 eCrime Researchers Summit. IEEE
17. Liaw A, Wiener M (2002) Classification and regression by randomForest. R News 2(3):18–22
18. Akinyelu AA, Adewumi AO (2014) Classification of phishing email using random forest machine learning technique. J Appl Math (2014)
19. DeBarr D, Ramanathan V, Wechsler H (2013) Phishing detection using traffic behavior, spectral clustering, and random forests. In: 2013 IEEE international conference on intelligence and security informatics (ISI). IEEE
20. Van Essen B et al (2012) Accelerating a random forest classifier: multi-core, GP-GPU, or FPGA?. In: 2012 IEEE 20th annual international symposium on field-programmable custom computing machines (FCCM), IEEE
21. Zhang N, Yuan Y (2012) Phishing detection using neural network. CS229 lecture notes (2012)
22. Martin A et al (2011) A framework for predicting phishing websites using neural networks. arXiv preprint arXiv:1109.1074

23. Mohammad RM, Thabtah F, McCluskey L (2014) Predicting phishing websites based on self-structuring neural network. Neural Comput Appl 25(2):443–458
24. Jameel NGM, George LE (2013) Detection of phishing emails using feed forward neural network. Int J Comput Appl 77(7)
25. UCI Machine Learning Repository https://archive.ics.uci.edu/ml/datasets/phishing+websites#
26. Paliwal P, Kumar D (2017) ABC based neural network approach for churn prediction in telecommunication sector. In: International conference on information and communication technology for intelligent systems. Springer, Cham
27. Singh A, Kumar D (2017) Novel ABC based training algorithm for ovarian cancer detection using neural network. In: 2017 international conference on trends in electronics and informatics (ICEI). IEEE

LMI and YALMIP: Modeling and Optimization Toolbox in MATLAB

Akhilesh Kumar Ravat, Amit Dhawan, and Manish Tiwari

Abstract In this paper, we present a MATLAB toolbox YALMIP and LMI. This paper narrated how YALMIP and LMI can be employed to model and solutions of the optimization problems arising in control systems. With the help of command of YALMIP, we can solve the optimization problem in control systems. The numerical examples also illustrate the success of the results presented in the paper.

Keywords LMI · YALMIP · MATLAB

1 Introduction

Semi-definite programming (SDP) and linear matrix inequality (LMI) are two important mathematical tools in control system and control theory during the past decade [1]. SDP combines a lot of control problems. It is operated in modern control theory, classical Lyapunov theory for linear systems, and H-infinity control theory. It has also given many results on stability analysis of uncertain systems, control of piecewise affine systems, model predictive control, and robust systems identification. The enormous number of uses of SDP has applied on research and growth of software for resolving the optimization issues [2–4]. There are almost ten public solvers accessible. Some solvers are free and simply accessible from Internet. Despite that, these solvers bring the problem explanation in complex format, time-consuming, and error likely. To remove this difficulty, modeling languages and interfaces are required [1].

A. K. Ravat (✉) · A. Dhawan · M. Tiwari
Department of Electronics and Communication Engineering, Motilal Nehru National Institute of Technology Allahabad, Allahabad, Uttar Pradesh 211004, India
e-mail: akravat1991@gmail.com

A. Dhawan
e-mail: dhawan@mnnit.ac.in

M. Tiwari
e-mail: mtiwari@mnnit.ac.in

D. Harvey et al. (eds.), *Advances in VLSI, Communication, and Signal Processing*, Lecture Notes in Electrical Engineering 683,
https://doi.org/10.1007/978-981-15-6840-4_41

In this Paper, the MATLAB toolbox YALMIP is introduced. This toolbox creates development of optimization problems and solves their optimization issue.

Initially, YALMIP was scratch for LMIs and SDP. In this way, YALMIP knew as Yet another LMI Parser. YALMIP supports quadratic programming (QP) and YALMIP certain commands are sufficient for mostly users to model and optimization.

2 Notations

$\boldsymbol{P}$ is symmetric matrix and it is indicated as positive semidefinite ($\boldsymbol{P} \geq 0$) if $z^{\mathrm{T}}\boldsymbol{P}\,z \geq 0\ \forall z$. For positive definite ($\boldsymbol{P} > 0$) if $z^{\mathrm{T}}\boldsymbol{P}\,z > 0\ \forall z \neq 0$.

3 Introduction to LMI

Linear matrix inequalities (LMIs) and LMI techniques have appeared as strong design tool used for system recognition and design in the area of control system. Numerous questions in control system can be prepared using LMI problems [5, 6].

These are clarifying with appropriate examples.

i. **Variables change:**
   This is frequently able to linearize nonlinear matrix inequalities with the help of new variables.
ii. **The Schur Complement:**
   Schur complement method is to be used in converting nonlinear inequalities of convex class into LMI.

Consider matrix $\boldsymbol{S} = \begin{bmatrix} \boldsymbol{S}_{11} & \boldsymbol{S}_{12} \\ \boldsymbol{S}_{12}^{\mathrm{T}} & \boldsymbol{S}_{22} \end{bmatrix}$, where $\boldsymbol{S}_{11}$ and $\boldsymbol{S}_{22}$ are symmetric matrix and invertible; and then the following three states are equivalence [4]:

(a) $\boldsymbol{S} < 0$;
(b) $\boldsymbol{S}_{11} < 0,\ \boldsymbol{S}_{22} - \boldsymbol{S}_{12}^{\mathrm{T}}\boldsymbol{S}_{11}^{-1}\boldsymbol{S}_{12} < 0$;
(c) $\boldsymbol{S}_{22} < 0,\ \boldsymbol{S}_{11} - \boldsymbol{S}_{12}\boldsymbol{S}_{22}^{-1}\boldsymbol{S}_{12}^{\mathrm{T}} < 0$.

Resolve LMI Utilizing MATLAB Toolbox
To solve LMIs, the LMI toolbox gives group of beneficial functions. Certain functions are mentioned with sample code.

First step: Initialization The LMI explanation with command setlmis ([ ]).
Second step: Identifying the decision variables it is required to defining the unknown variables of the LMI problem. For example, considering the LMI $F^{\mathrm{T}}\boldsymbol{X}\,F < 0$, where F and $\boldsymbol{X}$ stand for constant matrix and decision variables of matrix, respectively. This is identified with the help of function lmivar which is given as follows:

$$X = \text{lmivar(type, structure)}.$$

With the help of this command, various forms of decision matrices like rectangular matrices, symmetrical matrices or other type matrices are defined. Hence, first of all the type of the matrix and then structure of matrix which is dependent on type is defined.

For type = 1, the given matrix $\boldsymbol{X}$ is to be square and symmetrical.
For type = 2, the given matrix $\boldsymbol{X}$ is to be rectangular of size $m \times n$ as defined in structure = $[m, n]$.
For type = 3, the given matrix $\boldsymbol{X}$ is of different kind.

Third step: Defining the LMI One at a Time, it is done with the help of lmiterm command.

The syntax is lmiterm (termID, C, D, flag).

The lmiterm defines three or four arguments. The first argument of the above syntax is a 1×4 vector. An LMI is defined with help of first part of the vector. The second and third parts define the state or position of the term. The fourth part of vector shows the LMI decision variable is concerned. For left and right multipliers, we use second and third arguments of lmiterm function. If the flag is fixed to 's', then, it approves the term and its symmetrical which appear in LMI using a single command.

4 Introduction to YALMIP

In this paper, we introduce the basic commands of YALMIP.

a. **Identifying decision variables:**

The main element in any optimization problems is the decision variable. In YALMIP, the decision variable is indicated by **sdpvar** command. For rectangular matrix, m-rows and n-columns are denoted command as shown:

Q = sdpvar (m, n);

A square and symmetric matrix $\boldsymbol{Q} \in \boldsymbol{R}^{m \times m}$ is denoted as

Q = sdpvar (m, m 'symmetric', 'real');

Or it is described with the help of only dimension argument denoted as

Q = sdpvar (m, m);

We can also combine a third argument to clarify different type of variables such as diagonal, Hankel, and Toeplitz matrices utilizing through following commands:

$y =$ sdpvar $(m, 1)$;
$d =$ diag (y);
$h =$ hankel (y);
$t =$ toeplitz (y);

The various kinds of matrices with complex variable and fully parameterized are given as

$Y =$ sdpvar $(m, m$ 'full');
$Z =$ sdpvar $(m, m$, 'hermitian', 'complex');

b. **Defining constraint:**

The most usually applied constraints in YALMIP are equality, element-wise, and semi-definite constraints. For example:

$X =$ sdpvar (m, m);
$Y =$ sdpvar (m, m);
Constraint $= [X > 0, Y < 0]$;
$G =$ set $(X > 0)$;
$G = G +$ set $(P\ (:) > 0)$;
$G = G +$ set (sum (sum $(P)) = = m)$;

YALMIP also holds up suitable definition of integrity constraints, sum of square constraints, and second order cone constraints.

$G =$ set (integer (X));
$G =$ set (cone $(L * x + b, c^{\mathrm{T}} * x + d)$);
$G =$ set (sos $(1 + y + y^7 + y^8)$);

c. **Solving optimization problems:**

To solve optimization problem in YALMIP, the following command used. We assume that matrices $\boldsymbol{L}$, **b** and **c**. We would like to minimize $c^{\mathrm{T}}x$ subject to constraints $\boldsymbol{L}\,\boldsymbol{x} \leq b$ and $\sum x_i = 1$.

$X =$ sdpvar (length (c), 1);
$G =$ set $(L * x < b)$ + set (sum $(x) = = 1)$;
Solvesdp $(G, c^{\mathrm{T}} * x)$;

For illustration, a linear system

min $c'x$ subject to $\boldsymbol{Lx} \leq \boldsymbol{b}$ solved with code:
$x =$ sdpvar (length (c), 1);
constrt $= [L * x < = b]$;
$h = c' * x$;
solvesdp (constrt, h);
solve $=$ double (x)

YALMIP shall automatically divide as a linear programming problem and take appropriate solver. With the help of double (x) command find the optimal solution.

d. **Solving different control problem using YALMIP:**

We take an example related to control problem and solve using YALMIP.

To identify the stability of LTI systems, we assume the LTI system.

$$\dot{\boldsymbol{x}} = \boldsymbol{A}\,\boldsymbol{x} \tag{1}$$

The system is said to be stable if it satisfied the given below inequalities.

$$\begin{aligned} &\boldsymbol{P} > \boldsymbol{0}, \\ &\boldsymbol{A}^{\mathrm{T}}\boldsymbol{P} + \boldsymbol{P}\boldsymbol{A} < \boldsymbol{0}, \end{aligned} \tag{2}$$

We combined the two inequalities into singular LMI with the help of Schur complement,

$$\begin{bmatrix} \boldsymbol{A}^{\mathrm{T}}\boldsymbol{P} + \boldsymbol{P}\boldsymbol{A} & 0 \\ 0 & -\boldsymbol{P} \end{bmatrix} < 0 \tag{3}$$

For solving this stability problem, we take the value of $A = \begin{bmatrix} 0 & 1 \\ -2 & -3 \end{bmatrix}$.

YALMIP Code

clc;
clear all;
close all;
$A = \begin{bmatrix} 0\ 1;\ -2\ -3 \end{bmatrix}$; (system matrix)
$N =$ size (A, 1); (define unknown variable)
$P =$ sdpvar (N, N);
$M = \begin{bmatrix} A^{\mathrm{T}} * P + P * A & \text{zeros}(N,1) \\ \text{zeros}(N,1) & -P \end{bmatrix} < 0;$
Solvesdp (M)
$P =$ double (P)

Solution

Solvertime: 1.6332

$P = \begin{bmatrix} 84.4848\ 41.9548 \\ 41.9548\ 34.1295 \end{bmatrix}$

Applying LMIs to solve this problem MATLAB code given below:

$A = \begin{bmatrix} 0\ 1;\ -2\ -3 \end{bmatrix}$;

P = lmivar (1, $\begin{bmatrix} \text{size}(A, 1) & 1 \end{bmatrix}$); (for defining the structure and size of P)
lmiterm ($\begin{bmatrix} 1 & 1 & 1 & P \end{bmatrix}$, 1, A, 'S') (specify LMI)
lmiterm ($\begin{bmatrix} 1 & 1 & 2 & 0 \end{bmatrix}$, 1);
lmiterm ($\begin{bmatrix} 1 & 2 & 2 & P \end{bmatrix}$, −1, 1);
LMISYS = getlmis;
[tmin, Psol] = feasp (LMISYS);
P = dec2mat (LMISYS, Psol, P)

Solution

tmin = −2.615451 and $P = \begin{bmatrix} 65.992 & 12.8946; & 12.8946 & 15.1836 \end{bmatrix}$.

The main elements of an optimization are decision variables, objective, bounds, and constraints. Decision variable is a one-dimensional array whose value can be changed to determine the optimal solution. A solution is a set of values of decision variable. The objective is a function of the decision variables. It gives a single number solution in which optimizer efforts to minimize or maximize whatever you define in formulation.

Bounds define the set of possible solution. Each decision variable can be lower or upper bound. Constraints are bounds on functions of decision variable. It specifies which solution is feasible.

5 Numerical Examples

Problem 1 Consider the 2D discrete uncertain system

$$A = \begin{bmatrix} 1.05 & -3.0; & 0.1 & 0 \end{bmatrix}$$
$$B = \begin{bmatrix} 0.2; & 0.3 \end{bmatrix}$$

$$L_1 = \begin{bmatrix} 0 \\ 2 \end{bmatrix}$$
$$M_1 = \begin{bmatrix} 0.007 & -0.007 \end{bmatrix}$$
$$L_2 = 1$$
$$M_2 = \begin{bmatrix} 0.04 & -0.03 \end{bmatrix}$$

It is found using MATLAB YALMIP toolbox that the LMI [7]

$$\begin{bmatrix} -\boldsymbol{S} + \in_1 \boldsymbol{L}_1\boldsymbol{L}_1^{\mathrm{T}} + \in_2 \boldsymbol{B}\boldsymbol{L}_2\boldsymbol{L}_2^{\mathrm{T}}\boldsymbol{B}^{\mathrm{T}} & \bar{\boldsymbol{A}} & \boldsymbol{0} & \boldsymbol{0} \\ \bar{\boldsymbol{A}}^{\mathrm{T}} & -\boldsymbol{S} & \boldsymbol{S}\boldsymbol{M}_1^{\mathrm{T}} & \boldsymbol{S}\boldsymbol{M}_2^{\mathrm{T}} \\ \boldsymbol{0} & \boldsymbol{M}_1\boldsymbol{S} & -\in_1 \boldsymbol{I} & \boldsymbol{0} \\ \boldsymbol{0} & \boldsymbol{M}_2\boldsymbol{S} & \boldsymbol{0} & -\in_2 \boldsymbol{I} \end{bmatrix} < \boldsymbol{0}$$

is feasible for the given example and the feasibility answer is given below:

The best value of t must be negative for feasibility

Iteration : Best value of t so far

1	0.119737
2	0.011814
3	−0.016494

Result: best value of t: −0.016494

f-radius saturation: 0.000% of $R =$ 1.00e+09

yalmiptime: 1.8997

solvertime: 0.1013

info: 'Successfully solved (LMILAB)'

```
ans =
   0.9150    0
   0    0.8336

ans =
  -0.1628  -1.2799

ans =
    0.0484

ans =
    1.9106
```

Problem 2 Minimization the value of gamma

$$\min \gamma \begin{bmatrix} \boldsymbol{A}^{\mathrm{T}}\boldsymbol{P} + \boldsymbol{P}\boldsymbol{A} & \boldsymbol{P}\boldsymbol{B} & \boldsymbol{C}^{\mathrm{T}} \\ \boldsymbol{B}^{\mathrm{T}}\boldsymbol{P} & -\gamma \boldsymbol{I} & \boldsymbol{D}^{\mathrm{T}} \\ \boldsymbol{C} & \boldsymbol{D} & -\gamma \boldsymbol{I} \end{bmatrix} < 0$$

$$\boldsymbol{P} > 0$$

$$A = \begin{bmatrix} -4 & -3 & 0 & -1 \\ -3 & -7 & 0 & -3 \\ 0 & 0 & -13 & -1 \\ -1 & -3 & -1 & -10 \end{bmatrix}$$

$$B = \begin{bmatrix} 0 \\ -4 \\ 2 \\ 5 \end{bmatrix}$$

$$C = \begin{bmatrix} 0 \; 0 \; 4 \; 0 \end{bmatrix}$$

$$D = 0$$

Result Under LMI constraints, the solver for linear objective minimization

```
Iterations  :  Best objective value so far

   1          2.799901
   2          1.691414
   3          1.023500
   4          0.943996
   5          0.869379
   6          0.869379
   7          0.799262
   8          0.799262
   9          0.574049
  10          0.574049
  11          0.516617
  12          0.516617
  13          0.475772
  14          0.475772
***         new lower bound:   0.419195
  15          0.467467
  16          0.466237
***         new lower bound:   0.450554
  17          0.465347
***         new lower bound:   0.460304
  18          0.464993
***         new lower bound:   0.461488
  19          0.464163
  20          0.463996
* switching to QR
  21          0.463996
  22          0.463979
***         new lower bound:   0.463538

Result:  feasible solution of required accuracy

 Best objective value:    0.463979

Guaranteed absolute accuracy:  4.41e−04

 F-radius saturation:  0.002% of R = 1.00e+09

ans =  0.4640

ans = 1.0e+03 *
   1.5864   3.1002  −3.0683   3.7205
   3.1002   6.0597  −5.9990   7.2725
  −3.0683  −5.9990   5.9451  −7.2007
   3.7205   7.2725  −7.2007   8.7283
```

6 Conclusion

YALMIP tool can be widely used for optimization-based algorithm developed in MATLAB. The primary goal of this paper is intended for ease of the overall process as a powerful technical tool which can be used as a traditional framework for the optimization related to control in MATLAB. The numerical examples illustrate the performance of the results with the help of YALMIP.

References

1. Löfberg J (2004) YALMIP: a toolbox for modeling and optimization in MATLAB. In: Proceedings of the CACSD conference, vol 3, Taipei, Taiwan
2. Bemporad A, Morari M, Dua V, Pistikopoulos EN (2002) The explicit linear quadratic regulator for constrained systems. Automatica 38(1):3–20
3. Sturm JF (1999) A MATLAB toolbox for optimization over symmetric cones. Optim Methods Softw 11–12(1–4):625–653
4. Liao F, Li L (2017) Robust preview control for uncertain discrete-time systems based on LMI. Optimal Control Appl Methods 38(1)
5. Gahinet P, Nemirovaskii A (1995) LMI control toolbox: the LMI Lab. The Math Works Inc.
6. Boyd S, El Ghaoui L, Feron E, Balakrishnan V (1994) Linear matrix inequalities in system and control theory. SIAM studies in applied mathematics. SIAM, Philadelphia, Pennsylvania
7. Dhawan A (2012) Non-fragile controller design for 2-D discrete uncertain systems described by the Roesser model. J Sig Inf Process 3(2)

Improved Noise Margin and Reduced Power Consumption in Subthreshold Adiabatic Logic Using Dual Rail Power Supply

Sanjay Kumar Prajapati, Devarshi Shukla, and Santosh Kumar Gupta

Abstract Adiabatic logic has been found to be effective in achieving low power in VLSI circuits. In subthreshold region, the circuit conducts a small current and consumes lesser power. In this paper, the dual rail power supply with 0 and 180° phase difference has been used for subthreshold adiabatic logic (SAL). A priority encoder is designed using SAL with dual rail power supply. All the circuit simulations are done using HSPICE simulator at 22 nm PTM technology. In-phase dual supply offers low power consumption, whereas out-phase dual power supply offers an improved noise margin.

Keywords Subthreshold adiabatic logic (SAL) · Dual rail power supply · Leakage current

1 Introduction

Low power consumption is primary demand in these days. The advancement in battery technology is slow as compared to its applications; so, energy harvesting logic circuits are necessary to reduce power consumption. Moreover, with development in technology, transistors size has also reduced drastically over recent years. However, power dissipation gets increased due to reduction in size as well as integration density also increases. Nowadays, dynamic power dissipation has been the primary concern rather than static power dissipation to the designers.

Conventionally power dissipation occurs due to device switching from ON to OFF and vice versa. In an inverter circuit, when input is low (logic 0), then PMOS is ON and NMOS is OFF, therefore, output load capacitor starts charging. Similarly, when input is high (logic 1), then PMOS is OFF and NMOS is ON, due to this charge load capacitor starts discharging through ground. So, no energy recovery is possible in the conventional CMOS circuits and energy is dissipated in form of heat.

S. K. Prajapati · D. Shukla · S. K. Gupta (✉)
Department of Electronics and Communication Engineering, Motilal Nehru National Institute of Technology Allahabad, Prayagraj, Uttar Pradesh 211004, India
e-mail: sonu10396@gmail.com

D. Harvey et al. (eds.), *Advances in VLSI, Communication, and Signal Processing*, Lecture Notes in Electrical Engineering 683,
https://doi.org/10.1007/978-981-15-6840-4_42

To overcome the problem of power dissipation, adiabatic logic is adopted. In this technique, energy consumption can be reduced by suppressing the voltage applied to the resistance of the circuit. Furthermore, for low power operations, the circuit is operated in subthreshold region of operation. In subthreshold regime, the leakage current is used as its operating switching current.

In many research works, different types of adiabatic logic have been reported for low energy consumption, such as positive feedback adiabatic logic (PFAL) [1], modified positive feedback adiabatic logic (MPFAL) [2], efficient charge recovery logic (ECRL) [3], and energy-efficient adiabatic logic (EEAL) [3]. Subthreshold adiabatic logic (SAL) [4, 5] utilizes the advantage of both the adiabatic logic [6, 7] and subthreshold region of operation of CMOS logic. In adiabatic logic circuits, energy is recycled back to the source, so that to reduce total energy drawn from power source. Whereas, CMOS transistors are operated in subthreshold regime for low voltage operations. In this paper, the dual rail supply is incorporated in SAL technique to design the priority encoder and an invert circuit which shows reduced power consumption and improved noise margin.

2 Adiabatic Logic

In conventional CMOS logic, both pull up and pull down transistors have equal resistance for charging and discharging path as shown in Fig. 1. In adiabatic logic, a smooth changing power supply is used. The RC equivalent circuits and responses of (a) static CMOS and (b) adiabatic logic are shown in Fig. 2. From Fig. 2, it is evident that a sharp change in power supply dissipates more energy as compared to smooth changing power supply.

In CMOS, logic equal amount of energy dissipates during charging and discharging process which is given by [8, 9]

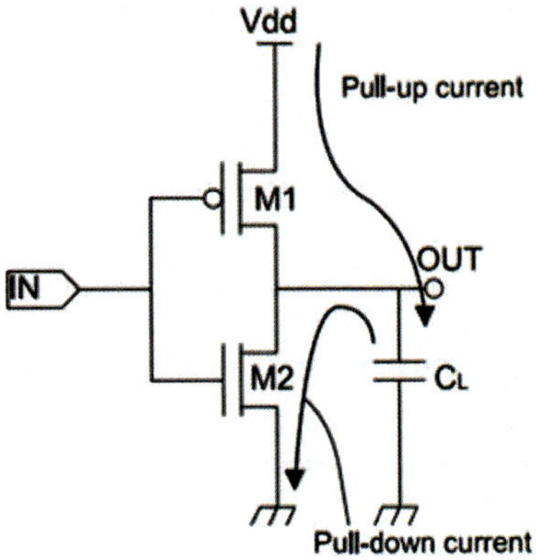

Fig. 1 Conventional CMOS inverter

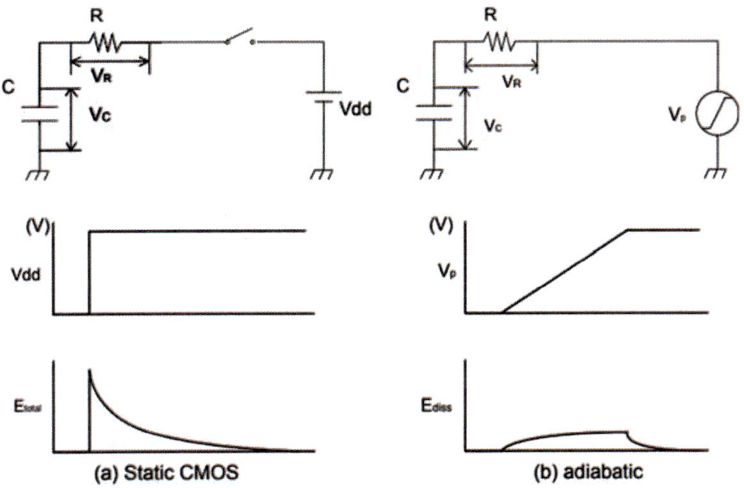

Fig. 2 RC equivalent and response of **a** static CMOS and **b** adiabatic logic [3]

$$E_{\mathrm{CMOS}} = (1/2)CV^2 \tag{1}$$

Also, the energy dissipation in adiabatic circuit is given by [3]

$$E_{\mathrm{adia}} = (RC/T)CV^2 \tag{2}$$

Equation (2) states that energy dissipation will be lower for higher values of T.

3 Subthreshold Adiabatic Logic

In subthreshold regime, circuits consume ultra-low power due to very small levels of ON state current. The drain current is assumed to be zero for V_{GS} (gate to source voltage) lower than V_{TH} (threshold voltage) during analysis of subthreshold circuits. But, practically, there exists a very small drain current in MOS transistor even when $V_{\mathrm{GS}} < V_{\mathrm{TH}}$.

The drain current expression for $V_{\mathrm{GS}} < V_{\mathrm{TH}}$ is given by [1]

$$I_{\mathrm{sub}} = I_0 \mathrm{e}^{\frac{V_{\mathrm{SG}} - |V_{\mathrm{TH}}|}{\eta_p V_{\mathrm{T}}}} \left(1 - \mathrm{e}^{\frac{-V_{\mathrm{SD}}}{V_{\mathrm{T}}}}\right) \tag{3}$$

where $I_0 = 2\eta_p \mu_p C_{\mathrm{ox}}(W/L)V_{\mathrm{T}}^2$; $V_{\mathrm{T}} = (kT/q)$ is thermal voltage, η_p is subthreshold slope factor, μ_p is the mobility holes, V_{SG} and V_{SD}, are source to gate and source to drain voltages of PMOS, respectively.

3.1 Subthreshold Adiabatic Logic (SAL) Circuit

SAL inverter circuit is represented by Fig. 3 in which a load capacitor is connected to the PMOS. In subthreshold adiabatic logic, a sinusoidal power supply is being used (Fig. 4a) which has peak value 0.4 V less than threshold voltage ($|V_{TH}| = 0.46$ V) of PMOS. The input voltage (Fig. 4b) applied to SAL inverter gives respective output (Fig. 4c). Output of the inverter has the high logic level (V_{OH}) equal to 0.4 V but has

Fig. 3 SAL inverter [1]

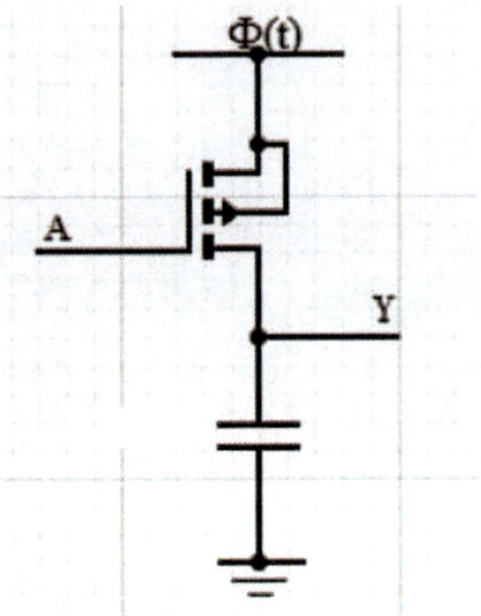

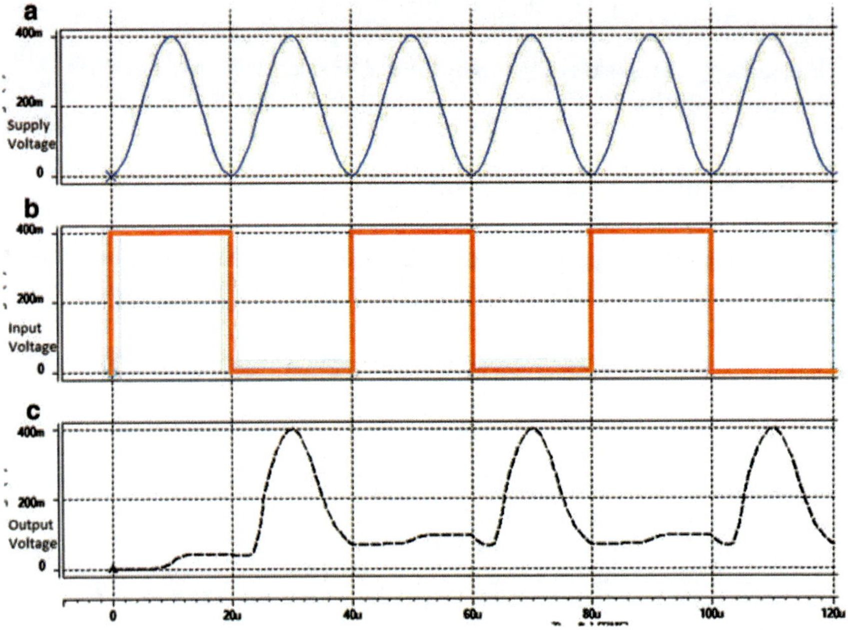

Fig. 4 SAL inverter **a** supply voltage, **b** input voltage, **c** output voltage

the low logic level (V_{OL}) higher than the ideal value (0 V). The higher value of low logic level reduces the noise margin of circuit which should be minimized to have a better noise margin.

3.2 Proposed SAL Circuit

In the proposed circuit, a dual rail power supply is being used as shown in Fig. 5. First power supply $\Phi_1(t)$ has higher magnitude connected to source of the pull-up network and the second power supply $\Phi_2(t)$ with lesser magnitude is connected to load capacitor. These power supplies can be used in two ways

1. In-phase dual rail power supply.
2. Opposite phase dual rail power supply.

In-phase dual rail power supply (Fig. 6) has 0° phase difference between the signals, whereas opposite phase dual rail power supply (Fig. 7) has 180° phase difference. V_{OH} of SAL inverter with in-phase and opposite phase dual rail power supply remains same as that of single rail power supply. V_{OL} lowest value of in-phase power supply is still high but power consumption is reduced to half of single rail power supply (Fig. 8c). Power consumption in SAL inverter with opposite phase dual

Fig. 5 SAL inverter with dual rail power supply

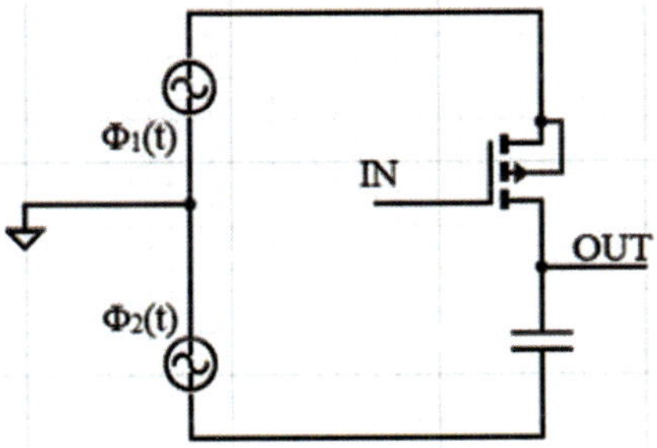

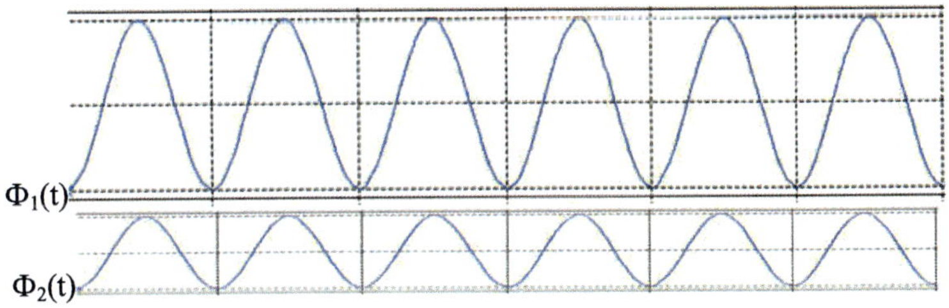

Fig. 6 In-phase dual rail power supply

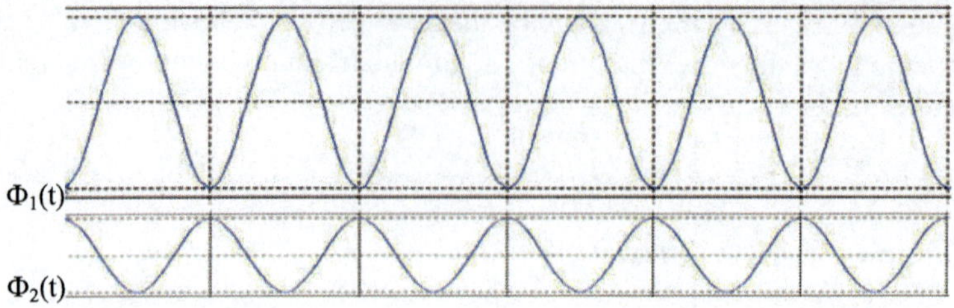

Fig. 7 Opposite phase dual rail power supply

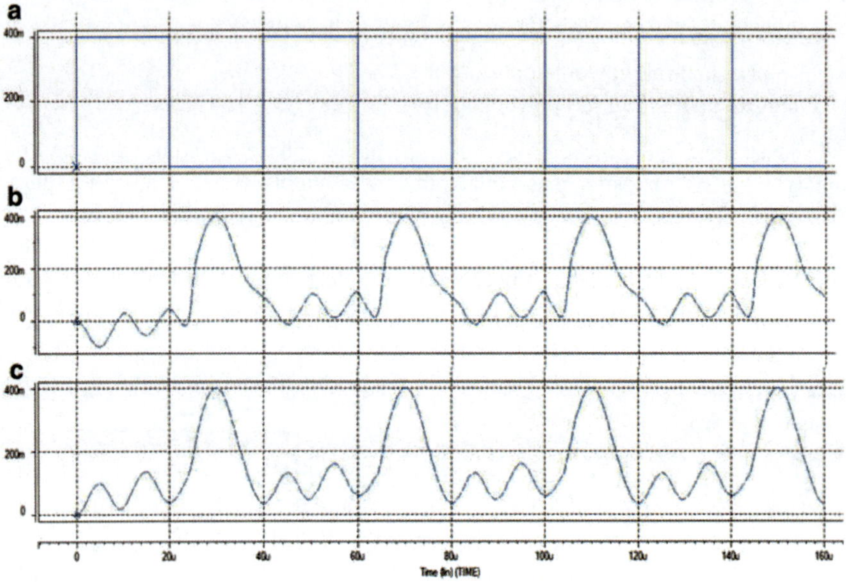

Fig. 8 SAL inverter **a** input signal, **b** output of opposite phase dual rail power supply, **c** output of in-phase dual rail power supply

rail power supply has increased but V_{OL} value has been reduced to 0 V (Fig. 8b). Table 2 shows different power supplies and their respective V_{OL} and average power consumption.

4 Results and Discussion

To evaluate circuit functionality and its performance, we design a priority encoder. This typical priority encoder uses 3 two-input NAND gate, 3 INVERTERS and 1 three-input NOR gates as shown in Fig. 9.

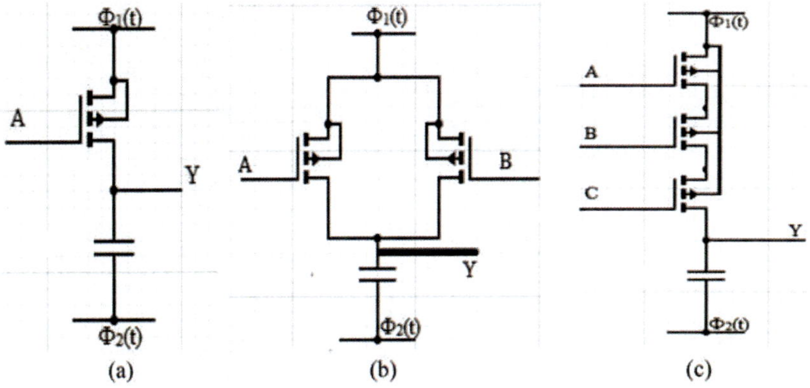

Fig. 9 **a** NOT gate, **b** NAND gate, **c** 3-input NOR gate

Figure 10 shows the block diagram representation of priority encoder. Table 1 shows the functionality of priority encoder. The output of a priority encoder is the binary representation of actual number starting from zero of the most significant bit (MSB) of input. The output V indicates if the input is valid. These find application in controlling of interrupt requests by pursuing on the highest priority. Figure 11 shows output waveform of priority encoder with opposite phase dual rail power supply, and Fig. 12 shows output waveform of priority encoder with in-phase dual

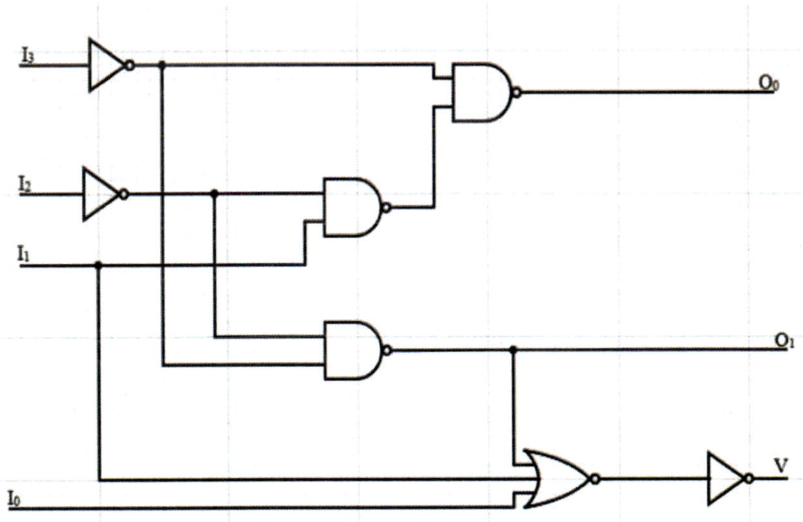

Fig. 10 Block diagram of priority encoder

Table 1 Truth table of priority encoder

I_3	I_2	I_1	I_0	O_1	O_0	V
0	0	0	0	X	X	0
0	0	0	1	0	0	1
0	0	1	X	0	1	1
0	1	X	X	1	0	1
1	X	X	X	1	1	1

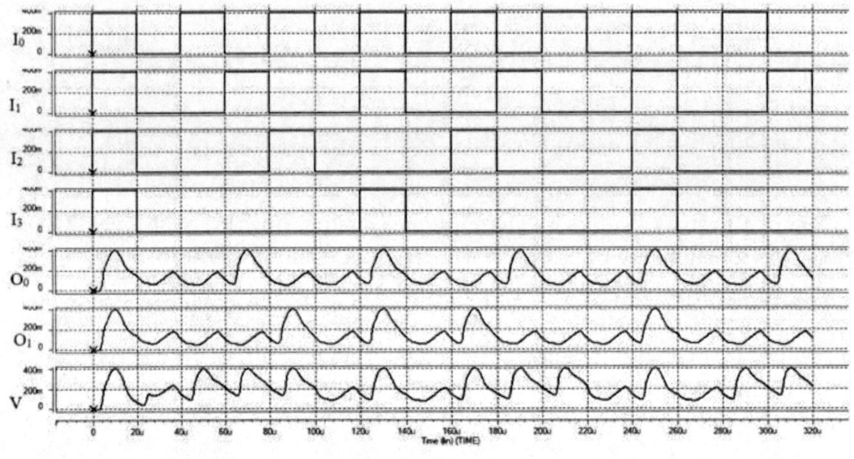

Fig. 11 Output waveform of priority encoder with opposite phase dual rail power supply

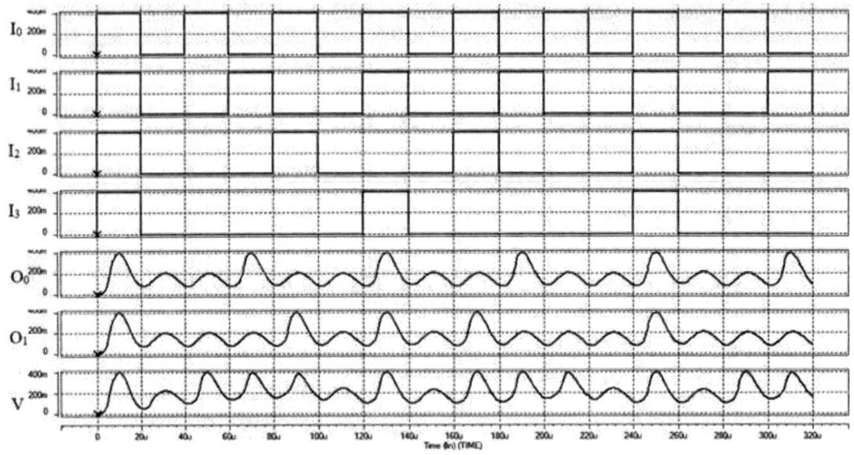

Fig. 12 Output waveform of priority encoder with in-phase dual rail power supply

Table 2 Power consumption and V_{OL} for different power supplies

S.No.	Power supply	V_{OL} (mV)	Average power (pW)	
			Inverter	Priority encoder
1.	Single rail	80	5.79	69.56
2.	In-phase dual rail	60	2.78	40.04
3.	Opposite phase dual rail	0	11	114.3

rail power supply. The values of V_{OL} and average power consumption for different power supplies are given in Table 2.

5 Conclusion

In this work, improved noise margin and low power consumption adiabatic logic have been presented. Further, for low supply voltage, the proposed design is operated in subthreshold region for different power supplies: single rail, in-phase dual rail and opposite phase dual rail. To verify the performance and functionality of the proposed methodology; an inverter and a priority encoder have been implemented. Among the three power supplies, in-phase dual rail power supply provides lowest average power consumption for inverter and priority encoder circuits, viz., 2.78 pW and 40.04 pW, respectively. Whereas, V_{OL} is lowest (=0 V) in opposite phase dual rail power supply for both inverter and priority encoder circuits. Therefore, for low power applications, we can use in-phase dual rail power supply, whereas, for improved noise margin, the opposite phase dual rail power may be preferred.

References

1. Vetuli A, Pascoli S, Reyneri LM (1996) Positive feedback in adiabatic logic. Electron Lett 32(20):1867–1869
2. Kushawaha SPS, Sasamal TN (2015) Modified positive feedback adiabatic logic for ultra-low power VLSI. In: 2015 International conference on computer, communication and control (IC4). IEEE, pp 1–5
3. Moon Y, Jeong D-K (1996) An efficient charge recovery logic circuit. IEICE Trans Electron 79(7):925–933
4. Chanda M, Jain S, De S, Kumar C (2015) Implementation of subthreshold adiabatic logic for ultralow-power application. IEEE Trans Very Large Scale Integr (VLSI) Syst 23(12):2782–2790
5. Ranjith KG, Chavan AKP, Ravish Aradhya HV (2017) Subthreshold adiabatic logic (SAL) based building blocks for combinational system design. In: 2nd IEEE international conference on recent trends in electronics information & communication technology (RTEICT), pp 96–100
6. Grover V, Gosain V, Pandey N, Gupta K (2018) Arithmetic logic unit using diode free adiabatic logic and selection unit for adiabatic logic family. In: 5th International conference on signal processing and integrated networks (SPIN), pp 777–781

7. Chanda M, Mal S, Mondal A, Sarkar CK (2018) Design and analysis of a logic model for ultralow power near threshold adiabatic computing. IET Circ Dev Syst 12(4):439–446
8. Kato K, Takahashi Y, Sekine T (2014) Two phase clocking subthreshold adiabatic logic. In: IEEE international symposium on circuits and systems (ISCAS), pp 98–601
9. Kalyan P, Satish Kumar P, Chandra Sekhar P (2017) Design of subthreshold adiabatic logic based combinational and sequential circuits. In: International conference on emerging trends & innovation in ICT (ICEI), vol 1, pp 9–14

Human Action Recognition Using a New Hybrid Descriptor

Om Mishra, Pranoti S. Kavimandan, M. M. Tripathi, Rajiv Kapoor, and Kalpana Yadav

Abstract Human action recognition is a very promising field for security, surveillance, elder care, etc. We proposed a new amalgamation of local and global features to describe the human action in a video. A new local spatiotemporal feature has been proposed in terms of fuzzy lattices of the frame in a video. These fuzzy lattices are described by the kinetic energy of the lattices. The kinetic energy of the lattices is found out by Schrödinger Wave Equation. Any changes in the body due to action motion change the kinetic energy of the lattices. Further to maintain the structural and shape information, most widely used global features descriptor motion energy image (MEI) and motion history image (MHI) are used. To validate the proposed methodology, three datasets Weizmann, KTH and IXMAS are used. The proposed methodology proved its superiority over other state-of-the-art methods.

Keywords Action recognition · Fuzzy lattices · Schrödinger Wave Equation

1 Introduction

Computer vision researchers have been working on human action recognition for a long time. Human action recognition is also used for security, surveillance, patient care, human--computer interactions, etc. The recently published reviews emphasize the challenges such as variation in viewpoint, occlusion, variation in execution rate and background cluttering. To address these challenges, generalized approaches of action recognition have been proposed by the researchers. These generalized approaches include global as well as local features to describe the actions. The global features represent the action globally that depicts the shape and the structure of the human body. The main disadvantages of the global features are that they are

O. Mishra (✉) · M. M. Tripathi · R. Kapoor
Delhi Technological University, New Delhi, Delhi 110042, India
e-mail: om.mishra83@gmail.com

P. S. Kavimandan · K. Yadav
Indira Gandhi Delhi Technical University for Women, New Delhi, Delhi, India

D. Harvey et al. (eds.), *Advances in VLSI, Communication, and Signal Processing*,
Lecture Notes in Electrical Engineering 683,
https://doi.org/10.1007/978-981-15-6840-4_43

very sensitive to the variations in the viewpoints, occlusion and background cluttering. The local features are very popular among the researchers because they show invariance against the challenges discussed above. But preserving the global features along with the local features makes human action recognition very efficient. In the proposed methodology, the global and the local features both are used for recognizing an action.

Many methodologies have been proposed by researchers to describe an action. The researcher used global features in [1–6]. Most of the global features descriptors are based on the silhouettes of the human body. These silhouettes are represented globally to describe the action. The appearance and structural features based on 3D shapes are used in [2]. The silhouette-based key poses of the actions are used in [3]. The features of these key poses are exploited by dividing the silhouettes into grids and cells. In the other work, 3D key poses are used in [6] for multi-view action recognition. The spatiotemporal volumes [7–10] are also used for action recognition. The space-time volume features are described by Poison's equation in [7]. The affine invariant moments are used as features to represent action volume in [8]. The action volumes are also described in terms of Hankel matrices. They require a very sophisticated technique of background subtraction. The disadvantage of this is that it cannot capture motion information properly which results in the misclassification of an action.

The local feature descriptors [11–15] give information on temporal motions. Accurate background subtraction is not required for these methods. They show viewpoints invariance. The Bag-of-visual-words model is a prominent methodology used by the researchers. The Histogram-of-Gaussian, Histogram-of-optical-flow, HOG3D, etc. are some major descriptors used by the researchers. The Bag-of-visual words are represented in terms of Bag-of-correlated poses in [11]. In [12], local features are extracted from negative space for each pose of the actions. In [13], local features are extracted in terms of the trajectories of the motion. They are divided into two trajectories: large and small. The different geometries of this trajectory described the action features. In other methodologies, the deep trajectory of the action motion is used in [15]. These methodologies cannot give the structural or shape information of the action and thus reduce the accuracy of the recognition.

In the proposed technique, we have used local features descriptors along with global descriptors to overcome the problems discussed above and to enhance the recognition rate of the action recognition. The diversities occur in the action videos are due to change in the background environment (illumination change or background cluttering). This results in a nonlinear relationship among pixels of the action frames. Thus, in the proposed methodology, a varying Gaussian fuzzy lattice [16, 17] is used. The other advantage of using a Gaussian fuzzy lattice is that they represent the complete information of the action frame because it can represent both low and high-frequency components of the frame. Another challenge in action recognition is a large dimension of features. To reduce features dimensionality, we used only those prominent fuzzy lattices that may carry important information for action recognition. The kinetic energy of the lattices is evaluated from Schrödinger Wave Equation [18].

2 Proposed Methodology

Figure 1 describes the proposed methodology in terms of the block diagram.

Local and global features are extracted from the input video. During local feature extraction fuzzy lattices are formed. These fuzzy lattices are expressed in terms of kinetic energy. Global features motion history image (MHI) and motion energy image (MEI) are also extracted. Further, these both local and global features are fused to recognize the action. Our proposed methodology is also summarized in subsequent Sects. 2.1 and 2.2.

2.1 Local and Global Feature Extraction

Local features have been extracted in terms of fuzzy lattices [16, 17] and the global features are extracted by the combination of MHI and MEI [11]. The features in action video may vary due to the motion. This develops a nonlinearity among the nearest neighboring pixels in the image. Thus, we used the Gaussian fuzzy membership function (G_{st}) described by Eq. (1) to capture this relationship among the pixels of an image.

$$G_{st}(p) = \exp^{\frac{-(p-m_s)^t}{2\mu_s^2}} \tag{1}$$

where m_s and μ_s are the mean and the variance of the Gaussian membership function and p is the gray-level intensity of pixels. Here, t is the total number of Gaussian membership functions having different means and variances. The fuzzy lattices [16, 17] are formed by comparing all the values of the pixel to its nearest neighboring pixels using these Gaussian membership function. These fuzzy lattices are formed in X and Y (spatial directions) and Z direction (gray-level value of the pixels). Gaussian lattice functions in X direction $L_{st}(X)$, in Y direction $L_{st}(Y)$ and in Z direction $L_{st}(Z)$ are given by Eqs. (2), (3) and (4), respectively.

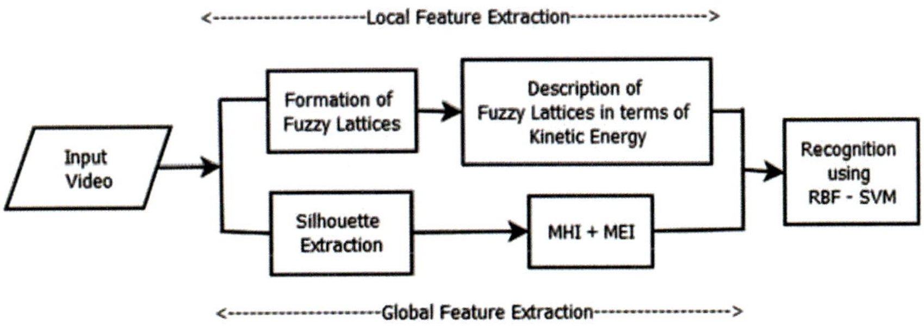

Fig. 1 Block diagram of the proposed methodology

$$L_{st}(x) = \exp^{-\frac{(x-x_{ms})^t}{2\mu_{xs}^2}} \tag{2}$$

$$L_{st}(y) = \exp^{-\frac{(y-y_{ms})^t}{2\mu_{ys}^2}} \tag{3}$$

$$L_{st}(z) = \exp^{-\frac{(p-p_{ms})^t}{2\mu_{zs}^2}} \tag{4}$$

where x_{ms} and μ_{xs} are means and variance of X coordinates of the pixels, respectively, y_{ms} and μ_{ys} are means and variance of Y coordinates of the pixels, respectively, and z_{ms} and μ_{zs} are the means and variance of the gray value of the pixels, respectively. These fuzzy lattices are described by kinetic energy using the solution of Schrödinger Wave Equation [18]. The kinetic energies of lattices are given by Eqs. (5), (6), and (7).

$$K_x = \frac{1}{L(x)} \frac{\partial^2 L(x)}{\partial x^2} \tag{5}$$

$$K_y = \frac{1}{L(y)} \frac{\partial^2 L(y)}{\partial y^2} \tag{6}$$

$$K_z = \frac{1}{L(z)} \frac{\partial^2 L(z)}{\partial z^2} \tag{7}$$

where K_x, K_y and K_z are the kinetic energies of the lattice in X, Y and Z directions. The addition of the kinetic energy of the lattice in X, Y and Z direction gives the total kinetic energy.

Any change in the human body will develop the change in the lattice kinetic energy. The change in the kinetic energy of the lattices offers a pattern for each action. We used MEI and MHIs [11] as global features to retain the structural information. All silhouettes in a video arrangement are anticipated onto one picture (MHI or MEI) as shown in Fig. 2a, b over a temporal axis. They extract the information about the location and how actions have been performed in a video.

The MHI and MEI give the structural description of the action where MHI described how the action performed and MEI described where the action is performed. Later, global and local features are fused to extract complete information of action.

2.2 Fusion and Classification

To fuse two modals information, two layers of RBF-SVM are used. The design of multimodal integration is done for a two-layer C-support vector machine fusion scheme [19] to discover the relation between global and local features modalities.

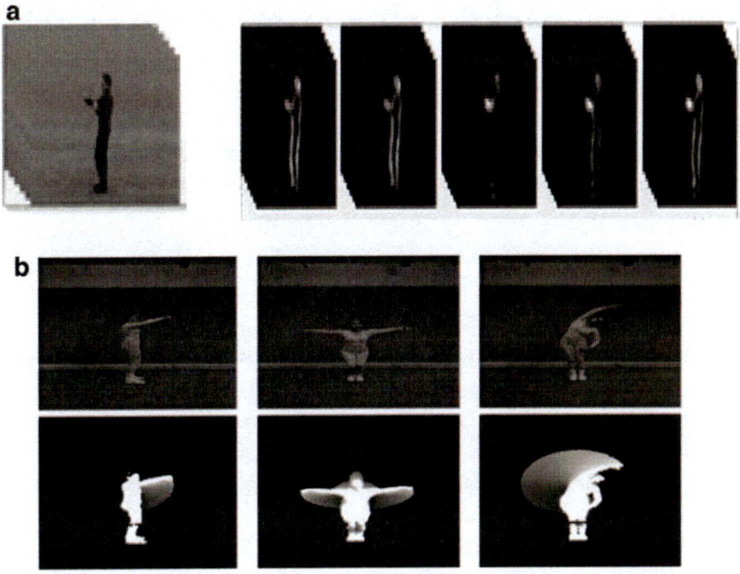

Fig. 2 **a** Illustration of motion energy image (MEI). **b** Illustration of motion history image (MHI)

The objective function for SVM is given in Eq. (8) as follows

$$f(x) = \sum_i \beta_i y_i \theta(x_i, x) - c \tag{8}$$

where $\theta(x_i, x)$ is a kernel function. We have opted for a radial basis function. In this experiment, during the first layer of classification, we train classifiers for global and local features respectively. Figure 3 shows the fusion model where the first level of fusion is done for local and global features, and again, they are fused at the second level.

To summarize, the algorithm of the methodology can be explained as follows:

1. The input action video is processed for local and global feature extraction.
2. The local features are represented in terms of fuzzy lattices formed in the frame of a video.

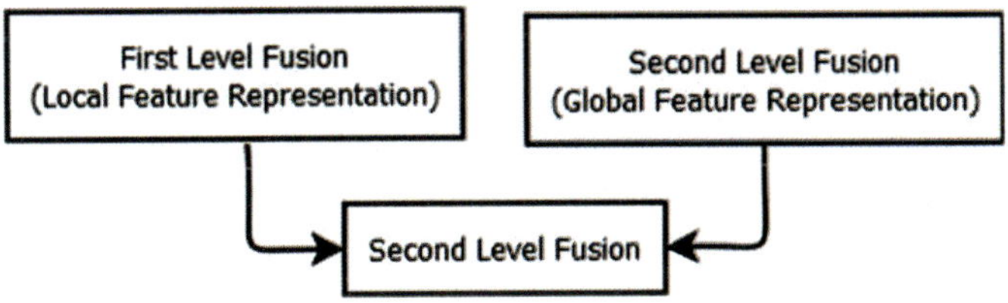

Fig. 3 Two-level C-support RBF-SVM

3. These fuzzy lattices are described by their kinetic energy calculated using the Schrödinger Wave Equation.
4. Dimensionality is reduced by the selection of only prominent lattices based on their kinetic energy.
5. Then these local features are combined with the global features MHI and MEI.
6. These features are classified into different actions using two levels (RBF-SVM).

3 Experimental Results and Discussion

The proposed method has been developed in MATLAB R2013a. The system configuration is as follows: Processor Intel(R) Core (TM) i5-6200U CPU @2.30 GHz 2.40 GHz with 8 Gb RAM and 64-bit operating system. The performance of the proposed method is measured using the parameter called accuracy. The proposed methodology is implemented on the three challenging action recognition datasets Weizmann [7], KTH [20] and IXMAS [21].

In Weizmann, the action dataset consists of ten actions such as walking, running, jumping, jogging, etc. They used 25 frames per second frame rate and 144×180 pixels resolution for the frames. The KTH dataset is more challenging than the Weizmann. It has different environment settings for action videos. Six different actions such as walking, jogging, running, boxing, hand-waving and hand-applauding are present in the dataset. It has four different environments and there are 100 recordings available for each action. The recording camera has a fixed position and frame resolution is 160×120 pixels. The IXMAS is the most challenging dataset out of three. They used five different viewpoints to capture the actions. Figure 4 shows the frames of these datasets. We compared the proposed methodology with other state-of-the-art methods. We used a leave-one-out testing strategy for the cross-validation. Figure 5 shows the results for kinetic energy for action walk for Weizmann, KTH and IXMAS datasets.

The sensitive parameter in our method is the number of fuzzy lattices. We tested the proposed method on 3, 5, 10, 15 and 20 fuzzy lattices. This is selected based on kinetic energy. We observed that the higher the number of fuzzy lattices, the higher is the accuracy. We got the best results with five fuzzy lattices. The number of fuzzy lattices higher than five shows only a 1–2% increase in accuracy. Figure 5 shows the kinetic energy of the fuzzy lattice having the highest kinetic energy for the action walk for all three datasets. The x-axis shows the number of frames and the y-axis shows the kinetic energy of these frames.

The proposed method is compared with other state-of-the-art methods. The combination of fuzzy lattices-based local features with the most popular global features MHI and MEI gave better results. Two-layer SVM is used to fuse those features. Because of this, the proposed method completely outperforms the other existing methods. Figure 6 shows the comparison for the Weizmann dataset. The methodologies [11, 22, 23, 24] achieved more than 95% accuracy, methodologies [25, 26]

Fig. 4 First row Weizmann, second-row KTH and third-row IXMAS dataset

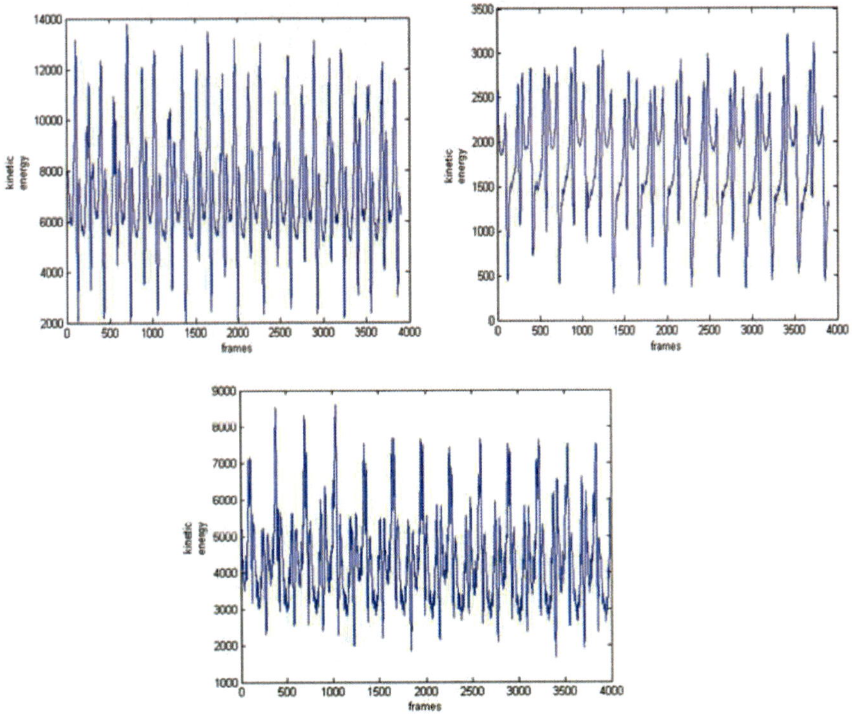

Fig. 5 Kinetic energy for action: walk for Weizmann, KTH and IXMAS datasets

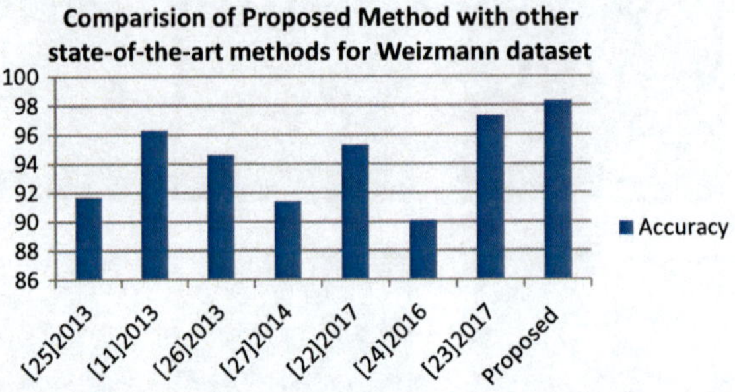

Fig. 6 Comparison of the proposed method with other state-of-the-art methods for Weizmann dataset

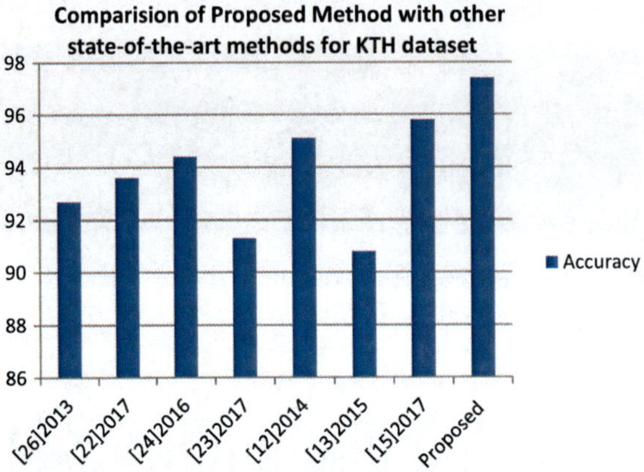

Fig. 7 Comparison of the proposed method with other state-of-the-art methods for KTH dataset

achieved around 92% accuracy but the proposed methodology achieved 98.3% accuracy which is better than these methods. Figure 7 shows the comparison among the methodologies for the KTH dataset. The methodologies [12, 15, 27] achieved better results in the comparison table. The proposed methodology shows comparable results by achieving an accuracy of 97.4%. For the IXMAS dataset, we used five different views for each action. Figure 8 shows the accuracy of all five camera views of the IXMAS dataset. The proposed method achieved accuracy of more than 90% for all the camera views and the average accuracy is 93.2%. It clearly shows its superiority over other state-of-the-art methods [27–30].

4 Conclusion

The proposed methodology used a new local feature descriptor fused with a widely used global feature descriptor for human action recognition. This local feature is extracted in the form of fuzzy lattices. These fuzzy lattices are described through kinetic energy. The proposed methodology is applied to the challenging datasets like Weizmann, KTH, and IXMAS and also compared with other state-of-the-art

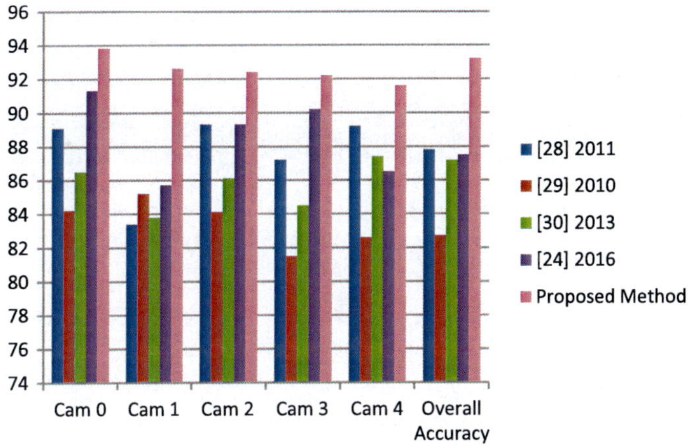

Fig. 8 Comparison of the proposed method with similar methods on IXMAS dataset

methods. The proposed method shows better accuracy than other existing methods. In the future work, a more sophisticated new global feature descriptor can be fused with the proposed fuzzy lattices-based local descriptor to recognize more complex actions.

References

1. Vishwakarma DK, Kapoor R, Dhiman A (2016) A proposed unified framework for the recognition of human activity by exploiting the characteristics of action dynamics. J Rob Auton Syst 77:25–28
2. Zhao D, Shao L, Zhen X, Liu Y (2013) Combining appearance and structural features for human action recognition. Neurocomputing 113(3):88–96
3. Vishwakarma DK, Kapoor R (2015) Hybrid classifier based human activity recognition using the silhouette and cells. Expert Syst Appl 42(20):6957–6965
4. Achar C, Qu X, Mokhber A, Milgram M (2008) A novel approach for action recognition of human actions with semi-global features. J Mach Vis Appl 19(1):27–34
5. Lin C, Hsu F, Lin W (2010) Recognizing human actions using NWFE-based histogram vectors. EURASIP J Adv Sig Process 9
6. Gu J, Ding X, Wang S (2012) Action recognition from arbitrary views using 3D-key-pose set. Front Electr Electron Eng 7(2):224–241
7. Gorelick L, Blank M, Shechtman E, Irani M, Basri R (2007) Actions as space-time shapes. IEEE Trans Pattern Anal Mach Intell 29(12):2247–2253
8. Sadek S, Hamadi A, Elmezain M, Michaelis B, Sayed U (2012) Human action recognition via affine moment invariants. In: International conference on pattern recognition, Tsukuba, pp 218–221
9. Presti L, Cascia M, Sclaroff S, Camps O (2015) Hanklet-based dynamical systems modeling for 3-D action recognition. J Image Vis Comput 44(C):29–43
10. Yilmaz A, Shah M (2008) A differential geometric approach to representing the human actions. Comput Vis Image Underst (CVIU) 335–351

11. Wu D, Shao L (2013) Silhouette analysis-based action recognition via exploiting human poses. IEEE Trans Circ Syst Video Technol 23(2):236–243
12. Rahman SA, Song I, Leung MKH, Lee I, Lee K (2014) Fast action recognition using negative space features. Expert Syst Appl 41(2):574–587
13. Conde IG, Olivieri DN (2015) A KPCA spatio-temporal differential geometric trajectory cloud classifier for recognizing human actions in a CBVR system. Expert Syst Appl 42(13):5472–5490
14. Li B, Camps OI, Sznaier M (2012) Cross-view activity recognition using Hanklets. In: International conference on computer vision and pattern recognition, Providence
15. Shi Y, Tian Y, Wang Y, Huang T (2017) Sequential deep trajectory descriptor for action recognition with three-stream CNN. IEEE Trans Multimedia 19(7):1510–1520
16. Kapoor R, Mishra O, Tripathi MM (2019) Anomaly detection in group activities based on fuzzy lattices using Schrödinger equation. Iran J Comput Sci
17. Kapoor R, Gupta R (2013) Non-linear dimensionality reduction using fuzzy lattices. IET Comput Vision 7(3):201–208
18. Schrödinger E (1926) An undulatory theory of the mechanics of atoms and molecules. Am Phys Soc Rev 28(6):1049–1070
19. Chang CC, Lin CJ (2011) LIBSVM: a library for support vector machines. ACM Trans Intell Syst Technol 2(3):1–27
20. Schuldt C, Laptev I, Caputo B (2004) Recognizing human actions: a local SVM approach. In: Proceedings of the 17th international conference on pattern recognition, Cambridge, UK
21. Weinland D, Ronfard R, Boyer E (2006) Free viewpoint action recognition using motion history volumes. Comput Vis Image Underst 104(2–3):249–257
22. Fu Y, Zhang T, Wang W (2017) Sparse coding-based space-time video representation for action recognition. Multimedia Tools Appl 76(10):12645–12658
23. Liu H, Shu N, Tang Q, Zhang W (2017) Computational model based on the neural network of visual cortex for human action recognition. IEEE Trans Neural Netw Learn Syst 29(5):1427–1440
24. Goudelis G, Karpouzis K, Kollias S (2013) Exploring trace transform for robust human action recognition. Pattern Recogn 46(12):3238–3248
25. Chaaraoui AA, Pérez PC, Florez-Revuelta F (2013) Silhouette-based human action recognition using sequences of key poses. Pattern Recogn Lett 34(15):1799–1807
26. Touati R, Mignotte M (2014) MDS-based multi-axial dimensionality reduction model for human action recognition. In: Canadian conference on computer and robot vision, pp 262–267
27. Lei J, Li G, Zhang J, Guo Q, Tu D (2016) Continuous action segmentation and recognition using hybrid convolutional neural network-hidden Markov model. IET Comput Vision 10(6):537–544
28. Wu X, Xu D, Duan L, Luo J (2011) Action recognition using context and appearance distribution features, In: IEEE conference on computer vision and pattern recognition (CVPR), pp 489–496
29. Weinland D, Özuysal M, Fua P (2010) Making action recognition robust to occlusions and viewpoint changes. In: European conference on computer vision (ECCV), pp 635–648
30. Mosabbeb EA, Raahemifar K, Fathy M (2013) Multi-view human activity recognition in distributed camera sensor networks. Sensors 13(7):8750–8770

Analog and Radio-Frequency Performance of Hetero-Gate-Dielectric FD SOI MOSFET in Re-S/D Technology

Nilesh Anand Srivastava, Anjali Priya, and Ram Awadh Mishra

Abstract Fully depleted silicon-on-insulator (FD SOI) technology has already been accepted for its short-channel immunity in nanoscale regime and the concept of metal-gate engineering has provided the flexibility to further scale-down the device dimensions. This paper presents the analog and radio-frequency behavior of hetero-gate-dielectric (HGD) triple-metal-gate (TMG) recessed-source/drain (Re-S/D) FD SOI MOSFET at 45-nm-technology node. Here, the analog performance has been investigated on the basis of numerical calculations of transconductance (g_m), output conductance (g_d), transconductance generation factor (TGF), and intrinsic gain (A_v). Further, radio-frequency performance has been monitored by analyzing parasitic capacitances (C_{gs} and C_{gd}) and cutoff frequency (f_T). All these studies have been performed using technology computer-aided design (TCAD) simulation tool Silvaco ATLAS. It is found that the proposed hetero-gate-dielectric-based TMG SOI MOSFET exhibits excellent transconductance behavior with higher gain and increased frequency of operation as compared to other state-of-the-art discussed in the literature. The proposed device offers cutoff frequency of 5.21×10^{11} Hz, which itself dictates the device immunity at high frequency. Simultaneously, the perspective of buried-oxide (BOX) thickness variation on the performance of HGD-TMG SOI MOSFET has also been investigated in order to tackle the challenges involved in nanoscale device design.

Keywords FD SOI · Hetero-gate-dielectric · Recessed-Source/Drain · Triple-metal-gate

---

N. A. Srivastava (✉) · A. Priya · R. A. Mishra
Motilal Nehru National Institute of Technology Allahabad, Prayagraj, Uttar Pradesh 211004, India
e-mail: rel1702@mnnit.ac.in

A. Priya
e-mail: rel1454@mnnit.ac.in

R. A. Mishra
e-mail: ramishra@mnnit.ac.in

D. Harvey et al. (eds.), *Advances in VLSI, Communication, and Signal Processing*, Lecture Notes in Electrical Engineering 683,
https://doi.org/10.1007/978-981-15-6840-4_44

1 Introduction

The recent advancement in complementary metal–oxide–semiconductor (CMOS) technology has provided the flexibility to accommodate billions of MOS transistors in an integrated circuit (IC). These high-density ICs led to the development of high-performance and low-power circuit and systems for analog/radio-frequency applications [1, 2]. However, the backbone of these high-speed ICs is metal–oxide–semiconductor field-effect transistors (MOSFETs) [3]. So, it is the prime requirement that the available nanoscale MOS technology should offer necessary and optimized performance for the design of such high-performance ICs. However, the endless scaling of MOS transistors beyond 100 nm is continuously deteriorating the device performance and results in various short-channel effects (SCEs), like drain-induced barrier-lowering (DIBL), threshold voltage roll-off, surface-scattering, mobility degradation. [3, 4]. These adverse effects are the main cause of off-state leakage and reduced circuit performance in high-density ICs.

Recently, various solutions have been proposed in standard literatures to off-flow these leakages in nano-scaled MOSFETs. The resulting technologies like FinFET [5], junctionless transistor (JLT) [6] and FD SOI [7] have been accepted worldwide due to their excellent immunity over short-dimensional effects. Also, various techniques like metal-gate-engineering [8], multi-gate (double-gate/gate-all-around) [9], hetero-gate-dielectric (HGD) [10] have further enhanced the device performance depending upon the requirements. As FinFETs offer better analog and radio-frequency performance. However, it results in disadvantageous due to complex structures.

Moreover, as per the industrial perspective and market demand, FD SOI technology has been attracting attention for low-power and high-performance constraints and having planer structure as compared to other technologies like FinFET [11–15]. Also, the analyzed channel density issues are almost overcome by the use of un-doped channel techniques in FD SOI MOSFETs. Young [11] has provided the very first mathematical analysis of FD SOI MOSFETs, and Suzuki et al. [12] have optimized the buried-oxide (BOX) thickness effects by proposing analytical model. Cheng et al. [13] have reviewed the features of FD SOI MOSFETs and recommended the back-biasing technique for improved radio-frequency performance. In continuation, the proposed source engineering technique in FD SOI MOSFET overcomes the parasitic effects [14, 15].

Metal-gate engineering in FD SOI MOSFETs has also been accepted to further reduce the threshold-voltage roll-off. Kumar et al. [16] have chosen the technique of dual-material-gate to excel in the performance of FD SOI MOSFETs by optimizing step-like potential profile. In next, Srivastava et al. [17] have utilized the concept of dual-metal-insulated gate in FD SOI technology along with the source engineering and analyzed the oscillation frequency of the designed ring oscillator circuit. Maity et al. [18] have discussed the impact of buried-oxide thickness on analog/RF and circuit performance in the design of FD SOI MOSFETs. However, due to the lesser thickness of source/drain regions, the nano-scaled thin-body FD SOI MOSFETs suffers from higher series resistance [19, 20]. Moreover, the suggested recessed-source/drain (Re-S/D) technology in FD SOI MOSFETs could be analyzed to reduce this resistance. Zhang et al. [19] had proposed the design of Re-S/D MOSFET and discussed its features. In continuation, the analytical model of this approach has been given by Sivilicic et al. [20]. Also, the metal-gate engineering has been utilized further to enhance the performance of Re-S/D MOSFETs [21]. Priya et al. [21, 22] had extensively analyzed the performance of multi-metal-gate FD SOI MOSFET in Re-S/D technology and also discussed the circuit-level performance. However, the scope and advantages of hetero-gate-dielectric (HGD) technique in the design of Re-S/D technology-based triple-metal-gate (TMG) FD SOI MOSFET have not been discussed yet.

In this contribution, the analog and radio-frequency performance of hetero-gate-dielectric-based TMG FD SOI MOSFET in Re-S/D technology has been analyzed for the first time. It has been assured that the studied HGD-TMG Re-S/D FD SOI MOSFET exhibits enhanced electric performance and could be suggested for analog and radio-frequency-based communication systems. This paper has been organized into five major sections. Section 2 discusses the device structure and specification used in the design of proposed SOI MOSFET, and the fabrication feasibility of the studied MOSFET is described in Sec. 3. Then, the TCAD simulation strategies have been explained in Sec. 4. The analog and radio-frequency performance of the device are analyzed in Sec. 5. Finally, the overall work is being concluded in Sec. 6.

2 Device Structure and Description

The proposed HGD-TMG Re-S/D FD SOI MOSFET is shown in Fig. 1, and the complete device dimensions and specifications are mentioned in Table 1. The device

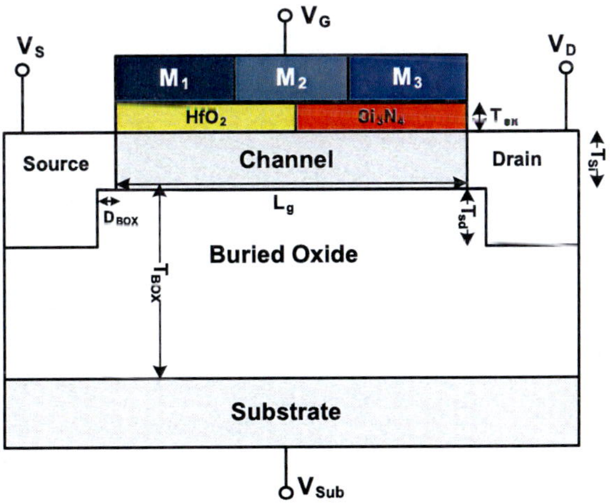

Fig. 1 Device structure of hetero-gate-dielectric triple-metal-gate Re-S/D FD SOI MOSFET

Table 1 Device design parameters and their specifications

Parameter	Value	Unit
L	45	nm
T_{ox}	1	nm
T_{BOX}	50–100	nm
T_{si}	10	nm
T_{rsd}	20	nm
D_{BOX}	3	nm
HGD ratio	1:1	–
$N_{S/D}$	1×10^{19}	cm^{-3}
$P_{Ch/Sub}$	1×10^{16}	cm^{-3}

is designed to operate at 45-nm-technology node such that all the three-metal-gates are of equal gate length ($L_1 = L_2 = L_3 = 15$ nm). The front and buried-oxide thickness is termed as T_{ox} and T_{BOX}, respectively, and the thickness of silicon film is T_{Si}. The extension of *S/D* depth is called as recessed-source/drain thickness (T_{rsd}), and the incremented thickness of overlapped *S/D* region at buried-oxide interface is named as D_{BOX}. This extension of S/D depth assures the effective reduction in series resistance as compared to other ultra-thin body FD SOI devices.

In proposed MOSFET, all three-metal-gates L_1, L_2, and L_3 are arranged form source-to-drain with materials aurum, molybdenum, and titanium with work functions 4.8 eV, 4.6 eV, and 4.4 eV, respectively. These metal gates are placed in such a manner that the material gate with higher work function is at source and lower work function metal is nearer to drain channel end. The metal gate with higher work function helps to off-flow the leakages in sub-threshold region, which results in increased gate controllability, so seen as control gate, whereas lower work function metal screens-off the changes at higher drain bias, so seen as screen gate.

Next, the hetero-gate-dielectric is featured with two different gate-oxide materials in symmetrical manner, such that the first HfO_2 ($\varepsilon_r = 32$) and second gate-oxide materials Si_3N_4 ($\varepsilon_r = 7.8$) are arranged form source-to-drain, respectively, as shown in Fig. 1. The hetero-gate-dielectric offers noticeable enhancement in short-channel performance as compared to single-gate-dielectric. In Table 1, the provided doping levels are used to examine the device characteristics, where source/drain doping is represented as $N_{S/D}$ and the channel/substrate doping as $P_{Ch/Sub}$.

3 Fabrication Feasibility of HGD-TMG Re-S/D FD SOI MOSFET

The FD SOI MOSFETs are experimentally feasible with reduced fabrication complexity as compared to FinFETs [23, 24]. Also, the process flow of triple-metal-gate re-S/D FD SOI MOSFET is already been discussed in our previous work [22].

However, the growth of hetero-gate-dielectric in the design of TMG Re-S/D FD SOI transistor could be targeted as discussed in [24], in which, the growth of hetero-gate-dielectric is described in detail and suggested using HV vapor method to properly etch the first deposited low-k dielectric then atomic layer deposition (ALD) to form high-k dielectric. Moreover, other fabrication steps remain the same as discussed in [22], like Si-wafer pattering, oxidation of BOX layer, SOI-layer deposition, Re-S/D formation, channel growth, oxide growth, S/D implantation, contact formation with proper etching.

4 Simulation Setup

The studied hetero-gate-dielectric-based FD SOI MOSFET has been designed and simulated using technology computer-aided design simulation tool Silvaco ATLAS [25]. Various simulation models have been used for the precise analysis of the studied device. As, for the analysis of majority carrier lifetime, recombination, and Shockley–Read–Hall model is taken into considerations. Similarly, Lombardi mobility and constant voltage, and temperature model are taken for the study of mobility variations due to temperature. Gummel–Newton along with drift-diffusion model is used to analyze the on–off switching of the device. Next, to effectively monitor the impact of transversal electric on-current conduction, field-dependant mobility model is chosen [25]. Also, the investigation of the quantum confinement effect is necessary below 50-nm-technology node. For this, Bohm quantum potential (BQP) model and quantum mechanical effect (QME) model have been used during TCAD simulations.

5 Performance Analysis of HGD-TMG Re-S/D FD SOI MOSFET

This section dictates the analog and radio-frequency performance of proposed FD SOI MOSFET on the basis of numerical simulations. The comparison of drain current characteristics of proposed FD SOI MOSFET with existing state of the art [21] at 45-nm-technology node is shown in Fig. 2. It is clear from the plot that the proposed HGD-based device offers lesser off-state-leakage current (I_{off}) and attains higher on-current (I_{on}) as compared to [21]. The proposed MOSFET and referenced devices show $I_{\text{off}} = 1.47 \times 10^{-13}$ A and 7.5×10^{-12} A and $I_{\text{on}} = 2.56$ mA and 1.12 mA, respectively. That results in switching ratio ($I_{\text{on}}/I_{\text{off}}$) of 10^{10} and 10^{9} for proposed and referenced device [21], respectively.

This increment in switching performance of proposed MOSFET is due to the use of hetero-gate-dielectric in place of single-gate-dielectric. As due to the inclusion of HGD, the effective electric-field under first gate-metal will get enhanced and that gives rise in gate control over device channel. Also, triple-metal-gate structure will

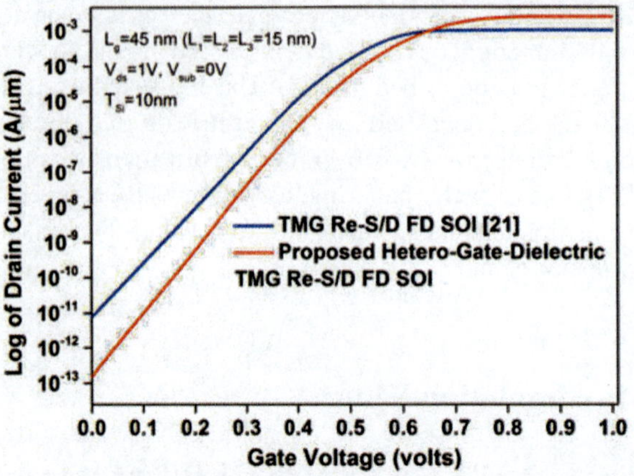

Fig. 2 Comparison of I_d versus V_{gs} characteristics of proposed HGD-TMG MOSFET at $T_{\text{BOX}} = 50$ nm with referenced FD SOI MOSFET [21] at $V_{ds} = 1$ V

provide step-like potential profile, which helps to reduce electric field penetrations due to applied drain voltage in short-dimensional MOSFET. This will further reduce drain-induced barrier-lowering effect.

5.1 Analog Performance of Studied FD SOI MOSFET

In this section, the analog performance of the studied FD SOI MOSFET has been monitored on the basis of numerical calculations of transconductance ($g_m = \partial I_d/\partial V_{gs}$), output conductance ($g_d = \partial I_d/\partial V_{ds}$), transconductance generation factor (TGF $= g_m/I_d$), and intrinsic gain ($A_v = g_m/g_d$).

The plot of transconductance as a function of gate voltage at different buried-oxide thickness is shown in Fig. 3 at V_{ds}= 1 V. The transconductance examines the carrier transport efficiency of the device, and it should be high for higher gain. It is depicted

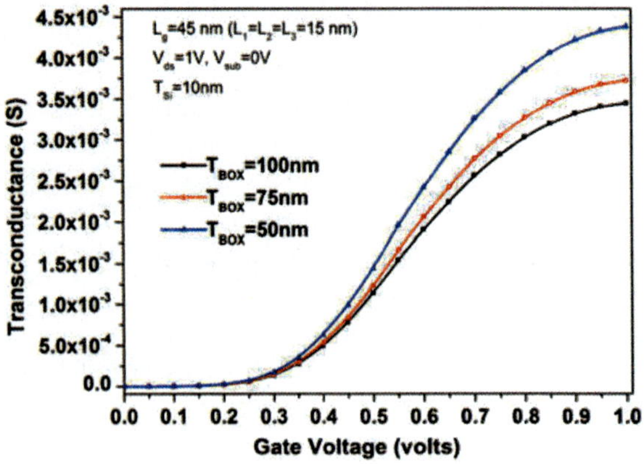

Fig. 3 Analysis of transconductance behavior versus gate voltage of proposed HGD-TMG MOSFET at $T_{\text{BOX}} =$ 50–100 nm

in Fig. 3 that the proposed HGD-based device excels the device transconductance and offers g_m of 4.38 mS, 3.72 mS and 3.45 mS at $T_{BOX} = 50$ nm, 75 nm, and 100 nm, respectively, with lower voltage of operation, and this leads to an improved performance when used with low-power analog circuits.

Also, the analysis of transconductance generation factor (TGF) at different T_{BOX} is shown in Fig. 4. The TGF investigates the effectiveness of transconductance toward the current conduction with thermal stability. One can get from Fig. 4 that the studied device offers reduction in TGF value with applied gate bias and better performance at $T_{BOX} = 50$ nm. This is needed for proper operation of highly linear analog ICs.

Similarly, output conductance behavior of the studied SOI MOSFET is plotted in Fig. 5. The output conductance must be as low as possible for higher gain in analog devices and circuits. The studied device exhibits lower output conductance at $T_{BOX} = 50$ nm and $L = 45$ nm. So, the studied MOSFET could be optimized at $T_{BOX} = 50$ nm for better analog performance. Now, the analyzed intrinsic gain (A_v)

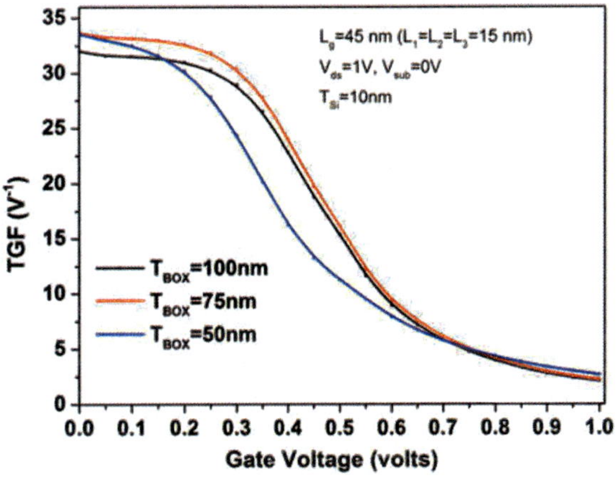

Fig. 4 Plot of transconductance generation factor (TGF) versus gate voltage with variation in T_{BOX} at $V_{ds} = 1$ V for proposed HGD-TMG MOSFET

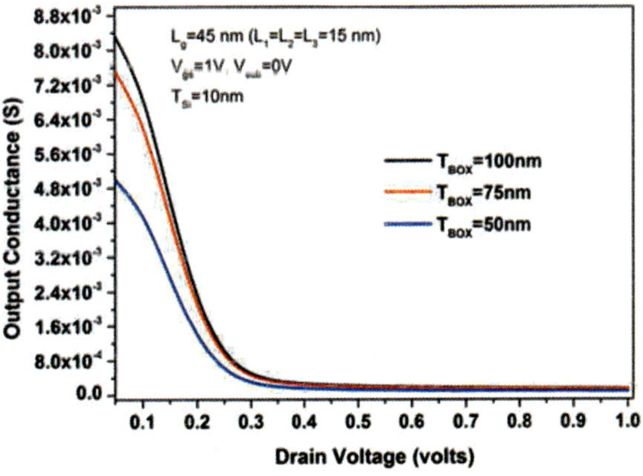

Fig. 5 Analysis of output conductance with applied drain bias at different T_{BOX} for proposed FD SOI MOSFET

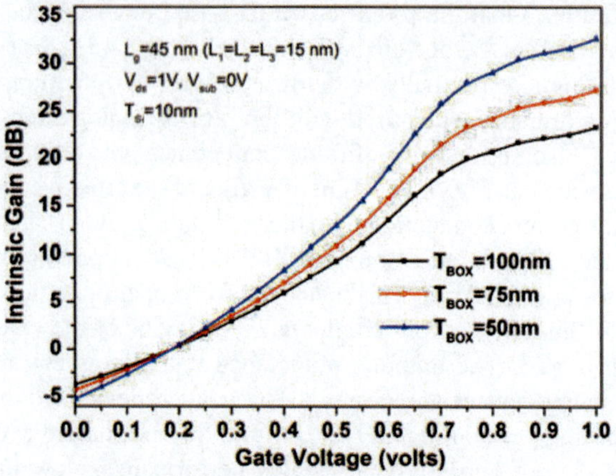

Fig. 6 Calculated intrinsic gain (A_v) of proposed FD SOI MOSFET at different T_{BOX}

of HGD-based SOI MOSFET is shown in Fig. 6. It is found from the plot that the device exhibits better intrinsic gain at optimized T_{BOX} and $L = 45$ nm.

5.2 *RF Performance Assessment*

For better RF characteristics, the device should attain excellent parasitic behavior along with higher cutoff frequency. Here, the radio-frequency performance has been investigated on the basis of *TCAD* numerical simulations. Firstly, the AC simulations have been performed to calculate the gate-to-source and gate-to-drain parasitic capacitances (C_{gs} and C_{gd}). The plot C_{gs} with variation in gate voltage is shown in Fig. 7. It is found that the studied device offers lesser value of parasitic capacitance at higher T_{BOX} and also considerable performance at lower T_{BOX}. This is due to the effective suppression of trans-capacitances due to the inclusion of HGD technique in

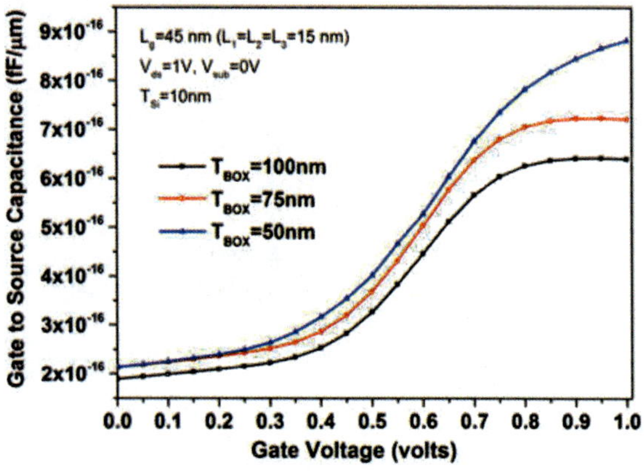

Fig. 7 Plot of gate-to-source parasitic capacitance (C_{gs}) versus gate voltage at $T_{\mathrm{BOX}} = 50$–100 nm and $V_{ds} = 1$ V for proposed MOSFET

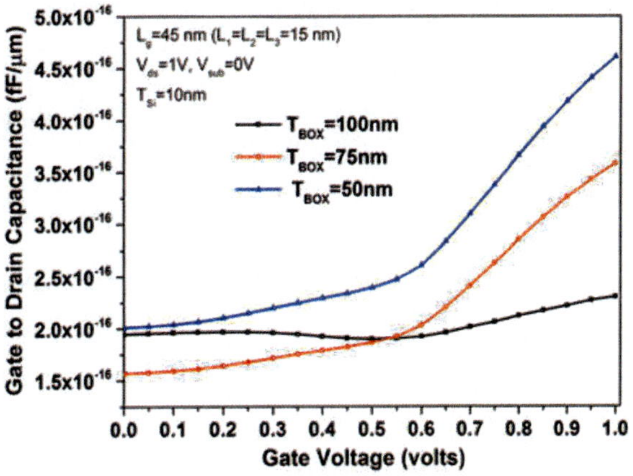

Fig. 8 Plot of gate-to-drain parasitic capacitance (C_{gd}) versus gate voltage at T_{BOX} = 50–100nm and $V_{ds} = 1$ V for proposed MOSFET

SOI MOSFETs. Similarly, plot of C_{gd} at different T_{BOX} is presented in Fig. 8, which also dictates the immunity of the studied device at 45 nm node. The device presents $C_{gs} = 0.639$ fF and $C_{gd} = 0.230$ fF at $T_{\text{BOX}} = 100$ nm and $L = 45$ nm.

Next the analysis of cutoff frequency $\left(f_T = g_m/2\pi\left(C_{gs} + C_{gd}\right)\right)$ is taken into consideration. The higher cutoff frequency is desired for the investigation of allowable bandwidth in radio-frequency integrated circuits (RFICs) communication systems. Figure 9 represents the plot of cutoff frequency with variation in gate voltage at different T_{BOX}. The device exhibits very close results at T_{BOX} ranging from 50 to 100 nm. However, the device offers better RF behavior at higher T_{BOX} and better transconductance at lower T_{BOX}. So, this leads to almost constant cutoff frequency for T_{BOX} range from 50 to 100 nm. The proposed hetero-gate-dielectric SOI MOSFET offers $f_T = 5.21 \times 10^{11}$ Hz at $T_{\text{BOX}} = 50$ nm and $L = 45$ nm.

Table 2 presents the exact analysis of analog and radio-frequency performance parameters of hetero-gate-dielectric TMG Re-S/D FD SOI MOSFET. One can depict

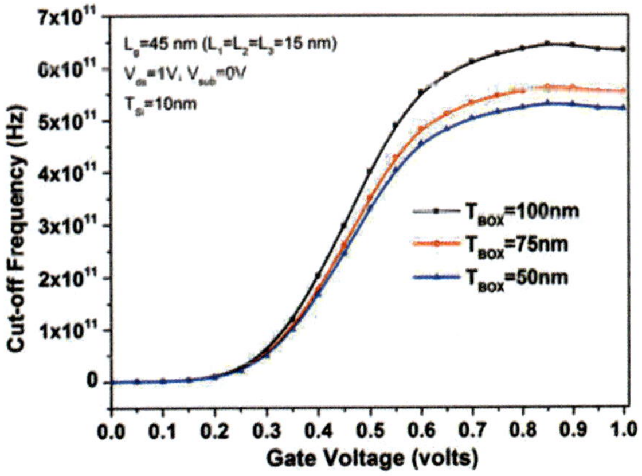

Fig. 9 Analysis of cutoff frequency at T_{BOX} = 50–100 nm and $V_{ds} = 1$ V for proposed MOSFET

Table 2 Analog and radio-frequency parameters of proposed HGD-based SOI MOSFET at different T_{BOX} for $L = 45$ nm

Parameters	BOX thickness (T_{BOX})		
	T_{BOX} = 100 nm	T_{BOX} = 75 nm	T_{BOX} = 50 nm
I_{on} (A)	3.30×10^{-3}	3.02×10^{-3}	2.56×10^{-3}
I_{off} (A)	4.1×10^{-14}	9.2×10^{-14}	1.47×10^{-13}
I_{on}/I_{off} ratio	8.1×10^{10}	3.2×10^{10}	2.24×10^{10}
g_m (mS)	3.45	3.72	4.38
TGF (V^{-1})	32.01	33.54	33.61
g_d (mS)	0.138	0.124	.083
A_v (dB)	23.34	27.23	32.68
C_{gs} (fF)	0.639	0.719	0.882
C_{gd} (fF)	0.230	0.357	0.460
f_T (Hz)	6.32×10^{11}	5.51×10^{11}	5.21×10^{11}

that the studied device is having better analog/RF performance. So, as per the above-discussion, it is clear that the utilization of hetero-gate-dielectric is advantageous in the design of triple-metal-gate Re-S/D FD SOI MOSFET. Also, the analysis over buried-oxide thickness provides the tradeoff to optimize the T_{BOX} according to the requirements in electronics circuit and systems. This type of performance itself dictates that the proposed device could be an alternative in RFICs-based communication systems.

6 Conclusion

This work presents the perspective of hetero-gate-dielectric technique on analog and radio-frequency performance of HGD-TMG Re-S/D SOI MOSFET. It is assured that the studied MOSFET exhibits significant features at 45-nm-technology node for low-power analog/RF applications. The proposed device exhibits off-state leakage of $I_{off} = 1.47 \times 10^{-13}$ A and on-current $I_{on} = 2.56$ mA at $T_{BOX} = 50$ nm, $L = 45$ nm, and $V_{ds} = 1$ V. This results in switching ratio $I_{on}/I_{off} = 10^{10}$, which is sufficient to reduce the leakage issues in off-state. So, it is advantageous to use the HGD technique in Re-S/D technology-based FD SOI MOSFET. This leads to a motivation to further investigate the device performance for analog/RF applications. Here, the analyzed results present a noticeable enhancement in transconductance behavior (g_m = 4.38 mS) with higher gain of 32.68 dB at optimized dimensions. Also, the device shows better g_m/I_d value of 33.61 V^{-1} in weak-inversion region, which is good for thermal stability of highly linear analog ICs. Further, the device shows a better parasitic immunity with C_{gs} and C_{gd} in order of femto-farads and the cutoff frequency of the device is found as $f_T = 5.21 \times 10^{11}$ Hz. Hence, the HGD-based SOI MOSFET could be suggested for the design of low-power analog ICs in radio-frequency-based communication systems.

References

1. Kumar M, Haldar S, Gupta M, Gupta RS (2014) Impact of gate material engineering (GME) on analog/RF performance of nanowire Schottky-barrier gate all around (GAA) MOSFET for low power wireless applications: 3D T-CAD simulation. Microelectr J 45(11):1508–1514
2. Trivedi N, Kumar M, Haldar S, Deswal SS, Gupta M, Gupta RS (2017) Charge plasma technique based dopingless accumulation mode junctionless cylindrical surrounding gate MOSFET: analog performance improvement. Appl Phys A 123(9):564
3. Troutman RR (1979) VLSI limitations from drain-induced barrier lowering. IEEE J Solid-State Circuits 14(2):383–391
4. Frank DJ, Dennard RH, Nowak E, Solomon PM, Taur Y, Wong HSP (2001) Device scaling limits of Si MOSFETs and their application dependencies. Proc IEEE 89(3):259–288
5. Saremi M, Afzali-Kusha A, Mohammadi S (2012) Ground plane fin-shaped field effect transistor (GP-FinFET): a FinFET for low leakage power circuits. Microelectr Eng 95:74–82
6. Khan U, Ghosh B, Akram MW, Salimath A (2014) A comparative study of SELBOX-JLT and SOI-JLT. Appl Phys A 117(4):2281–2288
7. Fossum JG, Choi JY, Sundaresan R (1990) SOI design for competitive CMOS VLSI. IEEE Trans Electr Devices 37(3):724–729
8. Yeo YC (2004) Metal gate technology for nanoscale transistors—material selection and process integration issues. Thin Solid Films 462:34–41
9. Lu H, Lu WY, Taur Y (2007) Effect of body doping on double-gate MOSFET characteristics. Semicond Sci Technol 23(1):015006
10. Hoi WY, Lee W (2010) Hetero-gate-dielectric tunneling field-effect transistors. IEEE Trans Electr Devices 57(9):2317–2319
11. Young KK (1989) Short-channel effect in fully depleted SOI MOSFETs. IEEE Trans Electr Devices 36(2):399–402
12. Suzuki K, Pidin S (2003) Short-channel single-gate SOI MOSFET model. IEEE Trans Electr Devices 50(5):1297–1305
13. Cheng K, Khakifirooz A (2016) Fully depleted SOI (FDSOI) technology. Sci China Inf Sci 59(6):061402
14. Mishra VK, Chauhan RK (2017) Performance analysis of modified source and TDBC based fully-depleted SOI MOSFET for low power digital applications. J Nanoelectr Am Sci Publisher 12(1):59–66
15. Srivastava NA, Mishra VK, Chauhan RK (2016) Analytical modelling of surface potential of modified source FD-SOI MOSFET. In: IEEE international conference on emerging trends in communication technologies (ETCT), pp 1–4
16. Kumar MJ (2004) Chaudhry, A: Two-dimensional analytical modeling of fully depleted DMG SOI MOSFET and evidence for diminished SCEs. IEEE Trans Electr Devices 51(4):569–574
17. Srivastava NA, Priya A, Mishra RA (2019) Design and analysis of nano-scaled SOI MOSFET-based ring oscillator circuit for high density ICs. Appl Phys A 125(8):533
18. Maity SK, Pandit S (2017) Effects of BOX engineering on analogue/RF and circuit performance of InGaAs-OI-Si MOSFET. Int J Electr 104(11):1777–1794
19. Zhang Z, Zhang S, Chan M (2004) Self-align recessed source drain ultrathin body SOI MOSFET. IEEE Electr Device Lett 25(11):740–742

20. Sviličić B, Jovanović V, Suligoj T (2009) Analytical models of front-and back-gate potential distribution and threshold voltage for recessed source/drain UTB SOI MOSFETs. Solid-State Electr 53(5):540–547
21. Priya A, Mishra RA (2016) A two dimensional analytical modeling of surface potential in triple metal gate (TMG) fully-depleted recessed-source/drain (Re-S/D) SOI MOSFET. Superlattices Microstruct 92:316–329
22. Priya A, Srivastava NA, Mishra RA (2019) Design of high speed and low-power ring oscillator circuit in recessed source/drain SOI technology. ECS J Solid State Sci Technol 8(3):N47–N54
23. Cho WJ, Ahn CG, Im K, Yang JH, Oh J, Baek IB, Lee S (2004) Fabrication of 50-nm gate SOI n-MOSFETs using novel plasma-doping technique. IEEE Electr Device Lett 25(6):366–368
24. Choi WY, Lee HK (2016) Demonstration of hetero-gate-dielectric tunneling field-effect transistors (HG TFETs). Nano Converg 3(1):13
25. ATLAS User Manual (2016) Device simulation software. Silvaco International, Santa Clara

Theoretical Analysis of Defected Ground Multiband Rectangular Shape Microstrip Patch Antenna

Salah Boubkar Salah Abdou, Anil Kumar, and Arvind Kumar

Abstract This research work presented and implemented a rectangular microstrip patched antenna executed at 2.4 GHz. The proposed antenna designed with defected ground structure. A 0.8-mm-thickness FR4 substrate material with size 60 mm × 60 mm and permittivity of 4.4 was utilized for the proposed antenna design. HFSS software was utilized for the simulation. The proposed antenna resonates at 2.4 GHz. The performance of the antenna was analyzed by changing the feed position.

Keywords Microstrip · Antenna · Multiband and defected ground

1 Introduction

Microstrip patch antennas (MPAs) have gained much attention in modern-day wireless communication technology or ultra-wideband (UWB) applications owing to their significant attributes say, thin profile together with lightweight, low-cost and can also be used on any host surfaces. MPA has limitations such as limited bandwidth along with lower gain and also poor efficacy [1]. Thus, there is a great need for designing multiband antenna solely for the purpose of diminishing the number of antenna's embedded in device by a single antenna amalgamating numerous applications frequency band [2, 3]. Several methods were presented in the past to achieve multiband with notches along with slots. A lot of papers on multiband antennas were offered in the literature. In [4], a range of papers are presented in the literature where numerous methodologies are used to accomplish improved radiation

S. B. S. Abdou (✉) · A. Kumar
Department of ECE, SHUATS, Prayagraj, India
e-mail: er.salah2018@gmail.com

A. Kumar
e-mail: anil.kumar@shiats.edu.in

A. Kumar
Department of ECE, Motilal Nehru National Institute of Technology Allahabad, Prayagraj, Uttar Pradesh 211004, India
e-mail: arvind@mnnit.ac.in

D. Harvey et al. (eds.), *Advances in VLSI, Communication, and Signal Processing*,
Lecture Notes in Electrical Engineering 683,
https://doi.org/10.1007/978-981-15-6840-4_45

pattern, UWB, and also current distribution over the radiating elements that include coplanar waveguide fed monopole antenna utilizing a defected substrate rhombus strip bounded annular ring antenna resulting from microstrip feedline using DGS [5]. Constructing slots on the patch might also be an easy approach for designing a multiband microstrip antenna [6]. The antenna should provide for a way to minimize the substrate size in order for the antenna size to be minimized [7]. A significant amount of miniaturization has been made possible by careful and meticulous investigation of slot insertion in patch and ground of MSA antenna [9], and the patch and ground plane are shorted for the further size reduction [10–15].

This research work presented proposed antenna design resonates at 2.4 GHz. The performance of the antenna was analyzed by changing the feed position. Parameters are simulated using AN SOFT HFSS v.13.

2 Antenna Design

Various designs for a variety of shapes small-size and wide-bandwidth of microstrip antennas are design through HFSS software. In the design process of the flat surface comprising of an infinite together with a finite ground plane, back lobes are offered for the ground plane. The blueprint structure could be executed on the infinite ground plane. An accurate value is required to plot the full structure. The value could be allotted onto three axes. The structures shape can be adjusted by altering the position value. Once the antenna design is complete, the next step is to allocate frequency sweep for the actual frequency range. If we want to utilize the antenna for specific application, the corresponding return loss along with VSWR together with radiation pattern can be acquired. For optimum antenna performance, a thick dielectric substrate with low dielectric constant is appropriate.

This offers better efficacy, greater bandwidth and also improved radiation. The elementary microstrip patch antenna is a strip conductor of size $L \times W$ and thickness hacked by aground plane. The substrate thickness is lesser on compared to the wavelength. While designing an antenna, rotation of the axis along different directions offers various antenna structures. These could be appraised using high-frequency structure simulator software (Table 1).

1. **Width of the Patch**

$$W = \frac{C}{2f_r\sqrt{\frac{2}{\varepsilon_r+1}}} \tag{1}$$

C Free space velocity of light.
f_r Resonating frequency.
ε_r Relative permittivity of substrate.

Table 1 Comparison with previously designed-type antennas

References	Dimension	Frequency of operation (GHz)	Gain (dB)	Feeding method	Substrate
[8]	52 × 58	$F1 = 1.87$	4	Coaxial feed	FR4
		$F2 = 2.25$	4.56		
		$F3 = 4.8$	5		
[7]	33.47 × 29.14 × 1	$F1 = 1.48$	4.11	Coaxial feed	Polyester
		$F2 = 2.24$	4.88		
		$F3 = 3.84$	6.47		
Proposed	60 × 60 × 1.5	$F1 = 8.7$	5.17	Microstrip line	FR4
		$F2 = 12.3$	3.16		
		$F3 = 14.6$	7.44		
		$F4 = 16.1$	11.22		

2. **Effective Length**

$$L_{\text{eff}} = \frac{C}{2f_r\sqrt{\frac{2}{\varepsilon_r+1}}} \tag{2}$$

3. **Effective Dielectric Constant**

$$\varepsilon_{\text{reff}} = \frac{(\varepsilon_r + 1)}{2} + \frac{(\varepsilon_r - 1)}{2}\left[1 + 12\frac{W}{h}\right]^{-\frac{1}{2}} \tag{3}$$

4. **Patch Length Extension**

$$\frac{\Delta L}{h} = 0.412\frac{(\varepsilon_{\text{reff}} + 1)\left(\frac{w}{h} + 0.264\right)}{(\varepsilon_{\text{reff}} - 0.258)\left(\frac{w}{h} + 0.8\right)} \tag{4}$$

5. **Length of the Patch**

$$L = L_{\text{eff}} - 2\Delta L \tag{5}$$

6. **Width of the Substrate**

$$L = 6h + L \tag{6}$$

7. **Length of the Substrate**

$$Wg = 6h + W \tag{7}$$

3 Geometry of the Antenna

Figure 1 gives an illustration of the rectangular-shaped microstrip patch antenna geometry. The substrate size is 60 mm × 60 mm and 1.5 mm thick. The patch is presented in a rectangular shape with an effective size of 35 mm × 18 mm. The antenna utilizes a microstrip line feeding method (Tables 2 and 3).

4 Result and Discussion

Simulation was executed utilizing HFSS Software. Subsequent sections offer the simulation result of the proposed antenna design (Fig. 2; Table 4).

5 Conclusion

A square-shaped antenna with defected ground structure was proposed and implemented. The suggested antenna's functional characteristics say, return loss together with its SWR along with the radiation pattern were examined. The designed antenna is resonating at 2.4 GHz. The proposed antenna's size is 60 mm × 60 mm. The offered antenna's size is small and meets the requirement.

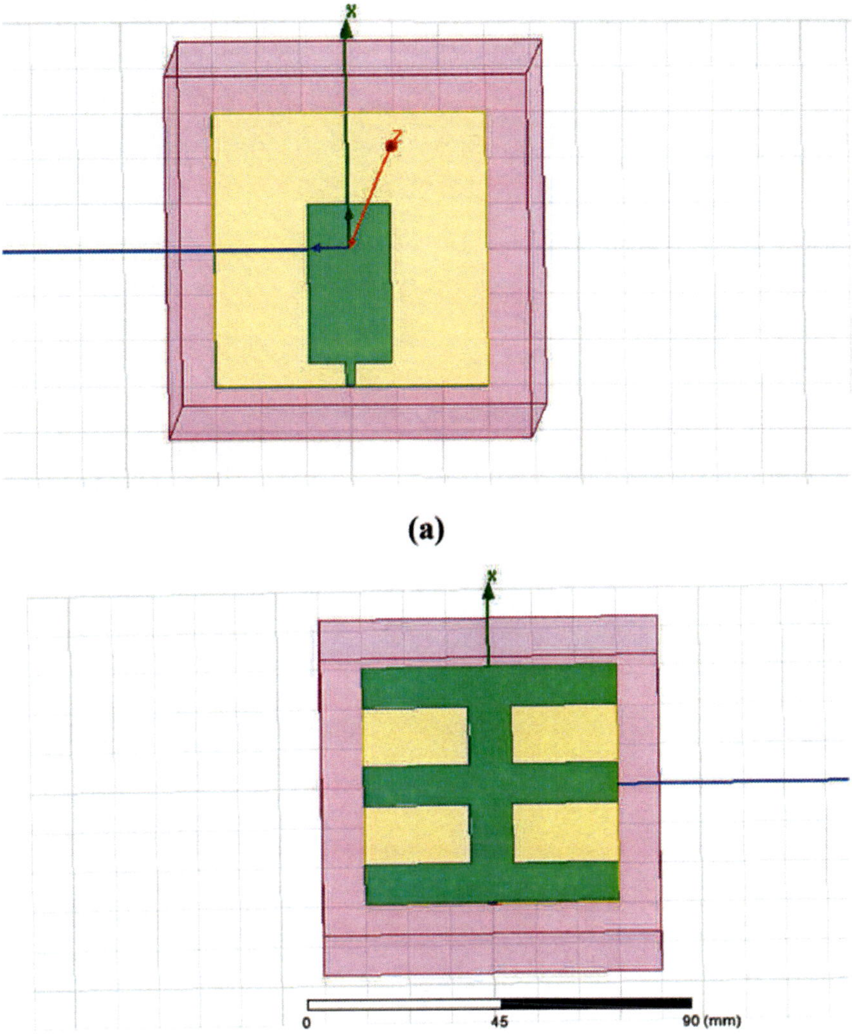

(a)

(b)

Fig. 1 **a** Rectangular-shaped microstrip patch antenna. **b** Ground defected structure

Table 2 Structural details of microstrip patch antenna

S. No.	Parameters	Value (mm)
1	Substrate width	60
2	Substrate length	60
3	Thickness of the substrate	1.5
4	Patch width	18
5	Patch length	35
6	Feed width	2
7	Ground width	60
8	Ground length	−60

Table 3 Structural details of different slot length and width

S. No.	Parameters	SLOT 1	SLOT 2	SLOT 3	SLOT 4
1	Slot width (mm)	10	−10	10	10
2	Slot length (mm)	−60	−60	−25	25

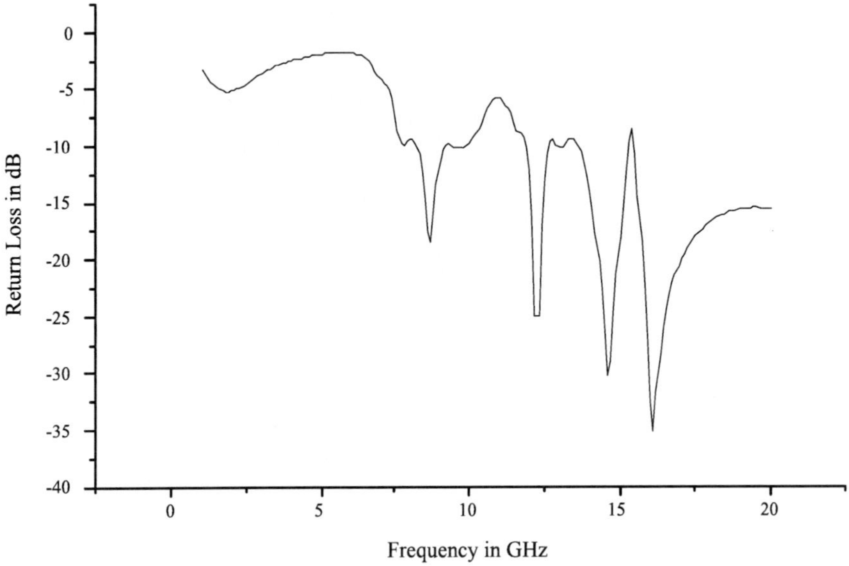

(a)

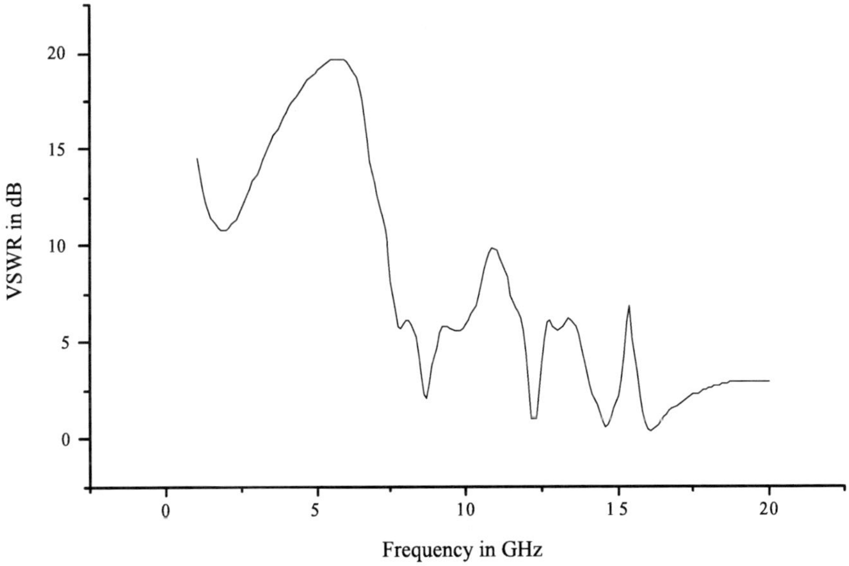

(b)

Fig. 2 **a** Return loss of rectangular shape microstrip patch antenna. **b** VSWR of rectangular shape microstrip patch antenna. **c** Gain of rectangular shape microstrip patch antenna. **d** Directivity of rectangular shape microstrip patch antenna. **e** Radiation efficiency of rectangular shape microstrip patch antenna. **f** Impedance of rectangular shape microstrip patch antenna. **g** Radiation pattern at frequency 8.7 GHz. **h** Radiation pattern at frequency 12.3 GHz. **i** Radiation pattern at frequency 14.6 GHz. **j** Radiation pattern at frequency 16.1 GHz

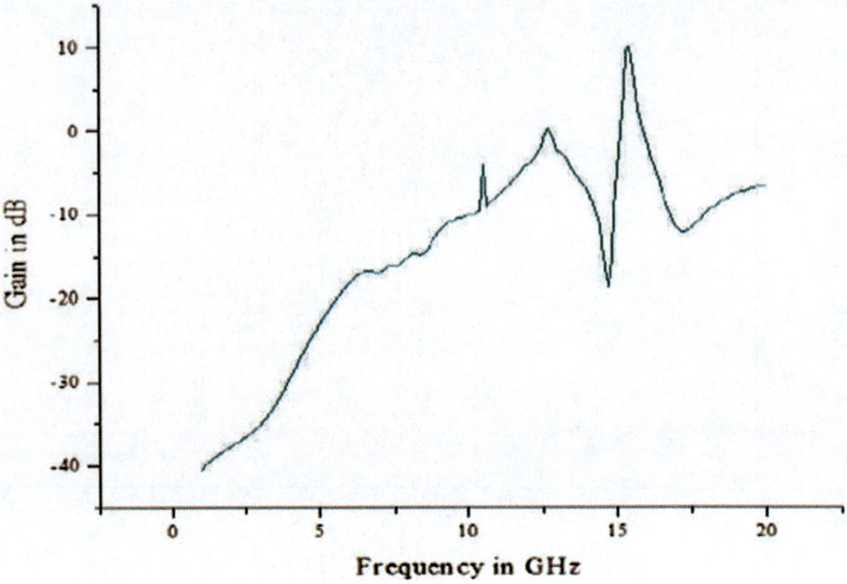

(c)

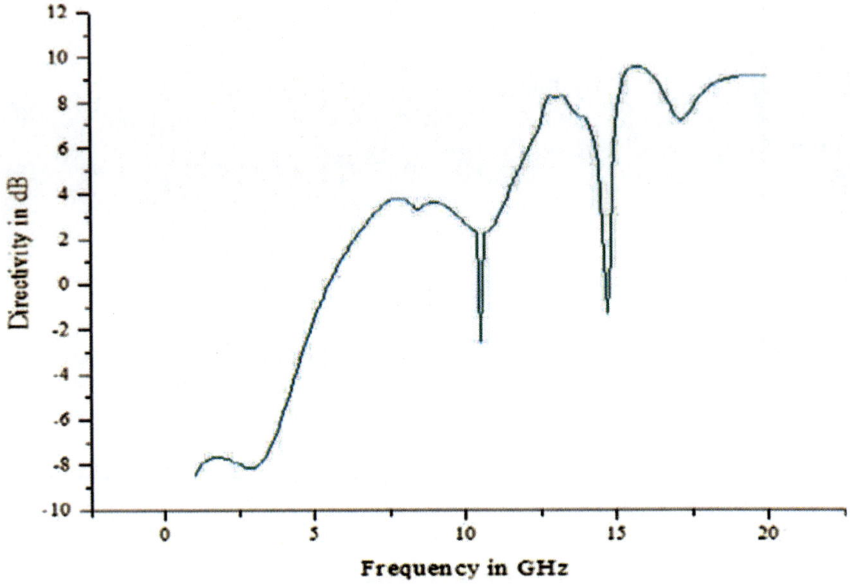

(d)

Fig. 2 (continued)

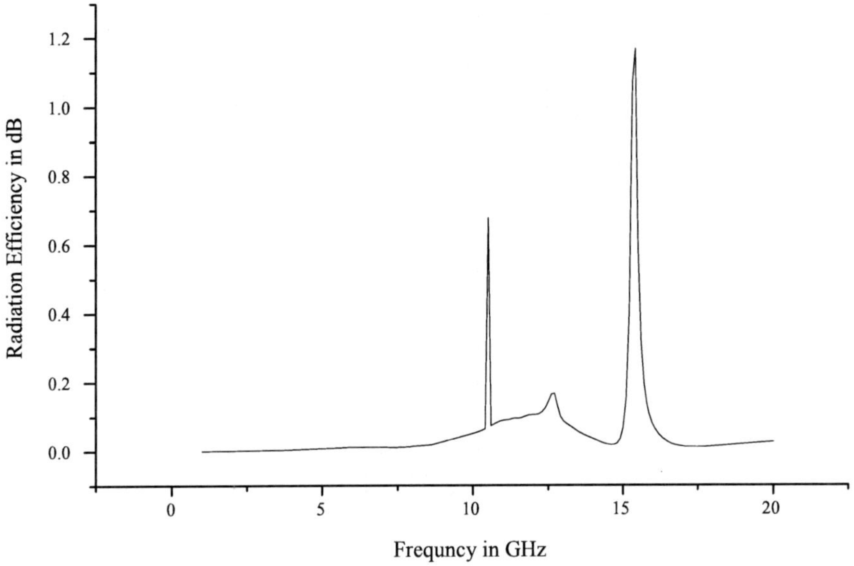

(e)

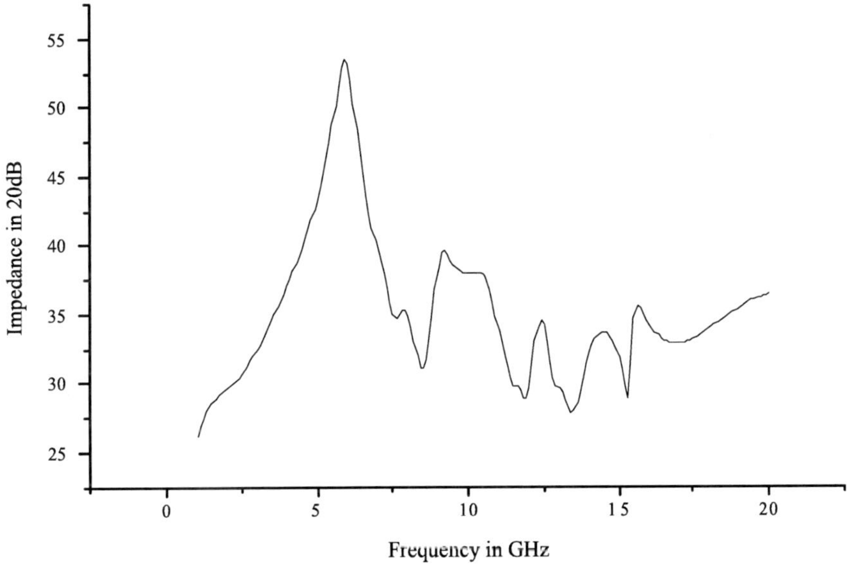

(f)

Fig. 2 (continued)

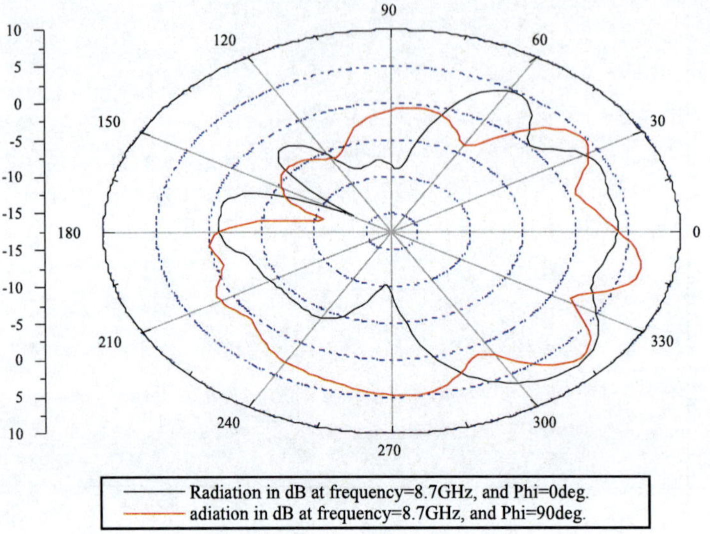

(g)

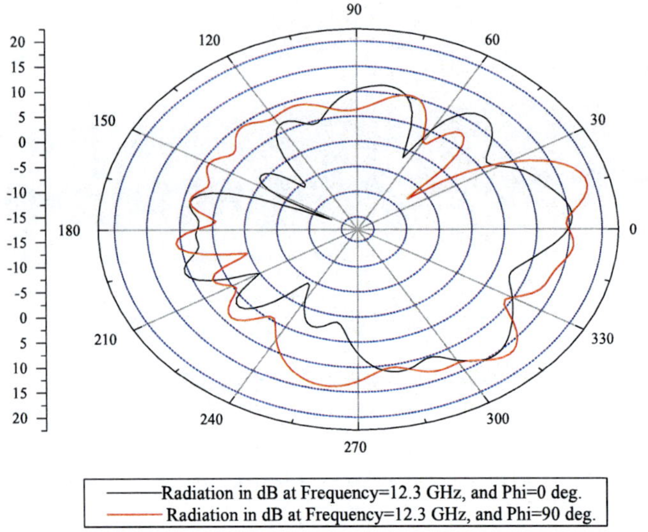

(h)

Fig. 2 (continued)

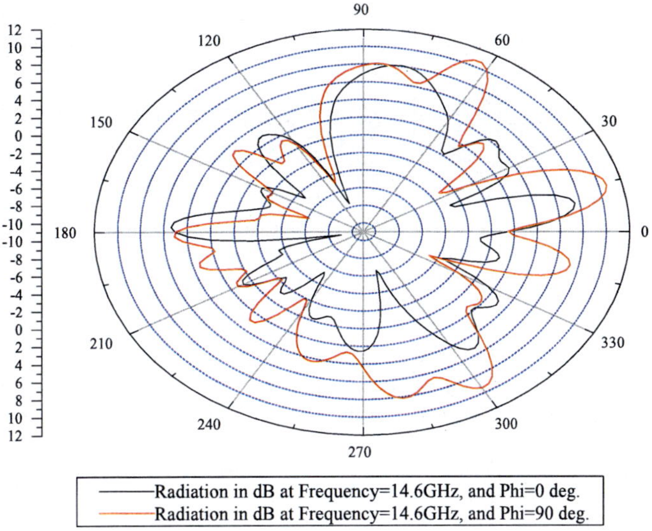

(i)

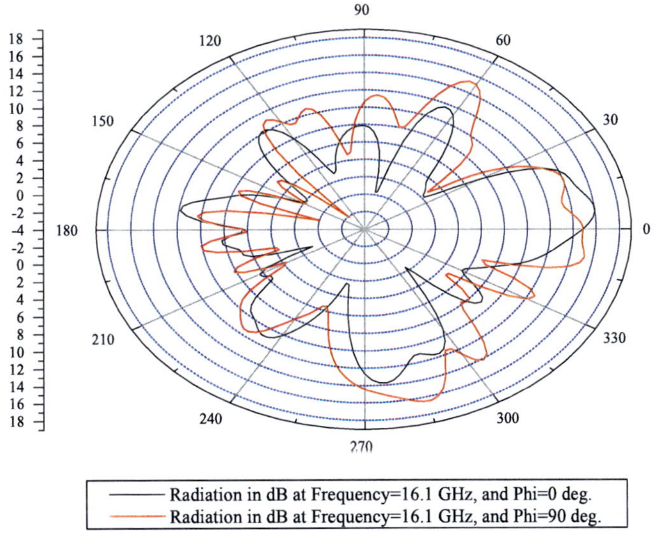

(j)

Fig. 2 (continued)

Table 4 Summary of the simulation results

Frequency of operation (GHz)	Return loss	Gain	Directivity	VSWR	Impedance in 20 dB
8.7	−18.467	5.179	6.6337	2.0824	33.1481
12.3	−24.975	3.161	5.0821	0.9807	34.1448
14.6	−30.223	7.441	8.2649	0.5356	33.5804
16.1	−35.244	11.226	19.6584	0.3004	33.8898

References

1. Kumar M, Ansari JA, Saroj AK (2019) A novel microstrip fed L-shaped arm slot and notch loaded RMPA with mended ground plane for bandwidth improvement. Electromag Res C 95:47–57
2. Chakraborty U, Kundu A, Chowdhury KS, Bhattacharjee KA (2014) Compact dual-band microstrip antenna for IEEE 802.11a WLAN application. IEEE Antenn Wirel Propag Lett 13
3. Chakraborty U, Chowdhury SK, Bhattacharjee KA (2014) Frequency tuning and miniaturizations of a bandwidth enhanced rectangular microstrip antenna. Microw Opt Technol Lett 56(7)
4. Kumar P, Dwari S, Pakariya SP (2017) Tripple-band microstrip antenna for wireless application. Springer Wireless Pers. Commun, Springer Science + Business Media, New York
5. Hashash AI, Bataineh MH, Ahmad I (2017) Using defected ground structure (DGS) to improve nonuniform microstrip bandpass filters' performance. 978-1-5090-4815-1/17/$31.00 ©2017 IEEE
6. Satish Jain K, Gautam S (2015) Miniaturization of microstrip antennas using defected ground plane. In: National conference on recent advances in electronics and computer engineering, RAECE-2015, liT Roorkee, India, 13–15 Feb 2015
7. Sharma A, Khanna P, Shinghal K, Kumar A (2016) Design of CPW-fed antenna with defected substrate for wideband applications. J Electr Comput Eng 2016:1–10, Article ID 6546481 (Hindawi)
8. Khajepour S, Ghaffarian SM, Moradi G (2017) Design of novel multiband folded printed quadrifilar helical antenna for GPS/WLAN applications. IEEE Electr Lett 53(2):58–60
9. Rashid M, Munir EM, Mahmood K (2018) Design of miniaturized multiband microstrip patch antenna using defected ground structure. (IJACSA) Int J Adv Comput Sci Appl 9(6)
10. Hussain SI, Bashir S, Hussain SSD (2014) Compact multiband microstrip patch antenna using defected ground structure (DGS). 978-88-907018-4-9/14/$31.00 ©2014 IEEE
11. Suganthi T, Robinson S, Kanimolhi G, Nagamoorthy T (2014) Design and analysis of rectangular microstrip patch antenna for GSM application. Int J Innov Sci Eng Technol 1(2)
12. Elftouh H, Touhami AN, Aghoutane M (2014) Miniaturized microstrip patch antenna with defected ground structure. Electromag Res
13. Shukla SS, Verma RK, Gohir GS (2015) Investigation of the effect of substrate material on the performance of microstrip antenna. 978-1-4673-7231-2/15/$31.00 ©2015 IEEE
14. Islam M, Shakib MN, Misran N (2009) Broadband E-H shaped microstrip patch antenna for wireless systems. Electromag Res, 163–173
15. Islam MT, Shakib NM, Misran N (2009) Multi-slotted Microstrip patch antenna for wireless communication. Electromag Res Lett 10:11–18

New Resistorless FDNR Simulation Configuration Employing CDDITAs

Kapil Bhardwaj and Mayank Srivastava

Abstract In this research work, a frequency-dependent negative resistance (FDNR) simulation circuit using a new active element, namely current differencing differential transconductance amplifier (CDDITA) has been designed. The developed configuration is an active-C circuit employing two CDDITAs along with only two capacitors. The presented configuration simulates the behaviour of a grounded FDNR without any requirement of passive/active element value matching and the working of proposed circuit found excellent under non-ideal considerations. To demonstrate the performance of presented simulator, an active second-order voltage mode band-pass filtering circuit is developed and simulated. All the theoretical results are confirmed by running PSPICE simulation using CMOS-based model of CDDITA.

Keywords CDDITA · Grounded FDNR · VMBPF

1 Introduction

Frequency-dependent negative resistance (FDNR), which is a very utilitarian element in high-order filter synthesis and various other signal processing applications, was firstly reported by Bruton in 1969 [1]. FDNR has been a very popular circuit concept among researchers and scientists since three decades. The conversion of conventional LCR circuits into FDNR-based circuits has been discussed in [1]. Various implementations of FDNR element employing different active devices have been available in open literature [2–20].

The grounded FDNR [12–20], which refers to an FDNR with one end connected to ground, is found very useful for analogue signal filtering applications. In [12], a grounded FDNR simulator using single second-generation current conveyor (CCII) and six passive components has been reported. But this configuration becomes a bulky configuration due to excessive no. of passive components. The configuration

K. Bhardwaj · M. Srivastava (✉)
Department of Electronics and Communication, National Institute of Technology, Jamshedpur, India
e-mail: mayank.ece@nitjsr.ac.in

D. Harvey et al. (eds.), *Advances in VLSI, Communication, and Signal Processing*, Lecture Notes in Electrical Engineering 683,
https://doi.org/10.1007/978-981-15-6840-4_46

developed in [13] employs three current-controlled current conveyors (CCCIIs). The use of large number of active components makes this realization unsuitable for on-chip implementation. Although the configuration of [14] uses only two CFOAs but the use of four passive components and also two of these are in floating, make this circuit also not suitable for monolithic integration. All the FDNR configurations of [15] exhibit the disadvantage of component value matching. Both the simulators of grounded FDNR presented in [16] are single current feedback operational amplifier (CFOA)-based. But these configurations employ floating capacitances and also the matched passive components. Similarly, the circuits reported in [17] which use single dual-X current conveyor (DXCCII) suffers with employment floating capacitances. Single operational transresistance amplifier (OTRA)-based grounded FDNR is given in [18], which is constructed by using single voltage buffer and four passive elements, and it also has the disadvantage of component matching. Two voltage differencing buffered amplifier (VDBA)-based FDNR simulation circuit with all passive elements in floating state is reported in [19]. The simulator presented in [20] uses five OTAs to simulate the FDNR behaviour. Hence, this circuit has the disadvantage of use of excessive number of active components.

Hence, from the above discussion, it can be said that the previously developed grounded FDNR simulation configurations have at least one of the following undesired features: (1) use of more than two active elements, (2) use of more passive elements, (3) constraint of component value matching, (4) use of all floating passive elements, (5) deviated behaviour under non-ideal considerations and (6) insufficient utilization of active elements.

Therefore, in this work, a new FDNR simulator has been presented employing two CDDITAs and two capacitances. Developed circuit is a purely active C configuration with no requirement of any external resistance. The presented configuration possesses other beneficial features like: low active and passive component requirement, no component matching constraint, good behaviour in non-ideal condition and low active--passive sensitivity index values and full utilization of both the CDDITAs, i.e. no terminal is left floating or connected to ground. As per the best knowledge of authors, there is no CDDITA-based FDNR configurations reported in open literature. Therefore, this work is also an attempt to fill this void.

2 CDDITA

CDDITA, which is an advanced version of conventional current differencing transconductance amplifier (CDTA), was first proposed by Biolek in [21] in 2008. It is a modern active building block used previously in several analogue signal processing circuits. Figure 1 shows the symbolic block representation of standard CDDITA. The CDDITA is a six terminal active circuit idea. The terminals P and N are low impedance current input ports while terminals $X+$, $X-$, Z and V are high impedance ports.

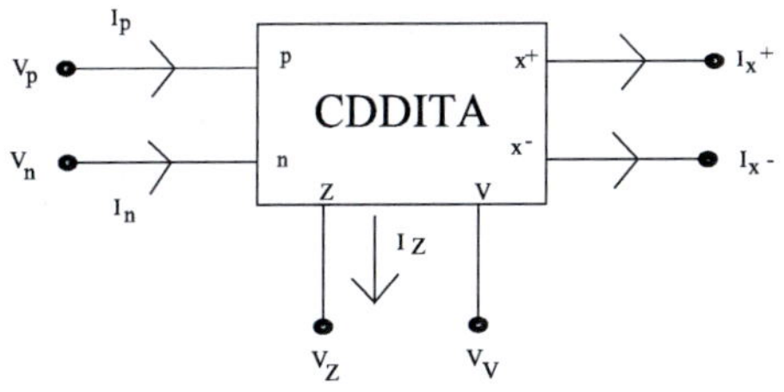

Fig. 1 Symbolic representation of CDDITA

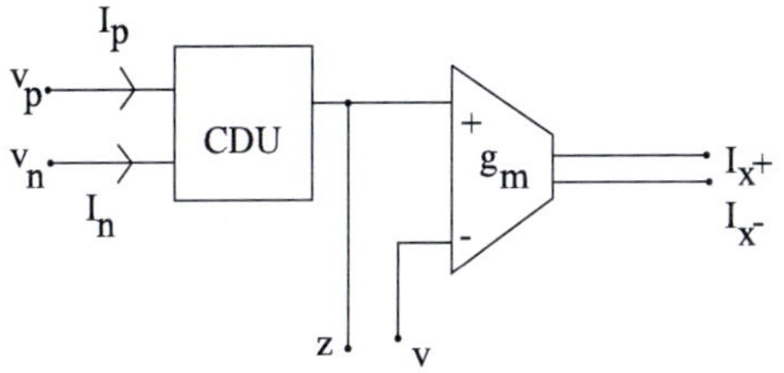

Fig. 2 Behavioural architecture of CDDITA

The behavioural architecture of CDDITA employing a current differencing unit (CDU) and an OTA is shown in Fig. 2.

Implementation of CDDITA active element using complementary MOS (CMOS) transistors is shown in Fig. 3.

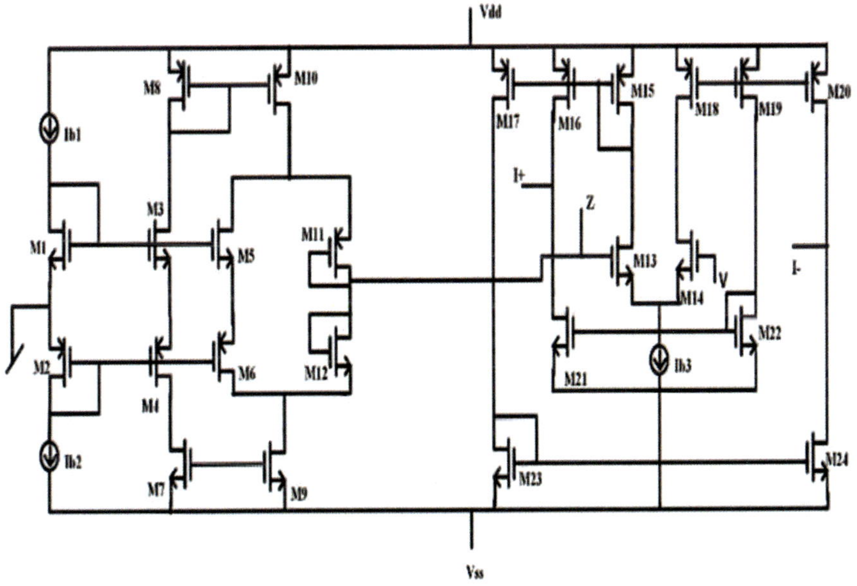

Fig. 3 CDDITA implementation based on CMOS transistors

The equations written below relate the current and voltages of different terminals of CDDITA, where g_m is transconductance of the second stage (OTA) of the CDDITA.

$$V_P = V_N = 0 \tag{1}$$

$$I_z = I_P - I_N \tag{2}$$

$$I_X+ = g_m(V_z - V_v) \tag{3}$$

$$I_X- = -g_m(V_z - V_v) \tag{4}$$

Some contemporary applications of CDDITA in analogue signal processing can be viewed in [22–24].

3 Proposed Configurations

The presented configuration of grounded lossless inductor using CDDITA is depicted in Fig. 4.

By simple mathematical analysis of circuit given in Fig. 4, the input impedance (Z_{in}) of this circuit can be obtained as:

$$Z_{in} = \frac{4g_{m1}}{s^2C_1C_2} + \frac{1}{2g_{m2}} \tag{5}$$

Or

$$Z_{in} = \frac{1}{s^2D_{eq}} + \frac{1}{2g_{m2}} \tag{6}$$

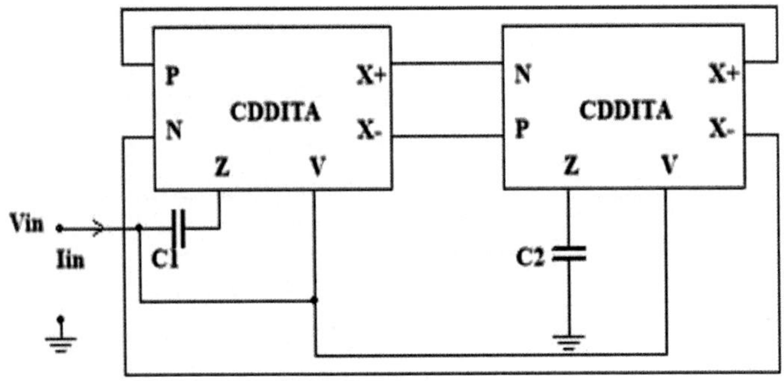

Fig. 4 Proposed CDDITA-based grounded FDNR simulation circuit

From Eqs. (5) and (6), it can be observed that the given circuit is representing the behaviour of a lossy FDNR with equivalent negative resistance value D_{eq} along with a lossy term $1/g_m$. For low and medium frequencies, the value of term $1/s^2\ D_{\mathrm{eq}}$ is much higher than 1/2 g_{m2} and in that case the expression of input impedance is:

$$Z_{\mathrm{in}} = \frac{1}{s^2 D_{\mathrm{eq}}} \tag{7}$$

Therefore, in such condition, the developed circuit simulates the behaviour of a lossless FDNR with negative resistance value D_{eq}:

$$D_{\mathrm{eq}} = \frac{C_1 C_2}{4 g_{m1}} \tag{8}$$

4 Non-ideal Analysis

When the non-ideal gain comes into picture, the relationship between different ports of CMOS CDDITA can be described as follows:

$$V_P = V_N = 0 \tag{9}$$

$$I_Z = \alpha_p I_P - \alpha_n I_N \tag{10}$$

$$I_X+ = \beta^+ g_m (V_z - V_v) \tag{11}$$

$$I_X- = -\beta^- g_m (V_z - V_v) \tag{12}$$

Here, α_p and α_n are current transfer errors, β_+ and β_- are voltage transfer errors.

Considering the equations from (9–12), non-ideal expression of admittance of lossless inductor shown in figure can be given by Eq.

$$Z_{\mathrm{in}} = \frac{4 g_{m1} \beta_1 \beta_2}{s^2 C_1 C_2} \tag{13}$$

Equation 13 clearly reflects that even under non-ideal circumstances the presented configuration simulates the behaviour of an FDNR with slightly deviated negative resistance value. Hence, the behaviour of proposed configuration is satisfactory under non-ideal environment.

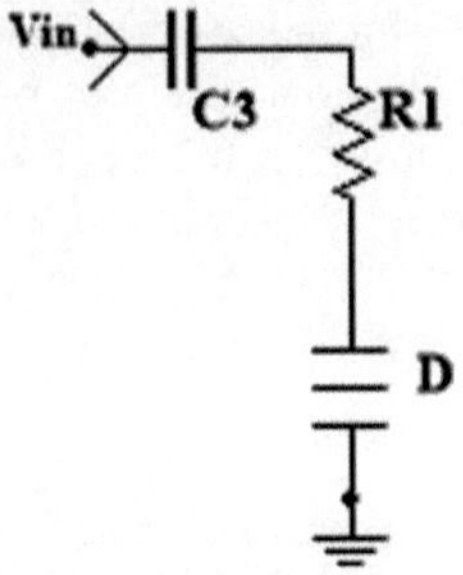

Fig. 5 Proposed passive element-based conventional band-pass filter

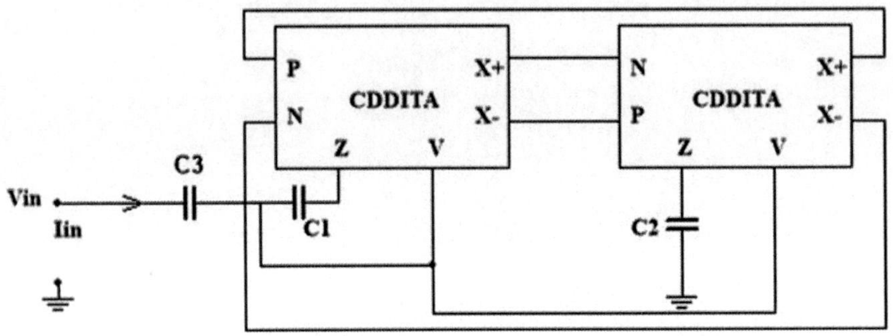

Fig. 6 Band-pass filter of Fig. 5 based on proposed inductance simulator of Fig. 4

5 Application Examples

To prove the usefulness of presented FDNR simulator, a voltage mode band-pass filter is constructed. The passive band-pass filtering network employing FDNR is shown in Fig. 5, and its implementation employing proposed simulator is given in Fig. 6.

6 Simulation Results

For validation of behaviour of designed grounded FDNR simulation configuration, simulations under PSPICE environment have been executed with CDDITA. Simulations were performed with ±3 V DC supply voltage and 100 uA CDDITA bias current. The external capacitance values were selected as $C_1 = C_2 = 0.01$ nF. The simulation generated magnitude response and the phase responses are depicted in Figs. 7 and 8, respectively. The magnitude response in Fig. 7 indicated the reduction of impedance with increasing frequency. Figure 8 clearly shows that the phase is almost −90° for a frequency range from 1.0 kHz to 10.32 MHz, which is acceptable for practical applications. To study the effect of C_1 and C_2, the magnitude responses

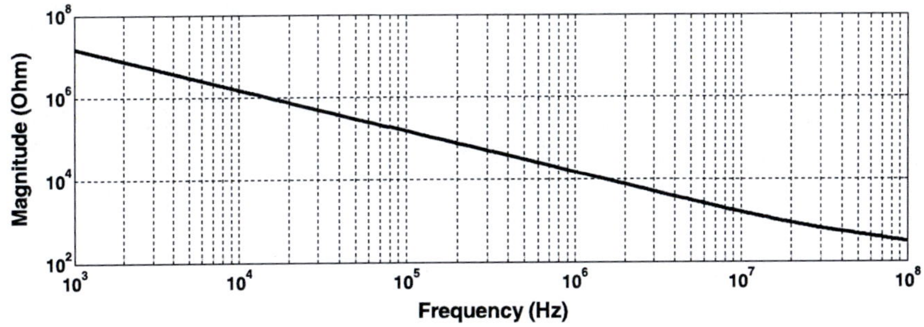

Fig. 7 Input impedance (magnitude) of FDNR simulator shown in Fig. 4

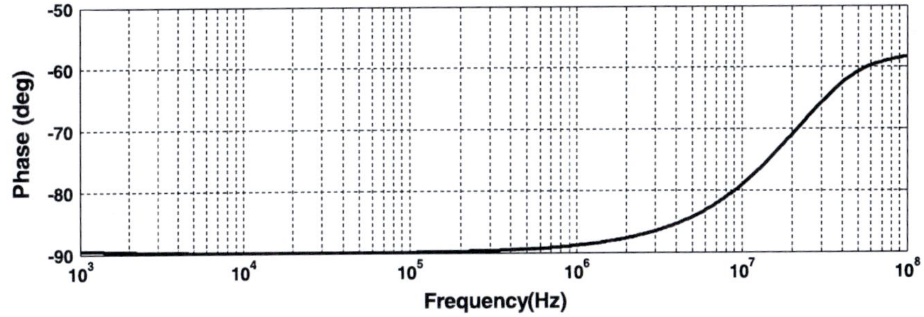

Fig. 8 Phase response of proposed FDNR simulation circuit

are plotted for different values of capacitances and are shown in Fig. 9. The PSPICE simulation of VMBPF given in Fig. 6 is executed with $C_1 = C_2 = 1$ nF, $R_1 = 10\,\mathrm{k\Omega}$ and $C_3 = 0.01$ nf. The frequency response of this filter is depicted in Fig. 10 for different values of C3. It can be observed that for $C = 0.01$ nf, value of Q factor is 0.25, for $C = 0.02$ nf the value of Q is 0.33 and for $C = 0.03$ nf the Q value is increased to 0.5. In Fig. 11, the frequency response plot is given for two different

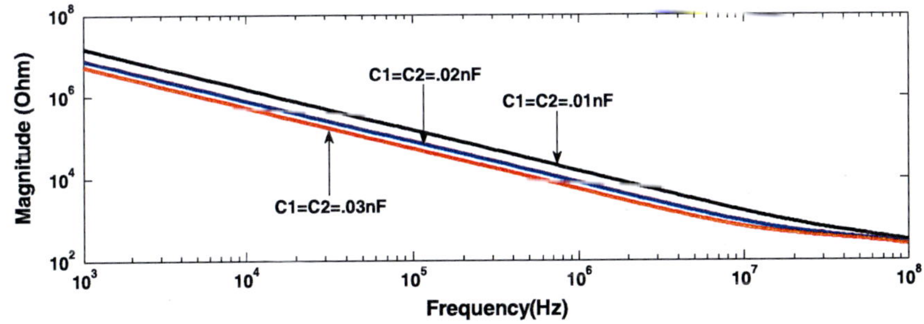

Fig. 9 Input impedance of realized negative resistance for different set of capacitance values

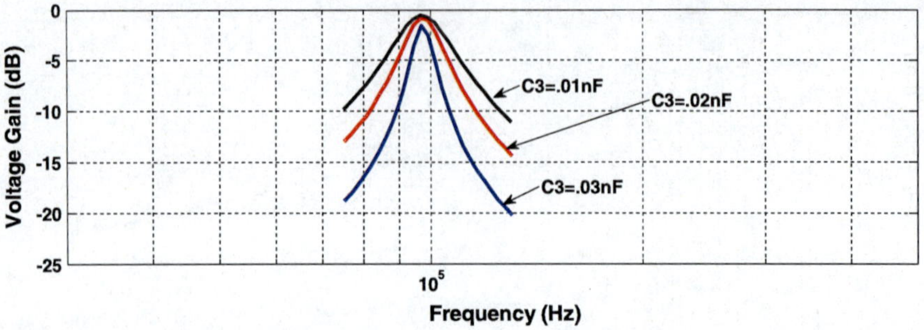

Fig. 10 Frequency response of developed band-pass filter of Fig. 6 for different Q values on varying C_3

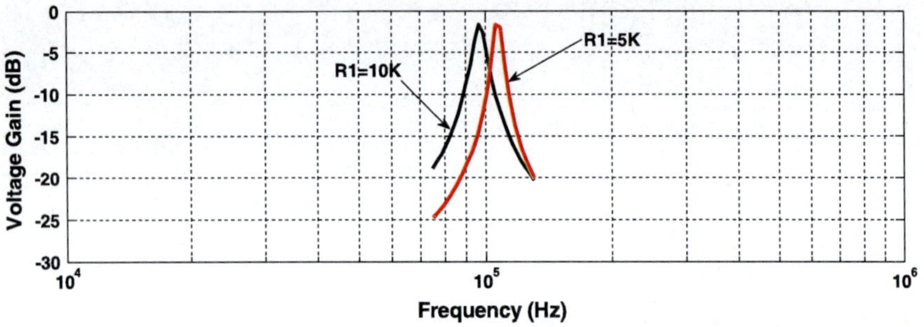

Fig. 11 Frequency response of designed band-pass filter for different centre frequency on varying R_1

values R_1. It can be seen that for low value of $R_1 = 10$ K the central frequency is found 1 MHz and for $R_1 = 5$ K the central frequency is 1.1 MHz.

7 Conclusion

The article presents novel simulation configurations of grounded lossless and lossy inductance based on CDDITA. Both circuits use only two grounded capacitors. The simulation circuits demonstrate merits like: electronic adjustment of realized inductance value, use of only grounded capacitors as external passive elements and behaviour is totally undeviated for non-ideal voltage and current errors. To justify the applicability of the designed circuits, a high-pass filter and voltage mode notch filer are derived from proposed lossless and lossy inductances, respectively. Presented simulation circuits and their application examples are verified under PSPICE environment using CMOS implementation-based model of CDDITA. Realization of lossless inductance simulators using single CDDITA is still an open research problem.

References

1. Bruton L (1969) Network transfer functions using the concept of frequency-dependent negative resistance. IEEE Trans Circuit Theory 16(3):406–408
2. Nandi S, Jana PB, Nandi R (1983) Floating ideal FDNR using current conveyors. Electron Lett 19(7):251–253
3. Senani R (1984) Floating ideal FDNR using only two current conveyors. Electron Lett 20(5):205–206
4. Higashimura M, Fukui Y (1986) Novel lossless tunable floating FDNR simulation using two current conveyors and a buffer. Electron Lett 22(18):938–939
5. Higashimura M, Fukui Y (1987) New lossless tunable floating FDNR simulation using two current conveyors and an INIC. Electron Lett 23(10):529–531
6. Toker A, Cicekoglu O, Kuntman H (1999) New active gyrator circuit suitable for frequency dependent negative resistor implementation. Microelectronics J 30(1):59–62
7. Yuce E (2006) On the realization of the floating simulators using only grounded passive components. Analog Intg, Cir Sig Proc 49, 161–166
8. Tangsrirat W (2013) VDTA-based floating FDNR simulator topology. Kmitl Sci Technol J 13(1):17–21
9. Jaikla W, Siripruchyanun M (2007) Realization of current conveyors-based floating simulators employing grounded passive elements. Proc ECTI conpp 89–92
10. Yuce E (2006) Floating inductance, FDNR and capacitance simulation circuit employing only grounded passive elements. Int J Electron 93(10):679–688
11. Yuce E (2007) On the implementation of the floating simulators employing a single active device. AEU—Int J Electron Commun 61(7):453–458
12. Soliman AM (1978) Realisation of frequency-dependent negative-resistance circuits using two capacitors and a single current conveyor. Proc IEE 125(12):1336–1337
13. Abuelma'Atti MT, Tasadduq NA (1999) Electronically tunable capacitance multiplier and frequency-dependent negative-resistance simulator using the current-controlled current conveyor. Microelectronics J 30(9):869–887
14. Nandi R, Sanyal S, Bandyopadhyay T (2006) Low sensitivity multifunction active circuits using CFA-based supercapacitor. Int J Electron 93(10):689–698
15. Kaçar F, Yeşil A (2012) FDCCII-based FDNR simulator topologies. Int J Electron 99(2):285–295
16. Kacar F, Kuntman H (2009) On the realization of the FDNR simulators using only a single current feedback operational amplifier. In: International conference on electrical and electronics engineering—ELECO 2009, vol 2, pp 223–226
17. Kacar F, Metin B, Kuntman H (2010) On the realization of simulated inductors with reduced parasitic impedance effects. Circuits Syst Signal Process 2:774–778
18. Gupta A, Senani R, Bhaskar DR, Singh AK (2012) OTRA-based grounded-FDNR and grounded-inductance simulators and their applications. Circuits Syst. Signal Process 31:489–499
19. Mongkolwai P, Tangsrirat W (2016) Generalized impedance function simulator using voltage differencing buffered amplifiers (VDBAs). In: Proceedings of the international multi-conference of engineers and computer scientists, vol 2, Hong Kong, 16–18 March 2016, pp 1–5
20. Theingjit S, Pukkalanun T, Tangsrirat W (2016) Grounded FDNC and FDNR realizations based on Gm-C technique and their applications to ladder filter design. Eng Lett 24:4–9
21. Biolek D, Senani R, Biolkova V, Kolka Z (2008) Active elements for analog signal processing; classification, review and new proposals. Radioengineering 17(4):15–32
22. Panwar K, Prasad D, Bhaskarand DR, Srivastava M (2017) Novel single resistance controlled oscillator employing MCDDITA. J Eng Technol 6:352–358
23. Prasad D, Panwar K, Bhaskar DR, Srivastava M (2015) CDDITA-based voltage mode first order all-pass configuration. Circuits Syst (USA) 6:252–256

24. Panwar K, Prasad D, Srivastava M, Haseeb Z (2018) New current mode Lossy integrator employing CDDITA. Cicuits Syst 9:117–123

Novel CDDITA-Based-Grounded Inductance Simulation Circuits

Kapil Bhardwaj and Mayank Srivastava

Abstract This article reports two current differencing differential input transconductance amplifier (CDDITA)-based configurations for simulating the behaviour of lossless- and lossy-grounded inductors. The first proposed configuration simulates the behaviour of a pure grounded inductor while the second realizes the working of a grounded series LC network. Both the proposed configurations are purely resistorless and exhibit several merits like: use of grounded capacitors only, electronic adjustment of realized inductance value, no requirement of matched capacitances and CDDITAs, un-deviated performance under nonideal conditions, full utilization of used CDDITA blocks and low active and passive sensitivities indexes. On detailed literature review, it has been found that there is no CDDITA-based pure inductor/series LC simulation circuits reported so far. Therefore, this work can be considered as the first attempt to design grounded pure inductor/series LC simulation circuits employing CDDITA active element. To ascertain the utility of proposed structures, a voltage-mode (VM) high pass filter and notch filtering circuit are realized using presented pure inductance simulator and LC network simulator, respectively. Validity of all the circuits is confirmed by executing PSPICE simulations based on CMOS CDDITA model.

Keywords CDDITA · Lossless inductor · Series LC

1 Introduction

There are several areas of electrical signal processing but particularly filter design, oscillation generation circuits, phase shifters, resonant circuits, tuned amplifiers and parasitic element cancellation are those areas where inductors and series LC networks are much needed. But conventional L/LC networks have some serious drawbacks, such as bulky nature, cause of unwanted harmonics and interference with electromagnetic waves present in surrounding atmosphere. And also dimensions of the

K. Bhardwaj · M. Srivastava (✉)
Department of Electronics and Communication, National Institute of Technology, Jamshedpur, India
e-mail: mayank.ece@nitjsr.ac.in

D. Harvey et al. (eds.), *Advances in VLSI, Communication, and Signal Processing*, Lecture Notes in Electrical Engineering 683,
https://doi.org/10.1007/978-981-15-6840-4_47

inductor present in the network need to vary proportionally to change the quality factor. That's why, simulating the behaviour of these inductive passive networks using active elements is a hot topic from several decades.

There are so many circuits for simulating the behaviour of pure/lossy-grounded inductors available in literature. These circuits are different active devices based like: Operational amplifiers (Op-amp) [1–6], current conveyors (CC) [7–15], current feedback operational amplifiers (CFOA) [16–19], four terminal floating nullors(FTFN) [20], current follower transconductance amps (CFTA) [21], differential voltage current conveyors (DVCC) [22], current differencing transconductance amplifiers(CDTA) [23], fully differential current conveyors (FDCCII) [24], operational trans-resistance amplifiers (OTRA) [25], dual-x second-generation current conveyors (DX-CCII) [26, 27], voltage differencing differential input buffered amplifiers (VDDIBA) [28], differential difference current conveyors (DDCC) [29, 30], voltage differencing transconductance amplifiers (VDTA) [31] and voltage differential buffered amplifiers (VDBA) [32].

On detailed investigation of configurations reported in [1–32], it has been observed that all the configurations undergo some or all of the following undesired features: (1) Use of enormous number of (three or more) building blocks, (2) Use of large value external resistances, (3) Use of one or more floating capacitances, (4) No provision for electronic controllability of realized inductance value, (5) Necessity of matched passive/active components, (6) Significant deviation in behaviour of circuits under nonideal considerations and (7) Wastage of circuit resources by partial utilization of employed active elements.

Therefore, the purpose of this research work is to report CDDITA-based pure grounded inductor simulator and grounded series LC network simulation circuit. The presented circuits are novel as there is no such configuration has been reported previously employing CDDITA block. Both the proposed configurations has many useful features like: resistorless implementation, use of grounded capacitors, facility of electronic adjustment of simulated inductances, free from matched requirement of matched components, no effect on behaviour under nonideal considerations, full utilization of employed CDDITA and low sensitivity of realized inductance with respect to passive and active components.

2 CDDITA

CDDITA, which is an advanced version of conventional current differencing transconductace amplifier (CDTA), was first designed by Biolek in [33] in 2008. Figure 1 shows the symbolic block representation of standard CDDITA. The CDDITA is a six-terminal active circuit idea. The current input ports P and N are of low impedance while terminal $X+$, $X-$, Z and V are high impedance ports.

The behavioural architecture of CDDITA employing a current differencing unit (CDU) and an OTA is shown in Fig. 2.

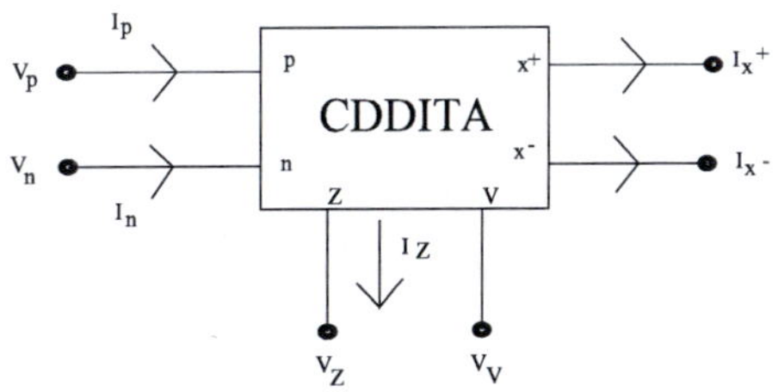

Fig. 1 Block form representation of CDDITA

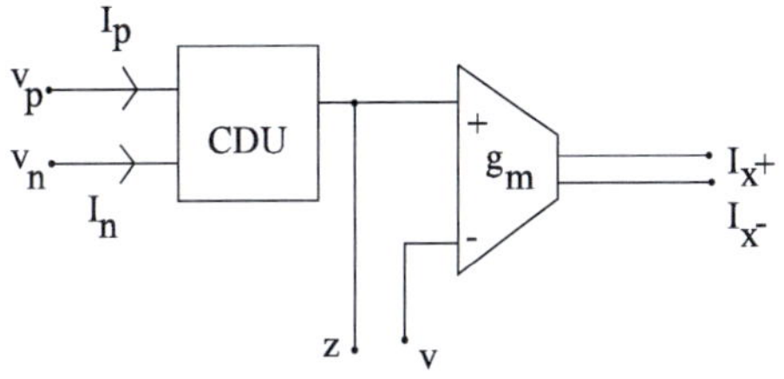

Fig. 2 Behavioural architecture of CDDITA

Implementation of CDDITA active element using complementary MOS (CMOS) transistors is shown in Fig. 3.

The equations written below relate the current and voltages of different terminals of CDDITA.

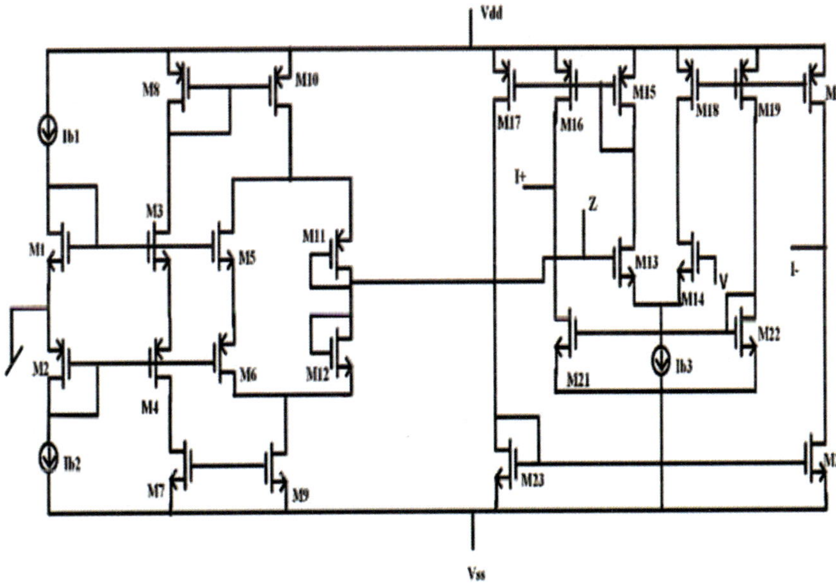

Fig. 3 CDDITA implementation based on CMOS transistors

$$V_P = V_N = 0 \tag{1}$$

$$I_z = I_P - I_N \tag{2}$$

$$I_X + = g_m(V_z - V_v) \tag{3}$$

$$I_X - = -g_m(V_z - V_v) \tag{4}$$

Some latest applications of CDDITA in analog signal processing can be viewed in [34–36].

3 Proposed Configurations

The presented configuration of grounded lossless inductor using CDDITA is depicted in Fig. 4.

Analysing this circuit using simple circuit mathematics, the admittance of the proposed circuit is obtained as:

$$Y_{\text{in}} = \frac{2g_{m1}g_{m2}}{s}\left(\frac{1}{C_1} + \frac{1}{C_2}\right) \tag{5}$$

Equation (5) clearly shows that the above circuit is realizing a grounded lossless inductor, whose inductance value can be given by Eq. (6).

$$L = \frac{1}{2g_{m1}g_{m2}}\left(\frac{C_1 + C_2}{C_1 C_2}\right) \tag{6}$$

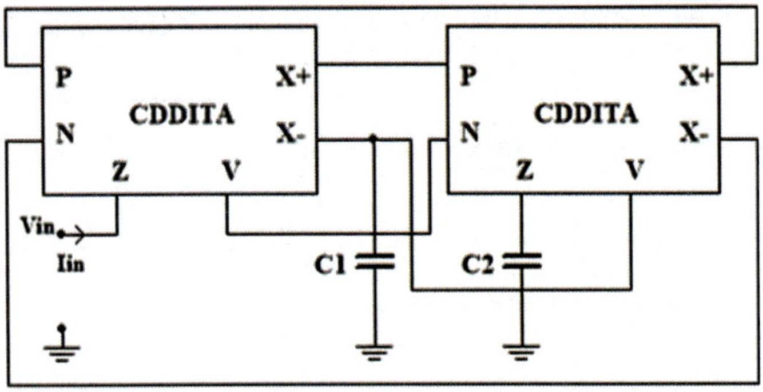

Fig. 4 Proposed CDDITA-based-grounded lossless inductance simulator

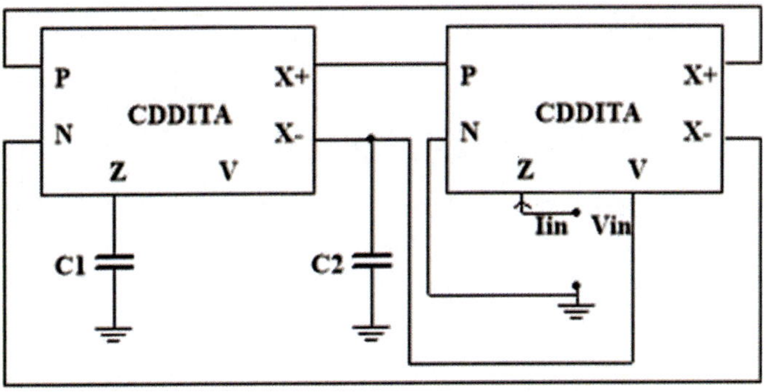

Fig. 5 Proposed CDDITA-based-grounded series RL network simulator

The proposed CDDITA-based-grounded series LC simulator is represented in Fig. 5.

The impedance of the circuit represented in Fig. 5 is found as:

$$Z_{\text{in}} = \frac{sC_1}{g_{m1}g_{m2}} + \frac{1}{sC_2} \tag{7}$$

The expression of Eq. (7) is clearly indicating that the configuration is simulating the behaviour of a grounded series LC network with equivalent inductance (L_{eq}) and capacitance (C_{eq}) described as:

$$L_{\text{eq}} = \frac{C}{g_{m1}g_{m2}} \tag{8}$$

$$C_{\text{eq}} = C_2 \tag{9}$$

From Eqs. 8 and 9, it can be noticed that the realized inductance terms in both the configurations can be adjusted by varying the transconductances g_{m1} and/or g_{m2} which confirm the availability of electronic tunability.

4 Nonideal Analysis

When the nonideal gain comes into picture, the relationship between different ports of CMOS CDDITA can be described as follows:

$$V_P = V_N = 0 \tag{10}$$

$$I_Z - \alpha_p I_P - \alpha_n I_N \tag{11}$$

$$I_X+ = \beta^+ g_m(V_z - V_v) \tag{12}$$

$$I_X- = -\beta^- g_m(V_z - V_v) \tag{13}$$

Here, α_p and α_n are current transfer errors, β_+ and β_- are voltage transfer errors.

Considering the equations from (10–13), nonideal expression of admittance of lossless inductor shown in Fig. can be given by Eq.

$$Y_{\text{in}} = \frac{g_{m1}g_{m2}}{sC[(\alpha_{P1}\beta_{X+2} + \alpha_{P1}\beta_{X-2})(\alpha_{P2} + \beta_{X-1})]} \tag{14}$$

where $C = C_1 = C_2$.

Similarly, impedance of series LC simulator shown in Fig. 5 under nonideal conditions is found as:

$$Z_{\text{in}} = \frac{sC_1}{g_{m1}g_{m2}\alpha_{P1}\alpha_{P2}\beta_{X+2}} + \frac{\beta_{X+2}}{s\alpha_{P2}\beta_{X-2}C_2} \tag{15}$$

From Eqs. (14) and (15), it can be cleducarly observed that behaviour of both the circuits is not changed for nonideal values of CDDTIA gains.

The active and passive sensitivities of L_{eq} from Eq. 15 are given in Eqs. 16 and 17.

$$S_{C_1}^{L_{\text{eq}}} = 0 \tag{16}$$

$$S_{g_{m1}}^{L_{\text{eq}}} = S_{g_{m2}}^{L_{\text{eq}}} = S_{\alpha_{P1}}^{L_{\text{eq}}} = S_{\alpha_{P2}}^{L_{\text{eq}}} = S_{\beta_{x2+}}^{L_{\text{eq}}} = -1 \tag{17}$$

Remaining all the sensitivity values are zero. Therefore, it is clear that all the sensitivity values are unity or zero.

5 Application Examples

To justify the usability of presented pure inductance simulation configuration shown in Fig. 4, a high pass filtering circuit has been constructed. The conventional-inductor-based HP filter is given in Fig. 6. On replacing conventional inductor of Fig. 6 by proposed pure active inductor, the resulting active HPF is developed in Fig. 7.

Similarly, to prove the usefulness of active series LC simulation configuration illustrated in Fig. 5, a voltage-mode notch filtering circuit has been designed. The conventional series RLC notch filter is given in Fig. 8. The complete grounded LC section of Fig. 8 can be replaced by the proposed LC simulator which results in an active notch filter shown in Fig. 9.

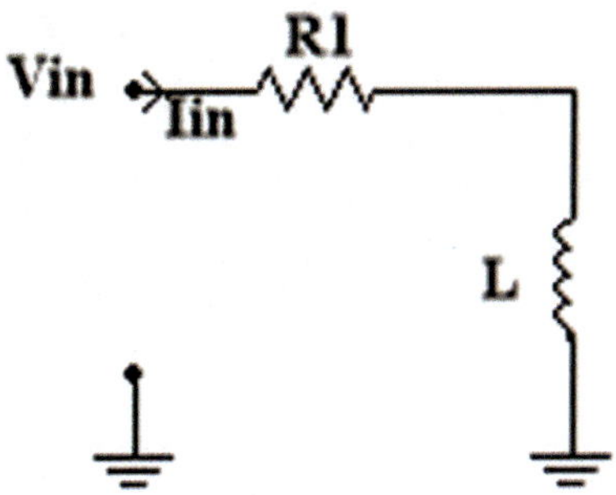

Fig. 6 Proposed passive element-based conventional VMHPF (voltage-mode high pass filter)

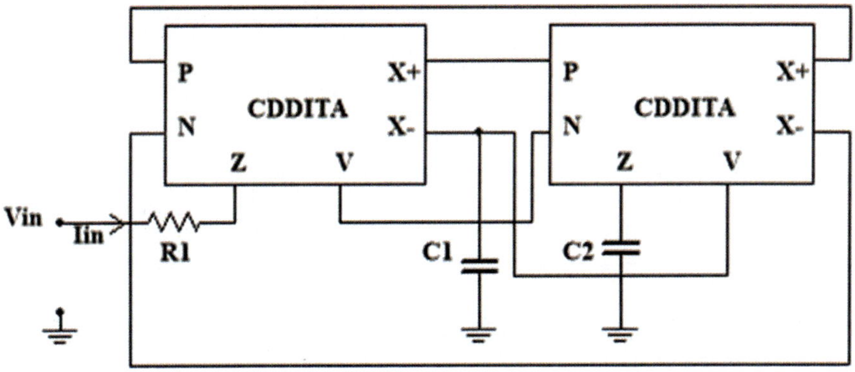

Fig. 7 High pass filter of Fig. 6 derived from proposed inductance simulator of Fig. 4

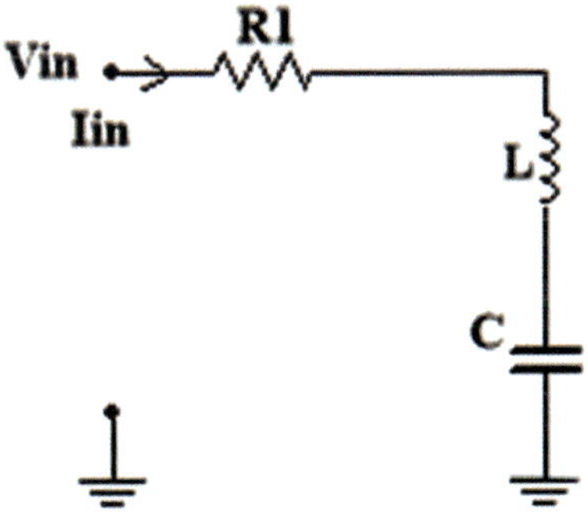

Fig. 8 Conventional passive elements-based RLC notch filter

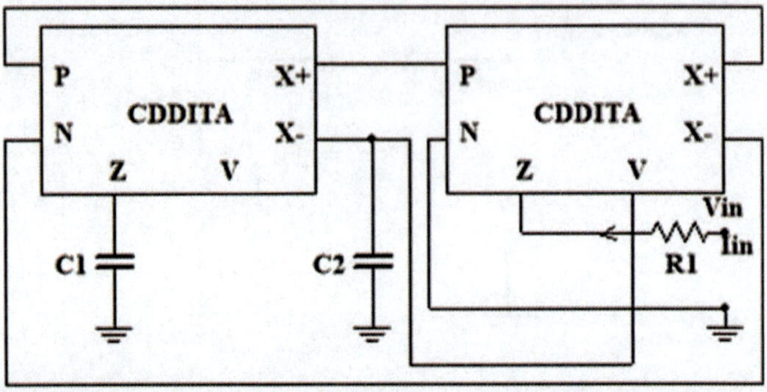

Fig. 9 RLC notch filter of Fig. 8 based on proposed series LC simulator shown in Fig. 5

6 Simulation Results

To verify the mathematical results obtained in previous section, PSPICE-based simulations of presented circuits have been carried out using CMOS-based CDDITA (shown in Fig. 3). For simulation purpose of presented pure inductance simulation configuration of Fig. 4, the capacitance values were chosen as $C_1 = C_2 = 0.01$ nf along with CDDITA bias current $I_b = 100$ uA. The simulation plots of magnitude of input impedance and phase are illustrated in Figs. 10 and 11, respectively, which verifies the behaviour of developed circuits as a pure grounded inductor approximately from 820 kHz to 2.23 MHz. The variation of simulated inductance with employed capacitances has been shown in Fig. 12. The feature electronic adjustability by changing the biasing current of used CDDITAs has been demonstrated in Fig. 13. Which clearly indicates the variation in realized inductance value for different sets of bias currents (Ib1 is bias current of first CDDITA and Ib_2 is for second CDDITA). The application filter of Fig. 7 was simulated for $C_1 = C_2 = 1$ nf, $R_1 = 0.2$ K and $Ib_1 = Ib_2 = 100$ uA and the simulation result shown in given in Fig. 14. Which is a high pass response of cut off frequency $f_c = 350$ kHz. The series

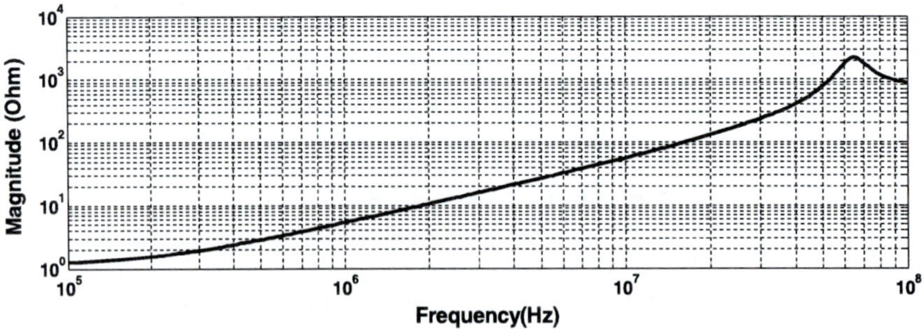

Fig. 10 Input impedance (magnitude) of inductance simulator shown in Fig. 4

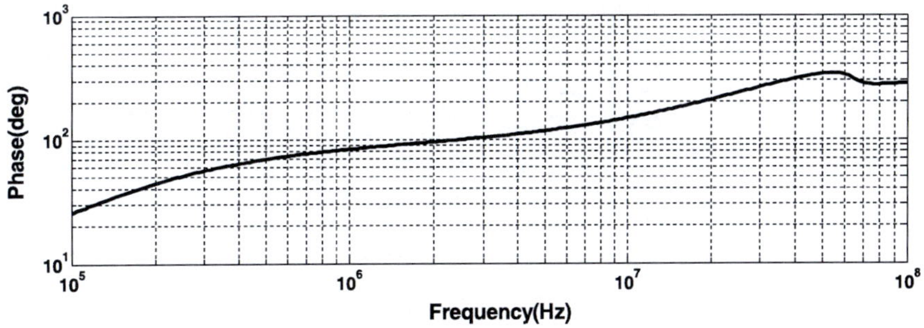

Fig. 11 Phase response of proposed lossless inductance simulation circuit

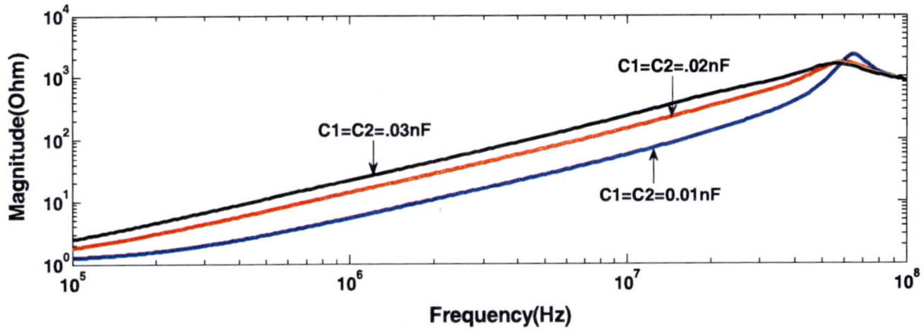

Fig. 12 Input impedance of realized inductance (pure L) for different set of capacitance values

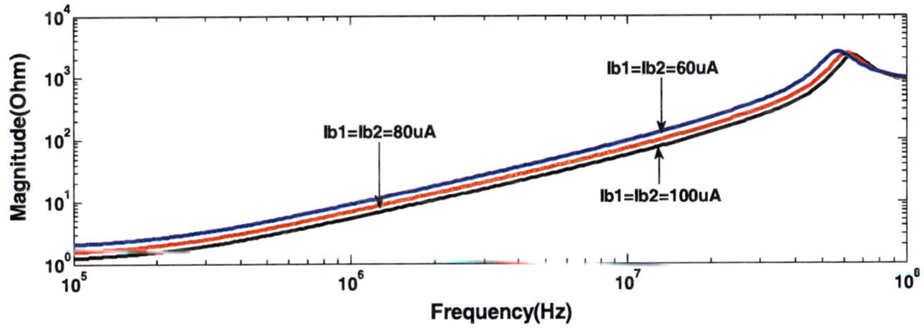

Fig. 13 Magnitude of impedance impedance for different CDDITA bias currents

LC simulator-based notch filter of Fig. 8 is simulated with element values $C_1 = C_2 = 5$ nf and $R_1 = 1$ K obtaining a notch filtering circuit response (as given in Fig. 15) with a value of notch frequency approximately $f_o = 700$ kHz.

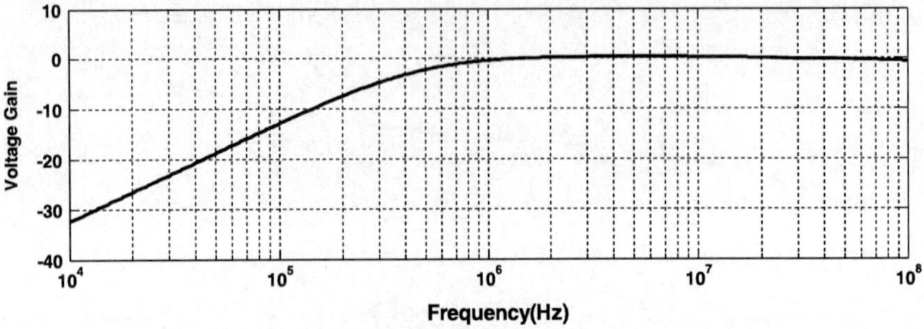

Fig. 14 Frequency response of developed high pass filter shown in Fig. 7

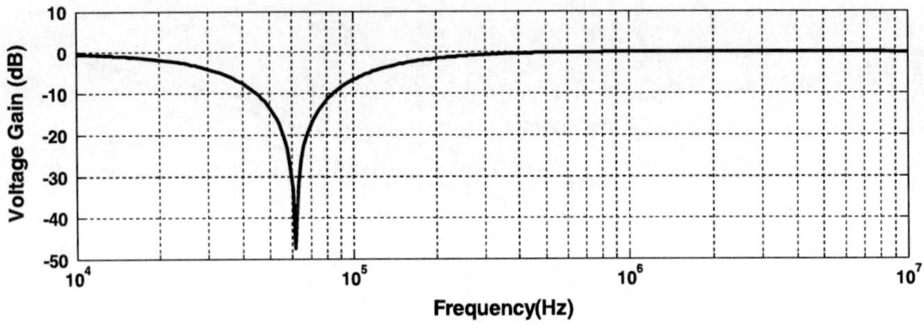

Fig. 15 Frequency response of designed notch filter presented in Fig. 9

7 Conclusion

The article presents novel simulation configurations of grounded lossless and lossy inductance based on CDDITA. Both circuits uses only two grounded capacitors. The simulation circuits demonstrate merits like: electronic adjustment of realized inductance value, use of only grounded capacitors as external passive elements and behaviour is totally un-deviated for nonideal voltage and current errors. To justify the applicability of the designed circuits, a high pass filer and voltage-mode notch filer are derived from proposed lossless and lossy inductances, respectively. Presented simulation circuits and their application examples are verified under PSPICE environment using CMOS implementation-based model of CDDITA. Realization of lossless inductace simulators using single CDDITA is still an open research problem.

References

1. Ford RL, Girling FEJ (1966) Active filters and oscillators using simulated inductance. Electron Lett 2(2):481–482
2. Prescott AJ (1966) Loss compensated active gyrator using differential input operational amplifier. Electron Lett 2(7):283–284
3. Orchard HJ (1974) New active gyrator circuits. Electron Lett 10(13):261–262
4. Roy SCD (1975) On operational amplifier simulation of grounded inductance. Archiv fuer Elektronik und Uebertragungstechnik 29:107–115
5. Nandi R (1980) Novel insensitive lossless inductor simulation through inverse function generation. Electron Lett 16(12):481–482
6. Fabre F (1992) Gyrator implementation from commercially available trans-impedance operational amplifiers. Electron Lett 28(3):263–264
7. Senani R (1978) Active simulation of inductors using current conveyors. Electron Lett 14:483–484
8. Paul AN, Patranabis D (1981) Active simulation of grounded inductors using a single current conveyor. IEEE Trans Circuits Syst 28:164–165
9. Arslan E, Cam U, Cicekoglu O (2003) Novel lossless grounded inductance simulators employing only a single first generation current conveyor. Frequenz; J RF Eng Telecommun 57:204–206
10. Yuce E, Minaei S, Cicekoglu O (2005) A novel grounded inductor realization using a minimum number of active and passive components. ETRI J 27(4):427–432
11. Parveen T, Ahmed MT (2006) Simulation of ideal grounded tunable inductor and its application in high quality multifunctional filter. Microelectron Int 23(3):9–13
12. Yuce ES, Minaei S, Cicekoglu O (2006) Limitations of the simulated inductors based on a single current conveyor. IEEE Trans Circuits Syst 53(12):2660–2667
13. Psychalinos C, Spanidou A (2006) Current amplifier based grounded and floating inductance simulators. Int J Electron Commun (AEU) 60:168–171
14. Yuce E (2008) Grounded inductor simulators with improved low frequency performances. IEEE Trans Instrum Measure 57(5):1079–1084
15. Pal K, Nigam MJ (2008) Novel active impedances using current conveyors. J Active Passive Electron Devices 3(29):29–34
16. Yuce E, Minaei S (2008) A modified CFOA and its applications to simulated inductors, capacitance multipliers, and analog filters. IEEE Trans Circuits Syst 55(1):254–263
17. Yuce E, Minaei S (2009) On the realization of simulated inductors with reduced parasitic impedance effects. Circuits Syst Signal Process 28:451–465
18. Yuce E (2009) Novel lossless and lossy grounded inductor simulators consisting of a canonical number of components. Analog Integr Circ Sig Process 59(1):77–82
19. Kaçar F, Kuntman H (2011) CFOA-based lossless and lossy inductance simulators. Radioengineering 20(3):627–631
20. Kumar P, Senani R (2010) New grounded simulated inductance circuit using a single PFTFN. Analog Integ Circuits Signal Process 62:105–111
21. Herencsar N, Koton J, Vrbra K (2010) CFTA-based active-C grounded positive inductance simulator and its application. Elcktrorevue 1(1):24–27
22. Nandi R (1980) Lossless inductor simulation: novel configurations using DVCCS 16(1):666–667
23. Prasad D, Bhaskar DR, Singh AK (2010) New grounded and floating simulated inductance circuits using current differencing transconductance amplifiers. Radioengineering 19(1):194–198
24. Kacar F (2010) New lossless inductance simulators realization using a minimum active and passive components. Microelectron J 41(2–3):109–113
25. Gupta A, Senani R, Bhaskar DR, Singh AK (2012) OTRA-based grounded-FDNR and grounded-inductance simulators and their applications. Circuits, Syst, Signal Process 31(2):489–499

26. Metin B (2011) Supplementary inductance simulator topologies employing single DXCCII. 20(3), 618–618
27. Myderrizi I, Minaei S, Yüce E (2011) DXCCII-based grounded inductance simulators and filter applications. Microelectron J 42(9):1074–1081
28. Prasad D, Bhaskar DR, Pushkar KL (2011) Realization of new electronically controllable grounded and floating simulated inductance circuits using voltage differencing differential input buffered amplifiers. Active Passive Electron Compon 8 p
29. Ibrahim MA, Minaei S, Yuce E, Herencsar N, Koton J (1974) Lossless grounded inductance simulation using only one modified dual output DDCC. 10(13):261–262
30. Ibrahim MA, Minaei S, Yuce E, Herencsar N, Koton J (2011) Lossy/lossless floating/grounded inductance simulator using one DDCC. In: Proceedings 34th international conference on telecommunications and signal processing (TSP2011), pp 261–264
31. Prasad D, Bhaskar DR (2012) Grounded and floating inductance simulation circuits using VDTAs. Circuits Syst 3(4):342–347
32. Yesil A, Kacar F, Gurkan K (2012) Lossless grounded inductance simulator employing single VDBA and its experimental band-pass filter application. Int J Electron Commun (AEU) 68(2):143–150
33. Biolek D, Senani R, Biolkova V, Kolka Z (2008) Active elements for analog signal processing; classification, review and new proposals. Radioengineering 17(4):15–32
34. Panwar K, Prasad D, Bhaskar DR, Srivastava M (2017) Novel single resistance controlled oscillator employing MCDDITA. J Eng Technol 6(13):352–358
35. Prasad D, Panwar K, Bhaskar DR, Srivastava M (2015) CDDITA-based voltage mode first order all-pass configuration. Circuits Syst (USA) 6(13):252–256
36. Panwar K, Prasad D, Srivastava M, Haseeb Z (2018) New current mode lossy integrator employing CDDITA. Cicuits Syst 3:117–123

New FDNR and FDNC Simulation Configurations Using Inverted VDDIBAs

Kapil Bhardwaj and Mayank Srivastava

Abstract In this research paper, two voltage differencing differential input buffered amplifier (VDDIBA)-based circuit configurations have been proposed. The first proposed circuit realizes the behaviour of grounded frequency-dependent negative resistance (FDNR), and the second circuit simulates the behaviour of a grounded frequency-dependent negative conductance (FDNC). As per the best knowledge of authors, there are no FDNR/FDNC configurations employing VDDIBA active elements that have been available in the literature and this work can be considered as the first attempt to bridge this gap. Both the developed circuit configurations employ two VDDIBA and three passive elements. The FDNC configuration employs only grounded capacitance and hence found fit for monolithic chip realization. Both the proposed configurations are free from component matching constraint and offer low sensitivity indexes. An active band-pass filtering circuit has been developed for justifying the utilization of one of the presented circuit. Employed VDDIBA model is based on the current feedback operational amplifier (CFOA) and CMOS OTAs. Hence, PSPICE models of AD844 (commercial IC of CFOA) and CMOS OTA are used for simulation purposes to validate all the circuits.

Keywords VDDIBA · Grounded FDNR · Grounded FDNC · VMBPF

1 Introduction

FDNR and FDNC are useful circuit concepts [1] for the synthesis of high-order active filters. The complete mathematical transformation of LCR circuits into FDNR/FDNC-based circuits has been reported by Bruton in [1]. Realization of FDNR behaviour employing active elements is a popular research domain, and various FDNR simulators using different types of active elements such as operational amplifier (OA) [2], current conveyer (CC) [3], second-generation current

K. Bhardwaj · M. Srivastava (✉)
Department of Electronics and Communication, National Institute of Technology, Jamshedpur, India
e-mail: mayank.ece@nitjsr.ac.in

D. Harvey et al. (eds.), *Advances in VLSI, Communication, and Signal Processing*, Lecture Notes in Electrical Engineering 683,
https://doi.org/10.1007/978-981-15-6840-4_48

conveyer (CCII) [4], operational transconductance amplifier(OTA) [5], operational transresistance amplifier (OTRA) [6], CFOA [7], fully differential-controlled second-generation current conveyer (FDCCII) [8], current floating transconductance amplifier (CFTA) [9], voltage differencing current conveyer (VDCC) [10], voltage differencing transconductance amplifier (VDTA) [11] and voltage differencing buffered amplifier (VDBA) [12] have been reported in the literature. Similarly, some FDNC simulators using modern active devices have been presented in [4, 13–15].

From detailed investigation of the literature available on FDNR/FDNC simulators, it has been observed that there is no FDNR/FDNC based on VDDIBA (−) active element which is available in the open literature. Therefore, this work is an endeavour to prove the versatility of VDDIBA (−) to realize both grounded FDNC and FDNR simulators and also without any component matching constraint.

2 VDDIBA (−)

VDDIBA (−) is an ABB of voltage differencing buffered amplifier (VDBA) family proposed in [16]. Being a voltage-mode (VM) device and having bias current controllability, VDDIBA (−) is a very useful active element. The electrical symbol of VDDIBA (−) is shown in Fig. 1. It has five ports, namely $V+$, $V-$, W, Z and V. The difference of voltages of terminal $V+$ and $V-$ is transferred to Z terminal in the form of current. The voltage difference between the V and Z terminals is transferred to the W terminal through a differential input unity gain buffer. The voltage–current relationships between different ports of VDDIBA have been given by Eqs. (1)–(3).

$$I_Z = g_m(V_{V+} - V_{V-}) \tag{1}$$

$$V_W = (V_V - V_Z) \tag{2}$$

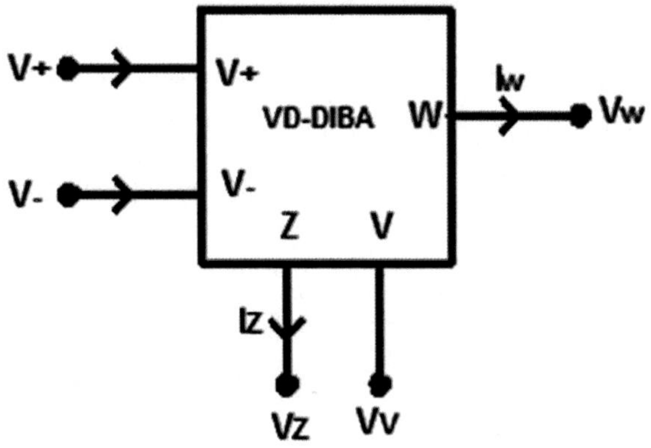

Fig. 1 Symbolic representation of VDDIBA (−)

$$I_{V+} = I_{V-} = I_V = 0 \quad (3)$$

The behaviour of the VDDIBA active element can be recognized from the behavioural diagram of Fig. 2.

The VDDIBA (+) and VDDIBA (−) are slightly different from each other. The voltage at *W* terminal of VDDIBA (−) is inverted into the voltage at *W* terminal of VDDIBA (+). In [17], the implementation of VDDIBA (+) is reported using three OTAs and AD844. On the basis of this realization of VDDIBA (+), the implantation of VDDIBA (−) employing two single output OTAs, one dual output OTA and single AD844 has been developed and shown in Fig. 3.

The 0.18 um MOS-based implementation of SO-OTA [18] is given in Fig. 4.

Fig. 2 Behavioural model of VDDIBA

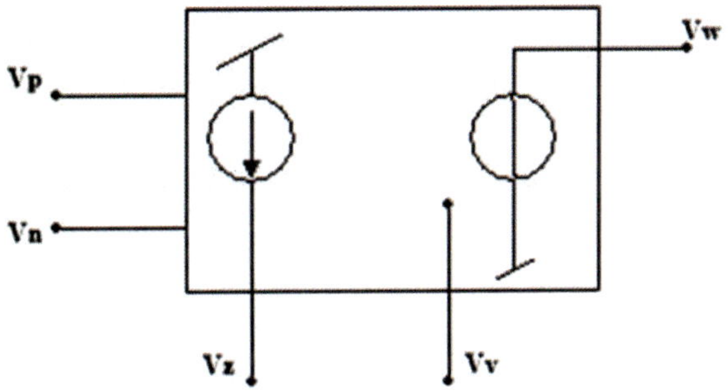

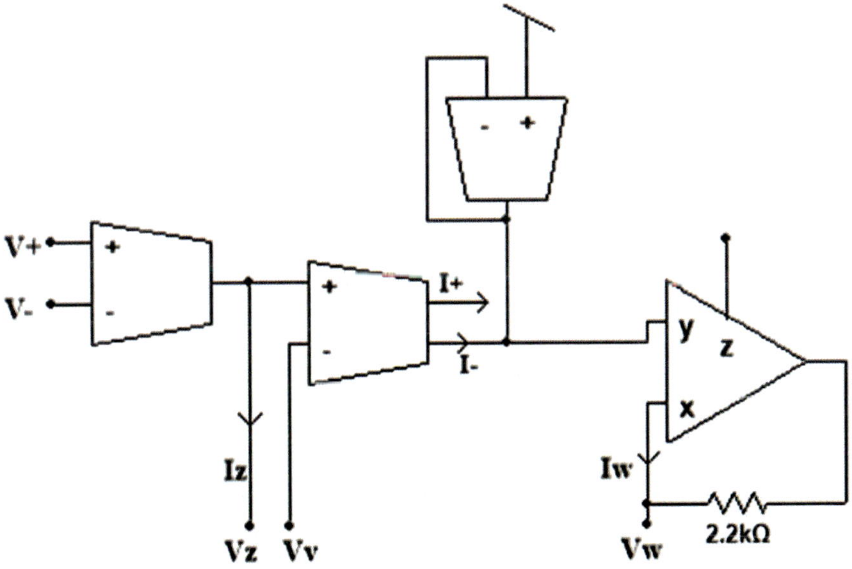

Fig. 3 Implementation of VDDIBA (−) using two SO-OTA, single DO-OTA and AD844

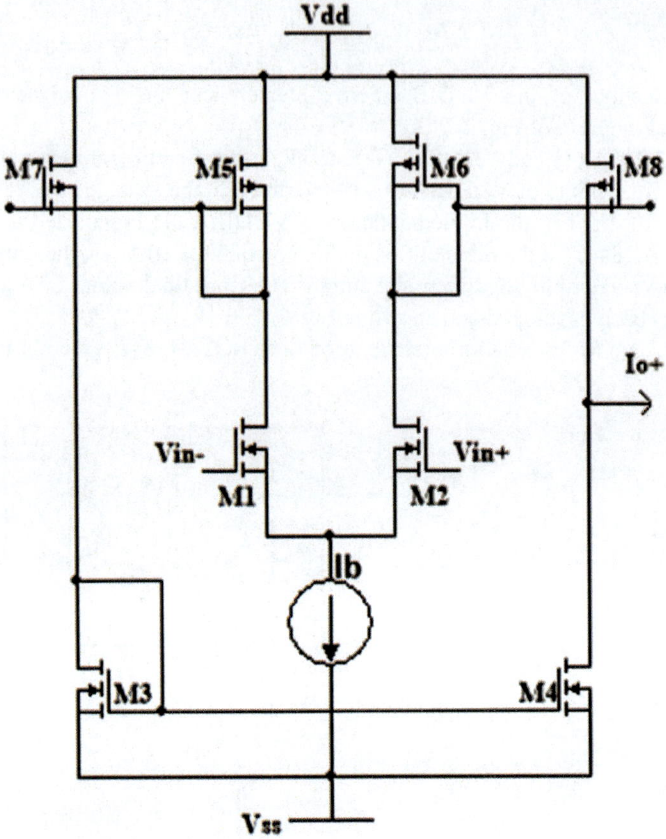

Fig. 4 MOS implementation of SO-OTA

Similarly, implementation of DO-OTA using the same 0.18 um technology-based MOS transistors is given in Fig. 5.

Some realization of the active filters and synthetic inductors using VDDIBA (−) are presented in [18, 19].

3 Proposed Configurations

The proposed VDDIBA (−)-based FDNR circuit is shown in Fig. 6.

On simple mathematical analysis of circuit of Fig. 2, the impedance value evaluated as

$$Z_{\text{in}} = \frac{g_{m2}}{s^2 g_{m1} R C_1 C_2} \tag{4}$$

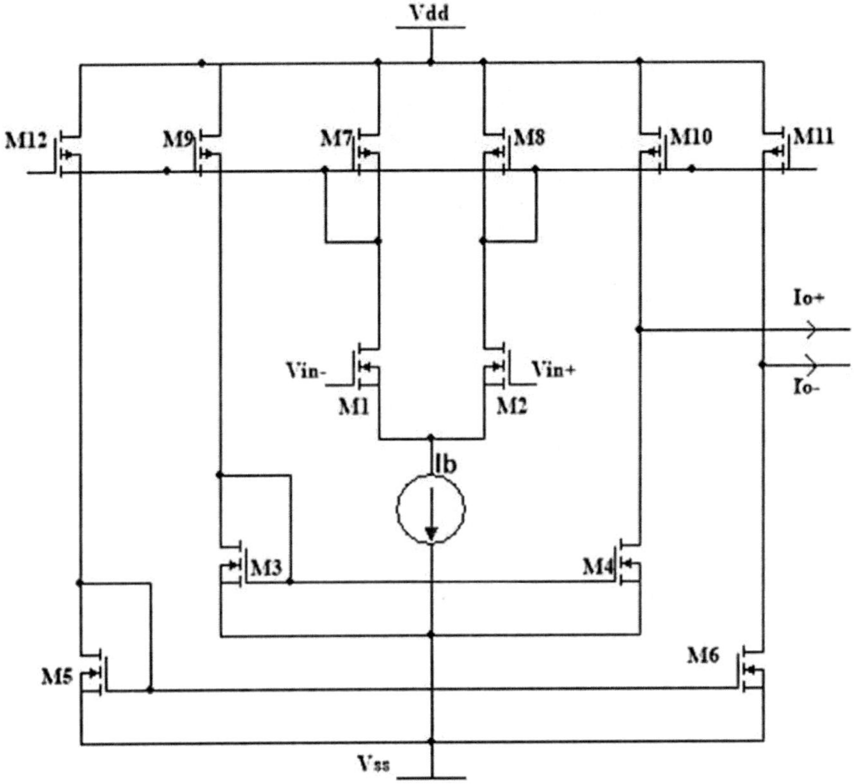

Fig. 5 MOS-based implementation of DO-OTA

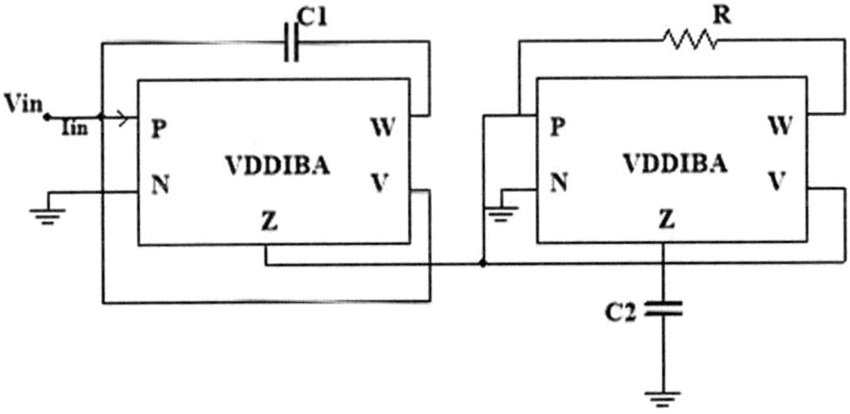

Fig. 6 Proposed grounded FDNR simulation circuit using VDDIBA (−)

where

$$D_{\mathrm{eq}} = \frac{g_{m1} R C_1 C_2}{g_{m2}} \tag{5}$$

Therefore from Eq. (4), it can be observed that the presented circuit simulates the behaviour of a grounded FDNR with parameter value given by Eq. (5).

Applying circuit mathematics for FDNC simulator of Fig. 4, the input impedance of presented FDNC:

$$Z_{\mathrm{in}} = \frac{s^2 R C_1 C_2}{g_{m1} g_{m2}} \tag{6}$$

Therefore, in such condition the developed circuit simulates the behaviour of a lossless FDNC with negative resistance value M_{eq}:

$$M_{\mathrm{in}} = \frac{R C_1 C_2}{g_{m1} g_{m2}} \tag{7}$$

4 Non-Ideal Analysis

On considering non-ideal voltage transfer errors of VDDIBA (−), the port relations of VDDIBA can be described as:

$$I_Z = \beta g_m (V_{V+} - V_{V-}) \tag{8}$$

$$V_W = (\beta^- V_V - \beta^+ V_Z) \tag{9}$$

$$I_{V+} = I_{V-} = I_V = 0 \tag{10}$$

where

β^- and β^+ are voltage transfer errors.

On using (8–10), the impedance expression for presented FDNR is found as

$$Z_{\mathrm{inFDNR}} = \frac{1/C_2 s(\beta^+ \beta_1 g_{m1} + s C_1 (1 - \beta^-))}{s\beta^+ \beta_2 g_{m2} R C_1 + \beta^+ \beta_1 g_{m1}(1 - \beta^-) + s C_1 (1 - \beta^-)^2} \tag{11}$$

The equivalent circuit developed from Eq. (11) is shown in Figs. 3, 7 and 8.
where

$$C' = (1 - \beta')C_2 \tag{12}$$

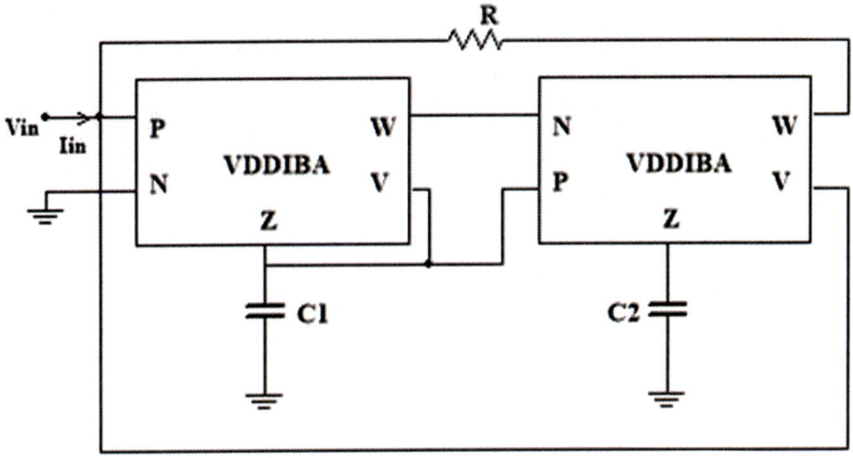

Fig. 7 Proposed grounded FDNC simulation circuit using VDDIBA (−)

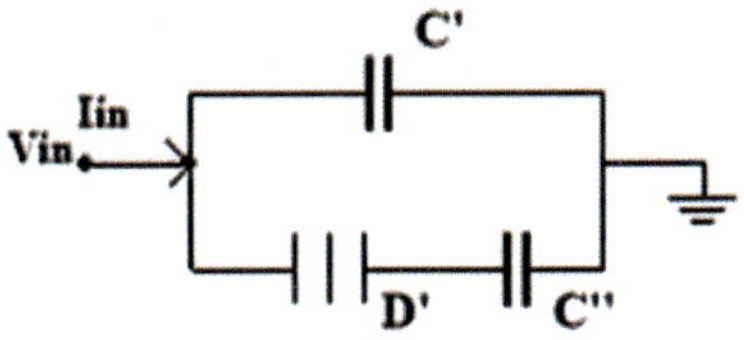

Fig. 8 Equivalent circuit of presented grounded FDNR of Fig. 6 based on Eq. 12

$$C'' = \frac{C_2\beta^+\beta_1 g_{m2}R}{(1-\beta^=)} \quad (13)$$

$$D'' = \frac{\beta_1 g_{m1}RC_1C_2}{\beta_2 g_{m2}} \quad (14)$$

Therefore on considering non-ideal constraints, the presented circuit behaves like a lossy FDNR with lossy capacitances C′ and C″.

Similarly, the impedance expression of the configuration shown in Fig. 7 considering non-ideal conditions can be obtained as:

$$Z_{\text{inFDNC}} = \frac{s^2R_1C_1C_2}{s^2C_1C_2(1-\beta^-) + \beta_1\beta_2 g_{m1}g_{m2}\beta^+[1-\beta^- - \beta^+]} \quad (15)$$

And the developed equivalent circuit from Eq. 16 is given in Fig. 5.
where

$$R' = \frac{R}{1-\beta^-} \quad (16)$$

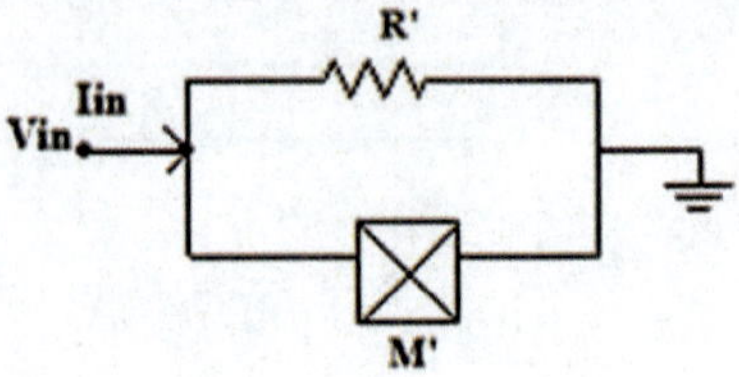

Fig. 9 Equivalent circuit of grounded FDNC presented in Fig. 7 from Eq. (16)

$$M' = \frac{CCR}{\beta_1 \beta_2 g_{m1} g_{m2} \beta^+ [1 - (\beta^- - \beta^+)]} \tag{17}$$

Clearly, we can observe from Fig. 9 that equivalent non-ideal model of grounded FDNC includes a lossy resistance R' along with the intended FDNC term M'.

5 Application Examples

To demonstrate the utility of developed FDNC simulator, a band-pass filtering circuit is constructed, whose passive element based realization is given in Fig. 10.

The active implementation of filter given in Fig. 10 employing proposed FDNC simulation circuit is shown in Fig. 11.

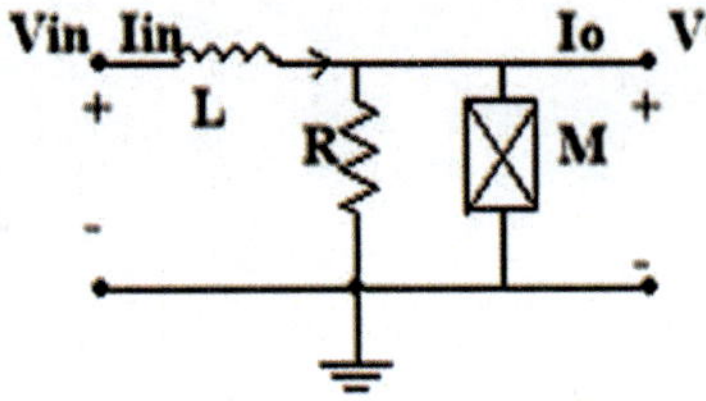

Fig. 10 Passive element-based realization of voltage-mode band-pass filter (VMBPF) using FDNC

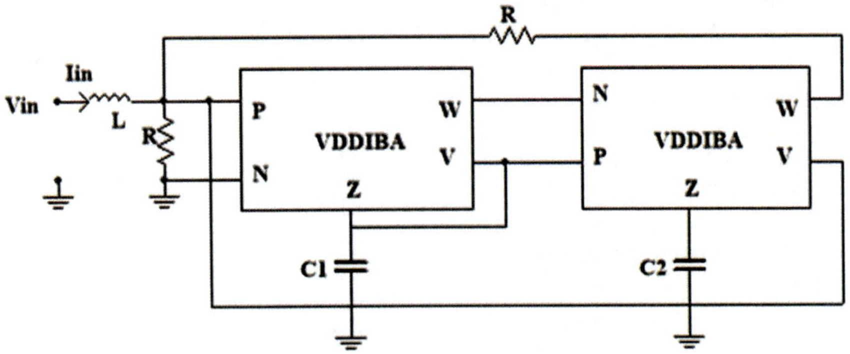

Fig. 11 Band-pass filter of Fig. 10 using presented FDNC simulator

6 Simulation Results

For the validation of the working of developed grounded FDNR circuit, simulations under PSPICE environment have been executed with VDDIBA (−). Simulations were performed with ±3 V DC supply voltage and 100 uA VDDIBA (−) bias current. The external passive element values were selected as $C_1 = 1$ nF, $C_2 = 2$ nF $R = 0.5$ K. The simulation generated magnitude response and the phase response of grounded FDNR simulator are shown in Figs. 12 and 13, respectively. The magnitude response in Fig. 7 indicates the reduction of impedance with increasing frequency. Figure 13 clearly shows that the phase is almost −90° for a frequency range from 60 to 100 MHz, which is acceptable for high-frequency applications. Similarly for presented grounded FDNC, the magnitude and phase plots are shown in Figs. 14 and 15. For the simulation of the presented FDNC simulator, the external capacitance values are taken as $C_1 = C_2 = 1$ nF and resistance value as $R = 0.5$ K. From the magnitude plot given in Fig. 13, it can be observed that the value of input impedance is getting increased for the increasing values of frequency. Similarly, the phase plot given in Fig. 14 is also indicating, that phase value is approximately 90° for 50–100 MHz. The band-pass filter based on presented FDNC given in Fig. 10 is verified

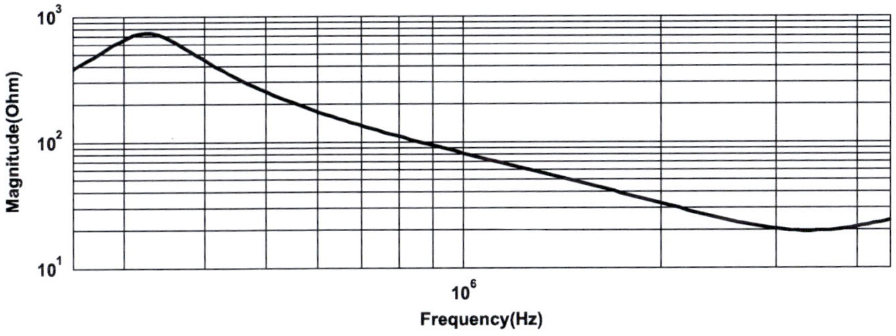

Fig. 12 Input impedance of presented FDNR simulator given in Fig. 6

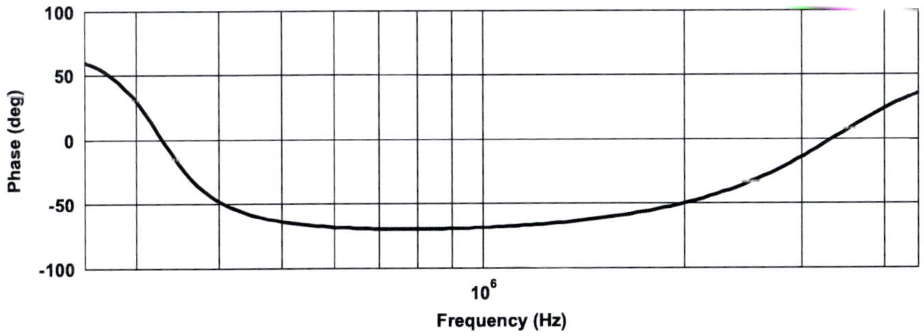

Fig. 13 Phase response of presented FDNR simulator

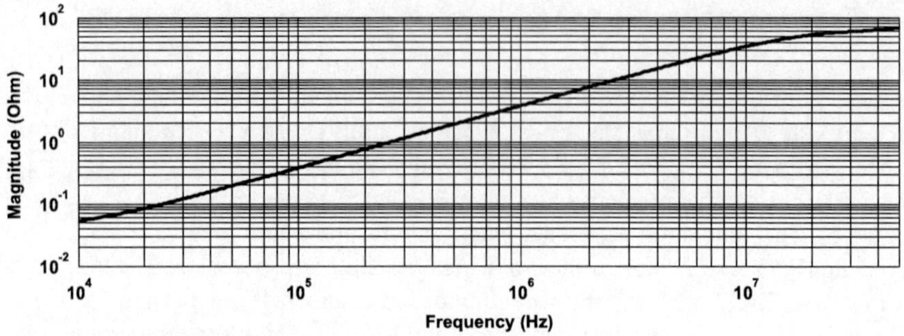

Fig. 14 Input impedance of presented FDNC simulator given Fig. 7

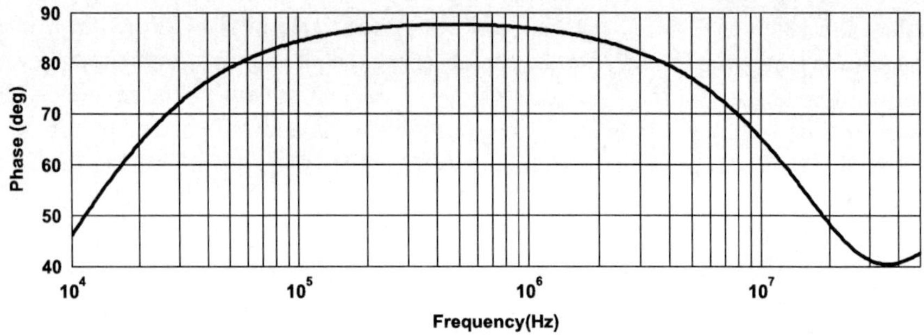

Fig. 15 Phase response of FDNC simulator

using PSPICE for external passive components values as $C_1 = C_2 = 1$ nF and $R = 0.5$ K, and the resulting frequency response plot is given in Fig. 16.

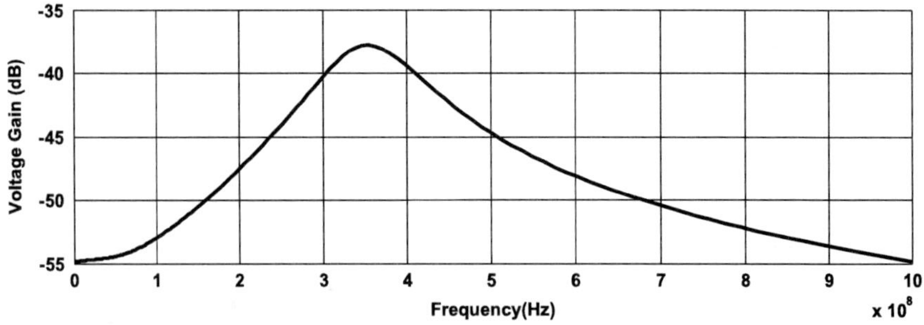

Fig. 16 Frequency response of developed band-pass filter of Fig. 11 using reported FDNC simulator

7 Conclusion

The work discussed the novel simulation configurations of grounded FDNR and FDNC simulators based on VDDIBA (−). Both circuits employ two VDDIBA (−)s along with three passive elements. Non-ideal equivalent model of both the circuits is also derived considering non-ideal gains of VDDIBA (−). To justify the applicability of the designed FDNC simulator, a VMBPF has been derived and simulated using this simulator. Presented simulation circuits and their application examples are verified under PSPICE environment using OTA and AD844-based implementation of VDDIBA (−). Realization of FDNR/FDNC simulators using VDDIBA (−) along with only grounded elements is still an open research problem.

References

1. Bruton L (1969) Network transfer functions using the concept of frequency-dependent negative resistance. IEEE Trans Circuit Theory 16(3):406–408
2. Nandi R (1979) Novel active-R ideal frequency-dependent negative-resistance simulation. Electron Lett
3. Soliman A (1978) Realisation of frequency-dependent negative-resistance circuits using two capacitors and a single current conveyor. PROC IEE 125(12)
4. Pal K Realization of ideal grounded inductances and frequency dependent negative resistances using current conveyors and without matched components. Int J Circuit Theory Appl
5. Theingjit S, Pukkalanun T, Tangsrirat W (2016) FDNC realization and its application to FDNR and filter realizations. In: Proceedings of the international multiconference of engineers and computer scientists 2
6. Gupta S, Senani R, Bhaskar DR, Singh AK (2012) OTRA-based grounded-FDNR and grounded-inductance simulators and their applications. Circuits Syst Signal Process 31:489–499
7. Kaçar F, Kuntman H On the realization of the FDNR simulators using only a single current feedback operational amplifier
8. Chung WS, Kim HJ, Cho KM (2004) Current-controllable FDNRs using linear transcapacitance amplifiers. Int J Electron 91(7):423–430
9. Li Y (2012) A series of new circuits based on CFTA. Int J Electron Commun pp 587–595
10. Srivastava, M (2016) New synthetic grounded FDNR with electronic controllability employing cascaded VDCCs and grounded passive elements. J Telecommun, Electron, Comput Eng 9(4)
11. Mongkolwai P, Tangsrirat W (2016) Generalized Impedance Function Simulator Using Voltage Differencing Buffered Amplifiers (VDBAs). In: Proceedings of the international multiconference of engineers and computer scientists 2
12. Tangsrirat W, Mongkolwa P (2013) VDTA-based floating FDNR simulator topology. KMITL Sci Technol J 13(1)
13. Senani R, Bhaskar DR (1979) Versatile voltage controlled impedance configuration. IEE Proc-Circuits Devices Syst 33:423–424
14. Soliman AM (2006) A new realization of FDNC using DVCCS/DVCVS. AEU—Int J Electron Commun 93(10):689–698
15. Pal K (2004) Floating inductance and FDNR using positive polarity current conveyors. Act Passive Electron Compon 27:81–83
16. Biolek D, Senani R, Biolkova V, Kolka Z (2008) Active elements for analog signal processing classification, review and new proposals. Radioengineering 17(4):15–32

17. Prasad D, Bhaskar DR, Pushkar KL (2011) Realizatio of new electronically controllable grounded and floating simulated inductance circuits using voltage differencing differential input buffered amplifier. Active Passive Compon
18. Biolek D, Biolkova V (2010) First-order voltage-mode all-pass filter employing one active element and one grounded capacitor. Analog Integr Circ Sig Process 65(10):123–129
19. Bhaskar DR, Prasad D, Pushkar KL (2013) Electronically-controllable grounded-capacitor-based grounded and floating inductance simulated. Circuits Using VD-DIBAs. Circuits Syst. http://dx.doi.org/10.4236/cs.2013.45055

A Low Power Approach for Designing 12-Bit Current Steering DAC

Abhishek Kumar, Santosh Kumar Gupta, and Vijaya Bhadauria

Abstract A low power 12-bit current steering DAC is designed using SCL 180-nm-technology. Various methodologies are considered to reduce the power consumption in current steering DAC. The current mode logic (CML)-based latch is incorporated to design latch cum-level shifters to reduce digital power consumption. A dummy switch compensation technique is also used at the output of switch to counter the effect of charge feedthrough from switch to the output of DAC. In addition, it also reduces the minimum allowable voltage headroom, thus reducing power consumption of analog block. The power consumption obtained is 12 mW for 12-bit design operating at 200 MHz frequency.

Keywords CML · LSB · INL · SFDR · Voltage headroom · Charge feedthrough

1 Introduction

The extensive use of battery-operated IOTs demands the use of low power DACs for transmitters; thus, DACs are one of the key blocks for portable IOTs. Lots of architectural and technology alternative available but CMOS technology DACs have the advantage of having high-speed and low power consumption [1–4].

CMOS current steering DACs may be classified based on architectural level as: (i) unary-weighted architecture (ii) binary-weighted architecture. Unary-weighted architecture consists of unit current element arranged in thermometer encoded

A. Kumar (✉) · S. K. Gupta · V. Bhadauria
Department of Electronics and Communication Engineering, Motilal Nehru National Institute of Technology Allahabad, Prayagraj, Uttar Pradesh 211004, India
e-mail: abhishekpintu4316@gmail.com

S. K. Gupta
e-mail: skg@mnnit.ac.in

V. Bhadauria
e-mail: vijaya@mnnit.ac.in

D. Harvey et al. (eds.), *Advances in VLSI, Communication, and Signal Processing*, Lecture Notes in Electrical Engineering 683,
https://doi.org/10.1007/978-981-15-6840-4_49

ways. However, the requirement of encoder–decoder increases area [5–10]. Binary-weighted architecture consists of weighted current cell arranged in binary-weighted form, due to which its architecture is simple and occupies less area. This architecture is preferred for low power solutions [4, 11–13]. Segmented architecture is hybrid between binary-weighted and unary architecture which also requires encoder–decoder and delay element; hence, it is not preferred over binary-weighted for low power DACs. The mismatch between weighted current sources limits the resolution of binary-weighted architecture up to 10 bits [14]. However, partial segmented architecture uses the advantage of both binary-weighted and segmented architecture and thus used for higher-resolution DACs [15–17]. In this, LSB current cell is connected in serial and parallel to form weighted array without the use of encoder and decoders.

There is very limited work done on the low power aspect of current steering DAC. Douglas A. Mercer [18] shows power consumption of different blocks used in current steering DAC with respect to clock frequency; however, the author used separate blocks for latch and level shifter which increases area as well as power consumption. The double cascode structures are also reported [12–18] in which higher supply voltage is required in turn adding a significant amount of power consumption.

A 12-bit partial segmented current steering DAC is proposed in this work, as shown in Fig. 1. The proposed architecture consists of CML latch which receives input digital signal, weighted current source array, differential switch (DS) and dummy switch (DUMS). First, power consumption of different blocks is explored targeting its low power implementation. A CML-based latch is used to reduce the power consumption. A dummy switch technique is used instead of using double cascode structure to reduce voltage headroom and to nullify the effect of charge feedthrough at the output of DAC.

Section 2 deals with the design and implementation of the proposed work. Results obtained are discussed in Sect. 3, and finally, conclusions are drawn in Sect. 4.

2 Power Consideration

Dynamic power consumption of a circuit is given by $P = C_{\mathrm{L}} V_{\mathrm{dd}}^2 f$, where C_{L} is load capacitance, V_{dd} is supply voltage and f is operating frequency. The power consumption of analog block, digital block and clock with respect to clock frequency is given in Fig. 2 of [12] which is reproduced here in Fig. 2 for ready reference. Digital block dynamic power consumption is dependent on the operating frequency.

The digital block becomes main contributor to total power at higher frequencies of operation. A CML-based latch [19] is used to design digital block replacing latch and level shifters of current steering DAC as shown in Fig. 3. The power consumption of digital block is 1.75 μW when operates at 200 MHz.

The output response of CML latch is shown in Fig. 4. The latch consists of differential pair MN_4 and MN_5, which used to track input signal and MN_6 and MN_7 for storing the data which changes with a compliment clock signal. The output swing of CML latch is 0.4–1.8 V; thus, it also serves the purpose of level shifter required

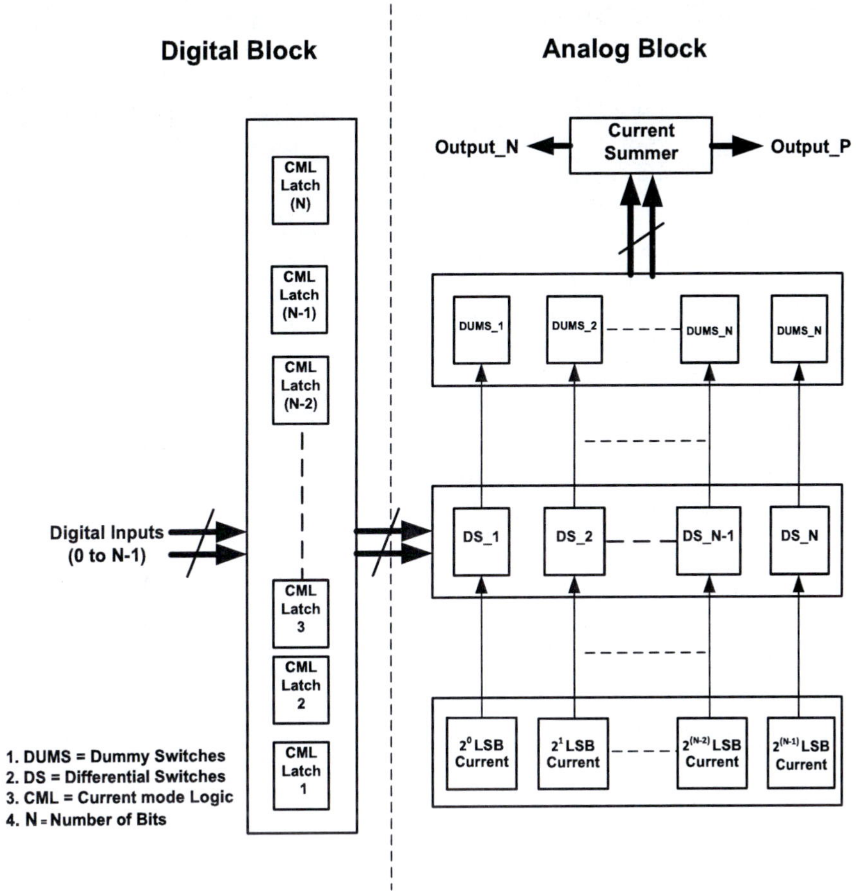

Fig. 1 Block diagram of the proposed architecture

in current steering DAC [14, 15, 17, 19]. The merging of both latch and level shifter blocks further reduces power consumption of digital block.

Analog block steers current and remains in "ON" state. Therefore, power consumption of analog block is independent of clock frequency as shown in Fig. 2. It may be seen that the analog block contributes significant amount of power consumption. Since the power consumption is independent of clock frequency, the best way to reduce the power consumption is to reduce the least significant bit (LSB) current, which reduces full-scale current of the current steering DAC. The LSB current is chosen to be 1 uA in this work, resulting full-scale current to be 4.096 mA

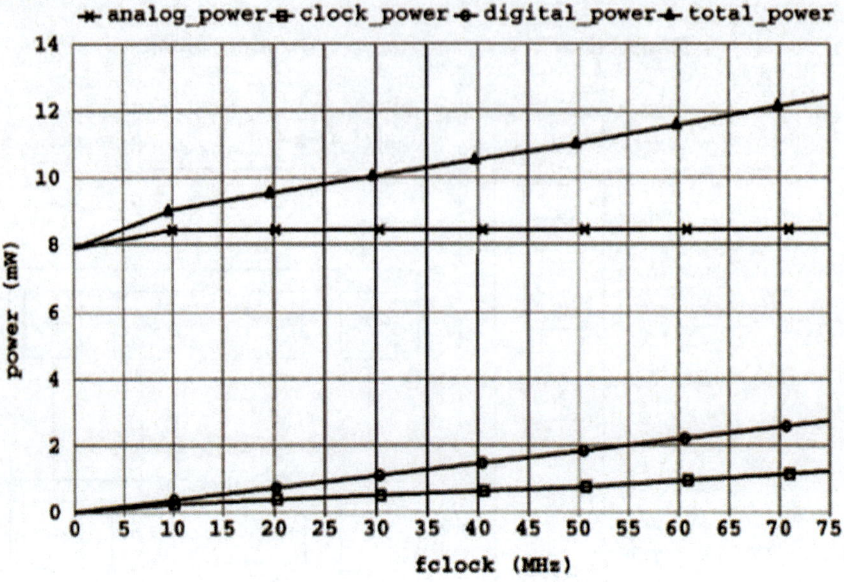

Fig. 2 Power consumption of different block in current steering DAC with frequency [12]

2.1 *Charge Feedthrough from Differential Switch to Output Node*

When output resistance is directly connected to drain of the switch, there is charge feedthrough from switch to output of DAC, which increases settling time of DAC, decreasing the operating frequency of DAC. Assume *Qch* be channel charge of M_3 transistor. When M_3 turns OFF from ON, equal charge goes from channel to source (*Qch*/2) and drain (*Qch*/2) shown in Fig. 5. The charge feedthrough from channel to drain of transistor M_3 disturbs output node of switch (*Outp*). Some author reported cascode transistor method [12–18] at the output to shield output terminal from charge feedthrough of channel to drain terminal of switch transistor (M_3). The cascode transistor is used in the past by researchers to countereffect of charge feedthrough increases voltage headroom [19], thus increases V_{dd} of the analog part of the circuit and thus increases power consumption.

The dummy switch compensation technique [20] (similar technique is used in sample hold circuit) is reported to countereffect of charge feedthrough at the output terminal of DAC as shown in Fig. 6. The source and drain terminal of M_5 and M_6 is shorted so that M_5 and M_6 act as voltage-controlled capacitor. The use of dummy switch compensation technique also reduces required V_{dd}, which reduces minimum allowable voltage headroom as compared to double cascode technique [12–18] and thereby further reduces power consumption of the analog block to 7.2 mW for 12-bit resolution.

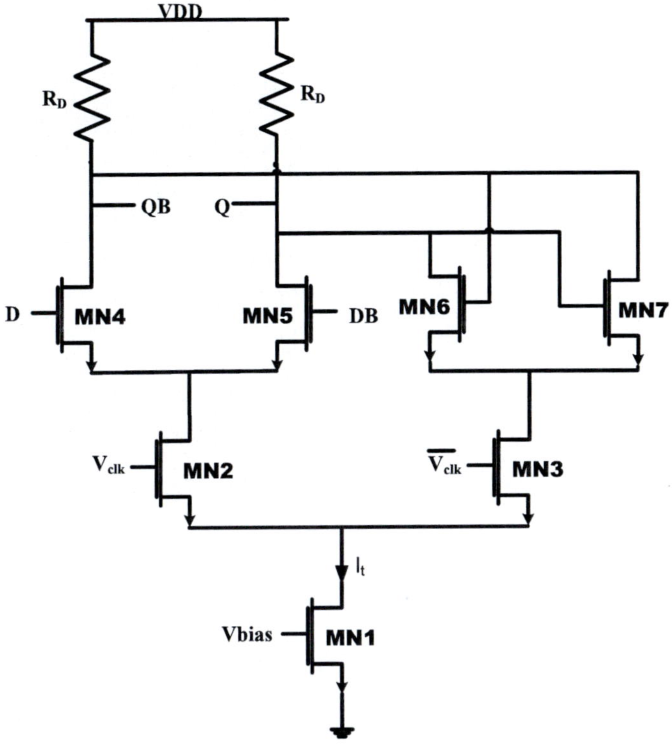

Fig. 3 Current mode logic (CML) latch

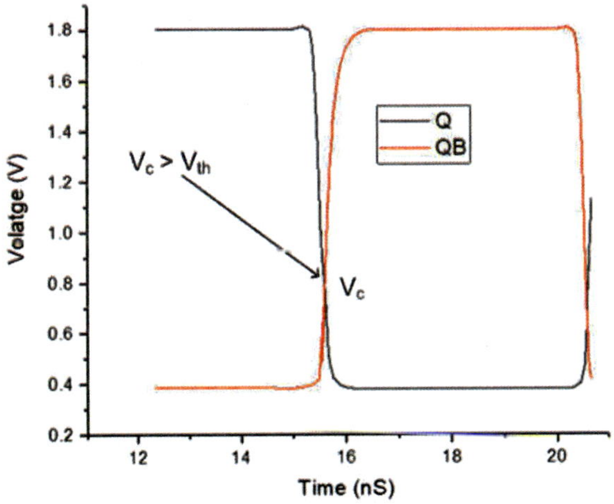

Fig. 4 Output response of CML latch

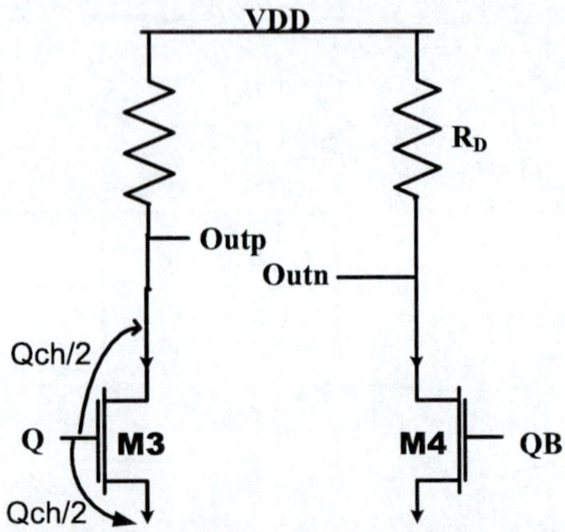

Fig. 5 Charge feedthrough from switch to output terminal of DAC

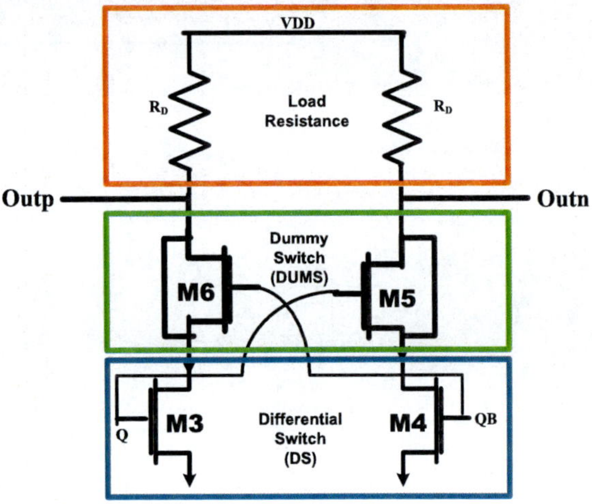

Fig. 6 Dummy switch technique to countercharge feedthrough from switch to output terminal of DAC

2.2 *Current Source Array*

See Fig. 7.

The current source array is formed by connecting two current source transistors in parallel. The LSB, 4 LSB, 16 LSB, 64 LSB, 256 LSB, 1024 LSB current sources are designed with high accuracy, and these current sources are connected in parallel to form 2 LSB, 8 LSB, 32 LSB,128 LSB, 512 LSB, 2048 LSB as shown in Fig. 6. The advantage of this architecture is that it requires only 6 precise current sources instead of 12 to reach up to 12 bits resolution without much increase in area overhead compared to binary-weighted architecture.

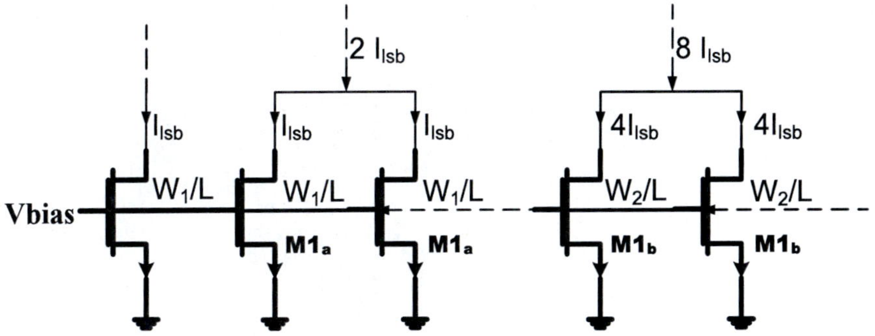

Fig. 7 Current source array in partial segmented architecture

3 Results and Discussion

The current steering DAC is designed using SCL 180-nm-technology library. The full-scale current obtained is 4.3 mA at both terminals *Output_P* and *Output_N* of DAC. The variation in current for different input bit is shown in Fig. 8. The INL of the proposed DAC is <0.02 LSB which can be shown in Fig. 9. Frequency spectrum at 1 GHz sampling rate is represented by Fig. 10. The spur-free dynamic range (SFDR) is 39 for input frequency 200 MHz at 1 GHz sampling frequency.

The results obtained are compared with various related works shown in Table 1. As the proposed circuit is using reduced full-scale current and less voltage headroom, the total power consumption of the circuit is found out to be 12 mW which is less than previously reported works as compared in Table 1. The SFDR is slightly reduced because of higher operating frequency and lowering full-scale current.

Fig. 8 Single-ended output of current steering DAC

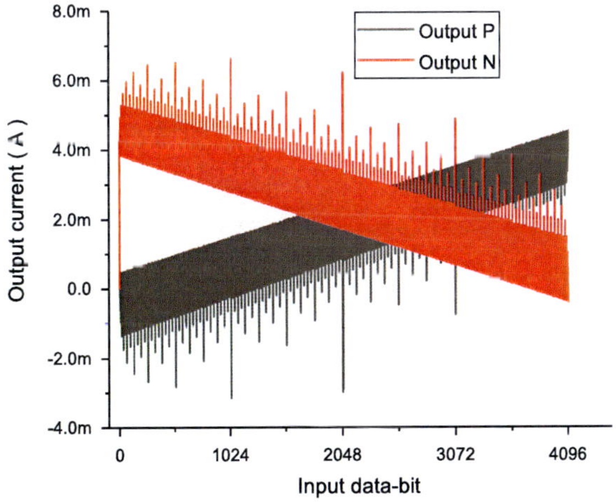

Fig. 9 INL error with input code

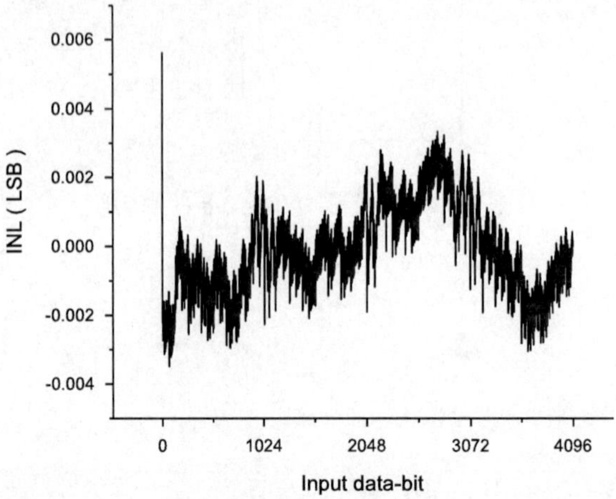

Fig. 10 SFDR obtained with input frequency 200 MHz with sampling frequency of 1 GHz

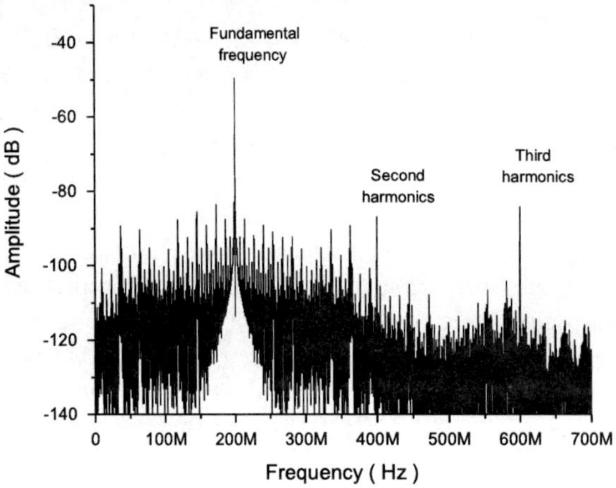

Table 1 Performance comparison with the reported work

Parameters	[8]	[11]	This work
Technology (μm)	1.8	1.8	1.8
Number of bits	14	10	12
SFDR	45 dB @25 MHz	49 dB @171 MHz	39 dB @200 MHz
INL (LSB)	<0.5	1.5	<0.02
Sampling rate	100 MS/S	400 MS/S	1 GS/S
Dynamic power dissipation (mW)	20 (analog)	20.7 (analog)	12 (analog + digital)
Architecture	Binary-weighted	Binary-weighted	Partial segmented
Supply voltage (V)	1.8	1.8	1.8

4 Conclusion

The paper presents a low power current steering DAC. The CML-based latch has been incorporated to reduce the power consumption of digital block. Further dummy switch compensation technique has been used for reducing power consumption of analog block which is 7.2 mW. The total power consumption has been reduced considerably to 12 mW using these techniques for 12-bit current steering DAC. The INL obtained is <0.02 LSB with the proposed partial segmented architecture which helps in decreasing power consumption. The proposed architecture is best-suited for battery-operated circuits.

References

1. Razavi B (2018) The current-steering DAC [a circuit for all seasons]. IEEE Solid-State Circuits Mag 10(1):11–15
2. Takakura H, Yokoyama M, Yamaguchi A (1991) A 10 bit 80 MHz glitchless CMOS D/A converter. In Proceedings of the IEEE Custom Integrated Circuits Conference. IEEE, pp 26.5/1–26.5/4
3. Bastiaansen CA, Groeneveld DWJ, Schouwenaars HJ, Termeer HA (1991) A 10-bit 40-MHz 0.8-μm CMOS current-output D/A converter. IEEE J Solid-State Circuits 26(7):917–921
4. Deveugele J, Steyaert MS (2006) A 10-bit 250-MS/s binary-weighted current-steering DAC. IEEE J Solid-State Circuits 41(2):320–329
5. Miki T, Nakamura Y, Nakaya M, Asai S, Akasaka Y, Horiba Y (1986) An 80-MHz 8-Bit CMOS D/A Converter. IEEE J Solid-State Circuits 21(6):983–988
6. Lin WT, Kuo TH (2012) A compact dynamic-performance-improved current-steering DAC with random rotation-based binary-weighted selection. IEEE J Solid-State Circuits 47(2):444–453
7. Chen T, Gielen GG (2007) A 14-bit 200-MHz current-steering DAC with switching-sequence post-adjustment calibration. IEEE J Solid-State Circuits 42(11):2386–2394
8. Lee DH, Kuo TH, Wen KL (2009) Low-cost 14-bit current-steering DAC with a randomized thermometer-coding method. IEEE Trans Circuits Syst II Express Briefs 56(2):137–141
9. Mao W, Li Y, Heng CH, Lian Y (2018) High dynamic performance current-steering DAC design with nested-segment structure. IEEE Trans Very Large Scale Integ (VLSI) Syst 26(5):995–999
10. Shen MH, Tsai JH, Huang PC (2010) Random swapping dynamic element matching technique for glitch energy minimization in current-steering DAC. IEEE Trans Circuits Syst II Express Briefs 57(5):369–373
11. Chou FT, Hung CC (2015) Glitch energy reduction and SFDR enhancement techniques for low-power binary-weighted current-steering DAC. IEEE Trans Very Large Scale Integr (VLSI) Syst 24(6):2407–2411
12. Van den Bosch A, Borremans MA, Steyaert MS, Sansen W (2001) A 10-bit 1-GS/s nyquist current-steering CMOS D/A converter. IEEE J Solid-State Circuits 36(3):315–324
13. Pelgrom MJ, Duinmaijer AC, Welbers AP (1989) Matching properties of MOS transistors. IEEE J Solid-State Circuits 24(5):1433–1439
14. Pal N, Nandi P, Biswas R, Katakwar AG (2016) Placement-based nonlinearity reduction technique for differential current-steering DAC. IEEE Trans Very Large-Scale Integr (VLSI) Syst 24(1):233–242
15. Tseng WH, Fan CW, Wu JT (2011) A 12-Bit 1.25-GS/s DAC in 90 nm CMOS With70 dB SFDR up to 500 MHz. IEEE J Solid-State Circuits 46(12):2845–2856

16. Chen T, Gielen G (2007) The analysis and improvement of a current-steering DAC's dynamic SFDR—II: the output-dependent delay differences. IEEE Trans Circuits Syst I Regul Pap 54(2):268–279
17. Bastos J, Marques AM, Steyaert MS, Sansen W (1998) A 12-bit intrinsic accuracy high-speed CMOS DAC. IEEE J Solid-State Circuits 33(12):1959–1969
18. Mercer DA (2007) Low-power approaches to high-speed current-steering digital-to-analog converters in 0.18-μm CMOS. IEEE J Solid-State Circuits 42(8):1688–1698
19. Heydari P Mohanavelu R (2004) Design of ultrahigh-speed low-voltage CMOS CML buffers and latches. IEEE Trans Very Large Scale Integr (VLSI) Syst 12(10):1081–1093
20. Eichenberger C, Guggenbuhl W (1989) Dummy transistor compensation of analog MOS switches. IEEE J Solid-State Circuits 24(4):1143–1146

Design and Implementation of an Efficient Mixed Parallel-Pipeline SAD Architecture for HEVC Motion Estimation

Mamidi Nagaraju, Santosh Kumar Gupta, Vijaya Bhadauria, and Devarshi Shukla

Abstract Video compression is a computationally intensive task that generally demands high performance for portable device application. This high performance is mainly attained by motion estimation (ME) in the video encoder. With the larger block size and flexible block, partitioning in High Efficiency Video Encoding (HEVC) makes ME block more complicated. The sum of absolute difference (SAD) is extensively used as a distortion metric in the ME process. It is highest computing task with more calculation time and hardware resource. In this paper, an efficient mixed parallel-pipeline SAD architecture is proposed for high frequency. The performance optimization techniques are used at different levels of abstraction: Architecture level (parallel and pipeline), RTL level (grouping and resource sharing) and Implementation level (retiming). The proposed architecture has been simulated, synthesized and prototyped using 28 nm Artix-7 Field Programmable Gate Array (FPGA). The result shows that the proposed design provides a significant increment in the maximum frequency of 498.4 MHz when compared with other existed work in the literature.

Keywords Sum of absolute difference · Video coding · HEVC · Motion estimation · FPGA

1 Introduction

To meet the continues growing demand of video, technologically, the design constraints such as performance, power and efficient memory at better transfer rate

M. Nagaraju (✉) · S. K. Gupta · V. Bhadauria · D. Shukla
Department of ECE, Motilal Nehru National Institute of Technology Allahabad, Prayagraj, Uttar Pradesh 211004, India
e-mail: rel1651@mnnit.ac.in

S. K. Gupta
e-mail: skg@mnnit.ac.in

V. Bhadauria
e-mail: vijaya@mnnit.ac.in

D. Shukla
e-mail: devarshi@mnnit.ac.in

D. Harvey et al. (eds.), *Advances in VLSI, Communication, and Signal Processing*, Lecture Notes in Electrical Engineering 683,
https://doi.org/10.1007/978-981-15-6840-4_50

are required for a wide range of applications such as high definition (HD) (1080 p resolution) and ultra-high definition (UHD) (4 K and 8 K resolution). To meet out this, robust and enhanced video compression efficient designs are required. HEVC or H.265 is the present popular video compression algorithm designed to increase the video coding efficiency significantly as compared to its previous standard called H.264 or Advance-Video-Coding (AVC) [1]. The HEVC is a video coding standard which supports higher resolutions and can achieve up to 50% bit-rate savings as compared to earlier H.264 for the same video quality [1, 2]. The considerable increase in throughput is due to the many enhanced tools, techniques and methodologies which has been introduced in HEVC. Some of this enhancements are included in the ME prediction process, which is the critical and complex block of video encoding. The experimental results show that the ME block is occupied approximately 70% of the total video encoding load [2].

The main objective of the ME process is to identify the suitable matching block in a search area (SA) of reference frame (RF) corresponding to a current frame (CF) and provides a map of the displacement directions using a motion vector (MV). Figure 1 shows the general motion estimation procedure.

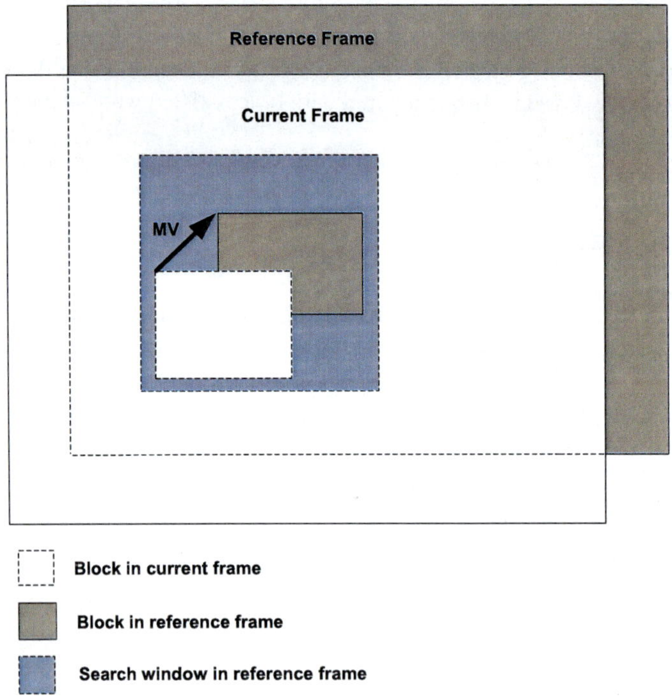

Fig. 1 Basic motion estimation process [3]

In the HEVC, a video frame encoding is performed using basic block called coding units (CU) of varied sizes (8×8 to 64×64) obtained by dividing the frames. The CU maximum size is 64×64 which increases complexity in the design of ME. In every level of CU, HEVC basically supports eight different partitions in the prediction block. As shown in Fig. 2, the block partitions from (a) to (d) are called symmetric mode (SM) partition and from (e) to (h) are called asymmetric mode (AM) partition.

For block-based ME, the SAD is the commonly used metric which perform addition of the absolute differences between corresponding elements in a current and reference block. The main bottleneck of the ME process lies in implementation with an appropriate and efficient SAD design. The SAD calculations are responsible around 65% of the total ME process [3]. The general prediction of SAD value between the current block (CB) and reference block (RB) is performed using the following equations [5].

$$\mathrm{SAD}(i, j) = \sum_{k=0}^{M-1} \sum_{t=0}^{N-1} |\mathrm{CB}(k, t) - \mathrm{RB}(i + k, j + t)| \tag{1}$$

$$\mathrm{SAD}_{\mathrm{min}} = \min(\mathrm{SAD}(i, j)) \tag{2}$$

where $N \times M$ is block size of the current and reference blocks, i and j are coordinates of the block and $\mathrm{SAD}_{\mathrm{min}}$ is the minimum SAD value. The HEVC defines different partitioning sizes for current blocks (CB) to reduce ME complexity. Among all, one of the constraint is disable the asymmetric motion partitioning (AMP) option with

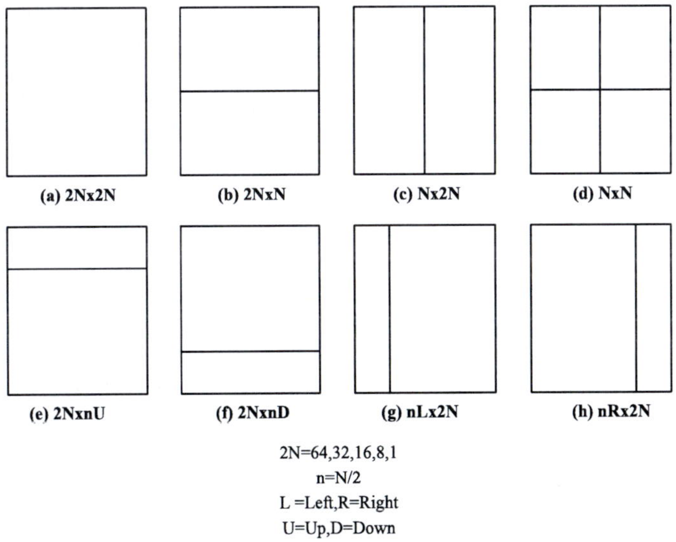

Fig. 2 Different block partitions and sizes for HEVC motion prediction unit [4]

Table 1 The sum of the total SADs required for every partition sizes in a 64 × 64 unit [6]

S. No.	Block sizes	No.of SAD's	S. No.	Block sizes	No. of SAD's
1	64 × 64	1	14	8 × 32 (left)	8
2	32 × 64	2	15	16 × 16	16
3	64 × 32	2	16	8 × 16	32
4	64 × 16 (up)	2	17	16 × 8	32
5	64 × 16 (down)	2	18	16 × 4 (up)	32
6	16 × 64 (right)	2	19	16 × 4 (down)	32
7	16 × 64 (left)	2	20	4 × 16 (right)	32
8	32 × 32	4	21	4 × 16(left)	32
9	16 × 32	8	22	8 × 8	64
10	32 × 16	8	23	4 × 8	128
11	32 × 8 (up)	8	24	8 × 4	128
12	32 × 8 (down	8	25	4 × 4	256
13	8 × 32 (right)	8	–		
Total No. of SADs		316 (Square) + 340 (Symmetric) + 168 (Asymmetric)			

fixed block size $N = 4$ [4]. Table 1 lists partitions for 64 × 64 CTB (coding tree block) and the number of SAD required for each partition.

By supporting variable block size, the compression coding performance can be improved significantly. Also, the encoding computational complexity is increased dramatically. The coding block size in HEVC is upto 64 × 64 used for HD/UHD resolutions as compared to 16 × 16 in H.264 makes the SAD process critical. Obviously, the number of SAD computations vastly increases resulting into substantial increase in the processing gate delay and hence a decrease in overall performance. For instance, the UHD video (4320 × 2160@30 frames/s) required the computation of 18,345,885,696,000 absolute differences (ADs) and 71,663,616,000 multi-value summation operations/s [5].

In the literature, several VLSI architectures for SAD computation are reported [4, 6–11]. These report a trade-off between speed, performance and power for the hardware implementation of SAD. The architectures considered are mainly either for low-power or high-performance implementation either on FPGA or on ASIC. The present available FPGAs contain a lot of dedicated hardware resources which are more suitable for high-performance optimization providing less opportunity for power optimization, whereas ASICs can be a good choice for both power and performance optimization. In this paper, FPGA is chosen for performance enhancement of SAD process. The main work contributions are outlined as follows.

1. On-chip memory and its efficient access architecture for SAD have been proposed and designed.
2. Performance optimization has been done at different abstraction levels, viz. **a** at architectural level by deploying mixed parallelism and pipelining, **b** at RTL level

with grouping and resource sharing and **c** at implementation level using retiming techniques.
3. Verified with test bench simulations, debug on FPGA with logic analyzer and constraint-driven synthesis for better performance.
4. Prototyped on FPGA using two consecutive video frames and calculated corresponding residual image and displayed output on monitor using display controller. The interfaces at input and output sides have also been developed.

For HEVC ME process, different SAD architectures were proposed in past targeting FPGA [4, 6–11] with various design and implementation approaches. Dinh Vu et al. [4] introduced a parallel SAD architecture to reduce the hardware cost and calculation time. They decreased the number of registers used but the speed degraded due to higher delay. Joshi et al. [7] reported a high-speed processing architecture with highest frequency of 475 MHz. Nalluri et al. [8] reported another high-speed architecture with trade-off between delay and hardware resources. Medhat et al. [6] reported a highly parallel architecture. It uses 64 PUs operating in parallel with two memory banks for each frame. The optimal data flow from the memory for SAD computation provides better speed and lower delay among above discussed works.

From the above brief literature survey, it is evident that there is a trade-off between speed and processing delay. So, the development of high-speed processing architecture for SAD on FPGA is of great interest. Hence, an efficient mixed parallel-pipelining architecture is proposed, designed and verified on FPGA and results compared with relevant SAD architectures. The implementation results show that the present architecture can process the video with a significant delay reduction in critical path and hence increased frequency which can efficiently meet the requirements of 4 K resolution at 30 frames/s for high-speed portable applications. The paper is organized as follow. Section 1 gives an introduction and overview of HEVC, ME and SAD process and provides literature review and related works reported till date. Section 2 explains the proposed SAD unit architecture and sub modules. Section 3 shows the hardware implementation. Section 4 presents result and comparison with existing works in terms of logic delay, hardware resource usage and operating speed. Finally, Section 5 concludes the findings and features of the proposed work.

2 Proposed Architecture and Design Process

The proposed architecture shown in Fig. 3 is mainly composed of: (i) External random access memory (RAM) for current and reference block data, (ii) on-chip memory register files (RF), (iii) processing element (PE), (iv) absolute difference (AD) circuit and (v) adder tree structure to calculate the SAD_{min}.

It receives the data related to reference and current pixels from external RAM. This data is pushed into on-chip memory RF. The data from on-chip memory RF is further fed to PE for calculation of ADs and not accessed from external memory directly. This reduces the access time of data, and hence, processing time is improved. Finally, ADs are fed to adder tree for calculating the SADs.

2.1 Memory Organization and Parallel Processing

The memory architecture for calculating all possible SADs value from 4×4 to 64 × 64 blocks of HEVC encoder has been considered.

At architecture level, 64 × 64 block unit of current and reference frames are fetched from external memory which are stored in the on-chip RFs. Each 64 × 64 block is divided into sixteen 32 × 32 blocks at level 1 and 16 × 16 blocks at level 2. The 16 × 16 block is divided into sixteen 4 × 4 block to be used for SAD calculations,

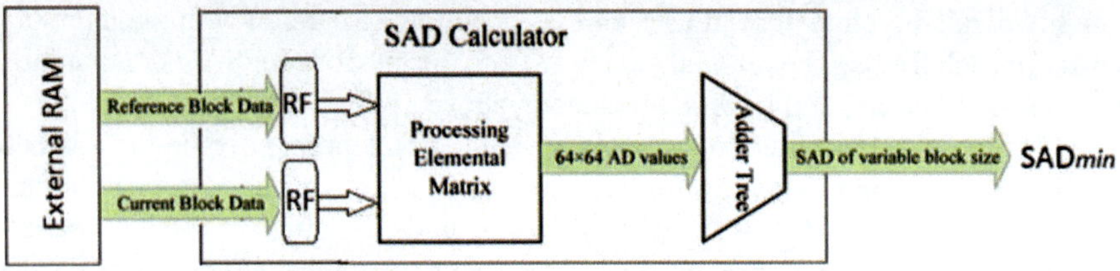

Fig. 3 Proposed SAD architecture

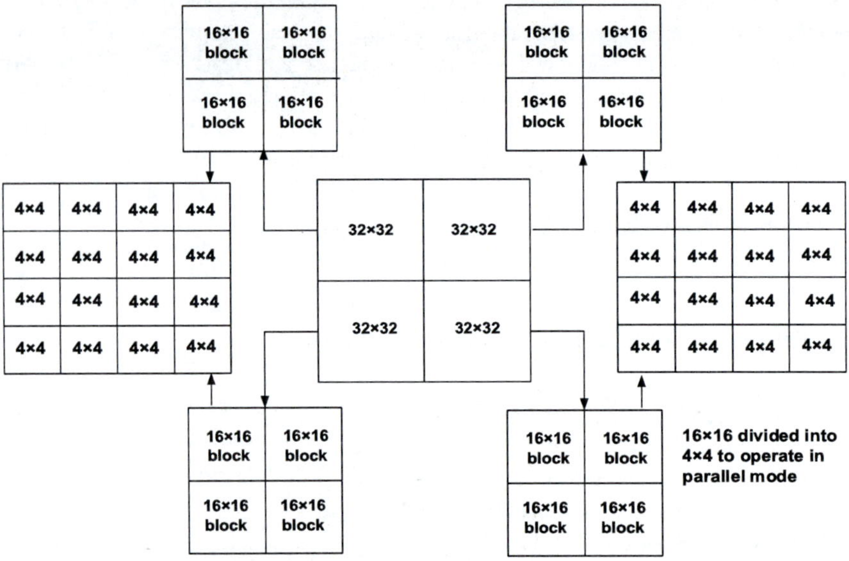

Fig. 4 Memory organization of 64 × 64 block unit

as shown in Fig. 4. Here, this process has been adopted to exploit more degree of parallelism.

2.2 Register File as On-Chip Memory

Register File (RF) stores the data of each current and reference block of 64 × 64 size. Each column and row data of 64 × 64 matrix are stored in associated memory address. The size of each RF requires approximately 4 KB of on-chip memory required. In FPGA, block random access memory (BRAM) is sufficient for storing the above mentioned data. However, for high resolution videos having higher number of bits per pixel requiring higher memory than 4 KB, look up tables (LUTs) can be used in the place of RF for on-chip data storage. In case for higher requirement, so, LUT can be used for RF.

2.3 Processor Element (PE)

The processor element (PE) block implementation is shown in Fig. 5, CP refers to current pixel and RP refers to reference pixels. It uses 4 ADs to calculate one row of 4 × 4 block in parallel. The AD values of each row are added to calculate 4 SAD values stored in next level RF memory which forms the primitive block for 64 SAD computations.

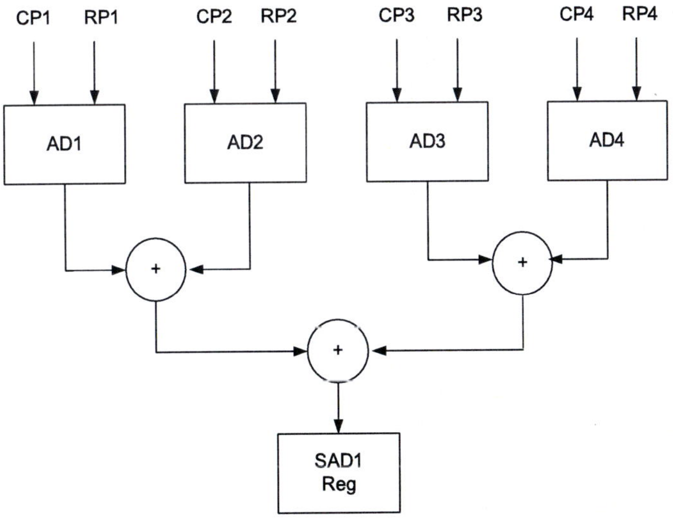

Fig. 5 The architecture of processor element

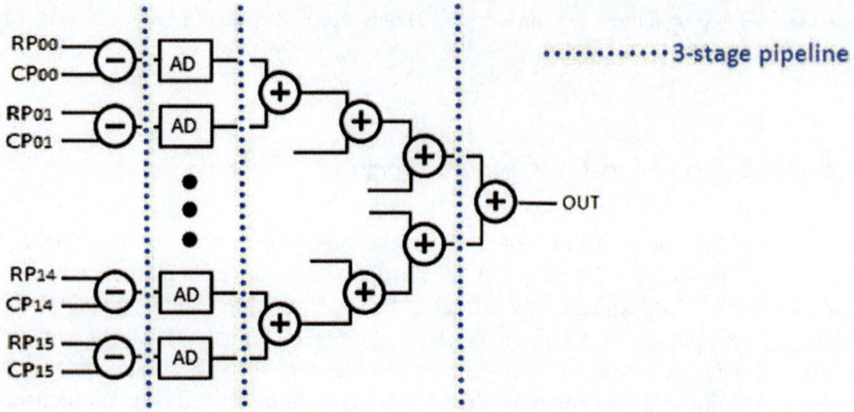

Fig. 6 16 × 16 pipelined architecture

The 16 × 16 array is implementing using a 4 × 4 block. The data block of 16 pixels is splitting into four groups and each group holds 4 pixels data and sent it to hierarchical SAD summation block.

The 16 × 16 size has implemented a three-stage pipeline structure as shown in Fig. 6, where all 4 × 4 blocks are operated in parallel. The critical path has a long path which is a tremendous amount of adder computation. The pipeline helps to decrease critical path delay.

2.4 Absolute Difference (AD)

An absolute difference (AD) circuit is 4 × 4 calculation of the absolute differences between current and reference pixels. Figure 7 shows AD block diagram implemented by using 8-bit comparator and subtractor. The comparator determines the higher value from current and reference pixel. The subtractor subtracts the lower value from higher to provide the AD value which is stored in an on-chip register file (RF) by 4-bit subtraction. Equation (3) represents AD operation [3].

$$AD = |C - R| = \begin{Bmatrix} C - R = C + \overline{R} + 1 \text{ if } C > R \\ R - C = R + \overline{C} + 1 \text{ if } R > C \end{Bmatrix} \tag{3}$$

where C is current pixel, R is reference pixel.

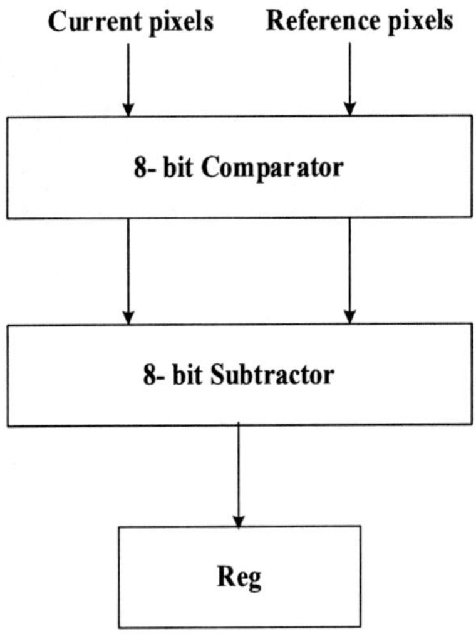

Fig. 7 Absolute difference circuit block diagram

2.5 *Adder Tree Structure*

The adder tree structure is used to calculate the SAD of different block sizes using SAD of 4 × 4 primitive block size. Figure 8 shows the hierarchical SAD adder tree structure. This is similar to the HEVC quad-tree CU structure in which each 2n × 2n blocks consisting of four $n \times n$ blocks.

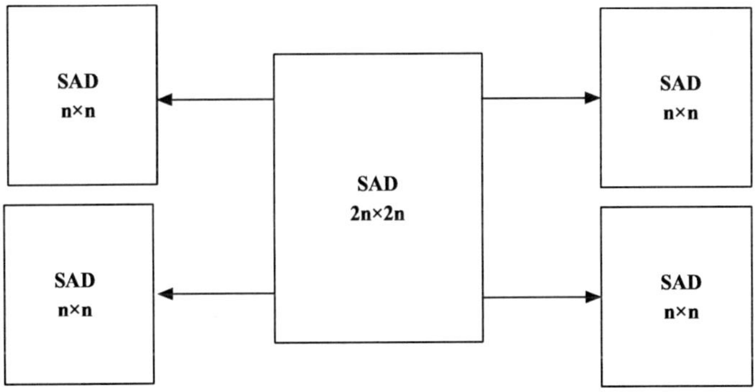

Fig. 8 The hierarchical view of SAD adder tree structure

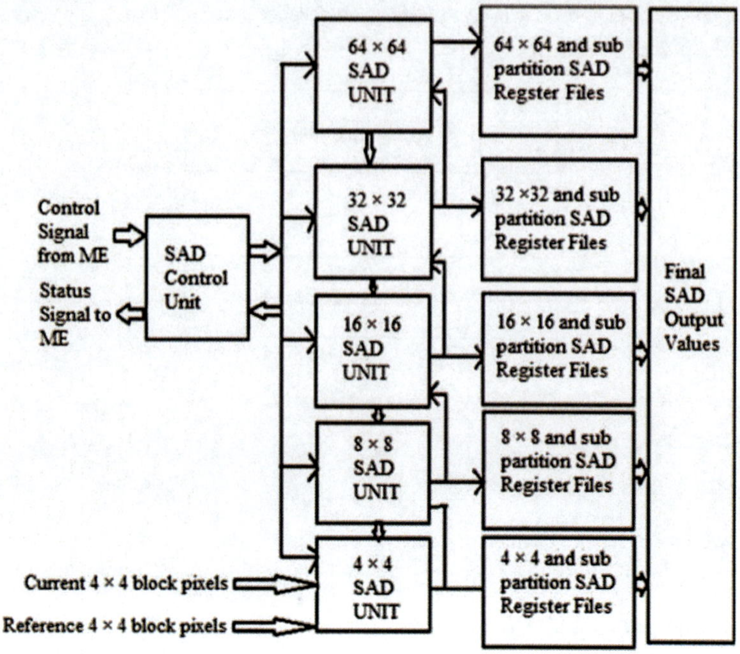

Fig. 9 SAD top-level architecture

2.6 *SAD Top Level*

The basic 4 × 4 block has been reused four times to obtain the SAD for 8 × 8 block size. Likewise, SAD of other block sizes viz. 16 × 16, 32 × 32 and 64 × 64 can also be calculated using reuse process. Figure 9 shows the top-level architecture of SAD calculation.

2.7 *Minimum Frequency Requirement*

In this design, each CTB of 64 × 64 size is decomposed into 256 number of (4 × 4) block size for SAD calculation. Therefore, it takes 2048 cycles (256 × 8 bit) to process each CTB. To achieve high throughput, for the application of 4 K video (3840 × 2160 at 30 frames/s) encoding, the following is minimum necessary operating frequency ($f_{\min}$) [12].

$$f_{\min} = (2048\ \text{cycle/CTB}) * \frac{(3840 * 2160 * 30\ \text{pixels/s})}{(64 * 64\ \text{pixels/CTB}\ )} = 124\text{MHz} \quad (4)$$

On the other hand, Fan and Yibo, et al. [13] reported the maximum necessary frequency for 4 K at 30 frames/s to be approximately 500 MHz. So, when a 4 × 4 SAD tree is adopted, the minimum and maximum frequency is 124 MHz and 500 MHz, respectively.

3 Hardware Implementation and Prototype

The proposed SAD architecture is designed in Verilog HDL and prototyped using Artix-7 FPGA. The test bench has been modeled for logic and functional verification with "foreman" video sequence of quarter common intermediate format (QCIF) 176 × 144 resolution [13]. The reason for choosing QCIF resolution is to keep it simple for simulation observations.

The simulation result of the proposed SAD architecture for two consecutive frames is shown in Fig. 10.

The two consecutive frames data are given to the test bench. It generates the corresponding SAD. The design is also verified with an on-chip debugging tool (chip-scope pro logic analyzer) as shown in Fig. 11.

Figure 12 presents the RTL schematic view of 4 × 4 SAD block in which PE and adders tree are operating in parallel. The synthesized logic can be further improved by (a) resource sharing, (b) grouping logic and (c) retiming pipeline.

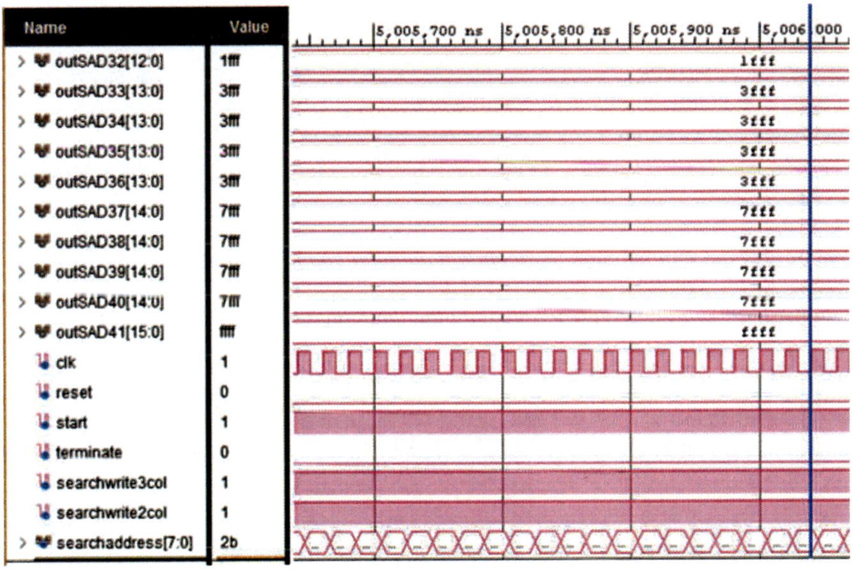

Fig. 10 Test bench simulation of the "foreman" test sequence for SAD calculations

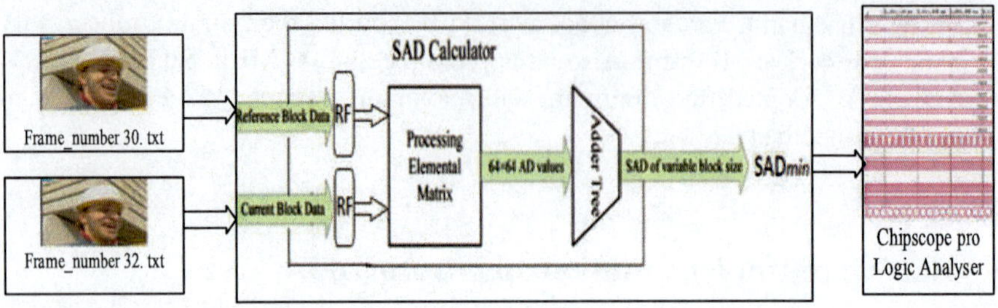

Fig. 11 Simulation and on-chip debug flow

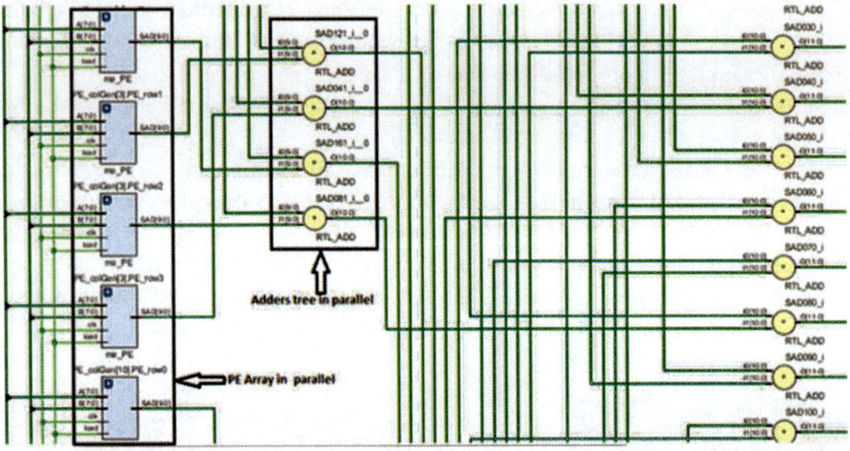

Fig. 12 RTL elaborated view of SAD 4 × 4 block

3.1 Resource Sharing [14]

At RTL level, resource sharing technique can be utilized for optimizing the performance parameter (such as speed). As shown in Fig. 13a and b, for some logics, if adders are used, it consumes more area. To overcome this problem, multipliers are used instead of adders to reduce the area. Moreover, we can use the optimized DSP48E1 slices for multipliers in FPGA.

This technique can also optimize the critical path of the logic designs which can decrease delay and increase processing speed.

3.2 Grouping Logic [14]

Similarly, the design performance can be improved by grouping the resources at RTL level, as shown in Fig. 14.

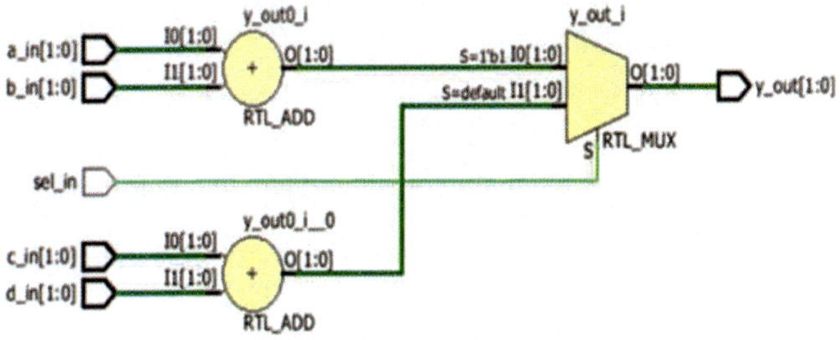

(a). without resource sharing

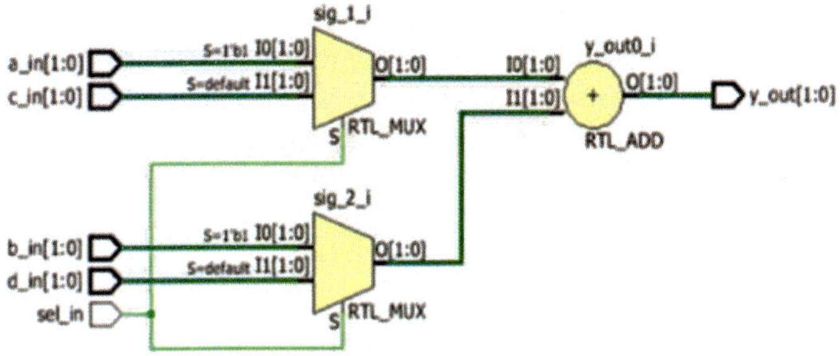

(b). with resource sharing

Fig. 13 Typical resource sharing techniques after synthesis

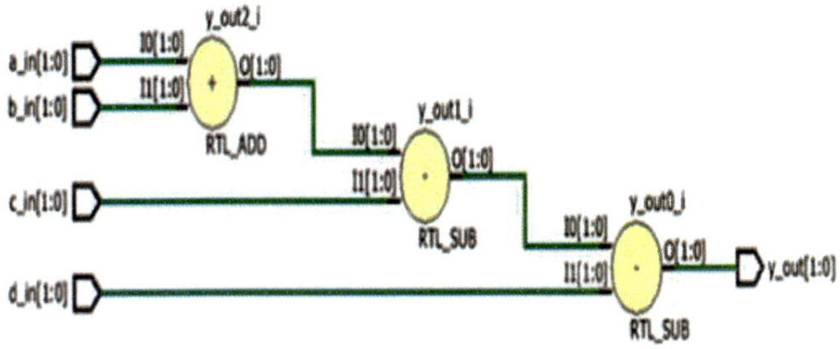

(a). without grouping

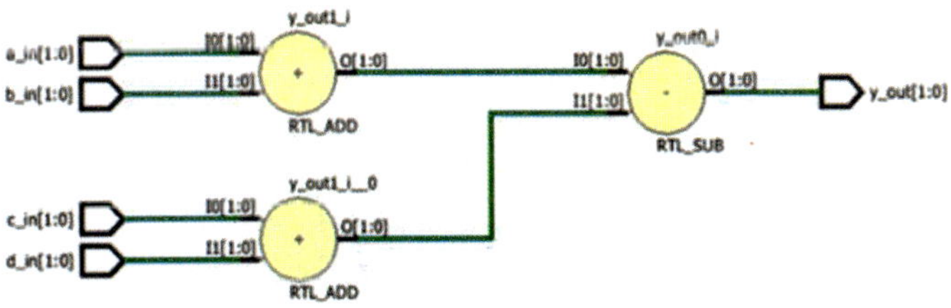

(b). with grouping

Fig. 14 Typical grouping techniques after synthesis

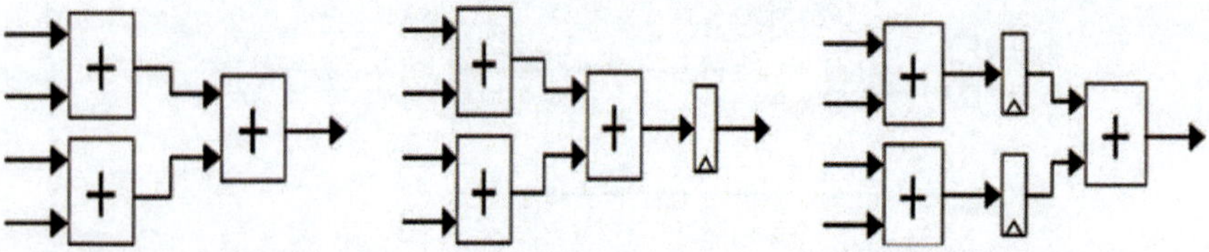

(a) Common logic inferred (b) Pipelined logic (c) Retiming logic

Fig. 15 Concept of retiming after synthesis

Figure 14a shown without grouping a cascaded logic comprising of one adder and two subtractors. The cascaded logic is a priority logic with estimated delay value of $n * t_{pd}$, where n *is* number of adders/subtractors with t_{pd} propagation delay. For the present case, the estimated delay is $3t_{pd}$, whereas in Fig. 14b shows the realization of same logic using grouping technique which utilizes two adders followed by one subtractor and estimated delay become $2t_{pd}$ only.

3.3 Retiming Pipeline [15]

Register timing concept is used at implementation level to improve timing by reordering the combinational and sequential logic in the data path. This technique can optimize registers and the balance pipeline to achieve higher operating frequency. Figure 15a–c shows the common logic inferred with combinational delay, pipeline register logic and retiming registers logic, respectively. For proposed SAD design at implementation level, retiming pipeline has been used to improve the speed.

3.4 Hardware Prototyping Setup

Represented in Fig. 16, the hardware was set up for prototyping on Xilinx Zybo kit (Artix-7 FPGA). The "foreman" video sequence has been prepared as data set for testing. The frame number 32 used as a current frame and frame number 30 used as a reference frame, which are stored in BRAM. These two data sets have been given as input to the SAD logic which computes the difference between the given two frames and stores it in flash memory.

Further, the flash memory data is provided as input to video controller unit to display the reconstructed residual frame on monitor as shown in inset of Fig. 16. Due to default tiles configuration of monitor (output display), output image of video controller unit displays all over the monitor screen.

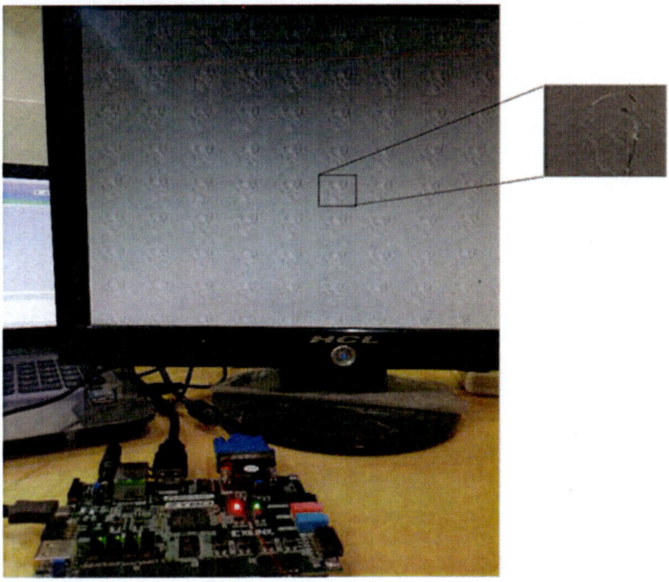

Fig. 16 Residual frame generation on FPGA

4 Results and Discussion

The design comparison of proposed work with existing design is shown in Table 2. More work has been reported on 65 nm technology node in past ten years [4, 7, 8], whereas very few work has been reported on 28 nm technology node [6]. The proposed design shows higher operating frequency because of the techniques employed as shown in Table 3, which optimizes the speed on the expense of higher number of LUTs and register as compared to 65 nm technology node.

Table 2 FPGA implementation results and comparison

	[4]	[7]	[8]	[6]	This work
FPGA	Virtex-5	Virtex-5	Virtex-5	Virtex-7	Artix-7
Technology node (nm)	65	65	65	28	28
No. of used registers	8841	4266	9180	39,901	22,880
No. of LUTs	17,992	14,761	15,453	24,957	25,072
Max. Frequency (MHz)	190.785	475.21	171.9	458.7	498.4
Block size (4 × 4 to 64 × 64)	Yes	Yes	Yes	Yes	Yes
Delay for every 64 × 64 block (ns)	167.73	44.19	372.2	45.98	32.06
AMP support	Yes	Yes	Yes	Yes	Yes

Table 3 List of optimization techniques employed

S. No	Technique	Levels
1	Parallel	Architecture
2	Pipeline	
3	Grouping logic	RTL modeling
4	Resource sharing	
5	Retiming	Implementation

The SAD design without optimization operates at 368 MHz with delay of 58.07 ns for each 64 × 64 block, whereas with optimization, it operates at 498.4 MHz with delay of 32.06 ns for each 64 × 64 block.

5 Conclusion

A mixed parallel-pipelined processing SAD architecture for ME has been presented. The architecture was synthesized and prototyped on Xilinx Artix-7 FPGA. The design uses 256 number of 4 × 4 SAD blocks which are operating in a mixed mode of parallel-pipeline to calculate SAD values of 64 × 64 block. The design operates at frequency of 498.4 MHz with the delay of 32.06 ns for each 64 × 64 block. The design performance has been improved at different levels viz. architecture, RTL modeling and implementation. This design can be used for any kind of ME processes including fast and full search algorithms. It can also be used in consumer electronics where the performance is most crucial.

References

1. Sullivan GJ, Ohm J-R, Han W-J, Wiegand T (2012) Overview of the high efficiency video coding (HEVC) standard. IEEE Trans Circ Syst Video Technol 22(12):1649–1668
2. Zheng, J, Lu C, Guo J, Chen D, Guo D (2019) A hardware efficient block matching algorithm and its hardware design for variable block size motion estimation in ultra-high-definition video encoding. ACM Trans Des Autom Electron Syst (TODAES) 24(2)15:1–21
3. Jia L, Tsui C-Y, Au OC, Jia K (2018) A low-power motion estimation architecture for HEVC based on a new sum of absolute difference computation. IEEE Trans Circ Syst Video Technol
4. Dinh VN, Phuong HA, Duc DV, Ha PTK, Tien PV, Thang NV (2006) High speed SAD architecture for variable block size motion estimation in HEVC encoder. In: IEEE sixth international conference on communications and electronics (ICCE). Vietnam pp 195–198
5. Silveira B, Paim G, Abreu B, Grellert M, Diniz CM, Ceśar da Costa EA, Bampi S (2017) Power-efficient sum of absolute differences hardware architecture using adder compressors for integer motion estimation design. IEEE Trans Circ Syst I: Regul Papers 64(12):3126–3137
6. Medhat A, Shalaby A, Sayed MS, Elsabrouty M, Mehdipour F (2014) A highly parallel SAD architecture for motion estimation in HEVC encoder. In: IEEE Asia Pacific conference on circuits and systems (APCCAS), Japan, pp 280–283

7. Joshi AM, Ansari MS, Sahu C (2018) VLSI architecture of high speed SAD for high efficiency video coding (HEVC) encoder. In: IEEE international symposium on circuits and systems (ISCAS). Italy, pp 1–4
8. Nalluri P, Alves LN, Navarro A (2014) High speed SAD architectures for variable block size motion estimation in HEVC video coding. In IEEE international conference on image processing (ICIP). France, pp 1233–1237
9. Abreu B, Santana G, Grellert M, Paim G, Rocha L, da Costa EAC, Bampi S (2018) Exploiting partial distortion elimination in the sum of absolute differences for energy-efficient HEVC integer motion estimation. In: 31st symposium on integrated circuits and systems design (SBCCI), Brazil, pp 1–6
10. Porto R, Agostini L, Zatt L, Roma N, Porto M (2019) Power-efficient approximate SAD architecture with LOA imprecise adders. In: IEEE 10th latin american symposium on circuits & systems (LASCAS), Colombia, pp 65–68
11. Vanne J, Viitanen M, Hamalainen TD, Hallapuro A (2012) Comparative rate-distortion-complexity analysis of HEVC and AVC video codecs. IEEE Trans Circ Syst Video Technol 22(12):1885–1898
12. He G, Zhou D, Li Y, Chen Z, Zhang T, Goto S (2015) High-throughput power-efficient VLSI architecture of fractional motion estimation for ultra-HD HEVC video encoding. IEEE Trans Very Large Scale Integ (VLSI) Syst 23(12):3138–3142
13. Foreman QCIF Video sequence: http://www2.tkn.tu-berlin.de/research/evalvid/qcif.html. Last accessed 18 July 2019
14. Vaibbhav T (2019) Advanced HDL synthesis and SOC prototyping, 1st edn. Springer Singapore, Singapore
15. Eguro K, Hauck S (2008) Simultaneous retiming and placement for pipelined netlists. In: 16th international symposium on field-programmable custom computing machines. California, pp 139–148

Minimization of Peak-to-Average Power Ratio in DHT Precoded OFDM System by A-Law Companding

Ajay Kumar Yadav, Dheeraj Dubey, and Y. K. Prajapati

Abstract In this paper, peak-to-average power ratio (PAPR) of orthogonal frequency division multiplexing (OFDM) is minimized by a new technique called as DHT precoder and A-law companding with variable parameters. This technique is implemented in two stages for PAPR reduction of OFDM system. In first stage, DHT precoder is used to spread the symbols and PAPR of this OFDM system is transformed into signal carrier modulation system which has a smaller amount of PAPR compared to multicarrier modulation system. In second stage, A-law companding is used to compressing large peaks and then, PAPR reduces. Therefore, proposed technique minimizes PAPR up to 4 dB ($A = 10$) and 2.98 dB ($A = 30$) which is better than DHT precoder, DFT precoder and A-law companding effectively and also preserves bit error rate (BER) performance.

Keywords OFDM · PAPR · DFT precoder · DHT precoder · IFFT · A-Law · AWGN channel

1 Introduction

OFDM is known for multicarrier modulation technique that has become high-speed digital communication. OFDM-based technologies are digital audio broadcasting (DAB), asymmetric digital subscriber line (ADSL), high definition television

A. K. Yadav (✉) · D. Dubey · Y. K. Prajapati
Electronics and Communication Engineering Department, Motilal Nehru National Institute of Technology Allahabad, Allahabad, India
e-mail: er.ajayyadavec@gmail.com

D. Dubey
e-mail: dubeydheeraj548@gmail.com

Y. K. Prajapati
e-mail: yogendrapra@gmail.com

D. Harvey et al. (eds.), *Advances in VLSI, Communication, and Signal Processing*, Lecture Notes in Electrical Engineering 683,
https://doi.org/10.1007/978-981-15-6840-4_51

(HDTV), terrestrial digital video broadcasting (DVB-T), and 4G wireless communication due to simple multi-path fading and robustness to narrowband interference [1].

In spite of having several benefits, OFDM signals are still considered as high PAPR that degrades performance of high power amplifier (HPA). As a result, HPA operates to nonlinear region from linear region. Due to nonlinearity HPA, distorted signals occur [2]. However, to overcome this problem, many PAPR minimization techniques are anticipated as partial transmit sequence (PTS) [3], selective mapping (SLM) [4], clipping and filtering [5], precoding [6], companding technique [7], and precoder and companding [8].

In this paper, a technique suggested based on the combination of DHT precoder with A-law companding to OFDM system for minimization of PAPR. A-law companding decreases PAPR but not up to a substantial level and also considered that it is responsible for distortion in amplitudes of the OFDM signals. When DHT precoder technique is associated with A-law in OFDM system, PAPR reduces up to significant level with minimum complexity by the adapting multicarrier OFDM into single carrier OFDM systems which have already low PAPR. Besides that, it is maintaining BER performance because DHT precoder is accountable for phase distortion.

This paper is arranged as follows: Original OFDM system model is described in Sect. 2. Section 3 is existing proposed system model for PAPR reduction. Simulation results and discussion are explained in Sect. 4, and Sect. 5 concluded this paper.

2 Original OFDM System

The OFDM signals with modulated symbols X_k can be expressed as:

$$s(n) = \frac{1}{\sqrt{N}} \sum_{k=0}^{N-1} X_k e^{j2\pi \frac{kn}{LN}}, \quad 0 \le n,\ k < (LN - 1) \tag{1}$$

where N is a number of subcarriers and L is oversampling factors. Symbols X_k are the statically independent and identically distributed symbols and being modulated by QPSK.

The PAPR of signals $s(n)$ is defined as:

$$\text{PAPR} = 10 \log_{10} \frac{\text{Max}\left[|s(n)|^2\right]}{E\left[|s(n)|^2\right]} \text{ dB} \tag{2}$$

The OFDM signal is recognized as high PAPR because of addition of same phase of subcarrier. Usually, complementary cumulative distribution function (CCDF) deals with performance of PAPR and can be defined as probability that PAPR of the signals exceeds threshold value. Let PAPR_0 assume to be threshold value, then

CCDF is as follows:

$$\text{CCDF} = \text{Prob.}(\text{PAPR} > \text{PAPR}_0) \tag{3}$$

3 Proposed System Model

3.1 DHT Precoded OFDM Signals

Figure 1 displays the proposed system model which is based on the combination of DHT precoder and A-law companding, and it is used as hybrid technique in OFDM system for PAPR reduction. The DHT precoder is employed before OFDM modulation whereas A-law after OFDM modulation. DHT precoder decreases PAPR by conversing multicarrier OFDM system into single carrier system, and PAPR of single carrier system is low in comparison to multicarrier OFDM system.

DHT is a real-valued function, which is associated to the Fourier transform, though it is simple kind of transform when compared to DFT. Let N real numbers x_0, x_1, …, x_{N-1} is input signals get transformed into N real numbers H_0, H_1, …, H_{N-1} as output. The real-valued function of DHT [6] is given by

$$H_k = \sum_{n=0}^{N-1} x_n.\left[\cos\left(\frac{2\pi nk}{N}\right) + \sin\left(\frac{2\pi nk}{N}\right)\right]$$

From Ref. [6], $\text{cas}\,\theta = \cos\theta + \sin\theta$

Therefore,

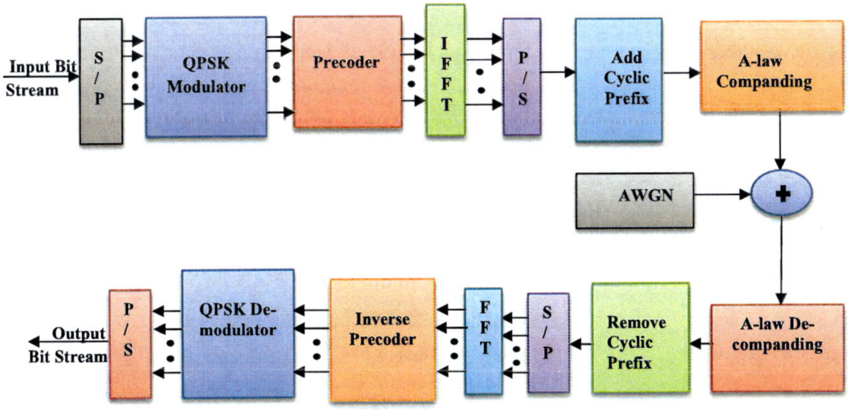

Fig. 1 Block diagram of proposed model for PAPR reduction

$$H_k = \sum_{n=0}^{N-1} x(n) \cdot \text{cas}\left(\frac{2\pi nk}{N}\right), \quad k = 0, 1, \ldots (N-1) \tag{4}$$

Now, inverse DHT is given by

$$x(n) = \frac{1}{N} \sum_{k=0}^{N-1} H_k \cdot \text{cas}\left(\frac{2\pi nk}{N}\right), \quad n = 0, 1, \ldots (N-1) \tag{5}$$

where $h_{l,k} = \text{cas}(\frac{2\pi lk}{N})$, $\forall l, k = 0, 1, 2, \ldots, (N-1)$ denotes DHT precoding matrix. The DHT precoding matrix $h_{l,k}$ of size $N \times N$, can be written as:

$$h_{l,k} = \frac{1}{\sqrt{N}} \begin{bmatrix} h_{0,0} & h_{0,1} \ldots & h_{0,(N-1)} \\ h_{1,0} & h_{1,1} \ldots & h_{1,(N-1)} \\ \vdots & \vdots \ddots & \vdots \\ h_{(N-1),0} & h_{(N-1),1} \ldots & h_{(N-1),(N-1)} \end{bmatrix} \tag{6}$$

Therefore, DHT precoded OFDM signals can be written as

$$S_h(n) = \frac{1}{\sqrt{N}} \sum_{l=0}^{N-1} Y_l \cdot e^{j2\pi \frac{kn}{LN}}, \quad \forall n, l = 0, 1, ..., (L \cdot N - 1) \tag{7}$$

where, $Y_l = \frac{1}{\sqrt{N}} \sum_{k=0}^{N-1} h_{l,k} \cdot X_k, \quad l = 0, 1, \ldots (N-1))$.

3.2 A-Law Companded Signals

A-Law companding technique decreases PAPR of OFDM signals excellently. This technique compresses high amplitudes as well as expands small amplitudes of OFDM signals so that reduces PAPR effectively. Companding function of OFDM signals [9] can be expressed by

$$F(x) = \begin{cases} \text{sgn}[S_h(n)] \cdot \frac{1+\log[A_s|S_h(n)|]}{1+\log A_s} & \frac{1}{A_s} \le |S_h(n)| \le 1 \\ \text{sgn}[S_h(n)] \cdot \frac{A_s|S_h(n)|}{1+\log A_s} & 0 \le |S_h(n))| \le \frac{1}{A_s} \end{cases} \tag{8}$$

where A_s denotes companding parameters. Decompanding function is used as inverse function of Eq. (8) to recover original signals and can be written as

$$F^{-1}(y) = \begin{cases} \text{sgn}[y] \cdot \frac{\exp[|y|(1+\log(A_s))-1]}{A_s} & \frac{1}{1+\log(A_s)} \le |y| \le 1 \\ \text{sgn}[y] \cdot \frac{|y|[1+\log(A_s)]}{A_s} & 0 \le |y| \le \frac{1}{1+\log(A_s)} \end{cases} \tag{9}$$

where y is received signal before companding function.

4 Results and Discussion

We performed MATLAB simulation to evaluate the performance of proposed system model that is based on the combination of DHT precoder and A-law companding technique. The input data streams are generated randomly which transformed into parallel. Then, modulated with subcarrier ($N = 256$) by QPSK and considered as cyclic prefix (CP) = 32, oversampling factor $L = 4$ and IFFT size = 1024.

As displayed in Fig. 2, the envelopes of the proposed technique ($A = 30$) are almost constant in comparison to original OFDM, DFT precoder, DHT precoder, and A-law companding. In Fig. 3, PAPR of the proposed technique ($A = 30$) is better than original OFDM signal by **8.1 dB**, DHT by **6.2 dB**, DFT precoder by **4.7 dB** and A-law companding (A = 30) by **0.65 dB** at same **CCDF** = 10^{-3}. However, The BER of proposed technique is preserved over A-law companding ($A = 30$) which has been shown in Fig. 4.

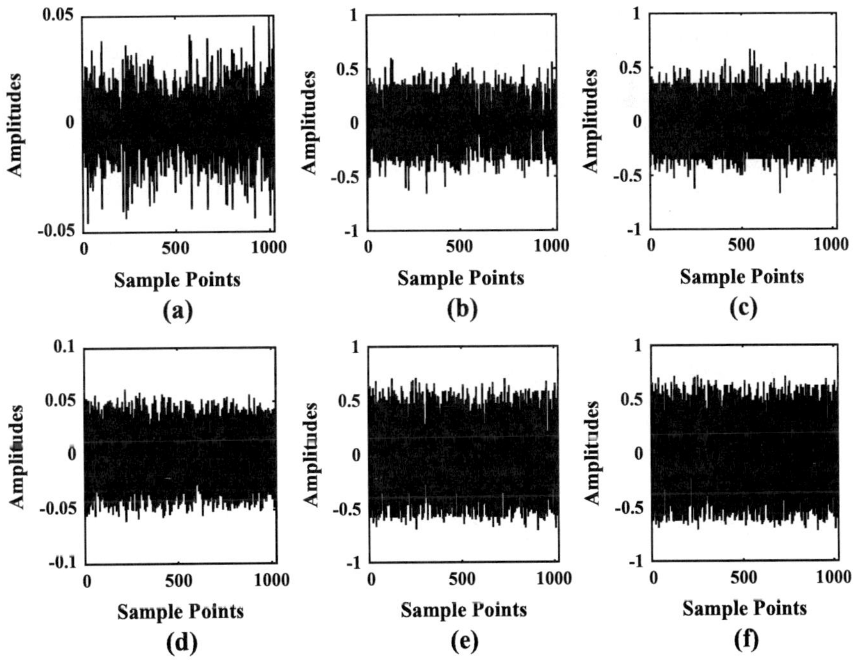

Fig. 2 Envelop of **a** original OFDM signals, **b** DHT precoded signals, **c** DFT precoded signals, **d** A-Law companded signals ($A = 30$), **e** proposed OFDM signals ($A = 10$), and **f** proposed OFDM signals ($A = 30$)

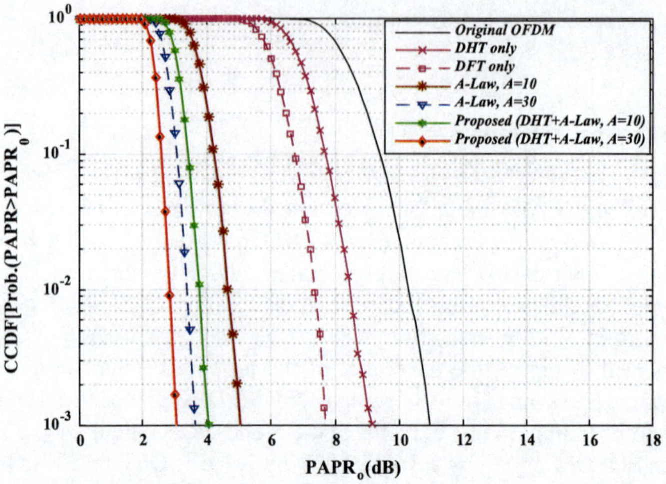

Fig. 3 Performance of PAPR of proposed technique and others

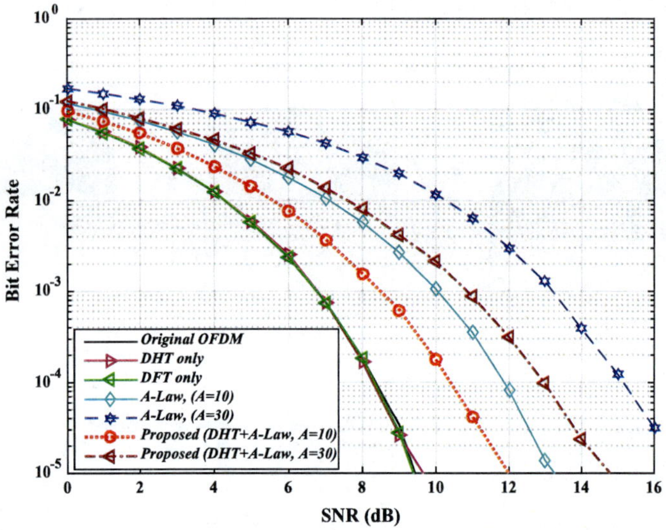

Fig. 4 BER performance of proposed technique and other techniques over AWGN channel

5 Conclusion

As far as PAPR minimization is concerned, the proposed technique is based on the combination of DHT precoder and A-law companding in OFDM system. Performance of BER of proposed technique is preserved in comparison to A-law companding ($A = 30$) over AWGN channel. Also, the computational complexity declines due to linearity of DHT precoder. By DHT precoder and A-law companding method, OFDM system has low PAPR. As shown in simulation results, the proposed technique reduces PAPR up to 4 dB ($A = 10$) and 2.98 dB ($A = 30$) and BER performance in comparison to other suggested techniques.

References

1. Jiang T, Zhu G (2004) 'Nonlinear companding transform for reducing peak-to-average power ratio of OFDM signals. IEEE Trans Broadcast 50(3):342–346
2. Nee RV, Prasad R (2010) OFDM for wireless multimedia communications. Artech House, Inc
3. Hung HL, Huang YF (2012) Peak-to-average power ratio reduction in orthogonal frequency division multiplexing system using differential evolution-based partial transmit sequences scheme. IET Commun 6(11):1483–1488
4. Li CP, Wang SH, Wang CL (2010) Novel low-complexity SLM schemes for PAPR reduction in OFDM systems. IEEE Trans Signal Process 58(5):335–341
5. Zhu X, Pan W, Li H, Tang Y (2013) Simplified approach to optimized iterative clipping and filtering for PAPR reduction of OFDM signals. IEEE Trans Commun 61(5):1891–1901
6. Ouyang X, Jin J, Jin G, Wang Z (2012) Low complexity discrete Hartley transform precoded OFDM for peak power reduction. Electron Lett 48(2):90–91
7. Yadav AK, Prajapati YK (2019) PAPR minimization of clipped OFDM signals Using tangent rooting companding technique. Wirel Personal Commun, 1–13
8. Yadav AK, Sahoo PK, Prajapati YK (2019) Peak-to-average power ratio reduction and improved bit error rate performance of orthogonal frequency division multiplexing system using discrete Fourier transform precoder and root-based nonlinear companding. Optical Engineering 76106:1
9. Wang X, Tjhung TT, Ng CS (1999) Reduction of peak-to-average power ratio of OFDM system using a companding technique. IEEE Trans Broadcast 45(3):303–307

Leakage-Tolerant Low-Power Wide Fan-in OR Logic Domino Circuit

Ankur Kumar, Pratosh K. Pal, Vikrant Varshney, Avaneesh K. Dubey, and R. K. Nagaria

Abstract A leakage-tolerant low-power wide fan-in OR logic domino circuit is presented to decrease the leakage current and to enhance the noise immunity. Primarily, an efficient switching control in keeper network is developed to reduce the switching of keeper transistor in both phases, so that dynamic power and noise immunity can be improved. Further, a diode-connected NMOS transistor in evaluation network is incorporated in series with the footer transistor of standard domino circuits. This significantly decreases the leakage current and charge sharing because of the stacking effect. This reduction in leakage and charge sharing ensures the improvement in the noise margin. Furthermore, a current mirror and feedback NMOS transistors are also employed in the evaluation network to improve the speed of the circuit and fully discharge the dynamic node. The simulation results of proposed domino and reported domino circuits are designed using Spectre simulator under cadence virtuoso models of 45-nm technology which shows the 31% reduction in power dissipation (PD) and 1.53 times improvement in noise immunity at the similar delay compared to the standard domino circuits.

Keywords Low-power dissipation · High-speed · Leakage current · Noise immunity · Wide fan-in OR logic domino

---

A. Kumar (✉) · P. K. Pal · V. Varshney · A. K. Dubey · R. K. Nagaria
Department of Electronics and Communication Engineering, Motilal Nehru National Institute of Technology Allahabad, Prayagraj, Uttar Pradesh 211004, India
e-mail: rel1604@mnnit.ac.in

P. K. Pal
e-mail: pratoshkumar1075@gmail.com

V. Varshney
e-mail: rel1551@mnnit.ac.in

A. K. Dubey
e-mail: akd8292@gmail.com

R. K. Nagaria
e-mail: rkn@mnnit.ac.in

D. Harvey et al. (eds.), *Advances in VLSI, Communication, and Signal Processing*, Lecture Notes in Electrical Engineering 683,
https://doi.org/10.1007/978-981-15-6840-4_52

1 Introduction

In today's technology progression, dynamic circuit is the first choice of the designer to implement the high-performance systems such as superscalar microprocessor, multiplexer, and comparator because of the improved noise immunity and reduced PDP [1–4]. However, threshold voltage along with power supply must be reduced to meet the desired performance of the circuit as per down scaling of technology [5]. Therefore, subthreshold current increases exponentially in the domino circuit which discharges the DN and makes the logic failure at the output [6–8]. A remedy of this problem is to incorporate a weak PMOS at the DN to improve the noise immunity against the noise and leakage [9, 10]. This representation of the transistor is called the conventional domino circuit. Hence, it can be stated that upsizing of the keeper transistor is a standard method [11–13].

The keeper ratio (KR) is defined as the ration of the current driving capability of the keeper transistor to the evaluation transistor [14–18]. Size of the keeper transistor is varied to enhance the noise immunity of the transistor. However, increasing the value of KR has two negative effects on PD [8–14]. First, it increases the contention current between the keeper and the PDN network which increases the short circuit PD [8–14]. Second, DN capacitance increases which increases switching PD. Moreover, upsizing of keeper ratio also increases the delay and process variability of the domino circuit [15–17]. This is the trade-off between the noise immunity and speed. Therefore, this method is not reliable solution because of the increased delay, PD, and process variability [15–18].

$$\mathrm{KR} = \frac{\mu_\mathrm{p} \times \left(\frac{W}{L}\right)_\mathrm{Keeper}}{\mu_\mathrm{p} \times \left(\frac{W}{L}\right)_\mathrm{Evaluation}} \tag{1}$$

where, μ and (W/L) stands for mobility and ratio of width to length of transistors, respectively.

As several techniques [6–17] have been developed to lessen the subthreshold leakage and to improve the noise immunity. Therefore, it is observed that subthreshold current is reduced by redesigning the evaluation network, and noise immunity is improved by redesigning the keeper network.

Conventional domino circuit (CDC) consists of precharge, keeper network, and evaluation network [3–13], in which, precharge and keeper transistor are for charging the DN and evaluation network for discharging the DN. Output logic is defined by the charge stored in the output capacitor due to charging and discharging. The power dissipation is obtained by using Eq. (2) for domino circuits [14–16]. The power dissipation is given as

$$\mathrm{PD} = P_\mathrm{Switcing} + P_\mathrm{Short} + P_\mathrm{lc} \tag{2}$$

where P_{switch} is the PD due to the charging and discharging of capacitor at output node, P_{short} is the PD due to the existence of short-circuit path between supply voltage and ground, and P_{lc} is the static PD due to the leakage current [17, 18].

This paper represents a proposed domino circuitry to decrease the subthreshold leakage current and improve the noise immunity. The organization of the rest paper is explained into the following sections. Literature review of the reported domino circuits is explained in Sect. 2. The working and explanation of the proposed domino are illustrated in Sect. 3. Simulation setup, results, and discussion are explained in Sect. 4. Conclusion of the research paper is explained in the last section.

2 Literature Review

2.1 Standard Domino Circuit (SDC)

CDC [3] is simply extension of dynamic gate to enhance noise immunity. In this circuit, a keeper along with static inverter is added to improve the performances as shown in Fig. 1. In evaluation phase, the keeper transistor keeps high voltage at DN when no input is applied in order that a correct logic can be gained at output node. However, keeper is enough to improve the noise immunity but at the expense of subthreshold leakage current, high power delay product (PDP) [6–10].

Fig. 1 Schematic of standard domino circuit (CDC) [3]

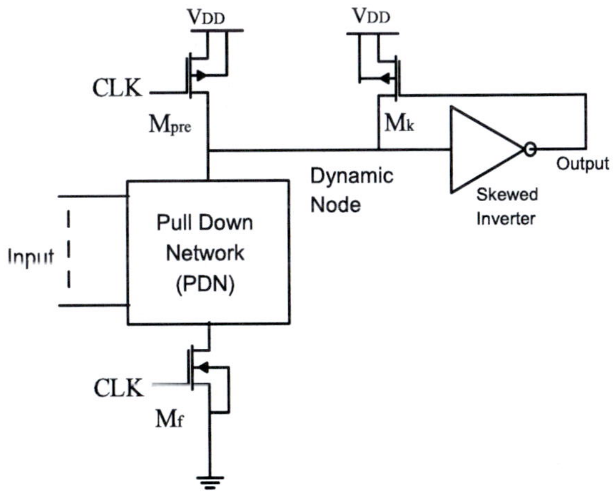

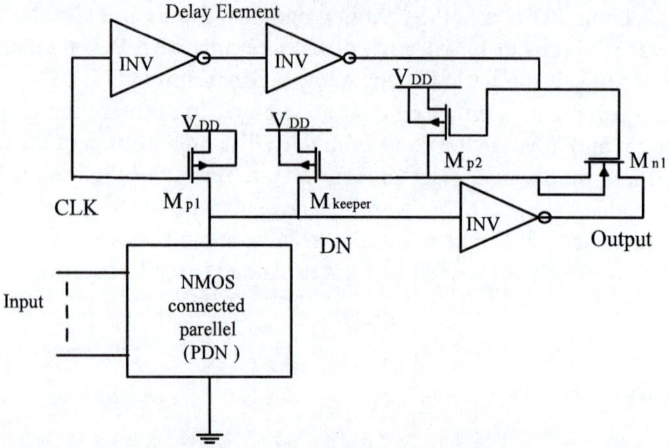

Fig. 2 Schematic diagram of high-speed domino circuit (HSDC) [6]

2.2 *High-Speed Domino Circuit (HSDC)*

Anis et al. have proposed HSDC [6] as shown in Fig. 2, in which performance and reliability both are optimized simultaneously. Keeper is controlled in well-mannered way because of two transistors (NMOS and PMOS transistor) to decrease the PC. Further, multi-threshold value transistors are used, so that leakage current and delay can be overcome during standby mode and active mode, respectively. This arrangement is inadequate to decrease the leakage current in case of large fan-in.

2.3 *Current Mirror Footed Domino Circuit (CMFDC)*

Moradi et al. have proposed another leakage tolerant domino called CMFDC [9] as given in Fig. 3. In which NMOS transistor is incorporated in evaluation network to decrease the leakage current by using stacking effect [5–15]. Further, mirror is incorporated in this circuit to reduce the evaluation delay. It is limited in terms of power dissipation and noise immunity.

3 Proposed Domino Circuit

It is observed though the literature review of reported domino circuits that wide fan-in OR logic domino circuits must have low-leakage, high-speed, and high noise immunity in contrast to conventional [6–15]. Therefore, a leakage-tolerant low-power domino wide fan-in OR logic domino circuit is proposed to decrease leakage current

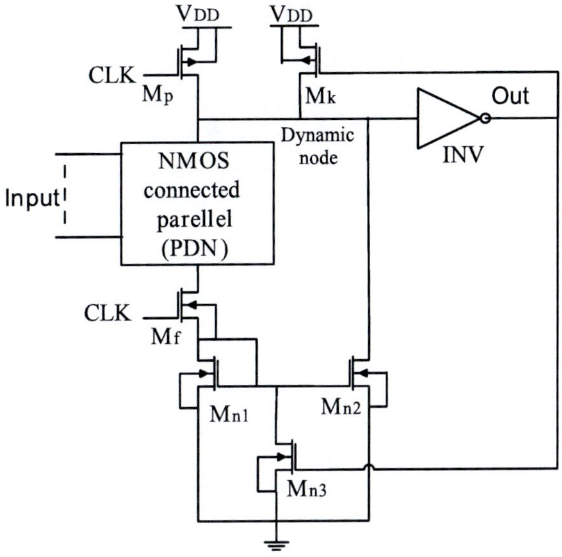

Fig. 3 Schematic diagram of the current mirror feedback domino circuit [9]

and improve the noise immunity as given in Fig. 4. In the proposed domino, an effective arrangement of transistors named as switching controlled network (SCN) is developed to control the switching of keeper transistor, which is the extension of the HSD [6]. Further, two transistors are added along with a parallel current mirror in the evaluation network to decrease the subthreshold leakage and delay across PDN as done in [9].

The working of the proposed domino circuit is illustrated in the two parts. In the first parts, keeper network is redesigned by adding SCN to reduce the dynamic power and enhance the noise immunity. The gate of the keeper transistor (M_k) is connected to output node and power supply through the TG and PMOS transistor (M_{pc}), respectively. Both M_{pc} and TG are driven by the delayed clock pulse. In precharge phase, M_{pc} transistor and TG turn ON and OFF, respectively, due to which gate of the M_k is connected to the supply voltage. This supply voltage switches OFF the M_k in the rest of this phase and reduce the dynamic power dissipation [6]. Moreover, M_{pc} and TG turn OFF and ON, respectively, in the evaluation phase and connected the gate to the output node. Now, M_k will be turned ON and OFF on the basis of the logic at output node in evaluation phase. If all inputs are at the lows logic, i.e., output node is at the high logic which turn ON the M_k transistor to recharge the dynamic node. And if at least one input is high, i.e., output node is at low logic which turns OFF the M_k transistor.

In the second phase, evaluation network is modified as in [9] to decrease the subthreshold leakage current. Primarily, M_{n1} transistor is stacked with the footer transistor (M_f) to increase the threshold voltage of the M_f and PDN transistor because of stacking effect [15, 17] which decreases the subthreshold current. However, delay in the evaluation phase is also increased at the same time because of the increased

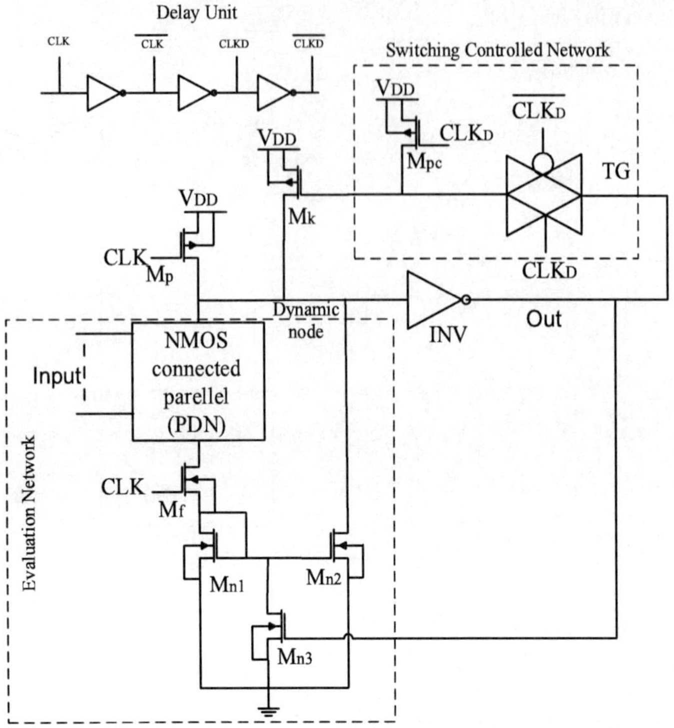

Fig. 4 Schematic diagram of the proposed domino circuit

length of critical discharging path (through PDN). Therefore, M_{n2} mirror transistor is added parallel to the M_{n1} transistor to decrease the evaluation delay. Now, dynamic node can also discharge through mirror transistor quickly. Further, transistor M_{n3} is added in the evaluation network which provides a feedback from the output to the dynamic node. This transistor is added to fully discharge the dynamic node in the evaluation phase if at least one of the inputs is high. Moreover, M_{n3} also helps to reduce the subthreshold current in standby mode. This combination of transistor reduces the charge sharing of the dynamic node which enhances the noise immunity of the proposed domino circuit. The proposed domino operations in two phases similar as CDC as explained below.

- **Precharge Phase**

In this phase, clock is low, and no input is applied resulting M_p turns ON and PDN transistors turn OFF. Hence, dynamic node gets charge because of the M_p transistor. Further, M_k transistor turns OFF due to the M_{pc} transistor after some certain delay. All the transistor in the evaluation network and TG are turned OFF which guide to increase the threshold voltage of the PDN transistor. Consequently, dynamic node

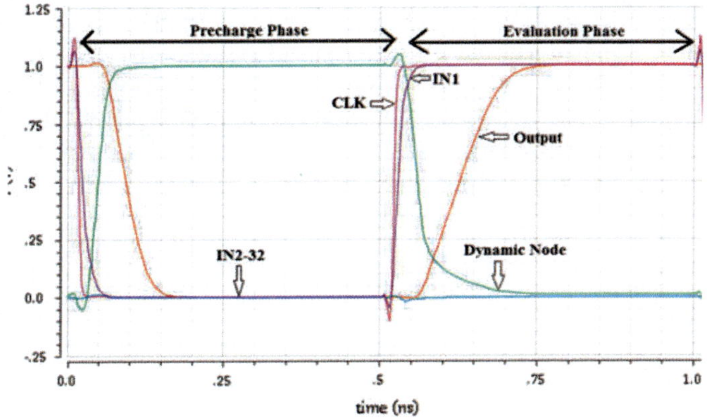

Fig. 5 Transient output of the proposed domino circuit at 32-bit fan-in OR logic

does not discharge. Finally, dynamic and output nodes are at the high and low logic, respectively.

- **Evaluation Phase**

In this phase, clock pulse (CLK = '1') becomes high resulting transistor M_p and M_{pc} turn OFF and applied inputs may be either at logic '1' or '0'. Now, two cases occur on the basis of applied inputs.

Case-I: In this case, no input is applied which means PDN transistors are off. After some certain delay, TG turns ON, and it passes the charge of output node to the M_k transistor. Thus, M_k turns ON and recharges the dynamic node if any charge sharing happens. Moreover, footer transistors work same as in precharge phase. Finally, dynamic and output nodes are at the high and low logic, respectively.

Case-II: At least one input is at logic '1' which means PDN is ON. Now, dynamic node starts to discharge through PDN and all footer transistor. And M_k turns OFF because of high logic at the output node. Finally, dynamic and output nodes are at the low and high logic, respectively.

Simulated output waveform for the 32-bit fan-in OR logic proposed domino circuit is shown in Fig. 5 when only single input is at high logic in the evaluation phase and no input is applied in precharge phase.

4 Simulation Results and Discussions

4.1 Standard Domino Circuit (SDC)

The proposed domino and reported domino circuits are designed and simulated for wide fan-in (8, 16, 32, 64 input) using the Spectre simulator of the cadence virtuoso.

Th initial values are fixed at 1 V, 1 GHz and 5 fF for the voltage, frequency, and load, respectively. A simulation setup [19] which is used to obtain all the simulation results for each circuit is shown in Fig. 5. Further, Monte Carlo simulations are performed to obtain the variations in the delay and power of proposed domino circuit and CDC by considering all the transistors between OUT and IN. All the simulation results are obtained under the same delay condition.

4.2 Design Metrics

- **Figure of merit (FOM)**

Figure of merit (FOM) [12–14] is taken as another parameter to evaluate the performance of proposed domino gate. FOM is calculated with the help of Eq. 3, in which each parameter is normalized by SFD gate.

$$\text{FOM} = \frac{\text{UNGM}}{A_{\text{norm}} \times P_{\text{avg-norm}} \times t_{\text{norm}}} \tag{3}$$

where UNGM, A_{norm}, PD_{norm}, and $\text{Delay}_{\text{norm}}$ stand for noise immunity, area, PD, and delay, respectively. Each factor is normalized by the SDC [3].

- **Noise Immunity**

Several metrics are developed to obtain the noise immunity of digital circuits. Unity noise gain margin (UNGM) [20] is one of them which is considered to observe the noise immunity in this work. The UNGM is the amplitude of input source due to which equal source can be obtained at the output.

$$\text{UNGM} = \{V_{\text{noise}} : V_{\text{noise}} = V_{\text{out}}\} \tag{4}$$

where, the noise pulse is considered similar to the real-time noise in terms of spike, glitches, and crosstalk.

The numerical values of normalized PD and UNGM are listed in Tables 1 and 2 for the proposed domino and some reported domino at different fan-in. These values

Table 1 Comparison of the normalized PD among the proposed domino and reported domino circuits at the different fan-in under similar delay condition

Fan-in	SDC [3]	HSDC [6]	CMFDC [9]	Proposed
8-bit (90 ps)	1.00	0.99	0.98	0.97
16-bit (100 ps)	1.00	0.92	0.84	0.82
32-bit (110 ps)	1.00	0.90	0.82	0.76
64-bit (120 ps)	1.00	0.86	0.76	0.69

Table 2 Comparison of the normalized UNGM among the proposed domino and reported domino circuits at the different fan-in under similar delay condition

Fan-in	SDC [3]	HSDC [6]	CMFDC [9]	Proposed
8-bit (90 ps)	1.00	1.17	1.14	1.34
16-bit (100 ps)	1.00	1.17	1.20	1.42
32-bit (110 ps)	1.00	1.16	1.15	1.53
64-bit (120 ps)	1.00	1.18	1.12	1.35

Table 3 Variation in the PD and delay with respect to voltage for the proposed domino and SDC [3] at the 32-bit fan-in

Node voltage	Normalized PD		Normalized delay	
	SDC [3]	Proposed	SDC [3]	Proposed
0.8	1.00	1.14	1.00	0.92
0.9	2.12	1.75	0.81	0.72
1.0	2.42	2.03	0.55	0.55
1.1	2.65	2.28	0.44	0.43
1.2	2.88	2.53	0.35	0.36

are normalized by the SDC [3] to show the improvement in the performance of proposed domino. Table 2 depicts that PD of proposed domino is reduced in contrast to the SDC [3]. Moreover, Table 3 depicts that noise immunity of proposed domino is improved 1.34–1.53 times to SDC [3]. Both tables show the reduction in PD and enhancement in noise immunity for the proposed domino in contrast to all other reported domino circuits. This reduction in power dissipation and improvement in noise immunity are owing to the effective controllability of keeper transistor and stacking effect.

Further, FOM is obtained by using Eq. (3) for the proposed domino and some reported domino circuits for the 32-bit fan-in OR logic. In this, area is obtained by the transistor count in which PMOS is given two count than NMOS because of the area and mobility. Figure 6 depicts that FOM is increased by 40% for the proposed domino in contrast to SDC [3].

It is necessary to include the process corner, voltage, and temperature (PVT) variations effect to determine the stability, robustness, and reliability of any domino circuits. Hence, the proposed domino circuit and SDC [3] for the 32-bit fan-in OR logic are simulated at different voltages supply, temperatures, and process corners. PD and delay are obtained at different temperature and the process corners in order to observe the effect of temperature and process corners, shown in Table 3 and Figs. 7, 8 and 9. It is observed through Table 3 and Figs. 7, 8 and 9 that stability, robustness, and reliability of proposed domino are more than the SDC [3] against the noise, leakage, and PVT.

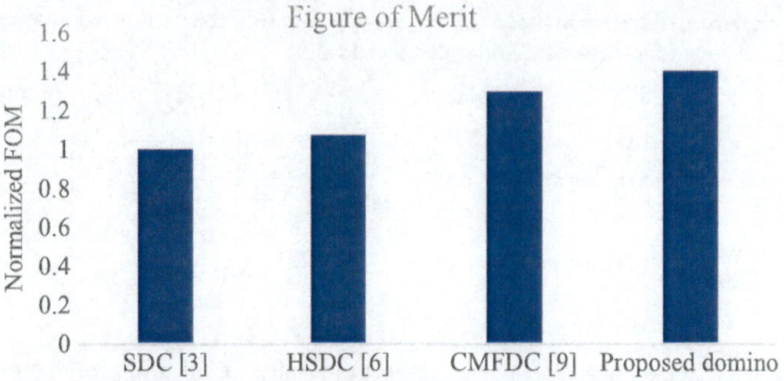

Fig. 6 Comparison of the FOM among the proposed domino and reported domino circuits at 32-bit fan-in under the similar delay condition

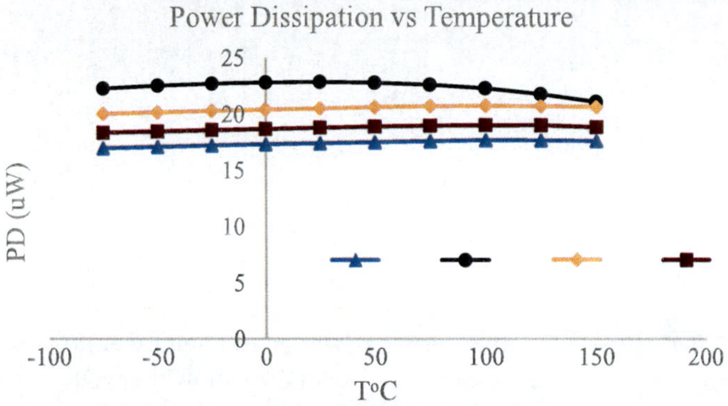

Fig. 7 Variation of the PD with respect to temperature for the proposed domino and SDC [3] at the 32-bit fan-in under the similar delay condition

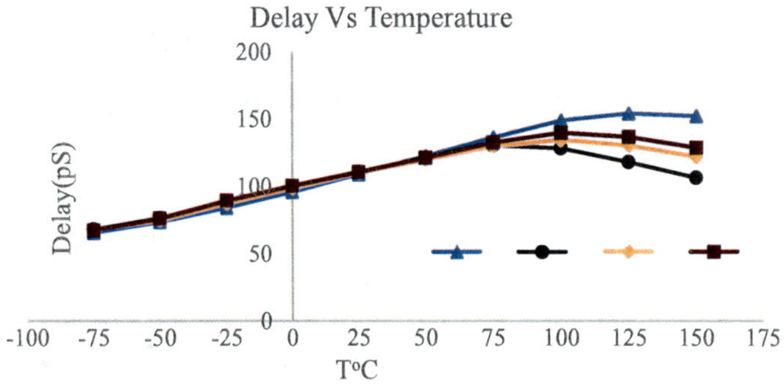

Fig. 8 Variation of delay with respect to temperature for the proposed domino and SDC [3] at the 32-bit fan-in under the similar delay condition

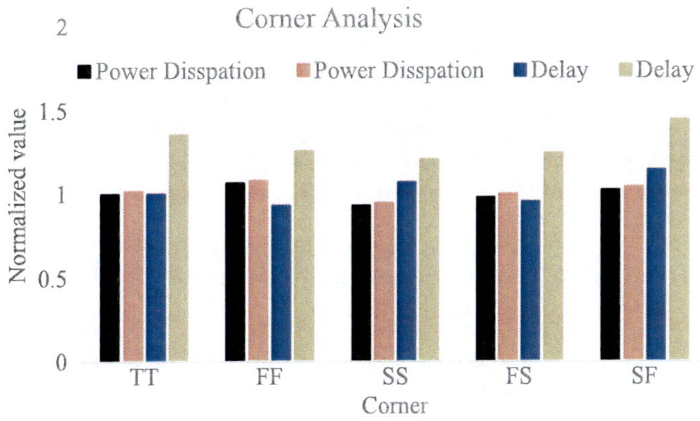

Fig. 9 Variation of the PD and delay at the different process corners and temperature for the proposed domino circuits at 32-bit fan-in

5 Conclusion

This paper introduced a leakage-tolerant low-power wide fan-in OR logic domino circuit to reduce the power dissipation and improve the noise immunity. First, keeper network is redesigned to make the efficient switching of the keeper network which decreases the dynamic power dissipation. Further, evaluation network is resigned to introduce the stacking effect which lessens the leakage current and overcomes the charge sharing. Both modifications in the circuit also improve the noise immunity. The simulation results confirm that PD is reduced by 31% and noise immunity enhanced by 53% of proposed in contrast with the standard domino circuit.

Acknowledgements The authors would like to thank Special Manpower Development Program Chip to System Design (SMDP-C2SD).

References

1. Gronowski P, Chandrakasan A, Bowhill WJ (2001) Issues in dynamic logic design. IEEE Press, Piscataway, NJ
2. Rabaey JM, Chandrakasan AP, Nikolic B (2002) Digital integrated gates, vol 2. Prentice Hall, Englewood Cliffs
3. Krambeck RH, Lee CM, Law H-FS (1982) High-speed compact circuits with CMOS. IEEE J Solid-State Circ 17:614–619
4. Alvandpour A, Larsson-Edefors P, Svensson C (1999) A leakage-tolerant multi-phase keeper for wide domino circuits. In: The 6th IEEE international conference on electronics, circuits and systems. Proceedings of ICECS'99, vol 1, pp 209–212
5. Roy K, Mukhopadhyay S, Mahmoodi-Meimand H (2003) Leakage current mechanisms and leakage reduction techniques in deep-submicrometer CMOS circuits. Proc IEEE 91:305–327

6. Anis MH, Allam MW, Elmasry MI (2002) Energy-efficient noise-tolerant dynamic styles for scaled-down CMOS and MTCMOS technologies. IEEE Trans Very Large Scale Integr (VLSI) Syst 10:71–78
7. Lih Y, Tzartzanis N, Walker WW (2006) A leakage current replica keeper for dynamic circuits. IEEE J Solid-State Circuits 42(1):48–55
8. Ye Y, Borkar S, De V (1998) A new technique for standby leakage reduction in high-performance circuits. In: 1998 symposium on VLSI circuits. Digest of technical papers, Cat. No. 98CH36215, pp 40–41
9. Moradi F, Peiravi A, Mahmoodi H (2004) A new leakage-tolerant design for high fan-in domino circuits. In: Proceedings. The 16th international conference on microelectronics, ICM 2004. IEEE, pp 493–496
10. Hua C-H, Hwang W, Chen C-K (2005) Noise-tolerant XOR-based conditional keeper for high fan-in dynamic circuits. In: IEEE International symposium on circuits and systems, ISCAS 2005, pp 444–447
11. Jeyasingh RGD, Bhat N, Amrutur B (2011) Adaptive keeper design for dynamic logic gates using rate sensing technique. IEEE Trans Very Large Scale Integr (VLSI) Syst 19(2):295–304
12. Peiravi A, Asyaei M (2012) Robust low leakage controlled keeper by current-comparison domino for wide fan-in gates. Integr VLSI J 45(1):22–32
13. Asyaei M (2015) A new leakage-tolerant domino circuit using voltage-comparison for wide fan-in gates in deep sub-micron technology. Integr VLSI J 51:61–71
14. Asyaei M, Moradi F (2018) A domino circuit technique for noise-immune high fan-in gates. J Circuits Syst Comput 27(10):1850151
15. Kumar A, Nagaria RK (2018) A new leakage-tolerant high speed comparator based domino gate for wide fan-in OR logic for low power VLSI circuits. Integr VLSI J 63:174–184
16. Kang S-M, Leblebici Y (2003) CMOS digital integrated circuits. Tata McGraw-Hill Education, New York
17. Kumar A, Nagaria RK (2019) A new process variation and leakage-tolerant domino circuit for wide fan-in OR gates. Analog Integr Circ Sig Process 1–17
18. Pal PK, Dubey AK, Kassa SR, Nagaria RK (2016) Voltage comparison based high speed & low power domino circuit for wide fan-in gates. In: IEEE international conference on electron devices and solid-state circuits (EDSSC), pp 96–99
19. Alioto M, Palumbo G, Pennisi M (2010) Understanding the effect of process variations on the delay of static and domino logic. IEEE Trans Very Large Scale Integr (VLSI) Syst 18:697–710
20. Ding L, Mazumder P (2004) On gate techniques to improve noise immunity of CMOS dynamic logic. IEEE Trans Very Large Scale Integr (VLSI) Syst 12:910–925

Virtually Doped Silicon-on-Insulator Junctionless Transistor for Reduced OFF-State Leakage Current

Prateek Kishor Verma and Santosh Kumar Gupta

Abstract In this work, a novel virtually doped silicon-on-insulator junctionless transistor (VD-SOIJLT) is analyzed and compared its DC characteristics with conventional SOIJLT. The VD-SOIJLT offers reduced I_{OFF} (leakage) and improved I_{ON}/I_{OFF} ratio over conventional SOIJLT having SiO_2 buried oxide (BOX) region. By 2D numerically calculated calibration, even for channel length of 20 nm, we have realized that virtually doped (VD) device delivers further reduction in I_{OFF} of the order of (~10^{-15} A/μm) which is ~10 order smaller than conventional SOIJLT and considerably enhanced I_{ON}/I_{OFF} ratio of the order of (~10^{10}) which is ~10 order higher than SOIJLT for BOX thickness (T_{BOX}) of 10 nm. It is noticed that a more perfect volume depletion can be achieved by applying charge plasma (CP) concept on conventional SOIJLT in thermal equilibrium and OFF-state conditions. The optimum DC characteristics, viz sub-threshold slope (SS), I_{OFF}, and I_{ON}/I_{OFF} ratio, can be obtained by $T_{BOX} = 10$ nm having channel length $(L) = 60$ nm for VD-SOIJLT. Therefore, the VD-SOIJLT is free from the doping issues, viz random doping fluctuations (RDF), doping activation, and threshold voltage variation, and it delivers a better agreement with DC characteristics over its conventional counterpart.

Keywords Charge plasma · Virtually doped · Silicon-on-insulator · Junctionless · OFF-state current · I_{ON}/I_{OFF} ratio

1 Introduction

In nanoscale regime, there are some complications that have arisen in the fabrication of junction MOSFETs due to the ultra-sharp doping profile of source/drain (*S*/*D*) and channel regions for p-type and n-type semiconductor separately. It is very hard

P. K. Verma (✉) · S. K. Gupta
Department of Electronics and Communication Engineering, Motilal Nehru National Institute of Technology Allahabad, Prayagraj, Uttar Pradesh 211004, India
e-mail: prateek.kishor05@gmail.com

S. K. Gupta
e-mail: skg@mnnit.ac.in

D. Harvey et al. (eds.), *Advances in VLSI, Communication, and Signal Processing*, Lecture Notes in Electrical Engineering 683,
https://doi.org/10.1007/978-981-15-6840-4_53

to maintain doping profile at the metallurgical junctions (ultra-shallow p-n junction) at sub-20 nm regime. The aforementioned problem can be resolved by a new device concept which does not involve any metallurgical junction but consists of uniformly distributed heavily doped (10^{19}–10^{20} cm^{-3}) *S*/*D* and the channel regions, better-known as a junctionless transistor (JLT) [1] which overcome the restriction related to ultra-sharp doping concentration, and it also reduces the higher thermal budget for activating the dopants. A JLT offers reduced short channel effects (SCEs), nearly ideal SS, low I_{OFF}, and less mobility degradation [2, 3] over junction-based transistors. Accordingly, JLT seems to be a capable substitute for the conventional junction based-MOSFETs. However, for highly doped n-channel devices to realizing volume depletion in the OFF-state, SOIJLT requires ultra-thin SOI films thickness ($\leq$5 nm) [4] and higher gate work function (~5.5 eV), but it is very unrealistic for the ultra-thin SOI substrate due to its fabrication's complexities. A bulk planner junctionless transistor (BPJLT) [5] was recently reported which solve some of these problems stated earlier, but it also has some limitations of choosing higher gate work function (5.5 eV) and process complexity of a trigate structure [6–10]. To overcoming these limitations, SOIJLT having high-k BOX along with highly doped ($10^{20}/cm^3$) ground plane is previously reported [8] which depleted the active device layer from the bottom that reduces the effective active device layer thickness. So, volume depletion can be realized even for a SOI film thickness of 10 nm, resulting in a significant reduction in I_{OFF} was obtained. However, due to very high doping concentration, JLT suffers from RDF which results in threshold voltage variation which significantly affects the overall device performances [11]. The charge plasma concept (dopingless) was reported to extricate the previously mentioned problems. For the first time, the CP concept has been fabricated and characterized by Rajasekharan et al., for p-n junction diode [12, 13]; the same concept has been investigated for bipolar junction transistor (BJT) [14], and recently dopingless tunnel FET [15], JLT [16], and gate-all-around (GAA) FET were reported [17] for enhancement of device performances.

In this paper, for the more perfect realization of the volume depletion, a virtually doped approach has been applied on SOIJLT. The numerical investigations of VD devices provide better DC characteristics over conventional SOIJLT in terms of SS, I_{OFF}, and I_{ON}/I_{OFF} ratio. ON-state current rises in conventional SOIJLT as compared to VD-SOIJLT due to enhanced electron concentration throughout the channel region. But here, the main intent of the work is not to improve I_{ON} but a more reduction in OFF-state current with the help of CP concept which makes net improvement in I_{ON}/I_{OFF} ratio. It is perceived that the effective active layer thickness in a VD-SOIJLT is reduced over actual physical device layer thickness in SOIJLT due to the more depletion of active device layer in the channel region above the BOX region which reduces the leakage current (I_{OFF}) of the order of 10^{-15} A/μm for $T_{BOX} = 10$ nm. Charge plasma concept reduces the fabrication complexity in terms of doping issues (ion-implantation) and is also more economical (reduces total cost of fabrication) than conventional junctionless devices.

2 Device Structures

Figure 1 demonstrates the 2D cross-sectional device structure of conventional SOIJLT and the virtually doped SOIJLT. *X*- and *y*-axis are taken along and across the channel of the devices, respectively. All the considered devices have identical dimension of the structure in terms of channel length (L) variations of 20–100 nm, channel thickness (T_{si}) of 10 nm, gate oxide thickness (T_{ox}) of 1 nm, gate to *S*/*D* spacer distance (L_S) of 25 nm each, BOX thickness (T_{BOX}) variations of 10–20 nm, and gate work function ($q\phi_M$) of 5.1 eV as given in Table 1. Here, the main differences in VD-SOIJLT as compared to previously reported work [8] are the dopingless (intrinsic) nature of the *S*/*D* and channel regions applied by the charge plasma concept and also an intrinsic (or lowly doped p^+) region of the ground plane. A majority of

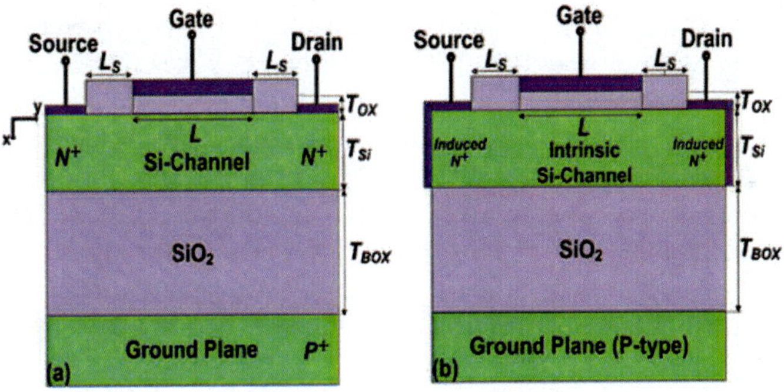

Fig. 1 2D cross-sectional view of **a** SOIJLT and **b** VD-SOIJLT

Table 1 Device parameters of SOIJLT and VD-SOIJLT

Device parameters	SOIJLT	VD-SOIJLT
BOX material	SiO_2	SiO_2
Channel length (L, nm)	20–100	20–100
Device layer thickness (T_{Si}, nm)	10	10
Gate to *S*/*D* distance (L_S, nm)	25	25
BOX thickness (T_{Box}, nm)	10–20	10–20
Gate oxide thickness (T_{ox}, nm)	1	1
S/*D* doping ($N_{S/D}$, cm^{-3})	10^{19}	Intrinsic
Channel doping (N_{Ch}, cm^{-3})	10^{19}	Intrinsic
Ground plane doping (N_A, cm^{-3})	10^{20}	Intrinsic
Gate work function ($q\phi_M$, eV)	5.1	5.1
S/*D* work function ($q\phi_{MSD}$, eV)	–	3.9
Gate voltage (V_{GS}, V)	0–1.0	0–1.0
Drain voltage (V_{DS}, V)	0–1.0	0–1.0
Temperature (T, K)	300	300

electrons and holes plasmas have induced in the undoped region of *S*/*D*, and it became N^+ (heavily doped) region due to the suitable adoption of *S*/*D* metal work function (Hafnium work function = 3.9 eV).

The VD concept follows two compulsory requirements which are: (i) The work function of *S*/*D* metal contacts must be less than undoped silicon, i.e., ($\phi_{\text{MSD}} < \chi_{\text{Si}} + (E_{\text{G}}/2)$), where χ_{Si} is electron affinity of silicon (=4.17 eV) and E_{G} is the silicon band gap (=1.12 eV), and (ii) device layer thickness (T_{Si}) must be less than Debye length (L_{D}), i.e., ($L_{\text{D}} = \sqrt{(\epsilon_{\text{Si}} \cdot (V_{\text{T}}/q) \cdot N)}$), where ϵ_{Si} is the silicon dielectric constant and V_{T}, q, and N are the thermal voltage, elementary charge, and carrier concentration, respectively. For the VD-SOIJLT, the first requirement has been fulfilled by using hafnium metal *S*/*D* contacts (3.9 eV < Si-work function) which accumulates the electron plasma near the metal-semiconductor interface and another requirement by keeping T_{Si} of 10 nm (<L_{D}).

The numerical investigations of 2D device structures have been performed by Silvaco TCAD (Atlas) [18]. Numerical calculations are based on physical models which include Shockley–Read–Hall (SRH) for carrier generation and recombination, Auger model for minority carrier recombination, and carrier statistics using the Fermi–Dirac distribution model. Lombardi mobility model (CVT) is for field- and doping-dependent mobility effects, CONMOB model is for low field mobility related to doping concentration, and FLDMOB is for lateral high field-dependent mobility model. Band gap narrowing (BGN) was preferred for high concentrations regions, and it also included interface trap charges and quantum confinement effects. The change in the lifetime of electron and hole were 10^{-7} s which corresponds to the carrier lifetime at highly doped regions. There is no tunneling models are used in the numerical calculations.

3 Numerically Calculated Results and Explanations

3.1 Electron Concentrations Profiles

Figures 2 and 3 depict graphical and contour representations of electron concentrations profiles of SOIJLT and VD-SOIJLT under thermal equilibrium ($V_{\text{GS}} = V_{\text{DS}} = 0$ V), OFF-state ($V_{\text{GS}} = 0$ V, $V_{\text{DS}} = 1$ V), and ON-state ($V_{\text{GS}} = V_{\text{DS}} = 1$ V), respectively. In Fig. 2a, under thermal equilibrium, the virtually doped device has highest electron concentration (~10^{20}–10^{21}/cm^3) at source and drain side of the junctionless transistor because of the induction of electron plasma near the metal-semiconductor interface. The reason behind higher electron concentration in *S*/*D* region is that when lightly doped n-type semiconductor having higher work function contacts with lower work function metal electrode (Hafnium = 3.9 eV), the electrons start to accumulate near the metal-semiconductor interface inside the semiconductor. At the channel region, SOIJLT has the highest electron concentration followed by VD-SOIJLT. In Fig. 2b, at OFF-state, the charge carriers redistribution occurs when applied a

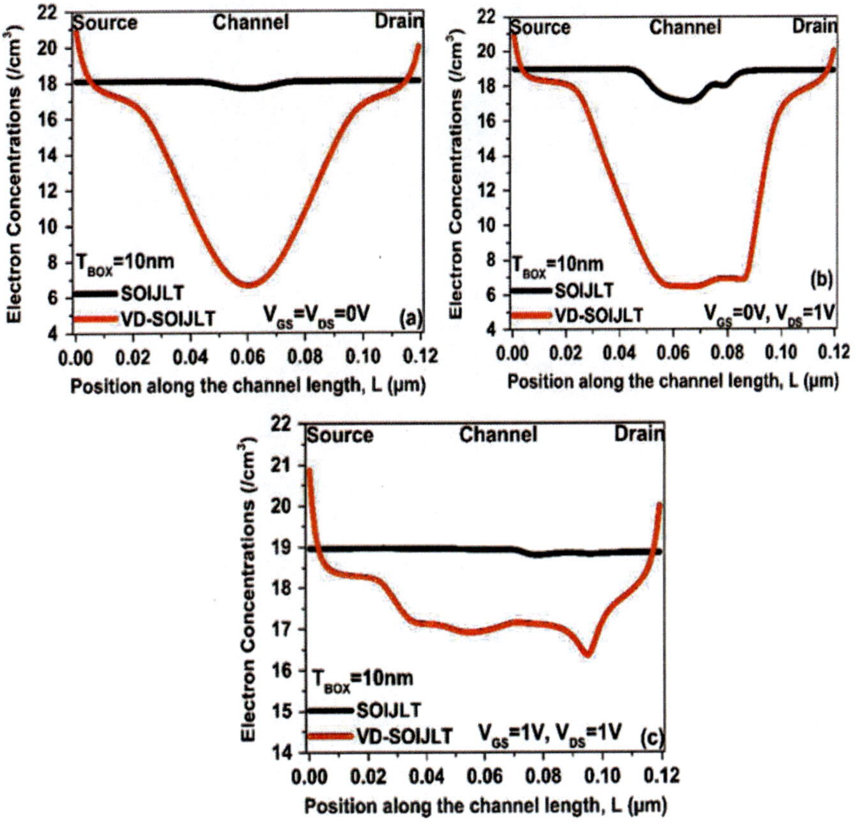

Fig. 2 Electron concentrations of SOIJLT and VD-SOIJLT under **a** thermal equilibrium, **b** OFF-state, and **c** ON-state

positive voltage (1 V) on drain terminal. For all devices, volume depletion of the active layer takes place in the channel region above the BOX region (Fig. 3). It is perceived that the effective active layer thickness in a VD-SOIJLT is reduced over actual physical device layer thickness in SOIJLT due to the depletion of active device layer in the channel region and above the BOX region which reduces the leakage current (I_{OFF}) of the order of 10^{-15} A/μm. It is noticed that VD-SOIJLT device has more volume depletion as compared to SOIJLT, so more reduction in I_{OFF} has been obtained in VD-SOIJLT due to the undoped region of *S*/*D* and channel over conventional SOIJLT. In Fig. 2c, for all devices, the electron concentration is highest for conventional SOIJLT, but VD-SOIJLT has lower electron concentration over virtually doped devices in the channel-drain interface to drain end. I_{ON} rises in SOIJLT as compared to VD-SOIJLT due to enhanced electron concentration throughout the channel region. The main intent of this work is to more reduction in OFF-state leakage current by using a virtually doped concept.

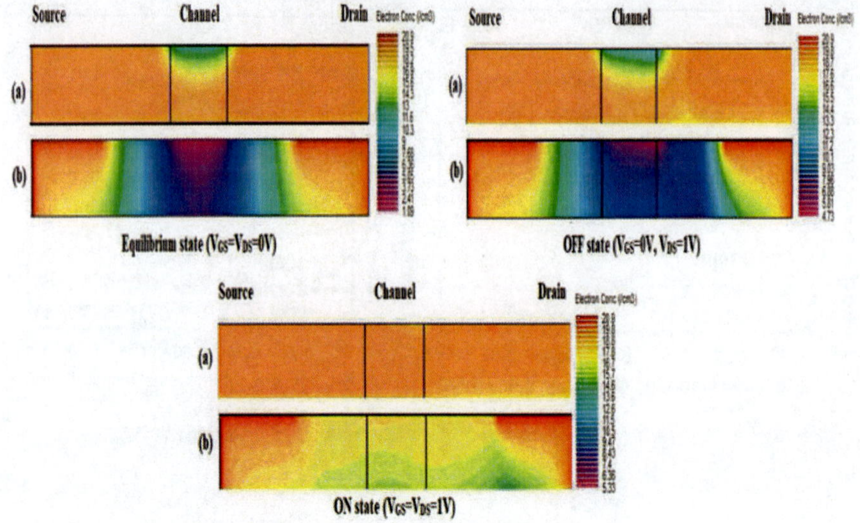

Fig. 3 Electrons contour plots of **a** SOIJLT and **b** VD-SOIJLT under thermal equilibrium, OFF-state, and ON-state

3.2 Hole Concentrations Profiles

Figures 4 and 5 illustrate graphical and contour representations of hole concentrations profiles of SOIJLT and VD-SOIJLT under thermal equilibrium, OFF-state, and ON-state, respectively. Thermal equilibrium condition, as shown by Fig. 4a, confirms the hole concentrations of VD-SOIJLT that is higher than SOIJLT inside the channel region (Fig. 5). SOIJLT has approximate equal hole concentrations throughout the S/D and channel regions. At the channel region, virtually doped devices have higher hole concentrations over its conventional counterpart under OFF-state condition which is responsible for lessening the I_{OFF} which is clearly depicted in Fig. 4b. In ON-state, again the hole redistribution occurred when applied a positive voltage (1 V) on a gate and drain terminals. Due to this, the hole concentration is higher for VD-SOIJLT over SOIJLT at the source and source-channel interface regions, but at channel-drain interface side, VD-SOIJLT has the lowest hole concentration due to maximum hole depletion at the channel-drain side which is shown by Fig. 4c.

3.3 Electrostatics Potential Profiles

Figures 6 and 7 illustrate graphical and contour representations of potential profiles of SOIJLT and VD-SOIJLT under thermal equilibrium, OFF-state, and ON-state, respectively. The thermal equilibrium state, exhibited by Fig. 6a, confirm the highest

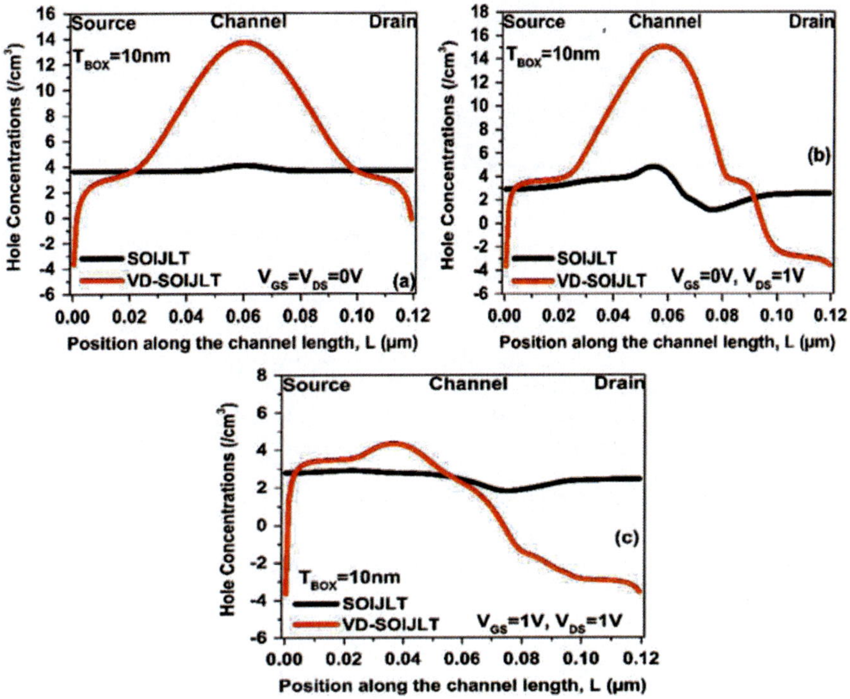

Fig. 4 Hole concentrations of SOIJLT and VD-SOIJLT under **a** thermal equilibrium, **b** OFF-state, and **c** ON-state

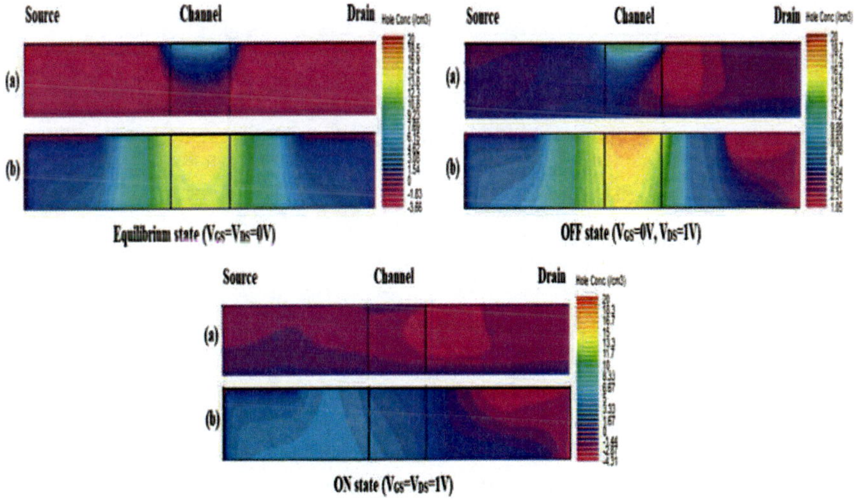

Fig. 5 Hole contour plots of **a** SOIJLT and **b** VD-SOIJLT under thermal equilibrium, OFF-state, and ON-state

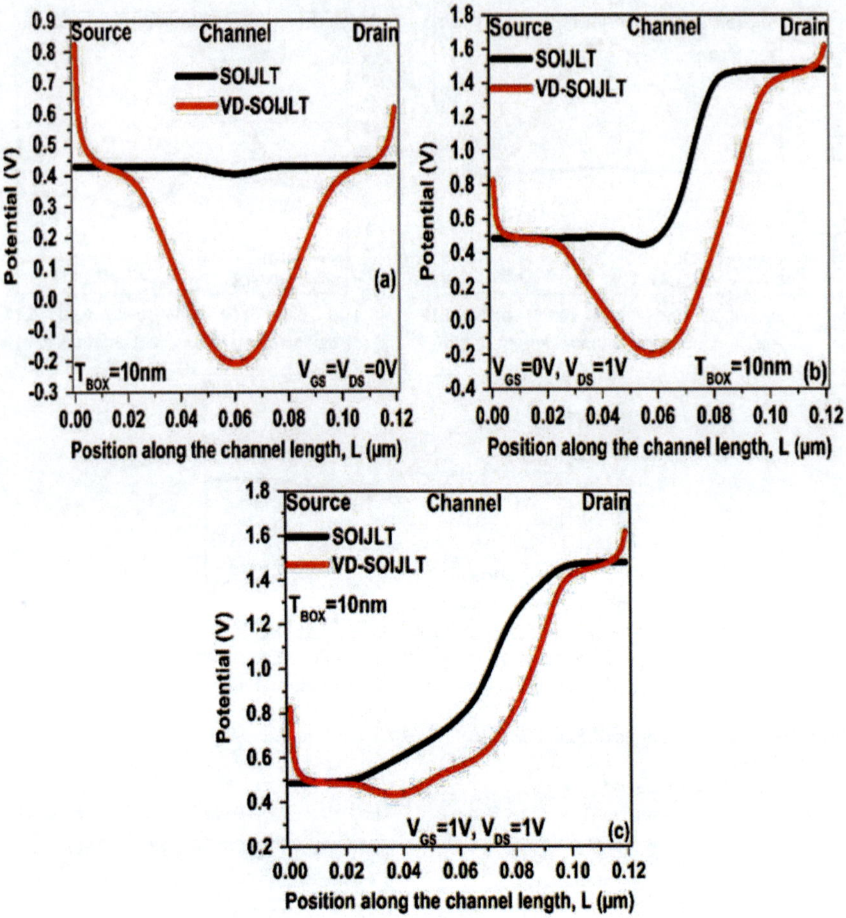

Fig. 6 Potential profiles of SOIJLT and VD-SOIJLT under **a** thermal equilibrium, **b** OFF-state, and **c** ON-state

S/D potential of dopingless device at *S/D* regions, but at the channel region, VD-SOIJLT offers minimum potential profile over SOIJLT at the center of the channel region (Fig. 7). In the OFF-state condition, at *S/D* regions, the potential of the dopingless devices increases, but shrinkage in the potential profile can be achieved at channel region under gate control of higher work function. The potential profile of all the devices increases toward drain side due to positive voltage applied on the drain terminal. VD-SOIJLT offers minimum potential over SOIJLT which is shown by Fig. 6b. In ON-state, at S/D regions, the electrostatic potential of VD device is moreover its conventional counterpart due to the use of lower work function metal electrode, but in the channel region, SOIJLT offers higher electrostatic potential, and VD-SOIJLT offers minimum potential. Potential profiles have increased and shifted

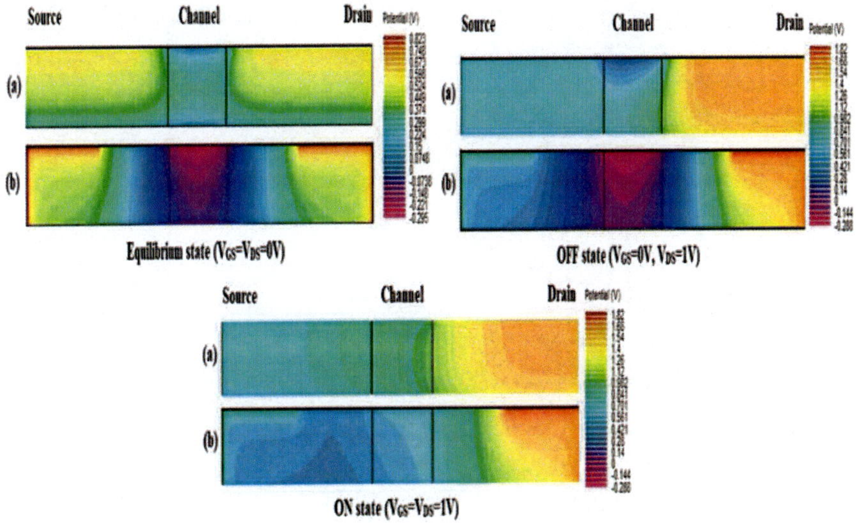

Fig. 7 Potential contour plots of **a** SOIJLT and **b** VD-SOIJLT under thermal equilibrium, OFF-state, and ON-state

toward the drain side due to the higher voltage applied at drain terminal which is shown by Fig. 6c.

3.4 *Transfer Characteristics ($I_D - V_{GS}$)*

Figure 8a demonstrates transfer characteristics of SOIJLT [8] and compared with the VD-SOIJLT. A considerably reduced OFF-state leakage current of VD-SOIJLT has achieved over conventional SOIJLT due to the efficient volume depletion at the bottom of the channel region. A charge plasma concept is employed on previously reported device structure [8], resulting in a more significant reduction in OFF-state leakage current over SOIJLT. Here, I_{ON} arises in SOIJLT as compared to VD-SOIJLT, but the main attentions of this work are to reduce the I_{OFF}, which ensuring net improvement in I_{ON}/I_{OFF} ratio. So, it enhances the switching capability in digital circuit applications. Here, virtually doped devices provide good drain current due to lower rates of Coulomb scattering because of undoped S/D and channel regions. VD-SOIJLT has lowest I_{OFF} (~10^{-15} A/μm) which is ~10 orders lesser than the conventional SOIJLT for $T_{BOX} = 10$ nm. For less OFF-state leakage current, VD-SOIJLT may be more preferable over SOIJLT. Maximum net improvement in I_{ON}/I_{OFF} ratio of the order of ~10^{10} has achieved by VD-SOIJLT which is ~10 order higher than SOIJLT. It is observed that at smaller L and smaller T_{BOX} based on VD-SOIJLT, deliver optimum DC characteristics in terms of SS, I_{OFF} and I_{ON}/I_{OFF} over conventional SOIJLT.

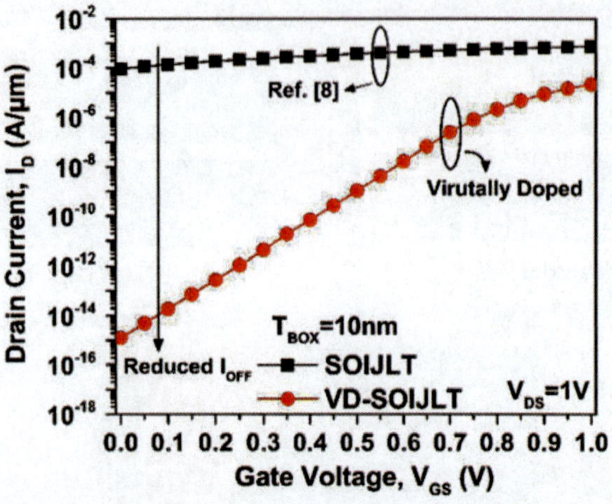

Fig. 8 Calibrated transfer characteristics (I_D-V_{GS}) of SOIJLT [8] with VD-SOIJLT; for $L = 20$ nm and $T_{BOX} = 10$ nm at $V_{DS} = 1$ V

3.5 *Transfer Characteristics of SOIJLT and VD-SOIJLT at* L = *20–100 nm and* T_{BOX} = *10–20 nm*

Figure 9a–c reveal I_D-V_{GS} of conventional SOIJLT and VD-SOIJLT as a variation in channel length (20–100 nm) and BOX thickness (viz 10, 15, and 20 nm). VD-SOIJLT has significantly reduced I_{OFF} and slightly reduced I_{ON} as compared to SOIJLT. As channel length has increased, the SS, I_{ON}, and I_{OFF} of both the devices are reduced, and I_{ON}/I_{OFF} ratio of both the devices increases. Here, for VD-SOIJLT, the optimum channel length of 60 nm is noticed due to the optimum DC characteristics over SOIJLT. For $L = 60$ nm with T_{BOX} variations, the DC characteristics in terms of optimum SS (nearly ideal), viz 61.084 mV/dec ($T_{BOX} = 10$ nm), 62.301 mV/dec ($T_{BOX} = 15$ nm), and 62.535 mV/dec ($T_{BOX} = 20$ nm) and optimum $I_{OFF} = 9.705 \times 10^{-18}$ A/μm ($T_{BOX} = 10$ nm) and optimum $I_{ON}/I_{OFF} = 1.619 \times 10^{12}$ ($T_{BOX} = 10$ nm), have obtained (Table 2). It is observed that at smaller T_{BOX} (10 nm), SOIJLT has reduced I_{OFF} and enhanced I_{ON}/I_{OFF} ratio over its larger T_{BOX} devices, and same trends are followed by the higher L devices over its lower L devices. For VD-SOIJLT having $L = 20$ nm, I_{OFF} is least (1.249×10^{-15} A/μm) for $T_{BOX} = 10$ nm, and I_{ON}/I_{OFF} ratio is highest (2.188×10^{10}) for $T_{BOX} = 10$ and 20 nm, respectively. For $L = 40$–100 nm variations, I_{OFF} is least (2.454×10^{-18} A/μm), and I_{ON}/I_{OFF} ratio is highest (9.912×10^{12}) for $L = 100$ nm and $T_{BOX} = 20$ nm.

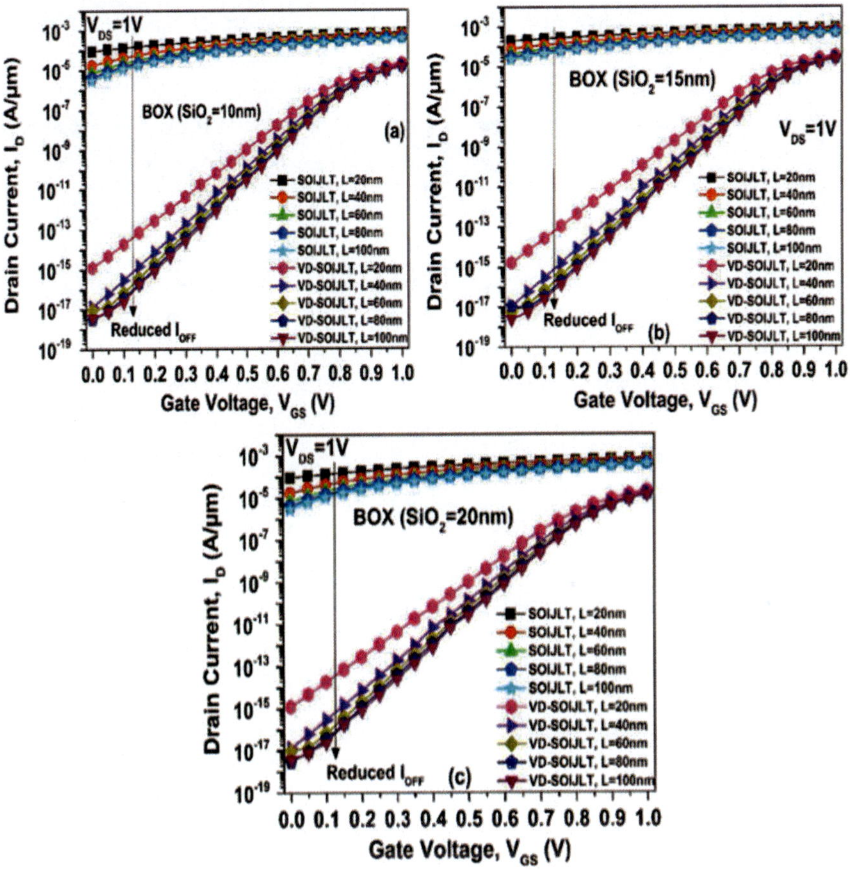

Fig. 9 Transfer characteristics of SOIJLT and VD-SOIJLT using SiO_2 BOX for different channel lengths, viz **a** $T_{BOX} = 10$ nm, **b** $T_{BOX} = 15$ nm, and **c** $T_{BOX} = 20$ nm at $V_{DS} = 1$ V

4 Conclusion

In this work, a virtually doped approach has been applied on SOIJLT for a meaningful reduction in OFF-state leakage current. Via 2D calibrated numerical calculations, it is realized that VD-SOIJLT has more perfect volume depletion over SOIJLT due to the depletion of active device layer from the bottom and inside the Si-channel, ensuring diminished effective active device layer over physical thickness. For the channel length of 20 nm, we have achieved that VD devices deliver further reduction in I_{OFF} of the order of (~10^{-15} A/μm) which is ~10 order lesser than I_{OFF} of conventional SOIJLT and considerably enhanced I_{ON}/I_{OFF} ratio of the order of ~10^{10} which is ~10 order higher than SOIJLT for $T_{BOX} = 10$ nm (Table 2). For VD-SOIJLT having $L =$ 60 nm with $T_{BOX} = 10$ nm, the DC characteristics in terms of optimum SS (nearly ideal), viz 61.084 mV/dec, optimum $I_{OFF} = 9.705 \times 10^{-18}$ A/μm, and optimum

Table 2 DC characteristics of SOIJLT and VD-SOIJLT (T_{BOX} = 10–20 nm and L = 20–100 nm)

T_{BOX}	Parameters	L = 20 nm	L = 40 nm	L = 60 nm	L = 80 nm	L = 100 nm
SOIJLT T_{BOX} = 10 nm	SS (mV/dec)	253.386	230.355	177.497	162.801	156.844
	I_{ON} (A/μm)	7.451×10^{-4}	6.049×10^{-4}	5.086×10^{-4}	4.377×10^{-4}	3.836×10^{-4}
	I_{OFF} (A/μm)	9.176×10^{-5}	1.716×10^{-5}	6.994×10^{-6}	4.210×10^{-6}	2.995×10^{-6}
	I_{ON}/I_{OFF}	0.081×10^{2}	0.352×10^{2}	0.727×10^{2}	0.103×10^{3}	0.128×10^{3}
VD-SOIJLT T_{BOX} = 10 nm	SS (mV/dec)	78.302	57.580	**61.084**	56.775	50.215
	I_{ON} (A/μm)	2.144×10^{-5}	1.721×10^{-5}	1.572×10^{-5}	1.489×10^{-5}	1.432×10^{-5}
	I_{OFF} (A/μm)	1.249×10^{-15}	1.324×10^{-17}	**9.705×10^{-18}**	2.904×10^{-18}	3.622×10^{-18}
	I_{ON}/I_{OFF}	1.716×10^{10}	1.298×10^{12}	**1.619×10^{12}**	5.129×10^{12}	3.955×10^{12}
SOIJLT T_{BOX} = 15 nm	SS (mV/dec)	82.036	39.091	30.428	27.637	26.379
	I_{ON} (A/μm)	8.482×10^{-4}	6.960×10^{-4}	5.907×10^{-4}	5.125×10^{-4}	4.519×10^{-4}
	I_{OFF} (A/μm)	1.657×10^{-4}	5.277×10^{-5}	2.744×10^{-5}	1.812×10^{-5}	1.344×10^{-5}
	I_{ON}/I_{OFF}	0.051×10^{2}	0.131×10^{2}	0.215×10^{2}	0.282×10^{2}	0.336×10^{2}
VD-SOIJLT T_{BOX} = 15 nm	SS (mV/dec)	76.971	55.520	**62.301**	55.024	51.769
	I_{ON} (A/μm)	2.905×10^{-5}	2.358×10^{-5}	2.162×10^{-5}	2.045×10^{-5}	1.959×10^{-5}
	I_{OFF} (A/μm)	1.376×10^{-15}	1.161×10^{-17}	5.587×10^{-18}	3.772×10^{-18}	2.884×10^{-18}
	I_{ON}/I_{OFF}	2.110×10^{10}	2.030×10^{12}	3.869×10^{12}	5.421×10^{12}	6.790×10^{12}
SOIJLT T_{BOX} = 20 nm	SS (mV/dec)	98.020	49.668	38.965	35.214	33.428
	I_{ON} (A/μm)	9.045×10^{-4}	7.464×10^{-4}	6.367×10^{-4}	5.547×10^{-4}	4.908×10^{-4}
	I_{OFF} (A/μm)	2.116×10^{-4}	8.057×10^{-5}	4.573×10^{-5}	3.137×10^{-5}	2.372×10^{-5}
	I_{ON}/I_{OFF}	0.042×10^{2}	0.092×10^{2}	0.139×10^{2}	0.176×10^{2}	0.206×10^{2}
VD-SOIJLT T_{BOX} = 20 nm	SS (mV/dec)	76.885	55.174	**62.535**	54.519	50.884
	I_{ON} (A/μm)	3.550×10^{-5}	2.920×10^{-5}	2.689×10^{-5}	2.544×10^{-5}	2.432×10^{-5}
	I_{OFF} (A/μm)	1.622×10^{-15}	1.105×10^{-17}	5.769×10^{-18}	9.268×10^{-18}	**2.454×10^{-18}**

(continued)

Table 2 (continued)

T_{BOX}	Parameters	$L = 20$ nm	$L = 40$ nm	$L = 60$ nm	$L = 80$ nm	$L =$ 100 nm
	I_{ON}/I_{OFF}	2.188×10^{10}	2.640×10^{12}	4.661×10^{12}	2.745×10^{12}	**9.912×10^{12}**

$I_{ON}/I_{OFF} = 1.619 \times 10^{12}$, have obtained. For $L = 100$ nm and $T_{BOX} = 20$ nm, VD-SOIJLT has least I_{OFF} (2.454×10^{-18} A/μm) and highest I_{ON}/I_{OFF} (9.912×10^{12}) (Table 2). Therefore, the VD-SOIJLT may be preferred as an alternative device over conventional SOIJLT with the advantage of minimum OFF-state leakage current and enhanced I_{ON}/I_{OFF} ratio, which may be beneficial in digital circuit applications.

References

1. Colinge JP, Lee CW, Afzalian A, Akhavan ND, Yan R, Ferain I, Razavi P, O'neill B, Blake A, White M, Kelleher AM (2010) Nanowire transistors without junctions. Nature Nanotechnol 5(3):225
2. Lee CW, Ferain I, Afzalian A, Yan R, Akhavan ND, Razavi P, Colinge JP (2010) Performance estimation of junctionless multigate transistors. Solid-State Electron 54(2):97–103
3. Gnani E, Gnudi A, Reggiani S, Baccarani G (2011) Theory of the junctionless nanowire FET. IEEE Trans Electron Devices 58(9):2903–2910
4. Lee CW, Afzalian A, Akhavan ND, Yan R, Ferain I, Colinge JP (2009) Junctionless multigate field-effect transistor. Appl Phys Lett 94(5):053511
5. Gundapaneni S, Ganguly S, Kottantharayil A (2011) Bulk planar junctionless transistor (BPJLT): an attractive device alternative for scaling. IEEE Electron Device Lett 32(3):261–263
6. Gundapaneni S, Ganguly S, Kottantharayil A (2011) Enhanced electrostatic integrity of short-channel junctionless transistor with high-k spacers. IEEE Electron Device Lett 32(10):1325–1327
7. Gundapaneni S, Bajaj M, Pandey RK, Murali KV, Ganguly S, Kottantharayil A (2012) Effect of band-to-band tunneling on junctionless transistors. IEEE Trans Electron Devices 59(4):1023–1029
8. Sahay S, Kumar MJ (2016) Realizing efficient volume depletion in SOI junctionless FETs. IEEE J Electron Devices Soc 4(3):110–115
9. Sahay S, Kumar MJ (2019) Junctionless field-effect transistors: design, modeling, and simulation. Wiley, Hoboken
10. Narula V, Agarwal M (2019) Enhanced performance of double gate junctionless field effect transistor by employing rectangular core–shell architecture. Semicond Sci Technol 34(10):105014
11. Leung G, Chui CO (2012) Variability impact of random dopant fluctuation on nanoscale junctionless FinFETs. IEEE Electron Device Lett 33(6):767–769
12. Hueting RJ, Rajasckharan B, Salm C, Schmitz J (2008) The charge plasma PN diode. IEEE Electron Device Lett 29(12):1367–1369
13. Rajasekharan B, Hueting RJ, Salm C, van Hemert T, Wolters RA, Schmitz J (2010) Fabrication and characterization of the charge-plasma diode. IEEE Electron Device Lett 31(6):528–530
14. Kumar MJ, Nadda K (2012) Bipolar charge-plasma transistor: a novel three terminal device. IEEE Trans Electron Devices 59(4):962–967
15. Kumar MJ, Janardhanan S (2013) Doping-less tunnel field effect transistor: Design and investigation. IEEE Trans Electron Devices 60(10):3285–3290

16. Sahu C, Singh J (2014) Charge-plasma based process variation immune junctionless transistor. IEEE Electron Device Lett 35(3):411–413
17. Singh S, Raman A (2018) Gate-all-around charge plasma-based dual material gate-stack nanowire FET for enhanced analog performance. IEEE Trans Electron Devices 65(7):3026–3032
18. Silvaco (2018) Atlas user's manual. Silvaco, Inc., pp 567–1000

Improved Store-Carry-Forward Scheme for Information Dissemination in Unfavorable Vehicular Distribution

Ankita Srivastava, Arun Prakash, and Rajeev Tripathi

Abstract Vehicular network assisting cooperative driving have gained much attention because they offer wide range of new possibilities to the modern intelligent transportation system (ITS). To enhance safety-related applications, vehicles conventionally broadcast warning message as soon as it finds hazardous situation. However, research work regarding data packet routing usually eliminates the effect of adverse density condition, i.e., density values that significantly varies from the average values, in spite such variations are quite common in urban environment. Conventionally, a store-and-carry forward scheme was utilized to handle such fragmented network issue, but with a large delay. In the work, a novel improved store-carry-forward scheme to overcome partitioned network is proposed. The simulation results shows that the modified scheme significantly reduces delay and improves delivery rate in low traffic density condition.

Keywords Vehicular network · Store-carry-forward scheme · Delivery rate · Delay · Partitioned network

1 Introduction

Modern intelligent transportation systems (ITS) are in progression by the influence of wireless communications and computation technologies, therefore enabling roads and highways to be efficient transportation platform [1]. A vehicular ad hoc network (VANET) is a wireless telecommunication network that permits cooperative driving

A. Srivastava (✉) · A. Prakash · R. Tripathi
Department of Electronics and Communication Engineering, Motilal Nehru National Institute of Technology Allahabad, Prayagraj, Uttar Pradesh 211004, India
e-mail: rel1708@mnnit.ac.in

A. Prakash
e-mail: arun@mnnit.ac.in

R. Tripathi
e-mail: rt@mnnit.ac.in

D. Harvey et al. (eds.), *Advances in VLSI, Communication, and Signal Processing*, Lecture Notes in Electrical Engineering 683,
https://doi.org/10.1007/978-981-15-6840-4_54

between vehicles. In VANET, vehicles functions as a communication nodes and relay, creating a dynamic network with other neighboring vehicles. VANETs have special features such as rapid variations, frequent fragmentation, and variable density [2].

Although VANET is recognized by various kinds of needful applications, it has few typical challenges yet to be resolved, such as connectivity. In VANET, the dynamic nature of vehicles creates unstable topology, and as a result, the connectivity is hardly steady in both time scale and space scale. Majorly in sparse network, limited numbers of vehicles are present in large space; therefore, the network gets disconnected more often [3]. Conventional routing methods are developed for fully connected network that mainly aims to maintain end-to-end connectivity among two desired vehicles [4]. Therefore, they fail to transmit data packets in sparse or partially connected network. In consideration of this disconnectivity issue, delay tolerant network was introduced where the delivery of packets was augmented by permitting vehicles to store the packets when no other option is available, and carry it for some distance until a feasible opportunity is available. This method is popularly known as carry-and-forward strategy [5–7]. Some of the routing protocols were developed to fulfill the necessities of sparse connectivity. Delay tolerant VANET (DTVANET) supports vehicular network that is specified by delay tolerance and asynchronous data traffic. Some DTVANET routing protocols are:

Vehicle-assisted data delivery (*VADD*) [8] reduces end-to-end delay by establishing connection from mobile vehicle toward the intended static destination. It brings into play carry-and-forward method based on predictable mobility of the vehicles that is restricted by traffic pattern and road layout. Generally, this method cannot assure multi-hop connectivity, mainly, if some abrupt variations in the road traffic flow occur [9]. *Reliable routing protocol* (*RRP*) [10] assumes that road side units (RSU) are present at each intersection which selects the road segment (block) so as to reach destination. Among the paths, the shortest one is preferred where no routing holes are present. RRP do not explores new route, instead reroutes the packets toward other block. *Reactive pseudo-suboptimal-path selection routing protocol* (*RSP*) [11] gives opportunity to the previously traversed intersection to renew its path selection decision in spite of following carry-and-forward mechanism. The simulation results depict that the overhead count is high in this method. GeOpps [12] takes benefit of the routes provided via navigation system that is embedded inside the vehicles to choose other vehicle that is likely to progress toward the destination. Inspired from GeOpps, GeoSpray [13] optimizes resource utilization such as storage, bandwidth, and energy. It follows spray phase and wait mechanism [14], where a small set of packet copies are sprayed to the vehicles which are found nearby destination. This method lacks simulation environment implementation which restricts its performance. *Opportunistic routing based on symmetrical traffic distribution* (*OSTD*) [14] utilizes vehicular traffic condition and drivers path to discover the most optimum route. GeoDTN + NAV [15] make use of vehicular mobility and onboard navigation system to work effectively even in fragmented network by utilizing store-carry-forward method whenever required. Limitation of this method is that the destination is assumed as static. However, carry-and-forward scheme may cause increased delay.

Therefore, to avoid frequent utilization of carry-and-forward scheme, *connectivity-aware routing* (CAR) [4] selects a road segment with less disconnection. As traffic lights play a vital role in creating clusters, the evaluation of cluster count can provide connectivity status of the road. Road-aware geographic routing protocol (RAGR) [16] handles efficient route formation and disconnectivity issue simultaneously. For network disconnectivity, store-carry-forward procedure is adopted simply. For efficient route formation, distance, direction, and traffic density metrics are used to make decision. The mechanism of reliable location-aware routing protocol (RLARP) [17] is similar to RAGR with a slight variation in route formation strategy. A two-level process is introduced to improve delivery of packets and reduce delay.

From the aforementioned routing protocols, majorly, simple store-carry-forward mechanism is used to get out of fragmented connectivity. However, it is quite challenging to obtain better delivery with less delay when the frequency of disconnectivity is high. As transmitting packets through wireless media is much faster to take it toward the destination as compared to vehicular speed [14], it is found that network disconnectivity in DTVANET is still a challenging task.

Due to the limited contribution mainly designed for adverse density condition, in this work, the main objective is to contribute a study on the transmission of data packets in fragmented network. The proposed method is an extension of RLARP, so as to reduce delay and packet drop rate in VANET. The main highlights of the paper are:

1. In this work, when the current vehicle is incapable to tracking the most suitable next hop, it adopts improved store-carry-forward (ISF) mechanism to increase efficiency. Suitability of next hop varies from protocol to protocol, like for greedy mechanism, the most suitable protocol is that which is closest to the destination.
2. In improved store-carry-forward (ISF) mechanism, primarily, the vehicle in the backward direction with high velocity is searched. If not found, the current forwarder vehicle increases its transmission range to discover the next hop. Preferably, vehicle that is predicted to move toward the destination in near future is selected as the next hop.
3. The performance is compared with existing RAGR and RLARP routing algorithms via network simulator (ns 2.34), and the results represents its effectiveness in terms of packet delivery ratio, throughput, and delay.

The rest paper is arranged as follows: Sect. 2 contains the proposed improves store-carry-forward algorithms. In Sect. 3 the performance evaluation of the protocol has been done with respect to RAGR and RLARP, and Sect. 4 concludes this work.

2 Proposed Work

2.1 Assumptions

Conventionally, vehicles periodically broadcast beacons to interchange its local information, which is utilized to gain knowledge about the surrounding area. To resolve network partition problem, this work makes the following assumptions:

- Global positioning system (GPS)-enabled vehicles.
- Depends only on one-hop neighbor status.
- Information exchange among vehicles involves location, velocity, moving direction, and previously obtained warning message.
- This scheme maintains a neighbor list that is revised periodically by means of beacon messages as a replacement of roadmap information.
- Signal transmission is not obstructed by obstacles such as trees and buildings.

2.2 Improved Store-Carry-Forward (ISF) Mechanism

With an aim to maximize the performance of conventional store-carry-forward scheme in sparse density distribution, this work presents improved store-carry-forward (ISF) mechanism. In this scheme, when a vehicle needs to transmit a warning message, but is incapable of finding the appropriate vehicle within its range in the forward direction, then in spite of blindly carrying the message, it first tries to find the better option. Primarily, to start the process, a vehicle must be well aware of its neighbors. For this, it is considered that all vehicles maintain updated neighbor list by means of beacons, avoiding any channel overhead. Whenever a beacon packet is received, the vehicle verifies its neighbor list to decide if the sender is new, in such condition, it will include this new information to the list. In case when no beacon packets are received from an already placed vehicle after 3 s from the time of adding, then that neighbor is removed.

A vehicle after obtaining a warning message first confirms from its neighbor list, whether it is a SCF vehicle of not. A vehicle assumes itself to be a SCF vehicle, when it resides at the edge of cluster, i.e., no vehicle is located within its range in its heading direction. Figure 1 illustrates the methodology adopted for assigning SCF vehicle as depicted by ISF mechanism.

Here, source S finds itself as the SCF vehicle because no other vehicle is found in its vicinity (shown by black dotted circle) in the forward direction. Now, if vehicle S will start carrying the message until it finds the better encountering opportunity, then the data dissemination will face high delay. Therefore, fundamentally the presence of any neighbor in the backward direction with higher velocity is verified from the neighbor table. If successfully found, then data packet is expected to be stored and carried by that neighbor. Otherwise, the current forwarder vehicle extends its transmission range to discover the appropriate vehicle that can be trusted to efficiently

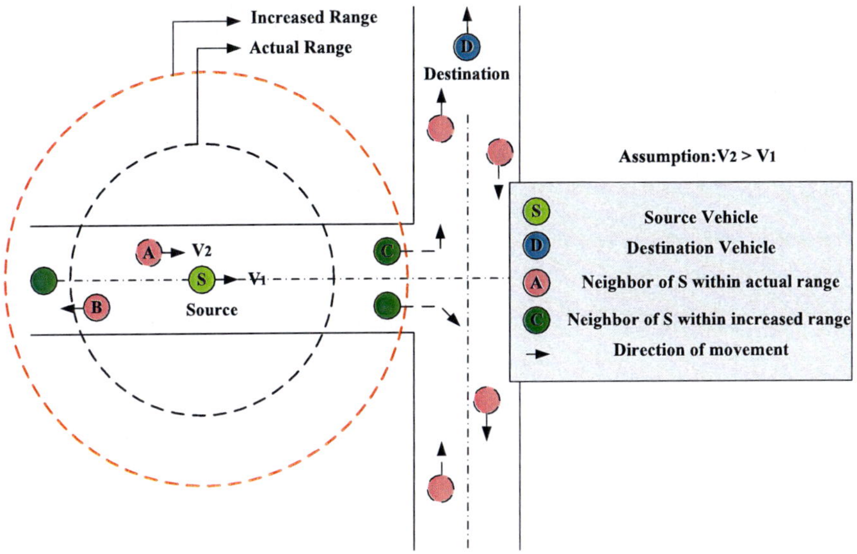

Fig. 1 Improved store-carry-forward (ISF) mechanism

carry the packets toward the destination. As it can be seen from Fig. 1, when the S increases its range, vehicle C is the most appropriate forwarder, because its direction of movement is expected to be toward the destination. The direction of movement of the given vehicle is obtained from its movement history. Figure 2 represents the flowchart of ICF mechanism.

3 Performance Evaluation

Under this section, the performance of the proposed algorithm will be estimated against the most recent routing protocol RAGR and RLARP by utilizing the metrics defined in Table 1. The simulation was carried using network simulator (NS) 2.34 over urban region that comprises nine intersections and twelve interconnecting roads as represented in Fig. 3. Additionally, Table 2 embraces other necessary parameters involved in the simulation.

3.1 Impact of Varying Vehicular Density on Packet Delivery Ratio

Figure 4 depicts PDR response of the simulated protocol for different vehicle densities. In all the three cases, PDR improves with increasing vehicle density, as higher

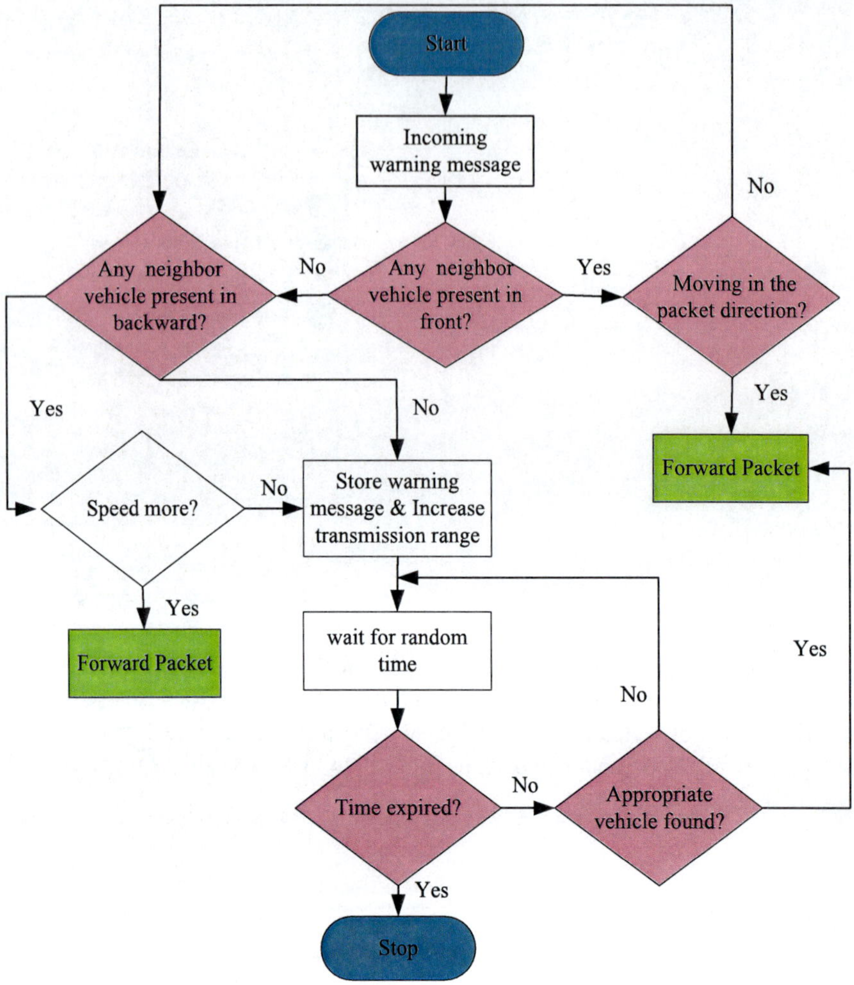

Fig. 2 ISF method working flowchart

amount of vehicles contribute better connectivity. In IFS, the packet drop rate due to carry-and-forward mechanism is generally lower, so it offers better delivery than the other two protocols. In the other two protocols, when the network encounters gap, the forwarder vehicle starts carrying the packet until a better forwarding opportunity is obtained. Therefore, in some situation, the vehicle might be incapable of tracking next hop for a longer time that exceeds from packet time-to-live (TTL), which drops the packet.

Table 1 Performance metrics

Metrics	Explanation
Packet delivery ratio (PDR)	It is defined as the data packet count that is delivered at the needful destination successfully. The PDR offered by the routing protocol can be evaluated by dividing the amount of packets collected at the destination to the total packets generated from the source
End-to-end delay	This parameter include the all feasible delays occurred as a result of buffering at the time of route discovery, queuing, retransmission, and propagation. It represents the time that data packets takes to move from source to the destination successfully
Throughput	It implies that in a certain amount of time, total number of useful packets obtained by the destinations. The unit of throughput can be Mb/s or kb/s

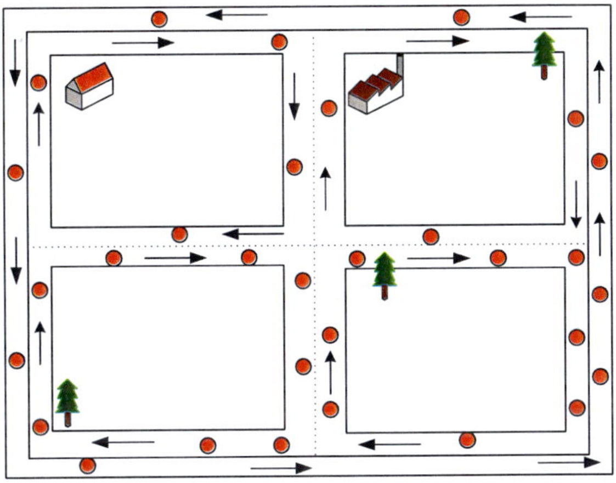

Fig. 3 Part of road scenario

Table 2 Simulation parameters

Parameters	Values	Parameters	Values
Simulation time	200 s	Physical layer	IEEE 802.11p
Scenario size	2×2 km^2	Propagation model	Nakagami radio propagation model
Packet transmission rate	1 packet/s	Data packet size	512 bytes
Transmission range	250 m	Vehicular density	60–120
CBR connection pairs	15	Speed of vehicles	10–20 m/s

However, in IFS, frequency of dealing network disconnectivity is fast, as the vehicle adjusts speed or transmission range to discover next hop quickly. The performance beyond 100 is almost constant because above that value the intermittent connectivity is always available.

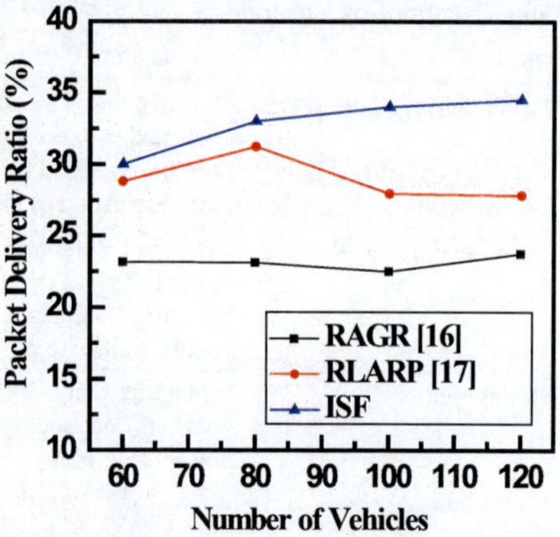

Fig. 4 Packet delivery ratio versus number of vehicles

3.2 *Impact of Varying Vehicular Density on End-to-End Delay*

The idea imposed behind IFS is to adopt a procedure that can tackle network gap issue quickly so as to reduce delay and packet drop rate. Both of these objectives are fulfilled when the next hop is discovered rapidly before the packet TTL expires. The proposed ICF routing protocol finds the next hop to meet the objective by adjusting speed and transmission range. For that reason, the end-to-end delay performance of ICF protocol is better which is shown in Fig. 5.

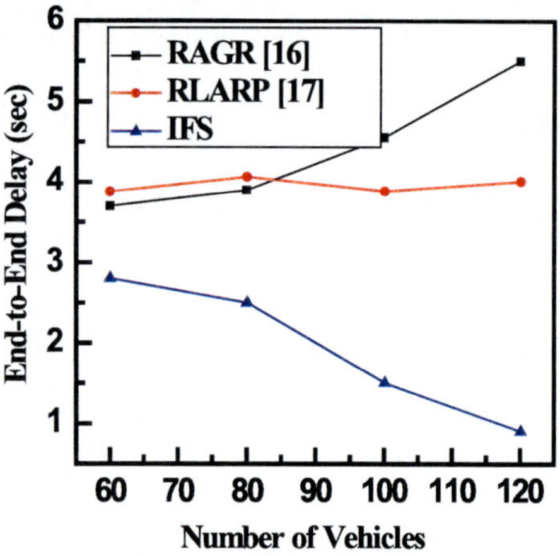

Fig. 5 End-to-end delay versus number of vehicles

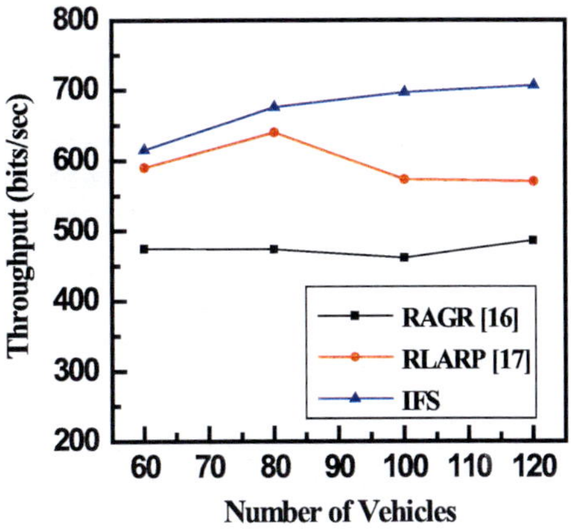

Fig. 6 Throughput versus number of vehicles

3.3 *Impact of Varying Vehicular Density on Throughput*

According to the definition of throughput, the protocol that can efficiently deliver packet at maximum rate in a given time has a high throughput. From the above discussion, it is obvious that the performance of IFS will be more efficient in terms of rate in comparison with the other two cases which can be seen from Fig. 6.

4 Conclusion

This paper presents improved store-carry-forward (ISF) protocol for packet forwarding in delay tolerant VANET network. In VANET, the previous work mostly inspects the heterogeneous traffic condition before transmitting the data packets between the road segments. Since, this estimation cannot guarantee connectivity, the existing work also utilizes store-carry-forward mechanism in such situation. However, delay offered by simple store-carry-forward is high when the routing protocol is incapable of tracking next hop for a long time. Also, sometimes, the next hop is selected blindly that instead of taking packets closer, the forwarder carry it away from the destination. In the proposed algorithm, the parameters such as velocity, transmission range, and movement history of the vehicles are changed accordingly to reduce delay and packet drop rate. The performance of the proposed protocols has been compared with the existing work in RAGR [16] and RLARP [17]. The results depict improvement in terms of packet delivery ratio, average end-to-end delay, and throughput with different vehicle density.

References

1. Nguyen TD, Le TV, Pham HA (2017) Novel store–carry–forward scheme for message dissemination in vehicular adhoc networks. ICT Express 3(4):193–198
2. Yousefi S, Mousavi MS, Fathy M (2006) Vehicular ad hoc networks (VANETs): challenges and perspectives. In: 2006 6th international conference on ITS telecommunications. IEEE, pp 761–766
3. Tonguz OK, Wisitpongphan N, Bai F (2010) DV-CAST: a distributed vehicular broadcast protocol for vehicular ad hoc networks. IEEE Wirel Commun 17(2):47–57
4. Yang Q, Lim A, Agrawal P (2008) Connectivity aware routing in vehicular networks. In: Wireless communications and networking conference
5. Lee KC, Lee U, Gerla M (2010) Survey of routing protocols in vehicular ad hoc networks. In: Advances in vehicular ad-hoc networks: developments and challenges. IGI Global, pp 149–170
6. Li F, Wang Y (2007) Routing in vehicular ad hoc networks: a survey. IEEE Veh Technol Mag 2(2):12–22
7. Oka H, Higaki H (2008) Wireless multihop communication in sparse vehicle ad-hoc networks. In: 2008 international conference on innovations in information technology. IEEE, pp 573–577
8. Zhao J, Cao G (2008) VADD: vehicle-assisted data delivery in vehicular ad hoc networks. IEEE Trans Veh Technol 57(3):1910–1922
9. Boussoufa-Lahlah S, Semchedine F, Bouallouche-Medjkoune L (2018) Geographic routing protocols for Vehicular Ad hoc NETworks (VANETs): a survey. Veh Commun 11:20–31
10. Kim JH, Lee S (2011) Reliable routing protocol for vehicular ad hoc networks. AEU-Int J Electron Commun 65(3):268–271
11. Wang X, Li C, Zhu L, Zhao C (2013) An effective routing protocol for intermittently connected vehicular ad hoc networks. In: IEEE wireless communications and networking conference (WCNC), pp 1750–1755
12. Leontiadis I, Mascolo C (2007) GeOpps: geographical opportunistic routing for vehicular networks. In: 2007 IEEE international symposium on a world of wireless, mobile and multimedia networks. IEEE, pp 1–6
13. Soares VN, Rodrigues JJ, Farahmand F (2014) GeoSpray: a geographic routing protocol for vehicular delay-tolerant networks. Inf Fusion 15:102–113
14. Mirjazaee NN, Moghim N (2015) An opportunistic routing based on symmetrical traffic distribution in vehicular networks. Comput Electr Eng 47:1–2
15. Cheng PC, Weng JT, Tung LC, Lee KC, Gerla M, Haerri J (2008) GeoDTN + Nav: a hybrid geographic and DTN routing with navigation assistance in urban vehicular networks. In: MobiQuitous/ISVCS
16. Qureshi KN, Abdullah AH, Altameem A (2017) Road aware geographical routing protocol coupled with distance, direction and traffic density metrics for urban vehicular ad hoc networks. Wireless Pers Commun 92(3):1251–1270
17. Srivastava A, Prakash A (2018) Reliable location-aware routing protocol for urban vehicular scenario. In: Advances in VLSI, communication and signal processing (VCAS 2018). https://doi.org/10.1007/978-981-32-9775-3_2

Information Theory-Based Defense Mechanism Against DDOS Attacks for WSAN

Jyoti Bhola and Surender Soni

Abstract In this paper, an energy-efficient secure protocol (ESP) for sensor and actor networks has been proposed to handle distributed denial-of-service (DDOS) attacks. The DDOS is a spy game among the detectors and attackers. It degrades services to legitimate users by exhausting the resources of the target node. In this work, information theory concepts are used to handle the DDOS attacks. In the proposed mechanism, when a large number of packets reaches to the sensors which are 1-hop (relay sensor's) away from the cluster head, they calculate the distance among the suspicious flows of different paths to the actor (victim). The suspicious flows can be confirmed as an attack if the distance among them are almost the same. NS2 is used for simulating the proposed protocol, and from the simulation results, it is evident that the proposed protocol performs better by discarding the attack packets from the network, hence reduces attack impact.

Keywords Actor · Sensor · Relay node · DDos · NS2

1 Introduction

The wireless sensor and actor (WSAN) network is an emerging area in the field of wireless communication to monitor, surveillance and acquisition for the critical as well as remote environments such as agriculture, health, home, military, and industry. [1–4]. The recently focused research in the field of miniaturization of electronic devices opens the door for new applications of WSAN such as body area networks (BANs), where the health monitoring of the patient is performed via sensor nodes, and if any discrepancy found, then immediate action is taken such as drug injection, break the blood clots, etc. Moreover, the small size of the sensor devices and battery-operated nature put a constraint on energy consumption. Therefore, energy-efficient

J. Bhola (✉) · S. Soni
Department of Electronics & Communication Engineering, National Institute of Technology Hamirpur, Hamirpur, India
e-mail: jyoti1may@gmail.com

D. Harvey et al. (eds.), *Advances in VLSI, Communication, and Signal Processing*, Lecture Notes in Electrical Engineering 683,
https://doi.org/10.1007/978-981-15-6840-4_55

WSANs are desired. Therefore, the energy-efficient nature of wireless sensor and actor network is also a milestone which restricts to implement the security techniques used in conventional networks [5, 6]. Most of the WSAN is implemented in hostile environments such as BAN and military environments where security is a prime concern. However, due to the broadcasting nature of WSAN, these are susceptible to several attacks. The key security concerns of the network are integrity, authentication, confidentiality, availability and access control, etc. [7]. The integrity ensures that the transmitted and received information are same and are not modified in the way. The authentication is regarding the receiving end and ensures that the receiver is authentic to whom the data is intended. The confidentiality refers to the ability of the network, so that contents of the message are understandable to only the intended node. The availability is defined to ensure that services need to be available for the authorized nodes. The access control is concerning providing the access only to the authorized nodes.

The attacks in the WSAN are generally of two types, i.e., passive and active attacks [7]. The passive attacks are defined as when the malicious/attacking nodes (MN) hear the message signal and do not affect the communication of authorized/legitimate nodes. These attacks affect the confidentiality of the WSAN since the information is accessible to the malicious nodes. Due to the unaffected nature of the authorized communication, passive attacks are challenging to detect since these do not affect the authorized communication. On the other hand, the active attacks affect the performance of authorized communication either by modifying or disrupting the information massage between transmitter and receiver. The Denial-of-Services (DoS) is the one among the primary active attacks, which refers to the failure of the communication nodes inadvertently and owing to malicious user attack [8–10]. This attack targets the availability of the WSAN by stopping communication between nodes or by avoiding a single node from sending traffic.

The DoS attacks are possible at several layers of the open system interconnection (OSI) model, as mentioned in the [11, 12]. Further, the distributed nature of WSAN to achieve energy efficiency by avoiding the information exchange with the centralized/controlling unit (CU) becomes a popular approach [13]. However, the distributed nature of WSAN makes the network vulnerable to new attack, i.e., distributed DoS (DDoS). In DDoS attack the compromised sensor nodes affected by the malicious attack act simultaneously and are controlled by a single attacking node via flooding the target nodes with false requests that exhaust the assets, and deny services of the legitimates sensor nodes forcefully defined as DDoS attack in WSAN. [14–18].

One popular way to perform the DDoS is to mimic the network traffic pattern by attacking node to hinder the detection algorithms [19]. The discrimination of the imitating DDoS attacks from massive legitimate network nodes is required, which is an exciting research area. In an emergency/ disaster scenario, the sensors will send a large volume of packets, if the detection algorithm blocks the communication as a DDoS attack, then it will cause more problems in the network. On the other hand, right attacks can access the victim by using many different paths with standard packet rate and distribution for every path. In this scenario, detection algorithms cannot detect the attack.

In this paper, we have put one step ahead to solve the above problem with the help of information theory. In the proposed approach, a large number of packets are reached to the sensors, which are 1-hop (relay sensor's) away from the cluster head (actor). The relay sensor's of the cluster calculates the distance among the chary data flows to the probable victim on various paths. If the distance among the chary flows is almost the same, then we can claim it as an attack.

2 Related Work

The DOS and DDOS attacks are the key milestones for the implementation of WSAN in the security-constraint applications such as military and body area networks which are vulnerable to the attacks intentionally. Various researchers have investigated the DDOS attacks in the wireless sensor networks and presented the possible way to diminish the effect of this attack [11, 14, 15, 20–22].

The authors in [11] have illustrated the various DDOS attacks and potential defenses in the wireless sensor networks (WSN). The key focus of the authors is to define the DOS attacks under the open system interconnection (OSI) model. The DOS service attacks at different layers are described, and prominent techniques to avoid or diminish the effect of these attacks are presented.

Yu et al. [20] have proposed a technique to discriminate the DDoS attacks from gush rightful accessing and detecting the attacks in the initial stage. The authors have exploited the various attack flows of one attack share a similar attack pattern that is not obsessed by rightful accessing flows in a small period. The potential benefit of the proposed scheme is that it avoids the storage load at the sensor node, which is the demand of WSNs. In [21], the authors have presented a profile-based protection scheme (PPS) next to the DDoS attack. The key concern of the security scheme is to visualize the effect of DDoS attack in WSN and recognize the sensor nodes which affect the WSN performance. This scheme verifies the profile of each node in WSN and detects the malicious sensor node by analyzing the extra flooding of message signal since only attacker/ MN transmits the flooded massage such hello messages, etc., to degrade the performance of the network. The performance metrics used to analyze the performance of the proposed security mechanism are routing load, throughput, etc.

In [15], the authors have illustrated the multilevel analysis of the DDoS attack problem and have presented eight scenarios where different security levels, i.e., encryption and no encryption and a diverse number of compromised devices are considered. Further, the performance and energy consumption of the sink node is analyzed under the DDoS attack. By examining the results gathered for prepared simulations, we identified a new kind of DDOS attack. It is reported that the security levels can be adjusted on the type of DDOS attack to avoid the various types of attacks. Moreover, in certain conditions, it is possible to prevent DDoS or delay its occurrence by lowering the security level.

Abidoye and Obagbuwa [14] have presented a security scheme to perceive and protect next to the DDoS attacks known as message analyzer scheme (MAS). This scheme has exploited the hash function and encryption to promise the data authenticity and integrity of the WSN and exploited the pre-shared keys to encrypt the message before it is passed to the receiving end. It is claimed and shown through simulations that the MAS outperforms to perceive and protect next to the DDoS attacks in WSANs.

Liu and Li [22] have investigated a protected distributed estimation problem for the WSNs by considering the malicious attacks on sensor nodes and communication. To tackle this attack, the authors have proposed a secure distributed estimation algorithm, known as "secure diffusion least mean squares" (S-dLMS). The S-dLMS is defined as a hybrid approach that comprises a "non-cooperative LMS" (nc-LMS) and "diffusion LMS" (dLMS) subsystems. The nc-LMS provides a reliable reference estimate; however, the dLMS subsystem is used to detect the protected network topology for attaining secure data fusion. Further, the authors have examined the performance of proposed S-dLMS algorithm and have developed an adaptive rule to select the threshold. Moreover, the S-dLMS is exploited to assess metric vectors from the various WSNs scenarios under a fixed, an elementwise as well as a time-varying attack. It is reported via simulation results that the proposed S-dLMS performs well for the malicious attacks. The scheme presented in [20] appears to be active and prominent technique due to since it avoids the storage load at the sensor node, which is prime requirements of the energy-efficient protocol designs for security systems. By keeping this in mind, we have further explored the concept of avoiding the DDoS attack by using the information theory perspectives.

3 Proposed Work

Distributed denial of service (DDoS) is a spy-on-spy game among the detectors and attackers. This attack exhausts resources of the target node by sending a large volume of dummy packets and does not allow services to the normal users. It is the advanced form of denial-of-service (DOS) attack. In a DoS attack, a single zombie host targets a legitimate node. On the other hand, in the DDOS attack, a large number of zombie hosts target a single node. Further, it is challenging to detect the DDOS attack as compared to the DOS attack. In a DDOS attack, the attackers try to mimic the traffic patterns to make it hard for a detection algorithm [23, 24]. Hence, a lot of research has happened on DOS attacks. However, research in the DDOS attack is still in an early stage. In WSAN, due to massive traffic in the network, it is a crucial problem to discriminate the DDOS attack with normal traffic.

In wireless sensor and actor networks, a huge number of sensors and few actors will be deployed in the target location. The sensors trace the events and send them to the nearest actor. The actor performs an action based on the information collected from the sensor. In WSAN, DDOS attack is hazardous as it exhausts the resources of sensors like the battery, computation and communication. The sensors need to send their data to the nearest actor as early as possible to perform actions on the target location effectively. An attacker can take control of a few sensors to target the actor. These zombie sensors can target an actor by sending a massive volume of data. Hence, the actor cannot communicate with legitimate sensors, which degrade the performance of the network.

It is essential to distinguish between real traffic and DDOS attack. In an emergency/ disaster scenario sensors will send a large volume of packets, if the detection algorithm blocks the communication as a DDoS attack, then it will cause more problems in the network. In the attack scenario, a large amount of data reaches the target node through various paths with standard packet rate. In this scenario, detection algorithms cannot detect the attack. In this work, we are concentrating on solving the above problem with the help of information theory. In the proposed mechanism, once a large number of packets are reached to the sensors which are 1-hop (relay sensor's) away from the cluster head (actor). "The relay sensor's of the cluster calculates the distance among the suspicious flows to the possible victim on different paths." If the distance between them is almost the same, then we can confirm it as an attack.

We have considered the following assumptions to solve the DDoS attack.

1. The attackers always use either Poisson or chi-square distribution to generate the packets at the malfunctioned sensors.
2. There is only one actor in the cluster that is under attack.
3. The network is linear and stable for the maximum amount of time. However, it is not stable when the actor performing actions in the cluster or leaves the cluster to help its neighbor actors.
4. The relay sensors are not malfunctioned by the attacker.

In this work, we denote the packets which share the same destination address as a flow on each relay node. Once the relay nodes receive the number of packets than usually, then it sample the number of packets of the suspicious flows in a time unit of approximately 0.15 s. In case of an attack, the attacker utilizes a random variable X to manage the packet generation speed of an attacker. The possible methods are as follows.

1. Generate the packets with a constant speed S $A(Y = S) = 1$.
2. Increase the packet generation speed at time t, $Y = at + b$.
3. Generate the packets using Poisson distribution $S(Y = k) = \frac{\lambda^k e^{-\lambda}}{k!}$, k=0,1..., and λ is a constant.

At any particular time, if we identify two apprehensive flows f_p and f_q, then the relay sensor will do the sampling for the two flows. Once the sampling job is finished, we get distributions of the two suspicious flows.

$$\begin{cases} D_1(Y) = d_1(y_1, y_2, y_3 \dots y_n) \\ D_2(Y) = d_2(y_1, y_2, y_3 \dots y_n) \end{cases} \quad (1)$$

The relative entropy (distance) of the two distributions is as follows.

$$D(D_1||D_2) = \sum_{x \in \chi} D_1 \log \frac{D_1}{D_2}, \tag{2}$$

where χ is the sample space of Y. If the two suspicious flows are attack flows, then they should be generated by the same function $f(x)$ at different zombies. Let us consider $g_p()$ and $g_q()$ are the functions for the suspicious flows f_p and f_q. Then, it can be denoted as

$$\begin{cases} d_1 = g_p(f(x)) \\ d_2 = g_q(f(x)). \end{cases} \tag{3}$$

If $g_{d_1}()$ and $g_{d_2}()$ are linear, then $D(d_1||d_2) = 0$. However, it is impossible in realistic scenarios. The distance of two attack flows should be minimal as compared to one attack flow and normal flow. The following equation is useful to identify whether two flows are attack flows or normal flows.

$$S(d_1, d_2) = \begin{cases} 1 & \text{if} \quad D(d_1||d_2) \leq \delta \\ 0 & \text{if} \quad D(d_1||d_2) > \delta, \end{cases} \tag{4}$$

where δ is a threshold value. Further, we can extend the number of attacks and suspicious flows means $d_{1i}, d_{2i}, i = 1, 2, 3 \ldots$. Hence, the $S_i(d_{1i}, d_{2i}$, i = 1,2,...m. Let us consider the probability of wrong judgement for each suspicious flows is p, then the true judgement for an attack is represented as,

$$S(d_{11}, d_{12} \ldots d_{1m}, d_{21}, d_{22} \ldots d_{2m}) = 1 - d_1{}^m \tag{5}$$

Using the above solution, we can identify the distributed denial-of-service attacks as early as possible. It is useful to discard the attack packets from the network to reduce attack impact on the network performance.

4 Experimental Setup and Analysis

The proposed protocol should be simulated in a proper simulation tool, and the results are to be compared with the existing competitive protocols for WSAN. NS2 is a popular simulator in the research world and which is the freely available source. Hence, we have simulated our proposed protocol in NS2 simulator. In our proposed mechanism, every sensor is embedded with a single radio and directional antenna. However, an actor is integrated with two radios for sensor–actor and actor–actor coordination. For simulations, the packet length is considered as 64 bytes. The number of channels in each radio is varied from 3 to 4 to analyze the performance of the protocol. In the 1000×1000 m^2 network area, the number of sensors is varied from 100 to 1000. Among the deployed sensors, 20% of sensors are malicious, which creates attacks in the network. The characteristics of WSAN are similar to nodes in the ad hoc network.

Table 1 Simulation parameters

Parameters	Values
Simulation duration	300 s
Actor's transmission range	300 m
Sensor's transmission range	100 m
Sensor's initial energy	2 J
Mobility pattern	Random waypoint
Packet size	64 B
Traffic flow	CBR
Number of channels	3–4
Data transfer rate	20–60 pkt/s

The nodes in ad-hoc have mobility. Thus, we used the random waypoint mobility model is used to consider the mobility of actors. Rest of the simulation parameters like sensor and actor transmission ranges, number of sensors and actors, etc., have been listed out in Table 1. The sensors are resource-constrained nodes; hence, energy is an important metric which is needed to be considered while designing any MAC protocol for WSAN. Hence, in the simulation, we have considered a standard energy dissipation mode, as shown in Fig. 1, which is used by several researchers to evaluate their proposed protocols. The radio model is used to compute how much energy is utilized in the network for a certain amount of time.

To analyze the simulation results of the proposed protocol and its competitive mechanisms, we have considered a few simulation metrics like packet delivery ratio, average energy dissipation, average end-to-end delay, and average throughput. "The packet delivery ratio is defined as the ratio of packets that are successfully delivered to a destination compared to the number of packets that have been sent out by the sender."

Figures 2 and 3 depict the packet delivery ratio versus data transfer rate and packet delivery ratio versus the number of sensors, respectively. Figure 2 indicates that the

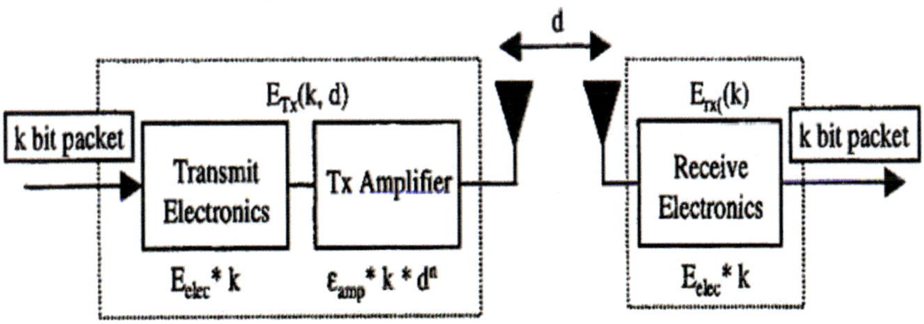

Fig. 1 Radio energy dissipation model

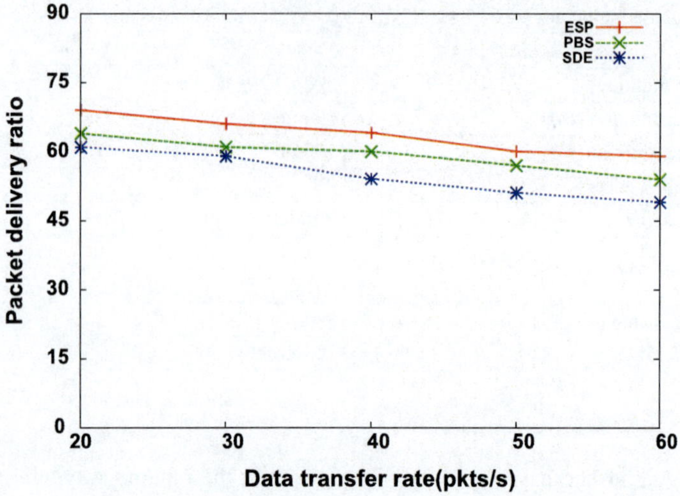

Fig. 2 Packet delivery ratio versus data transfer rate

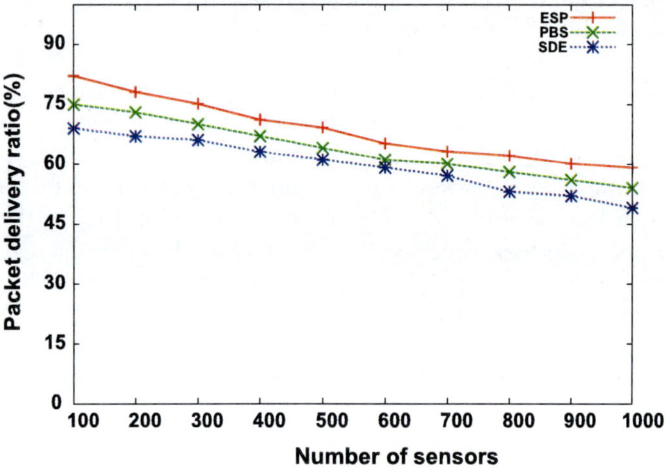

Fig. 3 Packet delivery ratio versus number of sensors

packet delivery ratio decreases with the increase in the number of packets transferred per second. Further, it also shows that the proposed protocol performs well as compared to the existing mechanism. Similarly, Fig. 3 depicts that the proposed protocol outperforms its competitive mechanisms like profile-based scheme against DDoS attack (PBS) and secure distributed estimation over wireless sensor networks under attacks (SDE).

The average energy dissipation is defined as the average amount of energy utilized in the network for network initialization, transversing the data, and receiving the data.

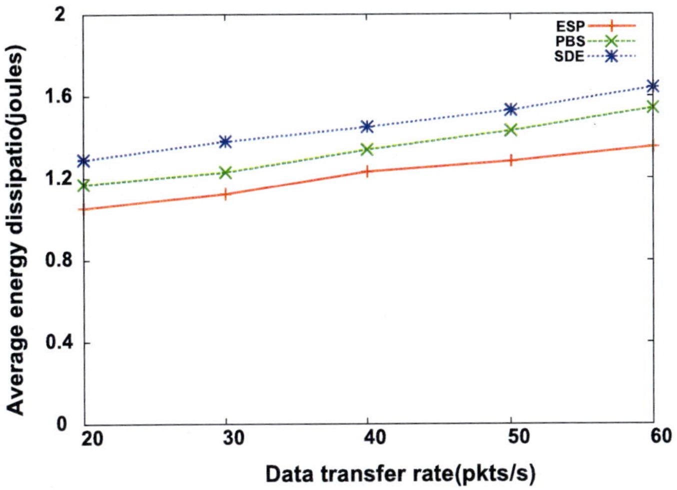

Fig. 4 Average energy dissipation versus number of sensors

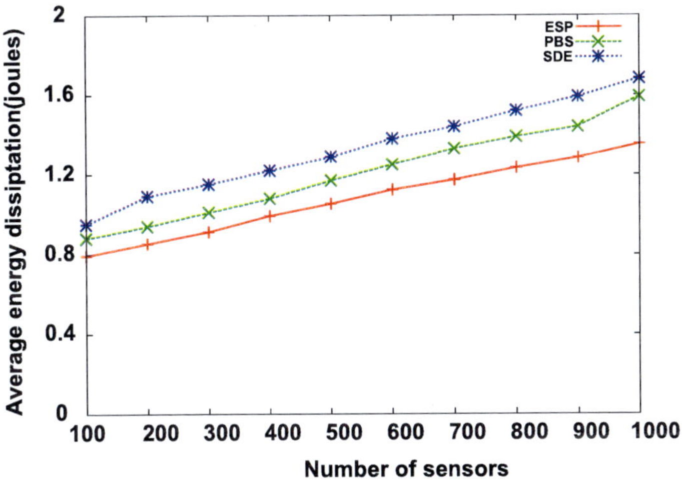

Fig. 5 Average energy dissipation versus data transfer rate

From the state of the art, the amount of energy is required to transfer the data is more as compared to receive the data.

Figure 4 shows the average energy dissipation of the proposed ESP protocol and existing mechanisms like PBS, SDE for variable data transfer rate. The figure indicates that the proposed protocol consumes less energy as compared to the existing protocol. Similarly, Fig. 5 depicts the average energy dissipation versus the number of sensors for all the three protocols. It also indicates that the average energy con-

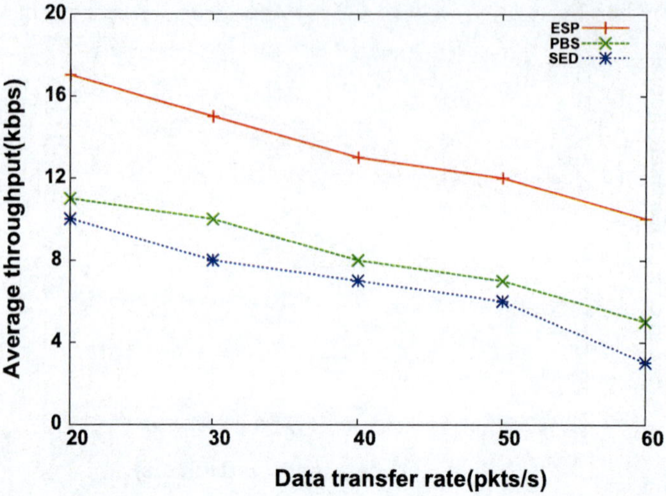

Fig. 6 Average throughput versus data transfer rate

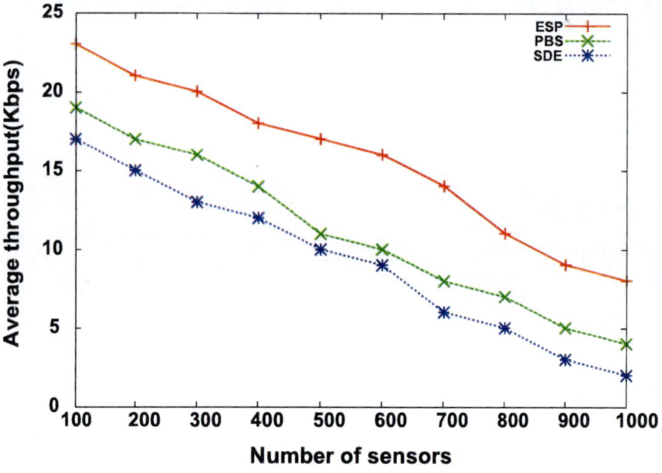

Fig. 7 Average throughput versus number of sensors

sumption increases with the increase in the number of sensors. Thus, the number of sensors in the network effects the average sensor lifetime in the network.

The throughput is defined as the number of bytes are transferred to the destination in a given amount of time. Figures 6 and 7 show the average throughput versus data transfer rate and average throughput versus the number of sensors, respectively. In both cases, the average throughput decreases with the increase in data transfer rate and the number of sensors. Further, Fig. 6 shows the proposed protocol achieves better throughput as compared to its competitive mechanisms. Similarly, Fig. 7 depicts that the proposed energy and secure protocol outperforms existing protocols like profile-based scheme against DDoS attack (PBS) and secure distributed estimation over wireless sensor networks under attacks(SDE).

5 Conclusion

In WSAN, it is important to distinguish between real traffic and DDOS attack. In disaster scenario, sensors will send a large number of packets; if the detection algorithm blocks the communication as a DDOS attack, then it causes more problems in the network. Hence, we have proposed an energy-efficient secure protocol for WSAN using information theory to handle these problems. It uses a distance metric among the suspicious flows to identify the attack as early as possible. Once we detect an attack, it is useful to discard the attack packets from the network to reduce attack impact. The proposed protocol is simulated in NS2, and its competitive MAC protocols are compated using various metrics like packet delivery ratio, average throughput and average energy dissipation. The results indicate that the proposed protocol performs better than its competitive mechanisms by reducing the impact of DDOS attacks.

References

1. Akyildiz IF, Kasimoglu IH (2004) Wireless sensor and actor networks: research challenges. Ad Hoc Networks 2(4):351
2. Guy C (2006) Sixth international symposium on instrumentation and control technology: signal analysis, measurement theory, photo-electronic technology, and artificial intelligence, vol 6357. International Society for Optics and Photonics
3. Zhou Y, Fang Y, Zhang Y (2008) Securing wireless sensor networks: a survey. IEEE Communications Surveys & Tutorials 10(3)
4. Sgora A, Vergados DD, Chatzimisios P (2016) A survey on security and privacy issues in wireless mesh networks. Secur Commun Networks 9(13):1877
5. Chen X, Makki K, Yen K, Pissinou N (2009) Sensor network security: a survey. IEEE Commun Surv Tutorials 11(2)
6. Perrig A, Stankovic J, Wagner D (2004) Security in wireless sensor networks. Commun ACM 47(6):53
7. Grover J, Sharma S (2016) 5th international onference on reliability, infocom technologies and optimization (Trends and Future Directions) (ICRITO) pp 397–404
8. Carl G, Kesidis G, Brooks RR, Rai S (2006) Denial-of-service attack-detection techniques. IEEE Internet Comput 10(1):82
9. Yusof MAM, Ali FHM, Darus MY (2017) Detection and defense algorithms of different types of ddos attacks. Int J Eng Technol 9(5):410
10. Masdari M, Jalali M (2016) A survey and taxonomy of dos attacks in cloud computing. Secur Commun Networks 9:3724–3751
11. Raymond DR, Midkiff SF (2008) Denial-of-service in wireless sensor networks: attacks and defenses. IEEE Pervasive Comput 7(1)
12. Hofmann S, Louizi M, Stoll D (2009) A novel approach to counter denial of service attacks against transport network resources. Bell Labs Tech J 14(1):219

13. Lameski P, Zdravevski E, Kulakov A, Davcev D (2011) International conference on distributed computing in sensor systems and workshops (DCOSS). IEEE, pp 1–3
14. Lameski P, Zdravevski E, Kulakov A, Davcev D (2018) IET wireless sensor systems, vol 8. IET, pp 52–59
15. Mazur K, Ksiezopolski B, Nielek R (2016) Multilevel modeling of distributed denial of service attacks in wireless sensor networks. J Sens 2016:1
16. Behal S, Kumar K (2017) Characterization and comparison of ddos attack tools and trac generators—a review. Int J Network Secur 19(3):383
17. Bedi H, Shiva S, Roy S (2014) A game inspired defense mechanism against distributed denial of service attacks. Secur Commun Networks 7(12):2389
18. Kaur T, Saluja KK, Sharma AK (2016) International conference on recent advances and innovations in engineering (ICRAIE). IEEE, pp 1–5
19. Xiang Y, Tian D, Zhou W (2009) A microscopic competition model and its dynamics analysis on network attacks. Concurrency Comput: Pract Experience 22(4):503
20. Yu S, Zhou W, Doss R (2008) Information theory based detection against network behavior mimicking ddos attacks. IEEE Commun Lett 12(4)
21. Nigam V, Jain S, Burse K (2014) Fourth international conference on communication systems and network technologies (CSNT). IEEE, pp 112–116
22. Liu Y, Li C (2018) Secure distributed estimation over wireless sensor networks under attacks. IEEE Trans Aerosp Electron Syst
23. Cover TM, Thomas JA (2006) Elements of information theory, 2nd edn. Wiley
24. Rahmani H, Sahli N, Kamoun F (2012) Distributed denial-of-service attack detection scheme-based joint-entropy. Secur Commun Networks 5(9):1049

Comparative Analysis of Channel Estimation Techniques in Vehicular Communication

Devesh Shukla, Arun Prakash, and Rajeev Tripathi

Abstract In the past few years, vehicular communication has gained a considerable interest in safety as well as the comfort of the users. IEEE 802.11p is been developed explicitly for providing a platform for the communication within the vehicles and between vehicle and surroundings. It utilizes a renowned methodology orthogonal frequency division multiplexing for implementing communication among vehicle and between vehicles to surrounding. In VANET scenario, an efficient channel estimation technique is crucial for the accuracy of the system due to time-variant channel characteristics. This paper presents an overview and comparative analysis of least square and minimum mean square error channel estimation techniques for vehicular communication. The two techniques are compared for several realistic channels in respect of bit error rate over varying signal-to-noise ratio. The estimation algorithm is substructured on the pilot-aided arrangement. The result indicates that the minimum mean square error achieves better performance with a higher complexity as compared to the least square due to reduced mean square error. Minimum mean square error shows an approximate gain of 15 dB signal-to-noise ratio relative to least square method. This work serves as a base for more advanced channel estimation schemes. Finally, this work has been concluded with some future scopes and open issues.

Keywords Vehicular communication · Channel estimation · IEEE 802.11p · Least square · Minimum mean square error

D. Shukla (✉) · A. Prakash · R. Tripathi
Department of Electronics and Communication Engineering, Motilal Nehru National Institute of Technology Allahabad, Prayagraj 211004, India
e-mail: rel1705@mnnit.ac.in

A. Prakash
e-mail: arun@mnnit.ac.in

R. Tripathi
e-mail: rt@mnnit.ac.in

D. Harvey et al. (eds.), *Advances in VLSI, Communication, and Signal Processing*, Lecture Notes in Electrical Engineering 683,
https://doi.org/10.1007/978-981-15-6840-4_56

1 Introduction

Vehicular communication much recognized as Vehicular Ad hoc Network (VANET) is emerging as a promising technology in combating with heavy traffic and reducing the risk of accidents. IEEE 802.11p standard has been developed mainly for Dedicated Short Range Communications (DSRC) or Intelligent Transportation Systems (ITS) applications. This technology provides a vast set of applications based on safety, efficiency, and comfort like collision avoidance, toll collections, gathering traffic updates, entertainment, etc. [1]. As per the survey made by the US Transportation Department, around 82% of losses caused by accidents could be addressed by the implementation of ITS technology [2]. VANET comprises of vehicle-to-vehicle (V2V) and vehicle-to-infrastructure (V2I) communications. DSRC operates in the range of 5.9 GHz licensed band having spectrum of 75 MHz providing a range of around 300 m in V2V scenario and 1000 m in V2I scenario and data rate varying from 3 to 27 Mbps.

Vehicular communication requires high-speed data transfer due to geographical change of vehicles and mobility. Vehicular communication generates or broadcasts the messages for other vehicles in the communication range isotropically. These warning messages contain vital information about vehicle speed and direction, traffic information and location which are conveyed in the form of alerts which may be visual or audible or combination of both. IEEE 802.11p standard being developed for ITS applications is obtained from much familiar IEEE 802.11a standard. It has been modified by amending the PHY and MAC layer of IEEE 802.11a considering the impact of the mobility environment [3]. It utilizes spectrally efficient digital modulation technique Orthogonal Frequency Division Multiplexing (OFDM) for data transmission.

The reliability of the VANET model is based on the accurate channel estimation determined from the preamble and pilots. In wireless LAN (WLAN) or indoor environments, the whole packet is equalized depending upon the estimated channel response for the entire packet duration due to relatively stable surrounding. While in vehicular communication channel estimation is a difficult task owing to the mobility of vehicles, the channel varies significantly over the packet duration. There are several other issues faced in 802.11p like multipath effects, frequency selective fading, rayleigh fading, delay spread and channel variation, which degrade the system performance. Channel estimation is an essential aspect for efficient and robust wireless communication systems. In VANET scenario, the accuracy of the channel estimation enhances the physical layer performance [4]. There are two basic methods for deriving the channel estimate in vehicular communication. Firstly by modifying the basic structure of the IEEE 802.11p and secondly by maintaining the structure of IEEE 802.11p standard [5]. However, later one achieves the target at the cost of spectrum bandwidth.

Channel estimation is based on pilot symbols where the receiver has the prior knowledge of these pilot symbols. Depending upon the pilot placement in the time--frequency grid, there exist block-type- and comb-type pilot-assisted estimation scheme [6]. In case of fast fading environment block type is preferred while comb-type placement is chosen for slow fading channel [7]. A lot of work has been done on channel estimation in OFDM and immobile systems whereas in vehicular communication where channel varies rapidly research is in phase to evolve an efficient scheme. In [8–11], several advanced channel estimation techniques for IEEE 802.11p standard are proposed such as Spectral Temporal Averaging, Decision Directed and Constructed Data Pilot where channel estimate is updated for every OFDM symbol with the help of previously updated symbol. This work deals in examining the attainment of Least Square (LS) and Minimum Mean Square Error (MMSE) channel estimation schemes for vehicular communication under several realistic scenarios and their comparison is shown.

The remainder of the paper is constructed as follows. Section 2 illustrates the system description for vehicular communication and channel characteristics. In Sect. 3, channel estimates for the two techniques have been derived. Section 4 furnishes simulation results and discussion for performance comparison. Finally, this work has been concluded in Sect. 5 providing some future scopes and issues.

2 System Description

This section presents the architecture of IEEE 802.11p physical layer to be used as a platform for obtaining channel estimate. Figure 1 shows the transmission architecture of vehicular communication which is very much similar to the parent IEEE 802.11a model. The processed incoming high data rate stream after being mapped by the defined mapping technique (BPSK, QPSK, 16-QAM, 64-QAM) is splitted into parallel low data rate streams that are multiplexed onto the subcarriers orthogonal to each other. OFDM uses guard band or cyclic prefix for removal of Intersymbol Interference (ISI). Inverse Fast Fourier Transform (IFFT) and Fast Fourier Transform (FFT) which are very swift and efficient algorithms are used for modulation and demodulation processes, respectively, in OFDM. In both the standards, it consists of 64 subcarriers out of which 48 are used as data, four as the pilot, and the rest as preambles.

The modulated data $\{S(k)\}$ having K number of subcarriers after IFFT is represented as:

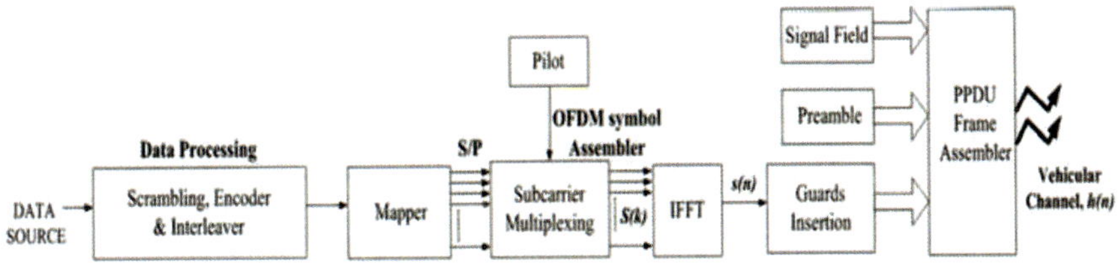

Fig. 1 IEEE 802.11p transmission process

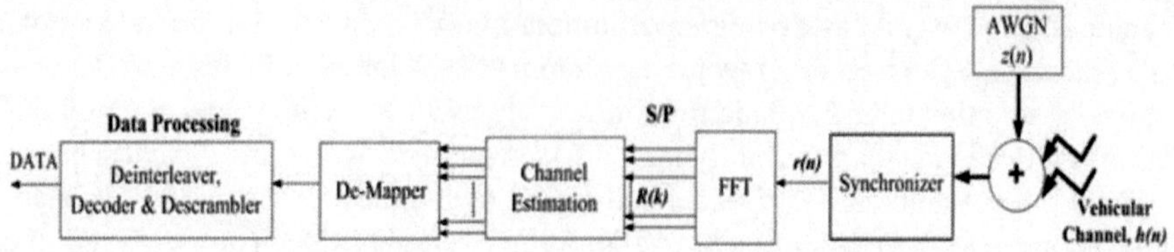

Fig. 2 Receiver architecture

$$s(n) = \text{IFFT}\{S(k)\} = \sum_{k=0}^{K-1} S(k)e^{j2\pi kn/K}, n = 0,1,\ldots,K-1 \tag{1}$$

where K represents the total number of subcarriers. As can be seen from Fig. 2, the signal received before frequency transform operation is represented by:

$$r(n) = h(n) * s(n) + z(n) \tag{2}$$

where $h(n)$ signifies channel impulse response and $z(n)$ as Gaussian noise and in case of wireless communication scenario, received signal is a composition of the number of transmitted signal each undergoing different attenuation effects resulting in either destructive or constructive interference. Assuming all the factors to be constant, channel can be considered as time-invariant and received signal is obtained as:

$$r(t) = \sum_{i=0}^{r-1} h_i s(t - \tau_i) \tag{3}$$

where h_i is time-invariant channel response, r represents propagation paths and τ_i is the delay of the i-th path. Due to the mobility of vehicles in case of VANET systems, the channel is found to be time-varying where output is represented as:

$$r(t) = \sum_{i=0}^{r-1} h_i(t)s(t - \tau_i(t)) \tag{4}$$

$$h(t) = \sum_{i=0}^{r-1} c_i(t)\delta(\tau - \tau_i) \tag{5}$$

where $h(t)$ signifies channel impulse response and $c_i(t)$ is time-dependent complex amplitude coefficients for multiple taps. Thus, radio channel can be interpreted as a combination of time-variant and time-invariant channel and Eq. (2) can be modified as:

$$r(n) = h(m, n) * s(n) + z(n) \tag{6}$$

Table 1 Channel model characteristics

Scenario	V (km/h)	d (m)	T_d (μs)	f_d (Hz)
V2V expressway same direction with wall	104	300--400	0.7	900--1150
V2V urban canyon oncoming	32--48	100	0.4	400--500
R2V expressway	104	300--400	0.4	600--700
R2V suburban street	32--48	100	0.7	300--500
R2V urban canyon	32--48	100	0.5	300
V2V expressway oncoming	104	300--400	0.3	1000--1200

where $h(n)$ and $h(m)$ are time-variant and time-invariant fading channel coefficients, respectively. Further received signal after FFT can be represented as:

$$R(k) = \text{FFT}\{r(n)\} = \frac{1}{K}\sum_{n=0}^{N-1} r(n)e^{-j2\pi kn/K}, k = 0,1,\ldots,K-1 \quad (7)$$

Further, to avoid ISI, it is assumed that the length of channel response is shorter than guard interval. Thus, received signal after de-multiplexing is further represented as:

$$R(k) = X(k)H(k) + Z(k), k = 0, 1, \ldots, K-1 \quad (8)$$

In this work, channel model has been adapted from the article [12] in which several realistic models have been proposed which has been accepted as a standard model for V2V channels. In [13], the tapped-delay line model has been considered where every taps is defined as having Rayleigh or Rician fading. The characteristics of the models are represented in Table 1, according to [12] in which six of the environments or scenarios have been introduced.

3 Channel Estimation

This section illustrates the channel estimators considered in this work. Pilot-assisted channel estimators are traditionally used widely for wireless communication systems. Training symbols are multiplexed together with data symbols at fixed location before signal transmission. Receiver has a prior knowledge of these training or pilot symbols. The channel estimation is carried out at pilot subcarrier positions based on the training symbols. Further, these training symbols assist in determining Channel State Information (CSI) correlating to its position. The CSI analogous to the information data location is achieved by interpolation among channel estimates derived earlier from the training symbols. Two estimation techniques, i.e., LS and MMSE channel estimation methods for vehicular communication implemented are discussed.

3.1 LS Channel Estimation

This is the elementary form of estimator utilized everywhere and also serves as a base for more advanced channel estimation schemes. This technique aims to minimize the squared error between the originally transmitted and received signal. In this technique channel estimate is determined based on log-likelihood function according to which LS estimate [14] is obtained as:

$$\hat{H}_{\mathrm{LS}}(k) = (X^H X)^{-1} X^H S(k) \tag{9}$$

In case of vehicular scenario, LS estimate is obtained further by averaging the individual estimates due to engagement of two training symbols. Thus, final estimate for the k-th subcarrier is:

$$\hat{H}(k) = \frac{1}{2}\left\{\hat{H}_1(k) + \hat{H}_2(k)\right\} \tag{10}$$

Now, estimate determined is equalized subsequently for the entire packet assuming constant channel. But in case of V2V communication, channel rapidly changes thus resulting in deteriorated system response. Though this scheme is the least complex, it suffers from higher MSE.

3.2 MMSE Channel Estimation

MMSE estimation applies second-order statistics of the channel condition for minimizing MSE in comparison to LS scheme. The channel coefficient estimate [15] is given by:

$$\hat{H}_{\mathrm{MMSE}}(k) = R_{\mathrm{HH}}\left(R_{\mathrm{HH}} + \sigma_n^2 (X^H X)^{-1}\right)^{-1} \hat{H}_{\mathrm{LS}}(k) \tag{11}$$

where $\hat{H}_{\mathrm{LS}}(k)$ is LS channel estimate, σ_n^2 is the noise variance and R_{HH} is auto-correlation matrix of the channel and is elucidated as:

$$R_{\mathrm{HH}} = E(\mathrm{HH}^T) \tag{12}$$

MMSE estimator performs efficiently having prior knowledge of channel statistics σ_n^2 and R_{HH}. This technique enhances the performance at the expense of higher complexity level because of the inverse operation performed every time data changes.

4 Results and Discussion

The above-stated estimation techniques are applied to the vehicular communication for performance evaluation in terms of BER and complex simulations. In this section, performance of the vehicular communication is compared using LS and MMSE channel estimation techniques depending upon the parameters given in Table 2. The channel model has been based on the realistic channel characteristics provided in Table 1. The simulation has been performed for QPSK & 16-QAM modulation scheme for Roadside-to-Vehicle (R2V) urban canyon, V2V urban canyon oncoming, and V2V expressway oncoming scenarios to cover high, medium, and low intensity channels. The packet length is kept smaller with number of OFDM data symbols to be 100 for better BER performance. The simulation result reflects the comparison between the two schemes is on the basis of BER versus varying SNR over different realistic channels.

Figure 3 shows the comparison between LS and MMSE estimation scheme using QPSK & 16-QAM constellation schemes for R2V urban canyon where vehicle speed is approximately 30--40 km/h with a Doppler shift of 300 Hz. A higher Doppler shift indicates stronger time-varying characteristic which is exhibited by V2V urban canyon oncoming and V2V expressway oncoming as demonstrated in Figs. 4 and 5, respectively. The result reflects that the MMSE estimation scheme performs relatively better in all scenarios. This is due to the reduced mean square error in the MMSE technique which results in an increase in system complexity. The result indicates trade-off between complexity level and channel estimation performance. Further,

Table 2 IEEE 802.11p Parameters

Parameters	Specifications
Modulation mode	BPSK, QPSK, 16QAM, 64QAM
Bit rate (Mbps)	3, 4.5, 6, 9, 12, 18, 24, 27
Code rate	1/2, 2/3, 3/4
Number of subcarriers	52
Data subcarriers	48
Pilot subcarriers	4
FFT period	6.4 μs
Guard time	1.6 μs
Symbol duration	8 μs
Preamble duration	32 μs
Subcarrier spacing	0.15625 MHz

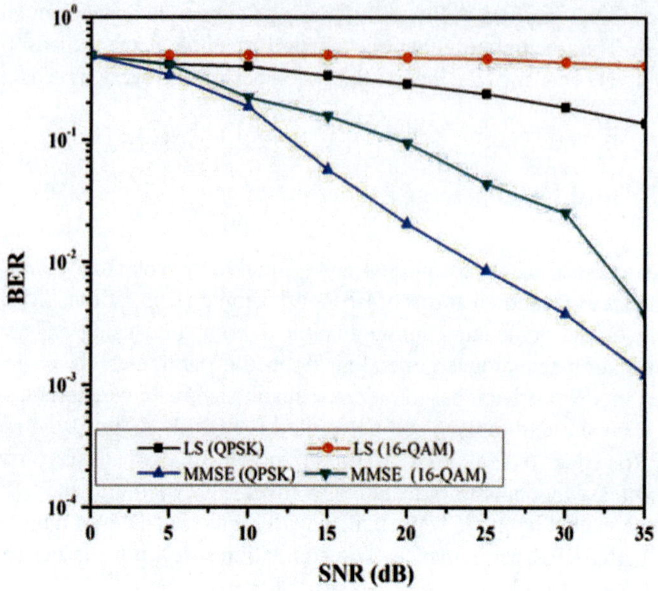

Fig. 3 Comparison of BER for R2V urban canyon

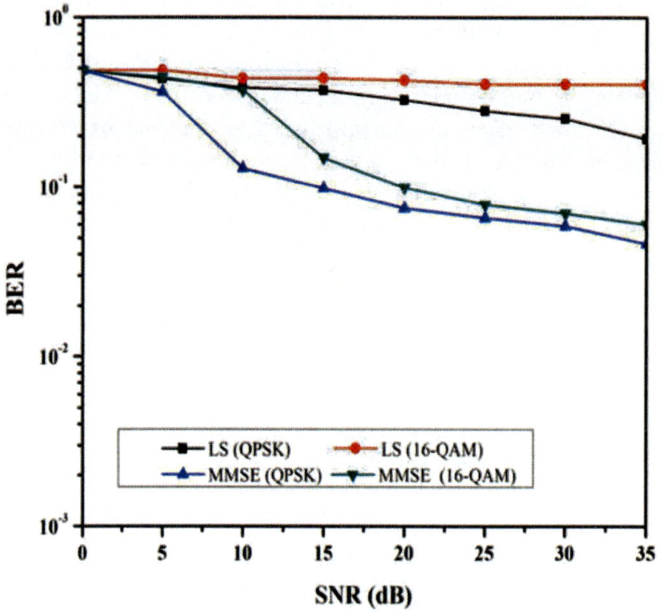

Fig. 4 Comparison of BER for V2V urban canyon oncoming

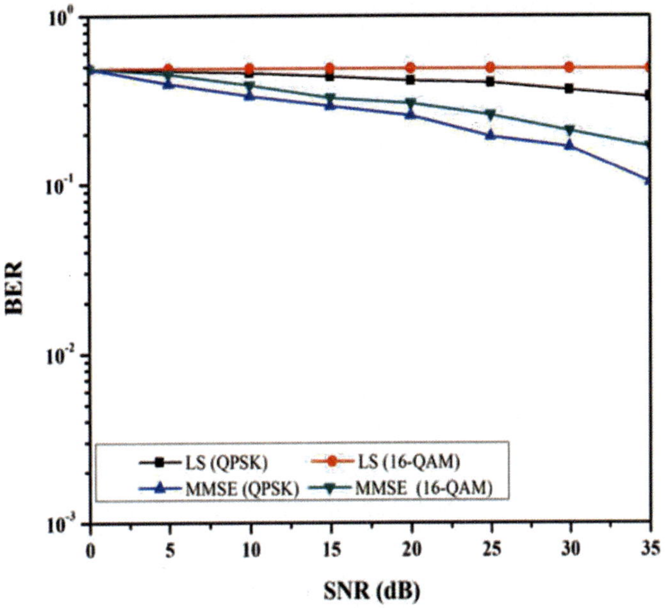

Fig. 5 Comparison of BER for V2V expressway oncoming

it is found that LS estimation is incapable of perfect or effective channel estimate. This is due to the mobility of vehicles in the V2V scenario where the channel varies rapidly.

5 Conclusion

Vehicular communication is evolving as an eminent research area that proves to be better in contributing toward safer, secure, and intelligent transportation systems. Channel estimation is vital in enhancing the reliability of the VANET system under highly time-variant channels. This work presented channel estimation schemes for vehicular communication, namely LS and MMSE estimation over for R2V urban canyon, V2V urban canyon oncoming, and V2V expressway oncoming channel scenario. The system and channel model has been developed and the estimate has been derived for both the schemes. The result shows that the MMSE technique outperforms LS channel scheme in the whole SNR regime for all the scenarios with increased complexity. Also MMSE technique works well where Doppler shift is in low range specifically for low mobility channel conditions. This work serves as a base for the proposal of a modified novel estimation which maintains a balance between complexity and BER performance. Although the implemented MMSE scheme handles the channel varying characteristics to some extent, still some challenges are open which can be resolved in the future work for better and accurate results.

References

1. Arslan S, Saritas M (2017) The effects of OFDM design parameters on the V2X communication performance: A survey. Vehicular Commun 7:1–6
2. Kenney JB (2011) Dedicated Short-Range Communications (DSRC) standards in the United States. U.S. Federal Communication Commission, vol 99, no 7, pp 1162–1182, July 2011
3. IEEE Std 802.11-2007 Part 11.: Wireless LAN Medium Access Control (MAC) and Physical Layer (PHY) specifications High-speed Physical Layer in the 5 GHz Band
4. Miao L, Djouani K, van Wyk BJ, Hamam Y (2012) Evaluation and enhancement of IEEE 802.11 p standard: a survey. Mobile Comput 1(1):15–30
5. Zhao Z, Cheng X, Wen M, Jiao B, Wang C-X (2013) Channel estimation schemes for IEEE 802.11p standard. IEEE Intell Transp Syst Mag 5(4):38–49
6. Shen Y, Martinez E (2006) Channel estimation in OFDM systems. Freescale semiconductor application note, pp 1–15
7. Coleri S, Ergen M, Puri A, Bahai A (2002) Channel estimation techniques based on pilot arrangement in OFDM systems. IEEE Trans Broadcasting 48(3):223–229
8. Fernandez JA, Borries K, Cheng L, Vijaya Kumar BVK, Stancil DD, Bai F (2012) Performance of the 802.11p physical layer in vehicle-to-vehicle environments. IEEE Trans Vehicular Technol 61(1):3–14
9. Bourdoux A, Cappelle H, Dejonghe A (2011) Channel tracking for fast time-varying channels in IEEE802. 11p systems. In: 2011 IEEE global telecommunications conference-GLOBECOM 2011, pp 1–6. IEEE
10. Zhao Z, Cheng X, Wen M, Yang L, Jiao B (2014) Constructed data pilot-assisted channel estimators for mobile environments. IEEE Trans Intell Transp Syst 16(2):947–957
11. Wang T, Hussain A, Cao Y, Gulomjon S (2019) An improved channel estimation technique for IEEE 802.11p standard in vehicular communications. Sensors 19(1):98
12. Acosta-Marum G, Ingram MA (2007) Six time-and frequency-selective empirical channel models for vehicular wireless LANs. IEEE Veh Tech Mag 2(4):4–11
13. Sen I, Matolak DW (2008) Vehicle-vehicle channel models for the 5-GHz band. IEEE Trans Intell Trans Syst 9(2):235–245
14. Van De Beek J-J, Edfors O, Sandell M, Wilson Sk, Ola Borjesson P (1995) On channel estimation in OFDM systems. In: 1995 IEEE 45th vehicular technology conference. Countdown to the wireless twenty-first century, vol 2, pp 815–819. IEEE
15. Edfors O, Sandell M, Vand de Beek J-J, Wilson SK, Ola Borjesson P (1998) OFDM channel estimation by singular value decomposition. IEEE Trans Commun 46(7):931–939

A Survey Study of Diseases Diagnosed Through Imaging Methodology Using Ultrasonography

Kumar Mohit, Jojo Johnson, Kumari Simran, Rajeev Gupta, and Basant Kumar

Abstract This paper is a work of survey to various diseases of human body being diagnosed through ultrasonography (USG) method, and later an introduction to modern medical database management system for the same. Human anatomy requires a confined methodology for its study and abnormalities identification. Diagnosis through imaging technology has been a revolutionary discovery in modern times. Among the various medical imaging technologies, USG is usually preferred. Ultrasound machines with the help of its transducers are able to image almost all body parts and hence diagnose almost all types of diseases. Starting with the introduction to USG, working principle, and advantages over other imaging methodologies, this paper covers near about hundreds of diseases being diagnosed through USG method, along with its diagnostic features, affecting body parts, and operating modes. Later in this paper, a modern algorithm, called MongoDB, is introduced for hospital information management system (HIMS) as an efficient database of patients and diseases.

Keywords USG · Modes · Transducer · MongoDB · NoSQL · GridFS

1 Introduction

Morphology of human being is very complex and so is diagnosing most of the diseases. Although many diseases affecting external body parts may be identified by just viewing through naked eyes, some external diseases and internal diseases required diagnostic test to be identified and cured. Diagnostic tests may include pathological tests, imaging technologies, and many other methods. The imaging methodologies for detecting some special features of particular disease being in practice are magnetic resonance imaging (MRI), computerized tomography (CT) scan, ultrasonography (USG) scan, X-Ray scan, positron emission tomography (PET), elastography, thermography, etc. [1]. One imaging technology is preferred over other

K. Mohit (✉) · J. Johnson · K. Simran · R. Gupta · B. Kumar
Motilal Nehru National Institute of Technology, Prayagraj, Uttar Pradesh 211004, India
e-mail: pmob3087@gmail.com

D. Harvey et al. (eds.), *Advances in VLSI, Communication, and Signal Processing*, Lecture Notes in Electrical Engineering 683,
https://doi.org/10.1007/978-981-15-6840-4_57

or in combination because of respective advantages and more helpful to particular region of interest. Among these imaging modalities, USG is more commonly used for diagnosis and therapeutic procedures because of involving no harmful ionizing radiations and no contrast enhancement, less cost, and fast procedure. Ultrasonography (USG) is term used for modern imaging technology for diagnosis of humans or animals body parts, organs and tissues, internally and externally. As the name suggest, this process utilizes high frequency sound waves, i.e., ultrasound waves to travel inside organs for its operation [2, 3]. For medical imaging process using USG, normally operating frequency lies between 2 and 10 MHz, which in some cases may go up to 40 MHz [1, 4]. Based on the operating frequencies and application to different organs, different transducers, or say probes, are designed, which is a major part of an ultrasound machine. They operate on phenomena of piezoelectric effect and hence transducers pressurized on organs/tissues generate electrical signals which are converted to sound signals and vice versa [1, 2, 5]. In this manner, it acts as transmitter and receiver both.

Diagnosis through any approach is done in two ways, either comparing recorded characteristics to stored characteristics of all diseases or by performing laboratory tests for some new one. In many diseases, like tumor, fast and correct diagnosis is very important, and therefore with the time many tools had been evolved and utilized for identification. In the same line, many medical image database algorithms are used for diseases related data collection, and quick detection purposes, one of which is MongoDB. Human physiology and anatomy database can be created through ultrasound medical imaging technology, which is also a step toward helping hospitals in identifying abnormalities. Some of the existing medical database are Marklogic, MySQL, PEDro, QResearch, EUROCAT, CPRD, PubMed, CHORUS, CouchDB, Solr, Elasticsearch, MedilinePlus, ENTREZ, MongoDB, etc. [6, 7]. MongoDB is new developed structure in this context which differs from all others in one way or other because of its improved architecture, framework, language, support features, simple syntax, etc. Based on NoSQL framework and including GridFS feature, it is becoming user-friendly platform and smart programming structure which can store, update, and archive large medical data of patients and diseases smoothly, even on cloud without causing load to servers [8, 9]. This will help in providing easy access to patient's related data whenever required.

2 Ultrasonography-Based Disease Detection

Ultrasonography(USG) is done through ultrasound machine which is a set of central processing unit (CPU), monitor, disk storage unit, keyboard and control knobs, printer and most important of all; transducer [5]. USG deals with ultrasound wave application in medical imaging and preclinical imaging technology, which can image almost all body organs, fluids, and tissues. The high frequency sound waves, in megahertz, are used to make images of the interior and/or external body parts for medical analysis and intervention purposes which further helps in diagnose and cure diseases

[2, 4, 5]. Depending upon application to imaging different body parts for clearer view and correct diagnosis, different modes of USG is available, such as amplitude mode (A-Mode), brightness mode (B-Mode), Doppler mode (D-Mode), time motion mode (M-Mode or TM-Mode), duplex mode presents B-Mode and D-Mode scan simultaneously, etc. [10] (Table 1).

Transducer probes are the most important equipment in ultrasound scanner because it is responsible for sending and receiving signals into/from body parts. Operating principle being piezoelectric phenomena, the pressure applied through transducers on body parts generates electrical signals which are converted into sound waves and reflected echoes from different organs or tissues are received for making images. Transducer probes are designed depending upon its application to different body parts and so, because different organs, tissues, and fluids required to be operated on different frequencies for clearer investigation [14]. Based on frequency and aperture area, it is classified as linear transducers, convex transducers, phased array transducers, single type transducers, etc. [5]. Different diseases being diagnosed through USG, their diagnostic features and working modes/types of scan have been enlisted in Table 2. Table has been grouped based on diseases having similar features.

3 Medical Database Management System

Medical images created through different imaging technologies of different patients and different diseases need to be saved at some places for future references and studies in order to get better and fast diagnosis. Toward this step, hospital information management systems (HIMS) with picture archiving and communication systems (PACS) have been a great help in health services field by storing medical data economically digitally in some form of database [1, 57]. But, as we know there is been numerous increase in number of patients and diseases daily which need database to be modify and update frequently, which is not easy task in early existing database, so some smart and easily accessible database management system needed to be employed; one of current applications which is being focused is MongoDB [7].

MongoDB is designed for storing, retrieving, and managing document-oriented information with the feature of scalability and indexing. It is a schema-less database management system (DBMS) which means; NoSQL database [9]. SQL database table has fixed database structures where similar data are stored as rows in tables, therefore once programmed it is difficult to improvise. Tables cannot be partially updated for storing any new data or files, rather all the fields need to be repost in order to update. On the other hand, MongoDB uses NoSQL and Java script object notation (JSON) database as the data exchange format whose structure is not fixed; it can be improvised easily whenever needed, any files can be updated without altering whole structure [57]. NoSQL database property is that it treats data as documents for storing and two documents in a collection need not have to be similar structure. In this way, it helps HIMS by changing format of data storing whenever needed or creating new files/information by not altering previously created database structure.

Table 1 Table represents brief introduction of USG modes in practice

Type of USG Modes	Mapping	Application	References
A-Mode	Information is displayed in one-dimensional map as amplitude versus time	Used for detecting depth or dimensions of organs in ophthalmology, neurology, and cardiology applications. Sometimes used in abdominal scan also	[4, 5, 10, 11]
	Echoes strength produces amplitude spikes on screen and gap between two amplitudes represents the time taken by sound to travel in human tissues; which helps in calculating tissues depth		
B-Mode	Information is displayed in two-dimensional map as brightness versus time	Represents body organs and tissues boundaries. Also called grayscale map where black represents fluid areas and white represents solid	[4, 5, 10, 11]
	The A-Mode scanning when rotated 90° then its amplitude will be shown as bright dots, called as B-Mode scanning		
	Pixel or dots sizes and brightness/gray level represents echoes strength, and gaps between them represents tissues depth	Mainly used in all types of abdominal and ophthalmology scanning	

(continued)

Table 1 (continued)

Type of USG Modes	Mapping	Application	References
B-Flow Mode	Special B-Mode scanning technique which is used to display intravascular echoes as two-dimensional view of moving dots on the screen	Displays flow information therefore represents blood flow, surrounding tissues vibrations, and vascular structures	[4, 5, 12, 13]
	Weak flowing blood cells are shown by brighter pixels		
D-Mode	It is an alternative to B-flow mode	Doppler's effect is used to image the movement of tissues and body fluids and their relative velocities to the probe. Mainly used in vascular and musculoskeletal applications	[4, 5, 10, 11]
	Two-dimensional plot of spectrum of velocity and frequency shift vs time. Spectrum lines or amplitude represents relative velocity, and brightness of lines represents quantity of blood cells causing frequency shift at a particular time		
M-Mode	It is one-dimensional plot of echoes from the moving body parts as amplitude versus time. Amplitudes are made by dots of different brightness levels moving at a pre-fixed speed, which in turn traces out the motion pattern of the moving body parts	Slope from M-Mode displays determines the velocity of moving body organs. Found its application mainly in cardiology for analyzing motion of interface	[4, 5, 10, 11]

Table 2 Table represents summary of the USG diagnosed diseases features

Diseases	Effected body parts	Diagnostic features	Modes/scan	References
Retinal detachment, retinal tears, vitreous detachment	Eye	A thin, continuous hyperechoic line separated from the wall of the globe	Brightness mode/amplitude mode	[15–17]
Carcinoma, sarcoma, lymphoma, leukemia, lipoma, fibroma, cirrhosi	Brain, lungs, skin, neck, stomach, liver, mouth, bone, breast	Isoperfusion or hypervascular and hyperperfusion pattern formation. Heterogeneous echotexture masses	Brightness mode, doppler mode	[18, 19]
Hyperplasia	Uterus, vagina, prostate, liver, throat, skin, breast	Mild proliferation of the tissues. Echogenic pattern	Brightness mode, doppler mode	[18, 20]
Emboli, thrombosis, nephrosis, ischemia, lupus, hematoma, hypoxia	Heart, lungs, brain, kidneys, arms, legs	Blood clotting in the veins or arteries. Hypoechoic heterogeneous oval shape elongagated masses	Doppler mode/B-flow imaging	[21, 22]
		Oscillating flow of blood		
Meningitis, encephalitis, hydrocephalus, ventriculomegaly	Brain, spine	Presence of the extra-axial fluid collection	Doppler mode	[23–25]

(continued)

Table 2 (continued)

Diseases	Effected body parts	Diagnostic features	Modes/scan	References
		Presence of the ventricular dilation and lesion		
Fistula, abscesses, sinus, phlegmon	Anal canal, vagina, spleen	Tract lined by granulation tissues with internal/external openings. Dimpling of the skin	Duplex mode	[26–29]
Hemorrhoid, teratoma, fissures, scleroderma	Rectum, anus	Swollen veins with surrounding tissues like "Mosaic Pattern" around dentate line. Mixed echo patterns	Brightness mode	[30, 31]
		Tear in the anus linings		
Polyps, cysts, nodule	Ear, cervix, colon, rectum, nose, stomach, uterus, knee, skin, bones, bronchi, kidney, liver, breast, buttocks, ovary, prostate	Presence of the twinkling artifact	Brightness mode, doppler mode	[18, 32, 33]
		Thickened fold of the linings or tissues. Isoechoic homogeneous lesion		

(continued)

Table 2 (continued)

Diseases	Effected body parts	Diagnostic features	Modes/scan	References
Goiter, dysphagia, hypothyroidism, hyperthyroidism	Neck, larynx, trachea, thyroid gland	Blood perfusion in the thyroid parenchyma is seen, increased nodule vascularity, multiple hyperechoic colloid nodules	Doppler mode	[34, 35]
		Affected maturation of bone		
Hemorrhage, aneurysms, floaters	Brain, eye, aorta, spleen, intestine, lungs, heart, other soft tissues	An excessive localized swelling of the wall of an artery or bulging of the artery showing hyperechoic image	Doppler mode/motion mode/amplitude mode	[36]
		Ballooning in the blood vessel		
Holoprosencephaly, meningomyelocele, cebocephaly, ethmocephaly, cyclopia	Brain, spine, eye.	Dilated monoventricular cavity and fused thalami. Absence of midline structure	Brightness mode	[37, 38]

(continued)

Table 2 (continued)

Diseases	Effected body parts	Diagnostic features	Modes/scan	References
Synovitis, arthritis, bursitis, tendinitis, enthesitis, spondylitis, fibromyalgia	Joints, tendons, ligaments, spine, eye	Presence of synovial thickness	Doppler mode/brightness mode	[39, 40]
		Hyperechogenicity and hypervascularity of the joints space		
Stenosis, vasospasm, occlusion, atherosclerosis, arteriosclerosis	Veins, stomach, vagina, spinal canal, larynx, liver, kidney, heart, penis	Narrowing of the arteries hypoechoic wall due to thickening of the wall near arteries. Region of high grayscale intensity	Duplex mode/motion mode/B-flow imaging	[35, 41, 42]
Torsion, hydrocele, epididymitis, orchitis, seminoma	Testis, scrotam	Differ in echogenicity between both testes. Also vascularity difference. Mixed echo patterns	Duplex mode. Brightness mode	[43–45]

(continued)

Table 2 (continued)

Diseases	Effected body parts	Diagnostic features	Modes/scan	References
Macrosomia	Abdomen	Abdominal circumference is more than normal. Increase in diagnose formula frequency with gestational age	Brightness mode	[46, 47]
Prolapse, rectocele, cystocele, endometriosis, enterocele, proctitis, adenomyosis	Vagina, rectum, anus, uterus, pelvis	Descent or bulging of the tissues. Separation between hyperechoic and hypoechoic line	Doppler mode	[48–50].
Leukomalacia, gliosis, necrosis, resorption, leukodystrophy, autolysis	Brain, spine	Coagulation of the white matter near lateral ventricles. Presence of increased echogenicity in periventricular areas	Doppler mode/motion mode/amplitude mode	[51, 52]
Appendicitis, pancreatitis, oedema, gastroenteritis, cholecystitis, hernia	Abdomen, bladder, colon, intestine, pancreas	Hypoechoic muscularis layer and fluid-filled lumen indicating heterogeneous mass	Doppler mode	[53, 54]

(continued)

Table 2 (continued)

Diseases	Effected body parts	Diagnostic features	Modes/scan	References
Hematochezia, hematuria, melena	Kidney, abdomen, ovary	Exophytic hypoechoic solid mass appearance	Brightness mode	[55]
Vasculitis, angiitis, arteritis, polyangiitis, cryoglobulinemia	Veins, arteries, kidney, lungs, sinus, head, nerve, skin, joints	Non-compressible, hypoechoic wall thickening of arteries	Doppler mode	[55, 56]

JSON employs human readable text for transmitting data objects, and also it can be run on any platform [7]. In this way, it can be said as document-based general purpose DBMS and cross-platform software which is easily accessible and understandable by anyone. Grid file storage (GridFS) system is included in MongoDB for file storing and archiving application, which is a salient feature of this platform because films obtained from imaging technologies are of big sizes, say in few MB's to GB's, whose storing of so many patients and retrieving whenever required is a big task today. GridFS helps by utilizing smaller file storage areas. It divides files into smaller parts or chunks, up to 255 kb smaller sizes, and save each of them as separate documents. In this way, it helps in load balancing and data compression [57, 58]. When original large file query is made, GridFS automatically reassembles the smaller files [7]; as shown in Fig. 1. MongoDB also supports cloud computing, hence, files can be stored on clouds; making it feasible for the hospitals to save as much files and providing accessibility anywhere in the world.

In the proposed database model, MongoDB cloud is being used for storing the records. All the patient's data and records are stored in individual document in JSON format [8]. The images are linked to the same file using GridFS. Searching and retrieving related queries can be written in any language which is linked to MongoDB using MongoDB connect. The database that we make exists in MongoDB atlas and the search and retrieval are done directly on MongoDB atlas interface which can further be extended to any Web sites [59].

4 Conclusion

This paper presented a survey of USG detectable human body diseases. USG was found to be a basic tool for scanning most of the body parts and first step toward many diseases identification. Not much survey-related research work had been done in this field to assemble almost all the body parts diseases with its diagnosis features and database framework on a single paper. In this paper, we have covered most of the diseases being diagnosed though USG based on diagnostic features associated with them, and suggested a modern database framework, MongoDB, for patient's information storing, retrieving and archiving purposes in efficient manner. Its role is helpful in making future references for diagnosis purposes, case study, and research work. With JSON database framework, GridFS system, and cloud computing, MongoDB is making it possible to store any size patient's data and simple access anywhere any platform.

This paper also provides a path to researchers to get through a more detailed study of some particular organ-related disease through the references, also working in the direction of evolving more efficient technique for diagnosis. Many diseases which have not been covered in this paper can be explored with operating frequencies information in diagnosis of diseases. MongoDB having some security-related limitations opens area of research toward improvements in this field.

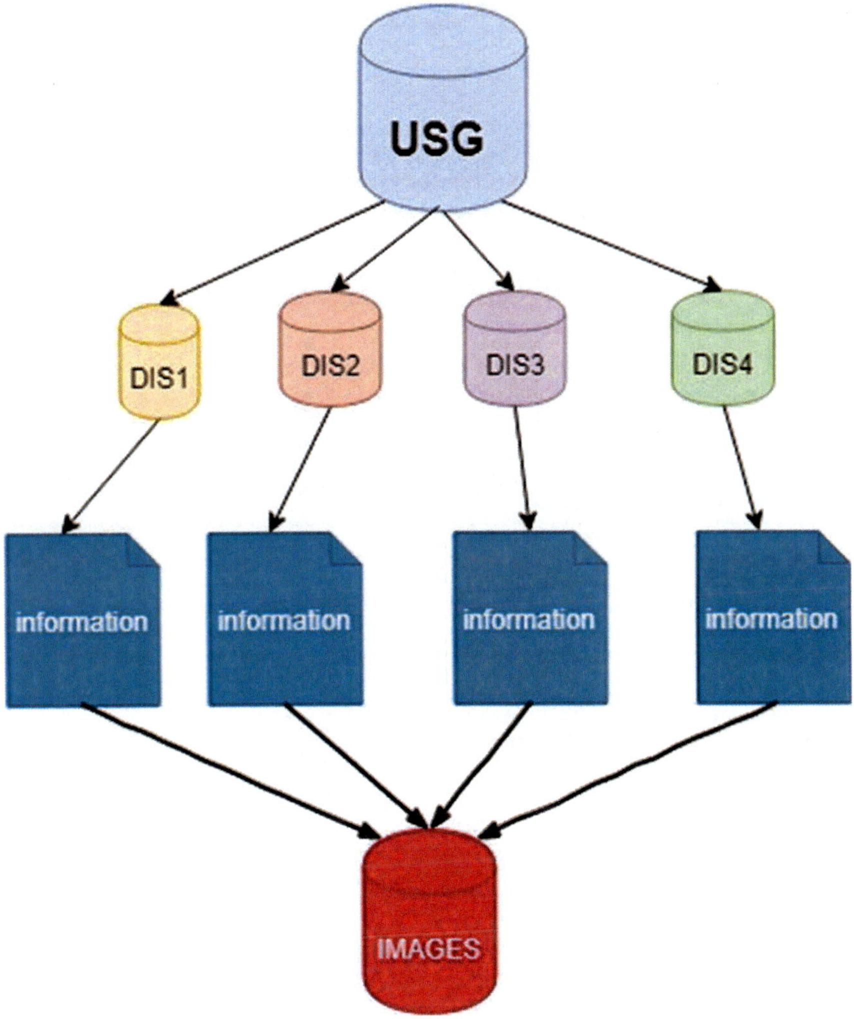

Fig. 1 Working model of GridFS

References

1. Lutz H, Buscarini E (2011) Manual of diagnostic ultrasound, 2nd edn. World Health Organization, Switzerland
2. Sprawls P (1995) Physical principles of medical imaging, 2nd edn. Medical Physics Publishing, USA
3. https://en.wikipedia.org/wiki/Medical_imaging. Last accessed 2019/07/25
4. http://www.fisme.science.uu.nl/woudschotennatuurkunde/verslagen/Vrsl2009/Haar_Romeny.pdf. Last accessed 2019/08/10
5. Khandpur RS (2004) Biomedical instrumentation: technology and applications, 1st edn. McGraw Hill Education, India

6. https://en.wikipedia.org/wiki/List_of_information_retrieval_libraries. Last accessed 2019/08/01
7. Erguzen A, (2018) An efficient middle layer platform for medical imaging archives. J Healthcare Eng
8. https://en.wikipedia.org/wiki/MongoDB. Last accessed 2019/07/29
9. https://www.mongodb.com. Last accessed 2019/07/29
10. https://www.flashcardmachine.com/d8-ultrasoundmodes.html. Last accessed 2019/08/10
11. https://en.wikipedia.org/wiki/Medical_ultrasound. Last accessed 2019/07/25
12. Wachsberg RH (2007) B-flow imaging of the hepatic vasculature: correlation with color doppler sonography. AJR 188(6):522–533
13. Hoskins P, Martin K, Thrush A (2010) Diagnostic ultrasound: physics and equipment, 2nd edn. Cambridge University Press, UK
14. https://en.wikipedia.org/wiki/Ultrasonic_transducer. Last accessed 2019/07/26
15. Gupta R (2018) A novel method for automatic retinal detachment detection and estimation using ocular ultrasound image. Multim Tools Appl, 1–19
16. Polo MDLH (2016) Ocular ultrasonography focused on the posterior eye segment: what radiologists should know. Insights Imaging 7(3):351–364
17. Aironi VD (2009) Pictorial essay: B-scan ultrasonography in ocular abnormalities. IJRI 19(2):109–115
18. Yang T (2013) Sonohysterography: principles, technique and role in diagnosis of endometrial pathology. WJR 5(3):81–87
19. Sofuni A (2005) Differential diagnosis of pancreatic tumors using ultrasound contrast imaging. J Gastroenterol 40(5):518–525
20. Mitterberger M (2010) Ultrasound of the prostate. Cancer Imaging 10(1):40–48
21. Lyaker MR (2013) Arterial embolism. IJCIIS 3(1):77–87
22. Ryu JK (2011) Sonographic appearances of small organizing hematomas and thrombi mimicking superficial soft tissue tumors. JUM 30(10):1431–1436
23. Gupta N (2017) Neonatal cranial sonography: ultrasound findings in neonatal meningitis-a pictorial review. QIMS 7(1):123–131
24. Yikilmaz A (2007) Sonographic findings in bacterial meningitis in neonates and young infants. Pediatr Radiol 38(2):129–137
25. Llompart-Pou JA (2013) Transcranial sonography and cerebral circulatory arrest in adults: a comprehensive review. ISRN 2013:1–6
26. Mentes O (2009) Ultrasonography accurately evaluates the dimension and shape of the pilonidal sinus. Clinics 64(3):189–192
27. Nuernberg D (2019) EFSUMB recommendations for gastrointestinal ultrasound part 3: endorectal, endoanal and perineal ultrasound. Ultrasound Int Open 5(1):34–51
28. Visscher AP (2015) Endoanal ultrasound in perianal fistulae and abscesses. Ultrasound Quart 31(2):130–137
29. Puranik CI (2017) Role of transperineal ultrasound in infective and inflammatory disorders. IJRI 27(4):482–487
30. Aimaiti A (2017) Sonographic appearance of anal cushions of hemorrhoids. WJG 23(20):3664–3674
31. Foxx-Orenstein AE (2014) Common anorectal disorders. Gastroenterol Hepatol 10(5):294–301
32. Chaudhary V (2013) Thyroid ultrasound. Indian J Endocrinol Metab 17(2):219–227
33. Clarke R (2016) Twinkle artefact in the ultrasound diagnosis of superficial epidermoid cysts. Ultrasound 24(3):147–153
34. Huel C (2009) Use of ultrasound to distinguish between fetal hyperthyroidism and hypothyroidism on discovery of a goiter. Ultrasound Obstet Gynecol 33:412–420
35. Mascia L (2009) Diagnosis and management of vasospasm. F1000 Med Rep 1
36. Intrapiromkul J (2013) Accuracy of head ultrasound for the detection of intracranial hemorrhage in preterm neonates: comparison with brain MRI and susceptibility-weighted imaging. J Neuroradiol 40(2):81–88

37. Chaudhari DH (2012) Prenatal ultrasound diagnosis of holoprosencephaly and associated anomalies. BMJ
38. Salama GSA (2015) Cyclopia: a rare condition with unusual presentation—a case report. Clinical Med Insights Pediatr 9:19–23
39. D'Agostino (2017) Scoring ultrasound synovitis in rheumatoid arthritis: a EULAR-OMERACT ultrasound taskforce-Part 1: definition and development of a standardised, consensus-based scoring system. BMJ 3(1)
40. Manik ZH (2016) Ultrasound assessment of synovial thickness of some of the metacarpophalangeal joints of hand in rheumatoid arthritis patients and the normal population. Scientifica
41. Hwang JY (2017) Doppler ultrasonography of the lower extremity arteries: anatomy and scanning guidelines. Ultrasonography 36(2):111–119
42. Wang HK (2005) B-flow ultrasonography of peripheral vascular diseases. J Med Ultrasound 13(4):186–195
43. Huang DY (2012) Focal testicular lesions: colour Doppler ultrasound, contrast-enhanced ultrasound and tissue elastography as adjuvants to the diagnosis. BJR 85(1):41–53
44. Bird K (1983) Ultrasonography in testicular torsion. Radiology 147(2):527–534
45. Kuhn AL (2016) Ultrasonography of the scrotum in adults. Ultrasonography 35(3):180–197
46. http://www.fetalultrasound.com/online/text/36–047.htm. Last accessed 2019/08/02
47. Mongelli M (2005) Ultrasound diagnosis of fetal macrosomia: a comparison of weight prediction models using computer simulation. UOG 26:500–503
48. Dietz HP (2007) Ultrasound assessment of pelvic organ prolapse: the relationship between prolapse severity and symptoms. UOG 29:688–691
49. Piloni VL (2007) Sonography of the female pelvic floor: clinical indications and techniques. Pelviperineology 26(2):59–65
50. Chamie LP (2011) Findings of pelvic endometriosis at transvaginal US, MR imaging and laparoscopy. Radio Graph 31(4):77–100
51. Schmidt WA (2018) Ultrasound in the diagnosis and management of giant cell arteritis. Rheumatology 57:22–31
52. Drera B (2014) Brain ultrasound in canavan disease. J Ultrasound 17(3):215–217
53. Lassandro F (2011) Abdominal hernias: radiological features. WJGE 3(6):110–117
54. Mostbeck G (2016) How to diagnose acute appendicitis: ultrasound first. Insight Imaging 7(2):255–263
55. O'Neill WC (2014) Renal relevant radiology: use of ultrasound in kidney diseases and nephrology procedures. CJASN 9(2):373–381
56. De Vries L (1993) Correlation between the degree of periventricular leukomalacia diagnosed using cranial ultrasound and MRI later in infancy in children with cerebral palsy. Neuropediatrics 24(5):263–268
57. Rebecca DR (2016) A NoSQL solution to efficient storage and retrieval of medical images. IJSER 7(2):545–549
58. https://docs.mongodb.com/manual/core/gridfs. Last accessed 2019/07/31
59. https://www.mongodb.com/cloud/atlas. Last accessed 2019/07/29

Sensor Localization in WSNs Using Rotating Directional-Antenna at the Base Station

Prateek Raj Gautam, Sunil Kumar, and Arvind Kumar

Abstract WSNs have found their application in all spheres of recent developments. WSNs are made up of hundreds of small battery-operated sensor nodes. These sensor nodes usually have limited, nonrechargeable battery, and they are deployed randomly. But the location of these nodes is often required. So, the localization process in WSNs must be energy-efficient. This paper further improves a new localization scheme named EE-LBRD (energy-efficient localization of the nodes in WSNs using beacons from rotating directional antenna). EE-LBRD is scalable, fast, and feasible at the nodes, and can be used in homogeneous or heterogeneous networks. The scheme is applicable for dense as well as sparse node deployment. The base station employs a rotating directional-antenna (DA) to transmits beacon toward nodes. The beacon contains information of DA's height and direction. The information from beacons received at a node is processed to estimate the location of the node. Simulation with MATLAB confirms that the proposed scheme minimizes the localization error and the energy consumption localization.

Keywords Localization · Energy efficiency · Directional-antenna · WSNs

1 Introduction

Recently, wireless sensor networks (WSNs) are used by many biomedical [2], industrial [6], and house-hold applications [18] due to the ease of deployment, cost-effectiveness, and adaptability [23], such as fire detection, gas leakage detection, an inspection of pipelines, security monitoring, health-care services, and industrial application [4, 22]. Reducing energy consumption at the nodes is a significant challenge in front of researchers while designing any application with WSNs. To prolong the stable life of the network, cluster-based routing schemes have been proposed

P. Raj Gautam (✉) · S. Kumar · A. Kumar
Motilal Nehru National Institute of Technology Allahabad, Prayagraj 211004, India
e-mail: prateekrajgautam@gmail.com; rel1601@mnnit.ac.in
URL: http://mnnit.ac.in

D. Harvey et al. (eds.), *Advances in VLSI, Communication, and Signal Processing*, Lecture Notes in Electrical Engineering 683,
https://doi.org/10.1007/978-981-15-6840-4_58

to save energy consumption while routing data [14, 15]. However, the nodes are deployed randomly, so the nodes do not know its location at deployment. Without the location of the node, information gathered sent by nodes in WSN is insignificant because proper control action can be taken only if the location of the triggering location is known. Hence, localization of nodes is necessary for WSNs even before routing, sensing, or communication. Due to the limited energy of sensor nodes, the localization must be energy-efficient, fast, and should be easy to implement and compute at the node. Localization techniques mainly focus on the accuracy of localization ignoring the computational capacity needed at the nodes, and its energy consumption is not compared.

1.1 Novelty

This paper further reduces the energy consumption of recent energy-efficient localization scheme EE-LBRD [8]. Transmission from nodes is not required, and computational capability required at the nodes is small. The proposed scheme is well suited for the dense or the sparse node deployment, as it does not depend on inter-communication among nodes.

1.2 Related Work

Present localization schemes are based on distance estimation using angle detection using antenna array or compass, received signal strength indicator (RSSI) [5],time difference of arrival (TDOA), time of arrival (TOA), angle of arrival (AOA), triangulation to find the position of unknown nodes with the help of beacon nodes, or using GPS signals [1, 3, 10, 11]. Some of these schemes require modifications in the hardware of the node that arises the cost, others increase energy consumption because of inter-communications among nodes, and some schemes need high computations capacity at the node. The use of TOA necessitates a compass and an antenna array results in increased cost of the sensor node [19].

Global positioning system (GPS) signal weakens in indoors, and it requires additional hardware at each anchor node [20]. Reference [13] proposed a beacon packet transmission with different power. Reference [12] aided localization that estimates the distance by using the power level of the received beacon. A three-dimensional localization based on beacons from GPS enabled flying beacon node is proposed [16], but the approach requires a well-controlled flying machine, computational complexity at the node is the high, and optimal path for a flying anchor is hard to detect. Mobile anchor nodes with GPS transmit beacons periodically, and nonanchor nodes can localize them self [17]. Ssu et al. [17] used mobile anchor points for location estimation; they have studied the accuracy and estimated computation cost of the

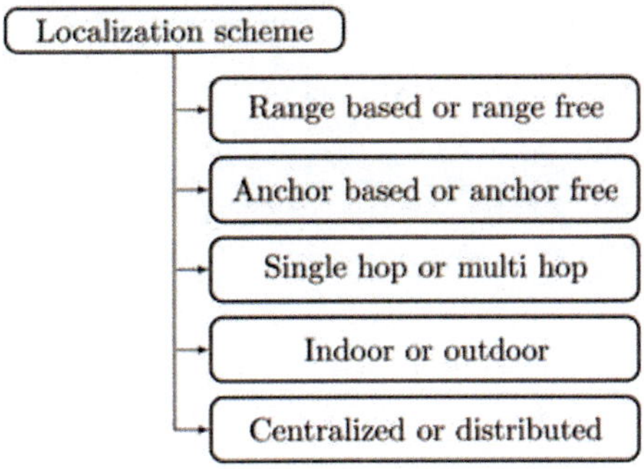

Fig. 1 Localization classification [8]

localization scheme. In [21] proposed the usage of multiple antennae for indoor localization and the estimated accuracy of the scheme. Brief classification of localization schemes is shown in Fig. 1 [7].

1.2.1 AREA Localization [9]

AREA localization is based on RLA [7], nodes receiving beacons from three or more beacon nodes can estimate their location. Beacon nodes are GPS enabled nodes or they are placed manual so they know their location. These nodes may serve as regular nodes to perform sensing and routing. In AREA localization, node select two closest nodes, as shown in Fig. 4, node $n1$ select two beacon nodes (b1) and (b2). Node $n1$ can estimate distance $b1n1$, $b2n1$, and also the distance between two beacons $b1b2$ from the coordinate transmitted by them. The line connecting $b1b2$ is referred as base-line. A perpendicular is dropped from $n1$ on the base-line (b1b2) and then it meets at point N. We can estimate the coordinate of $n1$ if the distance $n1 - N$ is known, later angle α and β can be determined. The area of triangle $n1b1b2$ is calculated as

$$\text{area} = \frac{1}{2}\text{base} \times \text{altitude} = \frac{1}{2}b1b2 \times n1N. \tag{1}$$

The heron's formula can be used to estimate the area of triangle, if all the sides of the triangle are known as

$$\sqrt{S(S-a)(S-b)(S-c)} \tag{2}$$

Where a, b, c represents sides of triangle and $S = \frac{a+b+c}{2}$. In Fig. 2 $a = b1n2, b = n1b2$, and $c = b1b2$. Let $n1N = d$ so from (1) and (2)

$$n1N = \frac{2 \times \sqrt{S(S-a)(S-b)(S-c)}}{b1b2} \tag{3}$$

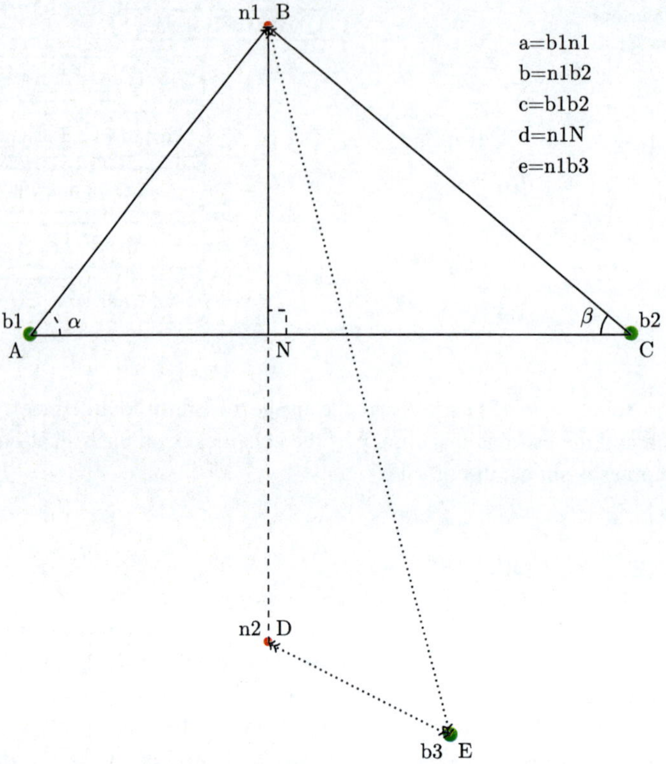

Fig. 2 Schematic diagram of localization using the area between a node and two beacons [9]

$$d = \frac{2 \times \sqrt{S(S-a)(S-b)(S-c)}}{c} \tag{4}$$

Now α and β can be estimated as

$$\alpha = \sin^{-1}\left(\frac{d}{a}\right) \tag{5}$$

$$\beta = \sin^{-1}\left(\frac{d}{b}\right) \tag{6}$$

and $b1N$ and $b2N$ and be determined as

$$b1N = a\cos\alpha \tag{7}$$

$$b2N = b\cos\beta \tag{8}$$

Now, the variables in triangle $b1n1b2$ are known; hence, coordinate of $n1$ can be determined using vector algebra by (9), where (x_A, y_A) represents $b1$ at point A, $\overrightarrow{AN}$ is vector in direction $b1$ to $b2$ with magnitude $b1N$ and $\overrightarrow{NB}$ is vector normal to direction of $b1b2$ with magnitude $n1N$. But this normal vector can also be in direction $\overrightarrow{ND}$ leading us to point D.

$$(x_A, y_A) + \overrightarrow{AN} + \overrightarrow{NB} \tag{9}$$

Solving this we end up two possibility in node $n1$ one at point B and anoth er at point D located equally opposite to the base-line.

So now, we select another beacon node, $b3$ in the region and distance of node from $b3$ is compared with two node points say $n1$ and $n2$, the point whose distance matches with the distance of beacon received is selected as estimated node point, means $|\overrightarrow{EB}|$ and $|\overrightarrow{ED}|$ is compared with distance estimated from node from beacon b3 located at E to select one point as estimated location.

In Sect. 2, LRDA scheme is described with the hardware modifications required at the BS. Error in localization, required time for localization, and the energy consumption at the nodes for localization with the proposed scheme are estimated with analytical expressions. Simulation results and paraeters user are presented with discussions in Sect. 3. Comparison with related work AREA [9] is presented in Sect. 4 followed by conclusion in Sect. 5.

2 Localization of Sensors in Wireless Sensor Networks Using the Beacons from Rotating Directional-Antenna (LRDA)

In the proposed scheme, nodes receive one or more beacons that are transmitted from BS using a directional-antenna. Each node estimates its location using the information present in the received in the beacon packets. The location of nodes is estimated in the polar form (ρ, ϕ) relative to the base station, here ρ represents the radial distance from the base station and ϕ is the angular position of the node.

2.1 Scheme Description

In Fig. 3, three-dimensional space is represented with Cartesian coordinates. The BS is placed at the center and a directional-antenna (DA) is kept at a height H and its direction can be changed. In Fig. 3, region in the XY-plane where the DA is pointing (ϕ, θ) is shown in green. The shaded region of Fig. 3 is shown in Fig. 4, and (10), (12), and (14) can be drawn from this figure .

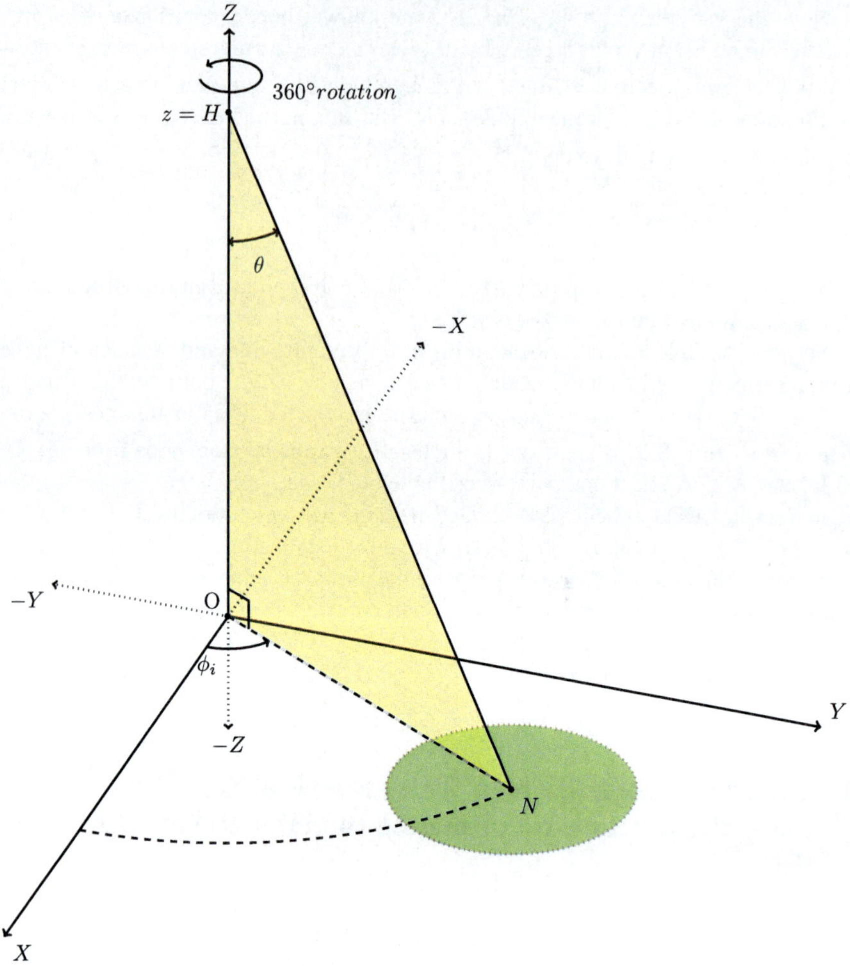

Fig. 3 Schematic diagram of LRDA[8]

The shaded (yellow) region (△ ONH) in Fig. 4 is a right-angled triangle. In ONH ∠O is 90° and height (=h) is represented as OH and $\angle \text{OHN} = \theta$) $\angle\theta$ is known at the node, ON= ρ, radial distance between node N and BS can be given by Eq. (10). After receiving one beacon, any node can estimate its location as (ρ, ϕ) in polar coordinates, and the location (x, y) coordinate of node can be obtained from Eq. (12) if required.

$$\rho = h \tan \theta. \tag{10}$$

$$x = \rho \cos \phi, \tag{11}$$

$$y = \rho \sin \phi. \tag{12}$$

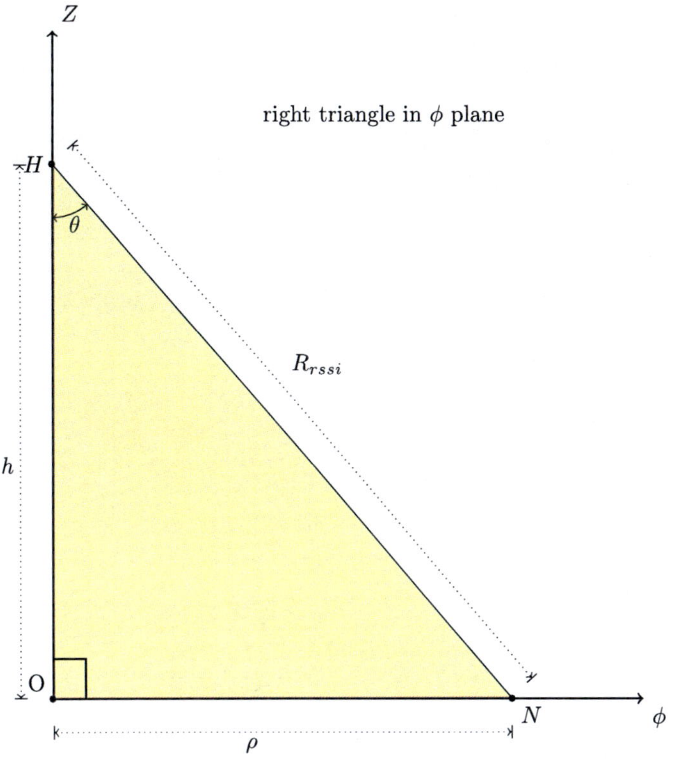

Fig. 4 Geometry of importance LRDA

A node after receiving one or more beacon signals averages the values of $\overline{\phi}$ and $\overline{\theta}$ and the location is re-calculated via Eqs. (10) and (12) using the updated value from Eq. (14) as ($\phi = \overline{\phi}$) and ($\theta = \overline{\theta}$). Here, number of beacon packets received by a node is represented by k.

$$\overline{\phi} = \sum_{i=1}^{k} \phi_i / k, \tag{13}$$

$$\overline{\theta} = \sum_{i=1}^{k} \theta_i / k. \tag{14}$$

The farthest node lies at a distance of $\rho_{\max}$ from BS. The upper value of θ can be calculated from Eq. (15), but $\theta_{\max} \leq \left(90^\circ - \frac{\theta_{\text{bw}}}{2}\right)$, leading to constrains on the height of the antenna h given by Eq. (16). Hence, the circular region around BS that can be localized depends on θ_{bw} and h, and it is given by Eq. (17).

Table 1 Simulation results

Parameter	Value
Maximum localization error e_{max}	4.2953 m
Average localization error e_{avg}	1.3101 m
Mean squared error MSE e^2	2.6743 m^2
Energy consumption of a node E_{loc}	0.0689 J
Energy consumption of n nodes $\overline{E_{loc}}$	0.07 J
Variance of energy consumption $\sigma^2(E_{loc})$	4.0 E−10 J
Localization efficiency η_{loc}	5.4281
Beacons received at node n_{rx}	16

$$\theta_{max} = \tan^{-1}\left(\frac{h}{\rho_{max}}\right). \tag{15}$$

$$h = \left\lceil \frac{\rho_{max}}{\tan\left(90 - \frac{\theta_{bw}}{2}\right)} \right\rceil. \tag{16}$$

$$\rho_{max} \le h \tan\left(90 - \frac{\theta_{bw}}{2}\right). \tag{17}$$

3 Simulation

Simulation is performed with MATLAB. Hundred nodes ($n = 100$) are randomly deployed in the test area of $100 \times 100\,\text{m}^2$. The BS is located at the origin, so ρ_{max} half the diagonal of the area, $\theta_{bw} = 20°$, hence, from Eq. (16) $h \ge 13$ m, height = 13 m. Table 2 lists simulation parameters. $T_a = mt_i - T_b$ where T_a is time for alignment, and T_b is time for beacon processing and transmission $m \in (1, 2, 3, \ldots)$.

3.1 Results and Discussions

Figure 5 shows the relationship between the required height of DA and θ_{bw} on the error in localization, when node received the single beacon. It shows that the error is proportional to the distance between a node and the BS. Figure 6 shows the required combination of h and θ_{bw} for any given ρ_{max}. For example, if $\rho_{max} = 100$ m and h is taken as 10 m, 20 m or 30 m then required θ_{bw} is nearly 13°, 23° or 35° respectively, which means highly directional-antenna (smaller θ_{bw}) requires smaller height of antenna (smaller h); however, broad beamwidth (large θ_{bw}) requires large height of antenna (large h). Figure 7 demonstrates the distance from BS and location

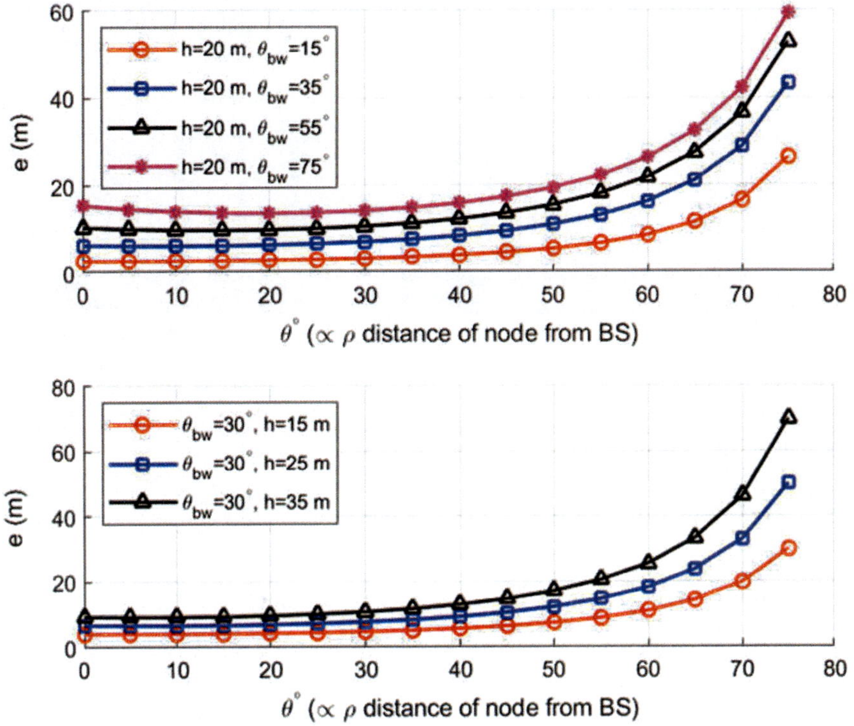

Fig. 5 Maximum error (with one beacon)

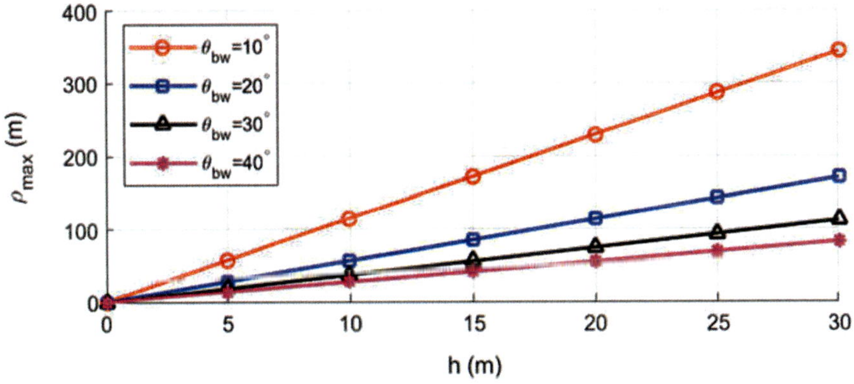

Fig. 6 Trade-off between h & θ_{bw} for $\rho_{\max}$ [theoretical]

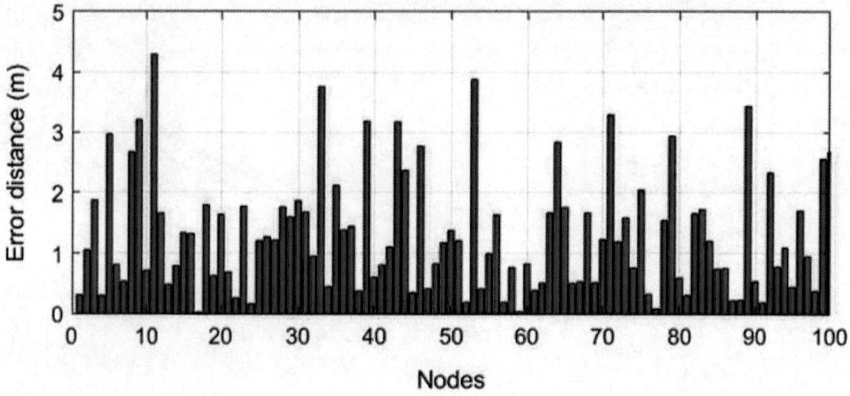

Fig. 7 Distance between node and BS and localization error

Table 2 Simulation parameters

Parameter	Notation	Value
Height of DA (m)	h	13
Beamwidth of DA	θ_{bw}	20°
Step size	$\delta\theta, \delta\phi$	5°, 4°
Starting direction of θ	θ_{init}	0°
Starting direction of ϕ	ϕ_{init}	$\frac{\theta_{bw}}{2}$
Alignment time (s)	T_a	$161.3E{-3}$
Time to transmit beacon (s)	T_b	$3.87E{-04}$
Beacon length (bytes)	L_{beacon}	8

error of each node. Radially outward lines indicate the distance of each node with its true location and estimated location (Table 1).

Figure 8 indicates the location error of different nodes on the XY-plane.

3.2 *Effect of Changing Step Size ($\delta\theta$, $\delta\phi$) on Accuracy, Energy Consumption, and Energy Efficiency of Localization*

Let the time required for a step change of DA remains same as T_a then, if $\delta\theta, \delta\phi$ is reduced, the accuracy of localization improves but T_{loc} and N_{rx} will increase resulting in an increase in total energy consumption at the nodes and vice-versa. It may be observed the small step size increases time required for localization and number of beacons received by nodes, together with, and increased transmissions form BS.

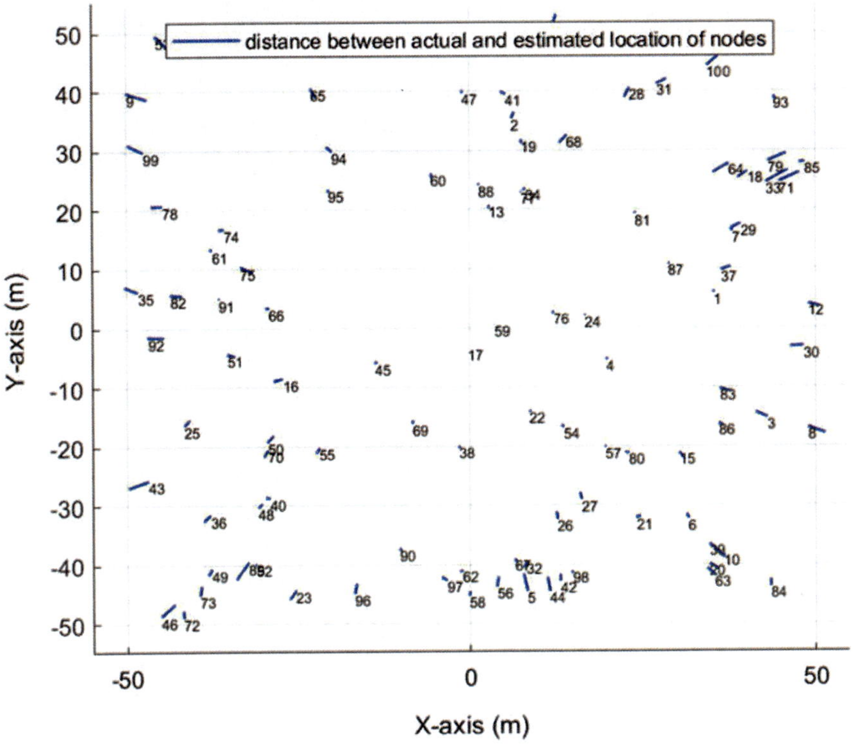

Fig. 8 Localization error on *XY* plane

Table 3 Trade-off between accuracy and energy consumption

$\delta\theta, \delta\phi$	N_{rx}	T_{loc}	MSE	η_{loc}
1,1	340	5.2E3	0.0638	11.4830
5,4	17	271	1.828	7.67
10,10	4	57	14.994	4.441

However, it reduces MSE and improves the efficiency of localization. Table 3 shows accuracy, energy consumption, and energy efficiency of localization.

4 Performance Evaluation

The performance of LRDA with AREA Localization [9] is presented in this section. In AREA, the dumb nodes in the network can estimate their location using the information present in the beacons. The number of beacons with different power

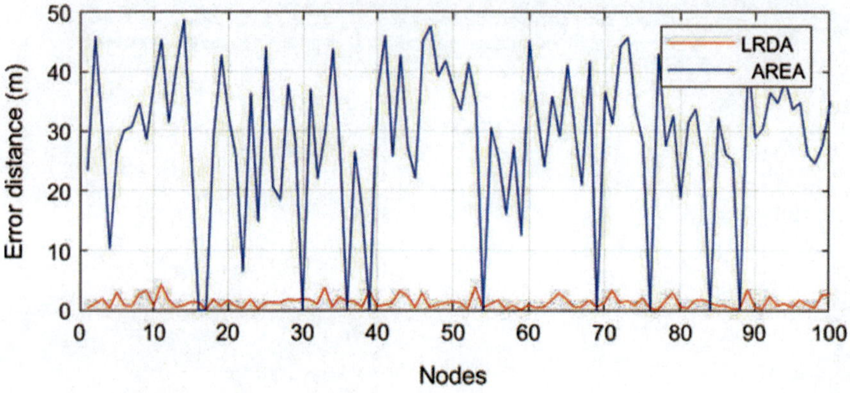

Fig. 9 Localization error in LRDA and AREA

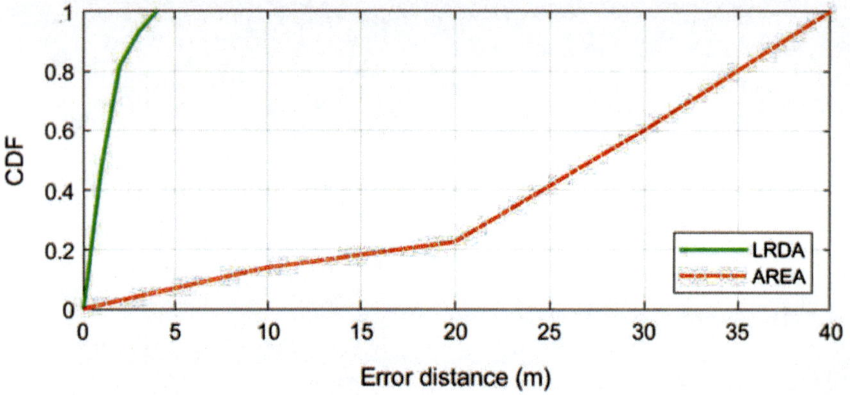

Fig. 10 CDF of LRDA and AREA

is transmitted from the anchor nodes (assuming 10% of nodes as beacon nodes). The parameters used in the simulation of AREA localization are radio communication range ($R_s = 100\,\text{m}$), the beacon signal step ($d_r = 10\,\text{m}$), $R_{\min} = 10\,\text{m}$ and $R_{\max} = R_s$.

Figure 9 shows LRDA is accurate in comparison to AREA. The beacon nodes in AREA show zero error. The CDF of Fig. 9 of LRDA and AREA is shown in Fig. 10. Maximum error magnitude in case of AREA is around 50m, whereas it is less than 4.9 m in LRDA. In LRDA, number of beacons transmitted from nodes is zero, and the average beacons received is around 18. Whereas, Nine beacons are transmitted from each beacon node and 49 beacons are received at dumb nodes in AREA. So, the energy consumption of LRDA is less compared to AREA.

In comparing to AREA, LRDA performs better on accuracy together with the reduced energy consumption at the node.

5 Conclusion

LRDA can localization sensor nodes without hardware modification at the nodes. The cost of the node remains the same. The small data scheming is needed at the node. Inter-communication among nodes is not needed for this scheme.

Energy consumption of localization is reduced. The small computations are required at the node.

The average accuracy of localization is ± 1.6 m and error less than 4.6 m. Energy consumption for localization of the nodes is less than 71 mJ.

Extended work can be found in [8].

References

1. Chelouah L, Semchedine F, Bouallouche-Medjkoune L (2017) Localization protocols for mobile wireless sensor networks: A survey. Computers & Electrical Engineering
2. Chiang PY, Chao PCP, Tarng DC, Yang CY (2017) A novel wireless photoplethysmography blood-flow volume sensor for assessing arteriovenous fistula of hemodialysis patients. IEEE Trans Ind Electron 64(12):9626–9635. https://doi.org/10.1109/TIE.2017.2701780 Dec
3. Chowdhury TJ, Elkin C, Devabhaktuni V, Rawat DB, Oluoch J (2016) Advances on localization techniques for wireless sensor networks: a survey. Comput Networks 110:284–305. https://doi.org/10.1016/j.comnet.2016.10.006
4. Chraim F, Erol YB, Pister K (2016) Wireless gas leak detection and localization. IEEE Trans Ind Inform 12(2):768–779. https://doi.org/10.1109/TII.2015.2397879 April
5. Cui W, Zhang L, Li B, Guo J, Meng W, Wang H, Xie L (2017) Received-signal-strength based indoor positioning using random vector functional link network. IEEE Trans Ind Inform 99:1–1. https://doi.org/10.1109/TII.2017.2760915
6. Derr K, Manic M (2015) Wireless sensor networks—node localization for various industry problems. IEEE Trans Ind Inform 11(3):752–762. https://doi.org/10.1109/TII.2015.2396007 June
7. Farooq-I-Azam M, Ni Q, Ansari EA (2016) Intelligent energy efficient localization using variable range beacons in industrial wireless sensor networks. IEEE Trans Industrial Inform 12(6):2206–2216. https://doi.org/10.1109/TII.2016.2606084 Dec
8. Gautam PR, Kumar S, Verma A, Rashid T, Kumar A (2019) Energy-efficient localization of sensor nodes in wsns using beacons from rotating directional-antenna. IEEE Trans Ind Inform 1–1. https://doi.org/10.1109/TII.2019.2908437
9. Gautam PR, Kumar S, Verma A, Rahsid T, Kumar A (2018) Localization of sensor nodes in wsn using area between a node and two beacons. In: Advances in VLSI, communication, and signal processing. https://doi.org/10.1007/978-981-32-9775-3
10. Han G, Jiang J, Zhang C, Duong TQ, Guizani M, Karagiannidis GK (2016) A survey on mobile anchor node assisted localization in wireless sensor networks. IEEE Commun Surveys Tutorials 18(3):2220–2243. https://doi.org/10.1109/COMST.2016.2544751
11. Haute TV, Verbeke B, Poorter ED, Moerman I (2017) Optimizing time-of-arrival localization solutions for challenging industrial environments. IEEE Trans Ind Inform 13(3):1430–1439. https://doi.org/10.1109/TII.2016.2550531 June
12. Huang JF, Chang GY, Chen GH (2015) A historical-beacon-aided localization algorithm for mobile sensor networks. IEEE Trans Mob Comput 14(6):1109–1122. https://doi.org/10.1109/TMC.2014.2346777

13. Jabbar S, Aziz MZ, Minhas AA, Hussain D (2010) A novel power tuning anchors localization algorithm for mobile wireless sensor nodes. In: 2010 10th IEEE international conference on computer and information technology, pp 2441–2446. https://doi.org/10.1109/CIT.2010.419
14. Kumar S, Gautam PR, Rashid T, Verma A, Kumar A (2018) ETDCC: energy-efficient transmission scheme for dynamic climatic conditions in WSN. TELKOMNIKA (Telecommunication Computing Electronics and Control) 16(3):1126. https://doi.org/10.12928/telkomnika.v16i3.8513
15. Luo J, Hu J, Wu D, Li R (2015) Opportunistic routing algorithm for relay node selection in wireless sensor networks. IEEE Trans Ind Inform 11(1):112–121. https://doi.org/10.1109/TII.2014.2374071
16. Ou CH, Ssu KF (2008) Sensor position determination with flying anchors in three-dimensional wireless sensor networks. IEEE Trans Mob Comput 7(9):1084–1097. https://doi.org/10.1109/TMC.2008.39
17. Ssu KF, Ou CH, Jiau HC (2005) Localization with mobile anchor points in wireless sensor networks. IEEE Trans Veh Technol 54(3):1187–1197. https://doi.org/10.1109/TVT.2005.844642
18. Wang JC, Lin CH, Siahaan E, Chen BW, Chuang HL (2014) Mixed sound event verification on wireless sensor network for home automation. IEEE Trans Ind Inform 10(1):803–812. https://doi.org/10.1109/TII.2013.2262283 Feb
19. Wang J, Ghosh RK, Das SK (2010) A survey on sensor localization. J Control Theory Appl 8(1):2–11. https://doi.org/10.1007/s11768-010-9187-7
20. Wang X, Liu Y, Yang Z, Lu K, Luo J (2014) Robust component-based localization in sparse networks. IEEE Trans Parallel Distrib Syst 25(5):1317–1327. https://doi.org/10.1109/TPDS.2013.85
21. Wen F, Liang C (2015) Fine-grained indoor localization using single access point with multiple antennas. IEEE Sens J 15(3):1538–1544. https://doi.org/10.1109/JSEN.2014.2364121
22. Wu D, Chatzigeorgiou D, Youcef-Toumi K, Ben-Mansour R (2016) Node localization in robotic sensor networks for pipeline inspection. IEEE Trans Ind Inform 12(2):809–819. https://doi.org/10.1109/TII.2015.2469636 April
23. Yan J, Zhou M, Ding Z (2016) Recent advances in energy-efficient routing protocols for wireless sensor networks: a review. IEEE Access 4:5673–5686. https://doi.org/10.1109/ACCESS.2016.2598719

A Survey on Proactive and Reactive Channel Switching Techniques in Cognitive Radios

Anushree Srivastava, Arun Prakash, and Rajeev Tripathi

Abstract Channel switching in cognitive radio networks (CRN) enables the switching of cognitive radio users to the idle channel for avoiding any disruption with the licensed users. The channel switching schemes provide an efficient utilization of the unused spectrum in a channel. In this survey paper, we have characterized the channel switching schemes into two main categories, proactive channel switching (PCS) and reactive channel switching (RCS). Proactive channel switching is based on predicting the channel first and then performing switching, whereas reactive channel switching is an on-demand switching method where switching request is made at the instant of channel sensing. There are some models based on which these switching schemes have been implemented. In the presented work, advantages and limitations of these schemes have been discussed for a better performance analysis.

Keywords Cognitive radios · Channel switching · Reactive channel switching · Proactive channel switching

1 Introduction

Cognitive radio is a software defined radio which has the potential to observe the environment, assess its parameters and adapt itself to those parameters so that the performance of the network can be enhanced. A CR has two main capabilities—cognition and reconfigurability. The concept of CR has been discussed in detail in [1–5]. A CR cycle is shown in Fig. 1. It consists of four steps which are sensing, analysis,

A. Srivastava (✉) · A. Prakash · R. Tripathi
Department of Electronics and Communication Engineering, Motilal Nehru National Institute of Technology Allahabad, Prayagraj 211004, Uttar Pradesh, India
e-mail: srivastava02bhagya@gmail.com

A. Prakash
e-mail: arun@mnnit.ac.in

R. Tripathi
e-mail: rt@mnnit.ac.in

D. Harvey et al. (eds.), *Advances in VLSI, Communication, and Signal Processing*, Lecture Notes in Electrical Engineering 683,
https://doi.org/10.1007/978-981-15-6840-4_59

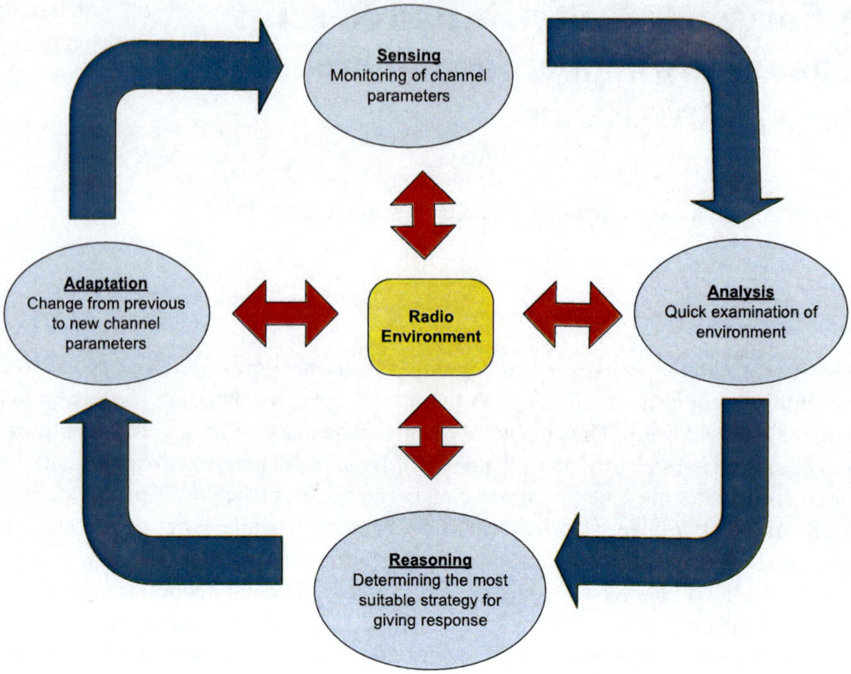

Fig. 1 A CR cycle

reasoning and adaptation. Problems of spectrum dearth as well as overutilization of a spectrum by PU can be dealt using CR technology. It provides an efficient way of utilizing spectrum by switching among the vacant channels depending on channel availability. A cognitive radio network (CRN) mainly consists of primary users (PUs)/licensed users and secondary users (SUs)/CR users. Through this CR technology, the SUs can switch in an opportunistic way to other channels which are currently not in use or they are being wasted by the PUs.

In [6], two types of spectrum allocation policies have been defined. Due to the fixed channel allocation policy, which is controlled by the government organization, some channels are highly utilized while others are lightly utilized or highly underutilized frequency bands. Hence, there exists, large number of spectrum holes known as white spaces in underutilized channels. These spaces or holes are nothing but slots left unexploited by the PUs. These holes can be exploited with the help of dynamic channel allocation policy where SU has to sense the PU traffic patterns in the channel and consequently occupy the channel when it is available. If the PU tries to reclaim the channel, the SU has to leave and switch to some other available vacant channel.

There are several channel switching techniques in CR viz. PCS, RCS, hybrid channel switching [7], smart switching [8], adaptive mode switching [9], fuzzy-based channel switching [10, 11], route switching [12], switch and stay; switch and

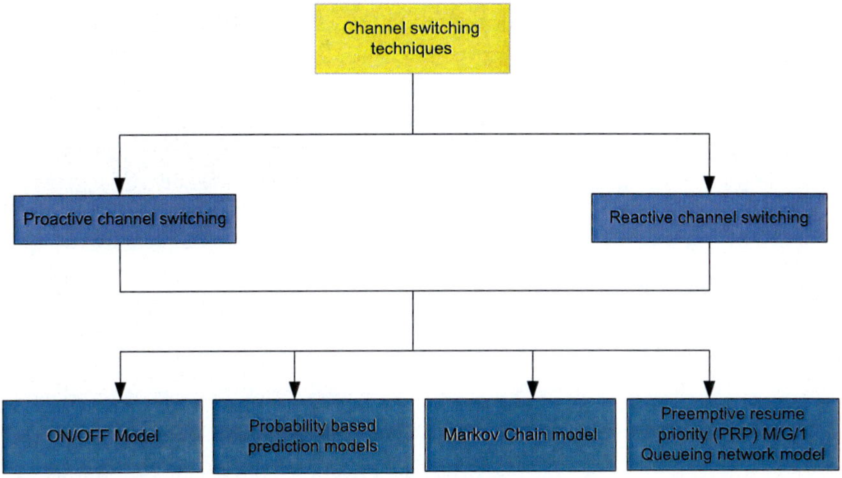

Fig. 2 Various models used to incorporate channel switching techniques

examine [13], channel switching in CRN based on reinforcement learning [14] and machine Learning [15].

In this paper, we are mainly going to discuss PCS and RCS in detail. The word proactive means to control a process by implementing it rather than waiting for the process to take place after some delay. Hence, in PCS mechanism, the future channel to which SU will switch to is sensed before the channel switching takes place. Another name for this method is active switching [16]. The word reactive means to show a response or some reaction in return. Hence, in RCS mechanism, the SU decides to switch to the target channel at the instant when PU makes an interrupt request. This mechanism is also known as passive switching. There are various models based on which these channel switching techniques are performed. The most commonly used models have been surveyed in further literature. They have been mentioned in the chart in Fig. 2.

1.1 *Purpose of This Paper*

We present a survey paper focusing on the recent enhancements made in the field of channel switching in CR networks. This paper analyzes the role of PCS and RCS in CRs and highlights the need of these switching techniques for avoiding interference, sustaining throughput, minimizing delay and minimizing switching overhead in a CRN.

It is equally important to discuss the significance of this paper in comparison to other existing survey papers. Till date, no such survey paper exists which specifically discusses the PCS and RCS methods in CRs. The most relevant survey paper is the comprehensive survey on spectrum handoff methods in CRNs provided in [6] which reviews all the existing handoff schemes like time-triggered handoff schemes, mobility-based handoff schemes, spectrum sensing based and operating mode-based handoff schemes and other miscellaneous spectrum handoff schemes.

Another survey paper has been presented by authors in [17] which focusses only on the timing-based handoff schemes. These handoff schemes are "time critical" because the entire process is based on continuous monitoring and sensing of the traffic patterns of a PU. Various types of time-triggered handoff schemes have been discussed in great detail. In [18] and [19], the authors have presented a brief survey on channel selection and switching strategies and have discussed the reactive and proactive sensing in detail.

1.2 Organization of the Paper

The paper is divided in five sections. Section 1 presents the introduction and a brief idea about the paper. Section 2 discusses the PCS techniques. Section 3 presents the RCS techniques. In Sect. 4, we have made a summary of all the existing schemes along with their advantages and limitations. Finally, the paper is concluded in Sect. 5.

2 Proactive Channel Switching Techniques

In proactive channel switching, before the actual switching process occurs, the future channel for transmission of data is pre-determined according to the previous traffic pattern or usage activities of a PU. The SU can shift to a new channel before the arrival of PU thereby reducing transmission delay. Figure 3 depicts a proactive channel switching mechanism.

The authors in [16] proposed a probabilistic model for channel selection and PCS. The model used is also popularly known as ON/OFF model [20] which is used to collect channel usage data of the PU. Probabilities of one state transition (OSTP) and two-state transition (TSTP) are calculated to obtain the idle states of the channel. Finally, channel ranking is done based on the values of OSTP and TSTP. Channel with highest TSTP is considered to be of highest rank and also the best idle channel for channel selection and switching.

The authors in [21] have presented a proactive spectrum access scheme in dynamic spectrum networks. The proposed architecture has two main parts: (a) proactive channel prediction—obtained using exponential and periodic models and (b) intelligent channel switching—SU can switch from one channel to the other using the predictions obtained from step (a) thereby avoiding collision with the PU. Also, they have made a comparison of results with the existing reactive schemes and finally concluded that proactive schemes give better channel utilization (here 5%).

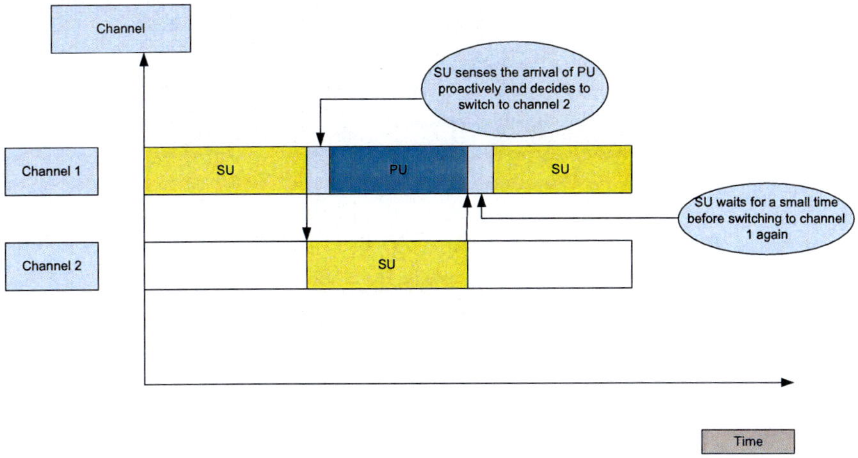

Fig. 3 PCS mechanism

In [22], a proactive spectrum handoff scheme using single rendezvous coordination (SRV) protocol has been proposed based on a multi-user greedy channel selection (GCS). The model used in this scheme is a Markov chain model which is used to evaluate the stochastic probability for channel selection of SUs and also for performing a theoretical analysis to obtain the aggregate throughput so that the performance of the proposed protocol can be analyzed.

In [23], the authors have proposed a PRP M/G/1 queueing network model for finding optimal target channel sequences for proactive spectrum handoff in CR network. Also, a GCS algorithm is proposed which can reduce total service time compared to other schemes.

The authors in [24] have presented a proactive switching scheme based on common hopping. Here, a communication link is established first over a common channel and then the packet transmission occurs between the nodes. The aim of the proposed proactive spectrum handoff protocol is to avoid the interference between SU and PU on PU arrival. Based on channel prediction, if the channel is about to be occupied by the PU, a channel switching (CSW) flag is set, i.e., SUs need to perform channel switching and leave the current channel to avoid any interference with the PU.

In [25], the authors have presented a prediction based proactive channel switching scheme in which time slots have been assigned to all PUs and SUs and transmission based on time slot takes place. The authors obtained results which show a better throughput and interference reduction than most of the existing reactive schemes.

The authors in [8] have proposed an intelligent channel switching algorithm, also known as smart channel switching algorithm in a CR network. It uses an ON/OFF model for prediction of channel usage. Smart channel switching is similar in approach to proactive channel switching technique. The only difference is that the smart switching algorithm takes into account factors such as channel switching cost, minimal switching overhead, etc.

The work presented in [20] and [15] is the most recent advancement in PCS methodology. In [20], the authors have presented a PCS scheme based on ON/OFF model. The basic concept involved is that before the actual transmission occurs, the PU channel list (PCL) is shared among the SUs. PCL contains the history of PUs traffic pattern. This scheme gives better throughput than RCS scheme and few disruptions to PUs. In [15], the authors have proposed a proactive spectrum switching scheme in CR using machine learning. Main focus of the authors is to increase the satisfaction of CR users by meeting the user requirements of delay, bandwidth and coverage. Two different ML strategies, i.e., ANNs and Q-learning, have been used to give a good sensing and switching performance.

3 Reactive Channel Switching Techniques

In reactive channel switching, the target channel is investigated by on-demand spectrum sensing after the channel switching request is made. Then, the interrupted secondary user can again continue its transmission on one of the vacant channel. Figure 4 depicts a reactive channel switching mechanism.

The authors in [26] have developed a Markov transition model in collaboration with the preemptive resume priority (PRP) M/G/1 queueing network model to illustrate the multiple handoff delay occurring as a result of the combined effects of the sensing time, handshaking time, channel switching time and the waiting time. A SUs connection might experience multiple interruptions from the PUs during its packet transmission period. Since these interruptions result in several switching, a

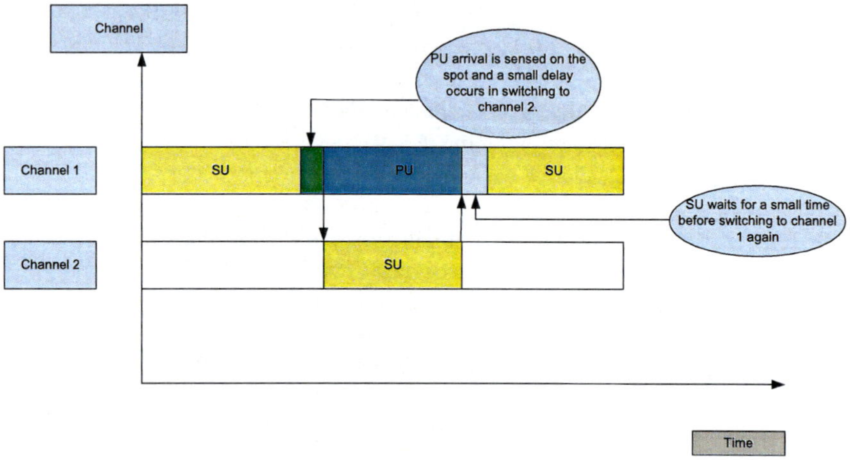

Fig. 4 RCS mechanism

sequence of target channels need to be selected, called the target channel sequence in this paper. Later, the analysis of prolonged data delivery time (total service time + cumulative handoff delay) is performed which is a significant performance parameter for analyzing the delay-sensitive traffic of SUs.

In [27], the authors have presented a model for analyzing reactive spectrum handoff in CRNs. The effect of reactive-decision channel switching on channel utilization has been evaluated by considering design features such as arrival rates of PUs, SUs and handoff processing time. A PRP M/G/1 queueing network model has been proposed to study the spectrum exploitation patterns between the PUs and SUs with multiple channel switching. The effect of delay incurred due to several handoffs on the prolonged data delivery time is examined using a state diagram. The results obtained in this paper can provide a deeper understanding of the effects of traffic patterns and service time on channel utilization and transmission delay.

The authors in [28] have presented a novel scheme based on RCS approach known as the movement and channel availability prediction (MCAP) scheme. In this scheme, two-state (idle state or busy state) Markov model is used to forecast the channel availability. The focus of this work is on reactive schemes because reactive schemes give a reduction in number of switches between the channels although some delay is observed.

Similarly, there are many other reactive spectrum handoff techniques presented in [29–32]. The goal in [29] is to achieve maximum throughput of SUs. Instead of utilizing one complete channel, several sub-channels are being utilized by SU for performing transmission. The approach is highly efficient in terms of throughput but it suffers from design complexity as a single channel is divided into multiple sub-channels, thereby avoiding disruption to PU on each sub-channel. The aim of the authors in [30] is to choose target channels on the arrival of PU on the basis of two criterions: the two-state prediction probability and the time duration of channel occupancy by PU. If the channel is empty and has a shorter busy period, then handoff occurs. This information is sent to all the neighboring SUs. This method leads to a reduction in number of handoffs but an increase in switching overhead because of communication among the SUs. In [31], a novel reactive handoff strategy based on realistic channel switching is proposed where "realistic" means that SUs can switch channels within the same frequency range. The authors in [32] have presented a delay-sensitive reactive strategy which is different from the above mentioned strategies. Here, SU is allowed to perform multiple transmission within a given time delay. Also, the authors have performed various calculations to experimentally show that less delay is incurred using this approach.

Based on the above discussion of proposed schemes by various authors, we can draw a comparison between PCS and RCS. In terms of switching time, PCS is better than RCS because sensing and switching time is comparatively less in PCS. But, in terms of accuracy, RCS is better than PCS because the target channel in PCS may not be available for switching even if it is pre-determined, whereas accuracy in selection of target channel is high in case of RCS.

4 Summary of All Switching Paradigms

A summary of all the switching techniques with their advantages and limitations has been compiled in Table 1.

5 Conclusion

Channel switching is an important phenomenon in CRNs. In this survey paper, we have discussed the proactive and reactive channel switching techniques in detail. Each of these techniques has their own advantages and limitations in a particular scenario which have been discussed in detail. The models presented in this paper have been assessed on basic performance criterions such as throughput, delay, jitter, switching overhead, number of SUs, channel efficiency, number of channel switches,

Table 1 Summary of all channel switching methods

Paper references	Channel switching technique	Advantages	Limitations
[8]	Proactive	• Reduction in channel switching by 20% • Reduction in interference	• Scheme is not valid for a scenario where multiple SUs exist
[25]	Proactive	• Less interference • Increased throughput	• Complexity is high
[24]	Proactive	• High packet delivery ratio • High throughput • Fewer collisions	• Model is quite complex
[16]	Proactive	• Improvement in overall throughput • Minimum channel switching	• Only single SU is taken into consideration.
[22]	Proactive	• Increase in throughput up to 38.7% in comparison to other similar studies	• Does not have a back-up channel mechanism
[23]	Proactive	• Reduction in total service time	• Model is complex despite of using low-complexity greedy algorithm
[21]	Proactive	• Reduces interference to PUs up to 30% • Decreases throughput jitter at SUs	• Imperfect predictions are observed called "dumb" channel switches

(continued)

Table 1 (continued)

Paper references	Channel switching technique	Advantages	Limitations
[26]	Reactive	• Decrease in cumulative handoff delay	• Restricted to only exponential distribution
[27]	Reactive	• Prolonged data delivery time is reduced even after multiple switching	• A research issue on how to use PRP repeat priority queuing still exists
[29]	Reactive	• Use of multiple sub-channels improves channel efficiency	• High complexity
[30]	Reactive	• Reduction in total number of handoffs	• Increase in switching overhead due to multiple SUs communication
[31]	Reactive	• Reliable and delay tolerant for a given transmission duration	• Total delay is not taken into consideration
[32]	Reactive	• Achievable and realistic results have been obtained in terms of channel utilization	• Limited to Markov model design only
[20]	Proactive	• High throughput • Few collisions • High packet delivery rate • Minimum switching overhead	• Limited to ON/OFF model design only but overall, it is a satisfactory model

packet delivery ratio and model complexity. The limitations of each model can be considered as a challenging research issue for future considerations.

References

1. Mitola J, Maguire GQ (1999) Cognitive radio: making software radios more personal. IEEE Pers Commun 6(4):13–18
2. Mitola JI (2002) Cognitive radio. An integrated agent architecture for software defined radio, pp 0474–0474
3. Giupponi L, Galindo-Serrano A, Blasco P, Dohler M (2010) Docitive networks: an emerging paradigm for dynamic spectrum management. IEEE Wirel Commun 17(4):47
4. Costlow T (2003) Cognitive radios will adapt to users. IEEE Intell Syst 18(3):7
5. Jayaweera S, Christodoulou C (2011) Radiobots: architecture, algorithms and realtime reconfigurable antenna designs for autonomous, self-learning future cognitive radios
6. Kumar K, Prakash A, Tripathi R (2016) Spectrum handoff in cognitive radio networks: a classification and comprehensive survey. J Netw Comput Appl 61:161–188

7. Karmokar AK, Senthuran S, Anpalagan A (2014) Physical layer-optimal and cross-layer channel access policies for hybrid overlay–underlay cognitive radio networks. IET Commun 8(15):2666–2675 (IEEE)
8. Xue F, Qu D, Zhu G, Li Y (2009) Smart channel switching in cognitive radio networks. In: 2nd international congress on image and signal processing, IEEE, pp 1–5
9. Ji X, Xu J, Che YL, Fei Z, Zhang R (2017) Adaptive mode switching for cognitive wireless powered communication systems. IEEE Wirel Commun Lett 6(3):386–389 (IEEE)
10. Arnous R, El-Desouky AI, Sarhan A, Badawy M (2018) ILFCS: an intelligent learning fuzzy-based channel selection framework for cognitive radio networks. EURASIP J Wirel Commun Netw, 247 (Springer)
11. Ali A, Abbas L, Shafiq M, Bashir AK, Afzal MK, Liaqat HB, Siddiqi MH, Kwak KS (2019) Hybrid fuzzy logic scheme for efficient channel utilization in cognitive radio networks. IEEE Access 7:24463–24476 (IEEE)
12. Liang Q, Wang X, Tian X, Zhang Q (2013) Route-switching games in cognitive radio networks. In: Proceedings of the fourteenth ACM international symposium on mobile ad hoc networking and computing, ACM, pp 249–252
13. Gaaloul F, Yang HC, Radaydeh RM, Alouini MS (2012) Opportunistic spectrum access in cognitive radio based on channel switching. In: 7th international ICST conference on cognitive radio oriented wireless networks and communications (CROWNCOM), IEEE, pp 270–274
14. Wang S, Liu H, Gomes PH, Krishnamachari B (2018) Deep reinforcement learning for dynamic multichannel access in wireless networks. IEEE Trans Cognit Commun Netw 4(2):257–265 (IEEE)
15. Ozturk M, Akram M, Hussain S, Imran MA (2019) Novel QoS-aware proactive spectrum access techniques for cognitive radio using machine learning. IEEE Access 7:70811–70827 (IEEE)
16. Shil S, Chauhan P, Deka SK, Sarma N (2017) Efficient proactive channel switching in cognitive radio networks. In: Conference on information and communication technology (CICT), IEEE, pp 1–6
17. Mir U, Bhatti ZA (2018) Time triggered handoff schemes in cognitive radio networks: a survey. J Netw Comput Appl 102:71–85
18. Grover A, Bali V, Singh S (2018) Channel selection and switching in cognitive radio networks: challenges and approaches. In: 8th international conference on cloud computing, data science & engineering (Confluence), IEEE, pp 14–15
19. Wang LC, Wang CW (2008) Spectrum handoff for cognitive radio networks: reactive-sensing or proactive-sensing? In: IEEE international performance, computing and communications conference, IEEE, pp 343–348
20. Rajpoot V, Tripathi VS (2019) A novel proactive handoff scheme with CR receiver based target channel selection for cognitive radio network. Phys Commun 36:100810
21. Yang L, Cao L, Zheng H (2008) Proactive channel access in dynamic spectrum networks. Phys Commun 1(2):103–111 (Elsevier)
22. Mehrnoush M, Fathi R, Vakili VT (2015) Proactive spectrum handoff protocol for cognitive radio ad hoc network and analytical evaluation. IET Commun 9(15):1877–1884 (IEEE)
23. Wang CW, Wang LC, Adachi F (2009) Modeling and analysis for proactive-decision spectrum handoff in cognitive radio networks. In: Proceedings of ICC, IEEE, 109(164), pp 13–18
24. Song Y, Xie J (2010) Proactive spectrum handoff in cognitive radio ad hoc networks based on common hopping coordination. In: INFOCOM IEEE conference on computer communications workshops, IEEE, pp 1–2
25. Devanarayana C, Alfa AS (2014) Proactive channel access in cognitive radio networks based on users' statistics. In: 1st international workshop on cognitive cellular systems (CCS), IEEE, pp 1–5
26. Wang CW, Wang LC, Adachi F (2010) Modeling and analysis for reactive-decision spectrum handoff in cognitive radio networks. In: IEEE global telecommunications conference GLOBECOM, IEEE, pp 1–6

27. Wang CW, Wang LC (2012) Analysis of reactive spectrum handoff in cognitive radio networks. IEEE J Sel Areas Commun 30(10):2016–2028
28. Kanan E, Husari G, Al-Ayyoub M, Jararweh Y (2015) Towards improving channel switching in cognitive radio networks. In: 6th international conference on information and communication systems (ICICS), IEEE, pp 280–285
29. Willkomm D, Gross J, Wolisz A (2005) Reliable link maintenance in cognitive radio systems. In: First IEEE international symposium on new frontiers in dynamic spectrum access networks, DySPAN, IEEE, pp 371–378
30. Tian J, Tian X, Qian L (2014) Secondary users operation strategies based on primary users activities. Sci J Inf Eng 4(1)
31. Premarathne US, Khalil I, Atiquzzaman M (2017) Reliable delay-sensitive spectrum handoff management for re-entrant secondary users. Ad Hoc Networks 66:85–94 (Elsevier)
32. NoroozOliaee M, Hamdaoui B, Cheng X, Znati T, Guizani M (2013) Analyzing cognitive network access efficiency under limited spectrum handoff agility. Trans Vehic Technol 63(3):1402–1407 (IEEE)

Controlling GIDL Using Core–Shell Technique in Conventional Nano-Wire

Abhishek Kumar, Sushmita Jaiswal, and Santosh Kumar Gupta

Abstract In this paper, detailed gate-induced drain leakage (GIDL) mechanism is analyzed in conventional nano-wire. It has been seen that conventional nano-wire suffers from both lateral and transversal band-to-band tunneling. The lateral component tunneling is more severe and active when device is in OFF condition. It is one of the contributors of leakage current in nano-wire transistors. SiO_2 as core–shell material is proposed for effectively controlling both lateral and transverse band-to-band tunneling (BTBT) current. The impact of core–shell on short channel parameters like I_{on}, I_{off}, I_{on}/I_{off} ratio, subthreshold swing and drain-induced barrier lowering (DIBL) are also analyzed. From the calibrated simulation, it has been found that the use core–shell significantly reduces I_{off} with only a minor reduction in I_{on}. On increasing core–shell radius, drain-induced barrier lowering (DIBL), I_{off} and subthreshold swing are also significantly and almost linearly decreases.

Keywords GIDL DIBL · Subthreshold swing · Tunneling · I_{off} · I_{on} · Nano-wire

1 Introduction

Continuous down scaling of device causes increase in leakage current due to gate inefficiency to control channel. Multi-gate technology is one of the efforts for effectively controlling channel by using multiple gates. According to ITRS report 2015 [1], gate all around MOSFETs are expected to drive the semiconductor technology in

A. Kumar · S. Jaiswal (✉) · S. K. Gupta
ECED, Motilal Nehru National Institute of Technology Allahabad, Prayagraj, Uttar Pradesh 211004, India
e-mail: ece.sush1014@gmail.com

A. Kumar
e-mail: abhishekpintu4316@gmail.com

S. K. Gupta
e-mail: skg@mnnit.ac.in

D. Harvey et al. (eds.), *Advances in VLSI, Communication, and Signal Processing*, Lecture Notes in Electrical Engineering 683,
https://doi.org/10.1007/978-981-15-6840-4_60

sub-nanometer regime. The use of high-k technology also has emerged as an effective tool to control gate leakage current [1–3].

A significant leakage current exists in short channel devices [4, 5]. Silicon-on-insulator (SOI) technology suffers from floating body effects [6] which increases bipolar effects thus suffers from increase in GIDL effects and decreases device performance [7–10]. There are a lot of solutions given like introducing under-lap area, spacer at overlapping region [11–15]. The use of spacer decreases coupling and gate capacitance (decreases gate to drain overlapping capacitance). Fan [15] shows from his experimental results that GIDL is one of the important parameters that seriously degrade static leakage power. They give various optimization techniques for reducing GIDL like doping variation, decreasing diameter and cross-section shaping. They also reported that on increasing diameter of a nano-wire, effects of GIDL increases. Kim et al. [16] showed gate stack configuration of a nano-wire for scaling at sub-10 nm range. Nah et al. [17] reported Ge–core–shell and P^{++} core–shell to sufficiently reduce GIDL but this is not valid in case of conventional nano-wire as current flow through center in case of junction-less MOSFET but it flows through surface in case of conventional MOSFETs [18–22].

In this paper, SiO_2 as core–shell material is proposed, which increases conduction band energy in leakage path, thus reduces overall GIDL in conventional nano-wire [14, 15, 17]. The effect of core–shell variation on device short channel parameters like drain-induced barrier lowering (DIBL), subthreshold swing, OFF current and ON current and their ratio are analyzed. It has been seen that there is minor reduction in ON current and large decrease in OFF current thus making I_{off} to I_{on} ratio increasing with increasing core–shell radius. There is also minor decrease in subthreshold swing and only decreases from 69 to 65 mV/dec. DIBL also decreases linearly with increasing core–shell radius. Overall device performance improves with minor decrease in ON current and subthreshold swing.

2 Device Structure and Simulation Calibration

2.1 Device Structure

Full three-dimensional view of a conventional nano-wire is shown in Fig. 1a. The proposed device structure is shown in Fig. 2b and its cross-sectional view is shown in Fig. 1c.

The nano-wire diameter (t_{si}) is taken as constant value of 8 nm. Radius of core (t_{core}) is taken as parameter. It's value is varied from 2 to 6 nm, while taking total diameter (t_{si}) as constant. To compare conventional with core–shell nano-wire, all parameters are taken as constant with variation in core–shell radius and its effect is studied. The simulation parameters taken are shown in Table 1.

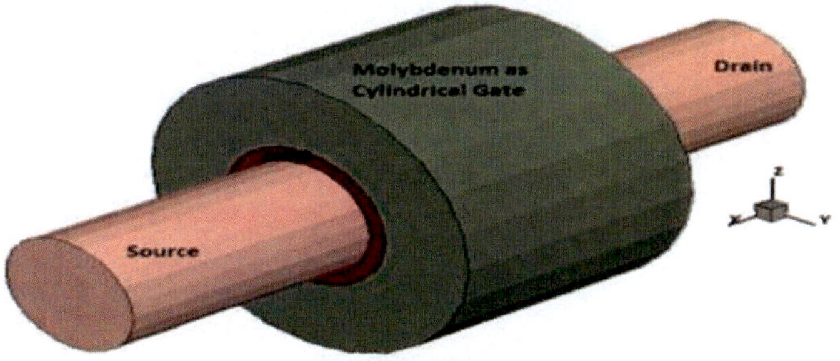

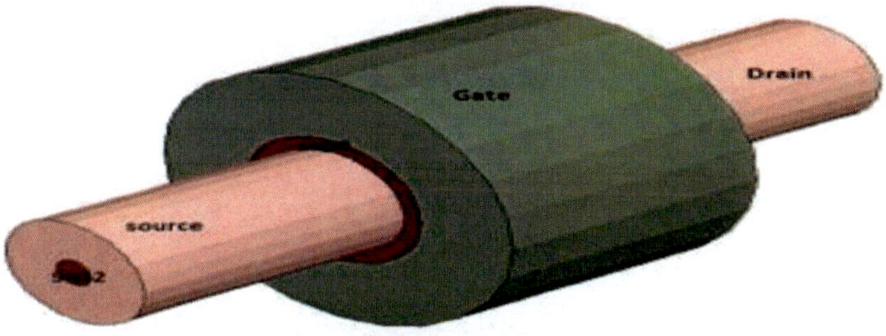

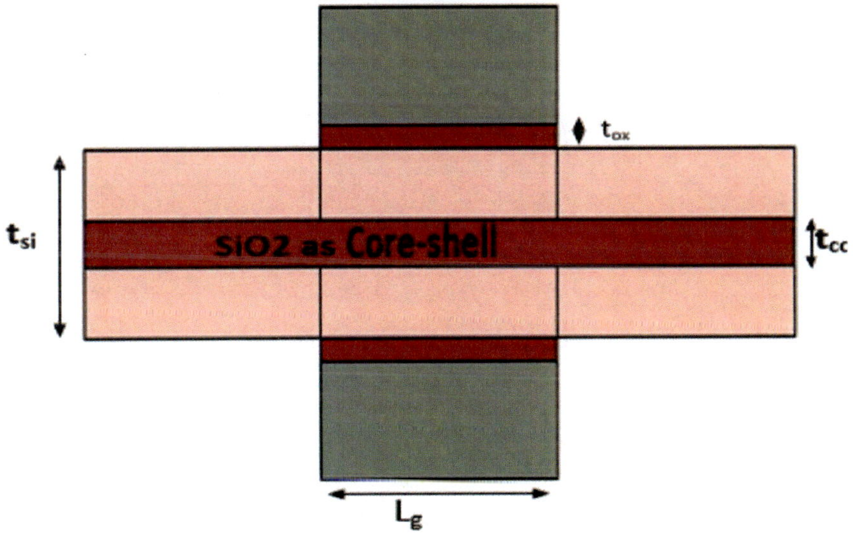

Fig. 1 **a** 3D view of conventional nano-wire, **b** 3D view of the proposed structure, **c** cross-sectional view of proposed structure

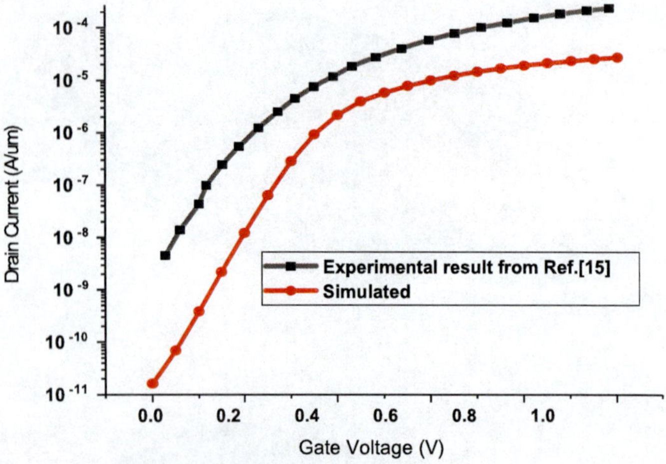

Fig. 2 Simulation models calibrated by reproducing the experimental results of [15]

Table 1 Parameters used for simulation

Parameters	Value
Nano-wire diameter (t_{si})	8 nm
Gate oxide thickness (t_{ox})	1 nm
Core–shell thickness (t_{core})	2, 4, 6 nm
Source/drain doping (N_d)	5×10^{19} cm^{-3}
Channel doping (N_a)	10^{16} cm^{-3}
Gate length (L_g)	14 nm
Gate work function	4.5 eV
Length of source/drain	28 nm
Electron tunneling effective mass (m_e)	0.4 m_o
Hole tunneling effective mass (m_h)	0.65 m_o

2.2 *Simulation Calibration*

The simulation is performed on 3D simulator Sentaurus *TCAD*. To consider the impact of majority and minority bulk carrier mobility, Philips mobility model is used and also Lombardi mobility model is used to consider its degradation at interface doping and field-dependent Shokley–Read–Recombination model is also activated. For highly doped source and drain region, Auger model and band gap narrowing along with slot boom model is used. Fermi model is also activated globally for higher carrier density. Full non-local BTBT model is activated in tunneling region, which includes tunneling with consistent parallel momentum integral with the direct non-local path to account the effect of lateral and transverse tunneling. The model is calibrated by reproducing the same result as reported in reference [15] as shown in Fig. 2.

3 Leakage Mechanism

Lateral band-to-band tunneling (L-BTBT) [23] and transverse BTBT (T-BTBT) [13, 22] are mainly responsible for leakage current. The details are explained as follow.

3.1 *Lateral Band-to-Band Tunneling (L-BTBT)*

L-BTBT is more pronounced along channel length, when gate voltage is 0. It is due to lowering of energy bands at channel when drain voltage is V_{dd}. The e-current density exists at center of device from source to drain. This is shown in Fig. 3.

It is clear from Fig. 3 that current density is maximum at the center of the device when device is in OFF state. The leakage path is from the body center of the devices. This is the main path for OFF current. To reduce contribution of L-BTBT to OFF current, this tunneling path must be blocked by using high band gap material, in this case, insulator is used.

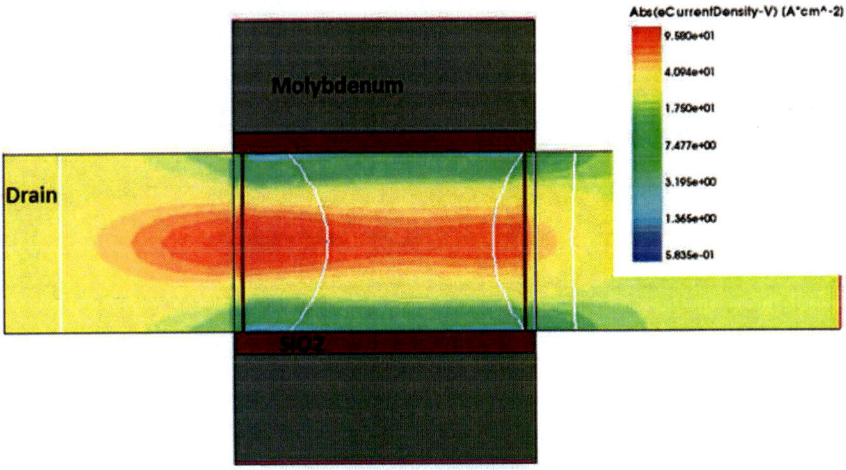

Fig. 3 E-current density along length of device in conventional nano-wire at $V_{gs} = 0$ V and $V_{ds} = 1$ V

3.2 *Transverse Band-to-Band Tunneling (T-BTBT)*

Transverse band-to-band tunneling is at overlapping region of source/drain and channel. This effect is more pronounced at high gate to drain voltage.

4 Optimizing GIDL Using Core–Shell Technique

From above discussion, it is clear that controlling lateral BTBT is more important than transverse BTBT. The detailed idea of optimizing GIDL using core–shell technique is explained below.

4.1 *Controlling GIDL Using Core–Shell Technique*

The core–shell technique as shown in Fig. 1b, c is used to effectively control GIDL. The band diagram of a nano-wire with and without core–shell is shown in Fig. 4.

As core–shell radius is increased from 0 (conventional) to 6 nm, the conduction band moves upward, indicating increasing in energy required for electron to tunnel to conduction band, thus, lateral tunneling is controlled. Figure 3 also indicates the current density to be maximum within device center at drain side, so, placing high

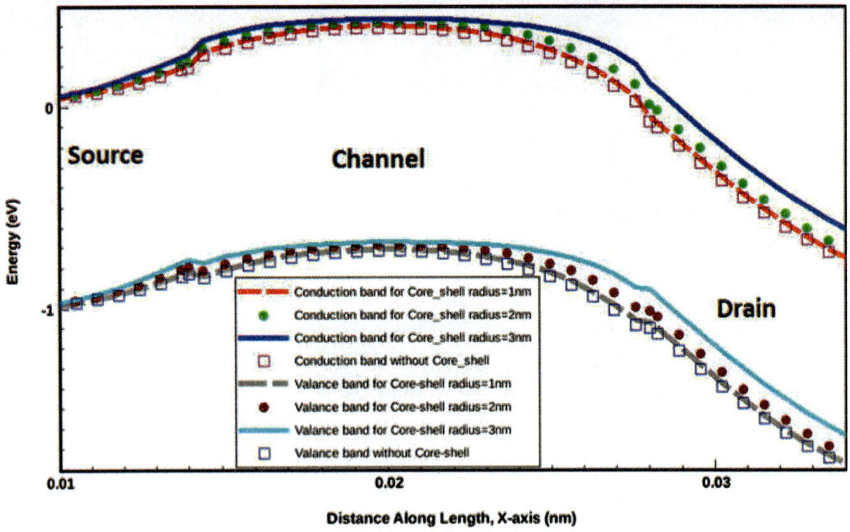

Fig. 4 Band diagram along lateral direction of the transistor with gate voltage (V_{gs}) = 0 V and V_{ds} (drain to source) voltage = 1 V

band gap material (insulator) at center controls L-BTBT. The electron current density contour with variation in core–shell radius from 0 to 3 nm is shown in Fig. 5a-c.

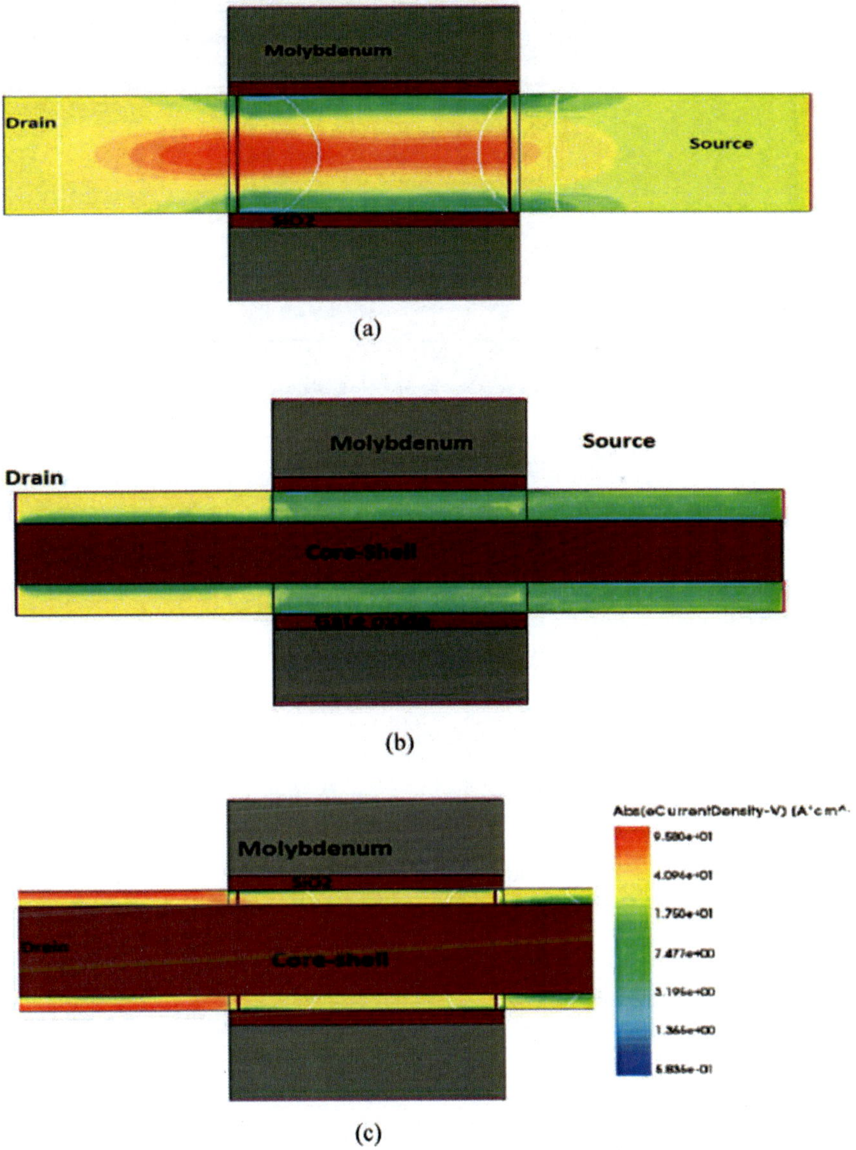

Fig. 5 **a** Electron current density without core–shell (conventional), **b** electron current density with core radius = 1 nm, **c** electron current density with core radius = 3 nm with gate voltage (V_{gs}) = 0 V and V_{ds} (drain to source) voltage = 1 V

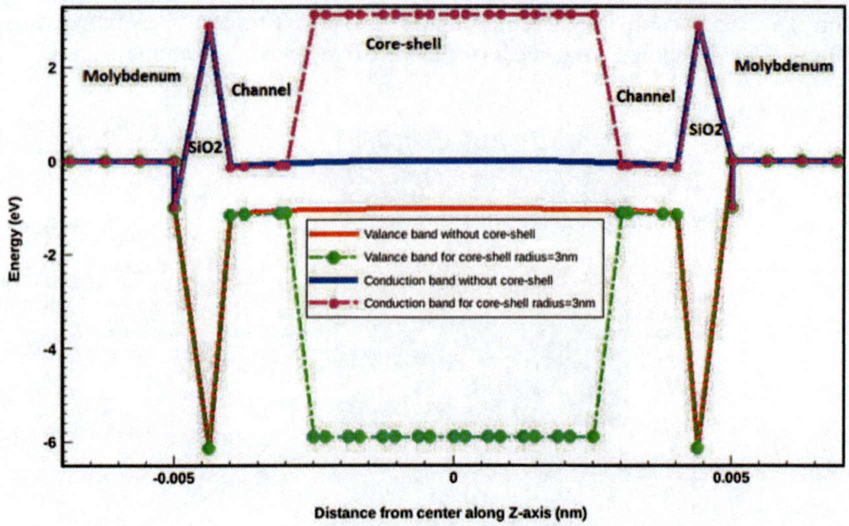

Fig. 6 Energy band diagram along transverse direction with $V_{gs} = V_{ds} = 1$ V

The transverse energy band diagram at drain and channel junction is shown in Fig. 6. The transverse band bending with core–shell is smaller as compared to conventional nano-wire. Thus, nano-wire with core–shell has better control over transverse band-to-band tunneling (T-BTBT) than conventional nano-wire.

From Figs. 4, 5 and 6, we can conclude that L-BTBT and T-BTBT can be controlled by using introducing core–shell of SiO_2.

4.2 *Effect of Core–Shell Technique on SCEs*

The effect of core–shell technique and radius on short channel parameter like drain-induced barrier lowering (DIBL), I_{off}, I_{on}, subthreshold swing and I_{off} to I_{on} ratio is being analyzed. The e-current flows from center of the device which is clearly visible in contour diagram of Fig. 5a. High bandgap material is used in form of SiO_2 as core–shell material. The decrease in e-current density is visible in Fig. 5b. On further increasing core–shell radius, results into decrease in conducting area, thus, the density at surface of silicon is increased. At the surface, the movement of electron is effectively controlled by all around gate. Variation of drain current when gate voltage is 0 V is large as compared to 1 as shown in Fig. 7a. Figure 7b shows variation of ON current with increasing core–shell radius. By increasing core radius from 0 to 1 nm, the decrease in ON current is 2 μA/μm but the decrease is much larger (10 μA/μm) when core-shell radius is increased from 2 to 3 nm. The decrease is 10 uA/um. Off current also decreases linearly up to core–shell radius = 2 nm. The decrease in I_{off} is 0.8×10^{-11} A/μm, but, the decrease increases to

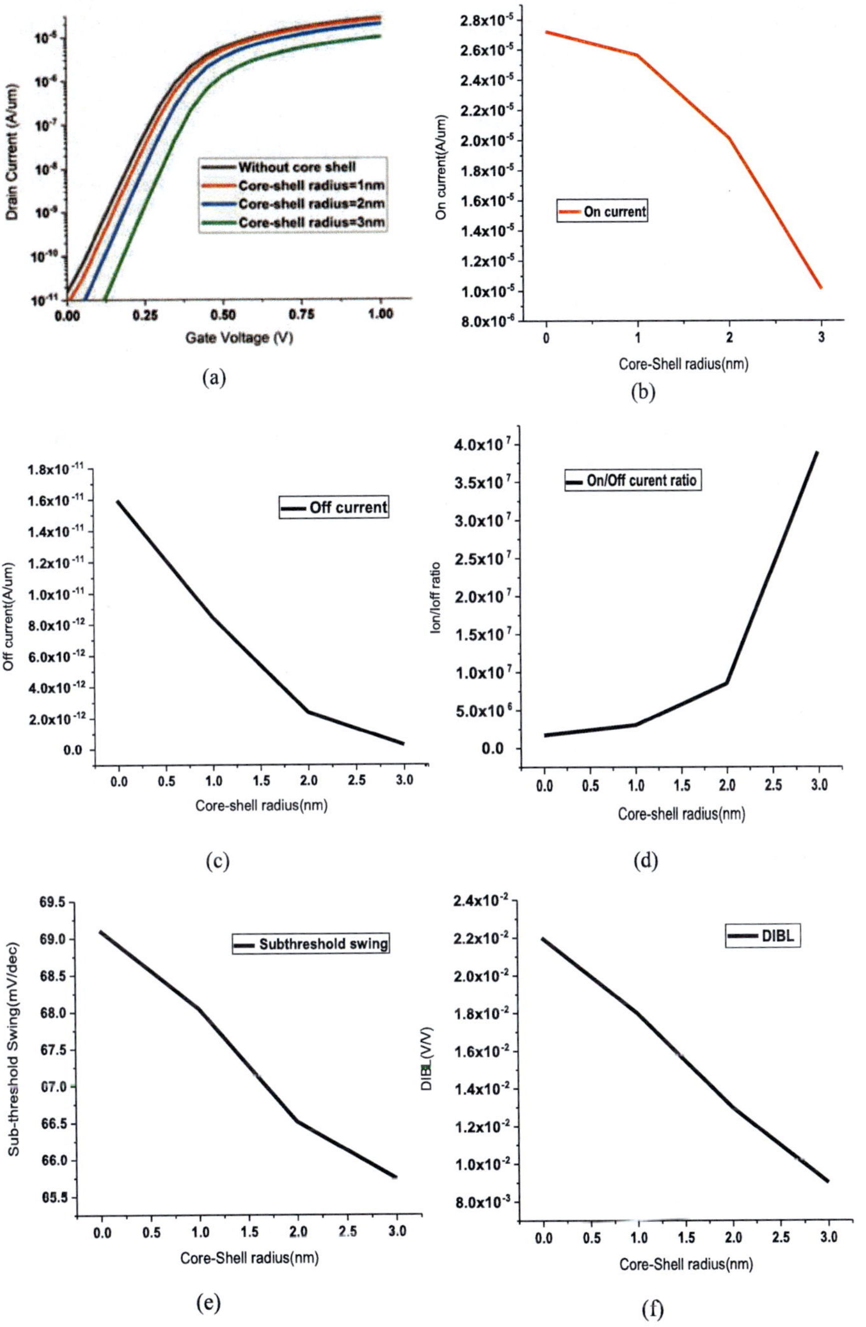

Fig. 7 **a** Transfer characteristics with $V_{ds} = 1$ V for different core–shell radius, **b** ON current variation with change in core–shell radius, **c** OFF current with changing core–shell radius, **d** OFF to ON current ratio for various core–shell radius, **e** subthreshold swing variation, **f** DIBL for changing core–shell radius

1.2×10^{-11} A/μm on increasing radius from 2 to 3 nm. The ratio of OFF to ON current is shown in Fig. 7d. The decrease in subthreshold swing (ss) varies almost linearly as seen from Fig. 7a and e. The least value of ss is 65.7 mV/dec for core radius = 3 nm, which is above silicon based MOSFET minimum limit of 60 mV/dec. Linear variation of DIBL is clearly visible in Fig. 7f. The value decreases from 22 to 9 mV/mV on changing radius from 0 to 3 nm. DIBL improves almost linearly on increasing radius.

5 Conclusion

In this paper, we have demonstrated that L-BTBT plays more significant role in transistor leakage current. Using SiO_2 as core–shell, effectively controls lateral (L-BTBT) and transverse tunneling (T-BTBT) current. The application of core–shell increases band energy in the tunneling path (center of device in case of conventional transistor), thus reduces L-BTBT and T-BTBT both. The reduction in BTBT also results in improving DIBL and OFF current with least affecting ON current and subthreshold swing.

References

1. ITRS (2015) Available: http://www.itrs2.net/itrs-reports.html
2. Chau R, Datta S, Doczy M, Doyle B, Kavalieros J, Metz M (2004) High-/spl kappa//metal-gate stack and its MOSFET characteristics. IEEE Electr Device Lett 25(6)
3. Mistry K, Allen C (2007) A 45 nm logic technology with high-k + metal gate transistors, strained silicon, 9 Cu interconnect layers, 193 nm dry patterning, and 100% Pb-free packaging. In: Electron devices meeting, IEDM 2007. IEEE international, 10–12 Dec 2007
4. Kumar A (2016) Analytical modelling of subthreshold characteristics of REGAA FinFET using center potential. Superlattices Microstruct
5. Kumar A (2016) Analog and RF performance of a multigate FinFET at nano scale. Superlattices Microstruct
6. Sahay S, Kumar MJ (2016) Realizing efficient volume depletion in SOI junctionless FETs. IEEE J Electr Device SOC 4(3):110–115
7. Wann H, Ko, Hu C, (1992) Gate-induced band-to-band tunneling leakage current in LDD MOSFETs. IEDM Tech Dig, 147–150
8. Chen J, Assaderaghi F, Ko P-K, Hu C (1992) The enhancement of gate induced drain leakage current in short channel SOI MOSFET and its application in measuring lateral bipolar current gain beta. IEEE Electr Device Lett 134(11):572–574
9. Hoffmann T, Doornbos G, Ferain I, Collaert N, Zimmerman P, Goodwin M, Rooyackers R, Kottantharayil A, Yim Y, Dixit A, De Meyer K, Jurczak M, Biesemans S (2005) GIDL and parasitic Schottky barrier leakage elimination in aggressive scaled HfO_2 TiN FinFET devices. In: IEEE international electron devices meeting, 2005. IEDM Technical Digest, pp 725–728
10. Fan J, Li M, Xu X, Huang R (2013) New observation on gate induced drain leakage in silicon nanowire transistor with epi-free CMOS compatible technology on SOI substrate. In: Proceedings of IEEE SOI-3D subthreshold micro electron technology unified conference, pp 1–2

11. Tanaka K, Takeuchi K, Hane M (2005) Practical FinFET design considering GIDL for LSTP devices. In: IEEE international electron devices meeting, 2005. IEDM technical digest, pp 1001–1004
12. Kuhn KJ (2012) Consideration for ultimate CMOS scaling. IEEE Trans Electr Devices 59(7):1813–1828
13. Chan TY, Chen J, Ko PK, Hu C (1987) The impact of gate induced band-to-band tunneling leakage current on MOSFETs scaling. IEDM Tech Dig, 718–721
14. Kerber P, Zhang Q, Koswatta S, Bryant A (2013) GIDL in doped and undoped FinFET devices for low-leakage applications. IEEE Electr Device Lett 34(1):6–8
15. Fan J (2015) Insight into gate induced drain leakage in silicon nanowire transistor. IEEE Trans Electr Devices 62(1)
16. Kim W, Chung TY, Oh KS, Lee W-S (2009) Sub 10 nm gate-all-around CMOS nanowire transistors on bulk Si substrate. In: VLSI technology symposium
17. Nah J, Liu E-S, Varahramyan KM, Tutuc E (2010) Ge-SixGe1-x core–shell nanowire tunneling field-effect transistors. IEEE Trans Electr Device 57(8):1883–1888
18. Sahay S, Kumar MJ (2016) Novel gate-stack engineering nano-wire fet for scaling to the sub-10 nm regime IEEE Trans Electr Devices 99
19. Colinge J-P, Lee C-W, Afzalian A, Akhavan ND, Yan R, Ferain I, Razavi P, O 'Neil B, Blake A, White M, Kelleher A-M, McCarthy B, Murphy R (2010) Nanowire transistor without junction. Nature Nanotechnol 5(3):225–229
20. Kranti A, Yan R, Lee C-W, Ferain I, Yu R, Akhavan ND, Razavi P, Colinge JP (2010) In: Proceedings of the European solid state device research conference, pp 357–360
21. Vijay Kumar MP, Hu C-Y, Kao K-H, Lee Y-J, Chao T-S (2015) Impacts of the shell doping profile on the electrical characteristics of junctionless FETs. IEEE Trans Electr Devices 62(11):3541–3546
22. Gundapaneni S, Bajaj M, Pandey RK, Murali KVRM, Ganguly S, Kottantharayil A (2012) Effect of band-to-band tunneling on junctionless transistor. IEEE Trans Electr Devices 59(4):1023–1029
23. Sahay S, Kumar MJ (2016) Controlling L-BTBT and volume depletion in nanowire JLFETs using core-shell architecture. IEEE Trans Electr Devices 63(9):3790–3794